Waste Biorefinery

Waste Biorefinery

Integrating Biorefineries for Waste Valorisation

Edited by

Thallada Bhaskar
Biomass Conversion Area,
Material Resource Efficiency Division,
CSIR-Indian Institute of Petroleum, Dehradun, India

Ashok Pandey
Centre for Innovation and Translational Research,
CSIR-Indian Institute of Toxicology Research, Lucknow, India

Eldon R. Rene
IHE Delft Institute for Water Education, Delft, The Netherlands

Daniel C.W. Tsang
Department of Civil and Environmental Engineering,
The Hong Kong Polytechnic University, Hong Kong

ELSEVIER

Elsevier
Radarweg 29, PO Box 211, 1000 AE Amsterdam, Netherlands
The Boulevard, Langford Lane, Kidlington, Oxford OX5 1GB, United Kingdom
50 Hampshire Street, 5th Floor, Cambridge, MA 02139, United States

Library of Congress Cataloging-in-Publication Data
A catalog record for this book is available from the Library of Congress

British Library Cataloguing-in-Publication Data
A catalogue record for this book is available from the British Library

ISBN: 978-0-12-818228-4

For information on all Elsevier publications visit our website at https://www.elsevier.com/books-and-journals

Publisher: Susan Dennis
Acquisitions Editor: Kostas KI Marinakis
Editorial Project Manager: Emerald Li
Production Project Manager: Selvaraj Raviraj
Cover Designer: Miles Hitchen

Typeset by TNQ Technologies

Contents

Section E: Sewage sludge biorefinery

Section F: Modelling and LCA studies

Section H: Country specific case studies

Contributors

Rabia Abad School of Applied Sciences, The University of Huddersfield, Huddersfield, United Kingdom

Abdelrahim Abusafa Chemical Engineering Department, An-Najah National University, Nablus, Palestine

Mortaza Aghbashlo Department of Mechanical Engineering of Agricultural Machinery, Faculty of Agricultural Engineering and Technology, College of Agriculture and Natural Resources, University of Tehran, Karaj, Alborz, Iran

Asam Ahmed Division of Systems, Power & Energy, James Watt School of Engineering, University of Glasgow, Glasgow, United Kingdom

Maher Al-Jabari Renewable Energy and Environment Research Unit, Mechanical Engineering Department, Palestine Polytechnic University, Hebron, Palestine

Maria Alexandri Leibniz Institute for Agricultural Engineering and Bioeconomy Potsdam, Potsdam, Germany

A.K.M. Kazi Aurnob Department of Chemical Engineering, Bangladesh University of Engineering and Technology, Dhaka, Bangladesh

Ayan Banerjee Material Resource Efficiency Division, CSIR-Indian Institute of Petroleum, Dehradun, Uttarakhand, India; Academy of Scientific and Innovative Research (AcSIR), New Delhi, India

Shishir Kumar Behera Industrial Ecology Research Group, School of Chemical Engineering, Vellore Institute of Technology, Vellore, Tamil Nadu, India

Laurent Belard NaturePlast, Ifs, France

Stella Bezergianni Chemical Process & Energy Resources Institute - CPERI, Centre for Research & Technology Hellas CERTH, Thessaloniki, Greece

Thallada Bhaskar Material Resource Efficiency Division, CSIR-Indian Institute of Petroleum, Dehradun, Uttarakhand, India; Academy of Scientific and Innovative Research (AcSIR), New Delhi, India

Nilutpal Bhuyan Department of Energy, Tezpur University, Tezpur, Assam, India

Hanif A. Choudhury Department of Chemical Engineering, Texas A&M University at Qatar, Doha, Qatar

Loukia P. Chrysikou Chemical Process & Energy Resources Institute - CPERI, Centre for Research & Technology Hellas CERTH, Thessaloniki, Greece

Caterina Coll Lozano Imecal S.A., Lálcudia, Valencia, Spain

Sutapa Das Department of Chemical Engineering, Indian Institute of Technology Guwahati, Guwahati, Assam, India

Antonio David-Moreno CIEMAT, Madrid, Spain

Francesca Demichelis DIATI, Politecnico di Torino, Torino, Italy

Chenyu Du School of Applied Sciences, The University of Huddersfield, Huddersfield, United Kingdom

Capucine Dupont Department of Environmental Engineering and Water Technology, IHE Delft Institute for Water Education, Delft, The Netherlands

Amer Elhamouz Chemical Engineering Department, An-Najah National University, Nablus, Palestine

Silvia Fiore DIATI, Politecnico di Torino, Torino, Italy

María García Torreiro AINIA-Centro tecnológico, Paterna, Valencia, Spain

Debashish Ghosh Material Resource Efficiency Division, CSIR-Indian Institute of Petroleum, Dehradun, Uttarakhand, India; Academy of Scientific and Innovative Research (AcSIR), New Delhi, India

Inmaculada González Granados Biomasa Peninsular S.A., Madrid, Spain

Vaibhav V. Goud Department of Chemical Engineering, Indian Institute of Technology Guwahati, Guwahati, Assam, India

Jasneet Grewal Enzyme and Microbial Biochemistry Laboratory, Department of Chemistry, Indian Institute of Technology Delhi, Hauz Khas, New Delhi, India

Natalia Herrero García Biomasa Peninsular S.A., Madrid, Spain

Homa Hosseinzadeh-Bandbafha Department of Mechanical Engineering of Agricultural Machinery, Faculty of Agricultural Engineering and Technology, College of Agriculture and Natural Resources, University of Tehran, Karaj, Alborz, Iran

Shu-Chien Hsu Department of Civil and Environmental Engineering, The Hong Kong Polytechnic University, Kowloon, Hong Kong, China

Kazi Bayzid Kabir Department of Chemical Engineering, Bangladesh University of Engineering and Technology, Dhaka, Bangladesh

Rupam Kataki Department of Energy, Tezpur University, Tezpur, Assam, India

Ravneet Kaur Dr. B.R. Ambedkar National Institute of Technology, Jalandhar, Punjab, India; Material Resource Efficiency Division, CSIR-Indian Institute of Petroleum, Dehradun, Uttarakhand, India

S.K. Khare Enzyme and Microbial Biochemistry Laboratory, Department of Chemistry, Indian Institute of Technology Delhi, Hauz Khas, New Delhi, India

Kawnish Kirtania Department of Chemical Engineering, Bangladesh University of Engineering and Technology, Dhaka, Bangladesh

Chor-Man Lam Department of Civil and Environmental Engineering, The Hong Kong Polytechnic University, Kowloon, Hong Kong, China

Marcos Latorre-Sánchez Imecal S.A., L'alcudia, Valencia, Spain

Raquel Lebrero Institute of Sustainable Processes, University of Valladolid, Valladolid, Spain; Department of Chemical and Environmental Engineering, University of Valladolid, Valladolid, Spain

Yize Li Division of Systems, Power & Energy, James Watt School of Engineering, University of Glasgow, Glasgow, United Kingdom

Diannan Lu Department of Chemical Engineering, Tsinghua University, Beijing, China

Mette Lübeck Department of Chemistry and Bioscience - Section for Sustainable Biotechnology, Denmark

Tiffany M.W. Mak Department of Civil and Environmental Engineering, The Hong Kong Polytechnic University, Kowloon, Hong Kong, China

Ritesh S. Malani Centre for Energy, Indian Institute of Technology, Guwahati, Guwahati, Assam, India

N. Arul Manikandan Department of Chemical Engineering, Indian Institute of Technology Guwahati, Guwahati, Assam, India

Vijayanand S. Moholkar Centre for Energy, Indian Institute of Technology, Guwahati, Guwahati, Assam, India; Department of Chemical Engineering, Indian Institute of Technology, Guwahati, Guwahati, Assam, India

Hamidreza Mojab Department of Water Resource Management, Faculty of Civil Engineering and Geoscience, Technical University of Delft, Delft, The Netherlands

Jose L. Molto Marin Exergy Ltd., Coventry, United Kingdom

Sidra Munir School of Applied Sciences, The University of Huddersfield, Huddersfield, United Kingdom

Raul Muñoz Institute of Sustainable Processes, University of Valladolid, Valladolid, Spain; Department of Chemical and Environmental Engineering, University of Valladolid, Valladolid, Spain

Hana Musinovic NATRUE, Brussels, Belgium

Ahaduzzaman Nahid Department of Chemical Engineering, Bangladesh University of Engineering and Technology, Dhaka, Bangladesh

M. M. Tejas Namboodiri Department of Biosciences and Bioengineering, Indian Institute Technology Guwahati, Guwahati, Assam, India

Abdul-Sattar Nizami Center of Excellence in Environmental Studies (CEES), King Abdulaziz University, Jeddah, Makkah Province, Saudi Arabia

Jose Miguel Oliva-Dominguez CIEMAT, Madrid, Spain

David Ovejero-Roncero Exergy Ltd., Coventry, United Kingdom

Santiago Pacheco-Ruiz Veolia Water Technologies Techno Center Netherlands B.V./Biothane, Delft, The Netherlands

Kannan Pakshirajan Department of Biosciences and Bioengineering, Indian Institute of Technology Guwahati, Guwahati, Assam, India

Hung-Suck Park Department of Civil and Environmental Engineering, University of Ulsan, Ulsan, Republic of Korea

Celia Pascual Institute of Sustainable Processes, University of Valladolid, Valladolid, Spain; Department of Chemical and Environmental Engineering, University of Valladolid, Valladolid, Spain

Andrés Pascual AINIA-Centro tecnológico, Paterna, Valencia, Spain

Víctor Pérez Institute of Sustainable Processes, University of Valladolid, Valladolid, Spain; Department of Chemical and Environmental Engineering, University of Valladolid, Valladolid, Spain

Greg Perkins Martin Parry Technology, Brisbane, QLD, Australia; School of Chemical Engineering, University of Queensland, Brisbane, QLD, Australia

Daniel Pleissner Sustainable Chemistry (Resource Efficiency), Institute of Sustainable and Environmental Chemistry, Leuphana University of Lüneburg, Lüneburg, Germany

G. Pugazhenthi Department of Chemical Engineering, Indian Institute of Technology Guwahati, Guwahati, Assam, India

Amélie Raingué Urbaser S.A., R&D and Innovation Department, Madrid, Spain

Eldon Raj Department of Environmental and Water Technology, IHE Delft Institute for Water Education, Delft, The Netherlands

Mohammad Rehan Center of Excellence in Environmental Studies (CEES), King Abdulaziz University, Jeddah, Makkah Province, Saudi Arabia

Eldon R. Rene Department of Environmental Engineering and Water Technology, IHE Delft Institute for Water Education, Delft, The Netherlands

Ali S. Reshad Department of Chemical Engineering, Indian Institute of Technology Guwahati, Guwahati, Assam, India

Alfredo Rodrigo AINIA-Centro tecnológico, Paterna, Valencia, Spain

Rocio Roldan-Aguayo Exergy Ltd., Coventry, United Kingdom

Ayesha Sadaf Enzyme and Microbial Biochemistry Laboratory, Department of Chemistry, Indian Institute of Technology Delhi, Hauz Khas, New Delhi, India

Megha Sailwal Material Resource Efficiency Division, CSIR-Indian Institute of Petroleum, Dehradun, Uttarakhand, India; Academy of Scientific and Innovative Research (AcSIR), New Delhi, India

Hassan Sawalha Renewable Energy and Environment Research Unit, Mechanical Engineering Department, Palestine Polytechnic University, Hebron, Palestine

Alba Serna-Maza Urbaser S.A., R&D and Innovation Department, Madrid, Spain

Izhar Hussain Shah Department of Civil and Environmental Engineering, University of Ulsan, Ulsan, Republic of Korea; Institute of Environmental Sciences and Engineering, School of Civil and Environmental Engineering, National University of Sciences and Technology, Islamabad, Pakistan

Shailendra Kumar Shukla Centre for Energy and Resources Development, Department of Mechanical Engineering, Indian Institute of Technology (BHU), Varanasi, Uttar Pradesh, India

Pushpendra Kumar Singh Rathore Centre for Energy and Resources Development, Department of Mechanical Engineering, Indian Institute of Technology (BHU), Varanasi, Uttar Pradesh, India

Mark Smith NATRUE, Brussels, Belgium

Debashis Sut Department of Energy, Tezpur University, Tezpur, Assam, India

Meisam Tabatabaei Faculty of Plantation and Agrotechnology, Universiti Teknologi MARA (UiTM), Shah Alam, Selangor, Malaysia; Microbial Biotechnology Department, Agricultural Biotechnology Research Institute of Iran (ABRII), Agricultural Research, Education, and Extension Organization (AREEO), Karaj, Alborz, Iran; Biofuel Research Team (BRTeam), Karaj, Alborz, Iran; Faculty of Mechanical Engineering, Ho Chi Minh City University of Transport, Ho Chi Minh City, Vietnam

Pankaj Tiwari Department of Chemical Engineering, Indian Institute of Technology Guwahati, Guwahati, Assam, India

Khanh-Quang Tran Department of energy and process engineering, Norwegian University of Science and Technology, Trondheim, Norway

Daniel C.W. Tsang Department of Civil and Environmental Engineering, The Hong Kong Polytechnic University, Kowloon, Hong Kong, China

Jack Van de Vossenberg Department of Environmental Engineering and Water Technology, IHE Delft Institute for Water Education, Delft, The Netherlands

Eric D. van Hullebusch Department of Environmental Engineering and Water Technology, IHE Delft Institute for Water Education, Delft, The Netherlands

Carol W. Wambugu Department of Environmental Engineering and Water Technology, IHE Delft Institute for Water Education, Delft, The Netherlands

Lei Wang Department of Civil and Environmental Engineering, The Hong Kong Polytechnic University, Kowloon, Hong Kong, China; Department of Materials Science and Engineering, The University of Sheffield, Sheffield, United Kingdom

Ian Watson Division of Systems, Power & Energy, James Watt School of Engineering, University of Glasgow, Glasgow, United Kingdom

Neerja Yadav Enzyme and Microbial Biochemistry Laboratory, Department of Chemistry, Indian Institute of Technology Delhi, Hauz Khas, New Delhi, India

Siming You Division of Systems, Power & Energy, James Watt School of Engineering, University of Glasgow, Glasgow, United Kingdom

Iris K.M. Yu Department of Civil and Environmental Engineering, The Hong Kong Polytechnic University, Kowloon, Hong Kong, China; Green Chemistry Centre of Excellence, Department of Chemistry, University of York, York, United Kingdom

Preface

Where there is righteousness in the heart, there is beauty in the character. When there is beauty in the character, there is harmony in the home. When there is harmony in the home, there is order in the nation. When there is order in the nation, there is peace in the world.

A. P. J. Abdul Kalam (1931–2015, Aerospace Scientist and the 11th President of India)

Rapid industrialization, population growth, unplanned expansion of urban zones and infrastructures, and inadequate policies have led to the mismanagement of solid waste in developed nations as well as poorer countries in the developing world. Solid and liquid waste, both the generation and disposal, is a topic of major public health and environmental concern. More often, these issues are engendered due to poor waste collection systems, lack of governmental or municipal services, limited budget, weak management policies, and lack of an efficient organizational infrastructure, among others. Therefore, solid waste piles up in streets, backyards, alleys, and illegal dumpsites; people scavenge them to earn a living. In many countries, these nonsanitary landfills have caused austere problems, including air, water, and soil pollution, and has induced the spread of disease-causing vectors. However, from a resource recovery viewpoint, solid waste can be considered a treasure house of enormous wealth, wherein electricity can be produced by combustion/incineration of the solid waste found in landfills. With the advent of advanced equipment, new processes, and better understanding of the mechanisms involved in biological and engineering sciences, solid waste can be efficiently transformed into energy, fuels, and value-added products. The solid wastes include a mixture of biological, combustible, and noncombustible materials such as biomass, grass clippings, wood, leaves, food waste, paper, cardboard, leather products, plastics, bedding materials, resins, metals, glass, etc.

By applying the concepts of pollution prevention, resource recovery, and cleaner production, a biorefinery can be defined as a facility that integrates different biomass conversion process and equipment to produce a wide range of biobased products such as biofuels, power, heat, and platform chemicals. A biorefinery can also be used to represent a stand-alone process, a plant or a group of synergistically linked facilities, e.g., ecoindustrial parks. The main aim of all these facilities are to integrate and apply the best engineering, biological, and management practices to minimize the impact on solid, liquid, and gaseous wastes on human health and the environment, convert waste into several value-added product streams, and sustainably manage the existing resources. Thus, the concept of a biorefinery has been constantly evolving, and a systematic transformation of the facilities has been envisioned in recent years. For example, the conventional biorefinery (first-gen) uses agricultural biomass to produce bioethanol or biodiesel, whereas the second and third Gen biorefineries uses advanced processes using lignocellulosic biomass, cereals, forestry biomass, algal biomass, waste gases, industrial sludges, oil residues, food waste, and high-strength wastewater streams to produce chemicals and energy. Depending on the source and characteristics of the raw materials, the processes can be either chemical, biological, thermochemical, and mechanical, or a combination of these processes.

Therefore, as citizens, we have to change our perspective to see how waste can be used as a secondary resource for the production of energy and other materials. In order to meet the growing demand of fuels, biofuels are emerging as an alternative clean fuel to replace the conventional fossil fuels. According to the European Union (EU) Energy Commission, by the year 2020, the EU aims to have 10% of the transport fuel of every EU country come from renewable sources such as biofuels. The fuel suppliers are also required to reduce the greenhouse gas intensity of the EU fuel mix by 6% by 2020 in comparison to 2010. Anew, due to the rising energy demand in the market, novel research areas have started to focus on resource recovery, and a galaxy of new technologies have been successfully tested, both at the lab and pilot-scale. Although all biorefinery-based processes are expected to produce fewer emissions and support sustainable local bioeconomy, the overall environmental implications and life-cycle impact analysis are still being studied. In this line of progressive research, there is still a lot to be done, and interestingly, standardization of protocols and methods should be documented clearly. Although regulations are well-established and implemented for biomethane and natural gas, the fuels, lubricants, and hydraulic fluids produced from mineral oil or biomass origin still does not have standardized methods of sampling, analysis, and testing, terminology, and specifications for application in the transportation, industrial, and domestic sectors.

To address some of the practical issues discussed above and to provide a general perspective of the different types of biorefineries, the first volume of the book entitled "Waste biorefinery: *Potential and perspectives*" was published in the year 2018. The book explored some of the recent developments in biochemical and thermochemical methods of waste-to-energy conversion and the potential generated by different kinds of biomass in more decentralized biorefineries. To address the most recent advancements made in the field of biorefineries, the second volume of this book series entitled *Waste biorefinery: Integrating biorefinery for waste valorization* has been compiled. This volume presents recent updates on the different types of biorefineries (e.g., solid waste, lignin residue, agroindustrial waste, lignocellulosic wastes, food waste, and nonedible oils), the application of multiscale modeling strategies, systems approach, life-cycle analysis (LCA), and carbon footprint estimation tools, and it presents different case studies related to the integration of biorefineries for waste-to-energy and fuels production. The volume comprises of twenty-five chapters, divided among the following eight thematic sections:

Session A: Municipal solid waste–based biorefineries
Section B: Lignocellulosic biomass-based biorefinery
Section C: Food waste and chitin-based biorefinery
Section D: Nonedible oils–based biorefinery and applications
Section E: Sewage sludge biorefinery
Section F: Modeling and life-cycle analysis studies
Section G: System dynamics and carbon footprints
Section H: Country-specific case studies

In Section A, the challenges and opportunities of applying gasification to municipal solid waste, its performance for the production of electricity and chemicals, economic considerations, and opportunities for the future development is presented in Chapter 1. The URBIOFIN demo-scale project presented in Chapter 2 explores the potential of the organic fraction of municipal solid waste (OFMSW) to produce bioblocks (bioethanol, volatile fatty acids (VFA), and biogas), biopolymers (short chain [scl-PHA]), medium chain polyhydroxyalkanoates (mcl-PHA), and additives (bioethylene and biofertilizers) using a battery of innovative and integrated physical, chemical, and biological processes.

In Section B, Chapter 3 highlights the working principle and concept of a nozzle reactor with countercurrent mixing for the upscaling of fast hydrothermal liquefaction (HTL) of solid biomass residues and wastes. Chapter 4 presents the advantages, limitations, and practical applications of an up-flow anaerobic sludge blanket (UASB) and expanded granular sludge bed (EGSB) for enhanced resource recovery (mainly biomethane) during wastewater treatment. Two case studies related to the application of UASB and EGSB systems in olive oil and the pulp and paper industries have also been discussed in this chapter. The valorization of agroindustrial wastes into platform chemicals (e.g. lactic acid, C_3) and its derivatives for applications in pharmaceutical, food, animal feed, dairy, detergent, and cosmetic industries is covered in Chapter 5. A similar approach has been demonstrated to convert lignocellulosic biomass for polyhydroxybutyrate (PHB) production in Chapter 6. Laboratory-scale and pilot-scale studies pertaining to the bioconversion of food waste, municipal solid waste, food processing waste, and agriculture residues to biofertilizers, including the practical field applications, has been reviewed in Chapter 7. In Chapter 8, the important role of trace elements (e.g., Fe, Ni, Co) in the methanogenesis step of anaerobic digestion has been discussed from a mechanism and metabolic engineering view point. The application of biochar for enhanced biogas production from the anaerobic digestion of food waste has been presented in this chapter as a case study.

Chapter 9 of Section C introduces the theory of planned behavior (TPB) that provides a theoretical framework to assist in our understanding of the factors influencing behavioral choices. In this chapter, the current implementation of TPB to predict food consumption pattern and to promote safe food handling and food-waste recycling in household and commercial sectors are discussed. In Chapter 10, an overview of chitin, chitosan, its properties and applications, metabolic pathway of chitin and chitosan, sources of chitin such as crustaceans, insects, and fungi, extraction methods and bioreactor configurations for chitosan production has been reviewed.

In Section D, the significant applications of castor plant (*Ricinus communis*) for the production of biofuels (bioethanol, biomethanol) and biochemicals (biophenolics) as well as the production of derivatives such as sebacic acid and ricinoleic acid from castor oil has been demonstrated in Chapter 11. In Chapter 12, the feasibility of biofuel production from nonedible rubber seed oil has been explained in detail. The useful properties of the rubber seed oil make it similar to well-known linseed and soybean oil. As the demand for biodiesel is increasing, the biorefinery approach in the field from rubber seed would be of added advantage. In another approach, the different waste carbon sources and related case studies for biodiesel production has been presented in Chapter 13. Meanwhile, in Chapter 14, the production and the application of biodiesel obtained from various plant species to run the engine and the effect of different biodiesel blends on the performance of the engine has been discussed. Additionally, the chapter also covers aspects related to the life cycle and cost-benefit analysis of biodiesel.

In Section E, Chapter 15 explores the possible application of sewage sludge for material and energy recovery through integrated thermochemical and biochemical conversion processes in a sewage sludge biorefinery. Section F covers chapters related to modeling and LCA. In this section, Chapter 16 highlights the application of multiscale models that range from molecular-level understanding of the biorefinery to a system-scale optimization of processes and product distribution. An overview of the different modeling approaches that shaped the current state of biorefineries, the procedure involved in selecting an appropriate model that is specific to the application, and a generic guideline has been presented in this chapter. In Chapters 17, 18, and 19, the application of LCA as a practical and methodological tool for the environmental characterization of a biorefinery has been presented. Accordingly, biorefineries present a favorable environmental profile in comparison with fossil-based reference systems, even though the results show great variability attributed mainly to the biorefineries configuration and complexity. Specifically, Chapter 18 also highlights the application of LCA, conventional macroscale management strategies, and laboratory-scale valorization techniques for a food-waste biorefinery. In Chapter 19, a summary of studies focusing on the LCA of waste biorefineries is presented.

In Section G, Chapter 20 provides information on the application of a systems dynamics approach to understand the relationship between the behavior of a system over time and its underlying structure. The chapter also addresses the various environmental issues and presents a comprehensive literature review on wood and yard waste management and the implementation of a systems dynamics approach in the stream of municipal solid waste and construction and demolition waste. In Chapter 21, the application of LCA in evaluating the carbon footprints of waste-to-biofuel systems has been explained in detail. The greenhouse gas emissions associated with the processes are also presented in this chapter with the identification of the carbon emission hotspots.

Section H deals with different case studies related to biorefineries. Chapter 22 presents case studies from Germany that are related to the simultaneous production of food and feed, materials, and energy in accordance to a cascading use of biogenic feedstocks as recommended by the German Bioeconomy Society. A pulp- and paper-industry case study from India has been discussed in Chapter 23, and the feasibility of integrating biochemical and thermochemical processes in a paper and pulp waste biorefinery to produce value-added chemicals, fuel, and energy has been demonstrated. In Chapter 24, several successful case studies such as landfill gas recovery from the retrofitted landfills, conversion of food waste and sewage sludge to biogas, and industrial symbiosis between a paper mill and zinc smelter have been demonstrated as pathways toward integrated biorefineries. Finally, in Chapter 25, the case study of a tannery is presented, and the most recent technologies to treat the wastewater discharged from tanneries is discussed. Options for resource recovery (e.g., by composting of solid wastes) and substitution of chromium and sodium sulfide are also presented as cleaner production options for tanneries.

The individual chapters of this book focus on the application of different biorefinery concepts in practice (i.e., at the lab, pilot, semiindustrial, and industrial scales), provide options for enhanced resource recovery from wastes (solid, liquid, and gaseous forms), and analyze the supporting tools and techniques for monitoring the performance of biorefineries. This book will serve as a useful resource for chemical engineers, environmental engineers, biotechnologists, researchers, and students studying biomass, biorefineries, and biofuels/products/processes, as well as chemists, biochemical engineers, and microbiologists working in industries and government agencies. We strongly hope that readers enjoy reading this book and find it of immense use.

We wish to thank and express our appreciation to the multidisciplinary team of authors for discussion and communication—above all, for their scientific contribution to this book. We also thank reviewers whose suggestions greatly helped to improve the quality of chapters. Our sincere thanks are due to Elsevier team comprising of Dr. Kostas Marinakis, Senior Acquisition Editor; Emerald Li, Editorial Project Manager; Mr. Selvaraj Raviraj, Project Manager; and their production and typesetting teams for supporting us constantly during the editorial process. We firmly believe that the information contained in this book will enhance the interdisciplinary scientific skills of readers while also deepening their fundamental knowledge on waste biorefinery.

Editors

Thallada Bhaskar

CSIR-Indian Institute of Petroleum, India

E-mail: thalladab@yahoo.com

Ashok Pandey

CSIR-Indian Institute of Toxicology Research, India

E-mail: ashokpandey1956@gmail.com

Eldon R. Rene

IHE Delft Institute for Water Education, Netherlands

E-mail: e.raj@un-ihe.org

Daniel Tsang

Hong Kong Polytechnic University, Hong Kong

E-mail: dan.tsang@polyu.edu.hk

SECTION A

MSW based biorefineries

CHAPTER 1

Production of electricity and chemicals using gasification of municipal solid wastes

Greg Perkins[1,2]
[1]*Martin Parry Technology, Brisbane, QLD, Australia;* [2]*School of Chemical Engineering, University of Queensland, Brisbane, QLD, Australia*

1.1 Introduction

The world currently generates 2.0 billion tonnes of municipal solid waste (MSW) each year and this value is expected to increase to 3.4 billion tonnes annually by 2050 [1]. On average 0.75 kg of waste is produced per capita per day, with national values varying from 0.11 to 4.54 kg per capita per day. Recyclables such as paper, cardboard, plastic, glass, and metals constitute a substantial fraction of the waste generated, ranging from 16% in low-income countries to about 50% in high-income countries. About ~30% of the waste is organic and can potentially be composted. In accordance with the waste hierarchy it is desirable to first reduce, reuse or recycle waste streams. The degree and sophistication of reuse and recycling programs varies widely around the world, driven by government policy, available infrastructure, local attitudes and incomes. Most recycling technologies require the waste to be sorted, which can be labor intensive and expensive, though automated systems have improved significantly over the past 15 years. As a result, recycling systems are not always available and large volumes of waste are currently disposed of in landfill. Most landfill waste contains a considerable fraction of combustible residuals that may be converted into electricity, heat and chemicals using thermo-chemical processes. Even when advanced waste schemes that separate the recyclables and organics from MSW are applied, there is still a residual portion of ~30%, such as contaminated paper and plastics, that cannot be recycled, and are preferably converted into energy or chemicals rather than becoming landfill.

Fig. 1.1 shows the distribution of waste disposal and treatment technologies utilized in each region of the world and in Japan and Sweden. In low-income areas, open dumps are not uncommon, while in some developed countries like United States, Canada, and Australia landfills can take over 50% of the waste that is disposed of. Many Western

Waste Biorefinery. https://doi.org/10.1016/B978-0-12-818228-4.00001-0

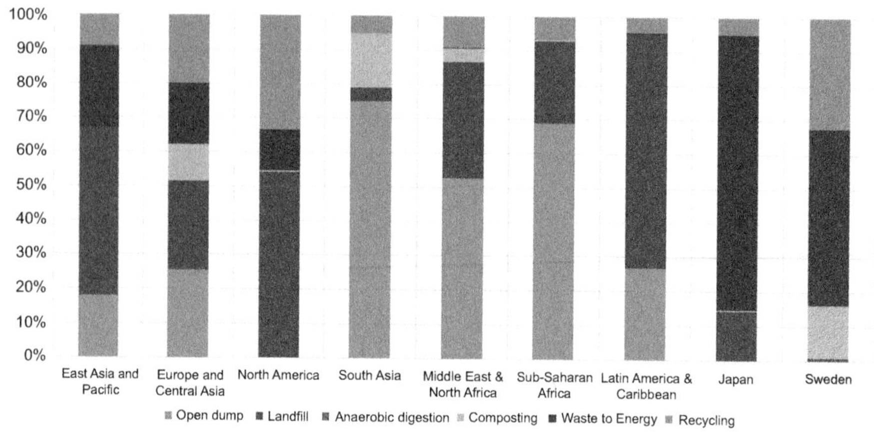

Figure 1.1

Distribution of waste disposal and treatment for selected regions and countries. *Data from Kaza S, Yao L, Bhada-Tata P, Van Woerden F. What a waste 2.0: a global snapshot of solid waste management to 2050. Washington, DC, USA: International Bank for Reconstruction and Development, The World Bank; 2018.*

European countries like Sweden have high rates of recycling and composting and send almost all the residual combustible material for incineration. Japan has a high reliance on waste-to-energy technologies such as incineration and gasification. In most jurisdictions there remains a significant opportunity to increase recycling rates, organic composting and deployment of waste-to-energy facilities as an integrated system to divert waste from landfill. Ideally, waste-to-energy facilities are used only to derive value from the waste components that cannot be recycled or used for composting (such as food organics). This includes the combustible residuals from MSW and also various commercial and industrial waste streams that are routinely sent to landfill. For example, fiber board from housing and automotive shredder residuals from used cars.

The most widely adopted waste-to-energy technology is incineration (combustion or mass burn), in which the waste is combusted in a boiler to generate steam which turns a turbine and electrical generator to make electricity. The principles of waste incineration are the same as conventional coal fired power plants, however the combustion methods, boiler configuration and flue gas treating systems are adapted for the properties of waste, namely its heterogeneous nature and the presence of a wide range of contaminants.

While waste incineration is mature and many technologies have been commercially proven, the process has some downsides. These include low power generation efficiency in comparison with conventional coal and biomass power plants, residual mineral matter and ashes which may not be suitable for landfill (depending upon environmental policies) and the requirement for large scale to achieve low operating costs. Waste incineration is only viable when plant capacities are significantly greater than ~150 ktpa, which requires large volumes of wastes to be transported to a central facility, sometimes creating logistic and contractual challenges.

Gasification is a thermochemical process like combustion undertaken at high temperatures, typically 800–1200°C. However, in gasification the amount of oxygen is controlled to be below the stoichiometric amount required for complete combustion of the fuel, thereby producing a synthesis gas (syngas) which contains up to 80% of the energy in the feedstock as chemical energy. The syngas being predominately CO and H_2 can be used in a range of applications including steam boilers, gas engines and gas turbines for electricity generation and for synthesis of chemicals, such as methanol, ethanol and jet fuels in catalytic reactors. Gasification is not a new technology and has been successfully deployed to produce syngas from coal, heavy oil, petroleum coke, natural gas and biomass for making electricity, hydrogen, fuels and chemicals as final products. While it is not well known, gasification of wastes has also been commercially proven, mostly in Japan and South Korea where diversion from landfill and generation of an inert vitrified slag from the waste are the main incentives for applying the technology. In the waste management literature, gasification is often classified as an advanced thermal technology (ATT).

To mitigate the impacts of climate change and reduce the impacts of waste disposal it is desirable to achieve a circular economy as shown in Fig. 1.2, in which products are preferentially reused, remanufactured and recycled and end of life materials are converted

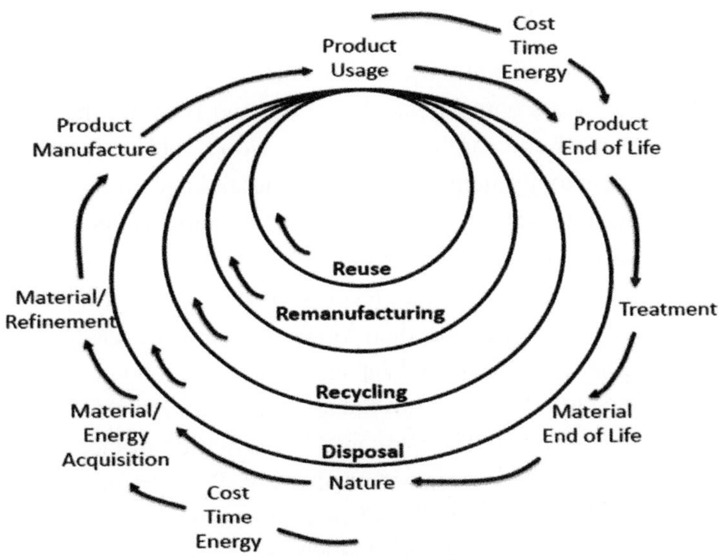

Figure 1.2
Schematic of the circular economy. *From Korhonen J, Honkasalo A, Seppälä J. Circular economy: the concept and its limitations. Ecological Economics 2018;143:37–46. doi:10.1016/j.ecolecon.2017.06.041.*

back into valuable products with a minimal release of waste to the environment. In the circular economy, gasification provides a flexible platform for recycling carbon and hydrogen molecules back into products, such as chemicals and plastics, the recovery of metals for reuse, and the transformation of inorganics into inert products for use in construction applications. This chapter provides an overview of gasification to produce electricity and chemicals from municipal solid waste (MSW), which can be considered the first steps toward the goal of using gasification to completely recycle waste atoms within the circular economy. Firstly, the fundamentals are summarized along with a review of waste gasification technologies. Commercial systems are described along with the material balances, economics and environmental impacts of several technologies. Finally, opportunities for using gasification to improve waste management are briefly described.

1.2 Fundamentals of MSW gasification

The main motivations for applying gasification for the conversion of wastes include generation of a high-quality energy carrier, ability to clean syngas prior to utilization, limiting dioxin/furan formation by operating under reducing conditions and flexibility to utilize syngas to produce electricity, hydrogen and chemicals. The major challenges with waste gasification have traditionally been associated with feeding materials of variable size and heterogeneity, achieving reliable gas clean up and overcoming the poor economics of projects at small scale (<100 ktpa of waste processed).

1.2.1 Characterization of MSW

The properties of MSW are significantly different from coal and biomass due to MSW being a mixture of various waste components. Fig. 1.3A and B shows the main constituents of MSW which includes paper and cardboard, plastics, wood, rubber, food and green waste, glass, metals and other commodity items from household garbage. The composition of MSW varies from one place to another, reflecting local consumption preferences, waste management policies and socioeconomic conditions. According to the World Bank, MSW from higher socio-economic areas has higher levels of paper, cardboard, plastics, metal and glass and lower amounts of food organics, than MSW from lower socioeconomic areas [1]. For example, in North America, Europe and Japan, most of the food consumed is supplied in packaging, whereas in South Asia and Sub-Saharan Africa, a smaller proportion of the food consumed comes packaged [1]. The per capita consumption of electronic appliances is another factor which impacts on the composition of waste. Waste regulations and collection services also impact on where waste is sent and what the composition of each stream is. For example, Sweden has implemented a scheme that requires residents to separate their waste by type, which maximizes reuse and recycling, with only residual wastes like diapers, nonpackaging materials and wood sent to energy from waste facilities.

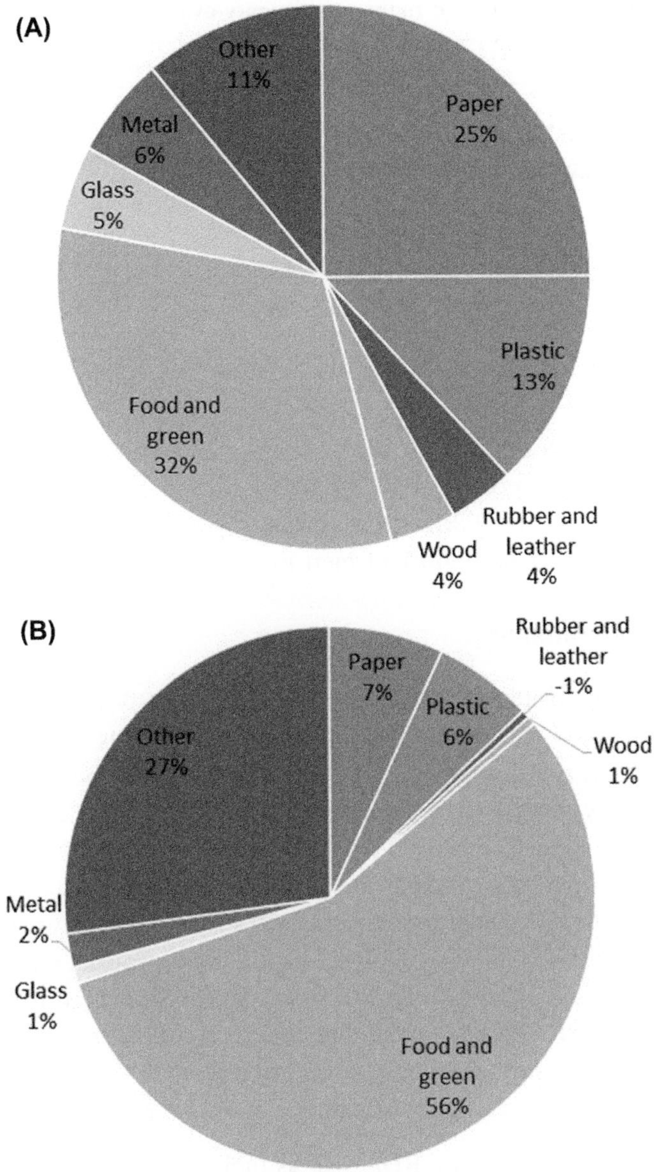

Figure 1.3

Typical composition of MSW by constituent for: (A) high and (B) low incomes. *From Kaza S, Yao L, Bhada-Tata P, Van Woerden F. What a waste 2.0: a global snapshot of solid waste management to 2050. Washington, DC, USA: International Bank for Reconstruction and Development, The World Bank; 2018.*

Table 1.1 shows the ultimate analysis of a typical MSW and its main constituents. The presence of PVC means that MSW has a much higher chlorine content than bio-derived materials, such as paper and food wastes. The amounts of sulfur in MSW can also be

Table 1.1: Ultimate analysis and main constituents of municipal solid waste (wt%).

Component	Paper	Plastic	Other	Food and green wastes	MSW
Carbon	27.40	63.57	24.92	27.00	29.60
Hydrogen	3.76	12.00	3.14	4.00	4.64
Oxygen	25.60	9.02	15.06	25.00	18.94
Nitrogen	0.16	0.90	0.73	1.00	0.67
Chlorine	0.27	3.38	0.41	0.40	0.78
Sulfur	0.16	0.34	0.18	0.02	0.12
Moisture	36.80	4.2	33.40	30.00	25.07
Mineral matter	5.85	6.59	22.16	12.58	10.19

From Mastellone ML. Waste management and clean energy production from municipal solid waste. New York, NY, USA: Nova Publishers; 2015.

significant, varying from about 0.1 to 0.5 wt%. The mineral matter from food and green wastes can have substantial amounts of alkalis. The moisture content of MSW is generally high compared to coal and can be difficult to reduce [4]. MSW may contain up to 40%−50% of biodegradables and 30%−40% of inert materials, but this depends on location.

1.2.2 Feedstock pretreatment

A variety of feedstock pretreatment steps may be applied to MSW to separate materials such as metals and glass for recycling and to produce residual fuels that have properties suitable for waste-to-energy technologies. The two most common approaches are (1) a material recovery facility (MRF) and (2) mechanical biological treatment (MBT). In both approaches, mechanical separation is used to remove metal, glass, paper and other recyclables and the carbonaceous residuals are shredded. Typically, it requires 80−100 kWh to process each tonne of MSW and a further 100−130 kWh to dry the waste [5]. In MBTs, the organic fraction is further processed by biological microorganisms via composting, anaerobic digestion and/or biodrying. Two fuel products can be derived from wastes - RDF (refuse derived fuel) and SRF (solid derived fuel). The main difference between RDF and SRF, is that SRF must meet stricter criteria on heating value (17−22 MJ/kg), moisture content (<15 wt%) and contaminant levels (chlorine < 0.9 wt%, sulfur < 0.5 wt%).

Most gasification technologies require RDF/SRF as fuel with well-defined properties, including size, morphology, composition and calorific heating value, though some can handle raw MSW without any pretreatment.

1.2.3 Gasification reactions

Waste gasification is usually undertaken auto-thermally, that is, oxygen is used in the gasification agent to generate exothermic heat release in order to reach temperatures sufficiently high (>800°C) that the gasification reactions convert the feedstock into CO and H_2. The amount of oxygen injected is limited to about 20%–40% of stoichiometric amount required to completely combust the feedstock, i.e., the equivalence ratio is ~0.2–0.4. The main reactions occurring in gasification are shown in Table 1.2.

Upon entering the gasification process the feedstock is dried by reaction (R1). If the feedstock has very high moisture it may be reduced in a separate drying stage prior to the gasification step. However, whatever moisture remains in the feedstock is driven off as the material is heated upon entering the gasification reactor.

At higher temperatures (300–800°C) the feedstock is thermally decomposed by pyrolysis reactions (R2), whereby some of the long chain molecules are broken to form smaller molecules, such as CO, H_2, CH_4, C_2H_6, along with oxygenated and nonoxygenated hydrocarbons called tars. For MSW, which is a mixture of paper, plastics, biomass and food organics, a very wide range of molecules are generated during the pyrolysis phase by a series of complex reactions. Generally, pyrolysis can be modeled simply with a two-step process—where the first step is the decomposition to form char, light gases and tars, and the second step is the conversion of tars into light gases and smaller tar molecules [8].

The tars play an important role in the design and operation of all practical gasification systems and may be characterized by a variety of methods, including by their dew point. Tar concentration can vary from ~1–180 g/Nm^3 [9]. Formation of tars is undesirable

Table 1.2: Main reactions involved in combustion and gasification reactions.

Reaction		Stoichiometry	ΔH^0_{298} (kJ/mol)
R1	Drying	Feedstock → Dry Feedstock + H_2O	+20
R2	Pyrolysis	Dry Feedstock → Char + Volatiles	~0
R3	Char combustion	$C + \frac{1}{2}O_2 \rightarrow CO$	−111
R4	Char combustion	$C + O_2 \rightarrow CO_2$	−393
R5	Steam gasification	$C + H_2O \leftrightarrow H_2 + CO$	+131
R6	Boudouard reaction	$C + CO_2 \leftrightarrow 2CO$	+172
R7	Hydrogasification	$C + 2H_2 \leftrightarrow CH_4$	−75
R8	Combustion	$H_2 + \frac{1}{2}O_2 \rightarrow H_2O$	−242
R9	Combustion	$CO + \frac{1}{2}O_2 \rightarrow CO_2$	−283
R10	Combustion	$CH_4 + 2O_2 \rightarrow CO_2 + 2H_2O$	−802
R11	Combustion	$C_nH_{2n+2} + 2nO_2 \rightarrow nCO_2 + 2nH_2O$	
R12	Water gas shift	$CO + H_2O \leftrightarrow CO_2 + H_2$	−41
R13	Steam-methane reforming	$CH_4 + H_2O \leftrightarrow CO + 3H_2$	+206
R14	Steam reforming	$C_nH_{2n+2} + nH_2O \leftrightarrow nCO + (2n+1)H_2$	
R15	Dry reforming	$C_nH_{2n+2} + nCO_2 \leftrightarrow 2nCO + (n+1)H_2$	
R16	Tar reforming/cracking	$Tar + H_2O \rightarrow \alpha CO + \beta CO_2 + \gamma H_2 + \delta CH_4$	

Adapted from Perkins G. Underground coal gasification – part II: fundamental phenomena and modeling. Progress in Energy and Combustion Science 2018;67:234–74. doi:10.1016/j.pecs.2018.03.002. Perkins G. Mathematical modelling of in situ combustion and gasification. Proceedings of the Institution of Mechanical Engineers, Part A: Journal of Power and Energy 2017. doi:10.1177/0957650917721595. pii:095765091772159.

since upon cooling they can severely foul up equipment and must be removed before the syngas can be utilized in engines, turbines or to synthesize other chemicals using catalysts. At high temperatures and sufficient residence times, the tars can be cracked and reformed into smaller molecules which do not cause problems. Each gasification system handles tars differently. In some designs, the syngas is produced at very high temperatures to avoid tar formation, while lower temperature processes will include specific unit operations to react or remove tars from the syngas. In many plants, the syngas is simply combusted in a boiler, which avoids the need to remove tars from the syngas. However, each of these "solutions" has pros and cons, which will be discussed throughout this chapter.

The oxygen injected into the gasifier reacts exothermically with char according to the reactions (R3 and R4):

$$C + \frac{1}{2}O_2 \rightarrow CO \tag{1.1}$$

$$C + O_2 \rightarrow CO_2 \tag{1.2}$$

At low temperatures (<1100°C), CO_2 is the primary product, while at high temperatures (>1600°C) CO is the primary product. At low pressures typical of waste gasification, the reaction is first order with respect to O_2. The steam gasification reaction (R5) is an important endothermic reaction:

$$C + H_2O \leftrightarrow H_2 + CO \tag{1.3}$$

and is also first order with respect to H_2O at low pressures, wherein the partial pressure of H_2O and H_2 are low. The Boudouard reaction (R6) is also endothermic:

$$C + CO_2 \leftrightarrow 2CO \tag{1.4}$$

The methanation reaction (R7) involves the reaction of char with H_2 to form CH_4. At high pressures (> ~30 bar), such as in coal gasification, the methanation reaction can become significant, however in waste and biomass gasification where the partial pressure of H_2 is low, the overall reaction rate is negligible. The majority of CH_4 present in the syngas generated from waste and biomass gasification is generated from the pyrolysis and thermal decomposition of the feedstock at lower temperatures by reaction (R2).

The products from the char gasification reactions may be combusted to form CO_2 and H_2O via reactions (R8)–(R11). The literature on the combustion of gaseous fuels is extensive (see for example Westbrook et al. [10,11], Jones and Lindstedt [12] and Ranzi et al. [13]). A generic one-step model of gaseous fuel combustion is:

$$C_nH_{2n+2} + 2nO_2 \rightarrow nCO_2 + 2nH_2O \tag{1.5}$$

where coefficient n is determined by the choice of fuel. The water gas shift reaction (R12) is important since it can control the ratio of CO and H_2 in the syngas product:

$$CO + H_2O \leftrightarrow CO_2 + H_2 \quad (1.6)$$

Thermodynamics favors the production of CO at high temperatures (>1000°C) and CO_2 at low temperatures (<1000°C). The reforming reactions (R13, R14, R15) break down larger molecules into CO and H_2 and are generally of less importance in waste and biomass gasification.

Complete conversion of the char formed from the pyrolysis of the feedstock into syngas is desired in the gasifier. The char reactions with H_2O (R5) and CO_2 (R6) are substantially slower than the combustion reactions and hence control the overall conversion. Thus, the chemical reactivity of the char formed from pyrolysis is an important parameter for the design of the gasification reactor. The most significant factors which control the overall reactivity of carbonaceous solids to H_2O, CO_2 and H_2 are known to be: (1) concentration of active sites, (2) presence of inorganic impurities which act as catalysts, and (3) diffusion limitations which control the rate at which reactive gases can reach the active sites [14]. The steam gasification (R5) and the Boudouard (R6) reactions are of similar magnitude, with steam gasification being approximately three times faster than the Boudouard reaction under the same conditions. Waste and biomass feedstocks contain mineral matter components that catalyze the gasification reactions. Most metals, metal oxides and salts act as catalysts, with the major catalysts being compounds of iron, magnesium, calcium and potassium.

1.3 Waste gasification technologies

1.3.1 Types of gasification reactors

Gasification has long been used for converting coal and biomass into synthesis gas and mature technologies exist for contacting the feed with the gasification agent using moving beds, fluidized beds and entrained flow reactors [15,16]. Fig. 1.4 shows schematics of the major types of conventional gasification reactors in use today, which can be classified based on how the solid feedstock moves through the reactor. In updraft and downdraft gasifiers a dense porous bed is formed by the feedstock, while in fluidized-bed gasifiers, the feedstock is partially entrained by the gasification agent (e.g., air) creating a turbulent bed of material. In entrained flow gasifiers the feedstock must be pulverized to a small size so that it can be conveyed into the reactor with the gasification agent.

Waste feedstocks, and MSW in particular, are very different to coal and biomass, being heterogeneous in size and composition, with generally higher mineral matter content, relatively low calorific value and a wide range of contaminant molecules such as sulfur, chlorine and heavy metals. The heterogeneous nature of waste feedstocks poses one of

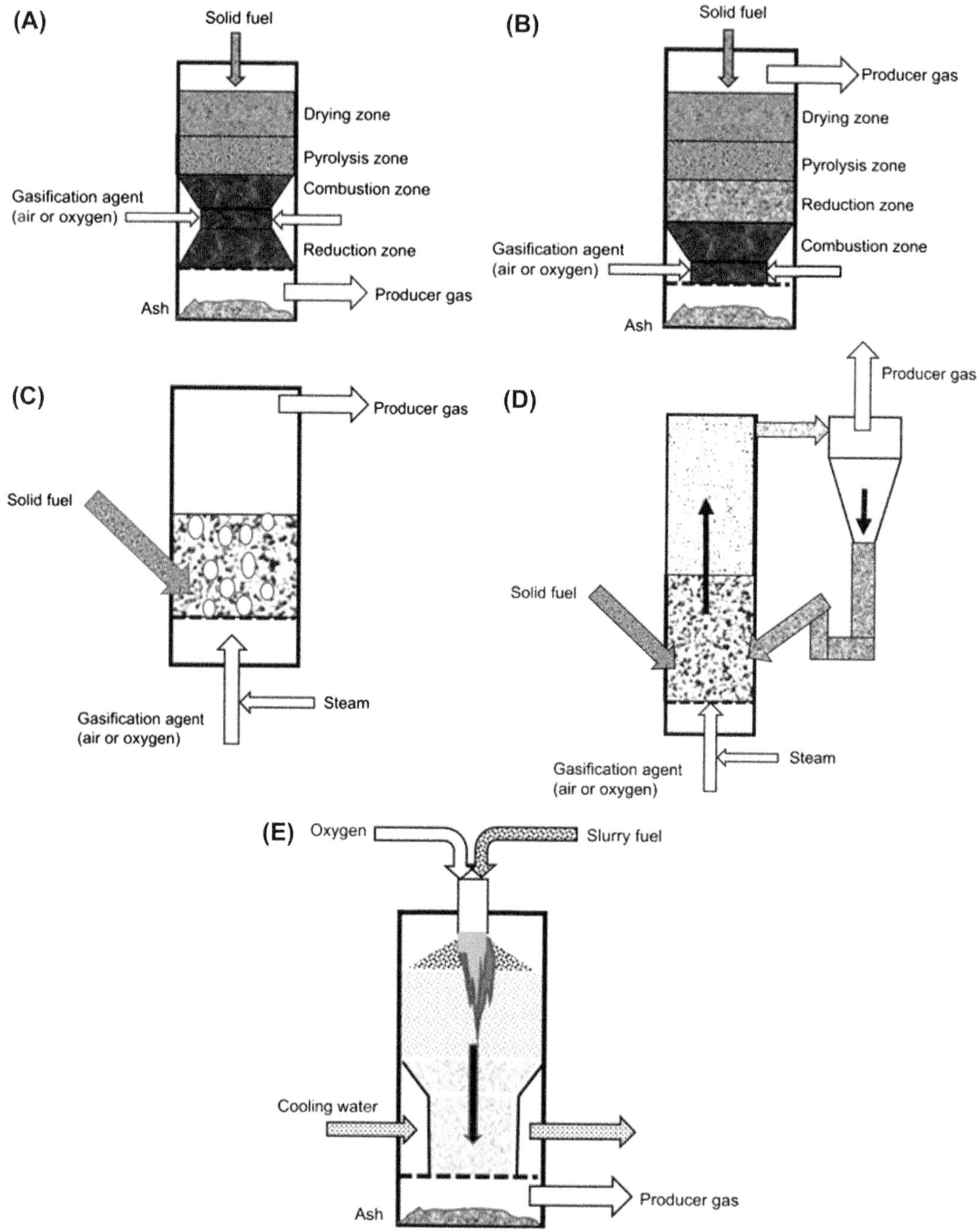

Figure 1.4

Major types of gasification systems for coal and biomass feedstocks: (A) downdraft moving bed, (B) updraft moving bed, (C) bubbling fluidized-bed, (D) circulating fluidized-bed and (E) entrained flow. *From Pang S. Fuel flexible gas production. In: Fuel flexible energy generation. Elsevier; 2016. p. 241–69. doi:10.1016/B978-1-78242-378-2.00009-2.*

the most difficult challenges for conventional gasification reactors which were originally developed for a well-defined particle size and only small variations in chemical composition. Since MSW is a mixture of different materials, including glass and metals, entrained flow reactors are not favored, because the waste cannot be easily milled to achieve the very small particle sizes required for entrainment in a gas. While some fluidized-bed technologies have been developed and adapted for processing wastes, a pretreatment step is invariably required to transform the raw MSW into RDF or SRF. Conventional moving bed reactors in which the ash is removed as a solid from a rotating grate have not been commercialized for raw MSW. However, some conventional moving bed reactor designs have been adapted to process RDF and SRF. Moving bed gasifiers which melt the mineral matter can convert raw MSW without any significant pretreatment step at all. The grate gasifier is a variant of the moving bed, adapted from combustion experience, which may also be used for the conversion of MSW.

Another important difference between waste and biomass gasification and the gasification of coal, petroleum coke and oil, is the operating pressure. Due to processing scale and the requirement for syngas at elevated pressure for chemicals synthesis, gasification of fossil fuels is usually undertaken at moderate to high pressures (20–80 bar). However, waste gasification is generally performed at low pressure (1–5 bar) since feeding heterogeneous feedstocks at pressure has not been commercially proven.

1.3.1.1 Moving bed reactors

In the moving bed variants designed for MSW, the feed is injected from the top or side and is gasified by air or oxygen injected via tuyeres at the side of the vessel. Syngas leaves from the top of the reactor at temperatures between about 900 and 1200°C. The bottom of the reactor is often further heated well above the ash fusion temperature to form a molten liquid from the mineral matter. Temperatures are above 800°C and the liquid slag is removed periodically from the reactor via tap holes. The moving bed waste gasifiers are therefore more similar to metallurgical reactors such as the blast furnace than to the traditional updraft fixed bed coal gasifiers, like the Lurgi gasifier [18]. To reach the high temperatures and achieve consistent operation with variable quality feedstocks, the waste may be complemented with a fossil fuel such as coke or natural gas. This of course aids process stability, but also incurs additional operational costs.

Fig. 1.5 shows the Nippon direct melting system moving bed gasifier for MSW. In this design oxygen-enriched air (36% O_2) is injected at the bottom of the gasifier via tuyeres and the waste feedstock slowly descends through the main zones of the reactor, being drying and preheating zone, thermal decomposition zone, combustion and melting zone. The combustible waste is gasified in the thermal decomposition zone and the syngas leaves from the top of the reactor. In the combustion zone added coke burns

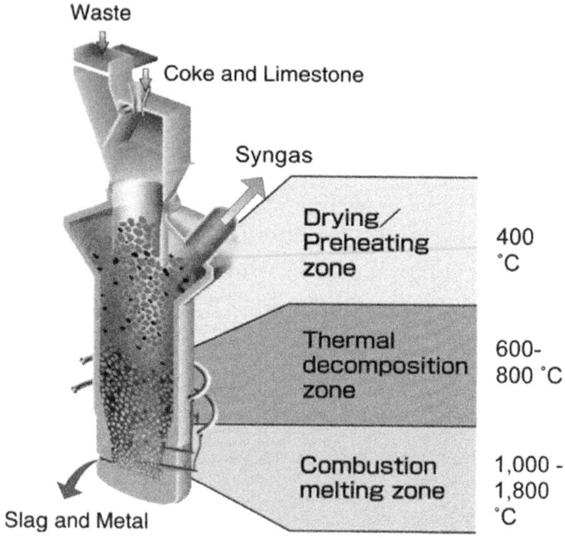

Figure 1.5

Nippon direct melting system moving bed gasifier. *From Tanigaki N, Manako K, Osada M. Co-gasification of municipal solid waste and material recovery in a large-scale gasification and melting system. Waste Management 2012;32:667–5. doi:10.1016/j.wasman.2011.10.019.*

and melts the incombustible portion of the waste feedstock at temperatures up to 1800°C, with molten slag being periodically discharged from a tap hole at the bottom of the reactor.

In some moving bed designs energy is added to the reactor using a plasma arc to reach very high temperatures so that the carbonaceous materials are decomposed predominately into CO and H_2. Plasma is fourth state of matter and is created by the exposure of gas to high temperatures to create ionization of gases [21]. The plasma is formed using torches and peak temperatures may reach over 5000°C in parts of the reactor. Plasma gasifiers are less dependent upon the composition of the waste feedstock, as additional energy is added via the plasma to regulate the operational conditions inside the reactor. The generation of the plasma requires electrical energy and presents an auxiliary load on the overall plant. One report states that the plasma torches require about 5%–10% of the energy in the feedstock, or equivalently ~15%–20% of the gross electrical output [22], while Alter NRG states that only 2%–5% of the energy in the feed is required [23]. Fig. 1.6 shows a schematic of a moving bed plasma reactor.

The consequences of the high temperatures and the presence of highly corrosive syngas and molten liquids means that the reactors must be refractory lined and many also require elaborate cooling systems to keep the steel parts of the reactor vessel within an acceptable temperature range. This makes the reactor vessels expensive to construct. Also, critical

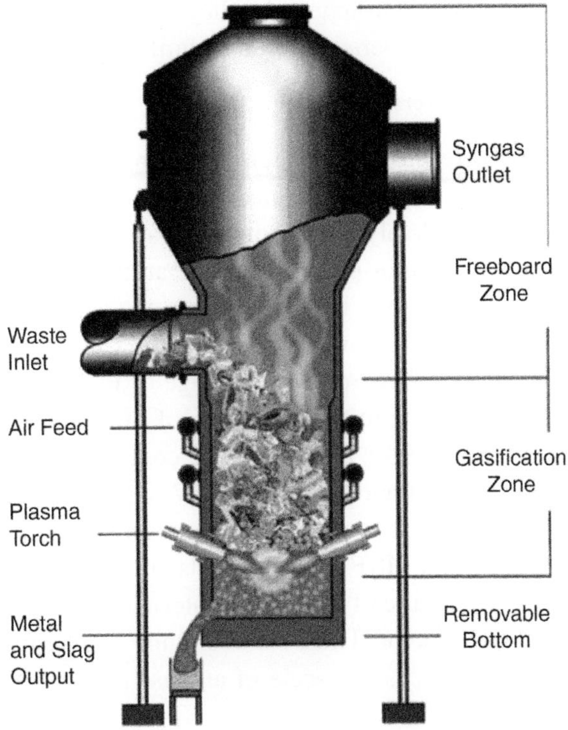

Figure 1.6
Plasma gasification reactor. *From van Nierop P, Sharma P. Plasma gasification — integrated facility solutions for multiple waste streams. In: Gasification technologies conference, Washington, DC; 2010.*

parts of the reactor may have relatively short lifetimes, which impacts on operating and maintenance planning and the configuration of the overall plant. For example, critical parts of the refractory lining in the Nippon direct melting system have a lifetime of about 12 months and requires a major shutdown of the gasifier to be replaced. Thus, to achieve high overall plant availability, multiple lines/trains of equipment are required to mitigate regular maintenance shutdowns.

All of the commercialized moving bed reactors for MSW gasification heat the mineral matter such that it forms a molten slag and is quenched and vitrified, rendering the ash inert and reducing the volume of residuals that need to be disposed of. This is a major advantage of these designs. However, reaching the high temperatures has a cost, either via the use of oxygen-enriched air and the addition of supplementary fuel such coke or natural gas, or the use of some of the generated electricity to create a plasma arc. The moving bed melting gasifiers have been commercialized almost exclusively in Japan, where the main drivers have been to reduce the quantity of residuals, vitrify the ash and recycle it for construction applications and where optimizing the efficiency of converting waste-to-energy has been a lower priority. Environmental policy has limited the recycling of bottom-ashes and fly-ashes from incineration without further treatment [24].

1.3.1.2 Bubbling fluidized-bed reactors

Bubbling fluidized-bed (BFB) reactors have been commercialized for a wide range of gas/solid reactions including combustion and gasification of coal, biomass and wastes; catalytic cracking of oil and the pyrolysis of biomass. Gas is injected at the bottom of a cylindrical vessel through a distributor plate to fluidize an inert bed of fine particulate material such as sand and the feedstock is injected from the sides or top of the reactor. When the gas flow reaches a superficial velocity of over ~1 m/s the bed materials are expanded forming a fluidized-bed. The major advantage of BFBs is that they achieve very uniform reaction conditions due to the high rates of mixing and heat and mass transfer. The feedstock particle size for BFBs must be controlled, therefore a pretreatment step is required to sort and size raw MSW to between 30–150 mm in size before it can be used as a feedstock for a fluidized-bed gasifier. Generally, the processing of MSW requires the building of a MRF so that metals and recyclables are removed from the feedstock, and the residual carbonaceous materials are reduced in size acceptable to the fluidized-bed reactor. Thus, for economic feasibility, the use of a BFB on MSW requires a significant processing scale of at least 100 ktpa, and preferably above 200 ktpa.

While bubbling fluidized beds enable good mixing of the feedstock and reactants, they generally have relatively low syngas production temperatures in the range of 750–950°C. As such, the lignocellulosic materials in the feed produce substantial amounts of tar in the syngas which must be removed for any downstream application other than combustion of the syngas in a boiler. Tar concentrations from BFBs are in the range of 10 g/Nm^3 [25]. Also, because of the low gasification temperature, the carbon conversion in the BFB is relatively low at about 70%–90%. Unconverted char must be separated from the gas flow and recycled back to the reactor to achieve an overall carbon conversion of >95%.

Enerkem has developed a BFB reactor using steam and oxygen for the gasification of sorted and shredded MSW [26], while Thermochem Recovery International has developed an indirectly heated steam reforming gasifier for converting biomass and sorted MSW wastes into a hydrogen rich syngas [27]. For unsorted waste feedstocks, Ebara has developed an internal circulating fluidized-bed which includes an ash melting furnace where the syngas is combusted at high temperature, as shown in Fig. 1.7. The major advantage of BFBs is that they are flexible, and the large thermal mass of the bed means that they can cope with a degree of feedstock variability. Pretreatment and feeding of the feedstock are the main challenges to achieve reliable operations.

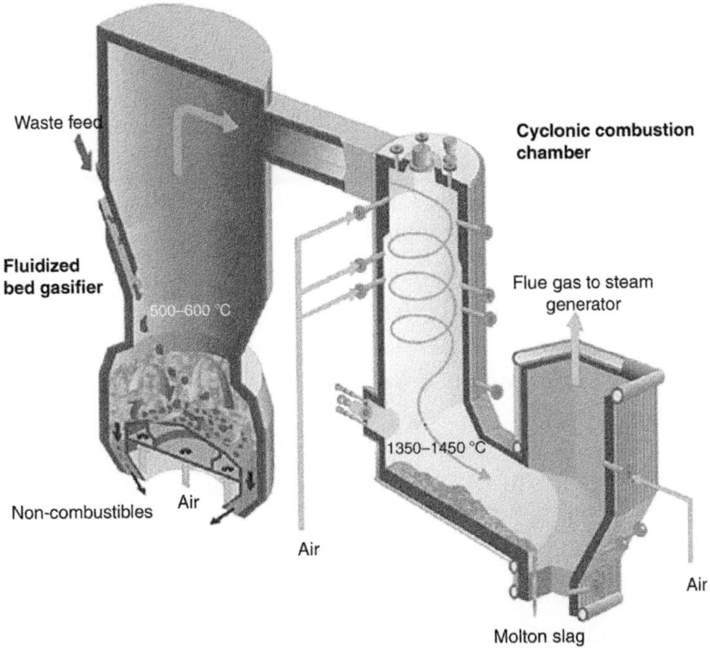

Figure 1.7
Ebara TwinRec internal fluidized-bed gasification process with ash melting furnace. *From Yoshikawa K. Gasification gasification and liquefaction alternatives incineration alternatives to incineration incineration in Japan. In: Kaltschmitt M, Themelis NJ, Bronicki LY, Söder L, Vega LA, editors. Renewable energy systems. New York, NY: Springer New York; 2013. p. 728–43. doi:10.1007/978-1-4614-5820-3_419.*

1.3.1.3 Circulating fluidized-bed reactors

In the circulating fluidized-bed and transport reactors the feedstock and/or a carrier solid are entrained in the flow in the riser section of the reactor, which necessities small particle sizes with a well-defined particle size distribution. Unconverted feed and heat carrier material is separated from the syngas in cyclones and recirculated back to the gasification section. The gas and solids are disengaged (using cyclones etc.) and the inert and unconverted solids are returned to riser via a return leg and loop seal arrangement. Circulating fluidized beds (CFBs) have been used for a variety of applications including coal combustion/gasification, catalytic cracking of oil and pyrolysis of biomass. Fig. 1.8 shows a schematic of a circulating fluidized bed gasifier, which uses air and steam as the gasification agent. CFBs have been developed for capacities spanning the range 20–300 MWth [30].

The advantages of CFBs are fuel flexibility, no moving parts in the reactor and high gasification efficiency. Fuels with calorific values in the range 9–20 MJ/kg including mixtures of origin sorted residential wastes, industrial waste, demolition wood and biomass have been successfully processed in CFBs. The major disadvantage of CFBs is that the feedstock needs to be pretreated to meet tight size specifications prior to use and the high velocities of the solids mean that wear and erosion can be significant. In the designs commercialized by the company Valmet, the product syngas is cleaned before being fed to a gas boiler [30]. In some cases, the gasifier is used to process biomass and waste into syngas to repower an existing coal combustion boiler. Since the syngas has been cleaned, this approach enables high temperature/pressure steam conditions in the boiler, which improves efficiency in comparison to technologies where the waste derived syngas is immediately combusted in a boiler.

1.3.2 Selection of gasification agent

The main oxidizing agents used in gasification systems are air, air enriched with oxygen, steam/air and steam/oxygen mixtures. The gasification agent will primarily be chosen based on the desired end-product, i.e., electricity or chemicals; and to create the right conditions for residual removal from the reactor, i.e., as dry or slagging mineral matter. In waste gasification the use of oxygen together with air or steam may be required to reach temperatures high enough to melt the ash.

Using air as the gasification agent produces syngas with a calorific value typically in the range of 4–7 MJ/Nm^3 and this is suitable for subsequent combustion in a boiler or when the syngas is suitably conditioned can also be used in gas engines and some gas turbines. When oxygen is used as gasification agent, the calorific value can reach 7–12 MJ/Nm^3 and an air separation unit (ASU) is required to separate oxygen from air. The ASU adds capital cost and consumes energy, reducing the net energy available for export in power plants [31,32]. However, the syngas is not diluted with N_2, leading to smaller equipment for gas conditioning and this makes synthesis into chemicals and synthetic fuels more feasible. Steam/O_2 may be used as the gasifying agent for biomass and wastes, though its use is generally restricted to when the syngas is required for making hydrogen or chemicals and/or when the project scales are relatively large. Indirect gasification using steam is also possible using a single or dual fluidized-bed gasifier [33]. The major advantage of indirect gasification is that a high heating value syngas can be made without the need for an ASU. A prominent example is the biomass gasification CHP plant in Güssing, Austria. However, currently the scale of indirect gasifiers is relatively small and in the context of waste gasification they are limited to processing RDF or SRF derived from MSW [34,35].

1.3.3 Synthesis gas processing

The main impurities present in syngas from biomass and waste gasification are soot, alkali compounds, nitrogen compounds, tars, light and heavy hydrocarbons, sulfur compounds, chlorine, fluorine, dioxins/furans and heavy metals. A major advantage of gasification over

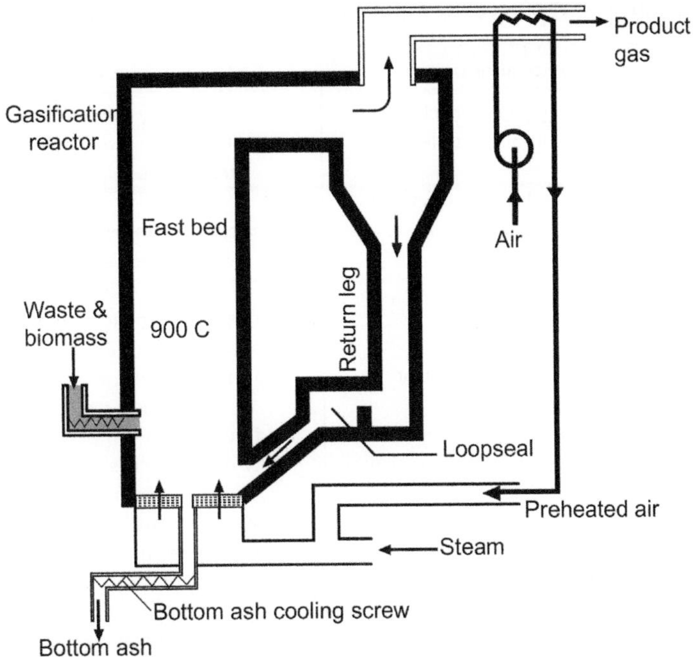

Figure 1.8
Circulating fluidized-bed gasifier. *From Basu P. Biomass gasification, pyrolysis and torrefaction. 2nd ed. London, United Kingdom: Academic Press; 2013.*

combustion is that the quantity of gas to be processed is much smaller and the waste experiences a reducing atmosphere which limits the oxidation of metals and the creation of dioxins. Table 1.3 provides a summary of the most common options for removing the main contaminants in syngas from the gasification of biomass and wastes.[1]

Solids in the form of soot and bed materials (in fluidized systems) can be removed using cyclones, though fine particulates will still remain entrained in the gas. These can be captured in liquid scrubbers if they are used. Filtering the syngas may be applied at high temperatures as is undertaken at the CFB biomass gasifier in Gussing, Austria [36], but is not recommended when the syngas is at low temperature due to tar condensation which can block filters.

Tars, all hydrocarbons with molecular weight above benzene, are one of the most troublesome contaminants in syngas and a range of options have been developed to remove them. Firstly, the tar content and the types of tars present in the syngas are influenced by the design and operation of the gasifier and also the feedstock being

[1] This section is focused on the conditioning of the syngas for beneficial use: the treatment of flue gas from combustion of syngas in a furnace is not discussed.

Table 1.3: Options for removing contaminants from syngas.

Contaminant	Removal options	Comments
Particles, soot	Cyclone (large particles)	Mandatory for fluidized-bed systems.
	Filters (small particles)	Only suitable when syngas at high temperatures, $>$400°C and higher.
	Aqueous/oil scrubber	
Tars	Aqueous scrubber	Aqueous scrubbers not very effective for tar removal and require additional water treatment.
	Oil scrubber	Oil scrubbing such as OLGA proven to be effective with multiple biomass gasifiers.
	Electrostatic precipitator	No moving parts, highly reliable. Wet or dry operation can be considered.
	Thermo-chemical conversion, i.e., combustion, cracking and reforming	Additional high temperature process unit with utility requirements. May be implemented with or without catalysts.
Sulfur (H_2S)	Physical absorption with liquids, e.g., methanol	Not economic at low sulfur capacity ($<$10 tpd) or low partial pressure of H_2S ($<$6.9 bar)
	Physical absorption with solids, e.g., $CaCO_3$	Bulk removal down to 50–100 ppmv.
	Physical absorption with solid metal oxides, e.g., ZnO, CuO, MnO	Fine removal down to ~0.1 ppmv.
	Chemical absorption with liquids, e.g., alkanolamine	Not economic at low sulfur capacity ($<$10 tpd).
	Membrane permeation	Small plant scale with syngas at high pressure.
Organics	Activated carbon	Relatively simple applied in fixed beds.
Chlorine	Dry removal with solid adsorbents	Use of $NaHCO_3$ at temperatures between 400–500°C can reduce HCl $<$ 1 ppmv
	Na-carbonate and Ca-oxide	
	Aqueous scrubber	Water treatment may be expensive.
	Caustic scrubber	Can remove HCl, COS and H_2S.
Ammonia (NH_3)	Aqueous scrubber	Water treatment costs can be prohibitive at small scale.

processed [36]. The tar content from biomass gasification varies from ~10–100 g/Nm^3 for updraft gasifiers, ~1–30 g/Nm^3 for downdraft gasifiers, ~1–50 g/Nm^3 for BFB gasifiers and ~1–10 g/Nm^3 for circulating fluidized-bed gasifiers [37]. For clean biomass feedstocks, such as wood chips, the downdraft gasifier is often preferred for small-scale projects as it can consume over 99% of the tars formed due to the configuration of reactants and products [38].

The tars may be removed with varying degrees of success using a variety of methods. Aqueous scrubbing is most successful for removing the water-soluble tars which are produced from primary pyrolysis in fixed bed gasifiers. However, aqueous scrubbing has the disadvantage of requiring tar separation from the water and further water treatment to meet disposal conditions.

Oil scrubbing is another option and Fig. 1.9 shows a simplified process diagram of the OLGA technology developed by ECN of The Netherlands [25,39]. The philosophy of OLGA is based on dew point control. Heavy tars are condensed in the collector vessel and together with fine solids may be recycled to the gasifier as a liquid. The syngas is then passed through an electrostatic precipitator (ESP) to remove fine aerosols and any remaining particulates. In the second stage, lighter tars are absorbed into the oil, which is regenerated by air or stream in the stripper. The stripping medium saturated with the light tars may also be used as part of the oxidant for the gasifier. The main advantage of the OLGA system over aqueous based systems is that water, tar and particulates are kept separate, the by-product streams may be recycled to the gasifier and the tar dew point of the clean syngas can be very low, below 0°C. The capital costs and operating costs are the main drawback of OLGA, particularly for small plants <5 MWth.

An alternative to water/oil scrubbing is to the selectivity react the tars to form smaller molecules at high temperatures using oxygen and/or steam. This approach is applied in Enerkem's fluidized-bed design, where the syngas from gasification at circa 850°C is oxidized in a tubular reactor at over 1000°C to remove the tar molecules. Catalytic cracking of tars has been studied extensively in the laboratory, but has scarcely been applied at larger scales in biomass or waste gasification systems [36]. Use of ESPs has been successfully applied to remove tars from the syngas in several biomass gasification plants.

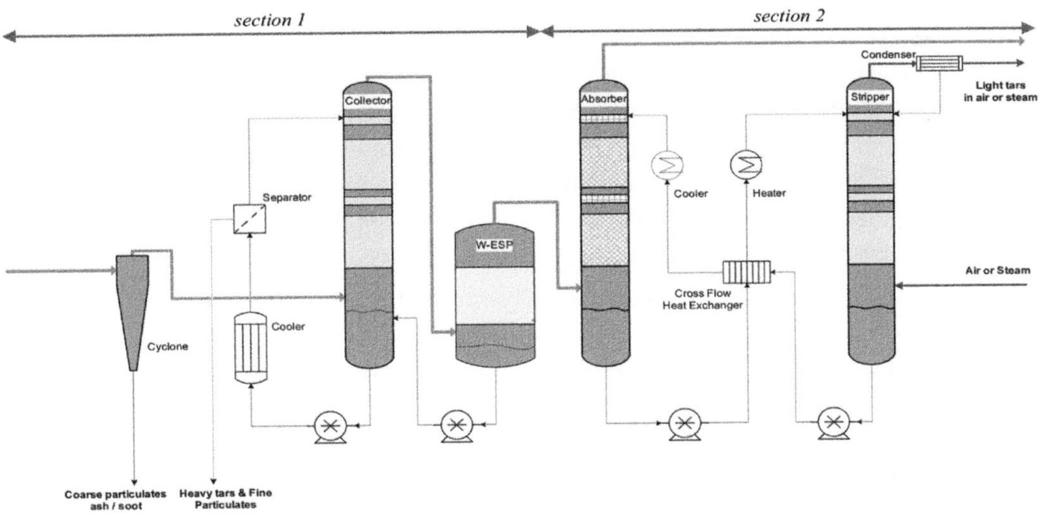

Figure 1.9

Simplified process flow diagram of the OLGA system. *From Panepinto D, Tedesco V, Brizio E, Genon G. Environmental performances and energy efficiency for MSW gasification treatment. Waste and Biomass Valorization 2015;6:123–35. doi:10.1007/s12649-014-9322-7.*

While tar content in the syngas depends heavily gasifier design and operation, the inorganic contaminants depend mostly on the composition of the feedstock being processed. Sulfur is another contaminant which can require special attention, especially if the syngas will be used for catalytic synthesis. Generally, proven approaches with liquids based on physical (e.g., methanol) and chemical (e.g., amine) absorption are not cost effective when the total amount of sulfur to be removed is small (<10 tpd) and the partial pressure of the H_2S is low. Solid adsorption is more economical at small scales and can be achieved using oxides of Ca, Fe, Mn, Zn, and Cu [40]. Activated carbon can be applied to remove any remaining organics and chlorine compounds (such as HCl) can be removed from syngas using aqueous and caustic scrubbers or a solid absorbent such as Na-carbonate.

Generally, successful design of the syngas processing steps, need to consider the feedstock, the gasifier operating conditions and the desired end-use for the syngas, and for this reason a variety of solutions are possible.

1.3.4 Electricity production

The production of electricity from the gasification of wastes is usually undertaken either by combusting the syngas in a boiler to produce steam for a steam turbine or directly combusting the conditioned syngas in a gas turbine or gas engine. Table 1.4 shows typical ranges for the net electrical efficiency of each of these configurations and the syngas conditioning required. The combustion of syngas in a boiler and the use of a steam turbine as the prime mover is the most common method of generating electricity from MSW gasification. The main advantage of this configuration is that the gas boiler does not require any syngas conditioning to remove tar, dust/soot, heavy metals and other contaminants and the steam turbine can have a high availability since it does not contact the combustion products of the syngas. However, this configuration can have low electrical efficiency at small scale and be rather costly due to the need for boiler, steam system, turbine and condenser systems. In addition, the flue gas from the boiler still needs to be cleaned to meet emissions requirements. Gas turbines may be used as the prime mover and potentially can lead to improved overall efficiency. However, as gas turbines operate at pressure, the syngas must first be compressed which adds costs and requires clean syngas. These factors combined with a relatively limited selection of small gas turbines warranted for syngas service in the desired range of 1–30 MWe, means that few waste gasification plants actually use gas turbines as the prime mover. Gas engines are commonly used for biomass and waste gasification plants. They are relatively inexpensive, durable and reliable and with a relatively high electrical efficiency. For projects in the 1–20 MWe range, gas engines are ideal. The major challenges are that the engine components are exposed to the syngas and combustion products, and this impacts on availability and maintenance. Ensuring the syngas is conditioned appropriately for use in

Table 1.4: Comparison of main electrical production devices that can be used in gasification-based waste-to-energy plants.

Electricity production device	Net electrical efficiency of gasification plant	Required level of syngas cleaning
Steam turbine	15%–24%	Tar: not limited. Dust/soot: not limited. Alkalis: not limited. Heavy metals: not limited. H_2S: not limited.
Gas turbine	20%–30%	Tar: 10 mg/Nm^3. Dust/soot: 5 mg/Nm^3. Alkalis: 0.1 ppm, wb. Heavy metals: 0.1 ppm wb. H_2S: 20 ppm, wb.
Gas engine	14%–26%	Tar: 100 mg/Nm^3. Dust/soot: 50 mg/Nm^3. Alkalis: 0.1 ppm, wb. Heavy metals: 0.1 ppm wb. H_2S: 20 ppm, wb.

Adapted from Arena U. Process and technological aspects of municipal solid waste gasification. A review. Waste Management 2012;32:625–39. doi:10.1016/j.wasman.2011.09.025.

the engine is critical as maloperation of the gasifier or syngas clean up units can have detrimental impacts on the gas engines.

1.3.5 Chemicals synthesis

A major advantage of biomass and waste gasification over combustion is that the syngas can be used to produce a wide range of chemicals. The most common of these products include methanol, Fischer-Tropsch liquids and synthetic natural gas (bioSNG). Prior to chemical synthesis the syngas must be properly conditioned: including removal of CO_2, sulfur (H_2S, COS, $Cs_2 < 10$ ppbv to <1 ppmv), Halogens (HCl, HBr, HF < 10 ppbv), nitrogen compounds (NH_3, HCN < 1 ppmv), heavy metals and other contaminants, adjustment of the syngas modulus, $M = \frac{H_2 - CO_2}{CO + CO_2}$ to the correct value and the gas pressurized. Due to the complexity of these steps, there are few plants in operation which transform biomass or wastes into chemicals.

Methanol synthesis is exothermic according to the reactions:

$$CO + 2H_2 \rightarrow CH_3OH \quad -91 \ \text{MJ/kmol} \tag{1.7}$$

$$CO_2 + 3H_2 \rightarrow CH_3OH + H_2O \quad -50 \ \text{MJ/kmol} \tag{1.8}$$

and occurs optimally at a pressure of between 50 and 100 bara with a syngas modulus of ~2. Multi-tubular reactors are often used to achieve the required heat removal by boiling steam and several vendors have methanol synthesis technology, including Johnson Matthey and Haldor–Topsoe. The raw methanol is separated from water in a methanol distillation unit and may be further purified if required. Methanol may also be converted into dimethyl ether (DME), ethanol and olefins. Enerkem is the only company in the world that has built a commercial waste gasification process which makes methanol and ethanol [41].

Long chain paraffinic hydrocarbons can be produced in a Fischer–Tropsch (FT) unit using syngas with H_2/CO ratio of ~2, according to:

$$CO + 2H_2 \rightarrow -[CH_2]- + H_2O \quad -159\ \mathrm{MJ/kmol} \tag{1.9}$$

Generally, the synthesis occurs at a pressure of 40–80 bara, with a cobalt or iron-based catalyst. The temperature is controlled by removing heat by boiling water for steam production. In coal applications multitubular reactors and very large bubble column reactors are often preferred, while for smaller plants microscale reactors have been developed. Since FT synthesis produces a wide range of hydrocarbon chain lengths, further processing via hydrocracking and hydrotreating is required to meet typical diesel, gasoline and naphtha specifications. These features mean that plant capital costs are high and the overall carbon efficiency to products is relatively low. While economic studies on using biomass for FT synthesis have not been encouraging [42], processing MSW into jet fuel via FT synthesis is receiving attention and Fulcrum Bioenergy is constructing a plant in Nevada, USA [43]. The combination of waste gate fee and high value of renewable jet fuels can make this complex endeavor economic.

Converting syngas to methane (bioSNG) can be achieved with a catalyst according to:

$$CO + 3H_2 \rightarrow CH_4 + H_2O \quad -206\ \mathrm{MJ/kmol} \tag{1.10}$$
$$CO_2 + 4H_2 \rightarrow CH_4 + 2H_2O \quad -165\ \mathrm{MJ/kmol} \tag{1.11}$$

The above reactions are highly exothermic and multiple stages of reaction with inter-stage cooling is required to convert all of the CO and H_2 in the feed to CH_4. Several vendors have substitute natural gas (SNG) synthesis technology, with the TREMP technology from Haldor–Topsoe being widely used [44,45]. The SNG synthesis occurs at temperatures of around 700°C (first stage) down to 250°C (third stage), with ~80% of the energy in the syngas being converted into methane, while ~20% is converted into high pressure steam. Production of bioSNG has proceeded to small demonstration scale in Europe. Studies show that large scale plants and either renewable subsidies for bioSNG or very high natural gas prices are required to make biomass and waste gasification to bioSNG feasible [46].

1.4 Commercial MSW gasification systems

Commercial gasification systems can be classified based on reactor type, mineral matter recovery and syngas use. Fig. 1.10 shows a graphical classification of the main gasification technologies for MSW, while Table 1.5 provides a summary of the MSW gasification facilities.

1.4.1 Nippon Steel direct melting system

Nippon Steel & Sumikin Engineering Co. Ltd. developed the direct melting system (DMS) waste gasification technology using a moving bed shaft-furnace type gasifier at

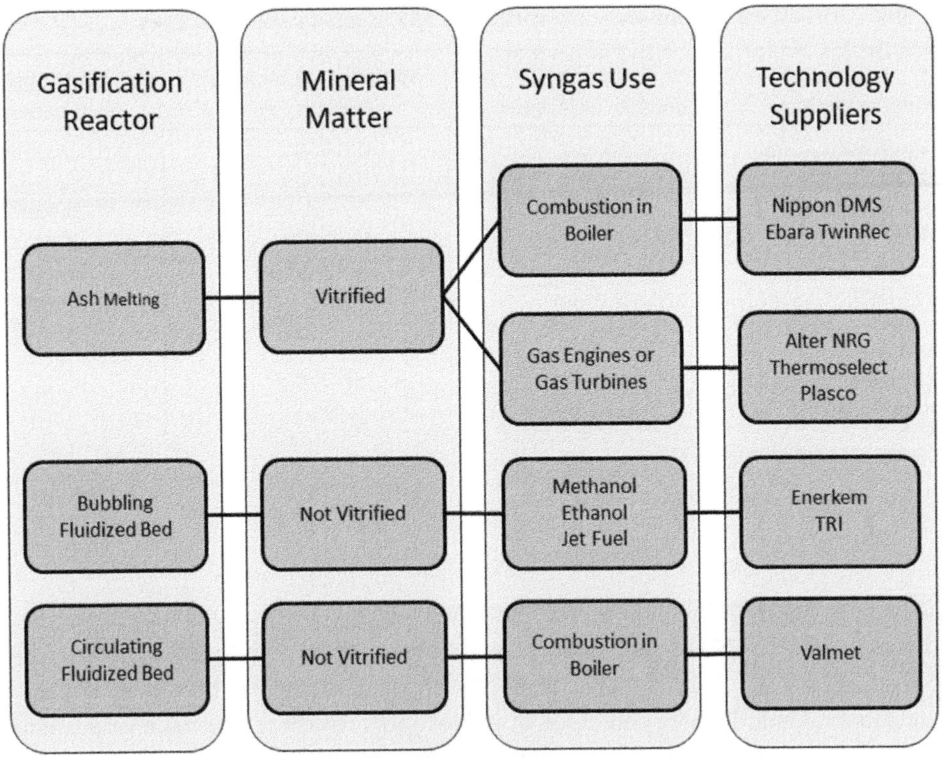

Figure 1.10
Classification of gasification technologies for processing MSW. *Adapted from Chromec P. Advanced technology: disparities between vision and reality. Nottwil, Switzerland: Hitachi Zosen INOVA Client Event; 2016.*

atmospheric pressure [24]. In this technology raw MSW at up to 800 mm in size can be processed or RDF can be used as a feedstock. The waste feed is charged into the direct melting furnace via a hopper system along with coke which functions as a reducing agent and limestone which functions as a viscosity modifier so that molten materials flow smoothly from the reactor. Fig. 1.11 shows a schematic of the process flow diagram of a typical DMS gasification plant. Japanese DMS plants are able to handle waste feedstock with calorific value in the range of 6–10 MJ/kg LHV, with up to 50% moisture [24,49].

The syngas from the direct melting reactor is fed through a cyclone to remove solid particulates and then combusted in a boiler to generate steam. Typical steam conditions are 400°C and 41 bar [50]. The flue gas is cooled and slaked lime is added upstream of the baghouse filter. Activated carbon is used to help gas cleaning. Ammonia is recovered in a selective catalytic reactor prior to the flue gas being released to atmosphere in a stack.

Over 38 DMS plants have been built in Japan and South Korea since 1979 with MSW capacities ranging from 30 to over 200 ktpa [48]. The principle driver for this technology

Table 1.5: Summary of worldwide MSW gasification facilities.

Company/technology	Approximate number of facilities	Approximate capacity range (tpd)	Locations of facilities
Nippon Steel	33	100–800	Japan, Korea
Thermoselect/JFE	5	100–600	Japan
Hitachi Zosen	6	30–300	Japan
Mitsui R-21	6	150–400	Japan
Ebara	7	100–400	Japan
Enerkem	1	275	Canada
Plasco	1	100	Canada
AlterNRG	2	150–200	China, Japan
Energos	7	100–250	Europe
Fulcrum Bioenergy	1[a]	480	USA

[a]Multiple plants under development in USA.

Adapted from Ciuta S, Tsiamis D, Castaldi MJ. Field scale developments. In: Gasification of waste materials. Elsevier; 2018. p. 65–91. doi:10.1016/B978-0-12-812716-2.00004-2.

has been to minimize landfill and recycle metal and slag products as much as possible. The main advantage of the DMS is that it can handle a wide range of different wastes, including raw MSW, RDF, sewage sludge, incineration bottom ash, automobile shredder residues (ASR), reclaimed landfill waste and medical wastes [50]. However, the main disadvantage is that the technology is relatively expensive and requires relatively large operating scale in most jurisdictions. Because of the high cost, the technology has not been adopted outside of Japan or South Korea. The key performance indicators for the DMS system are: carbon conversion of 91%–95%, cold gasification efficiency (CGE) of about 50%, equivalence ratio (ER) of 0.26–0.34 and a gross power generation efficiency of 19%–23%.

1.4.2 Thermoselect melting gasification

The Thermoselect gasification process transforms the waste feed into synthesis gas at ~1200°C and molten slag. Fig. 1.12 shows the basic process flowsheet and mass balance for the technology. Waste materials are stored in a bunker and transferred to a compaction press via grapple crane. The pressed waste has a density of ~1250 kg/m^3 and is pushed by hydraulic rams into a degassing channel (pyrolysis zone) which is externally heated to about 800°C [52]. The partially decomposed waste enters a vertical retort and is gasified with oxygen at about 1600°C and with a residence time >2 s. The mineral matter is heated to over 2000°C using oxygen and natural gas forming a high temperature melt. Further oxygen is added in a horizontal homogenization reactor, from which the melt flows into a quench system. The slag is granulated by water quenching and the metal and mineral granules are separated via magnetic separation. The syngas leaves the top of the vertical reactor at 1200°C and is quenched to 70°C using water sprays in the quench zone.

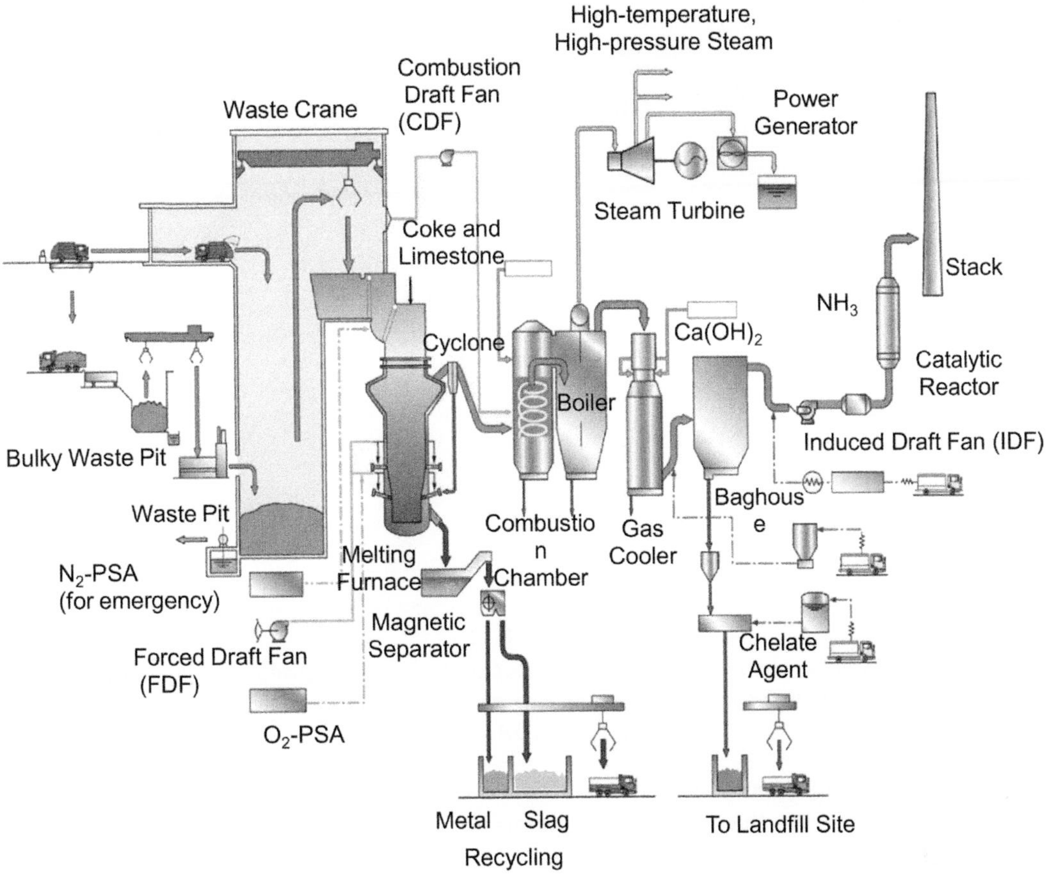

Figure 1.11

Schematic process flow diagram of the direct melting system. *From Tanigaki N, Manako K, Osada M. Co-gasification of municipal solid waste and material recovery in a large-scale gasification and melting system. Waste Management 2012;32:667–5. doi:10.1016/j.wasman.2011.10.019.*

The gas cleaning and conditioning system of the Thermoselect technology is involved and consists of acid scrubber, alkaline scrubber, de-dusting stage, desulfurization and gas drying. In the acid scrubber, HCl and HF acids are removed with water at pH ~ 3 and volatilized heavy metals are dissolved as metal ions. After the acid scrubber an alkaline scrubber using NaOH solution is applied to knock out any residual acid liquid droplets. De-dusting is achieved using a water/glycerine scrubber which reduces the surface tension of the liquid and enhances the capture of very fine residual dust particles from the syngas. Solids from de-dusting are removed from the liquid in a filter press and transferred back to the high temperature gasifier. The synthesis gas is passed through a desulfurization process, where the scrubbing liquid contains Fe-III complex to remove H_2S from the gas. The dew point of the gas is lowered by direct contact with cold water in a gas drying

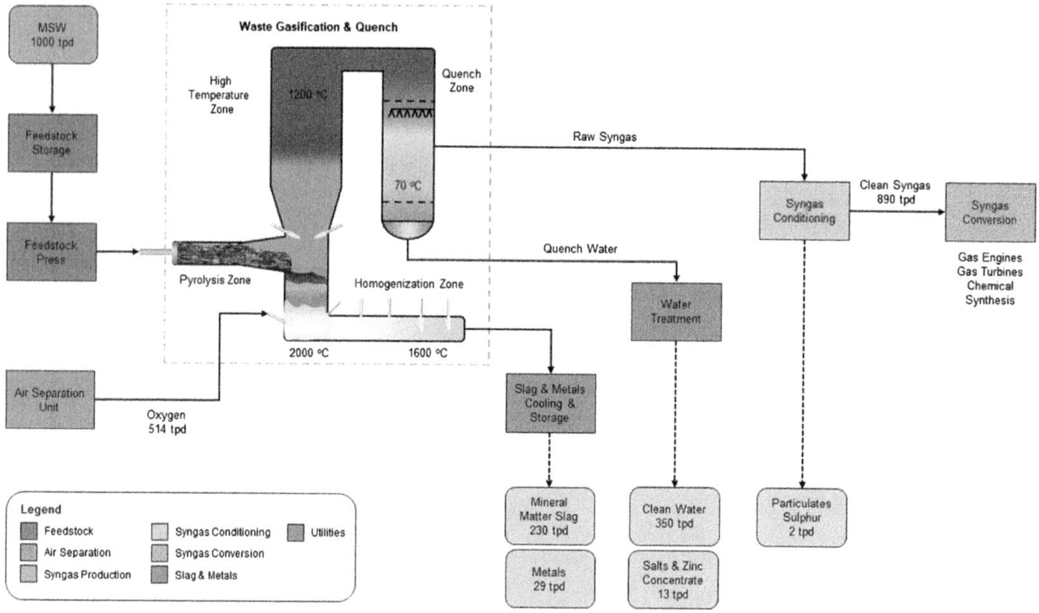

Figure 1.12

Schematic process flow scheme of the Thermoselect technology. *Adapted from Kais. Thermoselect – an advanced field proven high temperature recycling process. In: Gasification technologies council, San Francisco, CA, USA; 2003.*

scrubber. The resulting process water is treated in batches due to the relatively small quantities and returned to the gas cleaning system.

The typical mass balance for 1 tonne of MSW generates 0.89 tonne of purified syngas, 0.23 tonne of mineral matter and 0.03 tonne of metals. The cold gas efficiency for the gasification module is reported as 59% [53]. The net power generation efficiency using 12 MJ/kg MSW and gas engines is 19% and can reach about 23% with gas turbines [51,53]. Nine plants have been developed using the Thermoselect technology with seven constructed in Japan [53,54]. The processing capacity of the plants vary from 38 to 289 ktpa. The Japanese licensor of the technology, JFE, has stated that they no longer offer the technology in Japan because it is too expensive [55]!

1.4.3 Alter NRG plasma gasification

The Alter NRG plasma gasification technology has been developed over a period of 30 years and operates at temperatures up to 5000°C. The extremely high temperatures make the plasma gasification suitable for a wide range of waste feedstocks, such as MSW and hazardous medical wastes [21,56]. Waste feedstock is fed from the side of a large refractory lined vessel and is reacted with air and/or oxygen injected via tuyeres located in the side of the vessel. The plasma torches are directed to heat and melt the waste forming a molten liquid at ~1600°C from which slag is tapped periodically. The syngas exists the top of the reactor at ~950°C which means that tars are converted and reformed into smaller molecules such as CO, H_2, CH_4 and CO_2 [57]. The produced syngas is rapidly quenched with water to a temperature of 80°C. Fig. 1.13 shows the basic process flow sheet and mass balance for a plasma gasification plant configured to produce electricity from gas engines.

The advantage of plasma gasification is its extremely high temperatures which enable it to handle various types of waste materials, however, the high temperatures make design and operation challenging and while several commercial plants are successfully in operation there have also been notable failures. For example: the Ecovalley plant in Japan was shut down due to lack of feedstock [58], the Teeside plant in UK was abandoned before successful start-up and a plant in Sacramento, USA was shut down [59]. One major challenge is the reliable pretreatment and feeding of heterogeneous solid feeds into the plasma gasifier. The technology has high capital costs and high maintenance and operational costs [60,61]. While high CGE has been reported for plasma gasification [62], net electric efficiency is generally low due to the power consumption of the plasma torches and auxiliary loads. AlterNRG has four operating project references, two in China, one in India and one in Japan [57].

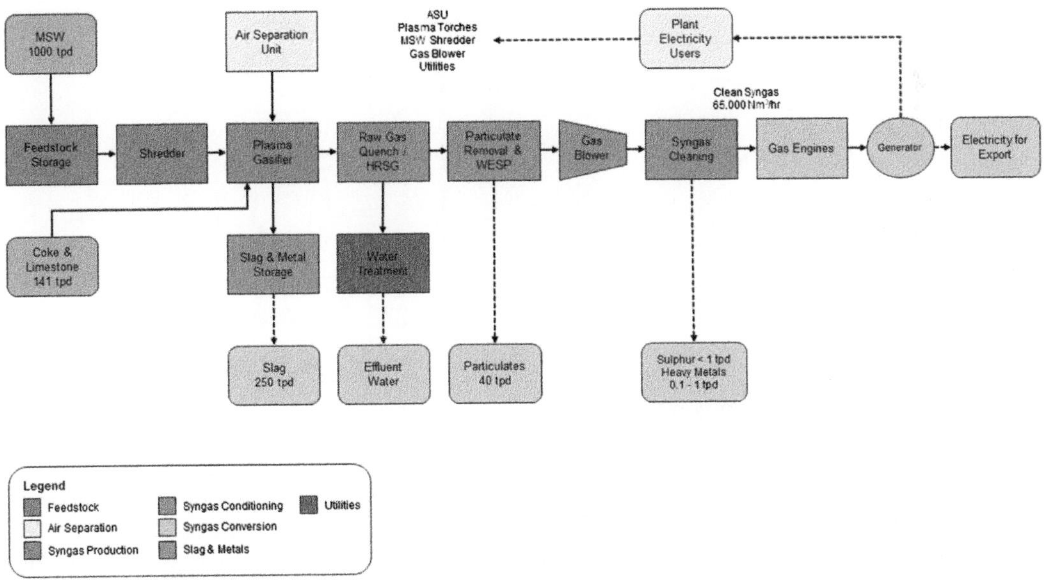

Figure 1.13
Schematic of the process flowsheet and mass balance for a plasma gasifier. *Adapted from Alter NRG. Summary of qualifications. Alter NRG; 2018.*

1.4.4 Ebara TwinRec fluidized-bed gasification

A fluidized-bed gasification and ash melting system has been developed and commercially deployed by Ebara Environmental Solutions of Japan. Fig. 1.14 shows a schematic of the process, which consists of a fluidized-bed technology, coupled with an ash melting furnace. The fluidized-bed design is a variant on Ebara's fluidized-bed combustion technology which is operated with limited air at relatively low temperatures of between 500 and 600°C. The pyrolysis and synthesis gases containing entrained char and ash from the fluidized-bed reactor are immediately passed to a cyclonic combustion chamber where temperatures reach over 1450°C, causing the ash to melt. The molten ash is removed from a special tapping point, while the combustion flue gases are sent to a heat recovery steam generator. Incombustibles like ferrous material and aluminum are recovered in an unoxidized state, further reducing final volumes disposed of in landfill. The hot flue gases from the combustion chamber immediately pass to a heat recovery steam generator (HRSG). Due to the absence of any gas treatment, the HRSG is exposed to risks of corrosion and erosion. The steam conditions are usually selected to be rather low, for example in the plant processing ASR, the steam conditions were 30 bara and 380°C [63]. Ebara has built five plants in Japan.

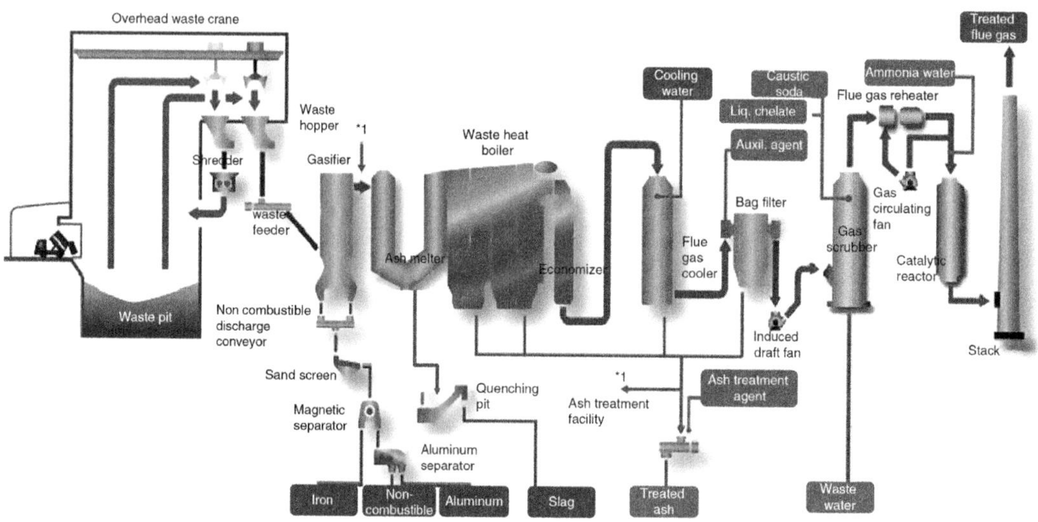

Figure 1.14

Schematic of the process flowsheet for the Ebara TwinRec internal circulating fluidized-bed gasification process with ash melting furnace. *From Yoshikawa K. Gasification gasification and liquefaction alternatives incineration alternatives to incineration incineration in Japan. In: Kaltschmitt M, Themelis NJ, Bronicki LY, Söder L, Vega LA, editors. Renewable energy systems. New York, NY: Springer New York; 2013. p. 728–43. doi:10.1007/978-1-4614-5820-3_419.*

1.4.5 Enerkem bubbling fluidized-bed gasification

Enerkem is a Canadian company that uses a BFB system to gasify nonrecyclable and noncompostable MSW to produce methanol and ethanol. The first commercial scale facility has been built in Edmonton, AB and processes up to 100 ktpa of RDF fluff. The RDF is produced in a MRF at the same site and ranges in size from about 120 to 230 mm and contains 60%–70% biogenic material and 30%–40% plastic. Fig. 1.15 shows the basic flowsheet of the Enerkem process. The BFB uses steam and oxygen as the gasification agent, with a small CO_2 purge gas and operates at 1–5 bar. The gasification temperature is reportedly about 750°C, so the syngas contains significant quantities of tars. The tars may undergo steam reforming downstream of the gasifier in a dedicated unit. After heat recovery, the gas is quenched and H_2S, heavy metals and CO_2 are removed in sequential steps. The syngas is pressurized in stages and used to produce methanol in a catalytic reactor. Enerkem is also aiming to produce ethanol by reacting methanol with CO to produce methyl acetate and then hydrogenating the methyl acetate to ethanol.

1.5 Process performance, economics and opportunities

To date, most of the commercial deployment of gasification for MSW has been undertaken in Japan where the primary drivers have been to minimize the amount of waste sent to

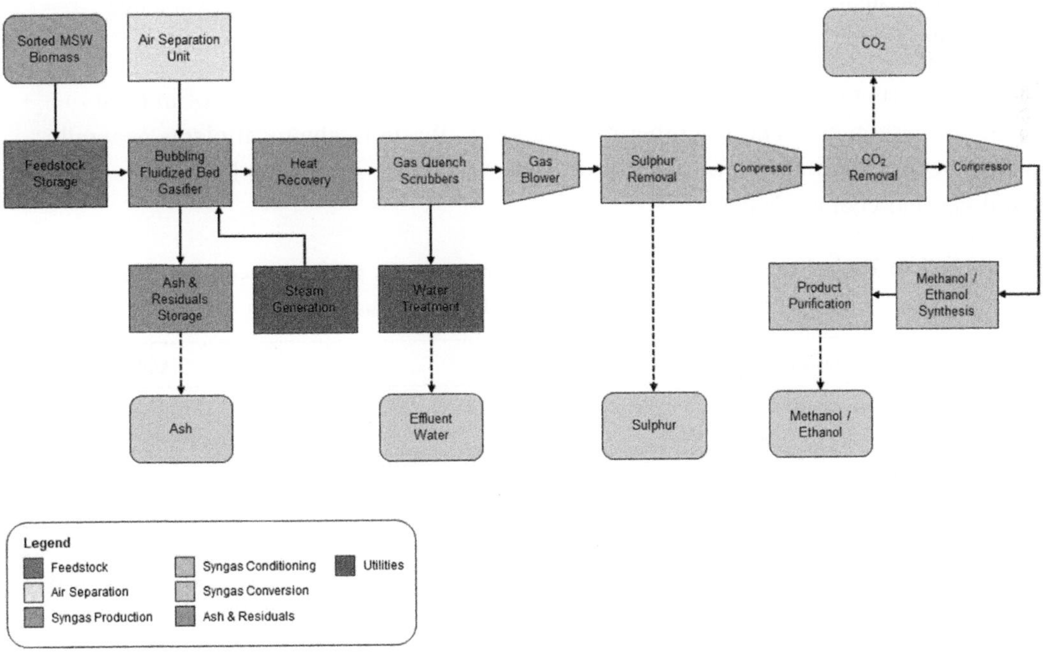

Figure 1.15
Schematic of the process flowsheet of the Enerkem process to convert waste into chemicals.

landfill. Environmental policy in Japan encouraged transformation of mineral matter into vitrified slag and hence most gasification systems in Japan have an ash melting function incorporated into the design. Outside of Japan, MSW gasification projects are less common and those that have been constructed have had mixed success, due to technical challenges or economics. In recent years there has been interest in converting waste to chemicals using gasification, as evidenced by the projects in operation and under construction by Enerkem and Fulcrum Bioenergy.

1.5.1 Process performance

When considering converting waste to electricity, the net electrical efficiency of gasification is similar to combustion (incineration). Consonni and Vigano compared the performance of two waste gasification systems using two-stage combustion (high temperature grate gasifier, HTGG and low temperature fluidized-bed, LTFBG) with a conventional grate combustion (GC) incineration process for a small-scale project of 50 MWth [63]. They found that the net electrical efficiencies were nearly identical at 19.7% for HTGG, 19.6% for LTFBG and 20.5% for GC. The trends were similar for larger scale plants, with GC having efficiencies about 1%–2% higher than two-stage gasification [63].

The net electrical efficiency of gasification coupled with gas engines can be very competitive at small scales, where combustion systems are prohibitively expensive. Table 1.6 provides typical ranges for the process performance of gasification.

When considering converting waste to chemicals, the process parameters of interest include the carbon efficiency and thermal efficiency. Waste to chemical plants generally have efficiencies similar to biomass based plants, which are somewhat lower than their natural gas and coal equivalents.

Table 1.6: Typical ranges of some operating and process performance parameters for gasification of municipal solid waste.

Operating parameters	Values
Equivalence ratio	0.25–0.35
Waste calorific value (MJ/kg, LHV)	7–18
Process performance parameters	
Carbon conversion efficiency (%)	90–99
Cold gas efficiency (%)	50–80
Syngas calorific value (MJ/Nm^3, LHV)	4–7 (air), 7–12 (oxygen)
Net electrical efficiency (%)	15–24
Specific net energy (kWh/tonne)	400–700

Adapted from Arena U. Process and technological aspects of municipal solid waste gasification. A review. Waste Management 2012;32:625–39. doi:10.1016/j.wasman.2011.09.025.

1.5.2 Air emissions

The major environmental concern of waste-to-energy plants is their air emissions. The major air pollutants that are regulated by standards on concentration include particulate matter, HCl, NOx, SOx, Hg and dioxin/furans. Table 1.7 shows the reported air pollutant emissions reported for five different gasification plants operating in Japan, Canada and Finland. It is observed that waste-to-energy plants using gasification can meet air pollution standards using appropriate gas cleaning and flue gas treatment technologies.

1.5.3 Economics of waste gasification

The economic feasibility of waste-to-energy plants is determined by a large range of factors, with the most important being government regulations and policies. The main revenue sources are: (1) waste gate fee, (2) product sales, and (3) byproduct sales, while the main operating expenses include: (1) labor, (2) plant maintenance, and (3) waste disposal costs. The conversion of waste into energy is capital intensive and so capital costs, financing costs and plant efficiency have a major role in feasibility. In addition, government subsidies and credit schemes also play a significant role. Generally, waste-to-energy plants require government policy to incentivize landfill diversion, since the

Table 1.7: Air pollutant emissions from waste gasification plants.

	PM	HCl	NOx	SOx	Hg	PCCD/PCDF
	mg/Nm^3					ng$_{TEC}$/Nm^3
Standard						
European Standard	10	10	200	50	0.05	0.1
Japanese Standard	10.1 −50.6	15.2 −50.6	30.3 −126.4	10.1 −30.3	0.03 −0.051	0.51
US Standard	24.3	25.3	151.7	30.3	0.03 −0.051	0.14−0.21
Plant						
Ebara TwinRec Kawaguchi Japan	1.0	2.0	29.2	2.8	4.99E-03	5.13E-05
JFE/Thermoselect—Nagasaki, Japan	3.3	8.3	—	—	—	1.78E-02
Nippon Steel DMS—Kazusa, Japan	10.0	8.9	22.2	15.6	—	3.21E-02
Plasco—Ottawa, Canada	9.1	2.2	106.8	18.5	1.42E-04	6.58E-03
Valmet—Kymijärvi II, Finland	—	1	161	7	1.00E-04	2.00E-03

From Pigneri A. Gasification technologies review – technology, resources and implementation scenarios. Sydney, Australia; 2014. Isaksson J. Commercial CFB gasification of waste and biofuels – operational experiences in large scale. In: Gasification technologies conference, Colorado Springs, CO, USA; 2015.

waste gate fee required is higher than the costs of simply putting the waste in landfill. Where, landfill taxes exist, waste-to-energy can be viable if the required waste gate fee is lower than the sum of the landfill tax and costs of operating landfills.

The UK has a landfill tax of 80 £/tonne (104 USD/tonne, 147 AUD/tonne), which has incentivized the construction of waste-to-energy facilities. About 4 million tonnes per annum are incinerated in the UK and the landfill diversion rate is around 75%.

Japan has installed the most waste gasification systems, largely driven by policies to minimize landfill and to ensure waste bottom ash materials are vitrified. Thus, in Japan gasification and melting can be more cost effective than conventional incineration and the separate melting of the bottom-ashes. Tanigaki et al. evaluated several waste management schemes relevant to the Japanese context and found that co-gasification of waste and bottom-ashes was most cost effective when landfill tax was above ~100 €/tonne (112 USD/tonne, 158 AUD/tonne) [65]. Higher landfill taxes are reportedly required when considering only the gasification of MSW.

In 2009, Alter NRG reported that a 750 tpd plasma gasifier would require a gate fee of ~80 USD/tonne for power prices of 50 USD/MWh and a 300 tpd plasma gasifier would require a gate fee of ~140 USD/tonne at the same power price [23]. Therefore, under current conditions, gasification of raw MSW using fixed bed plasma and melting reactors is expected to require waste gate fees in the range of 100–200 USD/tonne for projects >150 ktpa, with higher gate fees needed for smaller projects.

The economics of waste to chemicals are not yet clear, since only one plant is in operation and another is under construction. Certainly, the pioneer plants developed by Enerkem and Fulcrum Bioenergy are dependent upon government policies and also various subsidies and credit schemes. The high value and growing demand for renewable chemicals and fuels could make these plants economically attractive, despite the high initial capital costs.

1.5.4 Opportunities

The two main opportunities for the gasification of wastes are: (1) for small-scale (<150 ktpa) distributed waste to electricity projects and (2) for large-scale (>150 ktpa) waste to chemical projects. Two-stage gasification has no significant advantages over existing waste incineration, and in fact, capital cost is probably higher and net electrical efficiency is slightly lower.

At small scale, the main challenge for existing gasification technologies is the need for waste pre-treatment for fluidized-bed systems and the high costs of existing fixed bed melting systems. New gasification approaches will be needed to make small-scale waste-to-energy facilities viable. Wildfire Energy[2] of Australia has developed the MIHG technology (Moving Injection Horizontal Gasification) to pilot scale [66]. The main innovation in this technology, is the use of batch loading of the waste feed while enabling continuous production of high-quality syngas using a moving injection point for the oxidant.

Cleanup of syngas from gasification systems will continue to be an area for further development, especially to improve reliability and the costs of removing tars, sulfur, chlorine and other contaminants. The optimization of solid absorbents for small-scale facilities is another area of interest. Clean syngas is necessary to enable the use of high efficiency electrical generators like gas engines, gas turbines and fuel cells.

At large scale, conversion of waste to chemicals could have a bright future, if pioneer plants can operate reliably and overcome the challenges of gasifying heterogeneous wastes and conditioning the syngas to the very clean requirements demanded by catalytic synthesis. Unfortunately, most of the pioneer biomass to liquids (BTL) projects built to date have not succeeded due to technical issues and lack of capital. The experience from the oil and gas industry in processing natural gas and coal into synthetic fuels (GTL, CTL) indicates that success will require substantial capital and significant time to achieve high reliability and availability. However, once plant operations are optimized these facilities can be very profitable for many decades. Converting waste into hydrogen may also become of interest as the markets for hydrogen matures.

1.6 Conclusions and perspectives

This chapter has provided a review of the fundamentals and applications of producing electricity and chemicals from the gasification of municipal solid wastes. While it is not widely known, gasification of raw MSW has been proven with several technologies deployed in Japan. However, two-stage gasification does not offer enough advantages over conventional incineration to become widely adopted elsewhere. Therefore, further research and development to improve gasification technologies is needed. The main areas for investigation include (1) developing novel designs for smaller scale, distributed waste gasification, (2) reducing the cost of syngas cleanup, (3) improving methods to the utilize syngas (e.g., in fuel cells or chemical synthesis), and (4) research into using the residual ash and slag in higher value products such as building materials.

Many regions do not have sufficient waste volumes to warrant building large-scale incineration facilities, and one opportunity for gasification is in small-scale distributed waste to electricity plants where the use of gas engines can achieve efficiencies similar to

[2] The author is also a cofounder of Wildfire Energy.

combustion at larger scales. Converting waste to hydrogen is also of interest. Another opportunity lies in the conversion of wastes into fuels and chemicals at large scale (>150 kpta) as an alternative to incineration. At large scales, sorting and recycling of the waste in material recovery facilities and pretreatment into RDF is possible so that the gasifier has a well-defined feedstock to process. The complexity of waste to chemicals should not be underestimated and while most of the component technologies are proven, pioneer plants will likely face integration challenges and take a number of years to achieve high production availability. Nonetheless, the effort is warranted as waste to chemicals could become highly profitable and is a positive step toward the ultimate aim of building a circular economy.

References

[1] Kaza S, Yao L, Bhada-Tata P, Van Woerden F. What a waste 2.0: a global snapshot of solid waste management to 2050. Washington, DC, USA: International Bank for Reconstruction and Development, The World Bank; 2018.

[2] Korhonen J, Honkasalo A, Seppälä J. Circular economy: the concept and its limitations. Ecological Economics 2018;143:37−46. https://doi.org/10.1016/j.ecolecon.2017.06.041.

[3] Mastellone ML. Waste management and clean energy production from municipal solid waste. New York, NY, USA: Nova Publishers; 2015.

[4] Muthuraman M, Namioka T, Yoshikawa K. A comparative study on co-combustion performance of municipal solid waste and Indonesian coal with high ash Indian coal: a thermogravimetric analysis. Fuel Processing Technology 2010;91:550−8. https://doi.org/10.1016/j.fuproc.2009.12.018.

[5] Breeze P. Waste to energy technologies. In: Energy from waste. Elsevier; 2018. p. 29−37. https://doi.org/10.1016/B978-0-08-101042-6.00004-2.

[6] Perkins G. Underground coal gasification − part II: fundamental phenomena and modeling. Progress in Energy and Combustion Science 2018;67:234−74. https://doi.org/10.1016/j.pecs.2018.03.002.

[7] Perkins G. Mathematical modelling of in situ combustion and gasification. Proceedings of the Institution of Mechanical Engineers, Part A: Journal of Power and Energy 2017. https://doi.org/10.1177/0957650917721595. pii:095765091772159.

[8] Miller RS, Bellan J. A generalized biomass pyrolysis model based on superimposed cellulose, hemicelluloseand liqnin kinetics. Combustion Science and Technology 1997;126:97−137. https://doi.org/10.1080/00102209708935670.

[9] Ciuta S, Tsiamis D, Castaldi MJ. Fundamentals of gasification and pyrolysis. In: Gasification of waste materials. Elsevier; 2018. p. 13−36. https://doi.org/10.1016/B978-0-12-812716-2.00002-9.

[10] Westbrook C, Dryer FL. Chemical kinetic modeling of hydrocarbon combustion. Progress in Energy and Combustion Science 1984;10:1−57.

[11] Westbrook CK, Dryer FL. Simplified reaction mechanisms for the oxidation of hydrocarbon fuels in flames. Combustion Science and Technology 1981;27:31−43.

[12] Jones WP, Lindstedt RP. Global reaction schemes for hydrocarbon combustion. Combustion and Flame 1988;73:233−49.

[13] Ranzi E, Dente M, Goldaniga A, Bozzano G, Faravelli T. Lumping procedures in detailed kinetic modeling of gasification, pyrolysis, partial oxidation and combustion of hydrocarbon mixtures. Progress in Energy and Combustion Science 2001;27:99−139.

[14] Tomita A, Mahajan OP, Walker Jr PL. Reactivity of heat-treated coals in hydrogen. Fuel 1977;56:137−44.

[15] Gray D. Major gasifiers for IGCC systems. In: Integrated gasification combined cycle (IGCC) technologies. Elsevier; 2017. p. 305−55. https://doi.org/10.1016/B978-0-08-100167-7.00008-1.

[16] Breault RW. Gasification processes old and new: a basic review of the major technologies. Energies 2010;3:216–40. https://doi.org/10.3390/en3020216.
[17] Pang S. Fuel flexible gas production. In: Fuel flexible energy generation. Elsevier; 2016. p. 241–69. https://doi.org/10.1016/B978-1-78242-378-2.00009-2.
[18] Higman C, van der Burgt M. Gasification. 2nd ed. Burlington, MA, USA: Gulf Professional Publishing; 2008.
[19] Tanigaki N, Manako K, Osada M. Co-gasification of municipal solid waste and material recovery in a large-scale gasification and melting system. Waste Management 2012;32:667–75. https://doi.org/10.1016/j.wasman.2011.10.019.
[20] van Nierop P, Sharma P. Plasma gasification – integrated facility solutions for multiple waste streams. In: Gasification technologies conference, Washington, DC; 2010.
[21] Pourali M. Application of plasma gasification technology in waste to energy—challenges and opportunities. IEEE Transactions on Sustainable Energy 2010;1:125–30. https://doi.org/10.1109/TSTE.2010.2061242.
[22] Arena U. Process and technological aspects of municipal solid waste gasification. A review. Waste Management 2012;32:625–39. https://doi.org/10.1016/j.wasman.2011.09.025.
[23] van Nierop P. Alter NRG – a leading provider of clean energy solutions. In: Gasification technologies conference, Colorado Springs, CO; 2009.
[24] Tanigaki N, Ishida Y. Waste gasification technology with direct melting for energy and material recovery. Waste Management 2014;4:365–78. TK Verlag.
[25] Rabou LPLM, Zwart RWR, Vreugdenhil BJ, Bos L. Tar in biomass producer gas, the energy research centre of The Netherlands (ECN) experience: an enduring challenge. Energy & Fuels 2009;23:6189–98. https://doi.org/10.1021/ef9007032.
[26] Chornet E, Valsecchi B, Gagnon M, Nguyen B. Production and conditioning of synthesis gas obtained from biomass. WO 2009/132449 A1; n.d.
[27] ThermoChem Recovery International Inc. ThermoChem Recovery International Inc.; n.d. https://tri-inc.net/.
[28] Yoshikawa K. Gasification gasification and liquefaction alternatives incineration alternatives to incineration incineration in Japan. In: Kaltschmitt M, Themelis NJ, Bronicki LY, Söder L, Vega LA, editors. Renewable energy systems. New York, NY: Springer New York; 2013. p. 728–43. https://doi.org/10.1007/978-1-4614-5820-3_419.
[29] Basu P. Biomass gasification, pyrolysis and torrefaction. 2nd ed. London, United Kingdom: Academic Press; 2013.
[30] Isaksson J. Commercial CFB gasification of waste and biofuels – operational experiences in large scale. In: Gasification technologies conference, Colorado Springs, CO, USA; 2015.
[31] Ali OH, Hossein G, Hooman F. Energy efficiency improvement analysis considering environmental aspects in regard to biomass gasification PSOFC/GT power generation system. Procedia Environmental Sciences 2013;17:831–41. https://doi.org/10.1016/j.proenv.2013.02.101.
[32] Prabu V, Jayanti S. Integration of underground coal gasification with a solid oxide fuel cell system for clean coal utilization. International Journal of Hydrogen Energy 2012;37:1677–88. https://doi.org/10.1016/j.ijhydene.2011.09.132.
[33] Göransson K, Söderlind U, He J, Zhang W. Review of syngas production via biomass DFBGs. Renewable and Sustainable Energy Reviews 2011;15:482–92. https://doi.org/10.1016/j.rser.2010.09.032.
[34] Bolhàr-Nordenkampf M, Rauch R, Bosch K, Aichernig C, Hofbauer H. Biomass CHP plant Güssing – using gasification for power generation. In: 2nd regional conference on energy technology towards a clean enviornment, Phuket, Thailand; 2003.
[35] Gassner M, Maréchal F. Thermodynamic comparison of the FICFB and Viking gasification concepts. Energy 2009;34:1744–53. https://doi.org/10.1016/j.energy.2009.05.011.
[36] Anis S, Zainal ZA. Tar reduction in biomass producer gas via mechanical, catalytic and thermal methods: a review. Renewable and Sustainable Energy Reviews 2011;15:2355–77. https://doi.org/10.1016/j.rser.2011.02.018.

[37] Valderrama Rios ML, González AM, Lora EES, Almazán del Olmo OA. Reduction of tar generated during biomass gasification: a review. Biomass and Bioenergy 2018;108:345–70. https://doi.org/10.1016/j.biombioe.2017.12.002.

[38] Ciferno JP, Marano JJ. Benchmarking biomass gasification technologies for fuels, chemicals and hydrogen production. Morgantown, WV, USA: U.S. Department of Energy National Energy Technology Laboratory; 2002.

[39] Panepinto D, Tedesco V, Brizio E, Genon G. Environmental performances and energy efficiency for MSW gasification treatment. Waste and Biomass Valorization 2015;6:123–35. https://doi.org/10.1007/s12649-014-9322-7.

[40] van Paasen SVB, Cieplik MK, Phokawat NP. Gasification of non-woody biomass: economic and technical perspectives of chlorine and sulphur removal from product gas. The Netherlands: ECN; 2006.

[41] Enerkem. Enerkem; n.d. https://enerkem.com/.

[42] Trippe F, Frohling M, Schultmann F, Stahl R, Henrich E, Dalai A. Comprehensive techno-economic assessment of dimethyl ether (DME) synthesis and Fischer–Tropsch synthesis as alternative process steps within biomass-to-liquid production. Fuel Processing Technology 2013;106:577–86. https://doi.org/10.1016/j.fuproc.2012.09.029.

[43] Fulcrum Bioenergy. Fulcrum Bioenergy; n.d. http://fulcrum-bioenergy.com/.

[44] Haldor-Topsoe. From solid fuels to substitute natural gas (SNG) using TREMP™. Denmark: Haldor-Topsoe; 2017.

[45] Wix C, Dybkjær I, Udengaard NR. Coal to SNG – the methanation process. In: Twenty-forth annual international Pittsburgh coal conference, Johannesburg, South Africa; 2007.

[46] Aranda G, van der Drift A, Smit R. The economy of large scale biomass to substitute natural gas (bioSNG) plants. Petten, The Netherlands: ECN; 2014.

[47] Chromec P. Advanced technology: disparities between vision and reality. Nottwil, Switzerland: Hitachi Zosen INOVA Client Event; 2016.

[48] Ciuta S, Tsiamis D, Castaldi MJ. Field scale developments. In: Gasification of waste materials. Elsevier; 2018. p. 65–91. https://doi.org/10.1016/B978-0-12-812716-2.00004-2.

[49] Tanigaki N, Fujinaga Y, Kajiyama H, Ishida Y. Operating and environmental performances of commercial-scale waste gasification and melting technology. Waste Management & Research 2013;31:1118–24. https://doi.org/10.1177/0734242X13502386.

[50] Tanigaki N, Kashiwabara T. Operating experience from Japanese waste gasification plants with direct melting system. Waste Management 2017;7:379–88. TK Verlag.

[51] Kaiser W. Thermoselect – an advanced field proven high temperature recycling process. In: Gasification technologies council, San Francisco, CA, USA; 2003.

[52] Schilli J. The fourth dimension for waste management in the United States: thermoselect gasification technology and the hydrogen energy economy. In: Proceedings of NAWTEC12, Savannah, Georgia, USA. ASME; 2004.

[53] Campbell F. Overview of the history and capabilities of the thermoselect technology. Report by Interstate Waste Technologies. 2008.

[54] Gersham, Brickner & Bratton, Inc.. Gasification of non-recycled plastics from municipal solid waste in the United States. Fairfax, VA, USA: The American Chemistry Council; 2013.

[55] Whiting K, Wood S, Fanning M, Venn M. Review of state-of-the-art waste-to-energy technologies. Stage Two – Case Studies. London, UK. 2013.

[56] Favas J, Monteiro E, Rouboa A. Hydrogen production using plasma gasification with steam injection. International Journal of Hydrogen Energy 2017;42:10997–1005. https://doi.org/10.1016/j.ijhydene.2017.03.109.

[57] Alter NRG. Summary of qualifications. Alter NRG; 2018.

[58] Subramanyam V, Gorodetsky A. Municipal wastes and other potential fuels for use in IGCC systems. In: Integrated gasification combined cycle (IGCC) technologies. Elsevier; 2017. p. 181–219. https://doi.org/10.1016/B978-0-08-100167-7.00005-6.

[59] McGillivary D, Yi E, Chan F, Loboguerrero G, Lee K, Wei K. An analysis of the lack of investment and implementation of plasma gasification in the United States. USA: University of Chicago; 2015.
[60] Minutillo M, Perna A, Di Bona D. Modelling and performance analysis of an integrated plasma gasification combined cycle (IPGCC) power plant. Energy Conversion and Management 2009;50:2837–42. https://doi.org/10.1016/j.enconman.2009.07.002.
[61] Matveev IB, Washcilenko NV, Serbin SI, Goncharova NA. Integrated plasma coal gasification power plant. IEEE Transactions on Plasma Science 2013;41:3195–200. https://doi.org/10.1109/TPS.2013.2289908.
[62] Mazzoni L, Almazrouei M, Ghenai C, Janajreh I. A comparison of energy recovery from MSW through plasma gasification and entrained flow gasification. Energy Procedia 2017;142:3480–5. https://doi.org/10.1016/j.egypro.2017.12.233.
[63] Consonni S, Viganò F. Waste gasification vs. conventional Waste-To-Energy: a comparative evaluation of two commercial technologies. Waste Management 2012;32:653–66. https://doi.org/10.1016/j.wasman.2011.12.019.
[64] Pigneri A. Gasification technologies review – technology, resources and implementation scenarios. Sydney, Australia. 2014.
[65] Tanigaki N, Ishida Y, Osada M. A case-study of landfill minimization and material recovery via waste co-gasification in a new waste management scheme. Waste Management 2015;37:137–46. https://doi.org/10.1016/j.wasman.2014.07.024.
[66] Doucet D, Perkins G, Grant B. Moving injection horizontal gasification – a new process for waste-to-energy. In: Gasification and syngas technologies conference, Colorado Springs, CO, USA; 2017.

CHAPTER 2

Integrated innovative biorefinery for the transformation of municipal solid waste into biobased products

Víctor Pérez[1,2], Andrés Pascual[3], Alfredo Rodrigo[3], María García Torreiro[3], Marcos Latorre-Sánchez[4], Caterina Coll Lozano[4], Antonio David-Moreno[5], Jose Miguel Oliva-Dominguez[5], Alba Serna-Maza[6], Natalia Herrero García[7], Inmaculada González Granados[7], Rocio Roldan-Aguayo[8], David Ovejero-Roncero[8], Jose L. Molto Marin[8], Mark Smith[9], Hana Musinovic[9], Amélie Raingué[6], Laurent Belard[10], Celia Pascual[1,2], Raquel Lebrero[1,2], Raul Muñoz[1,2]

[1]Institute of Sustainable Processes, University of Valladolid, Valladolid, Spain; [2]Department of Chemical and Environmental Engineering, University of Valladolid, Valladolid, Spain; [3]AINIA-Centro tecnológico, Paterna, Valencia, Spain; [4]Imecal S.A., L'alcudia, Valencia, Spain; [5]CIEMAT, Madrid, Spain; [6]Urbaser S.A., R&D and Innovation Department, Madrid, Spain; [7]Biomasa Peninsular S.A., Madrid, Spain; [8]Exergy Ltd., Coventry, United Kingdom; [9]NATRUE, Brussels, Belgium; [10]NaturePlast, Ifs, France

2.1 Introduction

Globally, population tends to increase in urban rather than in rural areas. In 1950, 30% of the world's population was urban, while in 2018, 55% of the world's population is residing in urban areas and this figure is expected to grow until 68% by 2050 [1]. Thus, cities grow as major consumers of world's natural reserves and energy supplies, acting as concentrators of materials and nutrients, by aggregating inputs such as food from rural areas into a concentrated urban space. Considering that on average 483 kg of municipal solid waste (MSW) per capita are being produced in the EU, and the organic fraction of MSW (OFMSW) represents the largest fraction of the MSW (40%–50%), cities concentrate an enormous waste production as well, while the nutrients that return to the biosphere are nowadays considerably low [2]. Despite waste landfilling represents *linear economy*, still 24% of the MSW is managed in this way. At the expense of landfilling, other treatment methods such as recycling, incineration and composting or anaerobic digestion (AD) are being utilized for MSW treatment, managing 29%, 27%, and 16% of

Waste Biorefinery. https://doi.org/10.1016/B978-0-12-818228-4.00002-2

the generated MSW, respectively [3]. Waste composting and AD yield a fertilizer product and the possibility of energy recovery obtaining biogas, however, both are considered low-value products.

In this regard, new biobased building blocks, chemicals and materials like bioplastics will be required in the near future to face concerns about the environmental impact, availability and high cost of oil and its derivative products. Plans for a new biobased economy are being developed by the EU and by almost all countries worldwide. Biorefineries are considered a key instrument for achieving the bioeconomy goals and for promoting the linear-to-circular transition in the EU economy. The fundamental contribution of biorefineries to the concept of *circular economy* is based on their ability to transform biomass into different final products with high added value. Municipal biowaste (waste from food consumed in homes, restaurants, markets and commercial premises, as well as biodegradable waste produced in gardens and parks) are mainly composed of carbohydrates, proteins and lipids, all of which represent high potential raw materials for the creation of valuable products.

An integral management of the MSW focused on maximum efficiency in the production of valuable resources could contribute to the economic transformation toward the *circular economy* model, according to the EU strategy about circular economy, and to the environmental and economic sustainability of the MSW management model [4].

In this context, the URBIOFIN biorefinery (developed under the grant agreement No 745785—Biobased Industries Joint Undertaking; European Union's Horizon 2020 research and innovation program) represents a proof of concept of the techno-economic and environmental viability of the conversion OFMSW into:

- Chemical building blocks (bioethanol, medium or short volatile fatty acids (VFAs), biogas).
- Biopolymers (low- and medium-chain polyhydroxyalkanoates (PHA), composites combining different PHA).
- Additives (bioethylene, nutrients for fertilizers).

By using the biorefinery concept applied to MSW (urban waste biorefinery), OFMSW is exploited as feedstock, taking into account its heterogeneity and variable composition, to produce at semiindustrial scale different valuable marketable products for local consumption. The processes involved in the development of these bioproducts are be interconnected, thereby a versatile and efficient biorefinery for OFMSW can be created. Urban biorefineries offer a new feasible and more sustainable scenario alternative to the current treatment of the OFMSW. The production, collection, and management of the MSW should be analyzed in the European context as a starting point to estimate how the MSW could be directed to local management systems based on biorefineries.

In addition, waste biorefineries should aim at a versatile exploitation of OFMSW as feedstock according to their nature and composition. Thus, the most important industrial demands as well as the necessary synergies that allow the design of a versatile and efficient biorefinery must be identified.

The described biorefinery will validate the whole value chain at demonstration scale, including the implication of the waste management authorities and companies, technology developers and the validation of the final products by the end-users. The URBIOFIN biorefinery has been divided in three sections that include multiple improvements to the baseline treatment plant (Fig. 2.1). First, a fraction of OFMSW is pretreated, hydrolyzed and fermented into bioethanol, which will be catalytically converted into bioethylene to be used in fruit ripening applications. At the same time, the other fraction of OFMSW is treated in a two-phase AD unit, aimed at obtaining VFAs in the first stage and biogas in the second stage. Subsequently, the VFAs are biologically converted into short chain PHA or elongated with the bioethanol produced previously and further bioconverted into medium-chain PHAs to be used in packaging and agriculture. The VFA-free digestate can be mixed with the vinasse originated from

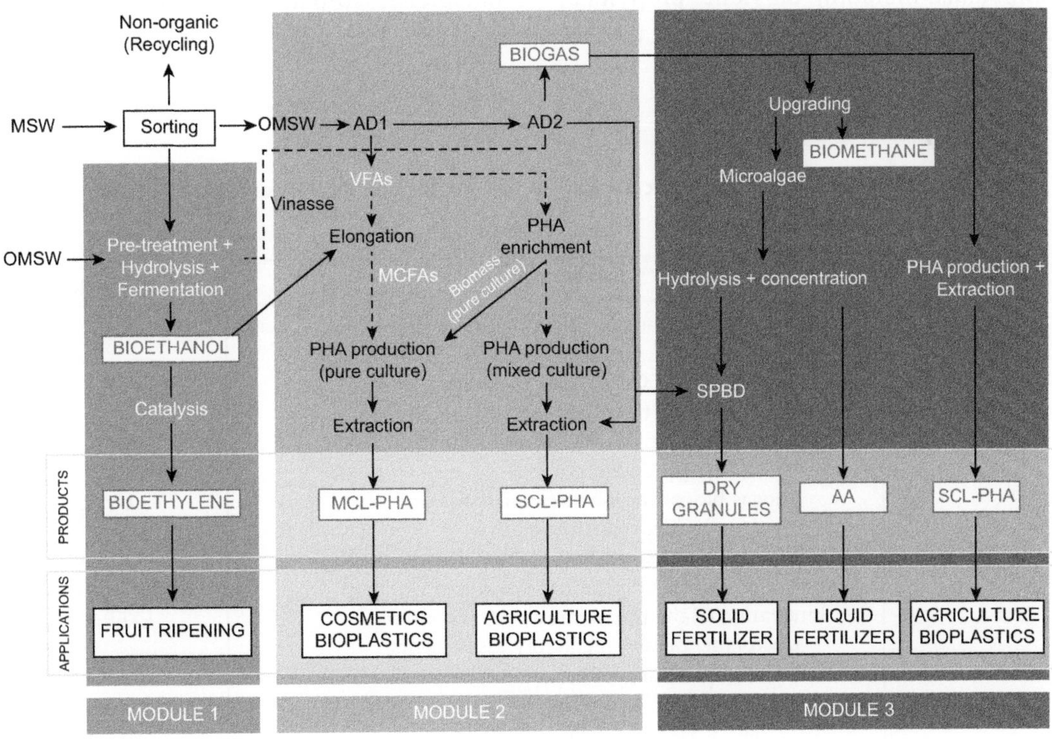

Figure 2.1

Overall URBIOFIN biorefinery diagram showing all biobased product lines.

OFMSW fermentation and anaerobically transformed into biogas. Downstream, the biogas produced in the second stage of AD can be photosynthetically upgraded to biomethane to be injected into natural gas grids or desulfurized and bioconverted into PHAs for packaging applications. The microalgae produced during photosynthetic biogas upgrading, together with the digestate and residual biomass from AD, is employed in the formulation of biofertilizers.

The URBIOFIN biorefinery was achieved by implementing eight different work packages (WP), led by the different partners of the consortium according to Table 2.1. The managerial structure of the consortium is depicted in Fig. 2.2.

2.2 Bioethanol from MSW as chemical building block

Bioethanol industry has rapidly increased worldwide, being the most used biofuel today globally. In 2017, the total bioethanol production was estimated in about 84 billion metric tons [5]. These figures have also brought up the huge market potential of bioethanol as feedstock for the production of alternative biobased chemicals [6,7]. As illustrated in Fig. 2.3, bioethanol can be catalytically converted into a wide range of compounds, including ethylene, propylene, 1,3-butadiene, iso-butylene, hydrogen, acetaldehyde, ethylene oxide, *n*-butanol, acetic acid, ethyl acetate, acetone and dimethyl ether [9].

Table 2.1: Distribution of the work packages and work package leaders in the URBIOFIN project.

WP No	WP title	WP leader
1	Preliminary actions for the urban biorefinery design	AINIA
2	Conversion of OFMSW to bioethanol as building block for the production of bioethylene	IMECAL
3	Conversion of OFMSW to VFAs for the production of PHA	URBASER
4	Biogas bioconversion to biomethane and added-value products	UVa
5	Final applications and industrial validation of the biobased products developed	BPE
6	Integration of the urban biorefinery. Economic, environmental and regulatory assessments	EXERGY
7	Communication, dissemination and exploitation activities	BCM
8	Project management	IMECAL

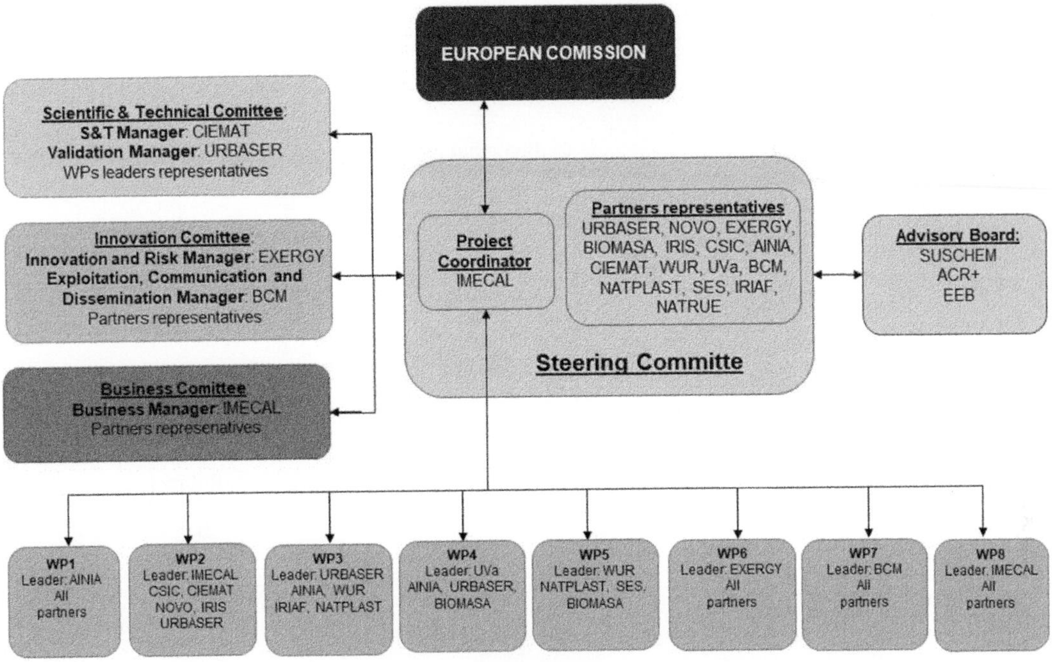

Figure 2.2
Scheme of the internal managerial structure of the URBIOFIN project.

Traditionally, bioethanol has been produced by converting edible sugar- and starch-based feedstock (e.g., sugarcane juice, molasses, corn) (Fig. 2.4). However, due to the competition with food and the concerns about ecological systems, recent research has shifted toward bioethanol production from nonfood biomass feedstock (lignocellulosic materials from agricultural, industrial and urban waste), which are low cost and abundantly available sources. In January 2013, bioethanol from lignocellulosic sources started to be produced at commercial scale [10]. Although its chemical composition greatly varies depending on factors such as culture, location, weather conditions, and degree of economy and development of a given society, the glucans contained in OFMSW have high potential as raw material for bioethanol production [11]. Starch and lignocellulosic carbohydrates (cellulose and hemicelluloses) from OFMSW can be converted into ethanol by either thermochemical or biochemical processing methods [12]. Bluefire, Coskata, New Planet Energy, Fulcrum and IMECAL (http://www.imecal.com/perseo/) are some examples of companies that have already developed technologies for converting OFMSW into ethanol at demonstration scale. Biotechnological OFMSW bioethanol production, such as the scheme proposed by IMECAL with its patented technology PERSEO Bioethanol, is preferred to thermochemical OFMSW bioethanol production due to the potential valorization of the

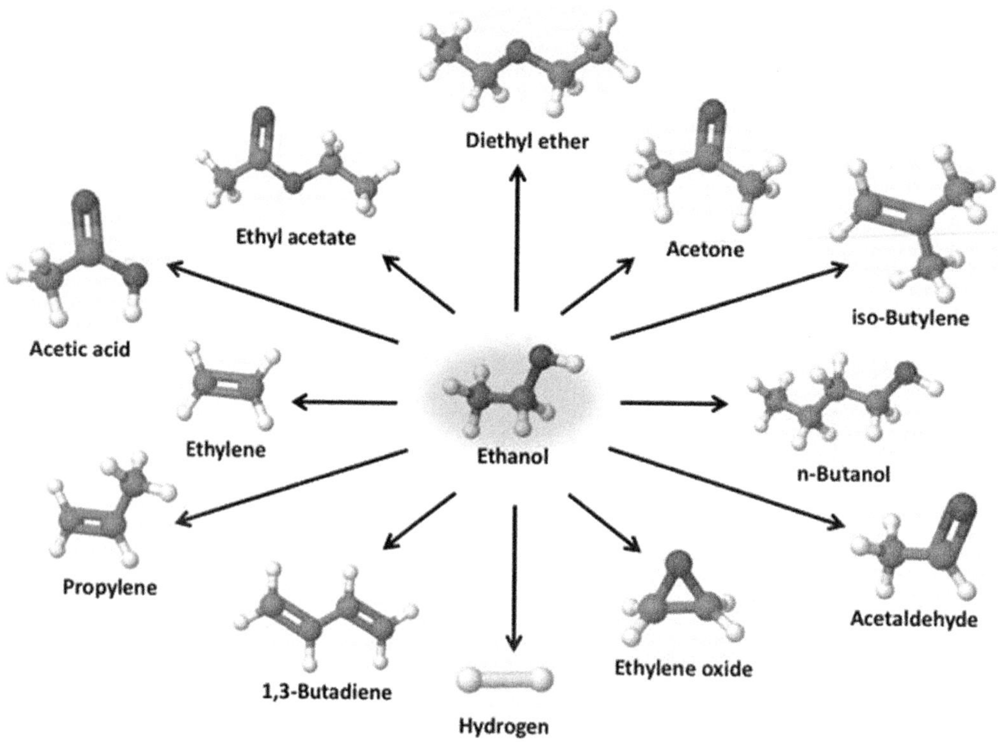

Figure 2.3
Ethanol as building block for the production of valuable chemicals. Carbon: dark gray; hydrogen: light gray; oxygen: red (black in print version). Molecule structures were obtained with JSME Molecular Editor [5,8].

remaining derived streams and its lower production costs. This process usually involves (1) a pretreatment step to ease the accessibility of hydrolytic enzymes to carbohydrates, (2) an enzymatic hydrolysis of starch, cellulose and hemicellulose, and (3) the subsequent microbial fermentation of the resulting sugars (mainly glucose) into ethanol. Compared to thermochemical processes, enzymatic processes avoid the presence of inhibitors to the fermentation stage. Saccharification and fermentation processes can either be performed separately, as separate hydrolysis and fermentation (SHF), or simultaneously, as simultaneous saccharification and fermentation (SSF). SSF strategies have shown superior process performance and better cost-effectiveness by reaching higher ethanol yields and reducing equipment needs by using a single reactor [13].

Current challenges for the biotechnological conversion of OFMSW into ethanol mainly include the development of improved process integration strategies, and increasing pretreatment and saccharification efficiencies. In URBIOFIN biorefinery, the pretreatment

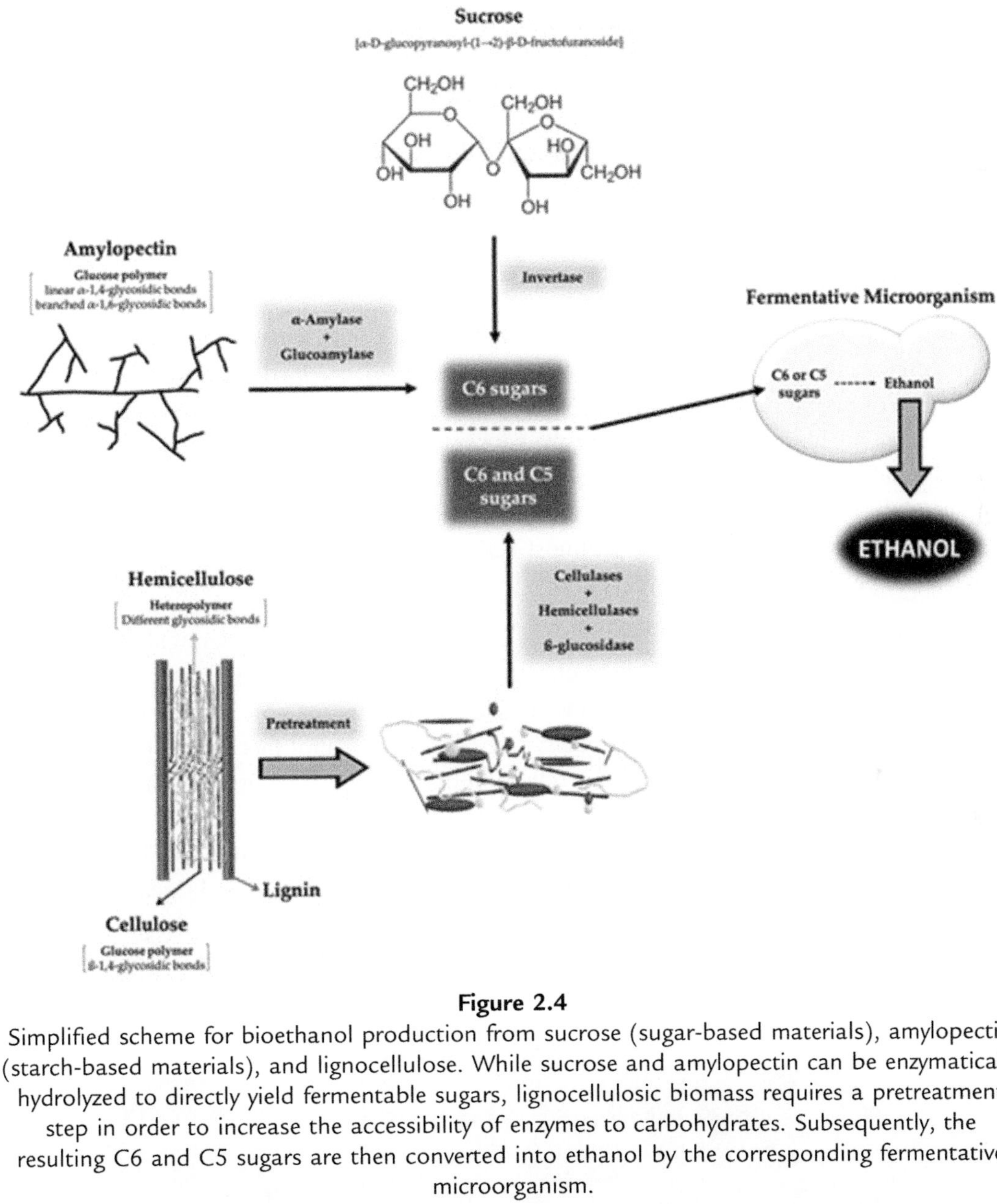

Figure 2.4

Simplified scheme for bioethanol production from sucrose (sugar-based materials), amylopectin (starch-based materials), and lignocellulose. While sucrose and amylopectin can be enzymatically hydrolyzed to directly yield fermentable sugars, lignocellulosic biomass requires a pretreatment step in order to increase the accessibility of enzymes to carbohydrates. Subsequently, the resulting C6 and C5 sugars are then converted into ethanol by the corresponding fermentative microorganism.

conditions (e.g., temperature, catalyst type and concentration) and the use of novel tailor-made enzyme cocktails to promote the effective depolymerization of cellulose and starch are being optimized to improve the OFMSW-based bioethanol economic competitiveness. Another critical aspect for the success of urban waste biorefineries is the development of the required process engineering for scaling-up technologies with such complex raw materials to a semiindustrial level. Finally, the technoeconomic improvements at the process engineering level determined in the previous stages of the project will be tested at the PERSEO Bioethanol semiindustrial plant (Fig. 2.5). The bioethanol produced in the URBIOFIN biorefinery presents the required quality as a chemical building block and can be further transformed into bioethylene.

2.3 Ethylene from OFMSW derived bioethanol

Ethylene is one of the most consumed chemicals worldwide, with a total production above 150 million tons in 2017 [14]. It is primarily a monomer used as a feedstock in the manufacture of plastic polymers (PE, PET, PEG, PVC and PS), fibers and other organic chemicals that are ultimately consumed in the packaging, transportation and construction industries, and in a multitude of industrial and consumer markets. Ethylene is also a gas plant hormone used for fruit ripening and degreening. During 2009–14, world ethylene consumption grew at an average rate of almost 4.5% per year and it is expected to grow at about 4% per year over the next 5 years [15].

Today, ethylene is mainly produced by cracking of naphtha in petrochemical industries [16]. However, concerns about climate change and global warming have encouraged the utilization of renewable materials and alternative energy sources instead of fossil resources. In conclusion, the catalytic dehydration of bioethanol has emerged as an alternative method to the fossil-based production of ethylene. The growing interest in this route to produce bioethylene is shown by the implementation of several commercial plants, which are currently in operation by companies such as Chematur, Braskem, TechnipFMC, or Axens, together with IFPEN and Total. The largest bioethylene production plant, with an annual production of 200,000 ton using sugarcane bioethanol as feedstock, is located in Brazil, [17]. In this context, bioethanol

Figure 2.5
PERSEO Bioethanol demonstration plant of IMECAL (L'Alcúdia, Spain).

can be catalytically dehydrated to produce bioethylene through chemical reaction (Eq. 2.1):

$$C_2H_5OH \leftrightarrow C_2H_4 + H_2O \tag{2.1}$$

The reaction of ethanol dehydration is endothermic, and ethylene yields are highly dependent on reaction temperature. The highest selectivity toward ethylene is obtained at 300–500°C. Higher temperatures shift the reaction toward acetaldehyde production, while lower temperatures result in production of diethyl ether [18]. In addition, the reaction is reversible, with the equilibrium being hindered by higher pressures and the presence of water vapor in the feed. The dehydration reaction occurs in the vapor phase inside the reactor in the presence of the corresponding catalysts. Maximizing ethanol conversion yield is crucial, since recycling of unconverted ethanol is a highly energy intensive process. By using specific acid catalysts designed for this reaction, a wide number of process schemes can be considered for supplying the required heat for the endothermic reaction. Four types of acid catalysts are mainly used in the ethanol dehydration reaction: phosphoric acid, molecular sieves (ZSM-5 type, Si–Al-phosphate (SAPO) type), oxides (activated-alumina based catalysts) and heteropolyacid catalysts [19].

Depending on the product application, the raw ethylene produced in the reaction needs to be further purified to reach different chemical grades of ethylene. Today, most part of the ethylene produced in the petrochemical industry is devoted to plastic production. To perform a successful polymerization reaction, a very high purity polymer-grade ethylene is required since the presence of impurities in ethylene has negative effects on the polymerization [20]. Typical byproducts formed during the catalytic dehydration of bioethanol include ethane, propane, propylene, butylene, or acetaldehyde. The ethanol feedstock can also contain some impurities such as methanol, propanol, or butanol [21]. These components are typically removed by additional distillation towers. Other mineral impurities such as salts in the form of cations, calcium, magnesium, sodium, potassium, iron and sulfate anions, chlorides and acidity of acetic acid, and even other possible contaminants present, can be retained by using ion exchange resins, where cations and anions are retained in two independent columns packed with resins, thus adapting the ethanol to the purity requirements.

In the URBIOFIN biorefinery, bioethylene is produced by catalytic dehydration of second-generation bioethanol derived from OFMSW. The produced bioethylene is used for citrus fruits degreening and banana ripening applications in postharvest fruit chambers. In this case, purity is not so critical compared to polymer application, where purities over 99% are required. When used as ripening gas, ethylene is usually sold mixed with other gases such as CO_2 or N_2 (with a concentration lower than 10%) to form a nonflammable mixture developed specifically to allow effective ripening without the high flammability risks of pure ethylene. The ethylene concentration necessary for ripening ranges from 5 to 150 ppm with exposure times between 12 and 72 h [22].

Considering bioethylene production as an independent process, the cost of this technology mainly depends on the feedstock price (i.e., bioethanol price). Thus, it is estimated that, depending on the region, bioethanol cost accounts for about 60%–75% of the bioethylene production cost [17]. Based on the fact that bioethanol production is a part of URBIOFIN biorefinery, the main cost of bioethylene production is associated to the energy consumption required to reach the high temperatures that shift the equilibrium toward ethylene production.

Another important aspect to consider during bioethylene production is catalyst deactivation. Although the catalyst is not consumed during the reaction, it can be deactivated, thus losing its ability to catalyze the reaction. The deactivation of the catalyst depends on: (1) the impurities contained in ethanol feedstock, (2) the conditions at which the reaction takes place, and (3) the catalyst nature. For instance, deposition of the coke formed during bioethylene production (depending on process conditions) is one of the main factors affecting catalyst deactivation. To prevent the conversion of ethylene into coke, the residence time for the ethylene product has to be kept as low as possible. Notwithstanding, such deactivation process is not fast, and the replacement of catalyst does not imply a critical cost in the process. During the plant operation, the possible impact of the associated impurities on the catalyst as well as on the chemical activity, selectivity and lifespan must be analyzed. The use of a modified catalyst could entail a reduction in the reaction temperature and therefore the energy consumption. However, changes in the catalyst have to be carefully evaluated because they can severely impact on the conversion yield, selectivity to the desired product and reaction stability. The chemical conversion, yield, and selectivity of the ethanol-to-ethylene catalytic reaction will be monitored by the installation of an in-line gas chromatograph in the pilot module. In the case that the catalyst lifespan under demo conditions is low, the cause of deactivation process will be studied and the required modifications of the catalyst to reduce the deactivation would be evaluated.

In conclusion, the catalytic dehydration of bioethanol to bioethylene is an alternative route to fossil-based production of ethylene that decreases the environmental impact of this chemical. Production of bioethylene from first generation bioethanol is a commercial process and several industrial plants are currently in operation. However, there is still no commercial process producing bioethylene from second generation bioethanol. An innovative bioethanol-to-bioethylene process for the production of bioethylene by catalytic dehydration of second-generation bioethanol derived from OFMSW is implemented in the core of the URBIOFIN biorefinery. This process constitutes a key market advantage for the potential customers of this biobased product. Particularly, the bioethanol-to-bioethylene demonstration module is located downstream in PERSEO Bioethanol pilot plant of IMECAL project coordinator (http://www.imecal.com/perseo/) located in L'Alcudia (Valencia, Spain).

2.4 VFA production from OFMSW

Traditionally, biodegradable waste has been transformed into biogas via AD to avoid the loss of its potential energy and to reduce the amount of waste going to landfill. In the process, anaerobic consortia conduct a series of syntrophic chemical reactions named hydrolysis, acidogenesis, acetogenesis and methanogenesis. Disposal of OFMSW currently faces environmental and economic challenges. Process efficiency, operational and equipment costs of biobased technologies need to be optimized to compete with fossil fuel-based commodities, e.g., natural gas and shale gas. Favoring the production of methane in the series of anaerobic reactions, the opportunity of obtaining a more profitable product is lost, since VFAs produced during the acetogenesis stage attain higher cost in the market [23].

Nowadays, the use of VFA produced by means of mixed culture fermentation through various fermentation pathways suffers from a severe handicap. The VFAs mixture produced has no clear economic value considering that the individual separation of VFA is not technically and economically feasible. An alternative in to surpass this economic issue in waste biorefineries is using the effluent from acidogenic AD directly in downstream processes, avoiding expensive purification processes, as nutrient and carbon source for biological processes for the production of both medium-chain fatty acids (MCFA) and PHA.

AD is still the core process technology of modern biorefineries, with the potential of waste stabilization, recovery of chemical building blocks for the carboxylate platform, renewable energy, and nutrients (Fig. 2.1). A two-phase AD system, comprised of a hydrolytic digester and a methanogenic digester, is operated to convert OFMSW in chemical building blocks (VFAs and biogas). Additionally, digestate is also used to generate solid and liquid biofertilizers, in order to valorize the whole OFMSW content. Mesophilic dry anaerobic digesters with VALORGA configuration are recommended due to their flexibility and ability to treat feedstock with high quantity of inert materials as it can be expected in mechanically sorted OFMSW (Fig. 2.6). It represents also an advantageous configuration due to the need for a low feedstock dilution, maximizing product and nutrients concentrations for its recovery in downstream processes.

The population of methanogens must be inhibited in the first hydrolytic digester of the 2-phase AD system in order to promote biological VFA production. This can be done by applying inoculation strategies of thermal pretreatment, pH shocks or inhibitor supplementation [24–27]. A most suitable approach to reduce operational costs and obtain an effective start-up of the hydrolytic digester is to inhibit the production of methane by overfeeding the methanogenic population. High organic loading rates (OLR)

Figure 2.6
Hydrolytic digester installed in URBASER's research center (CIAM, Zaragoza).

cause imbalance between the acidogenic and methanogenic populations, accumulation of VFA in the digester, a decline in buffer capacity, a pH decrease and the complete inhibition of the methanogenic stage [28]. The use of a high protein content feedstock can also generate an inhibitory ammonia concentration in the digester at high OLR, producing VFA concentrations as high as 100 g/L [29]. In the case of OFMSW, only thermophilic temperature can reach inhibitory ammonia concentrations [30].

Fig. 2.7 shows the general degradation route of particulate organic matter to VFA in the AD process. Anaerobic microorganisms compete for substrate in the complex metabolic pathways found in mixed culture fermentative systems. The activity of the enzymes involved in the hydrolytic step is vulnerable to decrease when the operational pH is not in the optimal range (5–7) [32]. This is the first essential and rate limiting stage in the anaerobic degradation of complex substrates such as OFMSW to VFA. Additionally, it is

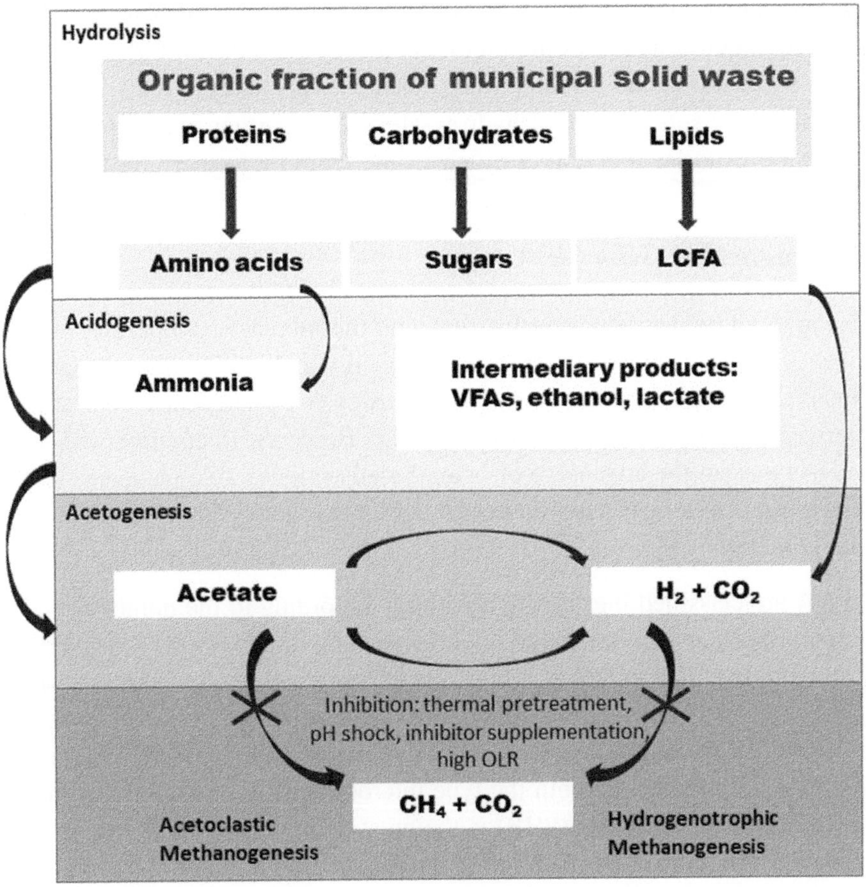

Figure 2.7

Anaerobic conversion of organic fraction of municipal solid waste to volatile fatty acids [31]. *Modified Arslan D., Steinbusch KJJ, Diels L, Hamelers HVM, Strik DPBTB, Buisman CJN, De Wever H. Selective short chain carboxylates production: a review on control mechanisms to direct mixed culture fermentation. Critical Reviews in Environmental Science and Technology 2016;46(6):592–634.*

key to maintain the operational pH value out of the optimum range for methanogenic *archaea* (6.8–8) in order to minimize CH_4 production. Therefore, the most advantageous pH interval to maximize acidification is 5–6, with or without pH control.

Several parameters such as the hydraulic retention time (HRT), OLR, pH, temperature and trace element concentration, can be controlled in order to alter the VFA yield and spectrum of individual VFAs produced. Additionally, the use of feedstock pretreatments, e.g., hydrothermal or alkaline, can stimulate the production of VFA at a restrictive cost [33,34]. Special attention needs to be paid when selecting the equipment for solid-liquid separation of the partially digested effluent. Dewatering represents a fundamental process in the treatment and valorization of waste. In particular, when the suspended biomass is still present in the liquid fraction of the digestate, it may compete with the specialized aerobic biomass that utilize the soluble molecules, i.e., VFA and nutrients, to produce PHA. Additionally, the solid stream of hydrolysate will be further treated in a conventional anaerobic digester for the production of biogas, which will be further bioconverted into PHA and biomethane.

2.5 PHA production from VFA

PHAs are a family of biodegradable polyesters with thermoplastic properties, produced by a wide variety of microorganisms as intracellular energy and carbon reserve material [35]. Industrial interest on PHA rely on its similar properties to common thermoplastics polymers derived from petrol such as polypropylene. However, in contrast with synthetic polymers, PHAs present the advantage of being based on renewable resources and completely biodegradable by a wide variety of microorganisms and under different environmental conditions [36,37].

Generally, PHA are classified into two main groups according to the number of carbon atoms that comprise their monomeric unit. Thus, short chain length PHA (scl-PHA) consists of three to five carbon atoms, whereas medium-chain length PHA (mcl-PHA) consists of 6−14 carbon atoms. PHA monomeric structure is directly related with the final physical properties of the material, determining its further applications. The monomer composition obtained depends on both the type microorganisms and nature of the carbon sources used. Polyhydroxybutyrate (PHB) is the most common scl-PHA and is characterized as a stiff and brittle material. From the incorporation of 3-hydroxyvalerate (HV) monomers into PHB, the resultant co-polymer becomes into a more flexible and easily processed by heat material proportionally to the amount of HV introduced. According to their intrinsic properties, scl-PHA are generally used for packaging material and disposable items, whereas mcl-PHA, which are characterized as elastomers, are suitable for film production and for high-value added applications such as surgical sutures, implants, drug delivery, etc. [38].

An environmental analysis of plastic production processes was carried out by comparing the petroleum-based polypropylene (PP) and the biological-based PHB production processes using LCA [39]. PHB was superior to PP in all the LCA categories. The energy requirements for PHB production are significantly lower than those of petroleum-based plastics. However, the environmental benefits that could be obtained by the replacement of conventional plastics by biopolymers may entail an important economic loss [40]. The main factors that increase the final costs of PHA are the price of the substrates used as a carbon source, bacterial productivity and downstream processing [41].

Several microorganisms with the ability to accumulate PHA at high yields have been studied in recent years, *Ralstonia eutropha* being one of the most referenced and well-known microorganisms due to its high productivities (up to 5 g/L h) [42]. Despite pure bacterial cultures are constantly investigated due to their high productivity, more efforts are necessary to develop bioprocesses for the valorization of waste streams and byproducts maintaining those high productivities. Since each type of waste or byproduct has different composition and characteristics, the selection of the appropriate microorganisms is of great importance.

An alternative scenario to pure cultures or engineered strains that would contribute to a reduction in production costs relies on the use of microbial mixed cultures (MMC). This approach is based on the utilization of mixed microbial consortia, operated under nonsterile conditions and ecological selection principles, where the operational conditions imposed on the biological system promote the development and maintenance of PHA-accumulating microorganisms. The efficiency of MMC for PHA production depends, to a great extent, on the microbial selection of the cultures by the application of different operational conditions on the bioreactor, known as enrichment step. Thus, PHA production with MMC is typically carried out in two different phases or stages, i.e., culture selection or enrichment and PHA accumulation. The separation of these two phases on two different bioreactors allows a better process optimization, since the optimal conditions and operational regimes of each stage are different. The main challenge of the enrichment step is the production of a culture highly enriched on PHA producing microorganisms and at high biomass concentrations. There are many methodologies to carry out the enrichment step, but one of the most employed is the classical aerobic dynamic feeding [43]. Enrichment process initiates with an aerobic sludge from a wastewater treatment plant, which is subjected to feast-famine cycles, where culture is first under an excess of carbon source and then under a limitation period of carbon source much longer than the feast phase. Microorganisms able to transform carbon source during feast period into PHA will have higher surviving rates during famine phase, since the stored PHA can be used as carbon and energy source, which allows them to outcompete the non-PHA-accumulating microorganisms. The enrichment step is the key reaction in the global scl-PHA production process. The purpose of the enrichment step is promoting the selection of those microorganisms able to accumulate PHA, and it determines the PHA-storing capacity of the MMC, but the operational conditions of this step also determine the biomass content on the system. Since PHA is an intracellular product, the biomass content reached in the enrichment phase is a rate limiting step of the process, similarly to the PHA-storing capacity of the system. In particular, the operational conditions applied during enrichment step, such as hydraulic retention time, solid retention time, organic loading rate and cycle length, together with the VFA stream composition, are directly related with the characteristics of the enriched biomass obtained after this process.

The biomass obtained in the enrichment step under stationary conditions is transferred to the accumulation step, where a continuous feeding of VFA stream is applied under nutrient (N and/or P) limitation conditions in order to obtain a biomass with a high PHA content. PHA production by MMC was successfully achieved for different type of industrial waste streams, which rendered this approach a promising option for cost reduction in PHA production [38,44,45].

In the URBIOFIN biorefinery, the valorization of waste byproducts as a feedstock to produce two different types of PHA using both pure and microbial mixed cultures has been implemented. On the one hand, the VFA present in the stream derived from the anaerobic hydrolyzation of OFMSW are biologically elongated to MCFA. Then, these MCFA are used as substrate to produce mcl-PHA using pure cultures (Fig. 2.8) of *Pseudomonas* strains, well-known mcl-PHA producers able to assimilate a wide range of carbon sources including fatty acids [46].

On the other hand, the VFA present in the stream derived from the anaerobic hydrolysis of OFMSW are used as direct substrate for scl-PHA using MMC following an aerobic

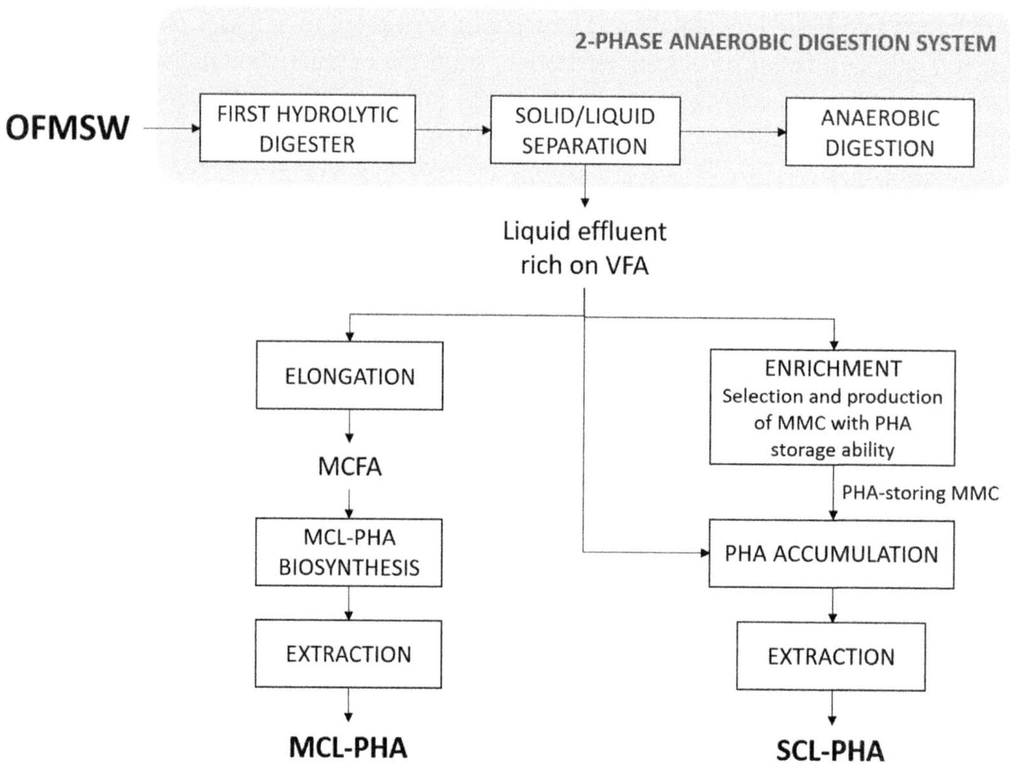

Figure 2.8
Block diagram for both scl- and mcl-PHA production in URBIOFIN.

dynamic feeding regime during enrichment step (Fig. 2.8). Finally, both types of biomass are subjected to an extraction procedure, adapted for each biomass requirements in order to obtain the final PHA products.

2.6 Biomethane production

The presence of pollutants such as CO_2, H_2S and volatile methyl siloxanes (VMSs) in biogas limits its use as a renewable alternative to fossil natural gas. The purification of raw biogas into biomethane, typically referred to as *biogas upgrading*, allows the injection of this green gas biofuel into natural gas grids or its use in automation [47]. Today, H_2S removal at commercial scale is carried out using adsorption-based technologies or *in-situ* chemical precipitation in the digester, while VMSs removal is conducted in activated carbon/silica gel filters or by cryogenic condensation [48,49]. On the other hand, water scrubbing and membrane separation rank among the most popular technologies for CO_2 removal, along with pressure swing adsorption and chemical scrubbing [47]. In this context, while research on biofiltration in the past decades has resulted in biological reactors such as biotrickling filters (BTF) for H_2S removal at commercial scale, biotechnologies for CO_2 abatement such as hydrogenotrophic or photosynthetic biogas upgrading are currently at a validation stage in demo-scale prototypes, and biological siloxane removal is still in a proof-of-concept stage. The simultaneous removal of H_2S and CO_2 from biogas in an algal-bacterial photobioreactor and siloxane removal in a two-phase partitioning bioreactor (TPPB) has been for the first time implemented at demo-scale in the URBIOFIN biorefinery.

The presence of CO_2 in biomethane decreases its specific calorific value, while H_2S is highly corrosive, toxic, and malodorous. Photosynthetic biogas upgrading is an emerging biotechnology engineered at Valladolid University (Spain) capable of removing simultaneously H_2S and CO_2 based on the symbiotic interactions between alkaliphilic microalgae and bacteria [50]. Microalgal-bacterial consortia are cultivated in a photobioreactor (typically an open high rate algal pond (HRAP), where dissolved CO_2 is photosynthetically fixed in the form of algal biomass in the presence of light) interconnected to a bubble absorption column (where CO_2 and H_2S are absorbed) via recirculation of the algal cultivation broth (Fig. 2.9). In the absorption column, H_2S is oxidized to sulfate by litoautothrophic bacteria using the CO_2 (as a carbon source) and the O_2 (as electron donor) present in the cultivation broth. O_2 stripping from the cultivation broth to the biomethane is considered as the rate limiting step of this upgrading technology. A recent optimization of this technology has shown that the implementation of a settler between the HRAP and the absorption column allows the operation of the column with a biomass-free scrubbing solution and the possibility to control biomass productivity by setting a fixed biomass wastage rate [51]. The nutrients (nitrogen and phosphorus) and

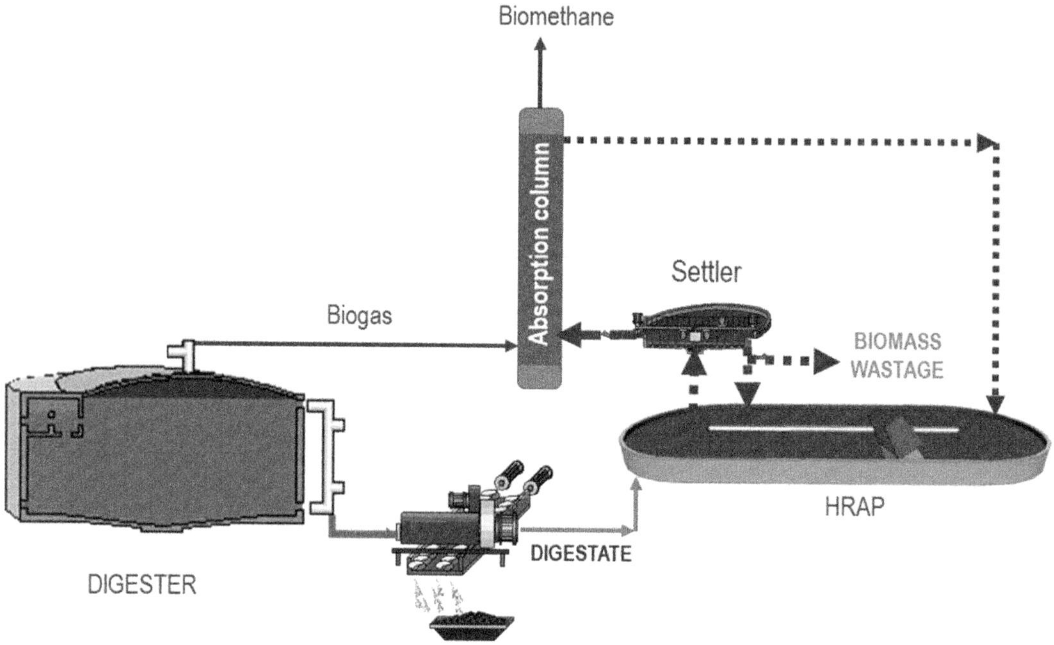

Figure 2.9
Schematic diagram of a photosynthetic biogas upgrading process coupled to nutrient recovery in the form of algal biomass.

water required to support microalgae and bacteria growth can be obtained from the digestate produced in the anaerobic digester, which entails an opportunity to produce microalgae-based biofertilizers while reducing the eutrophication potential of these high-strength wastewaters. This technology has been recently optimized in a 180 L HRAP indoors and outdoors in the context of domestic wastewater treatment, with a consistent production of a biomethane complying with most international regulations ($CH_4 > 95\%$, $CO_2 < 1\%$, $N_2 < 3\%$, $O_2 < 0.5\%$), and has been up-scaled in the URBIOFIN biorefinery in a 300 m^2 HRAP interconnected to a 0.5 m^3 absorption column fed with 12 m^3/d of biogas and 0.1 m^3/d of liquid fraction of digestate from the AD of OFMSW. The ratio between the biogas flow rate and the liquid recirculation flow rate in the absorption column, and the pH and alkalinity of the cultivation broth in the HRAP, have been consistently shown as the most relevant operational parameters determining biomethane quality in this innovative biotechnology [52]. Overall, photosynthetic biogas upgrading in algal-bacterial photobioreactors could decrease the operating costs of conventional biogas upgrading (H_2S adsorption by activated carbon filtration + CO_2 removal by water absorption) from 0.2 €/Nm^3 to 0.03 €/Nm^3, with a concomitant decrease in the energy demand from 0.3 kWh/Nm^3 to 0.08 kWh/Nm^3 [53].

The presence of VMSs in biogas represents an important handicap for the valorization of biogas as a renewable energy source, since VMSs are oxidized to silicon dioxide (SiO_2) during biogas combustion. These crystalline deposits of SiO_2 accumulate in the equipment, reducing the efficiency and increasing the maintenance costs due to corrosion, erosion and clogging of pipelines and engines [54,55]. Therefore, there is an increasing interest in VMSs removal from biogas. In this context, previous research has demonstrated that physical/chemical off-gas treatment technologies exhibit higher economic and environmental impacts than their biological counterparts [48]. In the last 10 years, biological technologies for VMSs removal have been studied, particularly in a BTF configuration. In a BTF, the VMS-laden biomethane passes through a packed bed of synthetic inert material where microbial communities responsible for VMS biodegradation are immobilized as a biofilm (Fig. 2.10). A continuous recirculation of a trickling liquid solution prevents drying of the packing material, provides nutrients for microbial growth

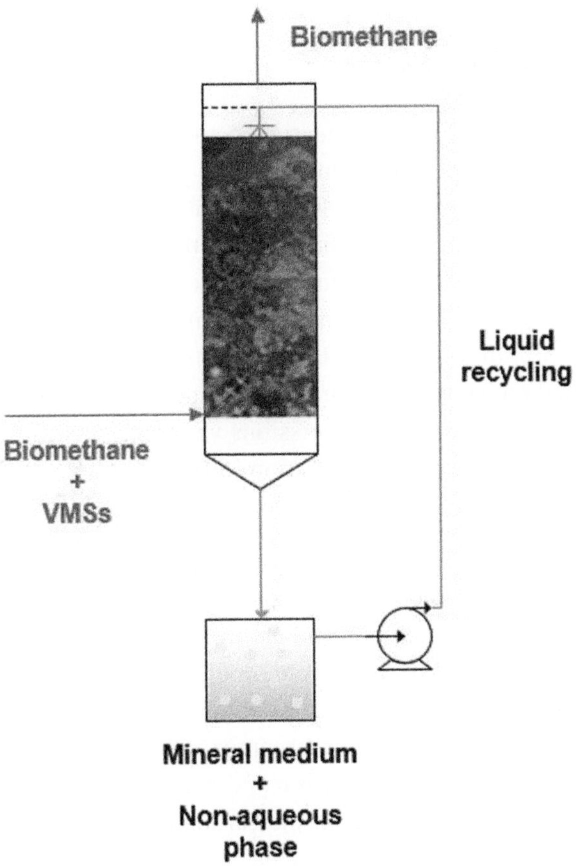

Figure 2.10
Schematic diagram of a two-phase partitioning biotrickling filter for VMSs treatment.

and wash-out from the biofilm any potential inhibitory metabolite produced during VMS biodegradation [56]. In this context, the operation of a BTF for the treatment of hexamethylcyclotrisiloxane (D3) under aerobic conditions has shown a removal efficiency (RE) of up to 20% at empty bed residence times (EBRT) of 2.1–3.6 min [54]. Likewise, the operation a BTF for octamethylcyclotetrasiloxane (D4) treatment under anaerobic and aerobic conditions resulted in REs of 43% at an EBRT of 19.5 min and 15% at an EBRT of 4 min, respectively [57]. These investigations suggested that an effective VMSs removal requires (1) high EBRTs, (2) the presence of an organic phase (nonaqueous phase) capable of enhancing the mass transfer of VMSs from biomethane to the microbial community, and (3) a microbial culture previously enriched with the ability to use VMSs as the only carbon and energy source. This technology is currently under optimization in the URBIOFIN biorefinery in a 2 L lab scale BTF operated with a nonaqueous phase supplemented to the mineral salt medium. The BTF will be inoculated with an enriched VMSs-degrading culture. This technology will be then validated in a 120 L BTF fed with 7.2 m^3/d of biomethane from the photosynthetic biogas upgrading unit.

2.7 PHA production from biogas

Methane (CH_4), with a global warming potential 85 times higher than that of CO_2 for a 20 years horizon and an atmospheric lifetime of 7–12 years, represents nowadays the second most important greenhouse gas (GHG). Its atmospheric concentration has steadily increased during the 20th century and the first part of the 21st century, achieving a value of $\approx$1835 ppb in 2015, an increase by a factor of 2.4 compared with preindustrial levels [58]. Methane concentration in anthropogenic emissions widely vary depending on the source, ranging from $<$3% up to 80% of CH_4 (i.e., land filling, composting, wastewater treatment). While diluted emissions are commonly vented to the atmosphere or flared as they are not suitable for energy recovery, emissions with CH_4 concentrations exceeding 20% are usually combusted for energy production [59]. Nevertheless, these current practices for the management of CH_4-loaded emissions entail a negative impact on the environment due to CH_4 and CO_2 release.

As opposed to traditional end-of-pipe physical-chemical methods for CH_4 abatement, biological technologies offer a sustainable and environmentally friendly alternative with lower associated operating costs. These biotechnologies are based on the biocatalytic action of aerobic methane-oxidizing bacteria called methanotrophs, able to degrade CH_4 into CO_2 and water using O_2 as the electron acceptor. Aerobic methanotrophic bacteria belong to the *Alphaproteobacteria*, *Gammaproteobacteria* and *Verrucomicrobia* phyla, and are classified into three groups depending on the carbon assimilation pathway: type I methanotrophs that use the ribulose monophosphate (RuMP) pathway; type II methanotrophs using the serine-pathway; and type X, a subset of type I that uses the

RuMP pathway for carbon assimilation, but also presents low levels of serine-pathway enzymes [60,61]. Moreover, methanotrophs have been reported to produce more than a dozen of different high added–value products using CH_4 as a feedstock (bioplastics (PHA), ectoine, biofuels, extracellular polysaccharides (EPS), single cell proteins (SCP), etc.). In this context, the implementation of a CH_4 biorefinery combining the abatement of this GHG contained in biogas and the production of multiple valuable byproducts has been recently investigated in order to boost the economic viability of biological CH_4 treatment technologies (Table 2.2).

Among these marketable products, PHAs represent one of the most profitable in terms of productivity, production-derived costs and proximity to market [63]. PHAs such PHB or poly-3-hydroxyvalerate (PHV) are biopolymers with comparable mechanical properties to those of polypropylene and polyethylene, besides being biodegradable and biocompatible. These properties make them attractive substitutes to oil-based plastics. These intracellular biopolyesters are produced under nutrient-limiting and CH_4-excess conditions by a wide range of microorganisms as carbon and energy storage resources. To date, the high costs of PHA compared to conventional petroleum-based plastics has limited their application, the feedstock cost (usually sugar from corn or sugar cane) accounting for more than 30% of the total production costs [62]. Therefore, CH_4 represents a low cost and environmentally friendly alternative substrate for the synthesis of PHA with a more competitive market price.

Production of PHA seems to be exclusively related to type II methanotrophs, since previous studies have suggested that PHA synthesis is linked to the serine cycle, whereas

Table 2.2: Potential added-value products obtained from methanotrophic strains using CH_4 as feedstock [59,62].

Product	Application	Market price (€/kg)	Genera	Production yield/ rate	Demo-scale
Ectoine	Medicine, cosmetology, dermatology, nutrition	900	*Methylomicrobium* *Methylobacter* *Methylohalobius*	70–230 (mg/ $g_{biomass}$)	Laboratory
PHAs	Substitutes of petroleum-based plastics	4–20	*Methylocystis* *Methylosinus* *Methylocella*	200–510 (mg/ $g_{biomass}$)	Industrial
Methanol	Biofuel	–	*Methylococcus* *Methylosinus* *Methylocystis*	0.13–25 μmol/ (mg_{cells} h)	Laboratory
EPS	Food, pharmaceutic, textile, oil industries	4–50	*Methylobacter* *Methylomonas* *Methylomicrobium*	300–1800 (mg/ $g_{biomass}$)	Laboratory

type I methanotrophs produce EPS instead of PHB under unbalanced growth conditions through the RuMP pathway. Moreover, at least three genes (phaCAB) are considered indispensable for PHB production in methanotrophs, encoding the condensation of two acetyl-CoA molecules to acetoacetyl-CoA (phaA), reduction of acetoacetyl-CoA to (R)-3-hydroxybutyryl-CoA (phaB), and polymerization of(R)-3-hydroxybutyryl-CoA monomer units into PHB (phaC) [60]. The synthesis of these genes has not been reported in type I methanotrophs. Accumulation of PHA is commonly performed in two stages, the first one aims at obtaining a high cell density culture under nutrient sufficient conditions, which is subsequently subjected to nutrient limitation to promote PHAs accumulation. The type of nutrient limitation applied will affect the maximum PHA content achieved and its molecular weight. Although N-limitation has been the most frequently studied, the influence of limiting other macro- and micronutrients such as Mn, Fe, P, K, or Mg on PHA yield deserves further investigation [64]. Besides, the addition of VFAs to the cultivation broth might increase PHA yield and tailor the composition of the composite. This option is particularly beneficial when CH_4 contained in biogas from AD is used as feedstock, since VFAs would be readily available during AD. In this context, Recent studies demonstrated that the addition of 10% extra carbon as VFAs resulted in an increase in PHA yield, and valeric acid supplementation supported a higher 3-hydroxyvalerate content in the final biocomposite [65].

Besides the lower production costs of biogas-derived PHA and the demonstrated high accumulation capacity of methanotrophic bacteria, the implementation of this technology is yet constrained by the low aqueous solubility of CH_4 and O_2 (necessary for CH_4 oxidation) which is considered, at this point, as the rate limiting step. Mass transfer of both compounds from the gas to the liquid interphase is limited by their low mass transfer coefficients (dimensionless Henry's law constant, $H = C_G/C_L$, of 29.0 and 31.3, respectively, at 25°C and 1 atm) [66]. Thus, reaching high CH_4 biodegradation rates and reasonable PHA productivities in the reactor require of prolonged contact times and improved mixing between liquid and gas phases. The increase in residence time is associated to higher reactor volumes that might lead to prohibitive investment costs, while a raise in the system turbulence is often related to excessive operational costs. Therefore, the cost-efficient implementation of this technology requires the development of innovative reactor configurations and smart operational strategies that maximize gas-liquid mass transfer and minimize reactor dimensions. In recent years, two different strategies have attracted the interest of researchers due to their improved performance in the removal of poorly water soluble compounds such as CH_4: (1) the addition of a secondary organic phase (NAP) and (2) the internal recycling of the gas stream [60,67]. NAPs are often nontoxic, nonbiodegradable and biocompatible compounds with high affinity toward CH_4 and O_2 [68]. These NAPs perform two actions simultaneously that facilitate the biodegradation of the target gas pollutant. On the one hand, NAPs act as pollutant reservoir allowing an additional mass transfer route from the NAP to the aqueous phase (in addition to the conventional gas-water mass transfer). On the other hand, the dispersion of these compounds has a positive effect on the hydrodynamics of the gas-liquid system by reducing the size of gas bubbles (which increases interfacial area) and reducing the gas-liquid interfacial tension [69,70]. Besides the addition of silicone oil as NAP at 5−10%*v/v* has shown an increase in CH_4 abatement in bioreactors by a factor of 1.4, the utilization of a secondary phase during the co-production of PHAs might complicate further downstream processes and deserves additional investigation [71].

Another promising strategy for enhancing gas-liquid mass transfer in bioreactors is the internal recycling of the gas mixture, which permits decoupling the system turbulence and the gas residence time, thus obtaining increased CH_4 biodegradation rates and avoiding an increase in reactor volume (or a decrease in overall residence time) [64,67]. This strategy was first employed for methane abatement in packed bioreactors such as BTFs, which are not recommended for co-production of bioproducts given the technical limitations of biomass harvesting [67]. However, in recent years, suspended growth bioreactors equipped with internal gas recycling as well as with other additional fittings for increasing system turbulence such as static mixers, external recycling pipelines, nozzles for gas side streams and additional liquid pumping systems have been used [64,72,73].

Suspended growth bioreactors (stirred tank reactors and bubble column bioreactors) have been traditionally used for fermentation processes due to the possibility of processing in a continuous or quasi-continuous regime at high biomass concentrations and convenient biomass harvesting [74]. However, most studies on PHA production by type II methanotrophs have used stirred tank reactors, which do not represent an economically feasible option at industrial scale due to the high cost dedicated to mechanical agitation of the culture broth. In this regard, bubble column bioreactors are the most promising bioreactors for PHA production given the minimized energy consumption for mixing in the absence of mechanical stirrers and their demonstrated high PHA accumulation capacity (40%−50%) [64,75,76].

In the URBIOFIN biorefinery, a two-stage system for the anoxic desulfurization and subsequent bioconversion into PHA of 60 m^3/d of biogas has been designed by the University of Valladolid. The bioconversion of the CH_4 contained in the biogas from the AD of OFMSW will take place in a bubble column bioreactor equipped with internal gas recycling. Biomass growth and the PHA accumulation occurs in a single stage. CH_4 removal efficiencies and PHA accumulation capacities over 80% and 40% are expected, respectively. H_2S must be removed from biogas to prevent clogging of the diffusers in the bubble column reactor. Due to the high operating cost of physical/chemical biogas desulphurization technologies, the biological oxidation of H_2S using nitrate as an electron

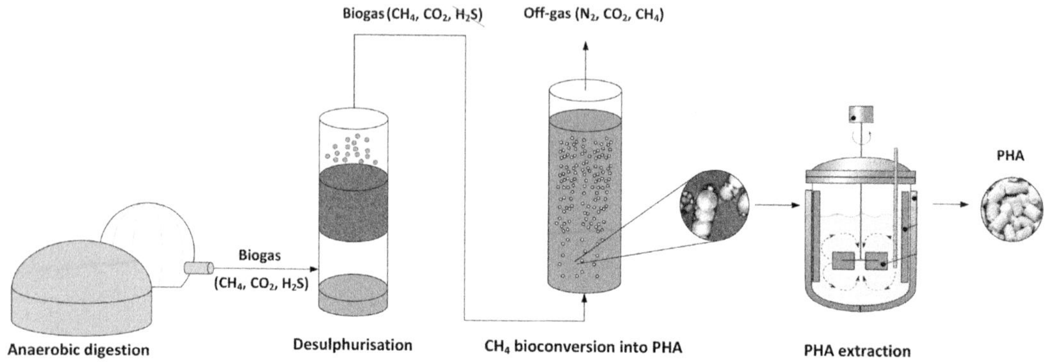

Figure 2.11
Schematic diagram of the biogas desulfurization, CH_4 bioconversion into PHA, and PHA extraction from the microbial cells.

donor in a biotrickling filter was selected as desulfurization technology in the URBIOFIN biorefinery (Fig. 2.11).

2.8 Biobased fertilizer production

Cities are major concentrators of materials and nutrients, contained in the urban waste streams, which are mainly extracted from agricultural soils. Nowadays, the return of these nutrients to soil is far from being achieved. At the same time, agricultural practices are leading to progressive soil degradation and mineral fertilizers are massively applied to increase crops productivity. European countries consume 11 Mton of nitrogen (N), 2.5 Mton of phosphate (P_2O_5) and 2.9 Mton of potash (K_2O) each year [77]. Current fertilizers regulatory framework (Regulation 2003/2003/EC) only covers mineral fertilizers [78]. As a part of the Circular Economy Action Plan, a new Fertiliser Products Regulation, FPR has been approved, opening the access to the EU market for biobased fertilizers. It is based on voluntary harmonization and set standards for marketing in Europe under CE mark. Moreover, it will promote the transformation of secondary raw materials into high-quality fertilizing products while reducing the dependency of the European Union with respect to the nutrients from third countries, meeting the challenges of sustainable agriculture and promoting a strong Circular Economy in Europe. FPR has been already published in *Official Journal of the European Union* on June 25, 2019 and will enter into force in 2022.

Organic fertilizer products are currently regulated only at regional and member state level, and its annual consumption (excluding manure) is estimated as 6% when compared to the inorganic fertilizers (332,800 and 549,800 ton of nitrogen and phosphate, respectively) [79].

Considering that mineral fertilizer supply is strongly dependent on nonrenewable resources (especially in the case of phosphorus), and that mines are often located in non-EU countries, Europe is making great efforts to promote partial replacement of mineral fertilizers by integrating organic fertilizers derived from recovered biowaste and biomass in the future fertilizers regulation (Proposal COM/2016/0157) [77]. It is expected that the integration of these organic fertilizers into the EU legal framework will also help to reduce both health and environmental risks derived from their application to agricultural soils.

It has been estimated that the total amount of nutrients present in food, animal and human waste on a global scale could represent nearly 2.7 times the nutrients contained on the chemical fertilizers currently used [80]. In particular, OFMSW represents an important source of nutrients. Each person generates annually 483 kg per capita of MSW, of which up to 50% in volume is represented by OFMSW [2]. OFMSW separation rates show significant differences between regions and member states. In this regard, with the aim to tackle the environmental impacts associated to landfilling of urban biowaste (i.e., GHGs, leachate formation, etc.), the new directive (UE) 2018/851 on waste commits all member states to separate at the source 55% of the urban waste for reuse and recycling by the year 2025 [81]. Today, more than half of total MSW generated in European countries is disposed of through landfilling or incineration [2], while the European Compost Network has estimated that 40% of source-separated municipal biowaste is composted or digested [82]. In total, the annual production of compost is 13 Mton, 79% coming from green waste and biowaste and 21% from sewage sludge and mixed waste [83]. It is worth highlighting that compost derived from nonseparately collected OFMSW presents a higher heavy metal concentration and presence of inert materials such as plastics and glass [84].

Urban solid water biorefineries such as URBIOFIN seek for the implementation of circular economy principles for MSW valorization and nutrient recovery through the integration of different biotechnological processes, as can be observed by analyzing the general nutrient flows of the URBIOFIN biorefinery (Fig. 2.12).

In this urban solid waste biorefinery, the carbohydrate fraction of the OFMSW is fermented and converted into bioethanol, which is then catalytically transformed into bioethylene. The residual stream, namely vinasse, is still rich on nutrients (N, P and K) and carbon (nondegraded OFMSW and yeast biomass). The vinasse is anaerobically digested together with fresh OFMSW in a 2-phase AD to produce biopolymers and biogas. Most of the N, P, K, and complex organic matter, such as lignin, will remain in the digestate. When digestate is applied as organic fertilizer, this organic carbon remains in the soil after 1 year of its application, thus contributing to hummus build-up. Therefore, the digestate provides a significant enhancement of soil properties [85].

During the AD, organically bound nitrogen is released as ammonium NH_4^+, becoming directly available for crop uptake. Hence, the efficiency of the digestate as N-fertilizer is

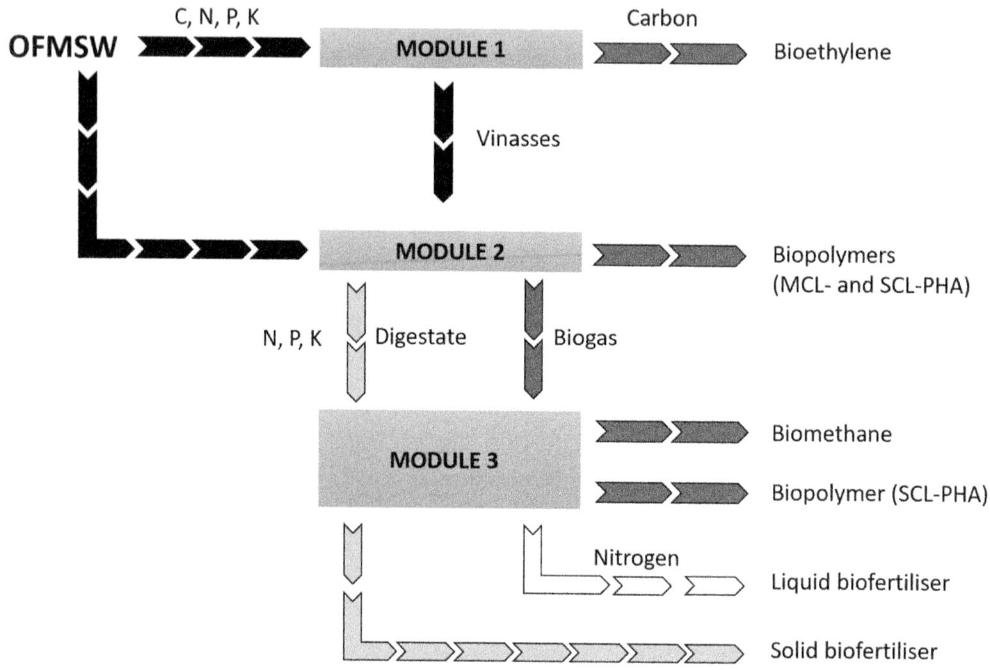

Figure 2.12

Basic nutrient fluxes in URBIOFIN biorefinery. Classified as streams containing only carbon, nitrogen, phosphorous and potassium (*black arrows*); only carbon (*dark grey*); mainly nitrogen, phosphorous and potassium (*light grey*); and mainly nitrogen (*white*).

highly dependent on $NH_4^+ - N$ concentration. OFMSW presents high organic nitrogen content, 7.6 ± 1.0 kg N/ton on average. The total phosphorus content does not change during AD, but the organically bound P becomes more available for crops after AD. Likewise, the total K, Ca, Mg and heavy metal content are not altered during AD, but some of them become soluble such as K, Ca and Mg. Potential toxic elements, such as Zn and Cu, may be critical in high concentrations despite being essential micronutrients for healthy plant growth [86].

Regarding biobased fertilizer production, the microalgae unit included in the biorefinery can simultaneously perform biogas upgrading into biomethane and the valorization of the nutrients contained in the liquid effluent streams. The amino acid concentration in the recovered liquid fraction must be determined after hydrolysis and solid-liquid separation of microalgae biomass and a concentration step is necessary to reach an amino acid concentration of ~10%*w/w*, for its final use as a concentrated amino acid-rich liquid biobased fertilizer with biostimulant properties (Fig. 2.13). The final products meet the end-of-waste status as well as technical specifications to fit future fertilizer regulations.

On the other hand, solid organo-mineral biobased fertilizers can be also developed using different waste streams coming from the biorefinery, i.e., solid residue from microalgae

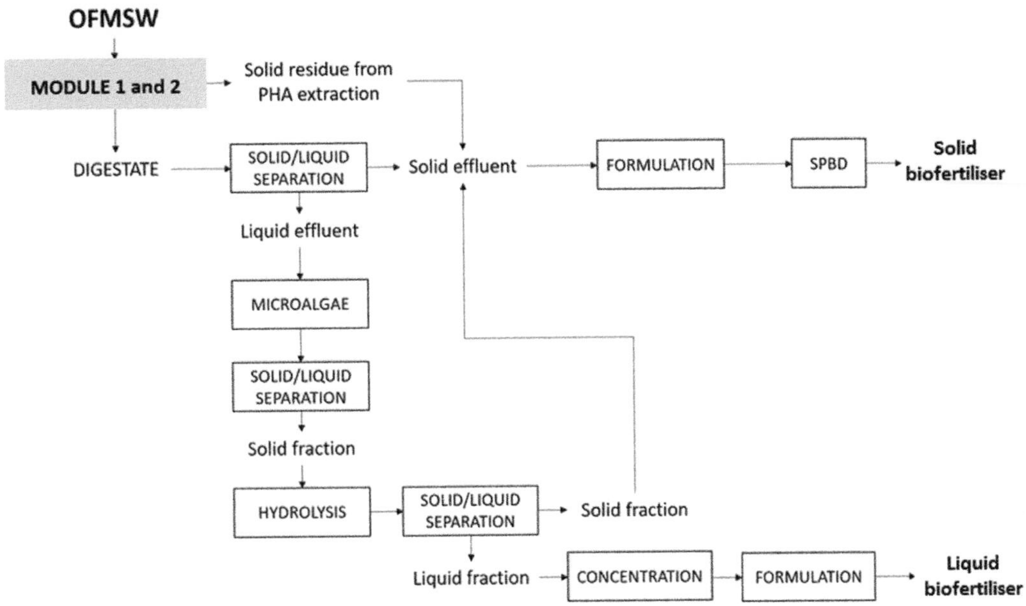

Figure 2.13
Block diagram for the biofertilizer production process in URBIOFIN for both liquid and solid biofertilizers.

hydrolysis, solid residue after PHA extraction, and solid fraction of the digestate (Fig. 2.13). This mixture must be fully characterized (nutrient composition, organic matter content, pH, conductivity and viscosity) in order to formulate the fertilizer product (adjustment of NPK balance, drying and granulation by spouted bed dryer (SPBD)) to accomplish specific requirements to reach the end-of-waste status as well as technical specifications to fit fertilizer legal standards.

2.9 Integrated URBIOFIN biorefinery: modeling, optimization, and environmental/economic assessments

Preliminary and subsequent assessments and updates are essential to define the boundaries, parameters, characteristics, and potential in any process engineering work. The modeling of future waste biorefineries is challenging and at the same time promising from a technical, economic, environmental and societal point of view [87]. The urban biorefinery scheme proposed (Fig. 2.1) might offer solutions to the current management of MSW, specifically to the current incineration and disposal of the OFMSW, to produce building blocks and final products that accounts for a higher value than the current end of life solutions [88,89]. Modeling work (Fig. 2.14) can offer alternatives to efficiently study, analyze, optimize, integrate, and up-scale current pilot plants. It can also assess the

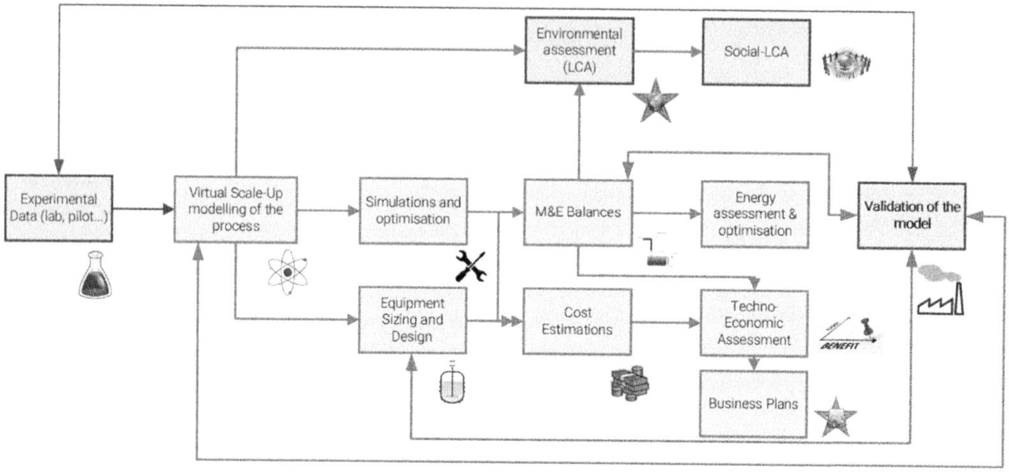

Figure 2.14
Biorefinery modeling and assessment stages. *Courtesy of Exergy Ltd. https://exergy-global.com/solutions/sustainable-processes/.*

future potential at larger scale, as well as being able to calculate material and energy (M&E) balances, study the sensitivity of critical factors or process conditions, and reach optimal results.

Modeling any processing plant is a complex task and some constraints and difficulties (i.e., in terms of calculations) can occur when biomass or nonconventional components, such as MSW, are involved. Decisions such as the thermodynamic model, experimental parameters and physical properties, equipment model selected in the simulator, possibility to integrate pretreatment, upstream and downstream processing, handling solids and accuracy of data (e.g., kinetics, exponential factors, process conditions), among others, will strongly influence the reliability of results from the simulation, and so will influence the behavior of the plant and the balances [90]. Starting from the software, the selection (and management) of the most appropriate software will be critical for the accuracy and reliability of the results to save time and costs. ASPEN Plus [91] software is considered one of the most thorough tools for the modeling and simulation (Fig. 2.1), and subsequently for its further integration and up-scaling of modern biorefineries. Some of the major criteria for the decision are based on (but not limited to) [92]: ASPEN Plus is a robust process simulator and includes tools to work in an environment with solids (critical for the modeling of OFMSW and some of the components). This software has been widely used for the modeling of some bioprocesses, for instance bioethanol production from corn, AD in wastewater or sludge treatment or biodiesel production from algae, is a reliable database, and exhibits the possibility to introduce new compounds and to change/introduce physical properties where appropriate. If an operation is not included in the software, it might be modeled and included using further tools such as Aspen Custom Modeler. In addition, ASPEN Plus it is among the most extended and applied software for the simulation of plants worldwide, allows the user to work with solids and includes an economic assessment tool, which can be ran in parallel to process simulations.

Simulations get started by introducing the list of components that are involved in the process. In respect of this, some problems may appear when nonconventional types of feedstock are added. OFMSW is made up of organic compounds that are not available in the ASPEN Plus database, such as glucan, xylan, pectin and lignin. The first three components can be modeled as compounds included in ASPEN Plus database, such as dextrose, xylose, and galacturonic acid, respectively. In the case of lignin, it can be modeled as solid component, however, no physical properties or information is available, which requires the use of the NREL database to complete the gaps [92]. Special mention deserves the option to introduce compounds as nonconventional, which can be the case of the biomass or other solid components that cannot be characterized by a molecular formula. These components are treated as pure components, though they are complex mixtures [93]. The data that is required to use this kind of component is the ultimate and proximate analysis and the enthalpy and density ASPEN models.

Another critical step to consider in a simulation is the selection of the most appropriate thermodynamic method. This essential choice will affect the estimation of accurate physical properties in the simulation. Several factors need to be considered prior to selecting the method: (1) the nature of the properties of interest. In this context, a collection of methods that will best predict the properties or results of interest should be selected. Some of the main properties of interest are enthalpy, heat capacities, viscosity, density, among others; (2) the composition of the mixture since the way to simulate the interactions among the components present in the mixture has an important influence in the results. In this regard, the polarity and molecular structure of the components and the whether the mixture is ideal or nonideal plays a key role in the final outcomes of the simulation; (3) the pressure and temperature range since each thermodynamic method has a certain operating condition where the model itself is valid. For example, activity coefficient methods are mainly described for subcritical systems, below 10 atm, and equation-of-state methods can operate in the critical region, high temperature and pressure; (4) the availability of parameters since the methods used by ASPEN Plus have collected in their database the necessary parameters to solve the equations involved in the models. These parameters are, for example, pure-component and binary parameters that relate the interactions of the components in a mixture [94]. Based on the type of processes involved in the URBIOFIN biorefinery, the previous mentioned considerations and different literature sources, the nonrandom two-liquid activity coefficient model (NRTL) is suggested as the most common method to carry out simulations for these processes, due to its wide application for low-pressure ideal and nonideal chemical systems. In addition, this route includes the NRTL liquid activity coefficient model, Henry's law for the dissolved gases and equation of state for the vapor phase, which could be desired to distill ethanol and to handle dissolved gases [18,92,95].

The (techno)economic assessment is revealed as one of the most important tasks to undertake in a project, based on its great influence on the feasibility evaluation. The American National Standards Institute (ANSI) defined three types of estimates: order-of-magnitude, budget and definitive. Order-of-magnitude estimates have an expected accuracy between +50% and −30%, are generally based on cost-capacity curves and cost-capacity ratios and do not require any preliminary design work. Budget estimates are based on flowsheets, layouts, and preliminary equipment descriptions and specifications and have an accuracy range of +30% to −15%. Design generally must be 5%–20% complete to permit such an estimate to be performed. Finally, definitive estimates require defined engineering data such as site data, specifications, basic drawings, detailed sketches and equipment quotations [96]. Design is generally 20%–100% complete and estimate accuracy should be within +15% to −5%. On the other hand, AACE International proposed an expansion of the ANSI estimate classifications to five types with expected accuracy levels based upon the amount of project definition available when the estimate is prepared as shown in Table 2.3 [97]:

Table 2.3: AACE cost estimate classification matrix for process industries.

Estimate class	Degree of project definition Expressed as % of complete definition	End usage Typical purpose of estimate	Methodology Typical estimating method	Expected accuracy range Typical variation in low and high ranges
Class 5	0%–20%	Concept screening	Capacity factored, parametric models, judgment, or analogy	L: −20% to −50% H: +30% to +100%
Class 4	1%–15%	Study or feasibility	Equipment factored or parametric models	L: −10% to −20% H: +20% to +50%
Class 3	10%–40%	Budget authorization or control	Semi-detailed unit costs with assembly level line items	L: −10% to −20% H: +10% to +50%
Class 2	30%–70%	Control or bid/tender	Detailed unit costs with forced detailed take-off	L: −5% to −15% H: +5% to +20%
Class 1	70%–100%	Check estimate or bid/tender	Detailed unit costs with detailed take-off	L: −3% to −10% H: +3% to +15%

The degree of accuracy is a function of the project development stage. For this reason, the economic assessment developed at early stages in a project of the nature of URBIOFIN can match with a −15% to −30% and +30% to +50% of accuracy. In the first stage, the equipment cost needs to be calculated by consulting both bibliography sources as well as potential suppliers quotations [98–100]. The next step comprises the development of the calculation methodology to estimate key parameters such as capital investment, utilities cost, etc. One method consists of the following conventional procedures for chemical plants based on initial factors due to the possibility to use it in parallel to the technical assessment/modeling of the plant [101]. Together with this methodology, some assumptions need to be taken into account with the aim of obtaining the best preliminary results. Based on the preliminary results, a couple of assertions can be made: (1) the feasibility of the project will strongly affect the sale price of some products such as PHA and liquid fertilizer; (2) the utility that has the largest impact is electricity, influenced by the electricity cost in Spain. The interpretation of these results is important to identify the strengths of the project, so do the availability to integrate the project in an MSW company to reduce equipment cost, the opportunities, production adjustment based on the market product price, plant location, etc., and the improvements, such as more techno-economic accuracy and plan optimization when higher project stages are reached.

The study of the environmental impacts and the comparison to current (incineration, disposal) and alternative (composting, anaerobic digestion) processes are important considerations in order to make final decisions, not only regarding the technical aspects of the project and products, but also from a policy making perspective [102]. Novel biorefinery schemes and bioprocesses are proposed and implemented, including the upgrading of fossil-based plants to biobased or environmentally friendly ones, with the objective to reduce the environmental impact and/or carbon footprint of the final product/s. However, this target is not always reached, and the LCA may show outstanding impacts at some points of the product or processes life cycle, which must be carefully investigated. A complete LCA (Fig. 2.15) is an unavoidable stage to make decisions from different points of view in regard to a novel biorefinery project, and it may influence the future definition of legislative and regulative aspects. A suitable definition of the processes/products boundaries, (harmonized) functional unit for all processes, scope, inventory, and impact assessment categories, as well as the appropriate interpretation and definition of results will be essential during the LCA [103] as shown in Fig. 2.16. Finally, the comparison to different scenarios will determine the best solution or combination of solutions for our biorefinery unit.

A preliminary LCA in the URBIOFIN biorefinery (which has not considered the integration of the plant as a whole) has reported the following points to be considered in future work: (1) it is necessary to decide whether the plant will be specifically located in a site (specific location) or the data will be used from general sources (i.e., EU data), which

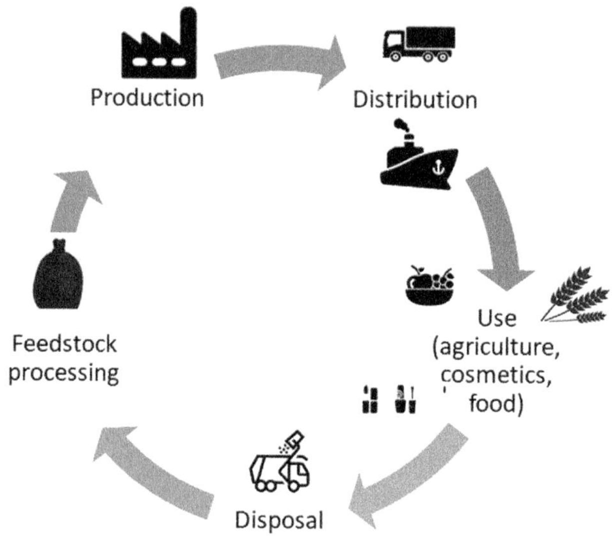

Figure 2.15
URBIOFIN's LCA process stages.

determines the impact of the electricity; (2) in some specific processes, the main impacts are due to other factors, especially the addition of chemicals (flocculants, solvents, sodium-based solutions and mineral salt mediums). These impacts will be further studied in detail associated to the potential plant integration after modeling, also considering a harmonized functional unit to study the process; (3) the potential use of biogas in the

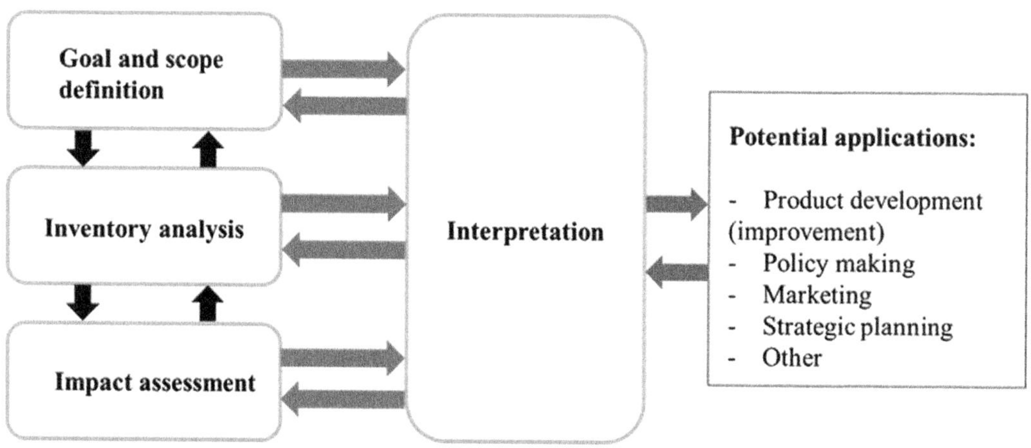

Figure 2.16
LCA stages based on standards ISO 14040, 14041, and 14044.

integrated plant as a source of energy could contribute significantly to the reduction of the environmental impact associated to the use of fuels and electricity; (4) a common functional unit based either on an integrated plant design or on a common reference point for the mass and energy balances is necessary to compare the different impacts associated to the products and processes.

2.10 Bioproducts downstream and applications

The term bioproducts or biobased products refers to products wholly or partly derived from biomass, such as plants, trees or animals (the biomass can have undergone physical, chemical or biological treatment). Some of the reasons of the increasing interest in biobased products lay in their benefits in relation to the depletion of resources and climate change. In this regard, biobased products could provide additional product functionalities, less resource intensive production and efficient use of all-natural resources.

Waste biorefineries such as URBIOFIN can produce a portfolio of different biobased products such as chemical building blocks (bioethanol, VFAs, biogas), biopolymers (PHA and biocomposites) or additives (bioethylene, microalgae hydrolysates for biofertilizers). Thus, URBIOFIN exploits the biowaste fractions of solid household waste as feedstock to transform it into building blocks that can, in turn, be used to produce different valuable marketable biobased products for different markets including agriculture and cosmetics. Beside the technical point of view, URBIOFIN also investigates the minimum requirements for the main bioproducts from a legal scope. Legal requirements are key aspects since it will be crucial to meet them in order to facilitate commercialization. An overview of the main bioproducts and their applications is described in this section.

2.10.1 PHA

In the past couple of years, the EU has adopted a number of strategies and regulations that aim to support the transition of Europe's economy from a linear fossil-based economy to a more renewable circular bioeconomy, including measures to improve waste management, increase recycling and usage of more biobased and biodegradable materials. The EU Commission's vision for improving the sustainability of the plastics industry in Europe is a key element of its Circular Economy Strategy. More specifically, EU Plastics Strategy aims to guide the transition of plastics economy using a 3R (reduce, reuse, recycle) strategy, thus ensuring that all plastic packaging in the EU market will be recyclable by 2030. EU Bioeconomy Strategy is focused on developing EU sustainable and circular bio economy. EU recognizes the importance of bioplastics (like PHA) in this transition, being a sustainable alternative to conventional petrochemicals plastics and thus a crucial component to create a strong and circular European bioeconomy.

PHAs are a group of biodegradable polymers of biological origin, polyesters produced by a number of bacteria mainly from saturated and unsaturated hydroxyalkanoic acids, and intracellularly deposited as energy storage or reserves. They have attracted considerable industrial interest by replacing the nonbiodegradable plastics derived from petroleum. In contrast to synthetic polymers, PHAs have the fundamental advantage of being based solely on renewable resources, and due to the fact that PHAs are completely digested and metabolized by a wide variety of bacteria and fungi, they are genuinely biodegradable. The variety of their microstructure creates potentially wide variety of biopolymers with different properties that are similar to conventional commodity plastics such as polypropylene, polyethylene and polystyrene. The variation in physical and mechanical properties of the different PHA-types offers a wide range of applications and it has been used to make various products, including films, coated paper, compost bags, disposable food service-ware, and molded products such as bottles and razors. The barrier properties of PHA are also important. Indeed, for products, notably cosmetics one, it is essential that the packaging has a very low permeability in the water and oxygen, in order to preserve or extend the life of the product. It allows a better conservation of the cosmetic properties of the product. Thus, target products for the use of PHA are cosmetic packaging, agricultural films and garbage bags. The PHAs produced in URBIOFIN biorefinery have been designed for molding into packaging by bioplastic companies and are under validation by cosmetic producers who are part of the NATRUE association.

2.10.2 Biobased fertilizers

Biofertilizers, also known as "plant probiotics" due to their favorable effects through the interaction between microorganisms, plants, and the environment, represent a segment of agricultural inputs (seeds, agri-chemicals, pesticides, etc.) with a great potential of substitution of the conventional agrichemicals (mineral) solid fertilizers. Their first value is the recovered/renewable origin and reduced environmental footprint of crop production: reduction of fossil fuel consumption, avoiding depletion of nonrenewable mineral fertilizers (P mainly), reduced NOx, and other GHG emissions. Furthermore, they improve soil quality and not only crop productivity. They have particular characteristics regarding the absence of involvement of chemical and/or synthetic products in their manufacture and consequently are suitable for its use in the organic or biological farming.

The stabilized organic matter solid fraction from end-of-waste separate collection can be an excellent raw material for elaboration of substrates, organic fertilizers and biofertilizers, with the requirements of no impurities and pollutants present, low conductivity, no refermentation, and temperature rise episodes. Modern biorefineries have the potential for developing a line of solid biobased fertilizers that will be produced, characterized, tested in laboratory, greenhouse and field and thoroughly evaluated. These segment of solid biofertilizers are more competitive with traditional mineral and organo-mineral fertilizers, with greater size but lower value added. In the context of biorefinery, solid and liquid biofertilizers are residuals and sidestreams of biowaste management. The connection between waste management and biotechnology companies should be adjusted and still work beyond the project duration to develop a long-term successful business strategy for the production and commercialization of biofertilizers.

2.10.3 Bioethylene

Bioethylene is produced via a biobased process such as dehydration of bioethanol unlike ethylene that in general can be processed with other processes such as steam cracking. Other than the way they are produced, there is not difference to bio ethylene and "normal" ethylene in terms of composition. Ethylene is a hydrocarbon which has formula C_2H_4. It's a natural gas without color or odor. Ethylene is the raw material for the most used plastic in the world, polyethylene, made of polymer chains of ethylene units and used as important building block for a broad range of end products. Beside that it is extensively used in agriculture for the ripening of fruits. Fruit ripening is a natural process in which the fruit goes through various chemical changes and gradually becomes sweet, colored, and gets soft and palatable. Often, ripening agents are used to speed up the ripening process. Ethylene is one of the most used agents. Externally applied ethylene is likely to trigger or initiate the natural ripening process of apple, avocado, banana, mango, papaya, pineapple and guava, and therefore, can be marketed before the predicted time.

2.11 Conclusions and perspectives

The URBIOFIN biorefinery demonstration plant, originated from the synergistic action of 16 European companies, universities and research centers, represents the most ambitious joint public-private research initiative to create a new model for the management of the organic fraction of MSW aligned with the EU strategy for promoting a circular bioeconomy. By using the biorefinery concept applied to MSW (urban waste biorefinery), the potential of OFMSW can be fully exploited as feedstock, taking into account its heterogeneity and variable composition, to produce at semiindustrial scale different valuable marketable products for local consumption. The project will optimise at a demo-scale the transformation of 10 ton/d of OFMSW into bioblocks (bioethanol, volatile fatty acids and biogas), biopolymers (short chain and medium-chain polyhydroxyalkanoates) and additives (bioethylene and biofertilisers) using a battery of innovative physical-chemical and biological processes configured in a three module approach. The technical, environmental and economical sustainability of these technologies will be assessed throughout the project by the whole value chain stakeholders such as waste management authorities and companies, technology developers and final products end-users. The main objective of this initiative is to provide an integrated marketable and scalable biorefinery able to obtain higher value bioproducts than biogas and compost than the current MSW treatment.

Acknowledgments

This publication is part of the project URBIOFIN, which has received funding from the Bio Based Industries Joint Undertaking under the European Union's Horizon 2020 research and innovation programme under grant agreement No 745785. The Regional Goverment of Castilla y León is gratefully acknowledged for the doctoral contract of Victor Perez.

References

[1] United Nations. World urbanization prospects: the 2018 revision. 2018. Available from:population.un.org/wup/Publications/Files/WUP2018-KeyFacts.pdf.
[2] Eurostat. Municipal waste statistics. 2018. Available from: ec.europa.eu/eurostat/statistics-explained/index.php/Municipal_waste_statistics.
[3] Eurostat. Recycling rate of municipal waste. 2018. Available from: ec.europa.eu/eurostat/tgm/table.do?tab=table&plugin=1&language=en&pcode=sdg_11_60.
[4] European Commission. Implementation of the circular economy action plan. 2018. Available from: ec.europa.eu/environment/circular-economy/index_en.htm.
[5] Statista. Global biofuel production from 2000 to 2017. 2018. Available from: www.statista.com/statistics/274163/global-biofuel-production-in-oil-equivalent/.
[6] Weusthuis RA, Aarts JMMJG, Sanders JPM. From biofuel to bioproduct: is bioethanol a suitable fermentation feedstock for synthesis of bulk chemicals? Biofuels Bioproducts and Biorefining 2011;5:486–94.
[7] Rass-Hansen J, Falsig H, Jørgensen B, Christensen CH. Perspective bioethanol: fuel or feedstock? Journal of Chemical Technology and Biotechnology 2007;82:329–33.
[8] Bienfait B, Ertl P. JSME: a free molecule editor in JavaScript. Journal of Cheminformatics 2013;5:24.
[9] Posada JA, Patel AD, Roes A, Blok K, Faaij APC, Patel MK. Potential of bioethanol as a chemical building block for biorefineries: preliminary sustainability assessment of 12 bioethanol-based products. Bioresource Technology 2013;135:490–9.
[10] Balan V, Chiaramonti D, Kumar S. Review of US and EU initiatives toward development, demonstration, and commercialization of lignocellulosic biofuels. Biofuels Bioproducts and Biorefining 2013;7(6):732–59.
[11] Alavi Moghadam MR, Mokhtarani N, Mokhtarani B. Municipal solid waste management in Rasht City, Iran. Waste Management 2009;29(1):485–9.
[12] Balat M, Balat H, Oz C. Progress in bioethanol processing. Progress in Energy and Combustion Science 2008;34:551–73.
[13] Sarkar N, Ghosh SK, Bannerjee S, Aikat K. Bioethanol production from agricultural wastes: an overview. Renewable Energy 2012;37:19–27.
[14] Merchant Research, Consulting Ltd. Ethylene (ET). World market outlook and forecast up to 2027. 2018. Available from: www.e4tech.com/project/from-the-sugar-platform-to-biofuels-and-biochemical/.
[15] Ring KL, deGuzman M. Ethylene. In: Chemical economics handbook. IHS Chemical; 2017. Available from: https://www.ihs.com/products/ethylene-chemical-economics-handbook.html.

[16] Matar S, Hatch LF. Chemistry of petrochemical processes. Houston, TX: Gulf Professional Publishing; 2001.

[17] Broeren M. Production of bio-ethylene: technology brief. Abu Dhabi: International Renewable Energy Agency (IRENA); 2013.

[18] Mohsenzadeh A, Zamani A, Taherzadeh MJ. Bioethylene production from ethanol: a review and techno-economical evaluation. ChemBioEng Reviews 2017;4(2):75–91.

[19] Zhang M, Yu Y. Dehydration of ethanol to ethylene. Industrial & Engineering Chemistry Research 2013;52(28):9505–14.

[20] Bergstra MF. Catalytic ethylene polymerization. Novel reactors for kinetics in gas, slurry and solution processes [Ph.D. thesis]. University of Twente; 2004.

[21] Almeida JRM, Runquist D, Sanchez Nogue V, Liden G, Gorwa-Grauslund MF. Stress-related challenges in pentose fermentation to ethanol by the yeast *Saccharomyces cerevisiae*. Biotechnology Journal 2011;6(3):286–99.

[22] TNAU Agritech Portal. Fruit ripening report. 2018. Available from: agritech.tnau.ac.in/horticulture/fruit_ripening.pdf.

[23] Calt E. Products from organic waste using managed ecosystem fermentation. Journal of Sustainable Development 2015;8(3):43–51.

[24] El-Bery H, Tawfik A, Kumari S, Bux F. Effect of thermal pre-treatment on inoculum sludge to enhance bio-hydrogen production from alkali hydrolysed rice straw in a mesophilic anaerobic baffled reactor. Environmental Technology 2013;34(13–14):1965–72.

[25] Xie L, Liu H, Chen Y-G, Zhou Q. pH adjustment strategy for volatile fatty acid production from high-strength wastewater for biological nutrient removal. Water Science and Technology 2014;69(10):2043–50.

[26] Dahiya S, Sarkar O, Swamy YV, Mohan SV. Acidogenic fermentation of food waste for volatile fatty acid production with co-generation of biohydrogen. Bioresource Technology 2015;182:103–13.

[27] Yu L, Bule M, Ma J, Zhao Q, Frear C, Chen S. Enhancing volatile fatty acid (VFA) and bio-methane production from lawn grass with pretreatment. Bioresource Technology 2014;162:243–9.

[28] Dogan E, Demirer GN. Volatile fatty acid production from organic fraction of municipal solid waste through anaerobic acidogenic digestion. Environmental Engineering Science 2009;26(9):1443–50.

[29] Plácido J, Zhang Y. Production of volatile fatty acids from slaughterhouse blood by mixed-culture fermentation. Biomass Conversion and Biorefinery 2018;8(3):621–34.

[30] Yirong C, Zhang W, Heaven S, Banks CJ. Influence of ammonia in the anaerobic digestion of food waste. Journal of Environmental Chemical Engineering 2017;5:5131–42.

[31] Demirel B, Scherer P. The roles of acetotrophic and hydrogenotrophic methanogens during anaerobic conversion of biomass to methane: a review. Reviews in Environmental Science and Bio/Technology 2008;7(2):173–90.

[32] Arslan D, Steinbusch KJJ, Diels L, Hamelers HVM, Strik DPBTB, Buisman CJN, De Wever H. Selective short chain carboxylates production: a review on control mechanisms to direct mixed culture fermentation. Critical Reviews in Environmental Science and Technology 2016;46(6):592–634.

[33] Yin J, Wang Y, Shen D, Wang M, Mo H. Improving production of volatile fatty acids from food waste fermentation by hydrothermal pretreatment. Bioresource Technology 2014;171:323–9.

[34] Pham TN, Um Y, Yoon HH. Pretreatment of macroalgae for volatile fatty acid production. Bioresource Technology 2013;146:754–7.

[35] Anderson AJ, Dawes EA. Occurrence, metabolism, metabolic role and industrial uses of bacterial polyhydroxyalkanoates. Microbiological Reviews 1990;54(4):450–72.

[36] Numata K, Abe H, Iwata T. Biodegradability of poly(hydroxyalkanoate) materials. Materials 2009;2:1104–26.

[37] Sashiwa H, Fukuda R, Okura T, Sato S, Nakayama A. Microbial degradation behavior in seawater of polyester blends containing poly(3-hydroxybutyrate co-3-hydroxyhexanoate) (PHBHHx). Marine Drugs 2018;16:34.

[38] Chen G-Q. A microbial polyhydroxyalkanoates (PHA) based bio- and materials industry. Chemical Society Reviews 2009;38:2434–46.
[39] Harding KG, Dennis JS, von Blottnitz H, Harrison STL. Environmental analysis of plastic production processes: comparing petroleum-based polypropylene and polyethylene with biologically-based poly-β-hydroxybutyric acid using life cycle analysis. Journal of Biotechnology 2007;130:57–66.
[40] Choi J-I, Lee SY. Process analysis and economic evaluation for poly(3-hydroxybutyrate) production by fermentation. Bioprocess Engineering 1997;17:335–42.
[41] Mozejko-Ciesielska J, Kiewisz R. Bacterial polyhydroxyalkanoates: still fabulous? Microbiological Research 2016;192:271–82.
[42] Wang F, Lee SY. Poly(3-hydroxybutyrate) production with high productivity and high polymer content by a fed-batch culture of Alcaligenes latus under nitrogen limitation. Applied and Environmental Microbiology 1997;63:3703–6.
[43] Kourmentza C, Plácido J, Venetsaneas N, Burniol-Figols A, Varrone C, Gavala HN, Reis MAM. Recent advances and challenges towards sustainable polyhydroxyalkanoate (PHA) production. Bioengineering 2017;4(55):1–43.
[44] Reis MAM, Serafim LS, Lemos PC, Ramos AM, Aguiar FR, Van Loosdrecht MCM. Production of polyhydroxyalkanoates by mixed microbial cultures. Bioprocess and Biosystems Engineering 2003;25:377–85.
[45] Kleerebezem R, van Loosdrecht MCM. Mixed culture biotechnology for bioenergy production. Current Opinion in Biotechnology 2007;18:207–12.
[46] Cruz M, Freitas F, Paiva A, Mano F, Dionísio M, Ramos A, Reis M. Valorization of fatty acids-containing wastes and byproducts into short- and medium-chain length polyhydroxyalkanoates. New Biotechnology 2016;33(1):206–15.
[47] Wall DM, Dumont M, Murphy JD. Green Gas. Facilitating a future green gas grid through the production of renewable gas. IEA Bioenergy Task 2018;37(2).
[48] Muñoz R, Meier L, Diaz I, Jeison D. A critical review on the state-of-the-art of physical/chemical and biological technologies for an integral biogas upgrading. Reviews in Environmental Science and Bio/Technology 2015;14(4):727–59.
[49] Shen M, Zhang Y, Hu D, Fan J, Zeng G. A review on removal of siloxanes from biogas: with a special focus on volatile methylsiloxanes. Environmental Science and Technology 2018;25:30847–62.
[50] Bahr M, Diaz I, Dominguez A, Sanchez-Gonzalez A, Muñoz R. Microalgal-biotechnology as a platform for an integral biogas upgrading and nutrient removal from anaerobic effluents. Environmental Science and Technology 2014;48(1):573–81.
[51] Toledo AL, Serejo M, Blanco S, Perez R, Lebrero R, Muñoz R. Photosynthetic biogas upgrading to bio-methane: boosting nutrient recovery via biomass productivity control. Algal Research 2016;17:46–52.
[52] Rodero MR, Posadas E, Toledo-Cervantes A, Lebrero R, Muñoz R. Influence of alkalinity and temperature on photosynthetic biogas upgrading efficiency in high rate algal ponds. Algal Research 2018;33:284–90.
[53] Toledo-Cervantes A, Estrada JM, Lebrero R, Muñoz R. A comparative analysis of biogas upgrading technologies: photosynthetic vs physical/chemical processes. Algal Research 2017;25:237–43.
[54] Accettola F, Guebitz GM, Schoeftner R. Siloxane removal from biogas by biofiltration: biodegradation studies. Clean Technologies and Environmental Policy 2008;10:211–8.
[55] Soreanu G, Béland M, Falletta P, Edmonson K, Svoboda L, Al-Jamal M, Seto P. Approaches concerning siloxane removal from biogas – a review. Canadian Biosystems Engineering 2011;53(8):1–18.
[56] Xue N, Wang Q, Wang J, Wang J, Sun X. Odorous composting gas abatement and microbial community diversity in a biotrickling filter. International Biodeterioration & Biodegradation 2013;82:73–80.
[57] Popat SC, Deshusses MA. Biological removal of siloxanes from landfill and digester gases: opportunities and challenges. Environmental Science and Technology 2008;42:8510–5.
[58] European Environmental Agency. Atmospheric greenhouse gas concentrations (CSI 013/CLIM 052). 2018.

[59] Lebrero R, Chandran K. Biological conversion of waste methane streams. Critical Reviews in Environmental Science and Technology 2017;47(22):2133–57.
[60] Pieja AJ, Rostkowski KH, Criddle CS. Distribution and selection of poly-3-hydroxybutyrate production capacity in methanotrophic proteobacteria. Microbial Ecology 2011;62:564–73.
[61] López JC, Quijano G, Souza TSO, Estrada JM, Lebrero R, Muñoz R. Biotechnologies for greenhouse gases (CH_4, N_2O, and CO_2) abatement: state of the art and challenges. Applied Microbiology and Biotechnology 2013;97:2277–303.
[62] Cantera S, Muñoz R, Lebrero R, López JC, Rodríguez Y, García-Encina PA. Technologies for the bioconversion of methane into more valuable products. Current Opinion in Biotechnology 2018;50:128–35.
[63] Strong PJ, Kalyuzhnaya M, Silverman J, Clarke WP. A methanotroph-based biorefinery: potential scenarios for generating multiple products from a single fermentation. Bioresource Technology 2016;215:314–23.
[64] García-Pérez T, López JC, Passos F, Lebrero R, Revah S, Muñoz R. Simultaneous methane abatement and PHB production by *Methylocystis hirsuta* in a novel gas-recycling bubble column bioreactor. Chemical Engineering Journal 2018;334:691–7.
[65] López JC, Arnáiz E, Merchán L, Lebrero R, Muñoz R. Biogas-based polyhydroxyalkanoates production by *Methylocystis hirsuta*: a step further in anaerobic digestion biorefineries. Chemical Engineering Journal 2018;333:529–36.
[66] Sander R. Compilation of Henry's law constants (version 4.0) for water as solvent. Atmospheric Chemistry and Physics 2015;15:399–4981.
[67] Estrada JM, Lebrero R, Quijano G, Pérez R, Figueroa-González I, García-Encina PA, Muñoz R. Methane abatement in gas-recycling biotrickling filter: evaluating innovative operational strategies to overcome mass transfer limitations. Chemical Engineering Journal 2014;253:385–93.
[68] Bruce LJ, Daugulis AJ. Solvent selection strategies for extractive biocatalysis. Biotechnology Progress 1991;7(2):116–24.
[69] Cantera S, Estrada JM, Lebrero R, García-Encina PA, Muñoz R. Comparative performance evaluation of conventional and two-phase hydrophobic stirred tank reactors for methane abatement: mass transfer and biological considerations. Biotechnology and Bioengineering 2016;113:1203–12.
[70] Stone KA, Hilliard MV, He QP, Wang J. A mini review on bioreactor configurations and gas transfer enhancements for biochemical methane conversion. Biochemical Engineering Journal 2017;128:83–92.
[71] Rocha-Rios J, Muñoz R, Revah S. Effect of silicone oil fraction and stirring rate on methane degradation in a stirred tank reactor. Journal of Chemical Technology and Biotechnology 2010;85:314–9.
[72] Petersen LAH, Villadsen J, Jørgensen SB, Gernaey K. Mixing and mass transfer in a pilot scale U-loop bioreactor. Biotechnology and Bioengineering 2017;114:344–54.
[73] Yazdian F, Shojaosadati S, Nosrati M, Pesaran, Mahdi P, Ebrahim V-F. Mixing studies in loop bioreactors for production of biomass from natural gas. Iranian Journal of Chemistry and Chemical Engineering 2012;31:91–101.
[74] Kraakman NJR, Rocha-Ríos J, van Loosdrecht MCM. Review of mass transfer aspects for biological gas treatment. Applied Microbiology and Biotechnology 2011;91:873–86.
[75] Rahnama F, Vasheghani-Farahani E, Yazdian F, Shojaosadati SA. PHB production by *Methylocystis hirsuta* from natural gas in a bubble column and a vertical loop bioreactor. Biochemical Engineering Journal 2012;65:51–6.
[76] Pieja AJ, Sundstrom ER, Criddle CS. Cyclic, alternating methane and nitrogen limitation increases PHB production in a methanotrophic community. Bioresource Technology 2012;107:385–92.
[77] Fertilizers Europe. Annual overview 2018. 2018. Available from: www.fertilizerseurope.com/uploads/media/Fertilizers_Europe_Annual_Overview_2018.pdf.
[78] Regulation (EC) No 2003/2003 of the European Parliament and of the Council of 13 October 2003 relating to fertilisers. November 21, 2003. Official Journal of the European Union.

[79] Proposal COM/2016/0157 final – 2016/084 (COD), for regulation on the making available on the market of CE marked products fertilising and Amending Regulations (EC) No 1069/2009 and (EC) No 1107/2009.
[80] Ellen MacArthur Foundation, Towards the circular economy, vol. 2.
[81] Directive 2018/851/CE of the European Parliament and of the Council of 30 May 2018, amending directive 2008/98/EC on waste. Brussels, Belgium.
[82] ECN 2017/01. European Compost Network Newsletter. Available from: www.compostnetwork.info/wordpress/wp-content/uploads/170713_ECN-NEWS-01_2017.pdf.
[83] ORBIT/ECN. Compost production and use in the EU. 2008. Available from: www.organicsrecycling.org.uk/dmdocuments/compostproduction_and_usein_EU.pdf.
[84] Cesaro A, Belgiorno V, Guida M. Compost from organic solid waste: quality assessment and European regulations for its sustainable use. Resources, Conservation and Recycling 2015;94:72–9.
[85] Vaneeckhaute C, Lebuf V, Michels E, Belia E, Vanrolleghem PA, Tack FMG, Meers R. Nutrient recovery from digestate: systematic technology review and product classification. Waste and Biomass Valorization 2017;8:21–40.
[86] Hillel D. Soil in the environment: crucible of terrestrial life. New York City: Academia Press; 2008.
[87] Urbiofin Urban Biorefinery: from waste to bioproducts through biorefinery. Available from: https://www.urbiofin.eu/.
[88] MAGRAMA. Management of biowastes of municipal competition. Guide for the implementation of selective collection and treatment of the organic fraction. 2013. Spain.
[89] Tyagi VK, Fernandez-Guelfo L, Zhou Y. Anaerobic co-digestion of organic fraction of municipal solid waste (OFMSW): progress and challenges. Renewable and Sustainable Energy Reviews 2018;93:380–99.
[90] Al-Malah KIM. Aspen Plus: chemical engineering applications. 1st ed. John Wiley & Sons; 2016.
[91] Aspentech. Aspen Plus. 2018. Available from: www.aspentech.com/en/products/engineering/aspen-plus.
[92] Wooley RJ, Putsche V. Development of an ASPEN PLUS physical property databased for biofuels components. United States: n.p.; 1996.
[93] Aspentech. Aspen physical property system V9 physical properties. 2016.
[94] Carson EC. Don't gamble with physical properties for simulations. Chemical Engineering Progress 1996;92(10):35–46.
[95] Liu F, Chen G, Yan B, Ma W, Cheng Z, Hou L. Exergy analysis of a new lignocellulosic biomass-based polygeneration system. Energy Journal 2017;140:1087–95.
[96] ANSI. Industrial engineering terminology: cost engineering. 1989.
[97] AACE. Skills & knowledge of cost engineering. 6th ed. 2015. 0086.
[98] Walas SM. Chemical process equipment: selection and design. 1st ed. Butterworth-Heinemann; 1988.
[99] Warren D, Seider JD, Lewin DR. Product and process design principles. 1st ed. John Wiley and Sons; 1998.
[100] National Energy Technology Laboratory; 2018. Available from: https://www.netl.doe.gov/.
[101] Peters MS, Timmerhaus KD. Plant design and economics for chemical engineers. 4th ed. McGraw-Hill Book; 1968.
[102] Shen Y, Chen TB, Gao D, Zheng G, Liu H, Yang Q. Online monitoring of volatile organic compound production and emission during sewage sludge composting. Bioresource Technology 2012;123:463–70.
[103] International Standards Organization. ISO 14040:2006 environmental management – life cycle assessment – principles and framework. 2006.

SECTION B

Lignocellulosic biomass based biorefinery

CHAPTER 3

Nozzle reactor for continuous fast hydrothermal liquefaction of lignin residue

Khanh-Quang Tran
Department of energy and process engineering, Norwegian University of Science and Technology, Trondheim, Norway

3.1 Introduction

Lignin is one of the three main components of lignocellulosic plant biomass, apart from cellulose and hemicellulose, accounting for 20–30 wt.% of the total mass of lignocellulosic material. Lignin is the main component of the residue from the fractionation step of bio23 ethanol production from lignocellulosic biomass (second generation bioethanol). Since lignin is not a carbohydrate, it cannot be converted to biogas by anaerobic digestion (AD), which is the technology commonly used today for energy recovery from wet biogenic residues such as sewage sludge and food wastes. The main today options for utilization of lignin from these wet waste streams are combustion, gasification, and/or pyrolysis. Apart from the high energy requirement for drying the fuel of wet lignin residue, major challenges associated with these options include high temperature corrosion and gas cleaning requirement to remove alkali metals and other impurities such as particulate matters, H_2S and HCl from flue and fuel gases. On the other hand, lignin is a complex 3-Dimensional aromatic polymeric material, consisting of an irregular array of hydroxyl- and methoxy-substituted phenylpropane units connected mainly through ether bonds [1–3]. From a thermochemical point of view, it is possible to decompose and breakdown lignin, producing valuable chemicals and liquid biofuel. However, unlike cellulose and hemicellulose, lignin conversion to chemicals and liquid biofuel is a big challenge. Research and technology development in this area is so far mostly limited to lab-scale tests and a few to pilot tests, including hydrothermal liquefaction (HTL). HTL oil produced by the state-of-the-art HTL technology is not analogous to petroleum oil. Along with the relatively low conversion efficiency, this makes today HTL technologies less economically competitive. Higher conversion efficiency and higher quality HTL oil with affordable energy supply and at reduced cost, as well as the ability for upscaling to commercial process are key issues, which are addressed in this chapter.

Waste Biorefinery. https://doi.org/10.1016/B978-0-12-818228-4.00003-4

3.2 Fast hydrothermal liquefaction

HTL involves the use of water in near-critical conditions (below 374°C, 22.1 MPa), as reaction media, mimicking the natural process converting ancient plant material into petroleum oil. It may become an important technology for valorization of wet biomass and biogenic residues, producing liquid biofuel and valuable chemicals. However, the technical hurdles have not been yet solved completely. The challenges are various including reactor material, conversion efficiency, slurry feeding, product separation, bio-oil stabilization, water management, process cost, and energy efficiency [4–7]. Major engineering challenges that hinder HTL from industrial applications are (1) corrosion; (2) precipitation of inorganic salts; (3) char and coke formation [8,9]. Among these, char and coke formation remains the most critical. It is because HTL is working in sub54 critical water conditions. The corrosiveness and pressure of HTL media are much lower than those of supercritical water (SCW) conditions for SCW gasification. More importantly, thanks to the technological advancements in metallurgical industry, it is possible to design and manufacture a suitable alloy that can stand and work effectively in hostile environments as such of subcritical water once the reaction conditions are defined [10,11]. Furthermore, the solubility of most inorganic salts in water only starts deceasing at the critical point of water. Since HTL is working in subcritical water conditions, the problem of salt precipitation is virtually minimized. Given that an HTL reactor is capable of tolerating char and coke particles, the reactor should also be able to tolerate salt particles since salt particles are formed in SCW via nucleation mechanism and therefore normally much smaller than char particles in particle size.

It is well known that char is formed by incomplete conversion of solid biomass, while coke is resulted from further thermal decomposition of HTL oil formed during HTL [12]. They are responsible for lower oil yields of HTL process and may cause serious deposition or even blockage in the relatively small pipes of a continuous flow HTL reaction system after a relatively long operation. Catalysts can be added to minimize the formation of these unwanted products [8,9,13]. However, the catalyst addition to HTL processes may be limited due to the harsh conditions of water in near-critical conditions and the cost of catalyst. Catalyst recovery may also be changeling. Alternatively, the problem of char and coke formation during HTL can be minimized via fast heating [14,15].

The effect of fast heating on char minimization during hydrothermal processing of biomass was noticed by Modell as early as in 1985 [16]. Later, a number of studies demonstrated that fast heating improved the yield of HTL oil. Brand et al. [14] have indicated that HTL consists of beneficial primary reactions (pyrolytic and& hydrolytic degradation) and non-beneficial secondary reactions (recombination and secondary cracking). In the case of slow heating, biomass undergoes step-by-step hydrolytic degradation, pyrolytic degradation and, if elevated temperatures are applied, recombination and secondary cracking reactions. In the case of fast heating (i.e., fast HTL) biomass is quickly heated to the desired temperature, which would lead to simultaneous hydrolytic and pyrolytic cleavage. Dominance of either pathway depends on the final reaction temperature. If higher final temperatures are applied, a long residence time leads to recombination and secondary cracking. Therefore, the yield of HTL oil can be maximized by combining fast heating with short reaction time, as well as relatively higher reaction temperatures. In addition to this, fast heating is also known to have effects on product characteristics, including the fuel properties and functional groups of HTL oil [14,17]. These effects are similar to that of catalyst addition to HTL process [17], which may beyond the state-of-the-art HTL technology with respect to the conversion efficiency, product selectivity and thus end-use compatibility.

Recently, the concept of fast HTL has been experimentally validated [17,18]. The possibility to improving the conversion efficiency and product selectivity of HTL by fast heating up to 585°C/min, using sealed capillary quartz reactors with a volume of approx. 0.5 cm^3 per reactor [17,18].

3.3 Nozzle reactor for upscaling fast HTL

3.3.1 The concept of nozzle reactor

A key question now to realizing the concept of fast HTL for sustainable production of liquid biofuel and chemicals from lignin residue is to develop a continuous flow tubular reactor, which can (1) achieve fast heating; (2) tolerate solid biomass particles suspended in a water stream flowing through the reactor; and (3) be up-scalable [19].

It has been identified that the concept of nozzle reactor may be suitable for fast HTL of biomass. It is because nozzle reactor virtually meets all the design criteria, discussed earlier for fast HTL of biomass, including:

- Instantaneous strong and uniform mixing of two fluid flows (the larger flow is pure water and preheated) of big difference in volumetric flow rate, to aid in achieving very fast heating for the smaller flow containing biomass particles;
- Short average residence time combined with a narrow residence time distribution, to minimize coke formation;
- Minimal heating of the slurry stream of biomass in water prior to the reactor, followed by immediate and rapid heating of the slurry stream within the reactor, to prevent side reactions;
- Strong net downstream flow/eddies for the rapid transport of product particles out of the reactor, to prevent particle accumulation within the reactor and to minimize subsequent coke particle formation.

Fig. 3.1 presents schematically the nozzle reactor and reaction system, which has been successfully developed for production of nanoparticles from aqueous solution of metal

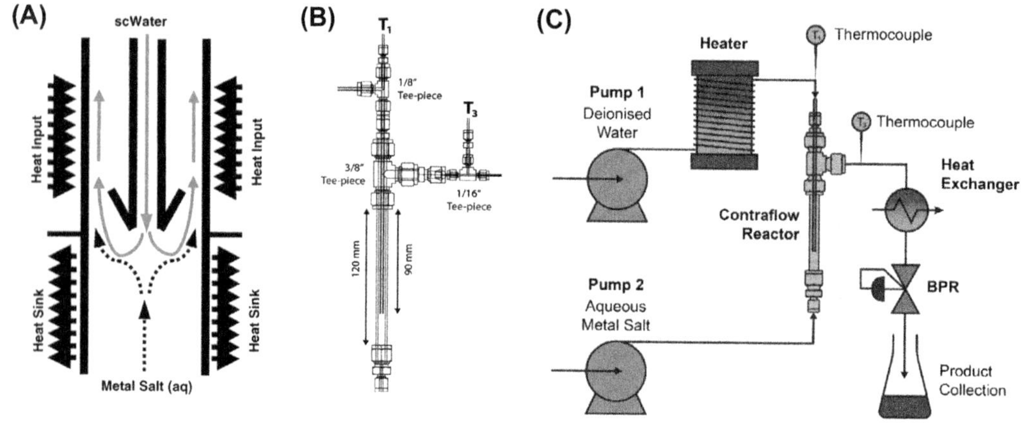

Figure 3.1
Schematic diagram of the nozzle reactor and continuous flow reaction system developed for hydrothermal synthesis of nanoparticles [20,21].

salts [20,21]. For fast HTL, the aqueous metal salt stream in Fig. 3.1A and C will be replaced with a lignin slurry stream.

The reactor in Fig. 3.1 is essentially a pipe-in-pipe concentric setup in which the internal pipe has an open-ended nozzle with a cone attached (optional). The optional cone would act in the same manner as a thin film reactor, spinning disc or spinning cone reactors for mixing intensification [22–24]. The "hot" stream (or flow) of preheated water is fed downwards through the internal pipe, reaching out at the exit end (nozzle) of the pipe. The "cold" stream is fed upwards through the outer pipe. The reactor outlet is situated on top of the outer pipe, leading the reactant mixture to a cooling unit, which in a full-scale industrial fast HTL system would be a heat exchanger (Fig. 3.1C) for heat recovery. Because of the impingement of the hot and cold streams, the forced countercurrent mixing process is enhanced by the natural convection due to the difference in density between the two steams. As a result, very good mixing and thus high heating rates can be achieved in the reactor. In addition, it has been proved that nozzle reactors developed for hydrothermal synthesis of nanoparticles are free from the blockage problem, even at a pilot scale of much higher concentrations of metal salt precursor than the laboratory-scale because of the higher velocities and larger physical dimensions of the nozzle and reactor [25].

3.3.2 CFD study of nozzle reactor for fast HTL assuming Newtonian fluid

3.3.2.1 Geometry and messing

Earlier, a computational fluid dynamics (CFD) study on nozzle reactor for hydrothermal synthesis of metal oxide nanoparticles, in combination with empirical measurement has been reported [20]. Although the geometry of nozzle reactor in that work has been extensively studied in comparison with the original [21], details of temperature profile in the reactor have not fully investigated. In addition, no study on the reactor for viscous cold flow has been reported so far. These two shortcomings of the previous study [20] are important for fast HTL of biomass. Therefore, a new study was performed as the first step of our project on application of nozzle reactor for fast HTL of biomass. The objectives of this study were to (1) develop and validate a CFD model of nozzle reactor in ANSYS FLUENT, and (2) predict process parameters suitable for fast HTL of biomass. For this purpose, the commercial CFD software ANSYS FLUENT, version 18.0 was used. Each analysis consisted of two steps; first a steady state run that established the flow and temperature fields of the entire reactor, and then a subsequent transient analysis to establish the residence time distribution (RTD) of the mixing zone. During the transient analyses, the flow and temperature fields were left unchanged at their steady state values; only the evolution of an inert scalar was simulated.

Since the focus of this study was on heating rate (mixing at the nozzle) and temperature profile in the reactor, and for the ease of modeling, it was assumed the reactor axisymmetric. This assumption is reasonable because a prestudy examination showed no existence of 3D flow structures. Thus, an axisymmetric model would suffice [26]. Furthermore, to simplify postprocessing of results as well as to simplify future modifications the decision was made to create a half-cylindrical 3D reactor. The reactor is consisted of two concentric cylinders: the inner (122.5 mm long) reaching down to approximately two thirds of the outer (203 mm long). Hot (preheated) water enters through the top end of the inner cylinder and mixes with a cold water/biomass solution entering through the bottom inlet of the outer cylinder. The streams will mix near the outlet (nozzle) of the inner pipe and will interact between the two cylinders until the top outlet. This reactor type is referred to as nozzle reactor, with an assumption of Newtonian fluids and dimensions similar those of the previous study [20].

The fluid volume and solid pipe walls were then meshed according to the standard demands of near-wall resolution of turbulence and heat transfer. A hexahedral/prismatic mesh of 960,244 cells was created with a nondimensional wall-adjacent cell height of less than 1 (scaled with laminar viscosity and wall friction velocity).

The 3D segregated double precision solver was used for the simulations. Residuals and global imbalances of energy were monitored in order to assure that the results were fully converged before interrupting the solver. For the transient analyses, a time-step of 0.01 s was used, which corresponds to a convective Courant number below unity. To validate the numerical setup, a comparison was made with RTDs found in literature [20]. Two sets of inlet flowrate were compared, 20:10 and 20:20 (mL hot water: mL cold water/biomass per minute).

3.3.2.2 Governing equations and turbulence model

The behavior of a fluid flow is governed by the fundamental set of Navier–Stokes equations for conservation of mass, momentum and energy. When modeling turbulent fluid flows, a suitable averaging method must be used [20]. For flow and mixing of fluids under supercritical conditions, it is convenient to use density or Favre averaging method, which applies fluctuation to average values instead of actual flow parameter in order to derive a time averaged solution [20,27]. The density averaged conservation equations for momentum, mass and energy are represented as

$$\frac{\partial \overline{p}}{\partial t}+\frac{\partial(\overline{p}\widetilde{u}_i)}{\partial x_i}=0 \tag{3.1}$$

$$\frac{\partial(\overline{p}\widetilde{u}_i)}{\partial t}+\frac{\partial}{\partial x_j}(\rho u_i, u_j)=-\frac{\partial \overline{p}}{\partial x_i}+\frac{\partial}{\partial x_j}\left(\widetilde{\sigma}_{ij}-\overline{p}\widetilde{u_i''u_j''}\right)+\rho g_i \tag{3.2}$$

$$\frac{\partial(\overline{p}\widetilde{e})}{\partial t}+\frac{\partial}{\partial x_j}\left(\overline{p}\widetilde{u}_j\left[\widetilde{e}+\frac{\overline{p}}{\overline{\rho}}\right]\right)=\frac{\partial}{\partial x_j}\left[\lambda\frac{\partial \overline{T}}{\partial x_j}+\widetilde{u}_j\left(\widetilde{\sigma}_{ij}-\overline{p}\widetilde{u_i''u_j''}\right)-\overline{p}\widetilde{e''u''}_j\right] \tag{3.3}$$

$$\frac{\partial\left(\overline{p}\widetilde{\phi}\right)}{\partial t}+\frac{\partial}{\partial x_j}\left(\overline{p}\widetilde{u}_i\widetilde{\phi}\right)=\frac{\partial}{\partial x_j}\left[\overline{p}D\frac{\partial\widetilde{\phi}}{\partial x_j}-\overline{p}\widetilde{\phi''u_j''}\right] \tag{3.4}$$

where u_j is the fluid velocity in direction x_j; p is the fluid pressure; ρ is the fluid density; e is the internal energy of the fluid; ϕ is a passive scalar; σ_{ij} is the viscous tress tensor of the fluid; and g_i is gravitational acceleration constant in direction x_i. If the viscosity fluctuation is neglected, the viscous tress tensor σ_{ij} will reduce to Eq. (3.5) where η is the molecular viscosity of the fluid and δ_{ij} is the unit matrix.

$$\sigma_{ij}=\eta\left[\left(\frac{\partial\widetilde{u}_i}{\partial x_j}+\frac{\partial\widetilde{u}_j}{\partial x_i}\right)-\frac{2}{3}\delta_{ij}\frac{\partial\widetilde{u}_k}{\partial x_k}\right] \tag{3.5}$$

The set of equations Eqs. (3.1)–(3.3) is not closed because the averaging procedure creates new variables for the Reynolds stress tensor $-\overline{p}\widetilde{u_i''u_j''}$. In order to close the equation set, the Reynolds stress tensors is modeled and calculated, adopting the Reynolds-averaged Navier–Stokes (RANS) method and the Realizable k-ε turbulent model, of which further details can be found in the literature [20].

3.3.2.3 Pure water simulations

To investigate the influence of the flow rate, eight analyses presented in Table 3.1 were conducted. The inlet flow rates were chosen so that the total flow rate and the ratio between them were varied. In each simulation, the material properties of the inlet streams were those of pure liquid water.

Table 3.1: Pure water simulation cases.

Case	Flow rate [mL/min^{-1}]				Temperature [°C]	
	Hot inlet	Cold inlet	Total	Ratio [−]	Hot inlet	Cold inlet
1	20	10	30	2.0	450	15
2	20	20	40	1.0	450	15
3	30	10	40	3.0	450	15
4	30	20	50	1.5	450	15
5	60	10	70	6.0	450	15
6	60	20	80	3.0	450	15
7	60	30	90	2.0	450	15
8	80	40	120	2.0	450	15

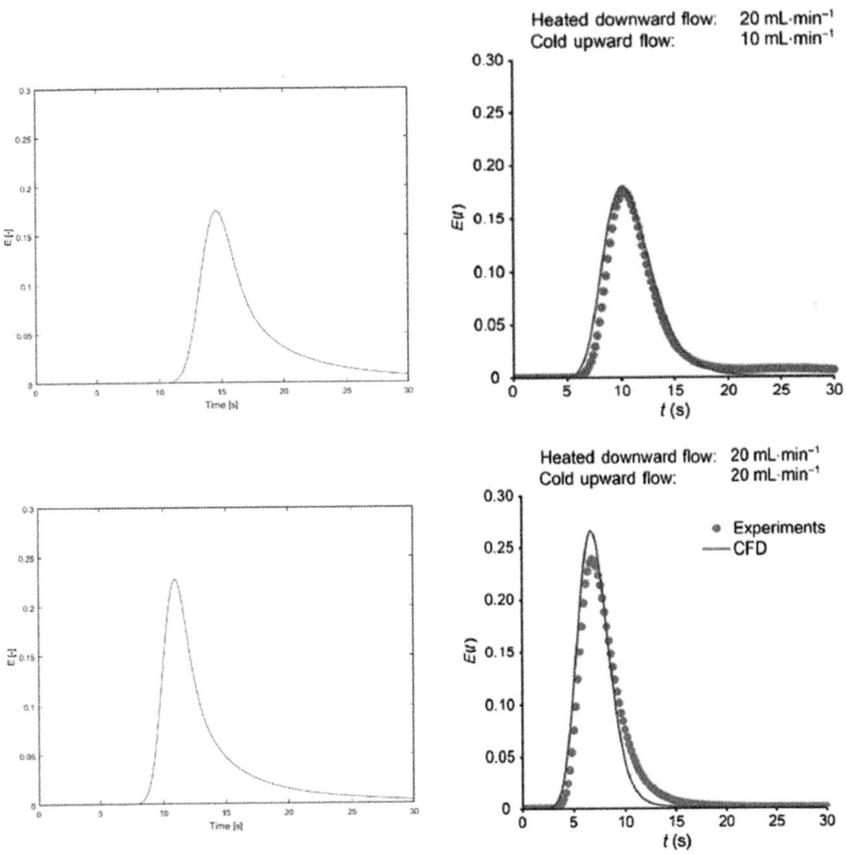

Figure 3.2
Comparison of RTDs. Upper left: our system (20:10), upper right: reference system (20:10), lower left: our system (20:20), lower right: reference system (20:20).

3.3.2.4 Model validation

The validation was made by comparing the RTDs obtained from our CFD model with the RTDs found in the literature (Fig. 3.2). The RTDs are very similar, although not identical. There is a time shift between our results and the literature. It is because the reactor volume is somewhat smaller in the compared literature than in our case. In addition, our RTDs have a more expressed tail. The tail indicates that some of the fluid elements remain inside the reaction zone longer. Overall, good agreements in the shape of the RTD curves obtained from two works are observed. This suggests that the validation of the CFD model reported in the literature is also valid for the present study.

3.3.2.5 Effect of mass flowrate on heating rate and temperature profile

3.3.2.5.1 Effect of mass flowrate ratio

The mass flow ratio was varied from 1.0 to 3.0 for this investigation, while the total mass flow rate was kept constant at 40 mL/min. These selections correspond to Case 2 and Case 3 in Table 3.1, respectively, with 450°C as inlet temperature of the hot stream and 15°C as inlet temperature of the cold stream.

The simulation results of these two cases are presented in Fig. 3.3, which shows that temperature of the hot streams decreases gradually when flowing from the right (the inlet) to the left (the mixing point). It is because heat is transferred, via thermal conductivity through the inner pipe wall, to the mixture steam from the hot stream flowing inside the inner pipe. That is the reason why the red curve decreases faster than the blue one, when move from the right to the left, co-considering the mass flowrate ratios. On the other hand, the temperature profiles of the mixture streams have a peak not at the nozzle exit (the

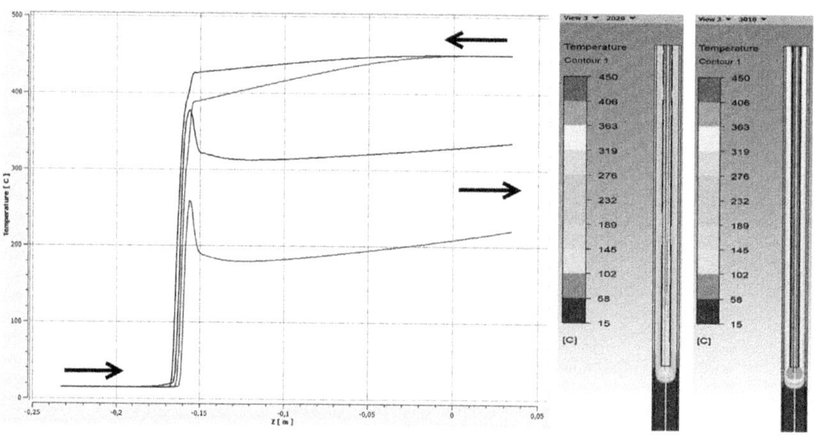

Figure 3.3
Temperature plots for different mass flow ratios. Left figure: Case 2 (red—top curves of higher temperatures; gray in printed version) and 3 (blue—bottom curves of lower temperatures; dark gray in printed version). The *arrows* indicate flow direction. The two plots to the right show the temperature distribution (Case 2 and 3).

inner pipe tip), but somewhere upstream of the reactor, with respect flow direction of the cold and mixture streams. This observation and the shape of the temperature profile of mixture streams are however in good agreement with the literature [28], which dealt with temperature measurements of a similar reactor in similar condition. The peak location is caused by the penetration of the hot stream in the cold stream, which is in turn resulted from the facts that the hot/cold mass flow ratio is normally larger than one and the smaller diameter of the inner pipe always lead to higher velocity of the hot stream compared to the cold stream. The peak of the temperature profile for Case 3 (the blue) is located slightly toward the left hand side compared to that of Case 2 (the red) confirms this observation, considering the mass flow rate of the hot streams (30 mL/min and 20 mL/min, respectively). After the peak, temperature of the mixture streams increase gradually, approaching above 340°C for Case 3 and 230°C for Case 2 at the reactor outlet. This suggests that mass flow ratio has very strong effects on the mixing and heat transferring.

3.3.2.5.2 Effect of total mass flowrate

The next comparison is between different cases (Case 1, 7, and 8) having the same flow rate ratio (2.0), but different total flow rates (30, 90 and 120 mL/min, respectively). Fig. 3.4A shows the temperature profile along the reactor and Fig. 3.4B presents the temperature contours of the cases under investigation. The temperature contours of Case 7 and 8 are similar to each other, but clearly distinguished from Case 1. The peak location of Case 1 (the red line) suggests that some heat exchange between the cold stream and the hot reactor wall (outer pipe) took place already before the actually mixing. This heat transfer is probably by conduction through the outer pipe. Comparing with the flow rate ratio, the total flow rate has weaker effects on mixing and heat transfer via both convection and conduction in the reactor.

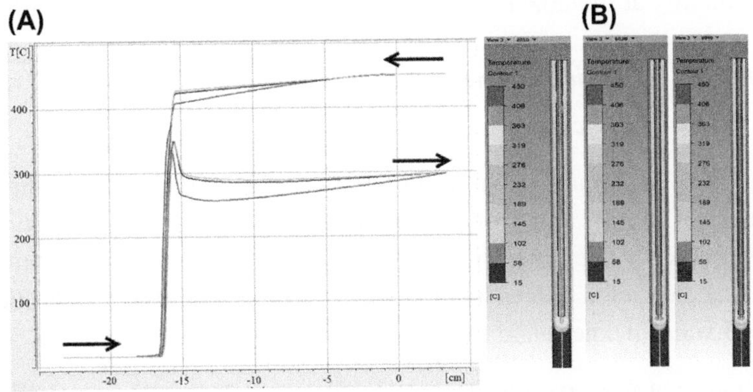

Figure 3.4
(A) Temperature profiles for Case 1 (red—bottom curves; gray in printed version), 7 (blue—middle curves; dark gray in printed version) and 8 (green—top curves; light gray in printed version); (B) Temperature contours for (left to right) Case 1, 7 and 8.

3.3.2.6 Variable viscosity simulations

To try resembling the cold flow consisting of biomass instead of pure water, a simulation was conducted where the viscosity of the cold water was significantly increased. The analysis conditions are presented in Table 3.2.

Table 3.2: Variable viscosity cases (viscosity of pure water is 0.001 Pa s).

	Flow rate [mL/min^{-1}]				Temperature [°C]		
Case	Hot inlet	Cold inlet	Total	Ratio [−]	Hot inlet	Cold inlet	Viscosity [Pa s]
6a	60	20	80	3.0	450	15	0.01
6b	60	20	80	3.0	450	15	0.05
6c	60	20	80	3.0	450	15	0.25

The temperature contours of this investigation were examined and analyzed, which indicates that the cold flow with viscosity higher than 0.01 Pa s would cause poor mixing at the nozzle. This suggests the need to increase the mass flowrates and the reactor diameters. On the other hand, it may be that biomass needs to be modeled differently at this point. Instead of a Newtonian assumption, maybe another viscosity model should be used.

3.3.2.7 Important remarks

A CFD model of Nozzle reactor has been developed and validated in ANSYS FLUENT for studying fast HTL of biomass. The results showed that the mass flowrate ratio between the hot and cold flows played the most important role in establishing high heating rates in the reactor, assuming an adiabatic system boundary. Tests for pure water with low viscosity (0.001 Pa s), the mass flowrate ratio of 60:20 mil/min (hot/cold) gave very good mixing and thus high heating rate. The effect of the total mass flow on the temperature profile in the reactor was not significant. Cold flows with viscosity higher than 0.01 Pa s resulted in poor mixing at the nozzle for the mass flowrate ratio of 60:20 mil/min.

3.3.3 Experimental validation of the Newtonian model

3.3.3.1 Nozzle reactor construction

A nozzle reactor, similar to that shown in Fig. 3.1B, has been construed in the Thermal Lab of Department of Energy and Process Engineering, Norwegian University of Science and Technology (NTNU). The dimension and materials reported in the literature have been adopted [21], using Swagelok pieces including 1/8″ tubing with 0.035 wall thickness, 3/8″ tubing with 0.065 wall thickness and two cross-pieces.

The fluid temperature at five positions (points 1−5) inside the reactor was measured, of which the longitudinal location relative to the exit end of the inner tube is −18; 1.5; 5; 63.5; and 125 mm, respectively. Four thermocouples, T1−T4 were used for every measurement. T1 were for monitoring the temperature of preheated water at source. T2 was used to measure the temperature at the outlet (125 mm). T3 was for measuring the temperature of the cold flow at the point it entered the reactor (−18 mm). T4 was a thin thermocouple (Ø = 0.5 mm) so that it can be introduced into the annulus between the 1/8″ and 3/8″ tubes for temperature measurement at the locations 1.5 mm; 5 mm; 63.5 mm.

3.3.3.2 Reaction system and experimental validation via temperature measurement

A reaction system similar to that in Fig. 3.1C has been developed by modifying an existing continuous flow tubular reactor system (Parr 5401) from Parr Instrument Inc. [29]. The electric furnace of Parr 5041 reactor system was used to preheat pure water fed by Pump 1 (High flow HPLC pump NJ2289HC, 0−100 mL/min, working pressures up to 4000 psi). Since the study employed pure water or aqueous glucose solution to simulate the cold flow, Pump 2 was another HPLC pump.

Seven experimental cases combining different ratios of hot/cold flows and temperatures of the preheated water (T1) presented in Table 3.3 were considered for the study. The group of cases 1−4 was for studying the behavior of the system with the hot flow preheated to the critical point of water (374°C), otherwise at supercritical conditions (450°C) for the group of case 5, 6 and 9.

Table 3.3: Experimental matrix.

Case	Hot flow (mL/min)	Cold flow (mL/min)	T1 (°C)
1	20	10	374
2	30	10	374
3	40	10	374
4	50	10	374
5	20	10	450
6	30	10	450
9	26	6	450

T1: Temperature of preheated water at source.

3.3.3.2.1 Temperature measurement for reactor fed with hot water preheated to 374°C

The investigation was first made for the hot flow preheated to the critical temperature of water, 374°C. It was because of the initial consideration over the reactor safety and economics. The result is presented in Fig. 3.5, in the form of temperature plotted versus axial locations relative to the nozzle (inner tube) exit. Numerical values of the measurement are tabulated in Table 3.2.

It is interesting to see in Fig. 3.5 in combination with data in Tables 3.3 and 3.4 that when the mass flowrate ratios (hot/cold stream) increased from two to five (case 1 to case 4), the heating rate decreased. This is obviously not reasonable, considering the driving force

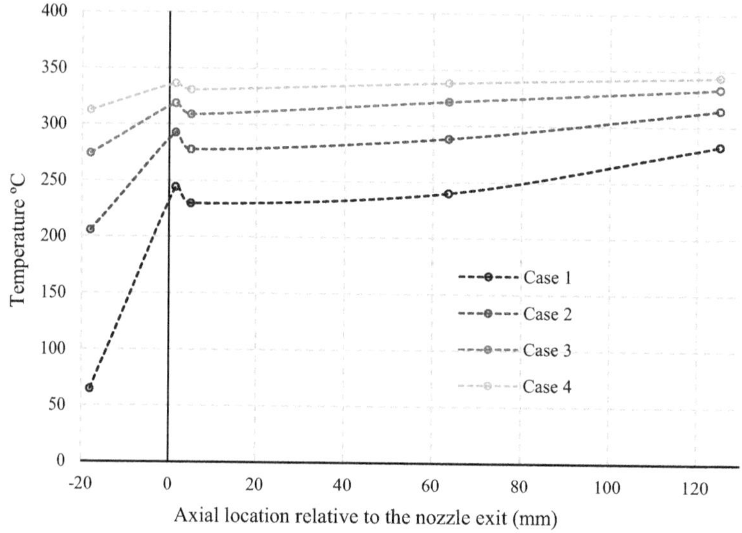

Figure 3.5
Measured temperatures for the reactor fed with hot water preheated to 374°C.

of the heat transfer. At this stage, it is not possible to fully explain this observation, although it is suspected that the thermocouple T3 was located too close to the mixing point (T – 18 mm) that the measurement result at this point was strongly affected by the heat transfer though the reactor wall via thermal conductivity. It is because the temperature of the cold flow at some point on the feeding line must be identical for all the case. For example, this temperature in pump 2 should be equal to ambient temperature. In fact, the temperatures measured along the reactor are always higher for higher mass flowrate ratio, which is reasonable but not for the first point.

Table 3.4: Measured temperatures (°C) for the reactor fed with hot water preheated to 374°C.

Case	T − 18 mm	T + 1.5 mm	T + 5 mm	T + 63.5 mm	T + 125
1	64.6	244.5	230.1	240.3	282.8
2	205.7	292.7	277.8	288.7	314.8
3	274.2	318.8	308.9	321.5	333.5
4	312.43	336.3	330.6	338	343.8

On the other hand, the temperature profiles of all cases share a common shape similar to that observed in the literature [28]. However, the temperature profiles in Fig. 3.5 peak at the second point, about 1.5 mm from the nozzle exit upward (T + 1.5 mm) downstream of the flow through the annulus between the two pipes. Among the measurement points within the annulus, temperature at point 2 (T + 5 mm) is the lowest. These observations are not in agreement with the literature, probably due to the difference in the measurement locations between the two works.

3.3.3.2.2 Temperature measurement for reactor fed with hot water preheated to 450°C

Similarly, the investigation was then made for the hot flow preheated to a temperature higher than the critical temperature of water, 450°C. It was performed because of the consideration over the heat loss during the process and the need to achieve relatively high temperature in the reactor, relevant for HTL. The result is presented in Fig. 3.6, in the form of temperature plotted versus axial locations relative to the nozzle (inner tube) exit. Numerical values of the measurement are also tabulated in Table 3.5.

Table 3.5: Measured temperatures (°C) for the reactor fed with hot water preheated to 450°C.

Case	T − 18 mm	T + 1.5 mm	T + 5 mm	T + 63.5 mm	T + 125
5	258.0	378.4	370.8	373.5	380.9
6	359.5	384.2	385.2	384.4	386.6
9	365.7	386.8	386.6	385.6	388.4

Similar to what observed for the case of hot stream of water, the temperature profiles of all cases share a common shape similar to that observed in the literature [28]. Similar disagreements with the literature are also observed. However, the differences between case 6 and 9 are not significant. This suggests the importance of initial temperature of the hot stream with respect to the heating rate as well, apart from the mass flow rate ratio.

3.3.3.2.3 Temperature measurement for reactor fed with hot water preheated to 450°C and cold flow of 10% w/w glucose solution

The experimental conditions and the cases of different mass flowrate ratios employed for this investigation are similar to that of the previous case presented in Section 3.2. The only difference is that the cold streams were 10% w/w glucose aqueous solution. Again, glucose solution was chosen to simulate the behavior of non-Newtonian fluid at the first approximation, considering the need of relatively long operation and practical limitations in pumping slurry biomass solution. The result is presented in Fig. 3.7, in the form of temperature plotted versus axial locations relative to the nozzle (inner tube) exit. Numerical values of the measurement are tabulated in Table 3.6.

Table 3.6: Measured temperatures (°C) for the reactor fed with hot water preheated to 450°C and cold flow of water containing 10% w/w glucose.

Case	T − 18 mm	T + 1.5mm	T + 5 mm	T + 63.5 mm	T + 125 mm
5 with glucose	213.2	378.0	368.2	374.1	380.3
6 with glucose	360.1	386.7	384.6	383.0	386.6
9 with glucose	358.3	387.1	386.9	387.1	388.8

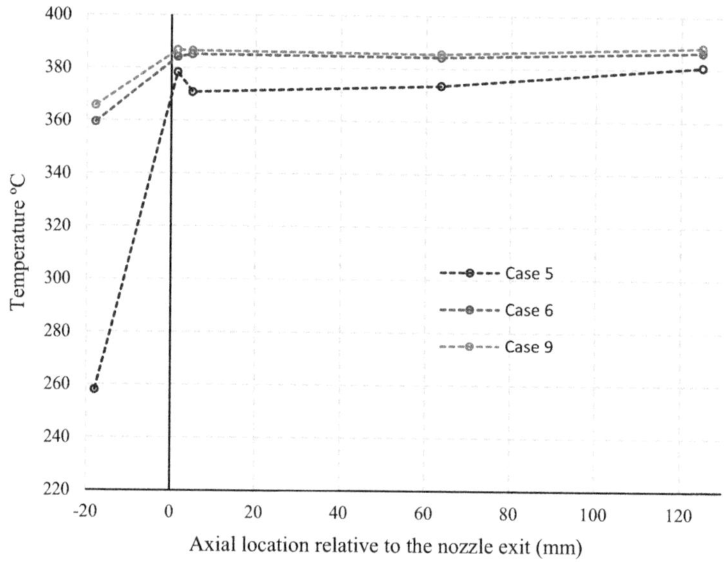

Figure 3.6
Measured temperatures for the reactor fed with hot water preheated to 450°C.

Comparing with data in Table 3.6 and Fig. 3.7 with Table 3.5 and Fig. 3.6, it is interesting to see that the differences between the two investigations are not significant. The shape of the temperature profiles and the numerical value are almost identical. This confirms that the assumption of Newtonian model for CFD modeling of biomass stream is not relevant.

3.3.3.2.4 Important observations and remarks

For all cases of temperature measurements including case 5, 6 and 9 with 10% w/w glucose solution, the curve of measured temperatures in the reactor exhibits the same trend with a peak at point 1.5 mm. The shape of the temperature curve constructed based on the measured points is similar to that reported in the literature [28] and predicted earlier in our CFD study. This suggests that the reactor can offer very high heating rates, suitable for continuous fast HTL of biomass. However, it is not relevant to indicate the measured temperatures as full temperature profile in the reactor, considering the fact that there is a large temperature gradient at the vicinity of the exit end (the nozzle) of the inner tube [28]. It means that the temperature peak might be at some point located at less than 1.5 mm downstream of the nozzle. The peak can even be located at negative location relative to the nozzle as observed and reported in the literature [28]. This was explained by the protruding of the hot flow into the cold flow for the case of unbalanced flow, i.e., for higher ratio of hot/cold mass flow. However, this explanation does not hold for the case with balanced flows.

On the other hand, it appears that the addition of 10% w/w of glucose into the cold flow did not have significant influence on the mixing and thus temperature profile in the reactor.

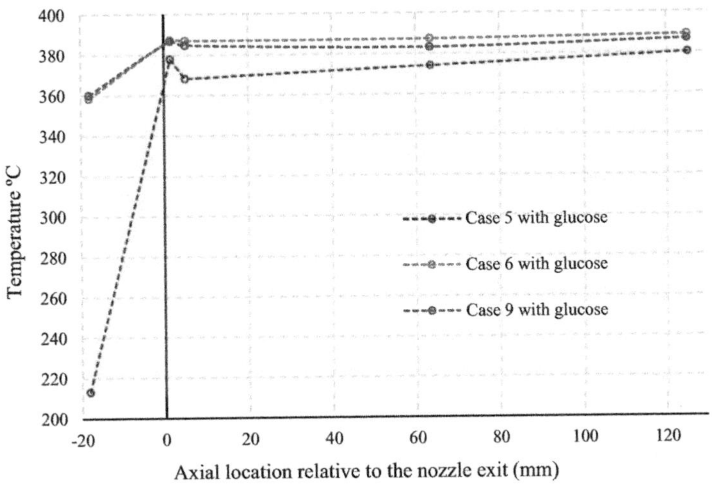

Figure 3.7
Measured temperature for the reactor fed with hot water preheated to 450°C and cold flow of 10% w/w glucose solution.

This confirms a conclusion remark from our previous work that the assumption of Newtonian fluid for CFD modeling of the nozzle reactor concept for fast HTL of biomass is not relevant.

3.3.4 CFD study of nozzle reactor for fast HTL assuming non-Newtonian fluid

In this study, a full geometry of nozzle reactor similar to that presented in Fig. 3.1B has been created. In the first instance, the thermodynamic and fluid dynamic setup was similar to that presented in the Newtonian simulation session. An RTD-based validation was performed, which gave similar results as for the simplified geometry assumption presented earlier. In addition, the mixing and thus reactor temperature profile is in good agreement with the literature (flow ratio of 20:10 mL/min) [20]. Then the power law non-Newtonian model was adopted to simulate the cold flow of water containing solid lignin particles (biomass slurry). Three investigations were performed using the developed non-Newtonian model. The first looked at the effect of the flow ratio on the mixing and thus temperature profile of the reactor. The second analyzed the effect of the viscosity change of the cold flow of biomass slurry. Third studied the effect of the total mass flow.

3.3.4.1 Effect of the flow ratio

Two simulations were performed and analyzed for the hot flow of the same pressure (25 MPa) and temperature (450°C) and a cold flow of 10%wt lignin slurry solution. The flows analyzed have been 20:10 and 30:10 mL/min. The results are shown in Fig. 3.8.

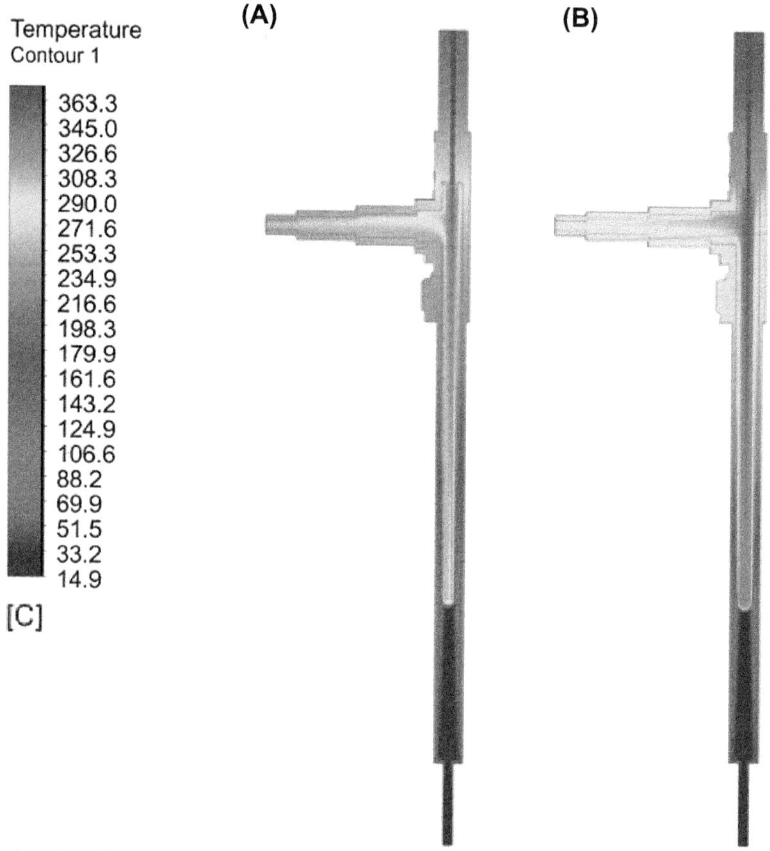

Simulated outlet temperature (°C)	248.4	286.3

Figure 3.8
Temperature profile of the 20:10 (A) and 30:10 (B) mL/min for non-Newtonian model.

It is clearly shown from the simulation result that the temperature at the outlet of the reactor is strongly dependent on the flow ratio. In addition, the temperature profile obtained from the non-Newtonian model is lower than that from the Newtonian. For example, comparing the 20:10 mL/min contours (Fig. 3.8A and B) it can clearly be seen that for non-Newtonian fluid, the temperature of the hot inlet arriving the mixing point has drastically been reduced to an around value of 280°C compared to the 340°C obtained from the Newtonian one. On the other hand, clear differences in the shape of the mixing between the Newtonian and non-Newtonian simulations were observed. Indeed, the mixing of the fluids occurs in a point nearer to the hot inlet tube outlet in the non-Newtonian model than in the Newtonian simulation. In addition, the heat transfer takes place in a smaller space although the mixture temperature is notably lower.

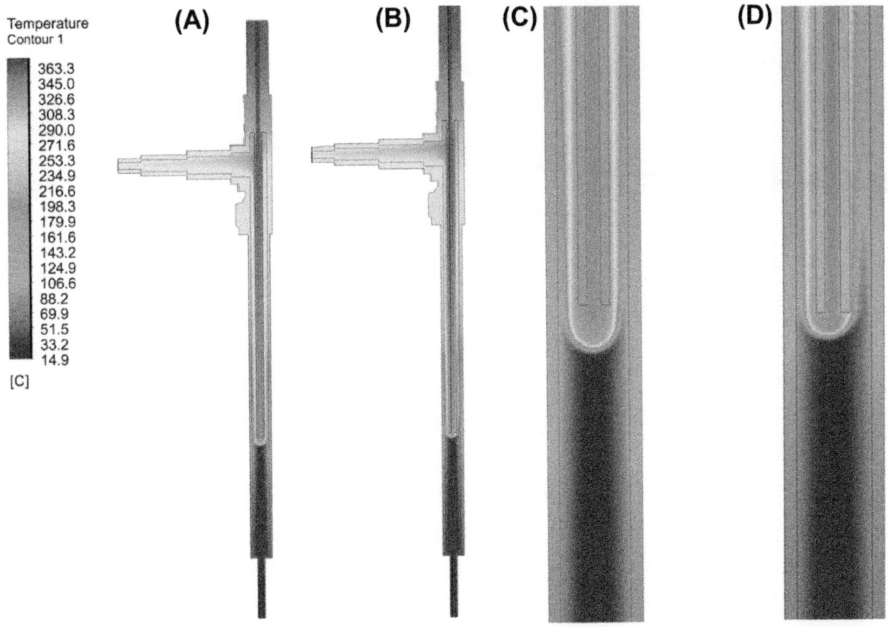

Figure 3.9
Temperature contour and (C) mixing point for 10%wt (A and C), and 12.5%wt (B and D) lignin.

3.3.4.2 Effect of viscosity of the cold flow

The effect of viscosity of non-Newtonian fluid was used to simulate the effect of lignin concentration and particle size in the cold flow. The investigation was performed for the flow ratio of 30:10 mL/min and the result is presented in Fig. 3.9.

It is interesting to see that the temperature profiles obtained from the simulations are pretty similar, with their outlet temperatures of 286.32°C and 288.3°C, respectively. The main difference between them resides in the asymmetry which appears along the reactor in the 12.5%wt case (Fig. 3.9D). Furthermore, the hot mixing zone appearing in the 10%wt is considerably longer than the one for 12.5%wt although being the same flow rates.

3.3.4.3 Effect the total mass flow rate

For this study, the total mass flow rate was varied within 30:10, 60:20, 90:30, and 120:40 mL/min to insure a constant flow rate ratio of 3:1. In addition, a cold flow containing 10%wt lignin was employed all cases. The obtained temperature contours for these flows and their outlet temperature were examined and analyzed. The temperature in the mixing zone increases and this mixing zone moves downwards with increased total mass flow rate.

3.3.4.4 Important remarks and implications

In this section, a full geometry of nozzle reactor was developed and modeled for Newtonian and non-Newtonian fluids. The Newtonian assumption was used for model validation. The non-Newtonian assumption was used for studying the effects of the flow rate ratio, total mass flow rate, and the viscosity of the cold flow on the mixing the thus temperature profile of the reactor. The results indicated that

- Non-Newtonian fluids behave in a different way of Newtonian ones in important terms such as viscosity, temperature, or flow.
- An important reduction in the temperatures along the reactor appears when working with non-Newtonian fluid compared to the results obtained with Newtonian one.
- The temperature profiles suffer very little change in value with the change of the viscosity.
- The mixing zone shape suffers a big change compared to the Newtonian fluid.
- Higher content of solid in the cold flow cause a bigger asymmetry in the temperature along the cross-sections of the reactor. This fact can be solved using higher hot:cold flow ratios.
- The outlet temperature increases with the increment of the total mass flow rate.

3.4 First test for fast HTL of lignin using nozzle reactor

The lab-scale continuous flow reaction system including a nozzle reactor developed at the Department of Energy and Process Engineering, Norwegian University of Science and Technology (NTNU) has been modified and developed further for fast HTL of lignin.

A preliminary study and validation of the reaction system for production of HTL oil from solid lignin via fast HTL has been performed. It has been demonstrated that the system works fine for fast HTL of lignin residue from second generation bioethanol production, producing HTL oil. Indeed, an aqueous slurry of lignin residue (from Sekab AB, Sweden) of 25 wt.%, with particle sizes not bigger than 125 μm and 2 wt.% emulsifier (a low cost vegetable cooking oil from Rema1000) on a dry basis was successfully fed into the system and converted to HTL oil in the conditions of 350°C with an estimated oil yield within 40–50 wt.% on a dry basis [30]. Continuous operations longer than 20 min for the system was not possible due to the limited volume (500 mL) of the high viscosity syringe pump. Two identical pumps in parallel are needed for fully continuous operations. This validation suggests the today fast HTL technology at NTNU is very promising but needs further investigation in order make it more relevant and more attractive to the industry.

3.5 Optimization needs

Optimization of the reactor design and process are crucial. The mixing geometry and dimension of the nozzle and the reactor, as well as the reactor orientation and streaming of cold and hot fluid flows would be optimized with to maximizing the mixing and thus heating rate, as well as to minimizing the risk of reactor blockage, and to improve the HTL oil yield and quality.

For the concept of nozzle reactor, there exist four different mixing geometries, of which three are presented in Fig. 3.10 (adapted from a study of the nozzle reactor concept for supercritical water oxidation (SCWO) of wastewater containing organic compounds [31]).

1 Countercurrent mixing with a *hot stream* of superheated water entering the reactor through the inner pipe (Fig. 3.10A)
2 Countercurrent mixing with a *cold stream* of water containing reactants entering the reactor through the inner pipe (not presented in Fig. 3.10)
3 Co-current mixing with a *hot stream* of superheated water entering the reactor through the inner pipe (Fig. 3.10B)
4 Co-current mixing with a *cold stream* of water containing reactants entering the reactor through the inner pipe (Fig. 3.10C)

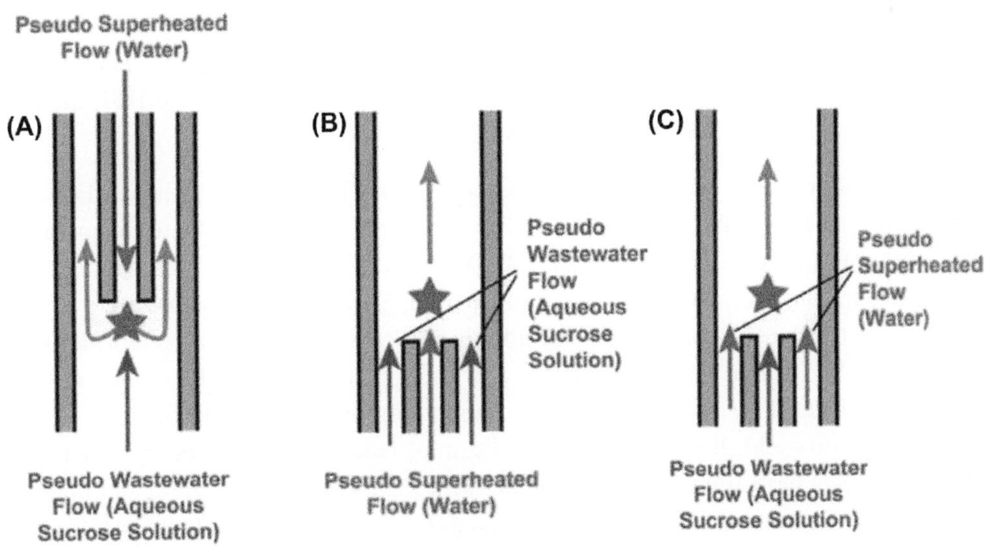

Figure 3.10
Scheme of three mixing geometries for nozzle reactor [31] (A) Countercurrent, (B) Co-current (C) Co-current protected wall geometry.

Among these four mixing geometries, the second (number 2, not presented in Fig. 3.10) and the fourth (number 4, Fig. 3.10C) may be more suitable for fast HTL of lignin residue because of two reasons. First, the flowrate ratio of hot stream over cold stream is normally larger than unity (the larger the ratio, the higher heating rate), thus it is more reasonable to feed the smaller cold stream (containing biomass) into the reactor through the smaller inner pipe. However, if the inner pipe is too small (in a laboratory-scale for example), feeding of concentrated biomass slurries through the inner pipe may be more difficult. In addition, the smaller cold stream may be no longer "cold" as it might be significantly heated by the larger hot stream of the geometry number 4 (Fig. 3.10C) or by the larger (hot) postmixing stream of the geometry number 2. Second, letting the hot stream of pure water entering the reactor through the outer pipe would help prevent solid particles, which have entered the reactor with the cold flow through the inner pipe, from depositing on the reactor wall and thus the blockage and corrosion risks are further minimized. These risks can even be minimized further by optimizing the reactor orientation, considering the influence of gravitational force and possible buoyancy mixing by natural convection as discussed earlier.

In addition, during fast HTL it is possible to add reducing gases such H_2, CO, or additives such as methanol or phenol into the process in order improve the fuel properties of HTL oil. H_2 addition would combine fast HTL with hydrogenation, a common technology in petroleum refinery for improving the fuel properties (e.g., heating value) of liquid fuels. The addition of methanol for example would (1) cool down the product stream to quench unwanted further reactions such as depolymerization or further thermal decomposition; and (2) make HTL oil more stable and less viscous by blocking (via transesterification/esterification) the reactive functional groups [32] (mainly guaiacol, methyl dehydroabietate, and ethyl- and methyl-substituted phenols [3,33,34]), and. Meanwhile, it has been demonstrated that a phenol addition of 3.4−9.7 wt.% increased the HTL oil yield from 33.9 wt.% to 77.3−102.3 wt.%, respectively [34].

On the other hand, catalyst addition into the reactor during HTL is also possible. For this purpose, various water-soluble catalysts (such as KOH and Na_2CO_3) and heterogeneous catalysts (Pd/C, Pt/C, Ru/C, $Ni/SiO_2{-}Al_2O_3$, $CoMo/Al_2O_3$, $NiMo/Al_2O_3$ and zeolite), which have been tested for catalytic HTL, can be tried. Since nozzle reactors are continuous flow tubular reactors, tuning the addition of catalyst along the reactor during the process of HTL (to find the right moment and right location) would be a good option of process optimization for in-process catalytic upgrading of HTL oil and further improving of the oil yield. It is because the addition of catalyst into the second reactor of a two-step HTL has been found more beneficial with regard to HTL oil yield and quality [35]. Finally yet importantly, (off517 process) catalytic upgrading of HTL oil would contribute to a full development of a novel process for sustainable production of higher grade diesel-like biofuel for transport sector.

3.6 Conclusions and perspectives

Nozzle reactor has been identified to be suitable for realizing continuous fast HTL of solid biomass, producing higher quality HTL that can be used directly as liquid fuel or upgraded further. The reactor is capable of very fast heating and tolerating solid particles suspended in the liquid stream flowing through the reactor. CFD (computational fluid dynamics) models have been developed in ANSYS Fluent for studying the use of the nozzle reactor concept for continuous fast HTL of solid biomass. A laboratory-scale setup has been developed, adopting the nozzle reactor concept, with modifications suitable for fast HTL. The reactor performance including the countercurrent mixing has been experimentally characterized by examining the temperature profile along the reactor. The characterization results are in agreement with the CFD predictions. The system has been successfully tested for fast HTL of lignin residues from steam explosion, which was prepared and continuously pumped in to the reactor in the form of a slurry solution of 25 wt.% lignin (125 μm in size) in water and 2 wt.% of cooking oil as emulsifier.

The nozzle reactor concept would make it possible to truly mimic the natural process producing biocrude analogous to petroleum crude, but in the time scale of seconds, leaving the regret to the Mother Nature. However, further investigations and optimizations of the reactor geometry and the process are needed.

Acknowledgments

The data presented in this paper is partly extracted from the master theses and project reports of my ex-students Manuel Villarreal Salcedo, Magnus Kyrkjebø, Arturs Curakovs, and Gastón Mauricio Cocco, whose efforts are highly acknowledged. The author would also like to thanks Love Håkansson and Jose Sierra-Pallares for discussions and sharing ANSYS FLUENT experiences in modeling of nozzle reactor.

References

[1] Yu Y, Lou X, Wu H. Some recent advances in hydrolysis of biomass in hot-compressed water and its comparisons with other hydrolysis methods. Energy & Fuels 2008;22(1):46–60.

[2] Fang Z, et al. Reaction chemistry and phase behavior of lignin in high-temperature and supercritical water. Bioresource Technology 2008;99(9):3424–30.

[3] Wahyudiono MS, Goto M. Conversion of biomass model compound under hydrothermal conditions using batch reactor. Fuel 2009;88(9):1656–64.

[4] Tekin K, Karagöz S, Bektaş S. A review of hydrothermal biomass processing. Renewable and Sustainable Energy Reviews 2014;40(0):673–87.

[5] Kruse A, Funke A, Titirici M-M. Hydrothermal conversion of biomass to fuels and energetic materials. Current Opinion in Chemical Biology 2013;17(3):515–21.

[6] Peterson AA, et al. Thermochemical biofuel production in hydrothermal media: a review of sub- and supercritical water technologies. Energy & Environmental Science 2008;1(1):32–65.

[7] Crocker M. Thermochemical conversion of biomass to liquid fuels and chemicals. Royal Society of Chemistry; 2010. p. 192.

[8] Matsumura Y, et al. Biomass gasification in near- and super-critical water: status and prospects. Biomass and Bioenergy 2005;29(4):269–92.
[9] Calzavara Y, et al. Evaluation of biomass gasification in supercritical water process for hydrogen production. Energy Conversion and Management 2005;46(4):615–31.
[10] Brunner G. Chapter 12 – corrosion in hydrothermal and supercritical water. In: Gerd B, editor. Supercritical fluid science and technology. Elsevier; 2014. p. 591–619.
[11] Marrone PA, Hong GT. Corrosion control methods in supercritical water oxidation and gasification processes. The Journal of Supercritical Fluids 2009;51(2):83–103.
[12] Wang Y, et al. Formation of coke during the pyrolysis of bio-oil. Fuel 2013;108(0):439–44.
[13] Sinag A, Kruse A, Rathert J. Influence of the heating rate and the type of catalyst on the formation of key intermediates and on the generation of gases during hydropyrolysis of glucose in supercritical water in a batch reactor. Industrial & Engineering Chemistry Research 2004;43(2):502–8.
[14] Brand S, et al. Effect of heating rate on biomass liquefaction: differences between subcritical water and supercritical ethanol. Energy 2014;68(0):420–7.
[15] Kamio E, et al. Effect of heating rate on liquefaction of cellulose by hot compressed water. Chemical Engineering Journal 2008;137(2):328–38.
[16] Modell M. Gasification and liquefaction of forest products in supercritical water. In: Fundamentals of thermochemical biomass conversion. Amsterdam: Elsevier; 1985. p. 95–119.
[17] Bach QV, Tran KQ, Lystad KQ. Fast hydrothermal liquefaction of macro585 alga: characterization of products. Chemical Engineering Transactions 2016:97–102.
[18] Bach Q-V, et al. Fast hydrothermal liquefaction of a Norwegian macro-alga: screening tests. Algal Research 2014;6(B):271–6.
[19] Tran K-Q. Fast hydrothermal liquefaction for production of chemicals and biofuels from wet biomass – the need to develop a plug-flow reactor. Bioresource Technology 2016;213:327–32.
[20] Sierra-Pallares J, et al. Prediction of residence time distributions in supercritical hydrothermal reactors working at low Reynolds numbers. Chemical Engineering Journal 2016:299–373.
[21] Lester E, et al. Reaction engineering: the supercritical water hydrothermal synthesis of nano-particles. The Journal of Supercritical Fluids 2006;37(2):209–14.
[22] Yatmaz HC, Wallis C, Howarth CR. The spinning disc reactor – studies on a novel TiO_2 photocatalytic reactor. Chemosphere 2001;42(4):397–403.
[23] Brechtelsbauer C, et al. Evaluation of a spinning disc reactor for continuous processing. Organic Process Research & Development 2001;5(1):65–8.
[24] Boodhoo KVK, Dunk WAE, Jachuck RJJ. Spinning disc reactor for photopolymerization. Abstracts of Papers of the American Chemical Society 2001:222.
[25] Gruar RI, Tighe CJ, Darr JA. Scaling-up a confined jet reactor for the continuous hydrothermal manufacture of nanomaterials. Industrial & Engineering Chemistry Research 2013;52(15):5270–81.
[26] Andersson B, et al. Computational fluid dynamics for engineers. 2011. p. 1–189.
[27] Versteeg HK, Malalasekera W. An Introduction to computational fluid dynamics: the finite volume method. Pearson Education Limited; 2007.
[28] Tighe CJ, et al. Investigation of counter-current mixing in a continuous hydrothermal flow reactor. The Journal of Supercritical Fluids 2012;62:165–72.
[29] Parr. 5400MB Series 5400 tubular reactor sales literature. 2018 [cited January 15, 2018]; Available from: http://www.parrinst.com/wp615-content/uploads/downloads/2016/05/5400MB_Parr_Tubular-Reactor-Systems-Literature.pdf.
[30] Curakovs A, Tran K-Q. On the application of Nozzle reactor for fast hydrothermal liquefaction of solid biomass. In: Sustainable energy symposium. Gothenburg, Sweden: Chalmers University of Technology; 2017.
[31] Huddle T, et al. Pseudo fluid modelling used in the design of continuous flow supercritical water oxidation reactors with improved corrosion resistance. The Journal of Supercritical Fluids 2017;120(Part 2): 355–65.

[32] Fushimi C, Umeda A. Comparison of biodiesel production by a supercritical methanol method from microalgae oil using solvent extraction and hydrothermal liquefaction processes. Energy & Fuels 2016;30(10):7916–22.
[33] Zhang B, Huang HJ, Ramaswamy S. Reaction kinetics of the hydrothermal treatment of lignin. Applied Biochemistry and Biotechnology 2008;147(1–3):119–31.
[34] Arturi KR, et al. Hydrothermal liquefaction of lignin in near-critical water in a new batch reactor: influence of phenol and temperature. The Journal of Supercritical Fluids 2017;123:28–39.
[35] Yang L, Li Y, Savage PE. Catalytic hydrothermal liquefaction of a microalga in a two-chamber reactor. Industrial & Engineering Chemistry Research 2014;53(30):11939–44.

CHAPTER 4

Granular sludge bed anaerobic treatment systems for resource recovery

Hamidreza Mojab[1], Eldon Raj[2], Santiago Pacheco-Ruiz[3]

[1]Department of Water Resource Management, Faculty of Civil Engineering and Geoscience, Technical University of Delft, Delft, The Netherlands; [2]Department of Environmental and Water Technology, IHE Delft Institute for Water Education, Delft, The Netherlands; [3]Veolia Water Technologies Techno Center Netherlands B.V./Biothane, Delft, The Netherlands

4.1 Introduction

For a long time, wastewater had been considered as a residual effluent that only needs to be treated in order to prevent environmental pollution. Although, the treatment had been done mostly by biological methods, but the fact that wastewater can be considered as a source of nutrients, chemicals and energy had been neglected. This was due to the lack of technological development from one side, and availability of resources and energy from the other side. Along with the expansion of industries, urbanization and population growth, concerns have been raised on the scarcity of resources and energy, and therefore, wastewater has gained more attentions, as a source of energy and materials. Anaerobic digestion was one of the first treating methods that was developed, which not only can treat wastewater with high efficiency, but also has the potential to produce energy during treatment processes. In this chapter, one of the most used anaerobic treatment systems, up-flow anaerobic sludge bed (UASB) and its improved version, expanded granular sludge bed (EGSB) will be introduced. These systems have been widely used, especially by the industries with high-strength wastewater in recent decades with high efficiency of treatment and energy recovery.

4.1.1 Sources of high strength wastewater

Industrial wastewater is one of the biggest sources of pollution of the environment. During the last century, due to the high expansion of industries, environmental problems has been increased significantly. As a result, more strict regulations and policies has been created in order to reduce the environmental problems, especially, pollution of the water bodies. Table 4.1 shows the maximum permissible concentrations (MPC) for surface water and sediment in The Netherlands [1]. Therefore, industries are forced to apply wastewater

Waste Biorefinery. https://doi.org/10.1016/B978-0-12-818228-4.00004-6

Table 4.1: Maximum permissible concentrations limits of major contents for surface water and sediments in the Netherlands [1].

Content	Maximum permissible concentration (MPC)
BOD_5	25 mg/L
COD	125 mg/L
TSS	35 mg/L
Total nitrogen	2.2 mg N/L
Total phosphate	0.15 mg P/L
Chloride	200 mg/L
Sulfate	100 mg SO_4/L

treatment systems to efficiently remove the pollutants from their effluent before discharging them to the environment. However, as the characteristics of these effluents vary largely from one industry to another, choosing the best treatment method is challenging.

Table 4.2 shows a general overview of the characteristic of some of the major industries. The wastewater from these industries also refer to what is called high-strength wastewater. High-strength wastewater, which is discharged from most of the industries, is generally described as a wastewater having pollutants at higher concentration than domestic wastewater. Although, these high concentration varies from one industry to another, but in general, wastewaters with biochemical oxygen demand (BOD) concentration between 100 and 4000 mg/L, oil and grease from 50 to 15,000 mg/L, COD from 800 to 15,000 mg/L and total suspended solids (TSS) from 140 to 4375 mg/L are considered as high-strength wastewater.

However, from another point of view, Hamza [3] states that wastewaters with COD concentration above 4000 mg/L, that conventional aerobic treatment methods are no longer

Table 4.2: An overview of the major water pollutants by some industries [2].

Sector	Pollutant
Chemicals	COD, organic chemicals, heavy metals, SS, cyanide
Pulp and paper	BOD, COD, chlorinated organic compounds
Iron and steel	BOD, COD, oil, metals, acids, phenols, cyanide
Petrochemicals and refineries	BOD, COD, mineral oils, phenols, chromium
Textile and leather	BOD, solids, chlorinated organic compounds
Non-ferrous metals	Fluorine and SS
Microelectronics	COD and organic chemicals
Mining	SS, metals, acids and salts

feasible, are identified as high-strength wastewater, which anaerobic treatment methods are suitable to apply.

According to Shi [2], industrial wastewater can be categorized into two groups; organic and inorganic industrial wastewater. Inorganic wastewater is mainly discharged from coal and steel industries as well as nonmetallic mineral industries and mining. These wastewaters usually identified by their high content of suspended matter which are inorganic materials such as sands and wasted materials, generated thorough production processes or washing step. On the other hand, organic wastewater is discharged from industries which use organic matter in their processes which make them suitable form biological treatment. Wastewater from pharmaceutical, cosmetic, pesticides and herbicides, tanneries, textile, pulp and paper, food, and oil-refining industries are examples of organic industrial wastewater [2].

4.1.2 UASB/EGSB systems for wastewater treatment and resource recovery

Anaerobic digestion has been considered as a solution for wastewater treatment for a long time in many countries. Simple septic tanks are a common use of anaerobic digestion which has been used by many households in a decentralized form in urban areas. This system has been able not only to treat wastewater, but also to produce biogas which has been used to generate energy and heat. A simple schematic of decentralized anaerobic digester is shown in Fig. 4.1.

In principle, an anaerobic process can degrade all organic compounds, while more readily biodegradable waste increases the efficiency and reduces the cost of such a system. Anaerobic digestion is defined as the biological degradation of organic matters into different compounds such as methane (CH_4) carbon dioxide (CO_2) and hydrogen (H_2), in different fractions, by a community of microbes in an oxygen-free environment. This biological process has been greatly applied not only for primary but also for secondary

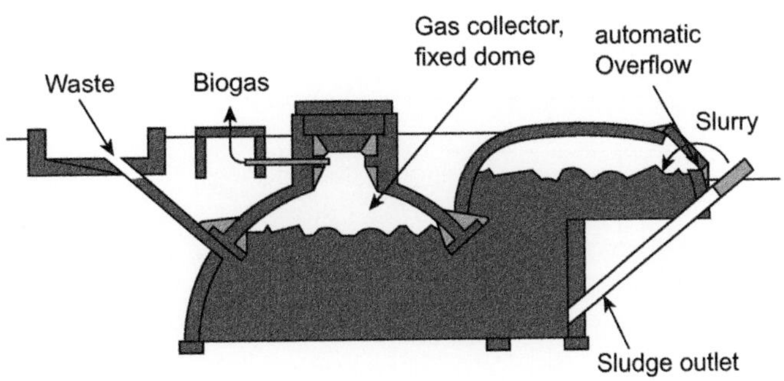

Figure 4.1
Schematic of a decentralized anaerobic digester [4].

treatment of numerous industrial wastewater streams [5]. This digestion occurs mainly in four steps which are shown in Fig. 4.2.

Anaerobic treatment of wastewater has several advantages compared to aerobic treatment. This technique has low energy nutrients requirements, while it generates biogas which can be used as an energy source. In theory, about 13.5 MJ is stored in 1 kg of COD, which according to the type of anaerobic process, it can be captured in form of gas (methane and hydrogen) or liquid (alcohol) [6]. Even at high loading rates, an excellent performance of the anaerobic system is expected, with high degradation efficiency and improved dewatering. Moreover, anaerobic microorganisms have the capacity to remain at storage without feed for several months. However, there are some disadvantages associated with anaerobic treatment such as process sensitivity, odor problems, long start-up time and in some cases, the need for posttreatment to comply with more strict discharge standards. Nevertheless, in a proper controlled environment, anaerobic digestion can be operated in stable condition [7].

UASB is one of the earliest developed systems for anaerobic treatment of industrial wastewater. UASB was developed by Dr. Lettinga and his colleagues in the late 1970s. From that time, this system has been widely used in industries with high-strength wastewater throughout the world. With its relatively simple and cheap design, this

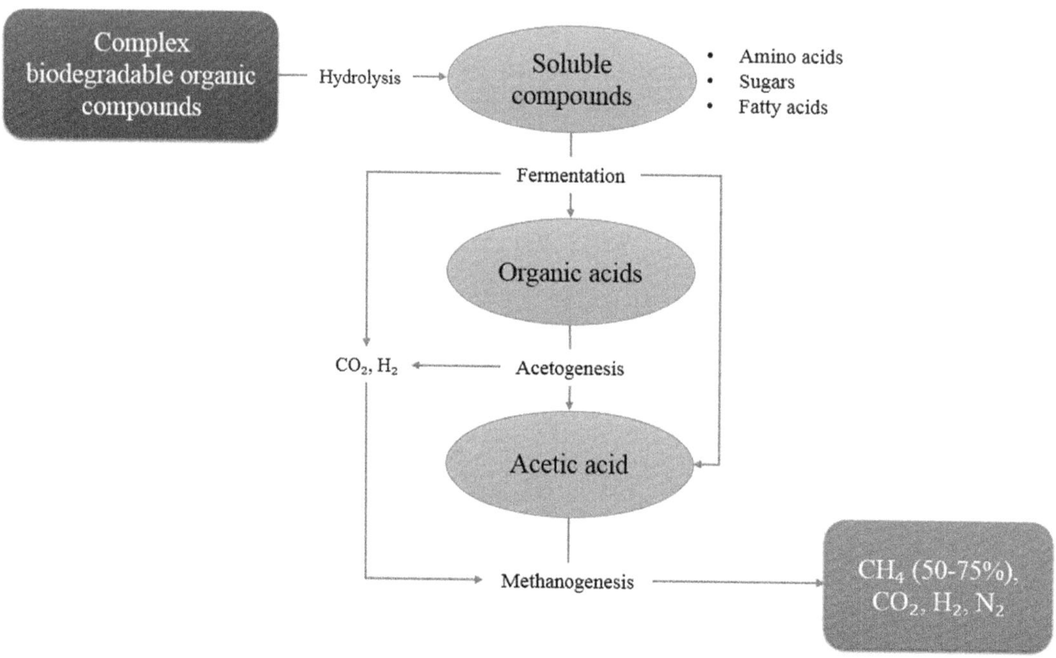

Figure 4.2
Process flow of methane production in anaerobic digestion.

bioreactor could treat wastewaters with high efficiency, and at the same time, produce energy, in form of biogas, in high volume. According to Khalil [8], costs related to operation and maintenance of an UASB reactor is 70% less than an aerobic reactor. The main feature of this system, which made it different than the simple septic tanks, was using a gas-liquid-solid separator (GLSS) inside the reactor, in order to separate different wastewater phases and especially, to prevent wash-out of the granules without help of any external energy and control device [9]. The GLSS can have different designs but its general configuration consists of a gas dome on top with an aperture opening part at the bottom. The opening part should be sufficient enough to avoid turbulence and provide return of the solids back to the reactor. Moreover, GLSS should be designed in a way to provide proper gas-water interface inside the dome and enough settling area outside the dome. Therefore, the geometry and hydraulics of the device play the main role in its design. Due to its reliable performance, application of UASB system for industrial wastewater treatment rapidly increased in different sectors, specially, in countries with warm climate which make it favorable for treatment under mesophilic condition. The early uses of UASB systems was mostly in agrofood industries during the 1980s, which then was expanded in a wide range of other industries, such as pulp and paper and chemical industries. Currently, about 60% of the total full scale anaerobic treatment plants are based on UASB system, which most of them are in tropic and subtropic countries [10]. High COD removal efficiency, considerable biogas production, and low capital and O&M costs are the main advantages of this system.

However, the need to cover the flaws of this system, such as low volumetric loading rate (VLR) and large required land, resulted in introduction of new granular-sludge-based systems. EGSB bioreactor was one of the first improvement of UASB. This system was developed by the Dutch company, Biothane Anaerobic Technologies (now as part of the Veolia Water Technologies). The improved configuration of EGSB allowed it to be operated under high up-flow velocity (4−6 m/h), which was not possible in UASB systems and could be built in smaller land size, due to its high height to width ratio. Moreover, the recirculation of the effluent resulted in better performance of the system due to expansion of the granules and better contact between the biomass and wastewater. These features made it possible to achieve high organic loading rate (20−40 kg COD/L/day) [11], which is a crucial parameter for industries, and thus, has made this system more attractive than UASB. Figs. 4.3 and 4.4, shows the number of different systems that has been installed by two of the major contractors, Biothane-Veolia, and Paques B.V., between 1981 and 2014 [11]. As it is illustrated in Figs. 4.3 and 4.4, the use of EGSB and IC reactors provided by the two company, has increased dramatically, from the time they were introduced in 90s, until recently that it has gained most of the market, which was held by UASB systems.

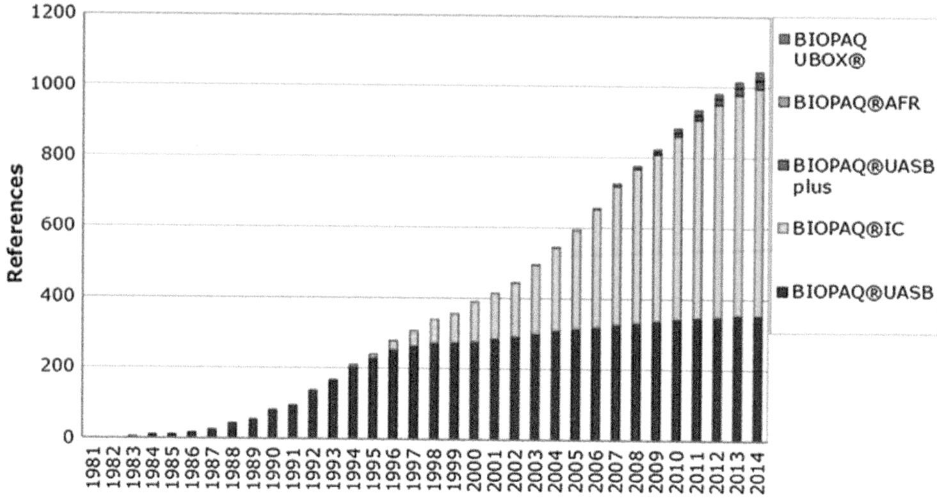

Figure 4.3
Sales of anaerobic high-rate reactors by Paques BV since the company's start-up (1981) [11].

4.1.3 Hybrid and coupled systems

In order to overcome the limitations of anaerobic treatment using UASB or EGSB reactors, the idea of coupling these systems with other treatment techniques has been

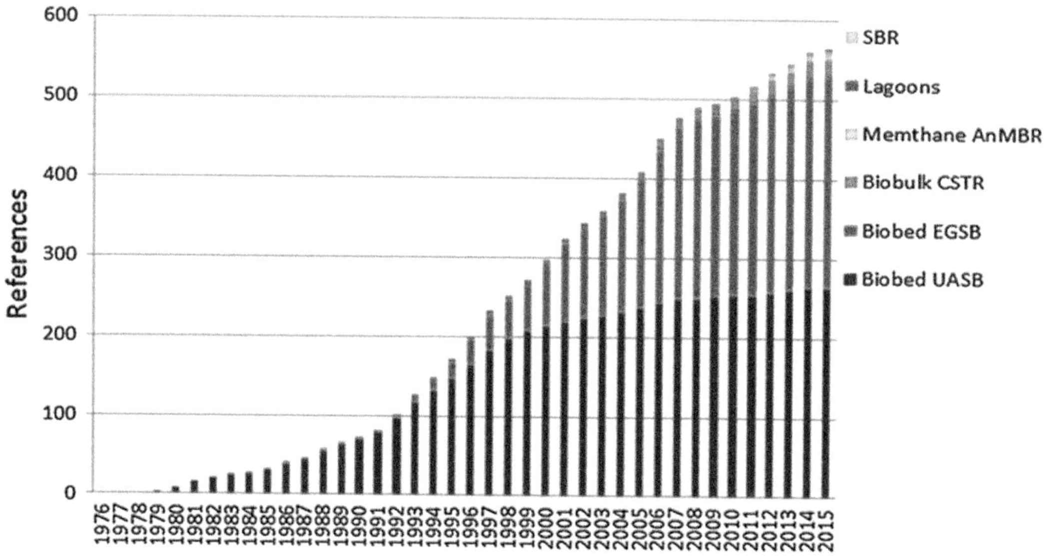

Figure 4.4
Sales of anaerobic high-rate reactors by Biothane-Veolia since the company's start-up (1981) [11].

developed in recent years. This improvement can be categorized in two different approach; combining the anaerobic reactor with an aerobic system, mostly as a posttreatment, or adding extra components to the configuration of a UASB or EGSB system, in order to enhance its performance.

A combination of aerobic and anaerobic system has gained more interests recently in order to reach higher quality of the final treated effluent from high-strength wastewaters. It has been suggested that the effluent of anaerobic treatment contains soluble organic materials which are suitable for postaerobic treatment. According to Hamza [1], the benefits of the anaerobic–aerobic process can be highlighted as follows:

- The aerobic posttreatment of the anaerobic effluent can highly increase the overall treatment efficiency, as it acts as a “polisher”.
- The excessive sludge production of aerobic treatment only, can be prevented by anaerobic pretreatment.
- Low energy consumption as high amount of energy is needed for aerobic treatment methods.
- The volatile organics which are present in the wastewater, are degraded anaerobically, lowering the chance of volatilization in the aerobic treatment.

Therefore, in terms of operation and feasibility, the combined aerobic-anaerobic system offers many advantages for treatment of high-strength industrial wastewaters.

On the other hand, the idea of combining an UASB reactor with an anaerobic filter (AF) reactor has resulted in development of a system called hybrid anaerobic reactor (HAR). As it is shown in Fig. 4.5, HAR is basically a UASB reactor, with a packing material (anaerobic filter) as an extra compartment to the top of the reactor. This filter can be used as a replacement of the GLSS, or together with it. This feature gives HAR systems more advantages than the normal UASB reactors, such as higher COD removal and biogas production. Moreover, the sludge wash-out, which is a common problem for UASB systems, can be reduced by 25% using HAR systems [13]. Also, this system shows a better performance for treating wastewater with complex organic compounds, such as phenols [12]. These advantages, together with possibility of operating under higher organic loading rate and less susceptibility to shock loads, has made this system attractive for many industries. However, due to limited number of full-scale application, and problems such as deterioration of the packing material, more research and study needs to be done on hybrid anaerobic reactors.

4.2 UASB/EGSB systems

In order to expand the use of anaerobic treatment systems, specially, in industries with high-strength wastewater, a major development was required on the configuration of

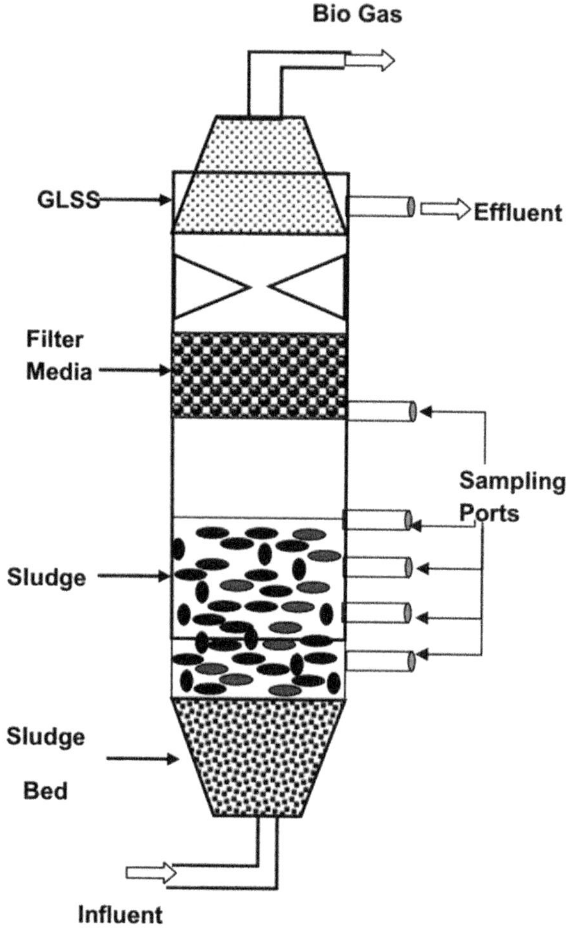

Figure 4.5
Schematic diagram of a hybrid anaerobic reactor (HAR) [12].

anaerobic digesters. This development, which started in early 1970s, allowed anaerobic technology to be used in a wide range of industries with high efficiency, treating high-strength wastewater with high flow rate and different characteristics. In this section, the mechanism and structure of two of the most widely used anaerobic treatment systems, will be studied. UASB reactor and EGSB reactor.

4.2.1 Definition and structure

UASB is one of the first major development in anaerobic treatment technologies. This system which was invented by Dr. Lettinga in 1976 in The Netherlands, dominated the market of anaerobic treatment plants in a short time. The concept of this system is based

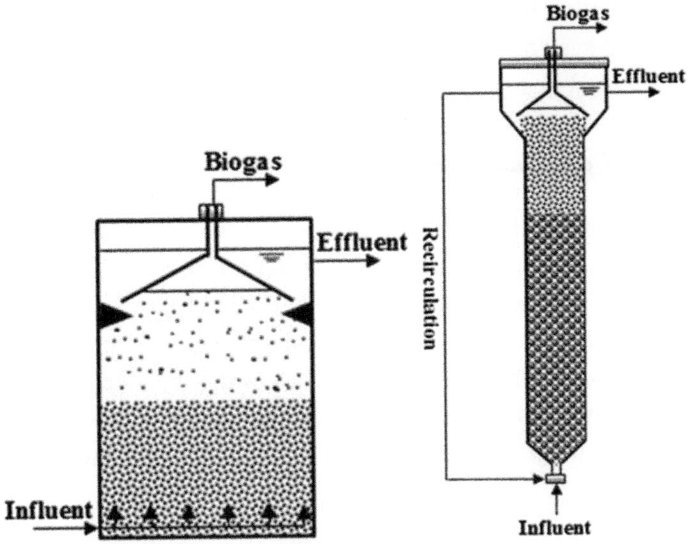

Figure 4.6
Schematic diagram of UASB (left) and EGSB (right) bioreactors.

on the conventional septic tanks, however, its special design made it suitable to be used in large scales with high-rate flows. Fig. 4.6, shows the structure of a UASB system. This reactor, which can be designed in rectangular or cylindrical shape, consists of mainly two parts. The bottom part is the location of the sludge bed and wastewater inlet. The sludge is in form of well settleable flocs and granules and the degradation of organic compounds occurs by flowing the wastewater thorough the sludge. Because of the high rate of the flow and the bubbles of the produced biogas, the sludge particles move upward along with flow, to the second part of the reactor, which is the main feature of this system. This part which is called GLS separator, divides the three phases of the system which allows the collection of the biogas, prevention of the sludge wash-out and leading the treated effluent outside of the reactor.

One of the most successful high-rate anaerobic reactor types are the EGSB reactors, which are advanced versions of the UASB reactor. In the late 1980s a new concept was introduced using granular sludge in anaerobic reactor. The EGSB system as it is shown in Fig. 4.6, is the developed version of UASB system, combining a GLSS and granular sludge. The main feature of this system is that much higher up-flow velocities for the liquid can be applied in the reactor which can be operated higher than 6 m/h. As a result, more expansion occurs in the sludge bed, with better hydraulic mixing which provides the granules to contact wastewater more efficiently and therefore, makes it possible to apply much higher VLR, around 20–35 kg COD/m^3/day. This feature makes this system suitable for treating high-strength organic wastewater [7]. Moreover, due to higher up-flow velocities that can be applied in the settler, the required settling surface area is much lower. Thus, EGSB type systems have smaller volumes than UASB systems and, therefore they are cheaper to build [14].

4.2.2 Advantages and disadvantages

In general, both UASB and EGSB systems have the common advantages of anaerobic treatment reactors, such as low sludge production, lower energy consumption than aerobic systems and generation of biogas which can be used as a source of energy. Moreover, reliable performance which has been gained through decades of application, have put them among top priorities of many industries in order to treat their wastewater. In addition, the relatively simple structure of these systems has made them easy to build with low capital cost and the possibility of using a wide range of materials such as steel, concrete and glass fiber.

However, due to its improved structure, EGSB offers more advantages than UASB. Because of its special design and recirculation of the effluent, it can be operated in extremely high VLR and up-flow velocity with high removal efficiency of COD, and therefore, suitable for treating of high-strength wastewaters. Moreover, due to its high height to width ratio, it requires less space for construction.

Despite all the advantages of UASB and EGSB systems, they are still facing with some major limitations. Both reactors need relatively long start-up time for the growth of proper microbial community. Presence of toxic compounds such as aromatics or high volatile fatty acids (VFAs) in wastewater can result in inhibition of microbial growth or acidification that leads to reactor failure [15]. However, EGSB system is less susceptible in this case, due to the recirculation of the flow, which results in exposure of biomass to a less concentration of toxic compounds [16]. Pathogens, coloring agents and nutrients are poorly removed in these systems and thus, a posttreatment facility is needed [17]. Wastewater with high concentration of suspended solids are not suitable for UASB and EGSB as it can affect the settle-ability of the sludge and wash-out. Finally, as anaerobic systems operate under mesophilic condition with temperature range of 25–40°C, a heating system is needed in case of application in cold climate regions.

Concerning its limitations, UASB reactor has poor removal ability for pathogens and similarly, elimination of nutrient by UASB reactors is low. Prolonged start-up period is also needed until steady state is reached due to the slow growth of methanogenic organism. Sometimes, industrial wastewaters treated with UASB reactor require posttreatment to fulfill discharge limits to surface water leading to additional treatment costs. The formation of hydrogen sulfide which cause bad odor and corrosion is another problem when using UASB reactor. Application of the system in the temperate region requires regulating the temperature between 16 and 35°C [17].

4.3 Operational parameters

In general, any treatment system are designed and operated under specific parameters. These parameters are important for a stable operation and avoiding system failure. When it comes to anaerobic treatment systems, these parameters become crucial due to the sensitivity of biological processes to unstable conditions. In this section, five of the most important parameters in UASB and EGSB systems are studied; Organic Loading Rate (OLR), Hydraulic Retention Time (HRT), Up-flow Liquid Velocity (ULV), pH, and temperature. These parameters are important for UASB and EGSB reactors as they directly affect the stability and performance of these systems and they need special care and attention during the design and operation procedure [18].

4.3.1 Organic loading rate

OLR or volumetric loading rate (VLR) is one of the main design criteria for UASB or EGSB systems. This parameter indicates the amount of gram COD per volume unit of the reactor per day that enters the reactor. This parameter is designed based on the capacity of the reactor to degrade COD in the wastewater, which depends on the total organic fraction in the sludge. OLR can be maintained by changing the inflow rate or COD concentration in the influent, for a fixed volume of reactor. Higher OLR results in higher production of biogas, until a maximum range which depends on the capacity of methanogens activity. However, it is important to operate the reactor with an optimum value of OLR. Excessive value of OLR can lead to high VFA production, if it is beyond the capacity of methanogenic activity. Moreover, biogas might be accumulated in the sludge and generate bubble pockets which will result in flotation of the sludge in some areas [9]. Eq. (4.1), shows how the OLR is calculated:

$$\mathrm{OLR} = \frac{Q_{\mathrm{in}} \times \mathrm{COD}_{\mathrm{in}}}{V} \tag{4.1}$$

OLR: Organic loading rate (g COD/L/day)
$\mathrm{COD}_{\mathrm{in}}$: Concentration of COD in the influent (g/L)
Q_{in}: Influent rate (L/day)
V: Reactor volume (L)

4.3.2 Hydraulic retention time

As it is demonstrated in Eq. (4.2), this parameter can be calculated by dividing the volume of the reactor to the inflow rate, therefore it is inversely related to inflow rate. HRT is basically the average time that a soluble compound spends in the reactor, before it flows out. For a fixed volume of reactor, the HRT is maintained by regulating the inflow rate, therefore, it can be easily controlled. Low HRT can decrease the performance of the reactor, as the contact time between biomass and organic compounds will be lowered [19]. However, specially, in EGSB systems, in order to have proper expansion of granules, high values for HRT should be avoided.

$$\mathrm{HRT} = \frac{V}{Q_{\mathrm{in}}} \tag{4.2}$$

HRT: Hydraulic retention time (day)
Q_{in}: Influent rate (L/day)
V: Reactor volume (L)

4.3.3 Up-flow liquid velocity

The up-flow liquid velocity is one of the main parameters that distinguishes EGSB systems from UASB. In UASB reactor, the optimum range of up-flow velocity is between 0.2 and 1 m/h, and higher velocity can result in wash-out of the suspended solids in the sludge [20]. However, in EGSB reactor, due to its special configuration and tall structure, it can be increased to 6 m/h. This high velocity allows the system to have higher inflow, and better expansion of the granules, without losing suspended solids. Eq. (4.3), shows the formula of the up-flow velocity.

$$V_{\mathrm{up}} = \frac{Q_{\mathrm{in}}}{A} \tag{4.3}$$

V_{up}: Up-flow liquid velocity (m/h)
Q_{in}: Influent rate (L/h)
A: Reactor cross-sectional area

4.3.4 pH

The importance of pH in UASB and EGSB reactors is due to the fact that methanogenic activity occurs in a specific range of pH. This range which is between 6.3 and 7.8, is required for the methanogenic bacteria to grow [10]. Therefore, it is a crucial parameter and needs to be controlled continuously. Controlling of pH is done by adding acid, mostly hydrochloric acid or caustic such as sodium hydroxide. However, pH values lower than this range is more toxic than higher values, as the methanogens are more susceptible to acidic condition, and they might not be able to recover even after adjusting the pH to its favorable range, resulting to failure of the reactor.

4.3.5 Temperature

One of the most important parameters in anaerobic digestion is temperature. Although anaerobic treatment can be done in all three ranges of temperature, but performance efficiency is hugely different in each range [21]. Temperature mainly affects the microbial community in terms of diversity and richness [22]. The appropriate temperature range for an anaerobic process is within mesophilic range, between 20 and 40°C, however, the optimum performance is achieved in anaerobic treatment plant between temperature of 35 and 40°C. Lower temperature decrease the activity of microbial community and therefore, higher SRT or HRT values are needed. Also, temperature above 45 can lead to inactivation of the microbes. For this reason, EGSB and UASB reactors operate with higher efficiency in tropic or subtropic regions, and in countries with cold climate, a heating system is needed to maintain the optimum temperature for the reactors.

4.4 Application in industry

Since the time of introducing UASB system as the first high flow rate anaerobic treatment reactor, which provided the application of anaerobic treatment in large scale, many industries have successfully applied this technology. However, agrofood and pulp and paper industries have had the highest contribution in UASB and EGSB market. Suitability of wastewater discharging from these industries has resulted in high stability, performance and efficiency of anaerobic treatment plants. In this section, two industries are studied, as case studies of using UASB and EGSB reactors.

4.4.1 Pulp and paper industry

The pulp and paper industry is one of the biggest water and resource consumers and at the same time, discharges high amount of waste in forms of liquid, gas and solids. Since the 1970s, concerns have been raised about sustainability of this industry, pointing on high consumption of natural resources and also environmental problems resulted from discharging toxic effluents.

Nowadays, this industry is facing big challenges to comply with strict environmental regulations and limits. Pulp and paper factories discharge huge amount of wastewater from their different types of processing units, which they can heavily affect the receiving environment. Scum and slim formation by microorganisms in water bodies, thermal impact, toxification of exposed community and color and odor problems are examples of the effects from discharging untreated wastewater from P&P factories [23].

Although due to the recent advances in the P&P wastewater treatment methods, concerns regarding toxicity of the disposed effluents, have been decreased, but pollutants are still likely to be found in the final treated wastewater. This is mainly because of technical and economical limitations which lead to insufficient degradation of pollutants in some P&P wastewater treatment methods. The characteristic and composition of wastewater in P&P industry varies largely not only from one mill to another, but also within a mill from different processes. Therefore, the treatment of these effluents become challenging due to this characteristic variation. For instance, in a study done by Zwain [24], the wastewater from recycled paper mill (RPM) was characterized based on different physicochemical contents, which is shown in Table 4.3, the results show the suitability of RPM wastewater for biological treatment as it consists organic and inorganic nutrients required for biological growth. Also, the pH is relatively neutral as it is in the range of 6.2–7.6 [24].

Since 1980s, anaerobic wastewater treatment has successfully been applied in the P&P industry. The relatively high amounts of organic compounds present in pulp and paper wastewater can easily be treated with anaerobic digestion and can be readily converted into biogas. Nowadays, after agro-food industries, pulp and paper industry is the second biggest market for anaerobic treatment plants, mostly UASB and EGSB systems, due to their high efficiency and feasibility. COD removal between 65% and 85% and high biogas production have been reported in many studies, using wastewater from different types of pulp and paper processes [25]. However, EGSB has showed better performance in terms of COD removal [5].

Nevertheless, to achieve higher efficiency and to meet discharge standards, anaerobic treatment plants are usually followed by a posttreatment facility. For pulp and paper industry, aerobic post-treatment is considered as one of the most suitable techniques in terms of performance and feasibility. Moreover, the effluent of the aerobic plant can be used to dilute the influent of the anaerobic plant, and thus, increase the COD removal efficiency.

Table 4.3: Physicochemical characteristic of recycled paper mill effluent [24].

Parameter	Value
pH	6.2–7.8
Floc size (μm)	208–300
Temperature (°C)	35–45
COD (mg/L)	3380–4930
BOD (mg/L)	1650–2565
BOD5/COD	0.488–0.52
Alkalinity (mg/L)	300–385
Volatile fatty acids (VFAs)	455–490
Ca (mg/L)	375–420
Mg (mg/L)	10–15
Total solids (TS) (mg/L)	3530–6163
Total dissolved solids (TDS) (mg/L)	1630–3025
Total suspended solids (TSS) (mg/L)	1900–3138
Total volatile solid (VSS) (mg/L)	840–292

4.4.2 Olive oil industry

Olive oil is produced from olive trees primarily from countries bordering the Mediterranean Sea. International Olive Oil Council's 2014/2015 report shows that the worldwide total olive oil production has reached 2,951,800 tons.

Production of olive oil in olive mills can be done by continuous centrifugation process or discontinuous press method. Depending on the type of the process used, it is believed that in every liter of olive oil production, up to 2.5 L of olive mill wastewater (OMW) will be generated. Globally, the production of OMW each year is predicted to be between 7 and 30 million m^3 [26].

OMW is a highly polluting wastewater. It is characterized by its high concentrations of BOD and chemical oxygen demand (COD). The composition of OMW varies both in quantity and quality depending on the olive tree type, climatic condition, soil type, and type of production. Generally, OMW composition has a total COD which corresponds to 30–320 g/L, total BOD of 25–135 g/L and TSS of 25–30 g/L. In addition, OMW contains a high amount of phenolic compounds and grease, which is in the range of 4–24 g/L and 4–25 g/L, respectively. It is also poor in nitrogen content wherein the total nitrogen contribution can be less than 0.6% of OMW and has relatively low pH values (2.4–5.9) [27].

Uncontrolled disposal of OMW causes severe environmental problems to the recipient water bodies. It is toxic to microorganisms, alters the color of water bodies, damages vegetation cover, causes odor and changes the pH and destroys the quality of soil [28]. Nevertheless, the nature of high-strength OMW makes it difficult to find affordable and efficient treatment method. In addition to the high organic load of the effluent, seasonal operation together with dispersed and localized production with low wastewater flow rates makes the treatment process difficult [29]. Thus, olive mill effluent is a main environmental concern in the Mediterranean area, particularly severe in developing countries like Palestine where OMW is disposed into waterways without treatment.

Biological treatment methods are considered to be environmentally friendly and cost-effective. It is widely applied for the treatment of olive mill wastewaters. The method requires low energy and there is also a production of biogas and the digestate can be used as a fertilizer for agriculture [30]. UASB reactors is the most popular anaerobic bioreactor since it has been proven to be effective and economical feasible for waste streams from both food industry and olive mills. Moreover, undiluted wastewater can be satisfactorily treated by UASB reactors under elevated organic loads. During the treatment of OM effluents, UASB reactors can generally be operated better than other reactors in addition to the possibility of energy recovery.

4.5 Conclusions and perspectives

Nowadays, UASB and EGSB systems are being used widely in many industries and they have become a proven technology for wastewater treatment and recovery of resources in form of energy. However, like other technologies, they are still facing with many limitations and challenges, which shows a big potential for research and development. In this section, some of the main challenges of these systems will be mentioned. Moreover, possible solutions and perspectives for future development will be introduced.

One of the main challenges of anaerobic reactors in general, is susceptibility of these systems to extreme conditions such as toxic organic compounds and high temperature. However, reactors with granular sludge are more sensitive to these conditions due to the fact that only specific microbial community can grow in the sludge. Toxic organic compounds such as aromatics that can be found in effluent from petrochemical industries inhibit the biological activity in UASB and EGSB systems, and therefore, result in failure of the system. As it was mentioned before, temperature is a crucial parameter for anaerobic process and optimum performance of anaerobic treatment systems is in mesophilic condition. Therefore, wastewaters with temperatures in thermophilic range need to be cooled down before entering the reactor, as they can be harmful to the bacteria. These two factors have resulted in limited use of UASB and EGSB systems in chemical and petrochemical industries. Moreover, treatment of wastewaters with high suspended solids (SS) is challenging for UASB and EGSB systems as they negatively affect the development of granular sludge. Also, in order to maintain a stable methanogenic activity, presence of nutrients should be controlled to prevent the starvation of bacteria. These nutrients are mainly nitrogen and phosphorous as macro nutrients, but also micronutrients such as Iron, that need to be added to the reactor.

Despite all the challenges and limitation, UASB and EGSB systems are still considered as the first option by many industries, for treatment of their wastewater. Stable performance, high efficiency in organic compounds removal, energy production and relatively low capital and O&M costs are the reason of the wide use of these systems. On the other hand, industrial growth, water scarcity and environmental problems, has resulted in more investment and focus on wastewater treatment by industries. Therefore, many research and studies are currently being done to cover the flaws and improve the performance of UASB and EGSB reactors. The result of these studies can lead to a wider use of these systems, specially, in industries with more extreme conditions in their effluent, which is limiting the use of UASB and EGSB reactors for their treatment. Use of additives such as trace metals for a better adaptation of microbial community to extreme conditions, modeling of the reactor behavior in different conditions in order to predict the performance of the system in different situations and combining UASB or EGSB systems with other treatment techniques are among the active research fields.

In conclusion, UASB and EGSB systems are playing an important role in sustainable development, by turning waste into energy and providing the possibility to reuse water in industries, and therefore, any further investment and research on these systems in the future, can maximize this role.

References

[1] Warmer H, van Dokkum R. Water pollution control in The Netherlands, policy and practice 2001. 2002.

[2] Shi H. Industrial effluents-types, amounts and effects, point sources of pollutions: local effects and its control 1. 2009.

[3] Hamza RA, Iorhemen OT, Tay JH. Advances in biological systems for the treatment of high-strength wastewater. Journal of Water Process Engineering 2016;10:128–42.

[4] Garfí M, Martí-Herrero J, Garwood A, Ferrer I. Household anaerobic digesters for biogas production in Latin America: a review. Renewable and Sustainable Energy Reviews 2016;60:599–614.

[5] Kamali M, Gameiro T, Costa MEV, Capela I. Anaerobic digestion of pulp and paper mill wastes – an overview of the developments and improvement opportunities. Chemical Engineering Journal 2016;298:162–82.

[6] Wan J, Gu J, Zhao Q, Liu Y. COD capture: a feasible option towards energy self-sufficient domestic wastewater treatment. Scientific Reports 2016;6:25054.

[7] Lim SJ, Kim TH. Applicability and trends of anaerobic granular sludge treatment processes. Biomass and Bioenergy 2014;60:189–202.

[8] Khalil N, Sinha R, Raghav A, Mittal A. UASB technology for sewage treatment in India: experience, economic evaluation and its potential in other developing countries. In: Twelfth international water technology conference, IWTC12, Alexandria, Egypt; 2008.

[9] Abdelgadir A, Chen X, Liu J, Xie X, Zhang J, Zhang K, Wang H, Liu N. Characteristics, process parameters, and inner components of anaerobic bioreactors. BioMed Research International 2014:10.

[10] Daud MK, Rizvi H, Akram MF, Ali S, Rizwan M, Nafees M, Jin ZS. Review of upflow anaerobic sludge blanket reactor technology: effect of different parameters and developments for domestic wastewater treatment. Journal of Chemistry 2018:13.

[11] van Lier JB, van der Zee FP, Frijters CTMJ, Ersahin ME. Celebrating 40 years anaerobic sludge bed reactors for industrial wastewater treatment. Reviews in Environmental Science and Bio/Technology 2015;14:681–702.

[12] Ramakrishnan A, Surampalli RY. Comparative performance of UASB and anaerobic hybrid reactors for the treatment of complex phenolic wastewater. Bioresource Technology 2012;123:352–9.

[13] Gupta P, Sreekrishnan TR, Shaikh ZA. Application of hybrid anaerobic reactor: treatment of increasing cyanide containing effluents and microbial composition identification. Journal of Environmental Management 2018;226:448–56.

[14] Perez KC, van Geest JF, Versprille A, Otten M, Heffernan B. Performance evaluation of a new (Biobed Advanced) EGSB settler. IWA Congress Anaerobic Digestion; 2008.

[15] Bobade V, Lomte A. Challenges in UASB reactor system design: a review [online]. In: Asia Pacific confederation of chemical engineering congress 2015: APCChE 2015, incorporating CHEMECA 2015. Melbourne: Engineers Australia; 2015. p. 318–25.

[16] Zoutberg G, Frankin R. Anaerobic treatment of chemical and brewery waste water with a new type of anaerobic reactor; the Biobed(R) EGSB reactor. Water Science and Technology 1996;34:375–81.

[17] Latif MA, Ghufran R, Wahid ZA, Ahmad A. Integrated application of upflow anaerobic sludge blanket reactor for the treatment of wastewaters. Water Research 2011;45:4683–99.

[18] Turkdogan-Aydınol FI, Yetilmezsoy K. A fuzzy-logic-based model to predict biogas and methane production rates in a pilot-scale mesophilic UASB reactor treating molasses wastewater. Journal of Hazardous Materials 2010;182:460–71.

[19] Lew B, Belavski M, Admon S, Tarre S, Green M. Temperature effect on UASB reactor operation for domestic wastewater treatment in temperate climate regions. Water Science and Technology: A Journal of the International Association on Water Pollution Research 2003;48:25–30.
[20] Zhang C, Wang A, Jia J, Zhao L, Song W. Effect of parameters on anaerobic digestion EGSB reactor for producing biogas. Procedia Engineering 2017;205:3749–54.
[21] Adrien NG. Processing water, wastewater, residuals, and excreta for health and environmental protection. Hoboken: An Encyclopedic Dictionary John Wiley & Sons, Inc; 2008.
[22] Gao WJ, Leung KT, Qin WS, Liao BQ. Effects of temperature and temperature shock on the performance and microbial community structure of a submerged anaerobic membrane bioreactor. Bioresource Technology 2011;102:8733–40.
[23] Kamali M, Khodaparast Z. Review on recent developments on pulp and paper mill wastewater treatment. Ecotoxicology and Environmental Safety 2015;114:326–42.
[24] Zwain HM, Hassan SR, Zaman NQ, Aziz HA, Dahlan I. The start-up performance of modified anaerobic baffled reactor (MABR) for the treatment of recycled paper mill wastewater. Journal of Environmental Chemical Engineering 2013;1:61–4.
[25] Ashrafi O, Yerushalmi L, Haghighat F. Wastewater treatment in the pulp-and-paper industry: a review of treatment processes and the associated greenhouse gas emission. Journal of Environmental Management 2015;158:146–57.
[26] Niaounakis M, Halvadakis C. Chapter 2: characterization of olive processing waste. In: Olive Processing Waste Management. vol. 5; 2006. p. 23–63.
[27] Dermeche S, Nadour M, Larroche C, Moulti-Mati F, Michaud P. Olive mill wastes: biochemical characterizations and valorization strategies. Process Biochemistry 2013;48:1532–52.
[28] McNamara CJ, Anastasiou CC, O'Flaherty V, Mitchell R. Bioremediation of olive mill wastewater. International Biodeterioration & Biodegradation 2008;61:127–34.
[29] Cañizares P, Paz R, Sáez C, Rodrigo MA. Costs of the electrochemical oxidation of wastewaters: a comparison with ozonation and Fenton oxidation processes. Journal of Environmental Management 2009;90:410–20.
[30] Paraskeva P, Diamadopoulos E. Technologies for olive mill wastewater (OMW) treatment: a review. Journal of Chemical Technology and Biotechnology 2006;81:1475–85.

Further reading

[1] Meyer T, Edwards EA. Anaerobic digestion of pulp and paper mill wastewater and sludge. Water Research 2014;65:321–49.

CHAPTER 5

Agroindustrial waste based biorefineries for sustainable production of lactic acid

Jasneet Grewal, Ayesha Sadaf, Neerja Yadav, S.K. Khare
Enzyme and Microbial Biochemistry Laboratory, Department of Chemistry, Indian Institute of Technology Delhi, Hauz Khas, New Delhi, India

5.1 Introduction

Currently, 90% of the chemicals and 80% of energy demands are met by fossil fuels and the consumptions are further increasing incessantly. The depleting fossil fuels, energy crisis, deteriorating environmental conditions and climate change have put impetus on developing sustainable technologies derived from renewable sources [1]. In this context, the plentiful available biomass has been perceived as a potential alternate to combat energy crisis and provide sustainable source of fuels and bio-based chemicals. The processing technologies for conversion of biomass into multitude of high-value products and fuels comes under the domain of biorefineries. The establishment of sustainable waste biorefineries for production of energy, platform chemicals, fuels and other value-added products is an effective way to reduce reliance on diminishing nonrenewable petroleum feedstocks [2,3].

The bio-based platform chemicals are the ones which originate from biomass and serve as building block for production of chemicals and other high-value products viz. polymers, fuels, solvents, plastics, pharmaceuticals etc. [4,5]. These are envisaged as the sustainable alternates to petro-based chemicals derived from finite fossil reserves. Thus, the sustainability and environmental challenges have focused colossal attention on waste biomass-based production of platform chemicals. Ethanol, glycerol, xylitol, sorbitol, furans (5-hydroxymethylfurfural, furfural, 2,5-furan dicarboxylic acid), hydrocarbons (isoprenes), 1,4-diacids (succinic, fumaric, and malic), 3-hydroxypropionic acid, levulinic acid, itaconic acid, glutamic acid, aspartic acid, and lactic acid are recognized among the top-value platform chemicals [6–8]. Except isoprene and glycerol, all others can be produced from biomass and the bio-based product sales are expected to reach 441 billion$ by 2020 [9].

Waste Biorefinery. https://doi.org/10.1016/B978-0-12-818228-4.00005-8

Lactic acid (2-hydroxypropanoic acid) is one such important C3 platform chemical, whose bio-based production has evoked great interest after the US Department of Energy identified it among the top platform chemicals which can be produced from biomass [10–12]. Conventionally, pure sugars are used for its production by fermentation but various agroindustrial wastes have better potential to serve as cheaper feedstocks. The challenges, however, are the highly complex and multiple processing steps required for its production from waste biomass. The present chapter highlights the valorization of both starchy and lignocellulosic agroindustrial wastes for production of lactic acid. The bioconversions mediated by both native and engineered microorganisms are discussed. The chapter also comprehensively analyses all the stages of production ranging from pretreatment of feedstock, fermentation mode, fermenting conditions, microbial strains and downstream processing to obtain the purified product.

5.2 Lactic acid and its application

Lactic acid, a highly abundant hydroxycarboxylic acid was first discovered in sour milk by Carl Scheele in 1780. Though its industrial production began in 1881 in Littleton, USA, the demand continues to rise and will escalate to 1960 kt by 2025 [13]. By the year 2025, its global market is expected to reach about 9.8 billion US$ [10]. Its annual global production was reported to be 260,000 metric tons in 2012 and predicted to reach 600,000 metric tons by 2020 [14]. The large demand is due to its versatile applications in range of industrial sectors viz. food, cosmetic, tanning, brewery, biomedical and pharmaceuticals [15].

The approval of Food and Drug Administration (FDA) and Generally Recognized as Safe (GRAS) status makes it highly acceptable in food industries, which share about 70% usage [16]. In food industry, the major applications of lactic acid and its derivatives are in buffering systems, as emulsifying, flavoring agents, dough conditioner in bakery goods and protection agents against bacterial spoilage.

Due to the reactivity of its two functional groups i.e., hydroxyl and carboxyl group, the most sought after chemicals viz. acrylic acid, lactate ester, propylene glycol, 2,3-pentanedione, acetaldehyde, ethyl lactate, pyruvic acid are produced from it, further adding to its market demands (Fig. 5.1) [17,19]. Most of the production technologies for above high-value chemicals are petro-based and thus, their production from bio-based lactic acid is forecasted as a sustainable alternative [18,20]. Further, the use of lactic acid and its derivative ethyl lactate as green solvent is being actively pursued as an alternate to conventional petrochemical solvents. Wang et al. [21] carried out ultrasound assisted synthesis of 18 pyrrole derivatives in lactic acid medium. Lactic acid was also found to be suitable as reaction media for various organic reactions involving styrenes,

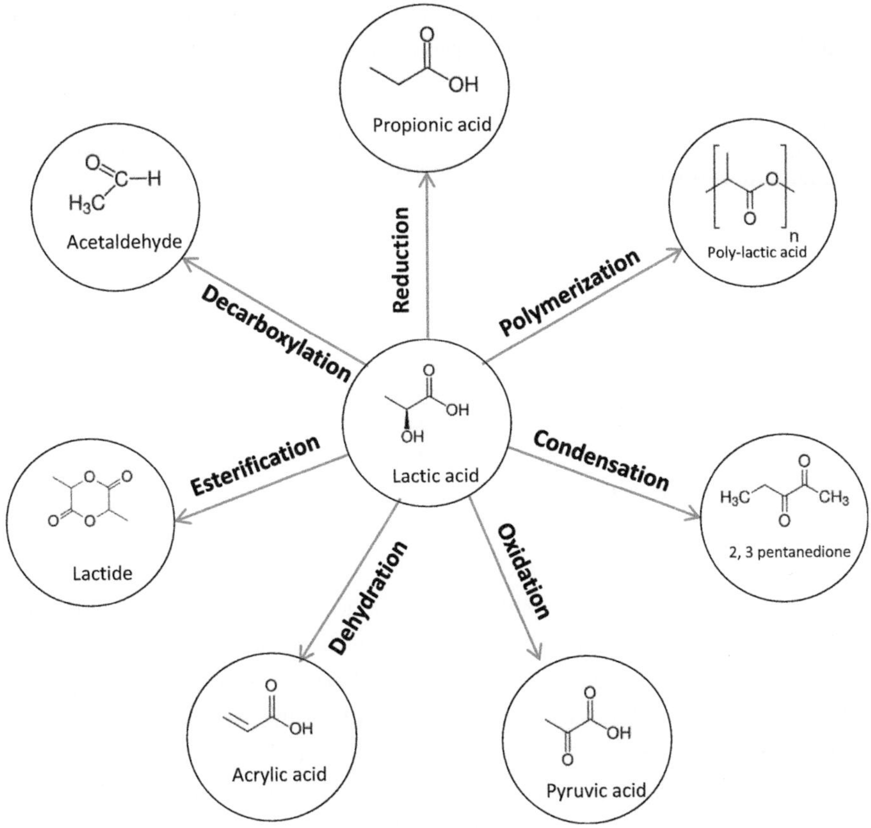

Figure 5.1
The role of lactic acid as platform chemical for production of various top-value commodities [17,18].

formaldehyde and N,N-dialkylacetoacetamides [22]. Solvents comprise 80% of the total volume of chemicals used in synthetic processes and thus, replacing them with nontoxic solvents like lactic acid, ethyl lactate is highly advantageous from environmental point of view [23].

5.2.1 Biopolymers synthesized from lactide monomer

The biopolymers synthesized using lactic acid are in high demand due to their biocompatibility, biodegradability and their ability to be customized for different biological applications by surface modification or tailoring the parameters of their synthesis. Moreover, the production of biodegradable polymers is an effective strategy to resolve environmental and marine pollution caused by the plastic wastes. The application of lactic acid for production of polylactide (PLA) based biopolymers used in textile, automotive and biomedical sector is on rise rapidly [15].

The global production capacity of bio-based plastics is expected to reach 7.85 million tons in 2019 and PLA will be among the leads [24]. PLA polymers with their distinctive thermal, mechanical properties and biodegradability serve as ideal base for fabrication of biomaterials such as tissue engineering scaffolds, absorbable sutures, coatings of implantable devices, orthopedic plates and stents. The current research efforts are extensively oriented toward development of PLA-based materials as a viable and sustainable alternate to fossil derived synthetic plastics [15].

Apart from PLA, synthesis of other lactic acid based polymers is also being actively pursued viz. lactic acid grafted gum-arabic copolymer was synthesized as bio-based adhesive for various structural applications [25]. Similarly, a copolymer of lactic and glycolic acid i.e., poly(lactic-co-glycolic acid) (PLGA) (Fig. 5.2) has gained noteworthy attention for targeted and sustained drug delivery besides its widespread biomedical applications such as use in biodegradable stents, sutures, bioabsorbable screws, tissue engineering scaffolds and other engineered nanodevices [26]. The fabrication of PLGA devices encapsulating therapeutic proteins or peptides for treatment of many diseases is an emerging research area [27]. The use of PLGA as theranostic agent is being extensively investigated to achieve molecular imaging and real-time monitoring of therapeutic response [28]. Radioiodinated PEGylated PLGA-indocyanine capsules were synthesized for in-vivo imaging and targeting cervical, breast and ovarian cancer cell lines [29]. Thus, for the fabrication of bio-based polymers, production of bio-monomers like lactic acid by microbial route becomes critically important.

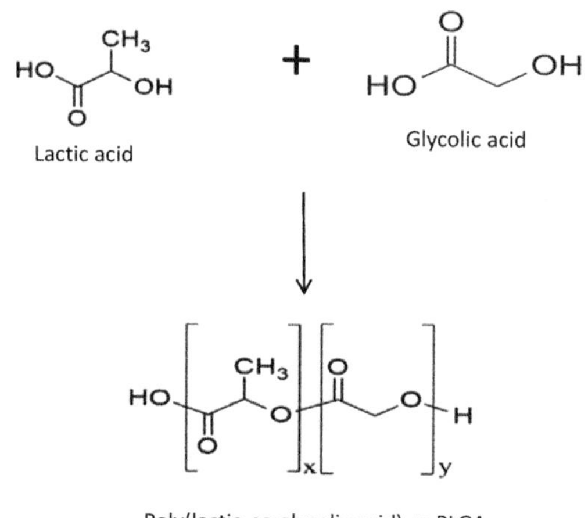

Figure 5.2
Copolymerization reaction of lactic and glycolic acid. The x and y represent the number of units of lactic acid and glycolic acid respectively [26].

5.3 Production of lactic acid

The lactic acid production can be carried out either by chemical synthesis or fermentation at commercial scale. The conventional chemical route for lactic acid synthesis utilizes petrochemical feedstock. It is based on hydrolysis of lactonitrile, which is generated by reaction of acetaldehyde and hydrocyanic acid [30,31]. In recent past, the chemical transformations involving renewable feedstocks such as cellulose, hexose sugars, trioses (glyceraldehyde), and glycerol have also been used for synthesis of lactic acid [18,20]. However, the use of different harsh homogeneous and heterogeneous catalysts in the nonfermentative chemocatalytic cascade along with stringent reaction conditions, high energy consumption and adverse environmental impact favor fermentation method. The production of racemic mixture of lactic acid by chemical synthesis is another bottleneck, which can be avoided by microbial fermentation as many microorganisms synthesize enantiomerically pure lactic acid [10,19].

Thus, the fermentative production route seems more advantageous for meeting lactic acid production demands. Juodeikiene et al. [32] compared sustainability metrics for lactic acid production from biotechnological and chemical processes and found 47% higher energy efficiency in fermentative route coupled with environmental friendliness and production of optically pure lactic acid. The bioprocesses involved in production of lactic acid from agroindustrial wastes and the downstream processing steps for obtaining pure bio-based lactic acid are discussed in subsequent sections.

5.3.1 Microorganisms utilized for fermentative production of lactic acid

The lactic acid can be produced by both fungal and bacterial fermentation. The fungi belonging to *Rhizopus* genus have been majorly used for lactic acid production due to release of extracellular hydrolytic enzymes and ease of separation of mycelium for downstream processing. *Corynebacterium glutamicum*, *Escherichia coli*, *Bacillus* strains and lactic acid bacteria (LAB) are the promising bacterial producers of lactic acid [30,33]. These microorganisms in both wild-type and engineered forms are being researched for cost-effective lactic acid production. Although many genera of bacteria produce lactic acid as a primary or secondary end-product of fermentation, the term LAB is conventionally reserved for seven genera, out of which *Enterococcus*, *Lactococcus*, *Pediococcus*, and *Lactobacillus*, have been extensively used for industrial production of lactic acid, nutraceuticals, antimicrobial products other high value-added metabolites [33,34]. They are emerging as major bioresource due to their GRAS status and production of many immunomodulatory compounds.

LAB comes from a highly heterogeneous group of Gram-positive, nonmotile, nonsporulating organisms that ferment a variety of sugars to produce lactic acid by either homofermentative or heterofermentative pathway resulting in different yields (Fig. 5.3) [30,34]. The homofermentative mode operates via Embden-Meyerhof-Parnas (EMP) pathway in which glucose is converted to pyruvate by glycolysis cycle and subsequently to lactic acid by lactate dehydrogenase. For every mol of glucose metabolized by EMP pathway, the maximum theoretical yield of 2 mol of lactic acid can be obtained. However, in the heterofermentative pathway the maximum theoretical yield drops to 1 mol of lactic acid per mol of metabolized sugar. It operates via phosphoketolase (PK) pathway by which both hexose and pentose sugars are fermented into lactic acid along with generation of acetic acid and ethanol [36,37].

5.3.2 Feedstocks used for fermentative lactic acid production

The raw materials constitute 60%–80% of the total cost of lactic acid production and hence, the availability of inexpensive, abundant renewable feedstock is essential to compete with chemical synthesis [38]. Replacement of traditional feedstocks such as refined sugars or edible starchy crops with agroindustrial wastes is a viable alternate. However, the utilization of agroindustrial waste substrates may require pretreatment and enzymatic hydrolysis for release of fermentable reducing sugars. The transition from linear

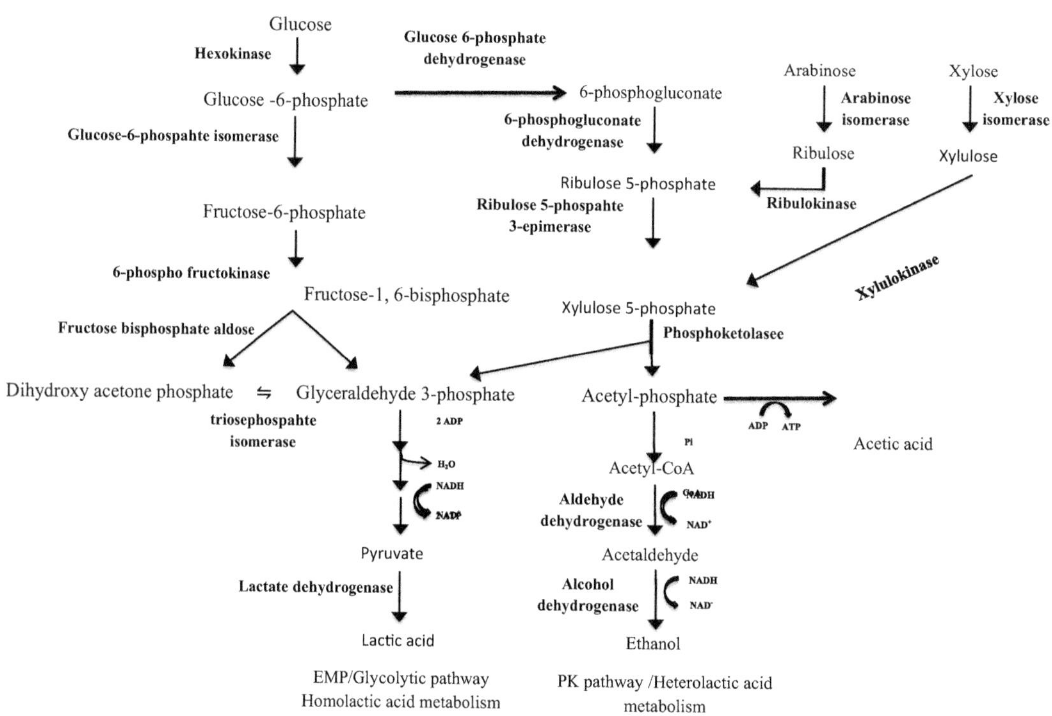

Figure 5.3
The metabolic pathway of homolactic and heterolactic fermentation in LAB [33,35].

fossil-based economy to bio-based circular economy relies on optimal harvesting of potential waste biorefineries [2,30].

5.3.2.1 Valorization of starchy agroindustrial wastes for lactic acid production

Starchy substrates derived from various plant sources are an attractive alternate to pure reducing sugars as feedstocks for lactic acid production. The use of edible carbohydrate substrates (potato, sorghum, cassava, rice and corn-starch) for lactic acid production can be challenged by food security issues, hence the valorization of starch-rich agroindustrial wastes seems better option. As starch is a complex carbohydrate composed of amylose and amylopectin, it needs to be hydrolyzed so as to release glucose for the next step of fermentation. Hence, enzymatic hydrolysis of complex starchy feedstocks by action of amylolytic enzymes viz. α-amylase, β-amylase, and glucoamylase becomes essential for utilization of starchy substrates [36,37]. This makes concomitant addition of saccharifying enzymes or secretory amylolytic activity of fermenting organism an important requisite. However, some starchy feedstocks may require pretreatment, which generally involves heating the starch substrate at 80–90°C for 10–30 min in a process called liquefaction, followed by enzymatic hydrolysis. For e.g., in a study, the potato peel waste was first gelatinized in boiling water for 30 min followed by amylase and glucoamylase treatment yielding 0.2 g/g of lactic acid [39]. In case of substrates containing resistant starches, alkali as well as acid pretreatments are also reported viz. Vavouraki et al. [40] showed the effect of acid (HCl) and alkali (KOH) pretreatment on the hydrolysis of kitchen waste. It was found that the use of both of these treatments along with enzymatic hydrolysis led to an increase in the glucose levels by 300%.

Most of the studies for lactic acid production from starchy feedstocks rely on use of commercial amylolytic enzymes for release of reducing sugars. The sugar rich hydrolysate generated after enzymatic hydrolysis is fermented to lactic acid in the subsequent second step and this approach is referred as separate hydrolysis and fermentation (SHF). With this approach, sweet potato waste, obtained as by-product after industrial processing was subjected to enzymatic hydrolysis by glucoamylase. The hydrolysate obtained after saccharification was fermented by *Lactobacillus rhamnosus* to achieve 10 g/L lactic acid production with 30% conversion yield [41].

For improving the economic feasibility, replacement of commercial enzymatic hydrolysis with in vitro fungal hydrolytic enzymes has also been advocated. With this approach, the valorization of mixed bakery waste (MBW) for lactic acid production was carried out by Yang et al. [42]. The nonsterilized solid-state fermentation of MBW was performed by *Aspergillus oryzae* and *Aspergillus awamori* producing proteolytic and amylolytic enzymes respectively. The resulting hydrolysate rich in sugars and other nutrients was used as feedstock for fermentation by engineered *Thermoanaerobacterium aotearoense* which resulted in overall lactic acid yield of 0.18 g/g raw waste. Similarly Kwan et al. [43] also reported the bioconversion of starch-rich food waste into lactic acid by fungal hydrolysis (*A. oryzae* and *A. awamori*) and subsequent fermentation by *Lactobacillus casei* Shirota, with overall yield of 0.27 g/g food waste.

In another study, the techno-economic evaluation of a food waste based biorefinery for lactic acid production indicated good economic feasibility in this sustainable technology, driving transition from current linear economy to circular bio-economy. The lactic acid was produced from valorization of food waste at a net production cost of 1066 US$/MT [44]. Their previous study also demonstrated the economic viability of lactic acid and plasticizer production from food waste with annual net profits of 422,699 US$ and 18.98% of rate of return [45].

As an alternate to two-step process i.e., SHF, the approach of simultaneous saccharification and fermentation (SSF) has also been used for bio-based production of lactic acid. Recently, the SSF of starch-rich restaurant food waste was carried out by *Streptococcus* sp. and 58 g/L L-lactic acid was produced from 20% (*w/w*) dry food waste [46]. Nair et al. [47] isolated *Streptococcus equinus* and used it for lactic acid production by SSF of Jackfruit seed powder, a starch-rich agrowaste. The maximum production of 109 g/L L-lactic acid was achieved and produced lactic acid was purified. The purified lactic acid was polymerized into PLA with yield of 62%

White rice bran, a by-product obtained from rice milling was used as feedstock in SSF for achieving 117 g/L L-lactic acid production by thermophilic *Bacillus coagulans* with yield of 98.75% [48]. Similarly, *L. rhamnosus* M-23 was able to produce 59 g/L L-lactic acid by SSF of nonsterilized mixture of rice bran and rice washing drainage. The lactic acid produced was 95% optically pure and the conversion yield was 0.85 g/g sugar [49]. Cassava bagasse is another starchy waste, whose disposal is a big environmental concern. John et al. [50] showed that the SSF of cassava bagasse by *L. casei* resulted in production of 83.8 g/L lactic acid with 96% conversion of starch to lactic acid. The SSF of starch-rich food wastes by *Lactobacillus delbrueckii* resulted in 27 g/L lactic acid production from 60 g/L of food waste [51].

The fermentation of organic wastes by mixed or undefined cultures is another cost-effective approach used for lactic acid production as it helps in carrying out the process under less stringent conditions and allows use of versatile feedstocks [52]. The fermentation of potato starch-waste by microbial inocula from rumen of Thai and Dutch dairy cows, resulted in lactic acid yield of 0.6 and 0.3 g/g starch respectively [53]. Nunes et al. [54] carried out fermentation of broken rice, a starch-rich by-product generated in rice milling by undefined mixed culture obtained from dewatered activated sludge. The lactic acid production (11.84 g/L) by mixed culture was much higher than production (3.19 g/L) by pure culture of *Lactobacillus amylovorus* used as control. Likewise, the fermentation of the gelatinized potato peel waste (PPW) by undefined mixed culture from wastewater treatment plant sludge resulted in 14.7 g/L lactic acid production with yield of 0.22 g/g PPW [39].

Table 5.1 summarizes the relevant studies on lactic acid production from starch-rich agroindustrial wastes.

As discussed previously, majority of lactic acid production process employed commercial enzymes for the hydrolysis of starch to glucose, which adds to the operational cost. Thus, fermentation by microorganisms possessing amylolytic activity is better as they can convert starch to lactic acid in a single step by simultaneously hydrolyzing and fermenting starchy feedstocks to generate lactic acid. Many wild-type microorganisms possessing amylolytic activity have been reported viz. *Lactococcus lactis*, *L. amylovorus*, *Lactobacillus amylophilus*, *Lactobacillus amylolyticus*, *Lactobacillus plantarum*, *Lactobacillus paracasei*, *Lactobacillus manihotivorans*, *Lactobacillus fermentum*, *Enterococcus faecium*, *Streptococcus bovis*, *Rhizopus oryzae* and *Rhizopus arrhizus*. They have been shown to produce lactic acid directly from conventional starchy substrates viz. potato, sorghum, cassava, rice, barley flour and corn-starch by using their amylolytic activity [58,59]. However, studies on amylolytic strains for lactic acid production from starch-rich agroindustrial wastes are rather scanty [60,61]. The screening of native amylolytic producers with potential for lactic acid production from different agroindustrial wastes is another hotspot research area for development of starch-waste based biorefineries.

On similar lines, the construction of recombinant amylolytic strains seems a promising approach to develop consolidated bioconversion process for lactic acid production. The fusant strain prepared by protoplast fusion of *Lactobacillus delbrueckii* and *Bacillus amyloliquefaciens* resulted in production of 40 g/L lactic acid from 42 g/L cassava bagasse starch [62]. The genetically engineered *A. oryzae* (*A. oryzae* LDHΔ871) produced 30 g/L L-lactic acid (99.9% optical purity) from 100 g/L starch [63].

L. plantarum was genetically modified to replace L-lactate dehydrogenase (L-ldh) gene with α-amylase secreting expression cassette and was used for production for D-lactic acid (99.6% optical purity) from corn-starch [64]. Briefly, a plasmid was designed to replace the L-ldh gene with amylase expression cassette. The disruption and substitution of L-ldh gene was done by using a suitable host plasmid based double cross over integration. The deletion and substitution of L-ldh gene by α-amylase secreting expression cassette was confirmed by PCR, as the ldhL1 substituted mutant showed a larger band (1621 bp) after agarose gel electrophoresis.

The development of these kind of recombinant strains will be highly advantageous as it will enable efficient and direct production of lactic acid from starchy feedstocks due to their indigenous amylolytic activity. However, there are many challenges in construction of

Table 5.1: Lactic acid production from starchy agroindustrial wastes as substrates.

Agroindustrial wastes used as substrate	Treatment	Fermenting microorganism	Fermentation mode	Lactic acid enantiomer	Lactic acid concentration (g/L)	Yield (g/g)[a,b,c]	References
Gelatinized starchy waste from rice noodle factory	–	*L. plantarum* S21	SSF	–	102.00	–	[55]
Potato starch-waste	–	Bovine rumen fluid from Dutch and Thai dairy cows	SSF	–	13.00 (Dutch), 25.00 (Thai)	0.30[c] (Dutch), 0.60[c] (Thai)	[53]
Starch-rich restaurant food waste	–	*Streptococcus* sp.	SSF	L	58.00	0.39[a], 0.81[b]	[46]
Broken rice	Gelatinization	Mixed culture from dewatered activated sludge	SSF	–	11.84	–	[54]
Mixed food waste	Fungal hydrolysis by *A. awamori* and *A. oryzae*	*L. casei* Shirota	SHF	–	94.00	0.27[a], 0.94[b]	[43]
Jackfruit seed powder	α-Amylase (5000 IU/mL) and glucoamylase (200 IU/mL)	*S. equinus*	SSF	L	109.00	0.55[a]	[47]
White rice bran	Fungal amylase and glucoamylase (1:1000 biomass)	*B. coagulans* LA-15-2	SSF	L	117.00	0.99[b]	[48]
Mixed bakery waste	Fungal hydrolysis by *A. awamori* and *A. oryzae*	*T. aotearoense* LA1002-G40	SHF	L	78.50	0.18[a], 0.85[b]	[42]

Sweet potato processing waste	Glucoamylase (80 U/100 g of waste)	*L. rhamnosus*	SSF	—	10.00	0.30^{b}	[41]
Potato peel waste	Gelatinization	Mixed culture from wastewater treatment plant sludge	SSF	—	14.70	0.22^{a}	[39]
Mixture of rice washing drainage and rice bran	—	*L. rhamnosus* M-23	SSF	L	59.00	0.85^{b}	[49]
White rice bran	Acid hydrolysis, amylase (1 mL/kg white rice bran)	*L. rhamnosus* LA-04-1	SHF	L	123.00	0.71^{b}	[56]
Cassava bagasse	α-Amylase (5000 IU/mL) and glucoamylase (2000 IU/mL)	*L. casei*	SSF	L	83.80	0.96^{c}	[50]
Defatted rice bran	Amylase and cellulase (5–50 g/m^3)	*L. delbrueckii* subsp. *delbrueckii* IFO 3202	SSF	D	28.00	0.28^{a}, 0.78^{b}	[57]

[a]Yield of lactic acid produced (g) to raw material/biomass used (g).
[b]Yield of lactic acid produced (g) to reducing sugar (g).
[c]Yield of lactic acid produced (g) to starch in raw material used (g).

recombinant amylolytic LAB, especially for gene manipulation of D-lactic acid producing LAB. The reports on D-lactic acid production by amylolytic LAB are scanty and need to be actively pursued [65]. The constraints in applying recombinant technology to LAB include difficulties in inserting gene construct into LAB, unpredictability of plasmid to replicate itself and express foreign genes, unavailability of efficient transformation systems and suitable cloning vectors, low expression and secretion of amylolytic activity due to incorrect folding of recombinant protein [65,66]. Thus, the engineering of fermenting microorganisms by using different biotechnological tools will play a critical role in production of high titers of lactic acid from starchy agroindustrial wastes.

5.3.2.2 Valorisation of lignocellulosic agroindustrial wastes for lactic acid production

Lignocellulosic biomass, comprising of cellulose (25%−55%), hemicellulose (11%−50%) and lignin (10%−40%) is an abundant renewable source which can serve as an ideal feedstock for production of bio-based chemicals [67]. Though abundant lignocellulosics are an attractive substrate for lactic acid production, the inability of lactic acid producing microorganisms to use complex carbohydrates (cellulose, hemicellulose) limits their exploitation. Summarily, the lactic acid production process from lignocellulosics is a multi-step process consisting of following stages: (1) pretreatment step for deconstruction of recalcitrance of lignocellulosic biomass; (2) saccharification by hydrolytic enzymes such as cellulases, xylanases and cellobiases; (3) microbial fermentation of released monomeric sugars into lactic acid; (4) downstream processing for recovery of pure lactic acid.

5.3.2.2.1 Pretreatment of lignocellulosic waste biomass

The pretreatment process is very critical as it requires high energy inputs for making rigid crystalline biomass amenable to enzymatic hydrolysis. The conventional processes employed for loosening of intertwined network of cellulose, hemicellulose and lignin include acid, alkali, steam, and ammonia fiber explosion (AFEX) pretreatment [68,69]. Most of the studies on lactic acid production from different lignocellulosic substrates viz. wheat straw, corn stover, sugarcane bagasse, corncob, paper sludge have used acid or alkali pretreatment [36,70]. The generation of inhibitory toxic compounds, extensive washing after pretreatment, disposal of residual effluents are the major drawbacks of these conventional pretreatment methods [68,71].

Ionic liquids (IL) and deep eutectic solvents (DES) have emerged as new class of green solvents for efficient solubilization and deconstruction of biomass under mild conditions [72,73]. The elimination of washing step to remove these residual green solvents may allow coupling of pretreatment and saccharification in a single step, provided the saccharifying enzymes i.e., cellulases and xylanases are stable in residual IL or DES. Therefore, development of compatible enzyme-IL/DES systems forms one of the thrust research areas for developing lignocellullosic biorefineries for lactic acid production. Recently, lactic acid itself has been used as one of the component of DES and these natural deep eutectic solvents (NADES) have been used for pretreatment of biomass viz. rice straw, xylose residue etc. [74,75]. Thus, use of bio-based lactic acid for pretreatment of agroindustrial wastes further potentiates efforts toward renewable circular economy.

5.3.2.2.2 Saccharification of waste biomass and fermentation of released sugars for lactic acid production

The enzymatic hydrolysis of pretreated biomass for release of free sugars is achieved by action of exoglucanases, endoglucanases, β-glucosidases, xylanases which are added in appropriate proportions to achieve high yields. The hydrolysate generated after saccharification is subjected to fermentation in subsequent step and this SHF approach has been investigated for lactic acid production from lignocellulosic waste substrates [35,36,76,77]. Nevertheless, the inhibitory effect of released sugars on hydrolytic activity of saccharifying enzymes has led to preferential use of SSF approach for lactic acid production, as the SSF process helps to minimize feedback inihibition by continuous microbial fermentation of released sugars. The techno-economic analysis of lactic acid production from lignocellulosic biomass revealed cellulose based lactic acid production is economically viable with 21.28% as internal rate of return (IRR) [78]. Similarly, the production cost from corn stover at 0.56 $/kg of 88% (*w/w*) L-lactic acid was competitive and comparable to commercial lactic acid produced from starchy feedstock and 24% less expensive than that of ethanol produced from corn stover [79]. The fermentative lactic acid production by wild-type as well as engineered microorganisms from lignocellulosic agroindustrial wastes, employing both SHF and SSF processes is summarized in Table 5.2. However, several challenges are encountered during lactic acid production from lignocellulosic wastes, which are addressed in next section.

5.3.2.3 Challenges hindering lactic acid production from lignocellulosic agroindustrial wastes

5.3.2.3.1 Release of inhibitors during pretreatment

The generation of various inhibitory compounds viz. furfural, 5-hydroxymethylfurfural, levulinic acid, vanillin, sulfate, acetate, formate, inorganic ions during pretreatment is one of the major constraint in fermentative lactic acid production. The presence of these compounds adversely affects the lactic acid yield by inhibiting the activity of saccharifying enzymes and interfering with growth/metabolism of fermenting microorganisms [70]. Thus, the pretreated biomass is subjected to detoxification for the removal of inhibitory compounds by activated charcoal treatment, vacuum membrane distillation, ion-exchange resins and electrodialysis [98]. Few lactic acid producing strains viz. *Lactobacillus brevis*, *Lactobacillus pentosus*, *L. plantarum*, *B. coagulans*, *Pediococcus pentosaceus*, *Pediococcus acidilactici* have been reported to withstand the presence of these inhibitors and residual IL after pretreatment [83,95,99,100]. Thus, a suitable

Table 5.2: Lactic acid production from lignocellulosic agroindustrial wastes as substrates.

Lignocellulosic agroindustrial wastes	Pretreatment	Enzymatic treatment	Fermenting microorganism	Fermentation mode	Lactic acid enantiomer	Lactic acid concentration (g/L)	Yield (g/g)^{a,b,c,d,e}	References
Corncob residue	–	Cellulase (15 U/g cellulose),β-glucosidase (15 U/g cellulose) and neutral protease (0.3 g/L)	*B. coagulans*	Batch SSF	L	68.00	0.85^{c}	[80]
Catfish waste	–	Cellulase (Cellic CTec2), 20 FPU/g biomass	*L. pentosus*	SSF	–	35.70	–	[81]
Wheat straw	Acid-pretreatment	Cellulase (Cellic CTec2), 10 mg total protein per gram of cellulose in the biomass	Engineered *P. acidilactici* TY112	SSF	L	130.80	0.67^{b}	[82]
Waste plywood chips	Acid-impregnated steam explosion	Cellulase (Cellic CTec3), 15 FPU/g of cellulose	*Enterococcus faecalis* SI	Batch SHF	L	59.81	0.95^{d}	[77]
Deoiled cottonseed cake	Ionic liquid [Emim][Ac] pretreatment	Immobilized cellulase (25 FPU/g)	*L. brevis*	SSF	L	–	0.22^{a}	[83]
Oil palm empty fruit bunch (EFB)	Acid-pretreatment	50 FPU/g cellulose of Cellic® CTec2 cellulase	*B. coagulans* JI12	SSF	L	120	0.49^{a}	[84]
Corn stover	Acid-pretreatment	Cellulase (Celluclast 1.5L), 30 FPU/g cellulose	*L. delbrueckii* sp. *bulgaricus*	SSF	D	18	0.18^{a}	[85]
Pulp mill residue	–	Cellulase (Cellic CTec2), 24.8 FPU/g cellulose	*Lactobacillus coryniformis* subsp. Torquens	SHF	D	57	0.97^{b}	[76]

Corn cob residue, cottonseed meal	—	Cellulase, 15 FPU/g dry CCR, Protease 810 U/g cottonseed meal	*Sporolactobacillus inulinus* YBS1-5	SHF fed batch	D	107.2	0.85[b]	[86]
Bagasse sulfite pulp	—	Cellulase 10 FPIU/g cellulose, Xylanase 120 IU/g hemicellulose	*B. coagulans* CC17	SSF	L	110	0.72[c]	[87]
Corn stover	Alkali pretreatment	Cellulase (Cellic CTec2), 5.6 FPU/g of corn stover	Engineered L. *plasntarum*	SSF fed batch	D	61.4	0.77[a]	[65]
Sugarcane bagasse	Acid-pretreatment steam explosion	Genencor GC220 enzyme cocktail, 15.8 FPU/g bagasse	*B. coagulans* DSM2314	SSF	L	70.4	0.83[b]	[88]
Tobacco waste extract	Acid-pretreatment	Cellulase 0.06 g/g and pectinase 0.01 g/g biomass	*R. oryzae*	SHF	L	173.5 g/L	0.86[b]	[89]
Coffee pulp	Acid-pretreatment	Cellulase Accellerase 1500 (0.3 mL/g pulp)	*B. coagulans*	SHF	L	45.3	0.78[b]	[90]
Corn stover	Alkali pretreatment	Cellulase (Cellic CTec2), 50 FPU/g stover	*B. coagulans* LA204	SSF	L	97.59	0.68[a]	[91]
Hardwood pulp	Mechanical milling	Cellulase, Cellic CTec2; (20 FPU/g pulp)	Engineered L. *plantarum*	SSF	D	102.3	0.88[b]	[92]
Waste wood chips	Acid-catalyzed steam explosion	Cellulase (Cellic CTec2), 15 FPU/g chips	Recombinant L. *paracasei* 7B (ldhD gene deficient)	SHF fed batch	L	99.2	0.96[b]	[93]
Soybean straw	Ammonia pretreatment	Cellulase, 50 FPU/g straw	*L. casei*	SHF	L	28	0.80[b]	[94]
Wheat straw	Acid-pretreatment	Cellulase (*T. reesei*), 20 FPU/g cellulose	*B. coagulans* IPE22	SSF	—	38.73	0.46[a]	[95]
Curcuma longa residue	—	Cellulase 179.2 FPU/L, cellobiase 870.8 U/L, α-amylase 1120 FAU/L and amyloglucosidase 420 U/L	*L. coryniformis*	SSF	D	91.61	0.65[a]	[96]

Continued

Table 5.2: Lactic acid production from lignocellulosic agroindustrial wastes as substrates.—cont'd

Lignocellulosic agroindustrial wastes	Pretreatment	Enzymatic treatment	Fermenting microorganism	Fermentation mode	Lactic acid enantiomer	Lactic acid concentration (g/L)	Yield (g/g)[a,b,c,d,e]	References
Curcuma longa residue	—	Cellulase 179.2 FPU/L, cellobiase 870.8 U/L, α-amylase 1120 FAU/L and amyloglucosidase 420 U/L	*L. paracasei*	SSF	L	97.13	0.69[a]	[96]
Rice straw	Acid-pretreatment	Cellulase, 10 FPU/g, cellobiase (Novozyme 188) 10 CBU/g	*L. brevis*	SSF	—	34.2	0.96[e]	[97]

[a]Yield of lactic acid produced (g) to raw material/biomass used (g).
[b]Yield of lactic acid produced (g) to reducing sugar (g).
[c]Yield of lactic acid produced (g) to cellulose in biomass (g).
[d]Yield of lactic acid produced (g) to reducing sugar consumed (g).
[e]Yield of lactic acid produced (mM) to reducing sugar consumed (mM).

pretreatment approach viz. use of compatible IL/DES along with selection of microbes tolerant to lignocellulose derived inhibitors and IL/DES is highly desirable for a productive bioprocess.

5.3.2.3.2 Fermentation of mixed sugars

The saccharification of cellulosic part of waste biomass generates glucose whereas hemicellulose part generates xylose and other pentoses. This presence of mixed sugars i.e., both hexose and pentose sugars in the lignocellulosic hydrolysate after saccharification is the next constraint, as majority of lactic acid producers are unable to ferment pentose sugars viz. xylose, arabinose etc. Screening of native strains capable of utilizing both hexose and pentose sugars and metabolic engineering for creation of tailored microorganisms have been suggested to overcome this challenge. Qiu et al. [101] engineered xylose-assimilating pathways in *P. acidilactici* to achieve 92.7% xylose conversion and produce 97.3 g/L of D-lactic acid from corn stover. In another approach, Zhang and Vadlani [102] advocated use of mixed cultures of *L. brevis* and *L. plantarum* for improving lactic acid yield by efficiently utilizing sugars released from both cellulose and hemicellulose components of biomass.

Further, the metabolism of sugars via heterofermentative (PK) pathway is another key constraint as it leads to lower yield of lactic acid due to by-product formation i.e., acetic acid, ethanol and carbon dioxide in addition to lactic acid. Thus, for achieving bio-based lactic acid production in high titers, the metabolic pathway of the fermenting microbial strains should be channeled toward homofermentative (EMP) pathway [19,30].

5.3.2.3.3 Carbon catabolite repression

The use of promising microorganisms with capability of consumption of both hexose and pentose sugars is further limited by their sugar utilization pattern. The presence of glucose generally represses utilization of other sugars, a well-known phenomenon called carbon catabolite repression (CCR). In context of lignocellulosic feedstocks, primarily abundant amounts of xylose remains unused in presence of glucose due to this CCR mechanism. Thus, selection of wild-type or engineered lactic acid producers with CCR negative phenotypes is very critical for optimising the fermentation of lignocellulosic substrates [30,70]. Yoshida et al. [103] integrated xylose-assimilating operon in *L. plantarum* and the engineered strain could produce D-lactic acid from mixed sugars without any CCR.

Similarly, *L. brevis and L. pentosus* are promising CCR negative strains for use in lignocellulosic biorefineries [83,104,105]. Kim et al. [106] reported putative mechanism of relaxed mode of CCR in *L. brevis*, enabling simultaneous utilization of secondary sugars (xylose, arabinose) along with glucose. In CCR sensitive LAB, the phosphoenolpyruvate-dependent sugar phosphotransferase system (PTS) favors glucose transport and inhibits the transport of other secondary sugars, a regulatory phenomenon called inducer exclusion.

However, in CCR negative phenotypes like *L. brevis*, the secondary sugars viz. xylose are transported by facilitated diffusion even in presence of glucose. Their transport further helps in relieving the inhibition of transcription of specific catabolic operons, usually mediated by carbon catabolite protein A (CcpA) complex in case of LAB with CCR. Summarily, deployment of CCR negative microbes is essential for efficient fermentation of lignocellulosic hydrolysates.

5.3.2.3.4 Enantiomeric purity: D and L lactic acid

The choice of fermenting microorganism and fermentation conditions needs to be critically evaluated for production of enantiomerically pure lactic acid. For wild-type strains producing racemic form of lactic acid, genetic engineering tools have been used to develop microorganisms producing enantiomerically pure form of lactic acid. Bhowmik and Steele [107] firstly reported a pure only L (+) lactic acid producing mutant of *Lactobacillus helveticus*, which was constructed by insertional inactivation of D(−) lactate dehydrogenase (ldhD) gene of wild strain. Similarly, Kuo et al. [93] interrupted ldhD gene of *L. paracasei*, to create a mutant strain, which could produce optically pure L-lactic acid from lignocellulosic agrowastes i.e., rice straw and wood waste chips.

Similar to the focus on pure L-lactic acid producing strains for commercial applications, the screening for pure D-form producing strains is also gaining attention. Few *Lactobacillus* and *Sporolactobacillus* strains in their native forms have been shown to produce D-lactic acid of high purity from hydrolysates of lignocellulosic agroindustrial wastes such as pulp mill residue, corncob residue etc. [76,86]. However, for biotechnological production at industrial scale, genetically engineered strains are being developed to produce optically pure D-lactic acid [108]. Table 5.2 also shows the current production of optically pure forms of lactic acid (both D and L) by wild and recombinant strains from lignocellulosic agroindustrial wastes.

5.3.2.3.5 Acid tolerance of fermenting microorganisms

Another challenging aspect in fermentative production of lactic acid is the selection of potent fermenting microorganisms that are resistant to acid stress. The lactic acid production leads to acidification of the fermenting medium which may lead to arrest of cell growth, denaturation of essential enzymes inside the cells and ultimately cell death [10,109]. To prevent this, addition of large amounts of neutralizing agents viz. gypsum, calcium carbonate is required which leads to complicated downstream processing and increased costs. Thus, selection of native acid-tolerant strains or improving the acid resistance of potent producers by genetic engineering is a desirable pursuit. Singhvi et al. [110] developed an acid-tolerant FM1 strain by protoplast fusion between *Acetobacter pasteurianus* and *L. delbrueckii*, which produced five-fold more lactic acid and also reduced the requirement of calcium carbonate by 50%. In another approach, adaptive evolution was used to create an *L. casei* mutant lb-2 which exhibited a 318-fold higher survival rate as compared to native parental strain [111].

5.4 Downstream processing for recovery of pure lactic acid

The purification of lactic acid from fermentation broth has several ecological and economic implications as it may constitute 50%−80% of the production costs [112]. Due to the complexity of fermentation media i.e., presence of biomass, microbial cells, metabolic by-products, residual sugars, inorganic ions and other organic acids along with lactic acid, the processes employed in the downstream processing become very complicated [113]. This influences the quality and price of lactic acid for commercial purposes. The purification procedures are further hampered by high affinity of lactic acid for water molecules, low volatility and tendency to self-polymerize.

Generally, centrifugation or filtration is performed as the preliminary downstream step to remove biomass or microbial cells and obtain clear supernatant for further processing. Since, majority of fermentation broths contain $Ca(OH)_2$ or $CaCO_3$ as neutralizing agents to prevent the pH drop due to lactic acid production, calcium salt precipitation process has been most widely used step in separation of lactic acid from fermentation broth [19]. The precipitated calcium lactate, recovered by filtration, is acidified with strong acid such as sulfuric acid to release free lactic acid along with low-value by-product calcium sulfate (gypsum). The released lactic acid is further purified by use of activated carbon, evaporation and crystallization processes. This conventional process based on neutralization and precipitation has been popularly used in industrial sector [114]. However, this separation method has several disadvantages such as environmental problems due to generation of one ton of gypsum per ton of pure lactic acid obtained, high cost of reagents, necessity of filtration and low purity of lactic acid [115]. Corbion, one of the global market leaders in lactic acid production claimed a patent for gypsum-free production of lactic acid in 2014 [19]. Daful and Görgens [78] also carried out cradle-to-gate Life Cycle Assessment (LCA) analysis of lignocellulosic lactic acid production from sugarcane bagasse and brown leaves and found that gypsum-free processes were attractive from both environmental and economic points of view.

The downstream processing by various other recovery techniques viz. solvent extraction, ion-exchange (adsorption), membrane separation by microfiltration, ultrafilitration, nanofiltration, electrodialysis and reverse osmosis, hybrid short path evaporation, reactive distillation and molecular distillation is also being actively pursued. Recently, the membrane-based technologies viz. electrodialysis, nanofiltration are gaining high attention for recovery of lactic acid especially from fermentation broths due to ease of scalability and environment friendliness [114,116,117]. Each of these processes offer their own distinct advantages and disadvantages and have been employed depending on the

Table 5.3: Different downstream processing methods used for purification of bio-based lactic acid produced via fermentation.

Feedstock used in fermentation broth	Fermenting microorganism	Downstream process for recovery/ purification	Lactic acid purity (%)	Recovery yield (%)	References
Dried distillers grains with solubles	*L. coryniformis* subsp. *torquens*	Ion-exchange (Amberlite IRA 120, Amberlite IRA 67)	91.80	80.40	[118]
Sweet sorghum juice	*Bacillus coagulans*	Ultrafiltration, mono- and bipolar electrodialysis, ion-exchange, vacuum distillation	99.80	85.60	[119]
Glucose	*Bacillus* sp.	Microfiltration, ultrafiltration, reverse osmosis membrane filtration	97.00	68.00	[113]
Mixed food waste hydrolysate	*L. casei*	Ultrasonic solvent extraction	98.00	83.89	[112]
Mixed restaurant food waste	*Streptococcus* sp.	Microfiltration, nanofiltration, softening electrodialysis, ion-exchange and distillation	99.70	38.00	[46]
Coffee pulp hydrolysate	*B. coagulans*	Microfiltration, nanofiltrations, softening, electrodialysis, ion exchange and distillation	99.70	23.00	[90]
Coffee mucilage	*B. coagulans*	Filtration, softening, electrodialysis, ion-exchange chromatography and distillation	99.80	38.20	[120]
Zizyphus oenophlia	*L. amylophilus*	Ultrafiltration, ion-exchange (Amberlite IRA 96 and Amberlite IR 120 resins)	99.17	98.90	[121]
Potato peel waste	Undefined mixed cultures	Ion-exchange (Amberlite IR 120), vacuum distillation	92.00	70.00	[122]

Table 5.3: Different downstream processing methods used for purification of bio-based lactic acid produced via fermentation.—cont'd

Feedstock used in fermentation broth	Fermenting microorganism	Downstream process for recovery/ purification	Lactic acid purity (%)	Recovery yield (%)	References
Sugarcane molasses	*L. plantarum*	Hybrid short path evaporation	89.70	–	[123]
Glucose	*P. pentosaceus*	Vapor permeation-assisted esterification, vacuum evaporation and distillation	–	95.00	[124]
Sugar cane juice	*L. plantarum*	Microfiltration, nanofiltration, electrodialysis	85.60	–	[125]
Cassava bagasse	*L. delbrueckii*	Ion-exchange chromatography (Amberlite IRA 67)	–	95.40	[126]
Paper sludge	*L. rhamnosus*	Ion-exchange (Amberlite IRA-92)	96.20	82.60	[127]

production process and the desired quality of final product [19,115]. Many studies concerning lactic acid purification from fermentation broth involves combination of aforesaid processes in multiple steps and summarized in Table 5.3.

The lactic acid separated by above processes is further subjected to rotary evaporation, distillation and crystallization procedures for concentration, while residual impurities are removed by chromatographic procedures such as ion-exchange. Summarily, multi-step downstream processing approach is generally applied for purification of lactic acid.

There are very few reports on downstream processing of lactic acid produced by lignocellulosic agroindustrial residues. Recently, Pleissner et al. [46] carried out the multi-step downstream processing for purification of lactic acid produced from restaurant food waste. The sequential use of micro and nanofiltration, electrodialysis, anion- and cation-exchange resins followed by distillation resulted in pure L(+)-lactic acid formulation with 99.7% optical purity. Neu et al. [120] also obtained 99.8% optically pure L(+)-lactic acid from coffee mucilage, a residue from coffee processing after employing similar multi-step downstream processing. However, for both the above studies about 38% of pure lactic acid could be recovered from the fermented broth, which necessitates effective optimization of processes especially for lactic acid produced from complex substrates.

Hu et al. [112] optimized a continuous ultrasonic solvent extraction method for lactic acid recovery from mixed food waste hydrolysate. The ethyl acetate mediated ultrasonic extraction led to separation of lactic acid of 98% purity with a recovery yield of 84%. Various solvents such as octanol, decanol, hexane, chloroform, diethyl ether, tertiary amines have been conventionally used as extractants for lactic acid, but this novel extractive method offered better sustainability and economic feasibility. The high recovery yield of 98.9% was also obtained by Bishai et al. [121] by use of ion-exchange chromatography for lactic acid purification. The two-step purification process involving anion (Amberlite IRA 96) and cation (Amberlite IR-120) exchangers led to recovery of 99.17% pure lactic acid from the fermentation broth. Similarly, lactic acid produced from jackfruit seed powder was purified by ion exchange employing Amberlite IRA 67 as the resin [47]. Extensive research efforts are still required to lower the operating costs and complexity of downstream processing so that production of lactic acid from agroindustrial wastes can be carried out at a competitive pace with chemical methods.

5.5 Conclusions and perspectives

The chapter encompasses the fermentative routes for the large scale production of lactic acid, which remains in high demand due to myriad industrial applications such as chemicals, pharmaceuticals, polymers, cosmetics etc. The industrial scale production requires the adoption of low cost as well as environment friendly technologies. The use of abundant and renewable agroindustrial wastes for sustainable production of high-value platform chemicals like lactic acid has garnered a lot of attention due to fast depleting fossil reserves. Due to the complexity of waste biomass, various challenges need to be addressed for their effective valorization. The use of native or genetically engineered CCR negative, acid-resistant microorganisms producing optically pure lactic acid via homofermentative pathway is being extensively researched. The optimization of simple and efficient downstream processing strategies will require high attention to implement lactic acid production via waste biorefineries at commercial scale.

Acknowledgments

The authors acknowledge the financial assistance provided by National Agricultural Science Fund (NASF), Govt. of India (Sanction No. NASF/AE-6017/2016-17/49) to carry out this study. JG, AS and NY duly acknowledge Ministry of Human Resource Development and Indian Institute of Technology Delhi for providing their research fellowship.

References

[1] Maity SK. Opportunities, recent trends and challenges of integrated biorefinery: part I. Renewable and Sustainable Energy Reviews 2015;43:1427–45.

[2] Demichelis F, Fiore S, Pleissner D, Venus J. Technical and economic assessment of food waste valorization through a biorefinery chain. Renewable and Sustainable Energy Reviews 2018;94:38–48.

[3] Nizami AS, Rehan M, Waqas M, Naqvi M, Ouda OKM, Shahzad K, et al. Waste biorefineries: enabling circular economies in developing countries. Bioresource Technology 2017;241:1101–17.

[4] Farmer TJ, Comerford JW, Pellis A, Robert T. Post-polymerization modification of bio-based polymers: maximizing the high functionality of polymers derived from biomass. Polymer International 2018;67(7):775–89.

[5] Jang YS, Kim B, Shin JH, Choi YJ, Choi S, Song CW, et al. Bio-based production of C2–C6 platform chemicals. Biotechnology and Bioengineering 2012;109(10):2437–59.

[6] Lee SY, Kim HU, Chae TU, Cho JS, Kim JW, Shin JH, et al. A comprehensive metabolic map for production of bio-based chemicals. Nature Catalysis 2019;2(1):18–33.

[7] Sheldon RA. The road to biorenewables: carbohydrates to commodity chemicals. ACS Sustainable Chemistry & Engineering 2018;6(4):4464–80.

[8] Chandel AK, Garlapati VK, Singh AK, Antunes FAF, da Silva SS. The path forward for lignocellulose biorefineries: bottlenecks, solutions, and perspective on commercialization. Bioresource Technology 2018;264:370–81.

[9] Takkellapati S, Li T, Gonzalez MA. An overview of biorefinery-derived platform chemicals from a cellulose and hemicellulose biorefinery. Clean Technologies and Environmental Policy 2018;20(7):1615–30.

[10] Singhvi M, Zendo T, Sonomoto K. Free lactic acid production under acidic conditions by lactic acid bacteria strains: challenges and future prospects. Applied Microbiology and Biotechnology 2018;102(14):5911–24.

[11] Becker J, Lange A, Fabarius J, Wittmann C. Top value platform chemicals: bio-based production of organic acids. Current Opinion in Biotechnology 2015;36:168–75.

[12] Choi S, Song CW, Shin JH, Lee SY. Biorefineries for the production of top building block chemicals and their derivatives. Metabolic Engineering 2015;28:223–39.

[13] López-Gómez JP, Alexandri M, Schneider R, Venus J. A review on the current developments in continuous lactic acid fermentations and case studies utilising inexpensive raw materials. Process Biochemistry 2019;79:1–10.

[14] Kim DH, Lee MK, Hwang Y, Im WT, Yun YM, Park C, et al. Microbial granulation for lactic acid production. Biotechnology and Bioengineering 2016;113(1):101–11.

[15] Djukić-Vuković A, Mladenović D, Ivanović J, Pejin J, Mojović L. Towards sustainability of lactic acid and poly-lactic acid polymers production. Renewable and Sustainable Energy Reviews 2019;108:238–52.

[16] Eş I, Khaneghah AM, Barba FJ, Saraiva JA, Sant'Ana AS, Hashemi SMB. Recent advancements in lactic acid production – a review. Food Research International 2018;107:763–70.

[17] Isikgor FH, Becer CR. Lignocellulosic biomass: a sustainable platform for the production of bio-based chemicals and polymers. Polymer Chemistry 2015;6(25):4497–559.

[18] Dusselier M, Van Wouwe P, Dewaele A, Makshina E, Sels BF. Lactic acid as a platform chemical in the biobased economy: the role of chemocatalysis. Energy & Environmental Science 2013;6(5):1415–42.

[19] de Oliveira RA, Komesu A, Vaz Rossell CE, Maciel Filho R. Challenges and opportunities in lactic acid bioprocess design—from economic to production aspects. Biochemical Engineering Journal 2018;133:219–39.

[20] Marszałek-Harych A, Jędrzkiewicz D, Ejfler J. Bio- and chemocatalysis cascades as a bridge between biology and chemistry for green polymer synthesis. Cellular and Molecular Biology Letters 2017;22:28.

[21] Wang SF, Guo CL, Cui KK, Zhu YT, Ding JX, Zou XY, et al. Lactic acid as an invaluable green solvent for ultrasound-assisted scalable synthesis of pyrrole derivatives. Ultrasonics Sonochemistry 2015;26:81–6.

[22] Yang J, Tan J-N, Gu Y. Lactic acid as an invaluable bio-based solvent for organic reactions. Green Chemistry 2012;14(12):3304–17.

[23] Gu Y, Jérôme F. Bio-based solvents: an emerging generation of fluids for the design of eco-efficient processes in catalysis and organic chemistry. Chemical Society Reviews 2013;42(24):9550–70.
[24] Klotz S, Kaufmann N, Kuenz A, Prüße U. Biotechnological production of enantiomerically pure D-lactic acid. Applied Microbiology and Biotechnology 2016;100(22):9423–37.
[25] Tripathi N, Katiyar V. Lactic acid oligomer (OLLA) grafted gum Arabic based green adhesive for structural applications. International Journal of Biological Macromolecules 2018;120:711–20.
[26] Swider E, Koshkina O, Tel J, Cruz LJ, de Vries IJM, Srinivas M. Customizing poly(lactic-co-glycolic acid) particles for biomedical applications. Acta Biomaterialia 2018;73:38–51.
[27] Lee PW, Pokorski JK. Poly(lactic-co-glycolic acid) devices: production and applications for sustained protein delivery. WIREs Nanomedicine and Nanobiotechnology 2018;10(5):e1516.
[28] Mir M, Ahmed N, Rehman Au. Recent applications of PLGA based nanostructures in drug delivery. Colloids and Surfaces B: Biointerfaces 2017;159:217–31.
[29] Akman L, Biber Muftuler FZ, Bilgi A, Yurt Kilcar A, Gokulu SG, Medine EI, et al. Synthesis of a theranostic agent: radioiodinated PEGylated PLGA-indocyanine capsules and in vitro determination of their bioaffinity on ovarian, cervical and breast cancer cells. Journal of Radioanalytical and Nuclear Chemistry 2016;308(2):659–70.
[30] Cubas-Cano E, González-Fernández C, Ballesteros M, Tomás-Pejó E. Biotechnological advances in lactic acid production by lactic acid bacteria: lignocellulose as novel substrate. Biofuels, Bioproducts and Biorefining 2018;12(2):290–303.
[31] Ghaffar T, Irshad M, Anwar Z, Aqil T, Zulifqar Z, Tariq A, et al. Recent trends in lactic acid biotechnology: a brief review on production to purification. Journal of Radiation Research and Applied Sciences 2014;7(2):222–9.
[32] Juodeikiene G, Vidmantiene D, Basinskiene L, Cernauskas D, Bartkiene E, Cizeikiene D. Green metrics for sustainability of biobased lactic acid from starchy biomass vs chemical synthesis. Catalysis Today 2015;239:11–6.
[33] Juturu V, Wu JC. Microbial production of lactic acid: the latest development. Critical Reviews in Biotechnology 2016;36(6):967–77.
[34] Sauer M, Russmayer H, Grabherr R, Peterbauer CK, Marx H. The efficient clade: lactic acid bacteria for industrial chemical production. Trends in Biotechnology 2017;35(8):756–69.
[35] Abdel-Rahman MA, Tashiro Y, Sonomoto K. Recent advances in lactic acid production by microbial fermentation processes. Biotechnology Advances 2013;31(6):877–902.
[36] Wang Y, Tashiro Y, Sonomoto K. Fermentative production of lactic acid from renewable materials: recent achievements, prospects, and limits. Journal of Bioscience and Bioengineering 2015;119(1):10–8.
[37] Hatti-Kaul R, Chen L, Dishisha T, Enshasy HE. Lactic acid bacteria: from starter cultures to producers of chemicals. FEMS Microbiology Letters 2018;365(20):fny213.
[38] Guo W, Jia W, Li Y, Chen S. Performances of *Lactobacillus brevis* for producing lactic acid from hydrolysate of lignocellulosics. Applied Biochemistry and Biotechnology 2010;161(1):124–36.
[39] Liang S, McDonald AG, Coats ER. Lactic acid production with undefined mixed culture fermentation of potato peel waste. Waste Management 2014;34(11):2022–7.
[40] Vavouraki AI, Volioti V, Kornaros ME. Optimization of thermo-chemical pretreatment and enzymatic hydrolysis of kitchen wastes. Waste Management 2014;34(1):167–73.
[41] Pagana I, Morawicki R, Hager TJ. Lactic acid production using waste generated from sweet potato processing. International Journal of Food Science and Technology 2014;49(2):641–9.
[42] Yang X, Zhu M, Huang X, Lin CSK, Wang J, Li S. Valorisation of mixed bakery waste in non-sterilized fermentation for L-lactic acid production by an evolved *Thermoanaerobacterium* sp. strain. Bioresource Technology 2015;198:47–54.
[43] Kwan TH, Hu Y, Lin CSK. Valorisation of food waste via fungal hydrolysis and lactic acid fermentation with *Lactobacillus casei* Shirota. Bioresource Technology 2016;217:129–36.
[44] Kwan TH, Hu Y, Lin CSK. Techno-economic analysis of a food waste valorisation process for lactic acid, lactide and poly(lactic acid) production. Journal of Cleaner Production 2018;181:72–87.

[45] Kwan TH, Pleissner D, Lau KY, Venus J, Pommeret A, Lin CSK. Techno-economic analysis of a food waste valorization process via microalgae cultivation and co-production of plasticizer, lactic acid and animal feed from algal biomass and food waste. Bioresource Technology 2015;198:292–9.
[46] Pleissner D, Demichelis F, Mariano S, Fiore S, Navarro Gutiérrez IM, Schneider R, et al. Direct production of lactic acid based on simultaneous saccharification and fermentation of mixed restaurant food waste. Journal of Cleaner Production 2017;143:615–23.
[47] Nair NR, Nampoothiri KM, Banarjee R, Reddy G. Simultaneous saccharification and fermentation (SSF) of jackfruit seed powder (JFSP) to L-lactic acid and to polylactide polymer. Bioresource Technology 2016;213:283–8.
[48] Wang Y, Cai D, He M, Wang Z, Qin P, Tan T. Open fermentative production of L-lactic acid using white rice bran by simultaneous saccharification and fermentation. Bioresource Technology 2015;198:664–72.
[49] Watanabe M, Makino M, Kaku N, Koyama M, Nakamura K, Sasano K. Fermentative L-(+)-lactic acid production from non-sterilized rice washing drainage containing rice bran by a newly isolated lactic acid bacteria without any additions of nutrients. Journal of Bioscience and Bioengineering 2013;115(4):449–52.
[50] John RP, Madhavan Nampoothiri K, Pandey A. Simultaneous saccharification and fermentation of cassava bagasse for L-(+)-lactic acid production using *Lactobacilli*. Applied Biochemistry and Biotechnology 2006;134(3):263–72.
[51] Kim KI, Kim WK, Seo DK, Yoo IS, Kim EK, Yoon HH. Production of lactic acid from food wastes. Applied Biochemistry and Biotechnology 2003;105(1):637–47.
[52] Liang S, Gliniewicz K, Gerritsen AT, McDonald AG. Analysis of microbial community variation during the mixed culture fermentation of agricultural peel wastes to produce lactic acid. Bioresource Technology 2016;208:7–12.
[53] Palakawong Na Ayudthaya S, van de Weijer AHP, van Gelder AH, Stams AJM, de Vos WM, Plugge CM. Organic acid production from potato starch waste fermentation by rumen microbial communities from Dutch and Thai dairy cows. Biotechnology for Biofuels 2018;11(1):13.
[54] Nunes LV, de Barros Correa FF, de Oliva Neto P, Mayer CRM, Escaramboni B, Campioni TS, et al. Lactic acid production from submerged fermentation of broken rice using undefined mixed culture. World Journal of Microbiology and Biotechnology 2017;33(4):79.
[55] Unban K, Kanpiengjai A, Khatthongngam N, Saenjum C, Khanongnuch C. Simultaneous bioconversion of gelatinized starchy waste from the rice noodle manufacturing process to lactic acid and maltose-forming α-amylase by *Lactobacillus plantarum* S21, using a low-cost medium. Fermentation 2019;5(2):32.
[56] Li Z, Lu J, Yang Z, Han L, Tan T. Utilization of white rice bran for production of L-lactic acid. Biomass and Bioenergy 2012;39:53–8.
[57] Tanaka T, Hoshina M, Tanabe S, Sakai K, Ohtsubo S, Taniguchi M. Production of D-lactic acid from defatted rice bran by simultaneous saccharification and fermentation. Bioresource Technology 2006;97(2):211–7.
[58] Reddy LV, Kim Y-M, Yun J-S, Ryu H-W, Wee Y-J. L-Lactic acid production by combined utilization of agricultural bioresources as renewable and economical substrates through batch and repeated-batch fermentation of *Enterococcus faecalis* RKY1. Bioresource Technology 2016;209:187–94.
[59] John RP, S AG, Nampoothiri KM, Pandey A. Direct lactic acid fermentation: focus on simultaneous saccharification and lactic acid production. Biotechnology Advances 2009;27(2):145–52.
[60] Huang LP, Jin B, Lant P. Direct fermentation of potato starch wastewater to lactic acid by *Rhizopus oryzae* and *Rhizopus arrhizus*. Bioprocess and Biosystems Engineering 2005;27(4):229–38.
[61] Smerilli M, Neureiter M, Wurz S, Haas C, Frühauf S, Fuchs W. Direct fermentation of potato starch and potato residues to lactic acid by *Geobacillus stearothermophilus* under non-sterile conditions. Journal of Chemical Technology and Biotechnology 2015;90(4):648–57.

[62] John RP, Gangadharan D, Nampoothiri KM. Genome shuffling of *Lactobacillus delbrueckii* mutant and *Bacillus amyloliquefaciens* through protoplasmic fusion for L-lactic acid production from starchy wastes. Bioresource Technology 2008;99(17):8008–15.
[63] Wakai S, Yoshie T, Asai-Nakashima N, Yamada R, Ogino C, Tsutsumi H, et al. L-lactic acid production from starch by simultaneous saccharification and fermentation in a genetically engineered *Aspergillus oryzae* pure culture. Bioresource Technology 2014;173:376–83.
[64] Okano K, Zhang Q, Shinkawa S, Yoshida S, Tanaka T, Fukuda H, et al. Efficient production of optically pure D-lactic acid from raw corn starch by using a genetically modified L-lactate dehydrogenase gene-deficient and α-amylase-secreting *Lactobacillus plantarum* strain. Applied and Environmental Microbiology 2009;75(2):462–7.
[65] Zhang Y, Vadlani PV, Kumar A, Hardwidge PR, Govind R, Tanaka T, et al. Enhanced D-lactic acid production from renewable resources using engineered *Lactobacillus plantarum.* Applied Microbiology and Biotechnology 2016;100(1):279–88.
[66] Serror P, Sasaki T, Ehrlich SD, Maguin E. Electrotransformation of *Lactobacillus delbrueckii* subsp. *bulgaricus* and *L. delbrueckii subsp. lactis* with various plasmids. Applied and Environmental Microbiology 2002;68(1):46–52.
[67] Kawaguchi H, Hasunuma T, Ogino C, Kondo A. Bioprocessing of bio-based chemicals produced from lignocellulosic feedstocks. Current Opinion in Biotechnology 2016;42:30–9.
[68] Khare SK, Pandey A, Larroche C. Current perspectives in enzymatic saccharification of lignocellulosic biomass. Biochemical Engineering Journal 2015;102:38–44.
[69] Sun S, Sun S, Cao X, Sun R. The role of pretreatment in improving the enzymatic hydrolysis of lignocellulosic materials. Bioresource Technology 2016;199:49–58.
[70] Tarraran L, Mazzoli R. Alternative strategies for lignocellulose fermentation through lactic acid bacteria: state-of-the-art and perspectives. FEMS Microbiology Letters 2018;365(15):fny126.
[71] Chen H, Liu J, Chang X, Chen D, Xue Y, Liu P, et al. A review on the pretreatment of lignocellulose for high-value chemicals. Fuel Processing Technology 2017;160:196–206.
[72] Zhang Q, Hu J, Lee D-J. Pretreatment of biomass using ionic liquids: research updates. Renewable Energy 2017;111:77–84.
[73] Satlewal A, Agrawal R, Bhagia S, Sangoro J, Ragauskas AJ. Natural deep eutectic solvents for lignocellulosic biomass pretreatment: recent developments, challenges and novel opportunities. Biotechnology Advances 2018;36(8):2032–50.
[74] Guo Z, Ling Z, Wang C, Zhang X, Xu F. Integration of facile deep eutectic solvents pretreatment for enhanced enzymatic hydrolysis and lignin valorization from industrial xylose residue. Bioresource Technology 2018;265:334–9.
[75] Kumar AK, Parikh BS, Pravakar M. Natural deep eutectic solvent mediated pretreatment of rice straw: bioanalytical characterization of lignin extract and enzymatic hydrolysis of pretreated biomass residue. Environmental Science & Pollution Research 2016;23(10):9265–75.
[76] de Oliveira Moraes A, Ramirez NIB, Pereira N. Evaluation of the fermentation potential of pulp mill residue to produce D(−)-lactic acid by separate hydrolysis and fermentation using *Lactobacillus coryniformis* subsp. *torquens.* Applied Biochemistry and Biotechnology 2016;180(8):1574–85.
[77] Yuan S-F, Hsu T-C, Wang C-A, Jang M-F, Kuo Y-C, Alper HS, et al. Production of optically pure L (+)-lactic acid from waste plywood chips using an isolated thermotolerant *Enterococcus faecalis* SI at a pilot scale. Journal of Industrial Microbiology & Biotechnology 2018;45(11):961–70.
[78] Daful AG, Görgens JF. Techno-economic analysis and environmental impact assessment of lignocellulosic lactic acid production. Chemical Engineering Science 2017;162:53–65.
[79] Liu G, Sun J, Zhang J, Tu Y, Bao J. High titer L-lactic acid production from corn stover with minimum wastewater generation and techno-economic evaluation based on Aspen plus modeling. Bioresource Technology 2015;198:803–10.

[80] Jiang S, Xu P, Tao F. L-Lactic acid production by *Bacillus coagulans* through simultaneous saccharification and fermentation of lignocellulosic corncob residue. Bioresource Technology Reports 2019;6:131–7.
[81] Shi S, Li J, Blersch DM. Utilization of solid catfish manure waste as carbon and nutrient source for lactic acid production. Applied Microbiology and Biotechnology 2018;102(11):4765–72.
[82] Qiu Z, Gao Q, Bao J. Engineering *Pediococcus acidilactici* with xylose assimilation pathway for high titer cellulosic L-lactic acid fermentation. Bioresource Technology 2018;249:9–15.
[83] Grewal J, Khare SK. One-pot bioprocess for lactic acid production from lignocellulosic agro-wastes by using ionic liquid stable *Lactobacillus brevis*. Bioresource Technology 2018;251:268–73.
[84] Juturu V, Wu JC. Production of high concentration of L-lactic acid from oil palm empty fruit bunch by thermophilic *Bacillus coagulans* JI12. Biotechnology and Applied Biochemistry 2018;65(2):145–9.
[85] Wang X, Wang G, Yu X, Chen H, Sun Y, Chen G. Pretreatment of corn stover by solid acid for D-lactic acid fermentation. Bioresource Technology 2017;239:490–5.
[86] Bai Z, Gao Z, Sun J, Wu B, He B. D-Lactic acid production by *Sporolactobacillus inulinus* YBS1-5 with simultaneous utilization of cottonseed meal and corncob residue. Bioresource Technology 2016;207:346–52.
[87] Zhou J, Ouyang J, Xu Q, Zheng Z. Cost-effective simultaneous saccharification and fermentation of L-lactic acid from bagasse sulfite pulp by *Bacillus coagulans* CC17. Bioresource Technology 2016;222:431–8.
[88] van der Pol EC, Eggink G, Weusthuis RA. Production of L(+)-lactic acid from acid pretreated sugarcane bagasse using *Bacillus coagulans* DSM 2314 in a simultaneous saccharification and fermentation strategy. Biotechnology for Biofuels 2016;9(1):248.
[89] Zheng Y, Wang Y, Zhang J, Pan J. Using tobacco waste extract in pre-culture medium to improve xylose utilization for L-lactic acid production from cellulosic waste by *Rhizopus oryzae*. Bioresource Technology 2016;218:344–50.
[90] Pleissner D, Neu A-K, Mehlmann K, Schneider R, Puerta-Quintero GI, Venus J. Fermentative lactic acid production from coffee pulp hydrolysate using *Bacillus coagulans* at laboratory and pilot scales. Bioresource Technology 2016;218:167–73.
[91] Hu J, Zhang Z, Lin Y, Zhao S, Mei Y, Liang Y, et al. High-titer lactic acid production from NaOH-pretreated corn stover by *Bacillus coagulans* LA204 using fed-batch simultaneous saccharification and fermentation under non-sterile condition. Bioresource Technology 2015;182:251–7.
[92] Hama S, Mizuno S, Kihara M, Tanaka T, Ogino C, Noda H, et al. Production of D-lactic acid from hardwood pulp by mechanical milling followed by simultaneous saccharification and fermentation using metabolically engineered *Lactobacillus plantarum*. Bioresource Technology 2015;187:167–72.
[93] Kuo Y-C, Yuan S-F, Wang C-A, Huang Y-J, Guo G-L, Hwang W-S. Production of optically pure L-lactic acid from lignocellulosic hydrolysate by using a newly isolated and D-lactate dehydrogenase gene-deficient *Lactobacillus paracasei* strain. Bioresource Technology 2015;198:651–7.
[94] Wang J, Wang Q, Xu Z, Zhang W, Xiang J. Effect of fermentation conditions on L-lactic acid production from soybean straw hydrolysate. Journal of Microbiology and Biotechnology 2015;25(1):26–32.
[95] Zhang Y, Chen X, Luo J, Qi B, Wan Y. An efficient process for lactic acid production from wheat straw by a newly isolated *Bacillus coagulans* strain IPE22. Bioresource Technology 2014;158:396–9.
[96] Nguyen CM, Kim J-S, Nguyen TN, Kim SK, Choi GJ, Choi YH, et al. Production of L- and D-lactic acid from waste *Curcuma longa* biomass through simultaneous saccharification and cofermentation. Bioresource Technology 2013;146:35–43.
[97] Kim J-H, Block DE, Shoemaker SP, Mills DA. Conversion of rice straw to bio-based chemicals: an integrated process using *Lactobacillus brevis*. Applied Microbiology and Biotechnology 2010;86(5):1375–85.
[98] Jönsson LJ, Martín C. Pretreatment of lignocellulose: formation of inhibitory by-products and strategies for minimizing their effects. Bioresource Technology 2016;199:103–12.

[99] de Oliveira RA, Rossell CEV, Venus J, Rabelo SC, Maciel Filho R. Detoxification of sugarcane-derived hemicellulosic hydrolysate using a lactic acid producing strain. Journal of Biotechnology 2018;278:56–63.

[100] Boguta AM, Bringel F, Martinussen J, Jensen PR. Screening of lactic acid bacteria for their potential as microbial cell factories for bioconversion of lignocellulosic feedstocks. Microbial Cell Factories 2014;13(1):97.

[101] Qiu Z, Gao Q, Bao J. Constructing xylose-assimilating pathways in *Pediococcus acidilactici* for high titer D-lactic acid fermentation from corn stover feedstock. Bioresource Technology 2017;245:1369–76.

[102] Zhang Y, Vadlani PV. Lactic acid production from biomass-derived sugars via co-fermentation of *Lactobacillus brevis* and *Lactobacillus plantarum*. Journal of Bioscience and Bioengineering 2015;119(6):694–9.

[103] Yoshida S, Okano K, Tanaka T, Ogino C, Kondo A. Homo-D-lactic acid production from mixed sugars using xylose-assimilating operon-integrated *Lactobacillus plantarum*. Applied Microbiology and Biotechnology 2011;92(1):67–76.

[104] Jeong KH, Israr B, Shoemaker SP, Mills DA, Kim J. Impact of lactic acid and hydrogen ion on the simultaneous fermentation of glucose and xylose by the carbon catabolite derepressed *Lactobacillus brevis* ATCC 14869. Journal of Microbiology and Biotechnology 2016;26(7):1182–9.

[105] Cubas-Cano E, González-Fernández C, Ballesteros M, Tomás-Pejó E. *Lactobacillus pentosus* CECT 4023 T co-utilizes glucose and xylose to produce lactic acid from wheat straw hydrolysate: anaerobiosis as a key factor. Biotechnology Progress 2019;35:e2739.

[106] Kim J-H, Block DE, Mills DA. Simultaneous consumption of pentose and hexose sugars: an optimal microbial phenotype for efficient fermentation of lignocellulosic biomass. Applied Microbiology and Biotechnology 2010;88(5):1077–85.

[107] Bhowmik T, Steele JL. Cloning, characterization and insertional inactivation of the *Lactobacillus helveticus* D(−) lactate dehydrogenase gene. Applied Microbiology and Biotechnology 1994;41(4):432–9.

[108] Zhang Y, Yoshida M, Vadlani PV. Biosynthesis of D-lactic acid from lignocellulosic biomass. Biotechnology Letters 2018;40(8):1167–79.

[109] Wu C, Huang J, Zhou R. Progress in engineering acid stress resistance of lactic acid bacteria. Applied Microbiology and Biotechnology 2014;98(3):1055–63.

[110] Singhvi M, Gurjar G, Gupta V, Gokhale D. Biocatalyst development for lactic acid production at acidic pH using inter-generic protoplast fusion. RSC Advances 2015;5(3):2024–31.

[111] Zhang J, Wu C, Du G, Chen J. Enhanced acid tolerance in *Lactobacillus casei* by adaptive evolution and compared stress response during acid stress. Biotechnology and Bioprocess Engineering 2012;17(2):283–9.

[112] Hu Y, Kwan TH, Daoud WA, Lin CSK. Continuous ultrasonic-mediated solvent extraction of lactic acid from fermentation broths. Journal of Cleaner Production 2017;145:142–50.

[113] Phanthumchinda N, Thitiprasert S, Tanasupawat S, Assabumrungrat S, Thongchul N. Process and cost modeling of lactic acid recovery from fermentation broths by membrane-based process. Process Biochemistry 2018;68:205–13.

[114] de Oliveira RA, Alexandri M, Komesu A, Venus J, Rossell CE, Filho RM. Current advances in separation and purification of second-generation lactic acid. Separation and Purification Reviews 2019:1–17.

[115] Komesu A, Wolf Maciel MR, Rocha de Oliveira JA, da Silva Martins LH, Maciel Filho R. Purification of lactic acid produced by fermentation: focus on non-traditional distillation processes. Separation and Purification Reviews 2017;46(3):241–54.

[116] Alexandri M, Schneider R, Venus J. Membrane technologies for lactic acid separation from fermentation broths derived from renewable resources. Membranes 2018;8(4):94.

[117] Mai TK, Rodtong S, Baimark Y, Rarey J, Boontawan A. Membrane-based purification of optically pure D-lactic acid from fermentation broth to poly(D-lactide) polymer. Journal of Membrane Science 2018;551:180–90.

[118] Zaini NAM, Chatzifragkou A, Tverezovskiy V, Charalampopoulos D. Purification and polymerisation of microbial D-lactic acid from DDGS hydrolysates fermentation. Biochemical Engineering Journal 2019;150:107265.

[119] Olszewska-Widdrat A, Alexandri M, López-Gómez PJ, Schneider R, Mandl M, Venus J. Production and purification of L-lactic acid in lab and pilot scales using sweet sorghum juice. Fermentation 2019;5(2):36.

[120] Neu A-K, Pleissner D, Mehlmann K, Schneider R, Puerta-Quintero GI, Venus J. Fermentative utilization of coffee mucilage using *Bacillus coagulans* and investigation of down-stream processing of fermentation broth for optically pure L (+)-lactic acid production. Bioresource Technology 2016;211:398–405.

[121] Bishai M, De S, Adhikari B, Banerjee R. A platform technology of recovery of lactic acid from a fermentation broth of novel substrate *Zizyphus oenophlia*. 3 Biotech 2015;5(4):455–63.

[122] Liang S, McDonald AG, Coats ER. Lactic acid production from potato peel waste by anaerobic sequencing batch fermentation using undefined mixed culture. Waste Management 2015;45:51–6.

[123] Komesu A, Martins PF, Lunelli BH, Oliveira J, Maciel Filho R, Wolf Maciel MR. Evaluation of lactic acid purification from fermentation broth by hybrid short path evaporation using factorial experimental design. Separation and Purification Technology 2014;136:233–40.

[124] Khunnonkwao P, Boontawan P, Haltrich D, Maischberger T, Boontawan A. Purification of L-(+)-lactic acid from pre-treated fermentation broth using vapor permeation-assisted esterification. Process Biochemistry 2012;47(12):1948–56.

[125] Sikder J, Chakraborty S, Pal P, Drioli E, Bhattacharjee C. Purification of lactic acid from microfiltrate fermentation broth by cross-flow nanofiltration. Biochemical Engineering Journal 2012;69:130–7.

[126] John RP, Nampoothiri KM, Pandey A. L(+)-Lactic acid recovery from cassava bagasse based fermented medium using anion exchange resins. Brazilian Archives of Biology and Technology 2008;51:1241–8.

[127] Tong W-Y, Fu X-Y, Lee S-M, Yu J, Liu J-W, Wei D-Z, et al. Purification of L(+)-lactic acid from fermentation broth with paper sludge as a cellulosic feedstock using weak anion exchanger Amberlite IRA-92. Biochemical Engineering Journal 2004;18(2):89–96.

CHAPTER 6

Value addition of waste lignocellulosic biomass through polyhydroxybutyrate production

N. Arul Manikandan[1], Kannan Pakshirajan[2], G. Pugazhenthi[1]

[1]Department of Chemical Engineering, Indian Institute of Technology Guwahati, Guwahati, Assam, India; [2]Department of Biosciences and Bioengineering, Indian Institute of Technology Guwahati, Guwahati, Assam, India

6.1 Introduction

Our continued reliance on nonrenewable energy sources originates from the widespread use and utilization of oil subordinates, which, along with the dwindling oil assets has led to ecological and political concerns. There is clear evidence that emission of greenhouse gasses (GHG), e.g., carbon dioxide (CO_2), methane (CH_4), and nitrous oxide (N_2O) emerging from massive utilization of nonrenewable energy source and human practices are aggravating the Earth's atmosphere [1]. The world's essential origin of energy for vehicle use and generation of chemicals is oil. Whereas vehicle use keeps on extending in the US and Europe, development in the rising economies of India and China is anticipated to be considerably more noteworthy. Concerning chemicals, their reliance is considerably more grounded on fossil assets. A large proportion of the present day commodities are derived from an oil refinery, out of which 4% of oil is overall utilized for plastic generation [2]. Keeping in mind the end goal to reduce the reliance on oil and relieve environmental change in polymer parts, flexible creation chains are essential. It is progressively recognized around the globe that biomass can possibly replace an enormous division of fossil assets as feedstocks for modern commodities, tending to the production of both chemicals and polymer [3].

At national, provincial and worldwide levels, there is a huge drive for utilization of biomass in biorefinery for creation of bioenergy, biofuels and biochemicals. Even though power and heat can be generated by a variety of sustainable options like wind, sun, ocean, etc., chemicals and polymers can be derived only by utilizing

Waste Biorefinery. https://doi.org/10.1016/B978-0-12-818228-4.00006-X

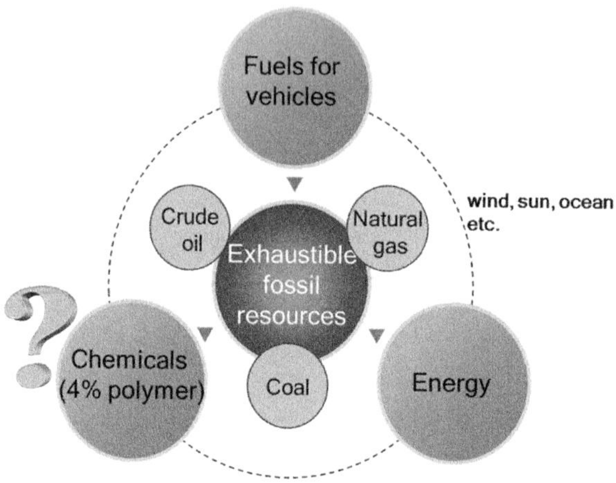

Figure 6.1
Fuels, energy and chemical dependency on exhaustible fossil resources.

biomass as the initial feedstock, as it contributes the main C-rich material source accessible on the Earth, other than fossils. Furthermore, biomass is a sustainable alternative to fossil assets for the creation of transportation fuels and chemicals (Fig. 6.1).

Polyhydroxybutyrate (PHB) is a biodegradable thermoplastic polyester, is one of the best alternatives to many conventional petrochemical-derived plastics which are currently in use. PHB-based bioplastics are currently produced by the process of microbial fermentation utilizing a two-stage cultivation process. The primary stage includes fermentation, which produces PHB intracellularly by utilizing sugars such as fructose, glucose, xylose, etc. as the carbon source. However, to keep the PHB production process cost low, cheaply available raw materials, for example, hydrolysate of lignocellulosic materials and starchy feedstocks were often examined [3,4]. Following the fermentation stage, PHB rich bacterial cells are collected and the product (PHB polymer) is isolated and recovered from the microbial cells by utilizing either a polymer dissolvable extraction procedure or an aqueous procedure in which the non-PHB part of the microbial cell is processed, either synthetically or enzymatically, leading to the separation of PHB polymer. The significant cost associated with this approach is majorly due to the high capital cost involved in establishing of efficient aerobic fermentation facilities as well as the huge cost involved in the purchase of sugar feedstock for production and organic solvents for polymer extraction. Hence, there is a need to evaluate cheaper PHB production methods utilizing suitable industrial waste/refuse as a potent source for biomass production and PHB extraction.

6.2 Polyhydroxybutyrate (PHB)

6.2.1 Properties of PHB

PHB application is expanding with support from purchasers of plastic thermoformed articles as it is seen as a sustainable and biodegradable plastic product obtained from nonpetroleum-based raw materials. Moreover, PHB is exceedingly crystalline because of its stereo consistency. PHB is water-insoluble and moderately impervious to hydrolytic degradation. In comparison with oil-based polymers (e.g., polypropylene), PHB has low oxygen permeability and great thermoplastic properties [5]. However, mechanical properties like Young's modulus and elasticity are considerably poor (Table 6.1). The densities of crystalline and amorphous PHB are 1.26 and 1.18 g/cm^3, respectively. The molecular weight of PHB obtained from wild type microbial strains is often in the range of 10–3000 kDa with a polydispersity index of around 2. PHB is optically pure and piezoelectric, which aids in the initiation of osteogenesis. PHB is exceptionally brittle and solid material. Increasing the molecular weight of PHBs enhances their physical properties.

Polylactic acid (PLA) is likewise not synthesized from petroleum derivatives, and it is biodegradable. However, PLA has a low mechanical strength when used at an elevated temperature. At a high temperature around 60°C, an article formed from PLA loses its ability to withstand twisting forces, which are often found during transportation. Moreover, PLA products may not be suitable for use in temperate regions [6].

PHB is a distinctive part of various life forms in a different environment and exists as both high and low molecular weight compounds. High molecular weight PHB greater than 60,000 Da is direct polyesters that accumulate in a wide variety of Gram-positive

Table 6.1: Physical properties of PHB in comparison with those of biopolymer (PLA) and other conventional polymers.

S. No.	Polymer	Young's modulus (GPa)	Tensile strength (MPa)	Elongation at break (%)	^aT_m (°C)	^aT_g (°C)
1	PHB	3.5–4	40	3–8	172–180	5–9
2	Isotactic polypropylene	1.0–1.7	29.3–38.6	500–900	170–176	−10
3	Polylactic acid	4	80	6	160	60
4	HDPE	0.4–1.0	17.9–33.1	12–700	112–132	−80
5	LDPE	0.05–0.1	15.2–78.6	150–600	88–130	−36
6	PS	3.0–3.1	50	3–4	80–110	21
7	Nylon-6,6	2.8	83	60	265	50
8	Polyethylene-terepthalate	2.2	56	7300	262	3400

^aT_m – Melting temperature; T_g – Glass transition temperature.

and Gram-negative microorganisms; in some Archaea, PHB is accumulated as an intracellular granular stockpiling material, when the organisms are exposed to harsh conditions for example, under nutrient starvation [7,8]. The stored polymeric granules are used for biomass growth when the condition becomes favorable. Low subatomic weight PHB (less than 15,000 Da) is essentially made out of the monomer 3-hydroxybutyrate, a characteristic ketone body found in human blood. Low atomic weight PHBs have been found in a wide variety of prokaryotes and eukaryotes, including humans, and are accepted to be a constituent of each living cell [7,9]. High subatomic weight PHBs have invoked a great enthusiasm from the industry since chemically, these are polyesters and, from physical properties viewpoint, these thermoplastics that can be dissolved in solvents and cast to form long-lasting structures. Therefore, these materials were often referred to as PHB bioplastics. PHB bioplastics are available all through nature and biocompatible materials. They are biodegradable in all bioactive environments and can be produced from various industrial sources. Microorganisms responsible for PHB production are found in a range of environment, but predominantly they can be isolated from root nodules as they lead a symbiotic relationship with nitrogen-fixing organisms found in the root nodules. Biocompatibility of PHB bioplastics in living creatures has been established in sheep, pig, and chicken [10]. Different areas in which PHB bioplastics can be used includes restorative applications in the human body where the material is implanted or used as a strong carbon substrate or as a support for aquaculture denitrification forms. Studies with the monomers of 3-hydroxybutyrate have recommended for their nutraceutical and other therapeutic applications [5,11]. In wild type microorganisms, PHB bioplastics are produced intracellularly as well-defined granules by various pathways that have been widely reported in the literature. This research focuses on the production of PHB as a large volume, value-added product in a closed-loop biorefinery set up.

6.2.2 Uses and applications of PHB

PHB bioplastics are forced into plastic pellets reasonable for further modification by equipment commonly used in the polymer processing industry. The physical properties of PHB bioplastics enable them as an excellent substitute for the oil-based plastics. These properties range from hard crystalline materials, closely resembling polypropylene or acrylonitrile butadiene styrene (ABS) gum, to milder, film-sort materials, for example, low-thickness polyethene, to elastomeric materials, for example, thermoplastic elastomer (TPE) mixes [12,13]. All these materials find an extensive array of end-use applications: blend formed materials from golf tees to hardware lodgings; films for diapers, packaging and agriculture; thermoforming for holders and espresso

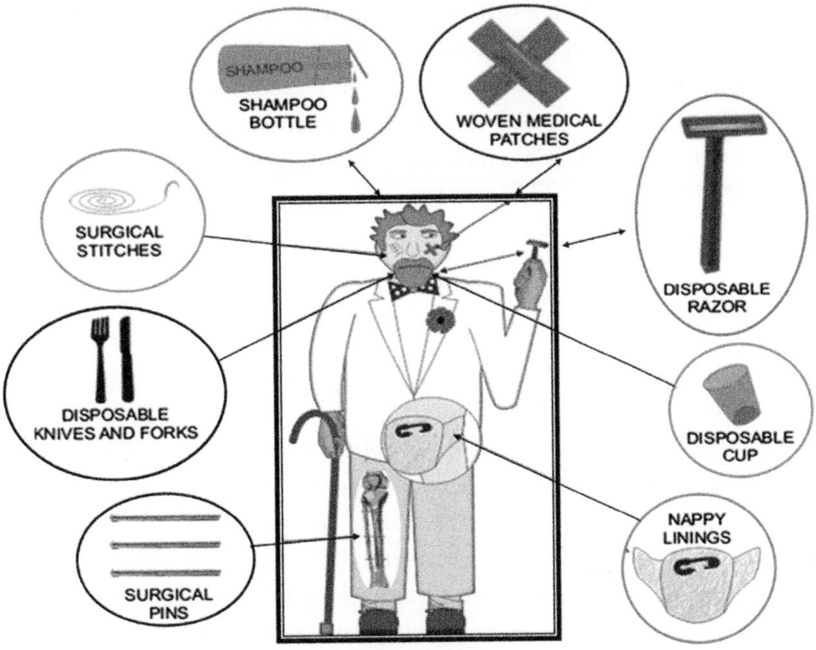

Figure 6.2
Various applications of PHB.

mug tops; strands for monofilament; nonwovens for use in everything from soundproofing in vehicles to nonwoven articles for wipes or diapers; froth for both packaging and food service wear; and coatings for paper for packaging and pharmaceutical products (Fig. 6.2). The properties and uses of PHB bioplastics can be improved by mixing PHB with other biobased/biodegradable materials and oil resins. Hence, PHB bioplastics can replace a critical extent of the oil-based polymers used worldwide. In any case, little is thought about the use of PHB for food packaging applications. One of the difficulties confronted by the food packaging industry is on its endeavors to create biodegradable and tough packaging film with an expanded time frame of realistic usability. The natural based packaging material must remain stable without changes in mechanical or potentially obstruction properties and must work appropriately without damage until disposal. Natural conditions prompting biodegradation must be abstained from the damage of the food item, though streamlined conditions for biodegradation must emerge in after the discarding [13]. Further, before the use of biodegradable material for essential food packaging, the safety must be analyzed.

Biodegradability of PHB bioplastics is another important feature that it offers completely new applications, for example, agricultural films, which is used as a sunscreen on the plants and these degrade toward the finish of utilization and doesn't require to be gathered again [14,15]. Of the billions of pounds of oil-based plastics created each year, 33% of the utilization is for packaging materials which unavoidably gets piled up in landfills. Finding appropriate biobased, sustainable, biodegradable choices have got colossal market potential.

6.2.3 PHB production pathway

The monomeric group of the polyhydroxyalkanoates (PHA) based polymer depends upon the type of sugar source present in the culture broth. For instance, polyhydroxybutyrate synthesis in *Cupriavidus necator* takes place as follows: two acetyl-coA generated from the tricarboxylic acid (TCA) cycle is condensed to form acetoacetyl-coA and this condensation reaction takes place in the presence of enzyme, β-ketothiolase (PhaA). Thereafter, acetoacetyl-coA is reduced to 3-hydroxybutyrylcoA by the action of acetoacetyl-coA reductase enzyme (PhaB). Finally, 3-hydroxybutyrylcoA is polymerized to form polyhydroxybutyrate by the esterification process that takes place in the presence of PHA synthase enzyme (PhaC) [16]. This polyhydroxybutyrate is reserved then as inclusion material in bacterial cells (Fig. 6.3).

6.3 Lignocellulosic biomass

Biomass and biomass derived from plant materials have been identified as the single most alternative to oil-based raw materials for polymer production. These materials are produced from CO_2 in the atmospheric air, water from the soil with the help of chlorophyll and sunlight through the process, photosynthesis. Accordingly, biomass has been considered to be the main practical source of natural carbon in the Earth and the ideal proportionate to oil for the generation of energies and fine chemicals with net zero carbon emission [3,4]. In this context, lignocellulosic biomass, which is the most abundant resource on the Earth, is of high volume. Many reports have demonstrated that lignocellulosic biomass holds massive potential for sustainable and large-scale production of chemicals and fuels. Also, it is an inexhaustible feedstock available throughout the world [17,18]. The use of lignocellulosic can further diminish CO_2 emission and environmental contamination. In this way, it is a promising alternative to conventional petroleum feedstock, for the production of biofuels, biomolecules and biomaterials. Lignocellulosic feedstocks are more interesting compared to other biomass varieties as these are a nonconsumable segment of a plant and, consequently, it doesn't compete with the food supplies [19]. Also, forestry, farming and agromodern lignocellulosic wastes are gathered every year in huge amounts; their disposal causes serious environmental

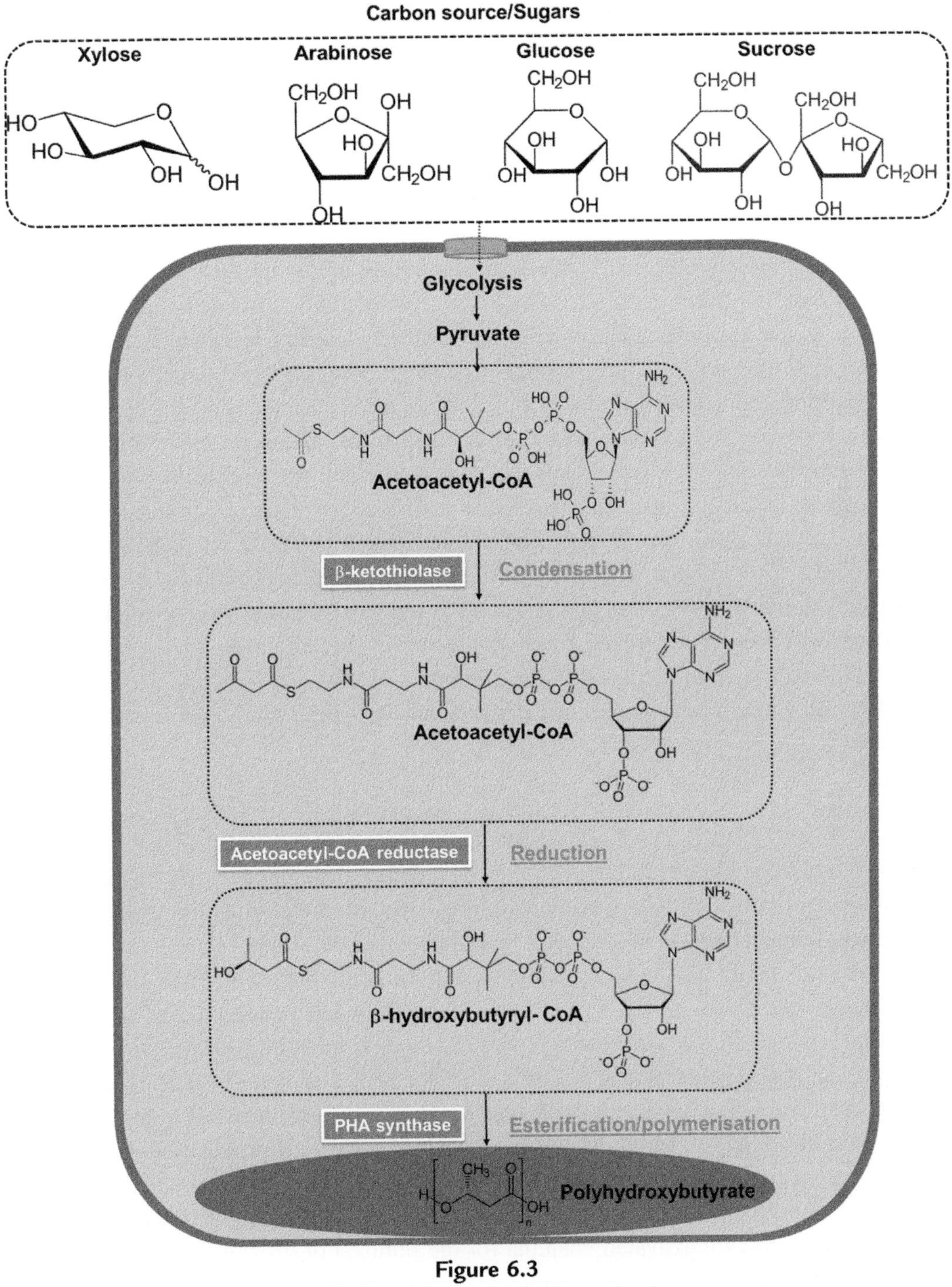

Figure 6.3
Metabolic pathway for PHB production.

concerns. Therefore, it could be beneficial to use such lignocellulosic biomass for the production of various value-added products [20]. From a financial perspective, lignocellulosic biomass can be produced rapidly and more cheaply than the other agronomically critical biofuel feedstocks, for example, corn starch, soybean, and sugarcane. Also, these feedstocks are less expensive than crude petroleum.

However, the advancement of lignocellulosic biomass conversion to fine chemicals and polymers remains a great challenge. Table 6.2 presents literature on biomass production and PHB accumulation by different microorganisms using industrial waste as feedstock, which reveals that the recent research is focused on value addition of such waste resources using suitable microorganisms. During the process of deriving key feedstock components from the raw agricultural plants, a vast number of residues was refused from these industries. Even though some of these refused sources were used as a cattle feed and in some cases with fewer processing, they were returned to the land as organic fertilizers. Much of these refused sources were disposed of as municipal, agroindustrial, and grassland wastes. Therefore, utilizing such waste feedstock for PHB production stands to be unique and an attractive topic of recent research. Since lignocellulosic material is a complex mixture of three polymers, viz. lignin, cellulose, and hemicellulose, it needs to be fractionated. The inhibitory portion (lignin) and the sugars (cellulose and hemicellulose) require for growth of microorganism. After fractionation, the cellulose and hemicellulose-containing portion are generally hydrolyzed using acid or enzyme to extract the sugars, which can be used as a ready source for PHB production in the presence of other mineral salts (Fig. 6.4).

6.3.1 Bagasse

Bagasse is a refused source from the sugar industry. Wherein sugarcanes are mainly processed for the production of sugar juice and the latter for the production of fine sugars and molasses. Though bagasse is used as a fuel for boilers, but many occasions, it is found to be less noticed and refused from the industry due to its lower fuel efficiency. Bagasse is fibrous in nature and many studies have reported the utilization of bagasse hydrolysate as a promising feedstock for PHB production. For instance, Silva et al. [32] used *Burkholderia* sp., bacterial strain for the production of PHB from the bagasse hydrolysate. In the particular study, the bagasse hydrolysate was processed by a three-fold pretreatment technique to enhance its fermentability and consecutively for PHB production. As a first step, the hydrolysate was concentrated to increase the sugar concentration and, thereafter, it is treated with lime to bring the hydrolysate to neutral pH; as a final step, the hydrolysate is treated with activated charcoal for the removal of growth inhibitors. Using the processed hydrolysate, among the different organisms *Burkholderia* sp. yielded a maximum biomass concentration of 4.4 g/L and PHB concentration of 2.7 g/L [32].

Table 6.2: PHB production by different microorganism using industrial refuse as cheap feedstock.

Biomass	Producer strain	Industrial source	Biomass (g/L)	PHB (g/L)	Yield	References
Bagasse hydrolysate	*Ralstonia eutropha*	Sugar mill	11.1	6.3	NR	[21]
Xylose	*Burkholderia sacchari*	Grassland	5.5	3.2	0.26	[22]
Coir pith	*Azotobacter beijerinickii*	Coir industry	5.0	2.4	NR	[23]
Xylose	*Burkholderia sacchari*	Grassland	5.3	2.7	0.17	[22]
Wood hydrolysate	*Burkholderia cepacia*	Sawmill	16.9	8.7	0.19	[24]
Hyacinth hydrolysate	*Ralstonia eutropha*	Municipal refuse	12	7.0	0.13	[25]
Grass biomass	*Pseudomonas* sp.	Grassland	0.9	0.3	NR	[26]
Rice straw hydrolysate	*Bacillus firmus*	Agroindustries	1.9	1.7	NR	[27]
Softwood hydrolysate	*Sphingobium scionense*	Sawmill	1.23	0.4	0.22	[28]
Spent coffee hydrolysate	*Burkholderia cepacia*	Instant coffee manufacturing	5.5	3.1	0.24	[29]
Sunflower husk	*Ralstonia eutropha*	Agroindustries	13.13	8.82	NR	[18]
Soybean straw			12.12	7.54		
Wood straw			11.42	6.79		
Rice paddy straw			15.50	10.87		
Fructose	*Cupriavidus necator*	Beverage industry	9.58	7.3	0.39	[30]
Date seed	*Cupriavidus necator*	Food industry	6.3	4.6	0.57	[31]
Parthenium						
Pentose	*Ralstonia eutropha*	Municipal	2.93	0.24	0.007	[17]
Hexose			3.35	0.60	0.017	
Hyacinth						
Pentose			3.70	0.30	0.007	
Hexose			4.44	0.96	0.36	

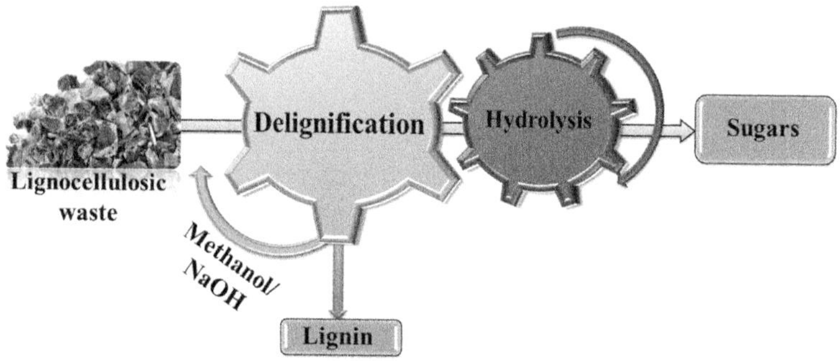

Figure 6.4
Fractionation and hydrolysis of lignocellulosic waste for the extraction of sugars.

In a study conducted by Yu and Stahl [21], *Ralstonia eutropha* grown on dilute acid pretreated bagasse yielded a biomass and PHB concentration of 11.1 g/L and 6.3 g/L, respectively. In another study carried out by Gowda and Shivakumar [33] growth of *Bacillus thuringiensis* on sugarcane bagasse hydrolysate resulted in the biomass and PHB concentration of 10.6 g/L and 4.2 g/L, respectively. In a prudent study carried out by Munoz and Riley [34], the authors used cellulosic fiber obtained from a tequila manufacturing industry and cultured *Saccharophagus degradans* without hydrolyzing the fibers but in the presence of minimal salt in the medium. The organism was able to degrade cellulose and directly utilize the sugars thereof for PHA production. This study is unique and interesting due to the fact that the upstream strategy involved no hydrolysis step for the fibers, which reduces the overall drastically. However, PHA yield obtained in the study could not be ascertained, and no progress seems to be made further following this strategy.

6.3.2 Spent coffee bean grounds

Next to bagasse, spent coffee bean grounds were observed to be the next promising feedstock for PHB production. For the past two century, coffee beans have gained commercial interest as it is one of the most widely used beverages around the world. As of the statistics in 2010, it was reported that there were more than 8 million tonnes of coffee beans produced worldwide annually. In the process of coffee drink preparation, the solid residues are rejected from instant coffee manufacturers, and these are commonly regarded as spent coffee bean grounds (SCBG). This SCBG comprises around 15% of oil, which can be extracted and used as a carbon source for PHB producing organisms, more specifically for *Cupriavidus necator* [29,35]. Even after the extracting oil from SCBG, the residue comprises primarily hemicelluloses and cellulose, which can be hydrolyzed further to convert to sugars. The sugars can be used for the production of PHB by using *Burkholderia cepacia.* It was reported that hexoses, i.e., galactose and mannose were the major sugars in the hydrolysate and these hexoses were found to be more favorable than the pentose sugars for PHB production. Furthermore, levulinic acid produced due to partial degradation of hexose during the acid pretreatment step gave rise to PHA production as it is considered a good precursor for the production of the copolymer poly-3-hydroxy-butyrate-valerate [29].

6.3.3 Coir pith

From the coir industry, a majority of coconut fibers are refused as coir pith, and, in many cases, due to their recalcitrant nature and high lignin content, it is considered a waste material. It is observed that on an average, the coir pith takes a decade to degrade, which gives rise to serious environmental threat and creates a further problem in solid waste management. Prabu and Murugesan [23] used this solid waste for PHB production by using *Azotobacter beijerinickii* by a multistep pretreatment procedure, wherein the coir pith was first delignified, and the cellulose fibers in it were treated with cellulase enzyme for the extraction of reducing sugars. Finally, these sugars were fed to *A. beijerinickii* that resulted in a maximum PHB titer value of 2.4 g/L [23].

6.3.4 Rice straw

Rice straw obtained from the vegetative part of paddy during its harvest is well-known as a cattle feed. However, in most cases, a large quantity of it is burnt or ploughed, which leads to air/land pollution. Sindhu et al. [27] utilized rice straw hydrolysate for PHB production by *Bacillus firmus*. The rice straw hydrolysate in the study was obtained by acid pretreatment, and the hydrolysate was reported to certain sugars and other sugar products such as acetic acid, furfuraldehyde, formic acid, and hydroxymethylfurfural (HMF). The authors cultured the organisms without removing the inhibitors, which lead to low biomass and PHB titer value of 1.9 g/L and 1.7 g/L, respectively. However, very high PHB content of about 89% was observed. This study suggests, that though the inhibitors reduced the biomass growth and PHB production, the PHB content was induced in the presence of these inhibitors [27].

6.3.5 Empty oil palm fruit bunches

In Southeast Asia, significantly high production of oil palm of about 15 million tonnes and even more is noticed annually. Oil palm is considered an industrial crop and is used as the feedstock for palm oil industries; Following the extraction of palm oil; the empty oil palm fruit bunches are refused from the industries. These oil palm empty fruit bunch were reported to comprise of cellulose, hemicellulose, lignin and ash content of 50.4%, 21.9% 10%, and 17.7%, respectively. Zhang et al. (2013) [36] in their study utilized this empty fruit bunches and hydrolyzed it first with chemical pretreatment, and after that, the authors used a cocktail of cellulase enzyme for the synthesis of reducing sugars. The reducing sugars in the hydrolysate in the presence of tryptone were identified to be a suitable source for the growth of *Bacillus megaterium* and PHB production. A maximum PHB content of 51.6%, PHB concentration of 12.48 g/L and a PHB productivity of about 0.260 g/L h was achieved [36].

6.3.6 Wheat straw

During wheat processing, wheat straw and wheat bran are refused as residues, which can serve as a potential feedstock for PHB production. In the year 2012–13, about 660 million tonnes of wheat were produced worldwide and of which about 15%–20% was wheat straw. The primary producers of wheat are Asia, Europe and North America, with a global share of about 43%, 32% and 15%, respectively [37]. In this regard, Van-Thuoc et al. [38] used wheat bran and hydrolyzed it by enzymatic hydrolysis and the sugars produced were used for PHB production using *Halomonas boliviensis*. However, to increase the PHB yield, the carbon content in the medium was increased by the addition of sodium acetate and butyric acid, which was obtained from anaerobic digestate of solid potato waste. The authors reported a maximum biomass and PHB concentration of 8.0 and 4.0 g/L, respectively [38]. Other than wheat bran, wheat straw has also been used for PHB production using *Burkholderia sacchari*. In a study by Cesário et al. [39], the authors hydrolyzed the wheat straw by ammonia fiber expansion (AFEX) process, and the cellulose and hemicellulose comprising mixture from the aforementioned process were further hydrolyzed using an enzymatic process. The final mixture comprises of hexose and pentose sugars, such as glucose, arabinose and xylose. In that study, the authors adopted a fed-batch strategy for PHB production, and a maximum PHB content of 72% was observed along with a PHB to carbon source yield of 0.22 g/g and PHB productivity of 1.6 g/L h.

6.3.7 Grassland refuse

Globally, nearly 69% of agricultural land accounts for about 3.4 billion hectares in the entire world is covered by grasslands. Particularly in Europe, out of 164 million hectares of agricultural area, 76 million hectares are permanently found to be the grassland. Some of the major advantages of grasslands are: unlike agricultural land, these grasslands do not require any fertilizers; secondly, there is no need of annual ploughing and, finally, instead of greenhouse gas emissions, they act as the sink to reduce the carbon pollution in the atmosphere [40,41]. In a study reported by Davis et al. [26], the authors used grass biomass as a feedstock for PHB production using *Pseudomonas* strains. The authors followed NaOH and hot water treatment for delignification, and, thereafter, the delignified biomass was treated with enzymes for sugar synthesis. The sugars were subsequently used for the growth of *Pseudomonas* strains, which yielded a maximum *mcl*-PHA content of around 20%–34% [26]. This study on grass biomass as the sole source of carbon and energy for PHB production was further developed by Koller et al. (2013) [42]; the authors used the juice extracted from the green grass and supplemented it with complex nitrogen and phosphate source for the enhancement of PHB production by *Ralstonia eutropha* [42]. In another study by Radhika et al. [25] reported on saccharification of water hyacinth; the authors reported that enzymatic hydrolysis is more preferred to acid hydrolysis for growth and PHB accumulation using *Ralstonia eutropha*. The authors optimized the process by using response surface methodology and found that with an initial sugar concentration in the hydrolysate of about 35 g/L and at the end of 72 h of fermentation, a maximum biomass and PHB concentration was estimated to be 12 g/L and 7 g/L, respectively were obtained [25].

6.3.8 Waste date seeds and citrus biomass

Annually, 7 million tonnes of dates are produced across 30 countries, and each fruit comprises of 10–15 wt% of seed in it; more than 1 million dates seeds are refused annually from the dates processing industries. Date seeds that are obtained after harvesting the date fruit are considered to be a waste; however, it has a high nutrient content that can be harnessed for PHB production. The date seed contains 50%–70% carbohydrates, 20%–40% proteins, and 10%–12% oil. Therefore, the carbohydrate can be readily hydrolyzed into sugars and protein, which can be used as a substitute for the growth of microorganisms. Yousuf and Winterburn [31] used date extract for PHB production using *Cupriavidus necator*, and reported a maximum biomass and PHB concentration of 6.3 g/L and 4.6 g/L, respectively, in the presence of an initial sugar concentration of 10.8 g/L. Furthermore, the citrus processing industry during the production of orange juice refuses orange peel as one of the main waste material. Every year it was perceived that 50 million tonnes of orange fruits are produced and out of which 3 million tonnes were considered to be the edible portion and 1.5 million tonnes were refused as waste from these industries. Lagunes and Winterburn [30] reported that the extract from the skin, seed and pulp of the orange resulted in a biomass and PHB concentration of 9.58 g/L and 7.8 g/L, respectively. In another study by Saratale et al. [43], utilization of kenaf fiber (a nonedible crop and lignocellulosic biomass) was reported for PHB production using *Ralstonia* eutropha. A maximum biomass and PHB concentration of 18.3 g/L and 13.12 g/L, respectively were reported.

Though lignocellulosic biomass offers a promising means for PHB production at reduced production costs, further studies are needed for the scale-up of the PHB production process from shake flask to bioreactors. Hence, appropriate bioreactor systems should be chosen with respect to the nature of feedstock used for PHB production. In connection with this the following section describes the different types of bioreactors used for PHB production.

6.4 Reactor considerations for upstream processing of PHB

Over the past decades, a variety of bioreactors have been studied for PHB production, and among which stirred tank bioreactor, airlift bioreactor, bubble column bioreactor and two-phase partitioning bioreactor stand out to be very common. Furthermore, these bioreactors were operated in various modes, viz. batch, continuous, continuous with cell recycle and fed-batch. All the aforementioned bioreactors used for PHB production are described briefly as follows:

6.4.1 Stirred tank bioreactor

Stirred tank bioreactors (STBRs) are most often used bioreactor for any biological processes and more specifically for aerobic fermentation. This bioreactor system comprises of a vessel (often made from glass) to hold the culture medium; air is sparged at the bottom of the vessel, and it is dispersed uniformly by an impeller connected to a rotating shaft (Fig. 6.5A). Baffles of varying number from four to six numbers are used to ensure sufficient mixing in the reactor. Increasing the impeller speed improves the oxygen transfer to the medium; however, beyond a certain speed, shear stress is imposed on the microorganism due to the sharp edges in the impeller and in the baffles. On increase in oxygen transfer also increases the biomass production and therefore PHB accumulation inside. Many studies have been carried out on PHB production using stirred tank bioreactors. Khanna et al. [44] employed stirred tank bioreactor for the cultivation of *Ralstonia eutropha* with an initial fructose concentration of 40 g/L. The authors observed a maximum biomass and PHB concentration of 20.73 g/L and 9.35 g/L, respectively. Furthermore, PHB to fructose yield of about 0.24 g/g was achieved [44].

6.4.2 Airlift reactor

Unlike stirred tank bioreactor, airlift reactors (ALRs) are unique in its design as there are no rotating parts inside, and air circulation inside serves to bring about the necessary mixing. Because of this feature, they are often classified as pneumatic reactors. In its basic design, an ALR comprises a draft tube in a long vertical cylinder with air sparger located at the bottom of the cylinder (Fig. 6.5B). However, ALR can be classified into two types: inner loop airlift reactor (ILALR) and external loop airlift reactors (ELALR). In the ILALRs, a partitioning wall is placed amid the vertical cylinder, thereby separating the

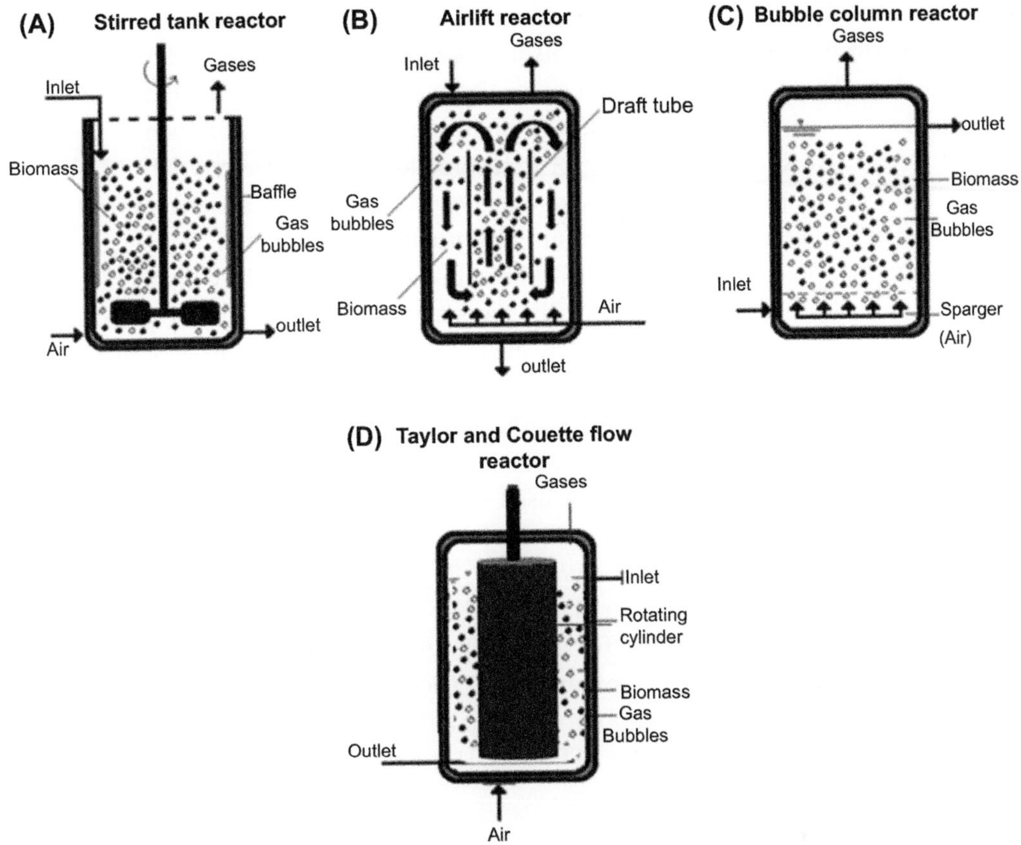

Figure 6.5

Different bioreactors: (A) Stirred tank reactor, (B) Airlift reactor, (C) Bubble column reactor used for PHB production along with the proposed, (D) Taylor and Couette flow reactor. *Taken with permission from* Espinosa-Ortiz EJ, et al. Fungal pelleted reactors in wastewater treatment: applications and perspectives. Chemical Engineering Journal 2016; 283: 553–571.

fluid medium into two regions, whereas one region is called as the upcomer and other one is called as the downcomer. These two regions give enough chance for the medium to circulate, thereby creating a well-defined fluid flow pattern. In the case of ELALR, a downcomer is placed as a separate arm to the airlift reactor and thereby, the fluid circulation is established in the reactor. Due to its simplicity in design and the absence of energy-intensive parts (e.g., impellers and shaft), both capital cost and operating cost are found to be relatively low. The modified flow of fluid introduced in the ALR ensures enhancement in oxygen and nutrient supply from liquid medium to solid biomass. Furthermore, better heat transfer, quick mixing time and cell retention are the added advantage of using ALRs. Compared to stirred tank bioreactors, ALRs are observed to maintain a high level of sterility inside the reactor.

Many studies compare the performance of stirred tank bioreactor (STBRs) with that of ALRs for the growth of various microorganism production of different metabolites and even for the degradation of different pollutants. These studies reported that though ALR operation is both simple and economical, it is slightly poor in terms of efficiency. In line with these studies, ALRs were also used for PHB production by numerous authors: for instance, Tavares et al. [45] cultivated *Ralstonia eutropha* in ALRs with a superficial gas velocity of 10 m/s and observed a maximum PHB productivity of 0.6 g/L h along with a PHB cell content of 50% [45]. Nevertheless, the authors reported slightly higher productivity of 0.82 g/L h with STBR. Finally, the authors concluded that the performance of ALR is better than STBR owing to its low energy demand. In another study by Da Cruz Pradella et al. [46], the authors cultivated *Burkholderia sacchari* in an airlift reactor and operated the reactor under high cell density cultivation mode. A maximum PHB productivity of 1.7 g/L h with a PHB to sucrose yield of 0.2 g/g was observed [46]. In another study by Du et al. [47], the authors used short-chain fatty acids obtained from an anaerobic digester as a substrate for the growth of *Ralstonia eutropha* in ALR. A maximum biomass concentration of 22.7 g/L with 72.6% PHB content present in the biomass was reported [47]. Gahlawat et al. cultured *Azohydromonas australica* using a modified airlift reactor, with provision for insitu cell retention. Performance of this novel airlift reactor was compared with that of an STBR. The results demonstrated that the maximum biomass and PHB concentration of novel airlift reactor (10.76 g/L; 7.81 g/L) is superior in comparison with that of the STBRs (8.31 g/L; 5.45 g/L) [48].

6.4.3 Bubble column reactor

Bubble column bioreactors (BLBRs) can be classified under the category of multi-phase bioreactors. BLBRs consists of a long vertical cylinder similar to that of ALRs, however without any draft tube or partition contained in it (Fig. 6.5C). These reactors are also not having any moving parts similar to ALR but comprise of air sparger, which is often called an air distributor. The sparger ensures uniform mixing in the rector by proper distribution of gas and liquid phase, thereby leading to excellent mass and heat transfer in the reactor system. The reactor design is simple, and its capital and operation costs are considerably low. BLBRs in PHB production has certain advantage such as they impart less shear stress on the PHB producing microorganisms, easy to maintain the sterility as there are no moving parts or sealing present in the reactor. Last few years, many studies have been conducted using BLBRs for PHB production. For instance, Rahnama et al. [49] utilized natural gas for PHB production using a bubble column bioreactor with *Methylocystis hirsute*. The authors observed a very low PHB concentration of 1.4 g/L; which was attributed to the source (natural gas) used that was not a favorable substrate for the microorganism [49]. Furthermore, in a study by García-Pérez et al. [50], the authors used the same strain and substrate as mentioned before; however, a slight modification in the bubble column bioreactor was brought by adopting an internal gas recycling strategy. Through this recycling strategy, the authors observed an enhanced delivery of methane from the gas phase to the liquid phase, which resulted in a maximum PHB productivity of 1.4 ± 0.4 kg/m^3 d and a PHB content of $34.6 \pm 2.5\%$ [50].

6.4.4 Two-phase partitioning bioreactor

In two-phase partitioning bioreactors (TPPBs), a nonaqueous phase is added along with the aqueous media in stirred tank bioreactor, which supports biological processes. The nonaqueous phase serves to overcome the limitation of the substrate or product toxicity experienced by the microorganisms [51]. In some other cases, when the substrate is insoluble in water, the nonaqueous phase is supplemented to increase the substrate bioavailability. Many organic solvents, polymer and co-polymers have been used as a nonaqueous phase. Initially, TPPBs was designed to enhance the oxygen transfer in the culture medium with a final goal of achieving a maximum product yield. These reactors were used for producing citric acid, and ethanol and for polyaromatic hydrocarbons (PAHs) degradation. Recently, TPPB has been employed for the cultivation of PHB producing *Methylobacterium organophilum* using methane as a substrate. PHB accumulation in the range of 34%–38% (w/w) was observed. However, this value was improved even up to 57% (w/w) in the limited nitrogen condition [51].

Thus, different types of bioreactor systems have been utilized for the production of PHB accumulating biomass: the stirred tank reactor being the most widely recognized one, despite the fact that bubble column and airlift reactors have likewise been investigated for PHB production. However, very recently, the Taylor and Couette bioreactor (TCBR) has different mechanical frameworks that permit efficient oxygen exchange without including any shear stress to the microorganisms (Fig. 6.5D). The TCBR has been used in a wide array of applications, such as water purification, emulsion polymerization, liquid-liquid extraction, pigment preparation, photocatalysis, a culture of animal cells, and cultivation of microalgae [52,53]. However, this reactor system has not been utilized for PHB production and hence using these bioreactors would result in the reduction of the net cost of PHB production.

6.5 Downstream processing for PHB recovery

Following biosynthesis of PHB, the separation of microorganisms from culture broth is the first step in the downstream operation, which is normally achieved by centrifugation, sedimentation and crossflow microfiltration [54,55]. Recently, Manikandan et al. [56] studied the application of low-cost ceramic membranes for the

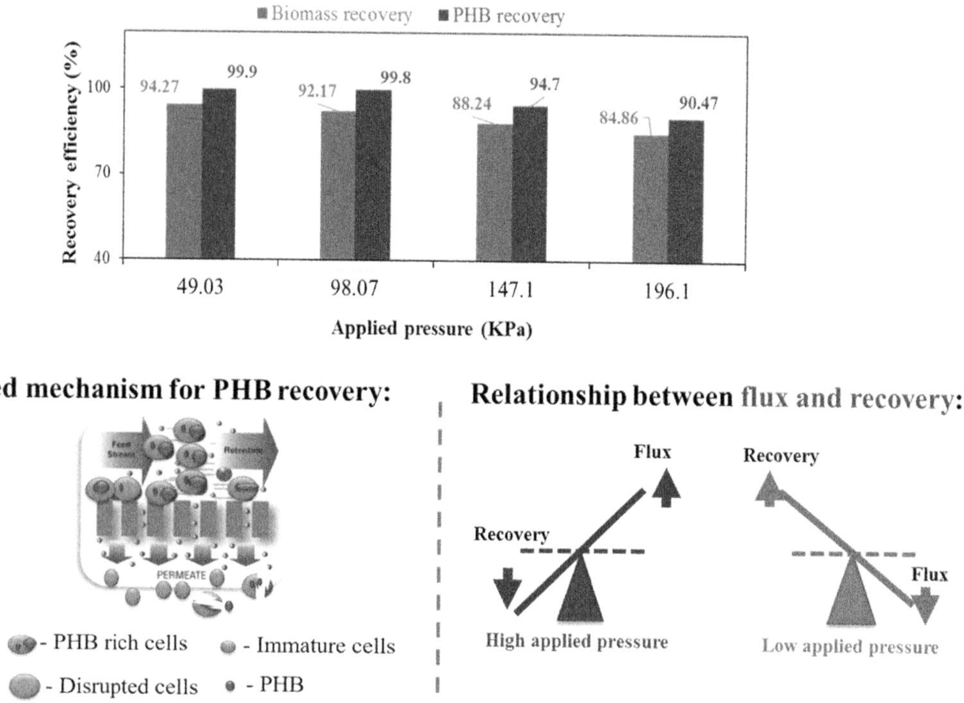

Figure 6.6

Microfiltration of bacterial biomass for PHB extraction. *Taken with permission from Manikandan NA, Pakshirajan K, Pugazhenthi G. A novel ceramic membrane assembly for the separation of polyhydroxybutyrate (PHB) rich Ralstonia eutropha biomass from culture broth. Process Safety and Environmental Protection 2019;126:106–18.*

separation of PHB rich *R. eutropha* biomass. It was reported that at the membrane-assisted separation process was highly efficient (<99%) in recovering the PHB rich bacterial biomass from the fermentation broth. The authors have reported that the increase in the applied pressure increased the flux greatly however at the cost of reduced biomass separation, even then the PHB recovery efficiency was found to be greater than 90% (Fig. 6.6).

Following biomass separation from the culture broth, solvent extraction for PHB recovery and purification is highly successful for large scale processes [42]. Modernly, high utilization of unstable solvents and toxic solvents are regular elements in PHB extraction (Table 6.3). The dissolvable blend remaining after PHB precipitation, comprising of ethanol and chloroform, is usually disposed off. Also, reutilization of solvents by distillation is not attractive from the financial perspective. This alienates the character of PHB biosynthesis as "green" innovation. Therefore,

Table 6.3: Comparison of various PHB extraction protocols as mentioned in the literature.

Method	Cost for chemicals	Time	Recovery yields	Scale-up is	Product purity	Impact on molar mass	References
Chloroform method	High	Medium	High	Not possible	High	Low	[57]
Hypochlorite digestion	Medium	Medium	Medium	Not possible	Medium–high	Medium–high	[18]
Cyclic carbonates	Medium	Medium	Medium	Not possible	High	High	[58]
Fusel alcohol	Surplus product	High	High	Possible	Low–medium	Medium–high	[59]
Lactic acid method	High	Medium–high	Low–medium	Possible	Medium	Medium–high	[42]
Enzymatic digestion	High	Low	High	Possible	Low before refining	No	[42]
Mechanical disintegration	Medium	Low	Medium–high	Possible	Medium	No	[17,60]

the use of green solvents such as makes the PHB production process both green as well as economical.

6.6 Strategy for PHB production using lignocellulosic waste

In all the aforementioned study on using lignocellulosic waste as a substrate for PHB production, it is clear that the lignin is either discarded or it is considered as an inhibitor of microorganism growth. Currently, lignin obtained from the delignification process is burnt to recover energy from it. However, this lignin discarded during the upstream processing of substrate preparation can act as a key component in the downstream processing to break the bacterial cell wall and to release PHB granules out of it. Such an approach would make the lignocellulosic waste to be a self-sufficient feedstock for upstream to downstream processing in PHB production. This approach of using lignocellulosic waste in a closed loop biorefinery set up (Fig. 6.7) does increase not only the value of lignin but also reduces the use of harsh chemical solvent. In this way, it makes the PHB production process to be an economical one. Further, the usage of lignin in the downstream operation would result in the formation of PHB/lignin composite, which would result in a polymeric film having exceptional properties such as antimicrobial activity, enhanced barrier property and so on. However, such a claim needs to be proved by carefully planning of bench scale experiments and demonstrating it in a pilot scale model.

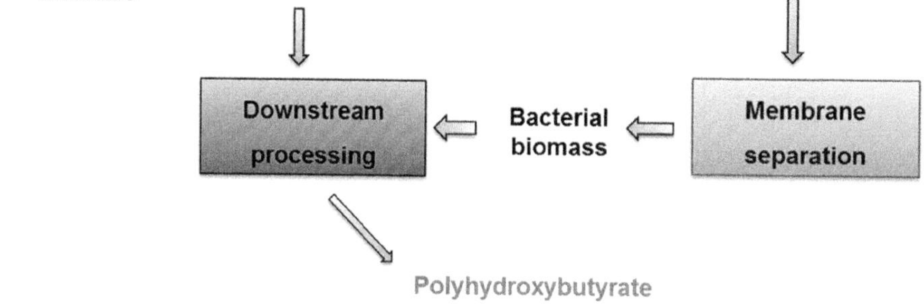

Figure 6.7
Proposed strategy on upstream to downstream processing for PHB production using lignocellulosic waste.

6.7 Conclusions and perspectives

PHB is found to be the best performing biopolymer currently available in the market, and its commercialization would greatly reduce the pressure on the depleting fossil petroleum sources. However, the significant cost associated with the PHB production keeps the PHB production process from actual commercialization. For instance, a high capital cost is involved in setting up of efficient aerobic fermentation facilities as well as the huge cost involved in the purchase of sugar feedstocks and organic solvents utilized for polymer extraction. However, the use of waste lignocellulosic biomass for commercial PHB production offers a great advantage. Furthermore, the use of novel reactors with modified reactor framework would bring down the cost incurred in the PHB production process significantly. Finally, replacement of harsh chemicals with a green solvent, such as lignin, which is a by-product from the waste lignocellulosic biomass, would lead to cheaper PHB production methods. Further studies on PHB production need to be directed not only toward efficient utilization of waste/refused sources as the feedstock, but also integrate into a closed loop biorefinery approach wherein every neglected component in the process would serve as a resource.

References

[1] Liguori R, Faraco V. Biological processes for advancing lignocellulosic waste biorefinery by advocating circular economy. Bioresource Technology 2016;215:13–20.

[2] Cherubini F. The biorefinery concept: using biomass instead of oil for producing energy and chemicals. Energy Conversion and Management 2010;51(7):1412–21.

[3] Kajaste R. Chemicals from biomass – managing greenhouse gas emissions in biorefinery production chains – a review. Journal of Cleaner Production 2014;75:1–10.

[4] Naranjo JM, Cardona CA, Higuita JC. Use of residual banana for polyhydroxybutyrate (PHB) production: case of study in an integrated biorefinery. Waste Management 2014;34(12):2634–40.

[5] Snell KD, Peoples OP. PHA bioplastic: a value-added coproduct for biomass biorefineries. Biofuels Bioproducts & Biorefining 2009;3(4):456–67.

[6] Arrieta MP, et al. Bionanocomposite films based on plasticized PLA-PHB/cellulose nanocrystal blends. Carbohydrate Polymers 2015;121:265–75.

[7] Valentino F, et al. Polyhydroxyalkanoate (PHA) storage within a mixed-culture biomass with simultaneous growth as a function of accumulation substrate nitrogen and phosphorus levels. Water Research 2015;77:49–63.

[8] Patnaik PR. Dispersion optimization to enhance PHB production in fed-batch cultures of Ralstonia eutropha. Bioresource Technology 2006;97(16):1994–2001.

[9] Rodriguez-Contreras A, et al. Modification of titanium surfaces by adding antibiotic-loaded PHB spheres and PEG for biomedical applications. Journal of Materials Science: Materials in Medicine 2016;27(8).

[10] Lizarraga-Valderrama L, et al. Biomedical applications of polyhydroxyalkanoates. In: Biomaterials from nature for advanced devices and therapies; 2016. p. 339–83.

[11] Borowitzka MA. High-value products from microalgae-their development and commercialisation. Journal of Applied Phycology 2013;25(3):743–56.

[12] Bugnicourt E, et al. Polyhydroxyalkanoate (PHA): review of synthesis, characteristics, processing and potential applications in packaging. Express Polymer Letters 2014;8(11):791–808.

[13] Savenkova L, et al. Mechanical properties and biodegradation characteristics of PHB-based films. Process Biochemistry 2000;35(6):573–9.
[14] Arrieta MP, et al. Plasticized poly(lactic acid)-poly(hydroxybutyrate) (PLA-PHB) blends incorporated with catechin intended for active food-packaging applications. Journal of Agricultural and Food Chemistry 2014;62(41):10170–80.
[15] Costa MME, et al. Use of polyhydroxybutyrate and ethyl cellulose for coating of urea granules. Journal of Agricultural and Food Chemistry 2013;61(42):9984–91.
[16] Anjum A, et al. Microbial production of polyhydroxyalkanoates (PHAs) and its copolymers: a review of recent advancements. International Journal of Biological Macromolecules 2016;89:161–74.
[17] Pradhan S, et al. Microbial production, ultrasound-assisted extraction and characterization of biopolymer polyhydroxybutyrate (PHB) from terrestrial (P. hysterophorus) and aquatic (E. crassipes) invasive weeds. Bioresource Technology 2017;242:304–10.
[18] Saratale GD, Oh MK. Characterization of poly-3-hydroxybutyrate (PHB) produced from Ralstonia eutropha using an alkali-pretreated biomass feedstock. International Journal of Biological Macromolecules 2015;80:627–35.
[19] Manikandan NA, et al. Biorefinery and possible deforestation. In: Platform chemical biorefinery. Elsevier; 2016. p. 307–22.
[20] Brar SK, Sarma SJ, Pakshirajan K. Platform chemical biorefinery: future green chemistry. Elsevier; 2016.
[21] Yu J, Stahl H. Microbial utilization and biopolyester synthesis of bagasse hydrolysates. Bioresource Technology 2008;99(17):8042–8.
[22] Lopes MSG, et al. PHB biosynthesis in catabolite repression mutant of Burkholderia sacchari. Current Microbiology 2011;63(4):319–26.
[23] Sathesh Prabu C, Murugesan A. Effective utilization and management of coir industrial waste for the production of poly-Î^{2}-hydroxybutyrate (PHB) using the bacterium Azotobacter beijerinickii. International Journal of Environmental Research 2010;4(3):519–24.
[24] Pan WY, et al. Production of polyhydroxyalkanoates by Burkholderia cepacia ATCC 17759 using a detoxified sugar maple hemicellulosic hydrolysate. Journal of Industrial Microbiology & Biotechnology 2012;39(3):459–69.
[25] Radhika D, Murugesan AG. Bioproduction, statistical optimization and characterization of microbial plastic (poly 3-hydroxy butyrate) employing various hydrolysates of water hyacinth (Eichhornia crassipes) as sole carbon source. Bioresource Technology 2012;121:83–92.
[26] Davis R, et al. Conversion of grass biomass into fermentable sugars and its utilization for medium chain length polyhydroxyalkanoate (mcl-PHA) production by Pseudomonas strains. Bioresource Technology 2013;150:202–9.
[27] Sindhu R, et al. Pentose-rich hydrolysate from acid pretreated rice straw as a carbon source for the production of poly-3-hydroxybutyrate. Biochemical Engineering Journal 2013;78:67–72.
[28] Bowers T, et al. Softwood hydrolysate as a carbon source for polyhydroxyalkanoate production. Journal of Chemical Technology and Biotechnology 2014;89(7):1030–7.
[29] Obruca S, et al. Production of polyhydroxyalkanoates using hydrolysate of spent coffee grounds. Process Biochemistry 2014;49(9):1409–14.
[30] Lagunes FG, Winterburn JB. Effect of limonene on the heterotrophic growth and polyhydroxybutyrate production by Cupriavidus necator H16. Bioresource Technology 2016;221:336–43.
[31] Yousuf RG, Winterburn JB. Date seed characterisation, substrate extraction and process modelling for the production of polyhydroxybutyrate by Cupriavidus necator. Bioresource Technology 2016;222:242–51.
[32] Silva L, et al. Poly-3-hydroxybutyrate (P3HB) production by bacteria from xylose, glucose and sugarcane bagasse hydrolysate. Journal of Industrial Microbiology & Biotechnology 2004;31(6):245–54.
[33] Gowda V, Shivakumar S. Agrowaste-based polyhydroxyalkanoate (PHA) production using hydrolytic potential of Bacillus thuringiensis IAM 12077. Brazilian Archives of Biology and Technology 2014;57(1):55–61.

[34] Alva Munoz LE, Riley MR. Utilization of cellulosic waste from tequila bagasse and production of polyhydroxyalkanoate (PHA) bioplastics by Saccharophagus degradans. Biotechnology and Bioengineering 2008;100(5):882–8.
[35] Cruz MV, et al. Production of polyhydroxyalkanoates from spent coffee grounds oil obtained by supercritical fluid extraction technology. Bioresource Technology 2014;157:360–3.
[36] Zhang Y, et al. Polyhydroxybutyrate production from oil palm empty fruit bunch using Bacillus megaterium R11. Bioresource Technology 2013;147:307–14.
[37] Kim S, Dale BE. Global potential bioethanol production from wasted crops and crop residues. Biomass and Bioenergy 2004;26(4):361–75.
[38] Van-Thuoc D, et al. Utilization of agricultural residues for poly (3-hydroxybutyrate) production by Halomonas boliviensis LC1. Journal of Applied Microbiology 2008;104(2):420–8.
[39] Cesário MT, et al. Enhanced bioproduction of poly-3-hydroxybutyrate from wheat straw lignocellulosic hydrolysates. New Biotechnology 2014;31(1):104–13.
[40] Fischer G, et al. Biofuel production potentials in Europe: sustainable use of cultivated land and pastures, part II: land use scenarios. Biomass and Bioenergy 2010;34(2):173–87.
[41] Fischer G, et al. Biofuel production potentials in Europe: sustainable use of cultivated land and pastures. Part I: land productivity potentials. Biomass and Bioenergy 2010;34(2):159–72.
[42] Koller M, Niebelschutz H, Braunegg G. Strategies for recovery and purification of poly[(R)-3-hydroxyalkanoates] (PHA) biopolyesters from surrounding biomass. Engineering in Life Science 2013;13(6):549–62.
[43] Saratale RG, et al. Pretreatment of kenaf (Hibiscus cannabinus L.) biomass feedstock for polyhydroxybutyrate (PHB) production and characterization. Bioresource Technology 2019;282:75–80.
[44] Khanna S, Srivastava AK. Statistical media optimization studies for growth and PHB production by Ralstonia eutropha. Process Biochemistry 2005;40(6):2173–82.
[45] Tavares LZ, da Silva ES, Pradella JGD. Production of poly (3-hydroxybutyrate) in an airlift bioreactor by Ralstonia eutropha. Biochemical Engineering Journal 2004;18(1):21–31.
[46] da Cruz Pradella JG, Taciro MK, Mateus AYP. High-cell-density poly (3-hydroxybutyrate) production from sucrose using Burkholderia sacchari culture in airlift bioreactor. Bioresource Technology 2010;101(21):8355–60.
[47] Du GC, Chen LXL, Yu J. High-efficiency production of bioplastics from biodegradable organic solids. Journal of Polymers and the Environment 2004;12(2):89–94.
[48] Gahlawat G, Sengupta B, Srivastava AK. Enhanced production of poly(3-hydroxybutyrate) in a novel airlift reactor with in situ cell retention using Azohydromonas australica. Journal of Industrial Microbiology & Biotechnology 2012;39(9):1377–84.
[49] Rahnama F, et al. PHB production by Methylocystis hirsuta from natural gas in a bubble column and a vertical loop bioreactor. Biochemical Engineering Journal 2012;65:51–6.
[50] Garcia-Perez T, et al. Simultaneous methane abatement and PHB production by Methylocystis hirsuta in a novel gas-recycling bubble column bioreactor. Chemical Engineering Journal 2018;334:691–7.
[51] Zúñiga C, et al. Production of poly-β-hydroxybutyrate (PHB) by Methylobacterium organophilum isolated from a methanotrophic consortium in a two-phase partition bioreactor. Journal of Hazardous Materials 2011;190(1–3):876–82.
[52] Ramezani M, et al. Experimental investigation of the effect of ethyl alcohol surfactant on oxygen mass transfer and bubble size distribution in an air-water multiphase Taylor-Couette vortex bioreactor. Chemical Engineering Journal 2017;319:288–96.
[53] Ramezani M, et al. Experimental measurement of oxygen mass transfer and bubble size distribution in an air-water multiphase Taylor-Couette vortex bioreactor. Chemical Engineering Journal 2015;279:286–96.
[54] Nath K. Membrane separation processes. PHI Learning Pvt. Ltd; 2017.
[55] Waszak M, Gryta M. The ultrafiltration ceramic membrane used for broth separation in membrane bioreactor. Chemical Engineering Journal 2016;305:129–35.

[56] Manikandan NA, Pakshirajan K, Pugazhenthi G. A novel ceramic membrane assembly for the separation of polyhydroxybutyrate (PHB) rich Ralstonia eutropha biomass from culture broth. Process Safety and Environmental Protection 2019;126:106–18.
[57] Ramsay JA, et al. Extraction of poly-3-hydroxybutyrate using chlorinated solvents. Biotechnology Techniques 1994;8(8):589–94.
[58] Lafferty RM, Heinzle E. Use of cyclic carbonic acid esters as solvents for poly-(β-hydroxybutyric acid). 1979. Google Patents.
[59] Nonato RV, Mantelatto PE, Rossell CEV. Integrated production of biodegradable plastic, sugar and ethanol. Applied Microbiology and Biotechnology 2001;57(1–2):1–5.
[60] Hwang K, You S, Don T. Disruption kinetics of bacterial cells during purification of poly-beta-hydroxyalkanoate using ultrasonication. Journal of the Chinese Institute of Chemical Engineers 2006;37(3):209.

CHAPTER 7

Valorization of organic waste into biofertilizer and its field application

Chenyu Du[1], Sidra Munir[1], Rabia Abad[1], Diannan Lu[2]
[1]School of Applied Sciences, The University of Huddersfield, Huddersfield, United Kingdom;
[2]Department of Chemical Engineering, Tsinghua University, Beijing, China

7.1 Introduction

Organic waste can be defined as waste materials that originated from living organisms, which includes mainly food waste, organic fraction of municipal waste, agriculture residue, forest waste, and organic wastewater. These materials are discarded during the food production, goods formation and human living activities. They have low economic value and have been a growing problem due to lack of suitable treatment method.

Organic waste accounts up to 70% of total waste stream [1]. Recently, reports on the amount of food waste attracted growing attentions worldwide. According to the report published by Food and Agriculture Organization (FAO) of the United Nations, around 1.3 billion tonnes of food was wasted every year, which is roughly 1/3 of the food produced worldwide. The wasted food could have a value worth 1 trillion US dollar [2]. Food is wasted in the food harvesting, preparation and distribution steps, such as discarded due to unpopular size or appearance. The loss of food during food production could account for up 50% of the available food [3]. Agricultural residues, such as wheat straw, rice husk are waste streams derived from crops harvesting. Give rice straw as an example, around 1.14 billion tonnes of rice straw could be generated based on the world rice production of 760 million tonnes [4]. In China around 580 million tonnes of various straw was produced annually [5], but no more than 5% of straw have been used in agriculture and in industry sectors, while majority of it was burnt in the field. Agriculture residue burning in rural area contributes significantly to the air pollution issues in China. In 2016, the China government set up a specific fund in six provinces to provide a stipend of RMB 300 per hectare to encourage alternative approaches of managing straws instead of direct burning of straw. To tackle food waste problem, the United Nations General Assembly set a target of reducing 50% food waste per capita by 2030. The European council published

Waste Biorefinery. https://doi.org/10.1016/B978-0-12-818228-4.00007-1

a directive on May 30, 2018 (DIRECTIVE (EU) 2018/851) to encourage its member states to promote prevention and reduction of food waste in line with the UN target [6].

To deal with the organic waste problems, growing amount of research have been carried out to explore various strategies to convert organic waste into value-added products, such as biofuel, biochemical and biopolymers [7]. Use of citrus peel is an example; the citrus peel can be treated using an acid for the production of D-limonene or as a feedstock for bioethanol fermentations [8]. Cashew nut waste can be subjected to solvent extraction, which generates a variety of phenols and acids, all of which can be used as catalysts or fine chemical production [9]. Liquid or semiviscous food waste such as rendered fat and grease have traditional outputs such as the production of candles, tallow, pet food, and cosmetic precursors, however, alternate handling of this waste toward biodiesel synthesis has been explored [10]. Valorization of organic waste not only reduces environmental pollution, but also has potential for providing significant amount of renewable energy, green chemicals and biodegradable polymers.

Unlike biofuel and biochemical production, there is a lack of investigation on the derivation of biofertilizer from organic waste. Fig. 7.1A shows the numbers of the published papers cited by Scopus related to food waste valorization and food waste derived biofertilizer/biofertilizer. In the past 10 years, there was a significant increase in research on food waste. Although the focus on food waste derived biofertilizer research increased in the past 5 years, it still cannot compare with food waste valorization. Research on agricultural residue valorization/biofertilizer showed a similarly trend (Fig. 7.1B). Actually, fertilizer has a significant market potential, with an estimated value of $150 billion per annum and is expected to grow at an annual rate of 5.9% in the following 5 years [11]. The conversion of organic waste into biofertilizer could not only avoid increasingly severe fees for waste disposal, but also potentially provide additional income for selling the products as a fertilizer/compost/soil conditioner. Environmental and health concern, including soil pollution, water pollution, and toxin accumulation limited fertilizer market. Using biofertilizer to replace synthetic chemical fertilizer could decrease the demand of synthetic fertilizers, alleviate pollution caused by synthetic fertilizers, and reduce the environmental impact of organic waste.

This review summarizes recent research into the conversion of organic waste into biofertilizer, including food waste, organic fraction of municipal solid waste (OFMSW) and agriculture residues. It also covers the characterization of biofertilizer produced from different feedstocks, as well as field trials of organic waste derived biofertilizer. The utilization of organic waste and forest waste for biofertilizer production is not included in this review due to the low availability of published reports.

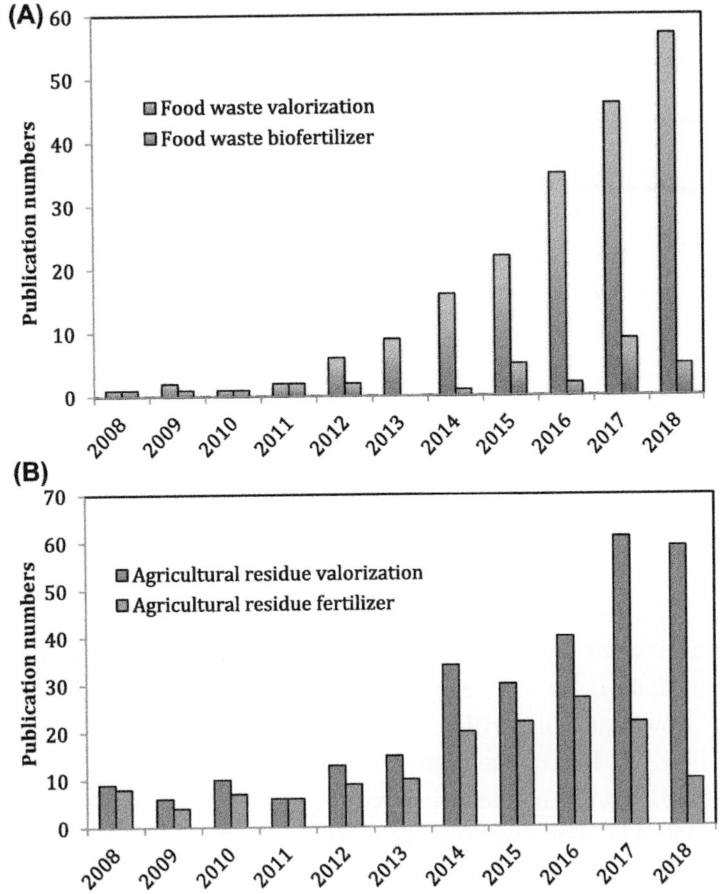

Figure 7.1

Publication trend for (A) food waste valorization and food waste derived biofertilizer; (B) agriculture residue valorization and agriculture residue derived biofertilizer in the past 10 years. *Data was collected in July 2018 in Scopus.*

7.2 Major technologies used for biofertilizer production

In the past decades, various technologies have been developed for the conversion of organic waste streams to biofertilizer either as the main product or as a by-product, as illustrated in Fig. 7.2.

7.2.1 Anaerobic digestion (AD)

Anaerobic digestion (AD) is a typical process for the treatment of organic waste, such as municipal solid waste, sewage wastewater, initially for the elimination of waste streams.

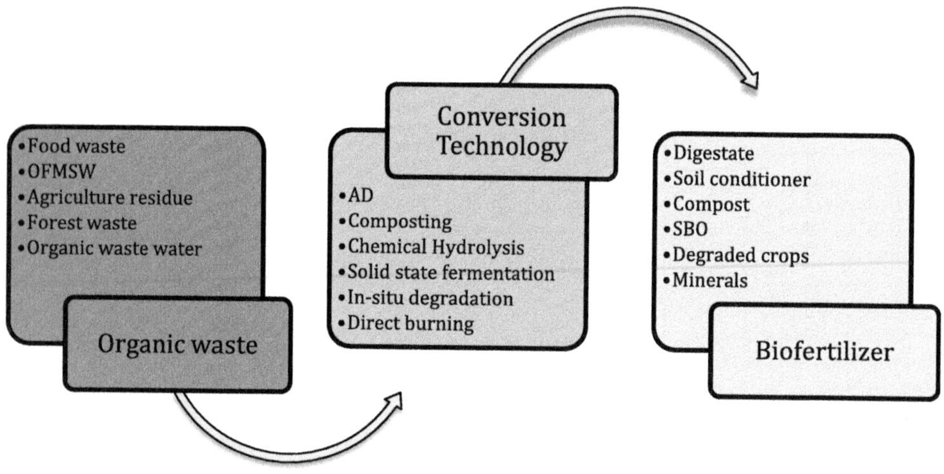

Figure 7.2
A simple diagram of the main technologies used for the conversion of organic waste into biofertilizer.

In the past decades, increasing numbers of AD plants have been built to convert organic waste to biogas (methanol) for bioenergy generation to reduce greenhouse gas emission. A combined heat, power and biofertilizer production via AD process is shown schematically in Fig. 7.3. In EU, the number of installed AD plant increase from 6227 in 2009 to 17,662 in 2016, with the electric capability of 9985 MW [12]. In the UK specifically, over 250 AD plants were in operation by March 2016, with over 40% using food waste as the primary feedstock [13]. In China, the AD plant number exceeded 38 million by 2010 [14].

Within an anaerobic digester, a consortium of microorganisms degrade the organic matter into biogas (a mixture of carbon dioxide and methane) in the absence of oxygen, while the remaining solid part is further process to a nitrogen-rich biofertilizer, digestate. Briefly, organic polymers, such as carbohydrate, protein, and fat/lipid, are hydrolyzed by extracellular enzymes into their respective monomers. The organic monomers are then absorbed by various microorganisms in the acidogenesis process to produce a mixture of volatile fatty acids, alcohols and other simple organic compounds. The higher volatile fatty acids are transformed to acetic acid, CO_2 and hydrogen in the acetogenesis stage, by acetogenic bacteria. Finally, acetoclastic methanogens, and hydrogen utilizing methanogens convert CO_2, H_2, and other compounds into methane and CO_2. The nitrogen compounds in the organic waste stream is concentrated and nitrogen content is enhanced via such processes [15]. The liquid fraction (liquor) following AD can also be processed to a biofertilizer as described by Tampio et al. [16].

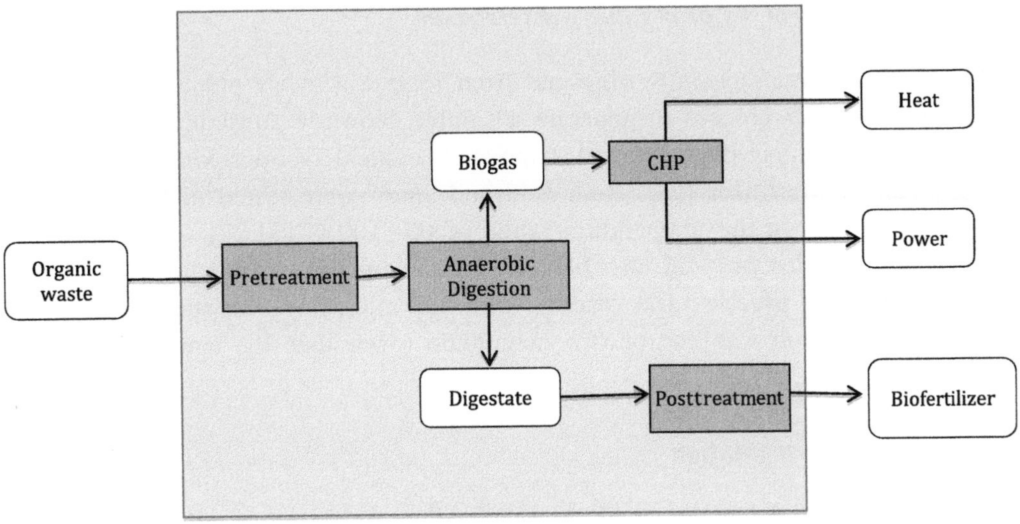

Figure 7.3
A schematic diagram of an AD process that coproduces heat, power, and biofertilizer.

7.2.2 Aerobic composting

Although organic waste is increasingly treated by AD with the aim of energy recovery, composting is still a viable process for the degradation of organic matter in the waste streams into a humus-rich compost. The compost contains high humus content, which can be used a fertilizer in garden and the field. Besides producing biofertilizer, composting treatment of organic waste also controls pathogens, controls germination of weeds, prevents undesirable odorous, Production scale compost is normally carried out in aerobic windrow or aerobic pile, e.g., in a shed, while designated container is commonly used for household composting of food and garden waste [17,18]. In composting of organic waste, natural microorganism community consumes the organic matter, breaking down them into short chain chemicals, such as humic acid. The heat release as a result of microbial activity plays a key role in aerobic composting. The peak temperature of composting process can reach to 75°C. As a consequence of a prolonged period of high temperature, the pathogen is killed and weed seed is deactivated [19].

The composting process has a long history of treatment of garden waste, farm slurry, agricultural residue, and wild grass. In the EU, approximately 66% of compost originated from farm slurry, animal manure, sewage sludge, and dedicated energy crops, 33% of compost obtained from food waste [20].

7.2.3 Chemical hydrolysis of organic waste stream

Organic waste stream, especially digestate from food waste AD plant can be treated via a thermochemical process to generate a soluble biowaste substance (SBO) [21,22]. This material is then dried to a moisture content of approximately 10% forming a solid biofertilizer [23]. Both acid and alkali were tested in the hydrolysis, normally in the temperature range of 60−100°C [21,22]. Thermochemical hydrolysis can also be achieved by microwave heating instead of conventional heating process. The reaction time for microwave assistant process completed the reaction was one or two magnitude lower than the control process [21].

7.2.4 Solid state fermentation

Solid state fermentation has been widely used for the production of industrial enzymes [24]. In comparison to submerged fermentation, solid state fermentation is operated in the absent or nearly absent of free water. It has several unique advantages for enzyme production, especially using solid organic waste as the substrate, including ease in operation, low equipment investment, particular suitable for solid substrate and potential high enzyme productivity [25,26]. Solid state fermentation of biofertilizer production is similar to aerobic compost in many ways, such as both have a high solid content, both require low maintenance, and both need a relatively long retention time. However, biofertilizer production via solid state fermentation requires external inoculation to start the fermentation, normally a pure microorganism [27,28]. Fermentation is normally carried out in mesophilic conditions. After fermentation, e.g., around 40 days, the fermented material can be used as biofertilizer directly [27,28].

7.2.5 In situ *degradation of agricultural residues*

Crop residue is major waste stream in agriculture practice. It occupies space and affects crop growth if it is left over in the filed after grain harvest. One of obvious choices is directly returning straw into soil, which simply involves burying the crop residue by soil after harvesting (Fig. 7.4). The target of returning straws/husks into soil is to use soil microorganisms to digest the crop residue to release nutritional compounds back to soil [29]. The microorganism community in soil breaks down the organic constituents into short chain organic compounds, containing carbon, nitrogen, phosphorus and potassium. These nutrients are then either transport to and absorbed by plants or stored in soil microbes for future plant usage [30].

Figure 7.4
A photo shows the activity of returning straw back to soil by a tractor.

7.2.6 Direct burning of biomass

Organic waste, especially crop residue can be burnt directly to generate a mineral rich ash, which can be used as fertilizer. Piekarczyk et al. analyzed the mineral compositions in the ash derived from wheat straw, rape straw, barley straw and hay [31]. On average, the ash contains 15.57% (w/w) potassium, 12.4% calcium, 1.51% phosphorus and 0.73% magnesium, together with track amount of essential micronutrients such as Fe, Mn, Zn and Cu. This is an old practice, which is mainly used to eliminate bulk, unwanted organic waste to free the field for the next round of plantation. However, during the burning process, organic compounds are oxidized to gaseous oxides and discharged to air without control in most cases. For the soil, the burning straw in the field resulted in loss in soil moisture and reduced organic matter content in the soil, therefore decreased the fertilities of soil [32,33]. The soil microorganisms can also be killed in the burning process, which leads to the soil vulnerable to plant pathogens. During the serious concerns on air pollution, soil erosion and nurturance loss, this practice is forbidden in the UK and many other developed countries [34].

7.3 Biofertilizer derived from food waste

7.3.1 Anaerobic digestion

In terms of biofertilizer generation using wasted food, food processing waste and OFMSW, AD is the most frequently used method. Many different types of food waste have been tested for biogas production at lab scale [35], as well as in the full production scale [36]. In industrial scale AD plants, biofertilizer production is an important byproduct to generate additional income to the plant.

Panuccio et al. [37] carried out AD of olive and citrus processing waste together mixed with animal manures. The derived biofertilizer (digestate) composition was strongly related to the composition of the feedstock. Tampio et al. [38] compared the AD of food waste, OFMSW and vegetable waste-activated sludge mixture at a scale ranging from lab scale to full production scale. The total nitrogen content ranged from 2.2 to 8.7 g/kg, with a NH_4−N percentage over 50% in four of five digestates analyzed. Regarding the heavy metal content, only the vegetable waste-activated sludge mixture derived biofertilizer exceeded the European legislation for Hg, Cu and Zn. Rigby and Smith [39] analyzed the composition of four biofertilizers, originated from (1) biosolids, (2) OFMSW, (3) food and farm waste, and (4) food waste containing bread, cooked meat, fruit, and vegetables. The total nitrogen contents of these four biofertilizers were 4.64%, 2.32%, 11.3%, and 3.5% (w/w, on the dry matter basis), respectively. The total phosphorus content and total potassium content both showed a wide variation depending on the origin of the biofertilizer. For the total phosphorus content, biosolids derived digestate contained 2.8% P (28 g/kg), while all other showed 0.38%−0.65%. Similarly, for the potassium content, food and farm waste derived digestate contained 3.76% K (37.6 g/kg), while all the other resulted in 0.2%−0.5%. Table 7.1 lists several recent case studies using food waste for biofertilizer production via AD process.

Food waste is normally codigested with farm waste, such as cow manure [40] and pig slurry [41] to boost the nutrient content, especially the nitrogen content. The food waste derived biofertilizer contains around 1.5−6.2 g/kg nitrogen (N), 0.2−2.6 g/kg phosphorous (P) and 1.2−11.5 g/kg potassium (K) [36]. However, the nutrient content of the digestate depends strongly on the composition of the food waste feed stream. Codigestion of food waste with farm slurry or green waste also facilitates the adjustment of C/N ratio to a preferable range to enable the efficient operation of AD plant [42].

Quality control in food waste derived biofertilizer is essential and challenging. Three main categories of unwanted contaminants could present in the biofertilizer, namely physical impurities (such as glass, plastic); chemical impurities (such as toxic compounds, heavy metals); and biological impurities (e.g., pathogens) [36]. These contaminations could be reduced/removed by pretreatment prior AD, or posttreatment after AD. Similar to lignocellulosic bioethanol production, various chemical, physical, and thermochemical hydrolysis processed have been proposed to treat food waste feed to AD plants, as reported by a recently review [43]. The benefits of pretreatment include screening out bulk size impurities, reducing particle size [44], kill plant and microbial pathogens [45], and reducing AD operation time [46]. When OFMSW is used in AD, the biofertilizer derived from the digestate is prone to heavy metal contamination. Abdullah et al. [47] analyzed the heavy metal content in a steam sterilized MSW, and found the present of cadmium (Cd), chromium (Cr), copper (Cu), lead (Pb), mercury (Hg), nickel

Table 7.1: Food waste–derived biofertilizer.

Feedstock	Conversion process	Total-N[a]	NH_4–N[a]	Total-P[a]/Total-K[a]	References
Wasted food and saw dust	Compost, 60 d, 30–65°C	1.6%	NA	0.6%	[58]
Winter wheat/ potatoes	Co-digestion, mesophilic	0.25% (FM)	0.16% (FM)	0.62 (P)	[59]
Olive oil husk and manure	Compost, 116 d, 30–65°C	1.4%–2.5%	NA	0.67%–0.71%	[51]
Wasted food, saw dust, rice husk and rice bran	Compost 6–15 d, 30–60°C	0.87%–1.59%	NA	NA	[53]
Ryegrass/sugar beet	AD, batch, mesophilic	6.2%	1.5 g/L	0.32 and 3.6 g/L	[60]
Food and farm wastes	Co-digestion	2.32%–4.64%	2.8–52.5 g/kg	3.81–28 and 1.94–37.6 g/kg	[39]
Wasted food	Compost, 30–33 d	1.73%–1.84%	1.3 g/kg		[61]
Olive mill waste and sheep/horse manure	Compost, 30 weeks, 20–70°C	1.47%–1.73%	NA	0.3%–0.4%	[55]
Palm oil mill waste	Compost, 35 d	NA	NA	NA	[62]
Fruit, vegetable waste and yard wastes	Compost, 15 weeks, 25–41°C	2.0%	NA	NA	[63]
Spent coffee grounds, spent tea leaves with yard wastes	Compost, 15 weeks, 25–44°C	0.3%–3%	NA	NA	[63]
Olive waste and citrus pulp	AD, batch, mesophilic	6.0%	149 mg/L	840 and 631 mg/L	[37]
Wasted food and saw dust	Compost, 35 d, 35–55°C	NA	1.6–6.0 g/kg	NA	[52]
MSW	Mesophilic wet AD	4.61% N	NA	3.33% P, 0.39% K,	[64]
Food waste with feldspar	SSF fermentation 7 days	NA	NA	NA	[65]

[a]Dry weight basis.

(Ni), and zinc (Zn) in higher concentrations that the standards for biofertilizer. The metal content in the biofertilizer may contaminate arable land and ground water; accumulate via food chain to human body. Heavy metal contamination in digestate/ biofertilizer is difficult to remove. Source separation is widely used to as an efficient approach to prevent heavy metals from entering the AD system [48]. Rigby and Smith characterized heavy metal contents in the food waste digestate [49]. It concluded that in the digestate derived from source separated food waste, heavy metal contents were within the limitations.

Postdigestion treatment, such as sanitization, chemical hydrolysis and pyrolysis was used to improve the nutritional levels of the biofertilizer. Opatokun et al. investigated the pyrolysis of industrial food waste derived digestate to a biochar like biofertilizer [50], which enhanced P and K contents with an average enrichment factor of 0.87. When the fertilizer was tested in sandy soil, seed germination and water retention capacity were improved.

7.3.2 Composting and chemical hydrolysis of compost

Similar to AD plant, food waste is commonly composted with farm waste [51], green waste [51], and bulking agents, such as sawdust [52] and rice husk [53]. This is mainly due to the high water content in the food waste [7] and unsuitable nutrient composition [54]. Cocomposting with other waste streams adjusted the C/N ratio of the substrate to a suitable range, e.g., 15–30 and led to a compost with related higher nitrogen content [51]. Compost should be dried to a moisture content of lower than 15% to prevent bacterial contamination and to facilitate transportation [55].

Tsai et al. explored food waste composting for the biofertilizer production [56]. Food waste that collected from a University restaurant mixed with saw dust was composted in a 250 L composter with agitation. Additional inoculation of thermophilic strain *Brevibacillus borstelensis SH168* improved the degradation rate. After 28 days composting, a biofertilizer containing 2.10% (w/w) nitrogen was obtained. Walker et al. reported a combined thermophilic AD with aerobic composting for the treatment mechanically sorted OFMSW [57]. After 21 days of treatment, a soil conditioner like biofertilizer was obtained, which contained 1.5% nitrogen and 0.314% (w/w) phosphorus. Table 7.1 summarized the composition of biofertilizer derived from several recent food waste compost studies.

7.3.3 Solid state fermentation

Several studies explored the possibility of using solid state fermentation to convert food waste into biofertilizer. Lim and Matu reported biofertilizer production via solid state fermentation of fruit wastes: watermelon, papaya, pineapple, citrus orange, and banana [28]. The first batch was carried out with the addition of water and the filtrate was then used as the inoculum for the following solid-state fermentation. The banana waste derived biofertilizer contained the highest potassium concentration of 3.9 g/L, but the nitrogen and phosphorus content were not reported. Application of the biofertilizers on the Mustard plant showed better biomass yield and fast growth rate than the control [28].

7.3.4 Field application of food waste derived biofertilizer

Rigby and Smith analyzed the physicochemical properties and microbial properties of the digestates from five food waste and animal slurry-based AD plants in Wales [41]. The digestates were then compared with nine market available garden fertilizers in the UK. The total Nitrogen (N), phosphorus (P), and potassium (K) contents of food waste derived digestate were 15%, 0.7%, and 4.7% respectively. The nitrogen content was higher than the average nitrogen content in the general purpose fertilizer (5.5%), vegetable fertilizer (5.46%), ericaceous plant food (7.8%), "root booster" (3.8%), rose and shrub feed (2.5%), tomato feed (3.8%), and turf fertilizer (12%). Furthermore, 62%–65% of nitrogen was in mineral forms, which could be easily accessed by plants. However, the phosphorus content was lower than any of these commercial fertilizers (1.0%–7.8%). The potassium content was comparable with some of these fertilizers (general purpose fertilizer 4.36%, tomato feed 4.6%, and turf fertilizer 6%), but was lower than ericaceous plant food (11.5%) and rose and shrub feed (17.5%). The heavy metal contents in the food waste–derived digestate were within the upper limit values specified in UK standards defined in the Quality Protocol PAS110 [41].

Mkhabela and Warman reported biofertilizer generated from source separated OFMSW for the application as the phosphorus fertilizer in the field for potato and sweet corn production [66]. Three different levels of OFMSW compost were used, as well mixture of biofertilizer and chemical fertilizer and chemical fertilizer alone as the control. The cultivation was carried out in Canada in 1996 and 1997. The results showed no statistically difference in tissue P concentration between chemical fertilizer and biofertilizer in both potato and sweet corn and for both years. However, the nitrogen content in the biofertilizer was not enough, which should be supplemented by inorganic fertilizer. The usage of similar OFMSW compost for strawberry cultivation also approved that nitrogen concentration in the biofertilizer was not sufficient [67]. Furthermore, the extractable soil mineral elements contents were increased. In a field trial of OFMSW derived compost for cultivation of forage and arable crops in New Zealand, a low nitrogen utilization ratio of only 13%–23% over 3–4 years was observed [68]. By contract, the nitrogen utilization ratio of c chemical fertilizer was normally in the range of 25%–50% [69]. Hargreaves et al. summarized around 30 reports on the utilization of MSW compost for cultivation of crops, vegetables, and fruits [70]. It concluded the heavy metal accumulation was an issue, but source separated MSW could derive a safe biofertilizer.

The postdigestation/composting treatment, such as chemical hydrolysis could improve nutrient level in the biofertilizer. Sortino et al. obtained a soluble biowaste substance via alkaline hydrolysis of food waste compost [23]. The Nitrogen (N), phosphorus (P), and potassium (K) contents were 5.1%, 0.37%, and 1.2% (w/w, db), respectively. Field test of the biofertilizer for red pepper production resulted in a 90% yield increase at a low dose of just 140 kg/ha [23]. Similar improvement of plant growth and reduction of plant pathogens were also observed in field test of chemical hydrolysis treated compost for beans [71] and radish [72] production.

7.4 Biofertilizer derived from agriculture residue

7.4.1 Biofertilizer production process

Agricultural residue is major organic waste stream from crop production. Using wheat straw as an example, around 1–2 kg of wheat straw is produced per kg of wheat grain is harvest, although this value varies due to the difference of cereal species, fertilizer condition, climate, and plantation practice. Agricultural residue contains high cellulose, hemicellulose, and lignin content, which has been widely investigated for the production of bioethanol [73]. Traditionally, agricultural residues were often directly returned to soil for recovering certain nutrients in the straw/husk/stalks/leaves. Glithero et al. collected data from farms in the UK for the wheat straw usage [74]. It revealed that over 36% of the straw was returned back to the soil. In a reported published by Yong et al. [75], around 68% of the straw generated in the USA were sent back to field. China produces around 20%–30% world straw, which is equivalent to around 580 Mt per year [5]. The Chinese government set a target to achieve converting 85% straw into valuable added products by 2020 [76]. *In situ* degradation of straw in the field as a biofertilizer was listed as one of the possible approaches.

In situ degradation of straw can be carried out using the endogenous soil microorganisms [77,78], or with the aid of additional microorganisms [79]. In European and North American where agricultural activities are well managed and the soil microorganisms show high degrading capacities, the degradation rate by endogenous soil microorganisms is sufficiently fast to avoid significant lignocellulosic residues [78]. In some countries, where crop rotation is high, chemical fertilizer is overused and straw is over produced, the loading of agricultural residue exceeds the degradation capacity of native soil microbes [79]. As a result, addition of external microorganism to facility soil microorganism degradation is utilized. This can be achieved by increasing the population of the soil microorganisms per unit area, or enhancing the average digestion capacity of individual microorganism, or the combination of these two approaches.

Many genera of microorganisms have been involved in degradation of agricultural residues. These strains include cellulose and hemicellulose degrading microorganisms, such as *Trichoderma*, *Aspergillus*, *Penicillium*, *Fusarium*, *Cytophaga*, *Sporocytophaga*, and *Polyangium* [80–83] and lignin-degrading microorganisms, such as various genera in *Ascomycota*, and species in *bacillus*, *Xanthomonas*, and *Pseudomonas* [84,85].

The degradation of agricultural residues occurs at both mesophilic and thermophilic environment. Bacterial strains usually dominate the population of soil microorganisms, which can reach 10,000 strains per gram of soil. Strom reported that in a thermophilic condition, strains in the genus of *Bacillus* accounted for 87% of randomly selected colonies [86]. Bacteria are smaller in size in comparison with fungi. As a result, the specific surface area is relatively high, and therefore allows faster mass transportation between cells and the environment [85]. Bacteria are less efficient in lignin degradation, especially in anaerobic condition [87]. Fungal strains secrete a wide range of enzymes for the degradation of lignocellulosic waste in nature. Brown-rot fungi degrade cellulose material effectively but cannot degrade lignin due to the lack of lignocellulolytic enzymes. However, brown-rot fungi are capable of modifying the lignin structure by demethylation. Usually, the depolymerization rate by brown-rot fungi is high due to the relatively smaller sizes of enzymes, which are relatively easier to penetrate into the plant cell wall matrix [88]. White-rot fungi degrade both cellulose and lignin [89]. They produce lignin-degrading enzymes that degrade the lignin to carbon dioxide and water, exposing the hemicellulose and cellulose in the wood matrix [90]. *Phanerochaete chrysosporium* was successfully used for biological pretreatment of cotton stalks under solid state cultivation and the results showed that 35.53% lignin was degraded within 14 days and carbohydrate was not consumed [91]. This strain was also studied to hydrolyze wheat straw. The result showed that 25% lignin was degraded within 1 week using a medium made up with wheat straw, inorganic salts and tween 80 [92]. Similar to white-rot fungi, soft-rot fungi can degrade cellulose and lignin simultaneously. One of the most extensively studied cellulolytic mechanisms in soft-rot fungi is *Trichoderma* sp., which has been widely used in industrial scale cellulase production [88].

In order to enhance agriculture residue hydrolysis, addition of a mixture of several microorganisms has been proposed [93]. Zhao et al. investigate the efficiency of using both cellulosic and lignin-degrading microorganisms to hydrolyze rice straw [93]. High degradation efficiency was achieved when *Cellulomonas flavigena* W9801 and *Trichoderma koningii* W9803 were inoculated together. Besides selection of microorganisms, the following factors should be optimized to facilitate in situ degradation: (1) the carbon to nitrogen ratio (C/N); (2) soil water content; (3) pH of the soil; (4) the depth of straws incorporation [36,94].

Agriculture residue could be converted to biofertilizer using solid state fermentation process as well. Chen et al. carried out solid sate fermentation using rice straw, cattle dung and vinegar production residue as the substrate by *Trichoderma harzianum* SQR-T037 [21]. The cultivation conditions were optimized for the production of fungal biomass for the degradation of lignocellulosic biomass as well as the production of 6-pentyl-a-pyrone (6PAP), an antimicrobial compound for the control of plant diseases.

7.4.2 Field test of biofertilizer derived from agriculture residues

During the in situ degradation of straw, the organic content such as lignocellulose and protein was hydrolyzed into short chain organic matters, mainly humic acid by soil microbes [95]. A stable particle cluster was then formed by the interaction between humic acid and divalent metal cations in the soil, which prevented soil erosion, increased soil permeability, and facilitated water transfer to the roots [96].

Several long-term field test experiments have demonstrated the above-mentioned benefits of in situ degradation of agriculture residue. Wei et al. reported that the soil receiving in situ degradation of wheat straw showed a better air permeability, heat preservation and water conservation [97]. Soil structure analysis demonstrated a significant increase of 202.9% in the number of soil particle cluster that is larger than 2 mm. The noncapillary porosity increased by 0.5%–3% as well [97]. Shen and Chen surveyed the impact of maize straw in situ degradation in Lingchuan, Shanxi province, China [98]. After 10 years of returning around 1/3 of the maize straw to the field, the soil permeability increased by 30%, soil erosion decreased by 60%–70% together with a 15% increase in crop yield. Lehtinen et al., reviewed 39 publications and concluded that soil organic carbon increased by 7% as a result of crop residue incorporation, but CO_2 and N_2O emissions were increased as well [99]. Wei et al. recorded a field test of in situ degradation of wheat straw carried out from 2007 to 2011 in a semiarid area of northwest China [100]. Three wheat straw loading ratios were investigated from 3000 kg per hectare to 9000 kg per hectare, with no in situ degradation as the control group. The results showed the high loading ratio group led to the best benefit. The N, P, K, and SOC contents in the soil were 9.1% –30.5%, 9.8%–69.5%, 10.3%–27.3% and 0.7%–23.4% higher than the control group. And the wheat yield increased by 26.75% as well [100].

The in situ degradation of crop residue also affected the soil microbial community by increase the strain numbers and enzyme activity [101]. A healthy microbial community facilitated the degradation of long chain carbon and nitrogen source, maintained a suitable carbon to nitrogen ratio, a suitable pH, improved pest resistance and degraded pesticide and chemical pollution [36,102,103]. Zeng et al. reported that in situ degradation of straw resulted in 142.9% and 115% increase of bacterial and fungal strain numbers, respectively [104]. Bandick and Dick analyzed urease, phosphatase and neutral phosphatase activities in the soil with in situ straw degradation [105]. It was found that these enzyme activities increased by 36.8%, 43.8% and 14.6% respectively in comparison to the control. Increase in other biomass hydrolysis related enzymes, such as cellulase, sucrose hydrolase, catalase, and ligninase was also observed [105]. Supplement of addition microorganisms is a common operation to improve in situ degradation of agriculture residue for biofertilizer production [106,107]. Cost effective nitrogen source could be added as well, which will be transformed into biological nitrogen and eventually stored in soil. In some cases, nitrogen fixing microorganism is considered as "biofertilizer" to add to soil to enhance absolute contents of nitrogen in soil [108].

7.5 Conclusions and perspectives

With the increase of world population, growing amount of organic waste has been generated. The treatment of organic waste became a challenge due to the lack of landfill places and increasingly strict environmental regulations. Utilization of organic waste for the production biofertilizer could be a viable choice to reduce environment burden and to generate income. Currently, biofertilizer derived from food waste via AD/composting has already been commercialized in certain countries. In situ degradation of agriculture residue to form a biofertilizer has been demonstrated to be a suitable approach to deal with the straw accumulation in the field. However, further improvement of the biofertilizer quality is required. Especially, a robust technology that is capable of handling waste streams with various compositions is desired. Further reduce production cost will be essential to enable biofertilizer to compete with synthetic fertilizer. With the continuous understanding of the mechanism of biofertilizer production process and further development of advanced technology development, the biofertilizer formation efficiency could be further improved.

Acknowledgments

This study was supported by the University of Huddersfield, under the program of URF (URF2015/24). The authors also thank the financial support provided by Ziguangtianhe Co. Ltd., Beijing China.

References

[1] Muralikrishna I, Manickam V. Chapter 16, Solid waste management. In: Muralikrishna I, Manickam V, editors. Environmental management; 2017. p. 431–62.
[2] http://www.fao.org/save-food/resources/keyfindings/en/.
[3] Hall KD, Guo J, Dore M, Chow CC. The progressive increase of food waste in America and its environmental impact. PLoS One 2009;4(11):e7940.
[4] Satlewal A, Agrawal R, Bhagia S, Das P, Ragauskas AJ. Rice straw as a feedstock for biofuels: availability, recalcitrance, and chemical properties. Biofuels, Bioproducts and Biorefining 2018;12:83–107.
[5] Wang Y, Bi Y, Gao C. The assessment and utilization of straw resources in China. Agricultural Sciences in China 2010;9:1807–15.
[6] https://eur-lex.europa.eu/legal-content/EN/TXT/PDF/?uri=CELEX:32018L0851&from=EN.
[7] Lin CSK, Pfaltzgraff LA, Herrero-Davila L, Mubofu EB, Abderrahim S, Clark JH, Luque R. Food waste as a valuable resource for the production of chemicals, materials and fuels. Current situation and global perspective. Energy and Environmental Science 2013;6(2):426–64.
[8] Marin FR, Soler-Rivas C, Benavente-Garcia O, Castillo J, Perez-Alvarez JA. By-products from different citrus processes as a source of customized functional fibres. Food Chemistry 2007;100(2):736–41.
[9] Attanasi OA, Berretta S, Fiani C, Filippone P, Mele G, Saladino R. Synthesis and reactions of nitro derivatives of hydrogenated cardanol. Tetrahedron 2006;62:6113–20.

[10] Woodgate S, van der Veen J. The role of fat processing and rendering in the European Union animal production industry. Biotechnology, Agronomy, Society and Environment 2004;8:283–94.
[11] Research market. 2017. https://www.researchandmarkets.com/research/fcrscs/fertilizers.
[12] http://european-biogas.eu/2017/12/14/eba-statistical-report-2017-published-soon/.
[13] Warp; 2017. http://www.wrap.org.uk/content/operational-ad-sites.
[14] Chen L, Zhao L, Ren C, Wang F. The progress and prospects of rural biogas production in China. Energy Policy 2012;51:58–63.
[15] Fuchs W, Wager F, Kirchmayr R, Braun R, Dorsg B. Digestate treatment: comparison and assessment of existing technologies. In: Third international symposium on energy from biomass and waste. Venice, Italy: CISA; 2010.
[16] Tampio E, Marttinen S, Rintala J. Liquid fertiliser products from anaerobic digestion of food waste: mass, nutrient and energy balance of four digestate liquid treatment systems. Journal of Cleaner Production 2016;125:22–32.
[17] Lim SL, Lee LH, Wu TY. Sustainability of using composting and vermicomposting technologies for organic solid waste biotransformation: recent overview, greenhouse gases emissions and economic analysis. Journal of Cleaner Production 2016;111:262–78.
[18] Pandey PK, Vaddella V, Cao W, Biswas S, Chiu C, Hunter S. In-vessel composting system for converting food and green wastes into pathogen free soil amendment for sustainable agriculture. Journal of Cleaner Production 2016;139:407–15.
[19] Miyatake F, Iwabuchi K. Effect of high compost temperature on enzymatic activity and species diversity of culturable bacteria in cattle manure compost. Bioresource Technology 2005;96:1821–5.
[20] Cesaro A, Belgiorno V, Guida M. Compost from organic solid waste: quality assessment and European regulations for its sustainable use. Resources, Conservation and Recycling 2015;94:72–9.
[21] Rosso D, Fan J, Montoneri E, Negre M, Clark J, Mainero D. Conventional and microwave assisted hydrolysis of urban biowastes to added value lignin-like products. Green Chemistry 2015;17(6):3424–35.
[22] Arshadi M, Attard M, Lukasik RM, Brncic M, Da Costa Lopes AM, Finell M, Geladi P, Gerschenson L, Gogus F, Herrero M, Hunt A, Ibáñez E, Kamm B, Mateos-Aparicio I, Matias A, Mavroudis N, Montoneri E, Morais A, Nilsson C, Papaioannoua EH, Richel A, Rupérez P, Skrbic B, Bodroza Solarov M, Švarc-Gajić J, Waldron K, Yuste-Córdoba FJ. Pre-treatment and extraction techniques for recovery of added value compounds from wastes throughout the agri-food chain. Green Chemistry 2016;18(23):6160–204.
[23] Sortino O, Dipasquale M, Montoneri E, Tomasso L, Avetta P, Bianco Prevot A. 90% yield increase of red pepper with unexpectedly low doses of com- post soluble substances. Agronomy for Sustainable Development 2013;33:433–41. https://doi.org/10.1007/s13593-012-0117-6.
[24] Thomas L, Larroche C, Pandey A. Current developments in solid-state fermentation. Biochemical Engineering Journal 2013;81:146–61.
[25] Pensupa N, Jin M, Kokolski M, Archer DB, Du C. A solid state fungal fermentation-based strategy for the hydrolysis of wheat straw. Bioresource Technology 2013;149:261–7.
[26] Hu Y, Du C, Leu SY, Jing H, Li X, Lin CSK. Valorisation of textile waste by fungal solid state fermentation: an example of circular waste-based biorefinery. Resources, Conservation and Recycling 2018;129:27–35.
[27] Chen L, Yang X, Raza W, Luo J, Zhang F, Shen Q. Solid-state fermentation of agro-industrial wastes to produce bioorganic fertilizer for the biocontrol of *Fusarium* wilt of cucumber in continuously cropped soil. Bioresource Technology 2011;102:3900–10. https://doi.org/10.1016/j.biortech.2010.11.126.
[28] Lim S, Matu S. Utilization of agro-wastes to produce biofertilizer. International Journal of Energy and Environmental Engineering 2015;6:31–5. https://doi.org/10.1007/s40095-014-0147-8.
[29] Gong C, Wang J, Chen X, Zhang Z. Comprehensive utilization of crop straw and returning technology. Science and Technology Innovation Herald 2008;9:251.

[30] Nie J, Zhou JM, Wang HY, Chen X, Du C. Effect of long-term rice straw return on soil glomalin, carbon and nitrogen. Pedosphere 2007;17:295–302.

[31] Piekarczyk M, Kotwica K, Jaskulski D. The elemental composition of ash from straw and hay in the content of their agricultural utilization. Acta Scientiarum Polonorum – Agricultura 2011;10:97–104.

[32] Rossi C, Pereira M, García A, Berbara R, Gazolla P, Perind A. Effects on the composition and structural properties of the humified organic material of soil in sugarcane strawburning: a chronosequence study in the Brazilian Cerrado of Goiás State. Agriculture, Ecosystems & Environment 2016;216:34–43.

[33] Ventrella D, Stellacci A, Castrignanò A, Charfeddine M, Castellini M. Effects of crop residue management on winter durum wheat productivity in a long term experiment in Southern Italy. European Journal of Agronomy 2016;77:188–98.

[34] UK Government; 2017. http://www.legislation.gov.uk/uksi/1991/1590/contents/made.

[35] Mao C, Feng Y, Wang X, Ren G. Review on research achievements of biogas from anaerobic digestion. Renewable and Sustainable Energy Reviews 2015;45:540–55.

[36] Du C, Abdullah JJ, Greetham D, Fu D, Yu M, Ren L, Li S, Lu D. Valorization of food waste into biofertiliser and its field application. Journal of Cleaner Production 2018;187:273–84.

[37] Panuccio MR, Attina E, Basile C, Mallamaci C, Muscola A. Use of recalcitrant agriculture wastes to produce biogas and feasible biofertilizer. Waste and Biomass Valorization 2016;7(2):267–80.

[38] Tampio E, Salo T, Rintala J. Agronomic characteristics of five different urban waste digestates. Journal of Environmental Management 2016b;15(169):293–302.

[39] Rigby H, Smith SR. Nitrogen availability and indirect measurements of greenhouse gas emissions from aerobic and anaerobic biowaste digestates applied to agricultural soils. Waste Management 2013;33(12):2641–52.

[40] Pivato A, Vanin S, Raga R, Lavagnolo MC, Barausse A, Rieple A, Laurent A, Cossu R. Use of digestate from a decentralized on-farm biogas plant as fertiliser in soils: an ecotoxicological study for future indicators in risk and life cycle assessment. Waste Management 2016;49:378–89.

[41] Koster JR, Dittet K, Muhling K-H, Kage H, Pacholkski A. Cold season ammonia emissions from land spreading with anaerobic digestates from biogas production. Atmospheric Environment 2014:8435–8.

[42] Yin Y, Liu Y, Meng S, Kiran EU, Liu Y. Enzymatic pretreatment of activated sludge, food waste and their mixture for enhanced bioenergy recovery and waste volume reduction via anaerobic digestion. Applied Energy 2016;179:1131–7.

[43] Zhang C, Su H, Baeyens J, Tan T. Reviewing the anaerobic digestion of food waste for biogas production. Renewable and Sustainable Energy Reviews 2014;38:383–92.

[44] Al Seadi T, Lukehurst C. Quality management of digestate from biogas plants used as fertiliser. IEA Bioenergy, Task; 2012.

[45] Evans TD, Boor M, MacBrayne D. Biofertiliser plant design – food waste to biofertiliser and biogas. In: Proc. 12th CIWEM European biosolids & biowastes conference, 13–15 November 2007 AquaEnviro, Manchester; 2007.

[46] Ma Y, Yin Y, Liu Y. New insights into co-digestion of activated sludge and food waste: biogas versus biofertilizer. Bioresource Technology 2017;241:448–53.

[47] Abdullah JJ, Greetham D, Pensua N, Tucker GA, Du C. Optimizing cellulase production from municipal solid waste (MSW) using solid state fermentation (SSF). Journal of Fundamentals of Renewable Energy and Applications 2016;6(3).

[48] Govasmark E, Stäb J, Holen B, Hoornstra D, Nesbakk T, Salkinoja-Salonen M. Chemical and microbiological hazards associated with recycling of anaerobic digested residue intended for agricultural use. Waste Management 2011;31(12):2577–83.

[49] Rigby H, Smith SR. New markets for digestate from anaerobic digestion. 2011. http://www.wrap.org.uk/sites/files/wrap/New_Markets_for_AD_WRAP_format_Final_v2.c6779ccd.11341.pdf.

[50] Opatokun SA, Yousef LF, Strezov V. Agronomic assessment of pyrolysed food waste digestate for sandy soil management. Journal of Environmental Management 2017:24–30. https://doi.org/10.1016/j.jenvman.2016.11.030.

[51] Montemurro F, Diacono M, Vitti C, Debiase G. Biodegradation of olive husk mixed with other agricultural wastes. Bioresource Technology 2009;100(12):2969–74.
[52] Wang X, Selvam A, Wong J. Influence of lime on struvite formation and nitrogen conservation during food waste composting. Bioresource Technology 2016;217:227–32.
[53] Chang J, Chen YJ. Effects of bulking agents on food waste composting. Bioresource Technology 2010;101(15):5917–24.
[54] Kiran EU, Trzcinski AP, Ng WJ, Liu Y. Bioconversion of food waste to energy: a review. Fuel 2014;134:389–99.
[55] Fernandez-Hernandez A, Roig A, Serramia N, Civantos CG, Sanchez-Monedero MA. Application of compost of two-phase olive mill waste on olive grove: effects on soil, olive fruit and olive oil quality. Waste Management 2014;34:1139–47.
[56] Tsai SH, Liu CP, Yang SS. Microbial conversion of food wastes for biofertilizer production with thermophilic lipolytic microbes. Renewable Energy 2007;32:904–15.
[57] Walker L, Cord-Ruwish R, Sciberras S. Performance of a commercial-scale DiCOM demonstration facility treating mixed municipal solid waste in comparison with laboratory-scale data. Bioresource Technology 2012;126:404–11.
[58] Lin C. A negative-pressure aeration system for composting food wastes. Bioresource Technology 2008;99:7651–6.
[59] Stinner W, Moller K, Leithold G. Effect of biogas digestion of clover/grass-leys, cover crops and crop residues on nitrogen cycle and crop yield in organic stockless farming system. European Journal of Agronomy 2008;29(2–3):125–34.
[60] Gunnarsson A, Bengtsson F, Caspersen S. Use efficiency of nitrogen from biodigested plant material by ryegrass. Journal of Plant Nutrition and Soil Science 2010;173:113–9.
[61] Sundberg C, Yu D, Franke-Whittle I, Kauppi S, Smårs S, Insam H, Romantschuck M, Jönsson H. Effects of pH and microbial composition on odour in food waste composting. Waste Management 2013;33:204–11.
[62] Mohammad N, Alam MZ, Kabashi NA. Optimization of effective composting process of oil palm industrial waste by lignocellulolytic fungi. Journal of Material Cycles and Waste Management 2015;17:91–8.
[63] Sangamithirai KM, Jayapriya J, Hema J, Manoj R. Evaluation of in-vessel co-composting of yard waste and development of kinetic models for co-com-posting. International Journal of Recycling of Organic Waste in Agriculture 2015;4:157–65.
[64] Paul S, Dutta A, Defersha F, Dubey B. Municipal food waste to biomethane and biofertilizer: a circular economy concept. Waste and Biomass Valorization 2017:1–11. https://doi.org/10.1007/s12649-017-0014-y.
[65] Xue S, Miao L, Xue J, Yan H, Li G. Conversion of food waste and feldspar into biofertilizer using a stress-tolerant keldspar-solubilizing *Bacillus subtilis* xue-113168. In: Liu H, Song C, Ram A, editors. Advances in applied biotechnology. ICAB 2016. Lecture notes in electrical engineering, vol. 444. Singapore: Springer; 2018.
[66] Mkhabela M, Warman P. The influence of municipal solid waste compost on yield, soil phosphorus availability and uptake by two vegetable crops grown in a Pugwash sandy loam soil in Nova Scotia. Agriculture, Ecosystems & Environment 2005;106(1):57–67.
[67] Hargreaves JC, Adl MS, Warman PR. The effects of municipal solid waste compost and compost tea on mineral element uptake and fruit quality of strawberries compost. Science & Utilization 2009;17(2):85–94.
[68] Horrocks A, Curtin D, Tregurtha C, Meenken E. Municipal compost as a nutrient source for organic crop production in New Zealand. Agronomy 2016;6(2):35.
[69] Hirel B, Tétu T, Lea PJ, Dubois F. Improving nitrogen use efficiency in crops for sustainable agriculture. Sustainability 2011;3:1452–85.

[70] Hargreaves JC, Adl MS, Warman PR. A review of the use of composted municipal solid waste in agriculture. Agriculture, Ecosystems & Environment 2008;123(1):1–14.

[71] Baglieri A, Cadili V, Mozzetti Monterumici C, Gennari M, Tabasso S, Montoneri E, Nardi S, Negre M. Fertilization of bean plants with tomato plants hydrolysates. Effect on biomass production, chlorophyll content and N assimilation. Scientia Horticulturae 2014;176(11):194–9.

[72] Monterumici CM, Rosso D, Montoneri E, Ginepro M, Baglieri A, Novotny EH, Kwapinski W, Negre M. Processed vs. non-processed biowastes for agriculture: effects of post-harvest tomato plants and biochar on radish growth, chlorophyll content and protein production. International Journal of Molecular Sciences 2015;16(4):8826–43.

[73] Mafe OAT, Davies SM, Hancock J, Du C. Development of an estimation model for the evaluation of the energy requirement of dilute acid pretreatments of biomass. Biomass and Bioenergy 2014;72:28–38.

[74] Glithero NJ, Wilson P, Ramsden SJ. Straw use and availability for second generation biofuels in England. Biomass and Bioenergy 2013;55:311–21.

[75] Yong H, Zhen R, Yong I. The effect of stubble return on agro-ecological system and crop growth. Chinese Journal of Soil Science 2001;32:209–13.

[76] Chinese Government; 2017. http://www.saic.gov.cn/ywdt/gsyw/zyxwbysj/gwywj/201701/t20170106_174029.html.

[77] Hong J, Ren L, Hong J, Xu C. Environmental impact assessment of corn straw utilization in China. Journal of Cleaner Production 2016;112:1700–8.

[78] Sun J, Peng H, Chen J, Wang X, Wei M, Li W, Mellouki A. An estimation of CO_2 emission via agricultural crop residue open field burning in China from 1996 to 2013. Journal of Cleaner Production 2016;112:2625–31.

[79] Zhao S, Li K, Zhou W, Qiu S, Huang S, He P. Changes in soil microbial community, enzyme activities and organic material fractions under long-term straw return in north-central China. Agriculture, Ecosystems & Environment 2016;216:82–8.

[80] Hyun HS, Chung JW, Lee HB, Youn JK, Lee CY, Kim DH, Cho KY. Isolation of cellulose-degrading myxobacteria *Sorangium cellulosum*. Korean Journal of Microbiology 2009;45:48–53.

[81] Eida MF, Nagaoka T, Wasaki J, Kouno K. Evaluation of cellulolytic and hemicellulolytic abilities of fungi isolated from coffee residue and sawdust composts. Microbes and Environments 2011;26(3):220–7.

[82] Wang X, Peng Z, Sun X, Liu D, Chen S, Li F, Xia H, Lu T. The FPase properties and morphology changes of a cellulolytic bacterium, *Sporocytophaga* sp. JL-01, on decomposing filter paper cellulose. Journal of General and Applied Microbiology 2012;58:429–36.

[83] Karpe AV, Beale DJ, Morrison PD, Harding IH, Palombo EA. Untargeted metabolic profiling of *Vitis vinifera* during fungal degradation. FEMS Microbiology Letters 2015;362(10).

[84] Odier E, Janin G, Monties B. Poplar lignin decomposition by gram-negative aerobic bacteria. Applied and Environmental Microbiology 1981;41:337–41.

[85] Tuomela M, Vikman M, Hatakka A, Itavaara M. Biodegradation of lignin in a compost environment: a review. Bioresource Technology 2000;72:169–83.

[86] Strom PF. Identification of thermophilic bacteria in solid- waste composting. Applied and Environmental Microbiology 1985;50:907–13.

[87] Moller J, Miller M, Kjoller A. Fungal-bacterial interaction on beech leaves: influence on decomposition and dissolved organic carbon quality. Soil Biology & Biochemistry 1999;31(3):367–74.

[88] Pensupa. The development of a biological pretreatment strategy for the conversion of wheat straw to biofuels or platform chemicals [PhD thesis]. University of Nottingham; 2015.

[89] Taherzadeh MJ, Karimi K. Pretreatment of lignocellulosic wastes to improve ethanol and biogas production: a review. International Journal of Molecular Sciences 2008;9:1621–51.

[90] Kumar P, Barrett DM, Delwiche MJ, Stroeve P. Methods for pretreatment of lignocellulosic biomass for efficient hydrolysis and biofuel production. Industrial & Engineering Chemistry Research 2009;48:3713–29.

[91] Shi J, Chinn MS, Sharma-Shivappa RR. Microbial pretreatment of cotton stalks by solid state cultivation of Phanerochaete chrysosporium. Bioresource Technology 2008;99:6556–64.
[92] Zeng J, Singh D, Chen S. Biological pretreatment of wheat straw by *Phanerochaete chrysosporium* supplemented with inorganic salts. Bioresource Technology 2011;102:3206–14.
[93] Zhao X, Lin Q, Sun Y, Wang Y, Zhang Y, Zhang M. Decomposition of different cellulose materials by some cellulose-decomposing microbes. Journal of Microbiology 2000;20(3):12–4.
[94] Zuo Y, Jia Z. Effect of soil moisture content on straw decomposing and its dynamic changes. Journal of Northwest Science-Technology University of Agriculture and Forest 2004;32(5):61–3.
[95] Song G, Novotny E, Mao J, Michael H. Characterization of transformations of maize residues into soil organic material. The Science of the Total Environment 2017;579:1843–54.
[96] Malik K, Azam F. Effect of humic acid on wheat (*Triticum aestivum* L.) seedling growth. Environmental and Experimental Botany 1985;25:245–52.
[97] Wei T, Cheng L, Zhu L. The research progress of improvement in soil after returning. Chinese Journal of Agricultural Mechanization Research 1990;2:48–52.
[98] Shen YY, Chen H. The progress of study on soil improvement research with straw stalk. Chinese Agriculture Science Bullet 2009;25(19):291–4.
[99] Lehtinen T, Schlatter N, Baumgarten A, Bechini L, Kruger J, Grignani C, Zavattaro L, Costamagna C, Spiegel H. Effect of crop residue incorporation on soil organic carbon and greenhouse gas emissions in European agricultural soils. Soil Use & Management 2014;30:524–38.
[100] Wei T, Zhang P, Wang K, Ding R, Yang B, Nie J, Jia Z, Han Q. Effects of wheat straw incorporation on the availability of soil nutrients and enzyme activities in semiarid areas. PLoS One 2015;10(4):e0120994.
[101] Marschner P, Kandeler E, Marschner B. Structure and function of the soil microbial community in a long-term fertiliser experiment. Soil Biology & Biochemistry 2003;35:453–61.
[102] Dong Y, Lang Z, Kong x, Lu D, Liu Z. Kinetic and multidimensional profiling of accelerated degradation of oil sludge by biostimulation. Environmental Science: Processes & Impacts 2014;17(4):763–74.
[103] Wu B, Lan T, Lu D, Liu Z. Ecological and enzymatic responses to petroleum contamination. Environmental Science: Processes & Impacts 2014;16(6):1501–9.
[104] Zeng G, Fu S, Jin P. The effect of organic materials on soil fertility improvement. Heilongjiang Agricultural Sciences 1988;3:35–9.
[105] Bandick A, Dick R. Field management effects on soil enzyme activities. Soil Biology & Biochemistry 1999;31:1471–9.
[106] Liu Y, Zhao S, Zhu Q, Ding W. The effect of microorganism inoculants on the color changing of the wheat straws after incorporated into soil. In: 2016 ASABE annual international meeting. American Society of Agricultural and Biological Engineers; 2016.
[107] Nilay B, Barua R, Nath D, Hazarika K, Phukon A, Goswami K, Barua DC. Low energy rice stubble management through *in situ* decomposition. Procedia Environmental Sciences 2016;35:771–80.
[108] Recous S, Aita C, Mary B. In situ changes in gross N transformations in bare soil after addition of straw. Soil Biology & Biochemistry 1999;31:119–33.

CHAPTER 8

Biochar from various lignocellulosic biomass wastes as an additive in biogas production from food waste

Carol W. Wambugu, Eldon R. Rene, Jack Van de Vossenberg, Capucine Dupont, Eric D. van Hullebusch

Department of Environmental Engineering and Water Technology, IHE Delft Institute for Water Education, Delft, The Netherlands

8.1 Introduction

Food waste can be described as the discarded food material that was intended for human consumption and is generated from a variety of sources such as households, institutions, restaurants and the food industries [1]. With the increasing population around the globe, the food waste generation has also increased. The traditional ways of disposal like the landfills and incineration are not suitable for its disposal due to the high moisture content (70%–80%) [2], low calorific value and high lability [1]. Anaerobic digestion (AD) has therefore been recommended as the most suitable method for dealing with the food waste as it can be used for biogas production, and the methane (CH_4) can be used for cooking at the domestic level or supplied to electrical grid to generate electricity at a large-scale level. Food waste is widely preferred for AD because of its high biodegradability and CH_4 yield [3], with a theoretical production rate ranging between 0.4 and 0.5 L CH_4 (g VS)$^{-1}$ [1]. It is composed of two parts: the solid food waste which includes vegetables, rice and meat, and the liquid food waste [4].

Depending on the diverse feeding habits around the globe, the characteristics of the food waste differ in terms of physical and chemical properties. Table 8.1 summarizes some characteristics of food waste from different sources and origin used in AD where the pH ranged from 3.9 to 6.5, carbon: nitrogen ratio (C:N) from 13.2 to 49.9, the total solids (TS) from 18% to 61%, and trace metals such as Ni 0.19 mg/L and Fe 766 mg/L.

Waste Biorefinery. https://doi.org/10.1016/B978-0-12-818228-4.00008-3

Table 8.1: Characteristics of food waste reported in the literature.

Parameters	Mixed cooked (Korean)	Raw food waste	Mixed cooked (California)	White bread from supermarket	Mixed from restaurant (China)	Mixed supermarket food waste
References	[28]	[4]	[29]	[20]	[30]	[13]
TS (%, w.b.)	18.1 (0.6)	23.1 (0.3)	30.90 (0.07)	61.2	-	*142 ± (20)
Volatile solids (%, w.b.)	17.1 (0.6)	21.0 (0.3)	26.35 (0.14)	59.5	-	*129 ± (20)
VS/TS (%)	0.94 (0.01)	90.9 (0.2)	85.30 (0.65)	-	0.97	-
pH	6.5 (0.2)	4.2 (0.2)	-	4.9	3.9	-
Carbon (% d.b.)	46.67	56.3 (1.1)	46.78 (1.15)	42.7	-	-
Nitrogen (% d.b.)	3.54	2.3 (0.3)	3.16 (0.22)	2.3	-	6.6 ± 0.0
C:N	13.2	24.5 (1.1)	14.8	18.7	49.9	-
Sodium (% d.b.)	0.84	3.45 (0.2)	-	-	-	-
Iron (mg/L, w.b.)	3.17	100 (23)	766 (402)	-	-	5.0 ± 1.0
Copper (mg/L, w.b.)	3.06	-	31 (1)	-	-	-
Zinc (mg/L, w.b.)	8.27	160 (30)	76 (22)	-	-	-
Aluminum (mg/L, w.b.)	4.31	-	1202 (396)	-	-	-
Manganese (mg/L, w.b.)	0.96	110 (95)	60 (30)	-	-	-
Chromium (mg/L, w.b.)	0.17	-	<1	-	-	-
Nickel (mg/L, w.b.)	0.19	-	2 (1)	-	-	0.1 ± 0.0

-, not analyzed; (), standard deviation; *d.b.*, on a dry basis; *w.b.*, on a wet basis; *., units in g/L.
Modified from Zhang C, Su H, Baeyens J, Tan T. Reviewing the anaerobic digestion of food waste for biogas production. Renewable and Sustainable Energy Reviews 2014;38:383–392.

One major issue in AD, and especially food waste AD, is inhibitors. A list of these inhibitors is proposed in Table 8.2 together with their major impact, and the counteraction possible. Direct inhibition (substrate-induced inhibition) occurs when the substrate

Table 8.2: Types of inhibitors and their effects.

Inhibitor	Inhibition action	Counteraction	References
	Direct		
Sulfate	- Competition for organic substrates between sulfate reducing bacteria and methanogens: Suppress CH_4 production - Toxicity of sulfide to other bacteria	- Dilution - Sulfate removal step (stripping, coagulation, precipitation) - Biological conversions - Acclimation by methanogens to hydrogen sulfide	[5]
Light metal ions (Na^+, K^+, Mg^{2+}, Ca^{2+}, Al^{3+})	- Al^{3+} competition for adsorption of Fe and Mn^{2+} - Precipitation of carbonates and phosphate by Ca^{2+} - Mg^{2+} restricts the formation of microbial double cells	- Acclimation of anaerobic bacteria	[6] [5]
Ammonia	- Intracellular pH change - Increased maintenance energy is required - Inhibition of enzymatic activity	- Air stripping - Chemical precipitation - Addition of adsorbents, e.g., biochar, zeolite	[5] [7]
Organic compounds	- Cause swelling and leaking of cell membranes - Disrupts the ion gradient - Causes cell lysis	- Acclimation of anaerobic microorganisms - Induction of some enzymes for the degradation - Genetic engineering	[5]
LCFA (Long Chain Fatty Acid)	- Adsorption on cell membrane (wall) - Interferes with the transport and protective functions - Sorption into biomass, leading to granular sludge flotation and wash out in UASB	- Codigestion with substrates having less lipid content - Addition of calcium	[5] [6]
VFA	- Decreases pH - Inhibits microbial metabolic activity	- pH adjustment with alkali - Adjusting the OLR - Supplementing trace elements - Addition of adsorbents (biochar)	[8] [9] [6]

contains compounds that could potentially inhibit the AD at the initial stages. They inhibit the growth of microbial cells by: diffusion through the bacterial cell membrane, increasing the cell membrane surface area and causes breakage of the cell membrane leading to leakage of its contents [6]. Indirect inhibition occurs from metabolic by-products and they cause potassium deficiency, proton imbalance, and inhibit the CH_4 catalyzing enzymes [6].

Another crucial aspect to be considered in food waste AD is trace elements. Trace elements are essential components of cofactors and enzymes and are present in minute amounts in a given sample or environment. Trace elements are a prerequisite in AD and are essential for microbial growth and metabolism [10]. The methanogenesis step involves the action of metal rich enzymes like acetyl-CoA synthase and coenzyme M reductase that require sufficient supply of Fe, Ni and Co to catalyze the key metabolic steps involved in AD [10,11]. The amount of trace element required depends on the bacterial species and their methanogenic pathway, but the most abundant is Fe, followed by Co and Ni, with trace amounts of Mo and Zn [10]. According to Ref. [10], the microorganisms uses the trace elements in two ways: fast, passive and unspecified; and slow, active and specified, which is driven by chemi-osmotic gradient across the cell membrane. A deficiency in trace element supply leads to reduced conversion of acetic acid into methane and an accumulation of volatile fatty acid (VFA) in the digester, hence creating an acidic environment which limits the microbial activity. The result is low methane yield and a subsequent AD system failure. The trace element can be supplemented by using a mix of the elements in solution form to improve the stability of the digesters [8]. Ref. [12] operated an anaerobic continuous stirred tank reactor (CSTR), at 55°C. The VFA concentration increased to 3.8 g/L on day 57. The authors added 10 mg/L Fe^{2+}, 1 mg/L Co^{2+} and 1 mg/L Ni^{2+} every 45 days during the second run of the CSTR and successful continuous operation was achieved by day 69 when the low pH, CH_4 concentration and reduced gas production was restored to optimum. They concluded that a deficiency in the three trace elements caused the VFA accumulation [12]. Carried out batch experiments at 36°C and used combinations of Mo, selenium (Se), Co, Ni, Fe and tungsten (W). The authors added Co to the digester with 12 g/L VFA accumulation and they observed a rapid fall of the concentration to less than 0.5 g/L. Table 8.3 shows a summary of the effects of trace element supplementation under different operating conditions. Recently, biochar has gained popularity in AD systems as it has been reported to reduce substrate-induced inhibition, increase the biogas yields and stability of the process. Biochar is a carbon-rich, porous solid formed as a result of heating biomass in absence of oxygen. The types and characteristics of biochar depend on the production temperature as summarized in Table 8.7.

The optimum trace elements amount required by the mesophilic methanogenic systems are summarized in Table 8.4 [14]. The authors also developed a correlation between the concentration of trace elements and the COD of the substrate (Eq. 8.1), to estimate the

Table 8.3: Supplementation of trace elements and its effects on the AD of food waste.

Food waste	Trace elements (mg/L)	Reactor type	Temperature (°C)	OLR	HRT days	Effect on AD	References
Source segregated food waste	Se (0.2), Mo (0.2), Fe (5.0), Ni (1.0), Co (1.0), W (0.2), Mn (1.0), Al (0.1), B (1.0), Cu (0.1) and Zn (0.2)	Semicontinuous	36	2–5 kg VS/ m^3/day	38 –95	- Se and Mo reduced the degradation time of acetic and propionic acids - CH_4 yields increased - Good AD stability	[12]
Potato 31%, vegetable patty 41%, bread powder 19% and onion 8%	Fe^{2+} (10), Co^{2+} (1) and Ni^{2+} (1)	CSTR	55	0.12 L/feed	100	- Addition of trace elements reduced the VFA concentration from 3800 to less than 1000 mg/L - pH increased to 7.0 and CH_4 production resumed after TE addition	[8]
Slaughterhouse waste	Ni (2.2, 0.7, 0.2), Co (1.6, 0.5, 0.2), Mo (22.3, 7.5, 2.5), Zn (7.4, 2.5, 0.8), Cu (27.6, 9.1, 3.0) and Se (197.4, 60.7, 19.7)	Semicontinuous	38	2.2 kg VS/ m^3/day	120	- High CH_4 yield (250–275 Nm^3/t COD) - Reduced VFA production from 10,000 to 700 mg/L when Ni (11.4 mg/L), Co (25.4 mg/L) and Mo (4.8 mg/L) were added - TE supplementation stabilized the performance of AD	[10]

Continued

Table 8.3: Supplementation of trace elements and its effects on the AD of food waste.—cont'd

Food waste	Trace elements (mg/L)	Reactor type	Temperature (°C)	OLR	HRT days	Effect on AD	References
Restaurant food waste	Ni (0.04), Co (0.04), Mo (0.03), Zn (0.02), Cu (0.01), Se (0.06), Al (0.01), B (0.008), Fe (0.56), and Mn (0.14)	Semicontinuous	37	1.0–2.3 g VS/L	20–30	- Reduced VFA concentration (<4000 mg/L) compared to control (15,000 mg/L) - Co, Mo, Ni, Fe were found to be important in VFA reduction	[3]
Mixed supermarket food waste	Fe (50, 100, 200, 400), Ni (2, 5, 10, 20), Co (0.5, 1, 2, 5), Se (0.1, 0.3, 0.6, 0.8), and Mo (2, 5, 10, 20), Fe-rich sludge (≥1000 mg Fe L^{-1})	Batch	37	2 g COD/g VSS	70	- High Fe in inoculum reduced bioavailability of added TE, combination of TE didn't improve CH_4 yield	[13]

Al, aluminum; *B*, boron; *CSTR*, continuous stirred tank reactor; *HRT*, hydraulic retention time; *Mo*, molybdenum; *OLR*, organic loading rate; *T*, temperature; *TE*, trace elements; *VS/L*, volatile solids per liter.

Table 8.4: Optimum trace elements concentration.

Trace element	Recommended value ($\mu g/g\ COD_{bio\text{-}influent}$)
Iron	84
Cobalt	5
Nickel	5.4
Zinc	19
Copper	5
Manganese	0.75
Molybdenum	2.1
Selenium	9.6
Tungsten	0.33
Boron	0.0114

Modified from Hendriks ATWM, van Lier JB, de Kreuk, MK. Growth media in anaerobic fermentative processes: the underestimated potential of thermophilic fermentation and anaerobic digestion. Biotechnology Advances 2018;36:1-13.

amount of trace element required per gram of biodegradable COD in the influent fed to the anaerobic digester.

$$C_E = COD_{bio-influent} \times E \tag{8.1}$$

where: C_E = concentration of the trace element in μg/L, $COD_{bio\text{-}influent}$ = biodegradable COD of the substrate in g/L and E = required trace element amount (μg/g $COD_{bio\text{-}influent}$).

However, the trace element supplemented in the AD systems may not always be bioavailable for uptake by the bacteria [15]. This is because they undergo complex physico-chemical reactions due to the influence of other key parameters such as; pH, alkalinity, the presence of sulfuric compounds and excreted microbial compounds which affect their occurrence and speciation as free ions, complex bound or precipitates [10]. The trace element concentration has widely been used to assess the effect of trace elements supplementation in AD, but Refs. [10,11,15] have reported it as an inadequate method of measuring their bioavailability. These authors have suggested several methods of measuring trace element bioavailability such as sequential extraction and metal fractionation, chromatography, spectroscopy and modeling.

8.2 Key parameters for performance of AD of food waste

The AD process occurs in four distinct steps [6]. The first step is the hydrolysis, where the macromolecules (proteins, fats and carbohydrates) are broken down into smaller molecules (peptides, fatty acids and saccharides). It is catalyzed by exoenzymes called hydrolyzes produced by the fermentative bacteria, as shown in Eq. (8.2) [16]. Example of these bacteria include *Bacterioides succinogenes* and *Clostridium thermocellum*. Hydrolysis is the rate-limiting step in AD [2]. This highly depends on the substrate particle size, which should be maintained at around 3 mm in large-scale reactors to prevent a very rapid process that would lead to accumulation of VFA.

$$nC_6H_{10}O_5 + nH_2O \rightarrow nC_6H_{12}O_6 \qquad (8.2)$$

The second step is acidogenesis, where the smaller molecules are converted into volatile fatty acids (VFA) such as propionic, acetic and butyric acid, and other byproduct gases like ammonia, carbon dioxide, hydrogen sulfide, alcohols, and aldehydes by acidogenic bacteria (*Clostridium butyricum*); Eq. (8.3) [2,16].

$$nC_6H_{12}O_6 \rightarrow 3nCH_3COOH \qquad (8.3)$$

The third and fourth steps, i.e., the acetogenesis and methanogenesis involve conversion of the acetic acid into acetate, which is then converted into carbon dioxide and methane (Eq. 8.4) by acetoclastic methanogens such as *Methanosarcina* and *Methanosaeta*. The hydrogenotrophic methanogens also produce CH_4 by using CO_2 as a carbon source and hydrogen as a reducing agent (Eq. 8.5) [7,16].

$$CH_3COOH \rightarrow CH_4 + CO_2 \qquad (8.4)$$

$$CO_2 + H_2 \rightarrow CH_4 + 3H_2O \qquad (8.5)$$

Some oxidation-reduction reactions involved in the anaerobic digestion (AD) process can be summarized as follows (Table 8.5):

The oxidation reactions have a $\Delta G_0 > 0$; hence, the degradation of propionate and butyrate cannot occur under standard conditions and the reactions cannot shift to the right due to high amount of hydrogen ions. The continuous removal of hydrogen from the AD

Table 8.5: Oxidation-reduction reactions in anaerobic digestion.

		ΔG_0 (kJ/mole)
Oxidation reactions		
Propionate → acetate	$CH_3CH_2COO^- + 3H_2O \rightarrow CH_3COO^- + HCO^- + H^+ + H_2$	+76.1
Butyrate → acetate	$CH_3CH_2CH_2COO^- + 2H_2O \rightarrow 2CH_3COO^- + H^+ + H_2$	+48.1
Reduction reactions		
Bicarbonate → acetate	$2HCO_3^- + 4H_2 + H^+ \rightarrow CH_3COO^- + 4H_2O$	−104.6
Bicarbonate → methane	$HCO_3^- + 4H_2 + H^+ \rightarrow CH_4 + 3H_2O$	−135.6

ΔG₀, Gibb's free energy; *kJ*, kilojoules.

Modified from Marcus Von Sperling and Carlos Augusto de Lemos Chernicharo (2005), Table 24.1 (p.670) from Biological Wastewater Treatment in Warm Climate Regions, ISBN 9781843390022, © IWA Publishing.

system will ensure efficient completion of the process [17]. This can be achieved by the addition of alkalinity (electron acceptors) in the form of bicarbonate to effectively complete the methanogenesis step. The stability of the AD depends on several key parameters that must be maintained in order to provide a favorable environment for the maximum metabolic activity of the anaerobic bacteria.

8.2.1 Nature of the substrate

The physical and chemical nature of the input material determines the success of the AD process. Some of the chemical characteristics of food waste are summarized in Table 8.1. Trace elements (Fe, Ni, and Co) concentration in the substrates and the inoculum should be sufficient to start off the digestion process; however, supplementation might be required if the levels are very low to avoid acidification. The organic loading rate (OLR) differs depending on the substrate and the inoculum being used. According to a review by Ref. [16], the optimal OLR in a two stage AD was found to be 22.6 kg $VS/m^3/day$ for hydrogen and 4.6 kg $VS/m^3/day$ for CH_4, with a solid retention time (SRT) of 160 h and 26.7 days, respectively. Ref. [16] indicated that certain water-soluble substances like amino acids, proteins and sugars decreases with the age of plants, whereas lignin, hemicellulose and polyamides increase with the age of plants. This suggests that young vegetative materials produce more biogas than the older ones. Ref. [16] also reported that biogas production from animal products waste could be affected by the type, age, feeding and living conditions of the animals, and storage of the waste.

The physical size of the substrates could also limit the hydrolysis step, which is usually the rate-limiting step in AD [2]. Large materials are not easily hydrolyzed by the anaerobic bacteria and hence pretreatment is necessary to reduce the volume and the particle size. This will eventually increase the surface area for contact and microbial activity. Some common pretreatment methods include (1) mechanical (e.g., grinding, shredding and sieving): where the substrate particle size is reduced to less than 3 mm [8]; (2) thermal (i.e., heat treatment, greater than 120°C for 30 min), where the degradation of lipids increased from 67% to 84% with 190°C and methane production increases by 25% [2,16]; and (3) chemical, where the destruction of organic compounds occurs by the use of strong acids, alkali, and oxidants [18]. The effectiveness of the treatments depends on the composition of the substrates.

8.2.2 Temperature

This is a very important parameter in AD because it influences the activity of enzymes, methane yield, digestate quality and coenzymes [2]. Ref. [16] elaborated that it influences AD microbial systems that can affect the metabolic rate, bioavailability of trace elements and substrate solubility. Anaerobic bacteria can thrive in psychrophilic (<25°C), mesophilic (25–40°C) and thermophilic (>50°C) conditions [19]. Most research done so far on AD of food waste has been under mesophilic conditions [4,6,20] and all the authors have reported sufficient CH_4 yields of up to 581 mL/g-VS. However, Ref. [2] reported a higher CH_4 yield at 55°C than at 15°C. The authors also reported that thermophilic operating conditions can be advantageous for the degradation of organic nitrogen and phosphorous assimilation, but these systems are very sensitive to temperature fluctuation and are highly unstable as compared to mesophilic systems.

8.2.3 pH and volatile fatty acids (VFAs)

VFAs are the main byproducts of AD, with the main ones being acetic, propionic, valeric, and butyric acids [2], which are transformed into CH_4 and CO_2, respectively. At high OLR, the rate of VFA production overwhelm the metabolic activity of the microbes (indirect substrate-induced inhibition), thereby leading to acidification and low pH of the digesters, and finally the failure of the AD system. Table 8.6 shows the pH change in AD of food waste from previous studies. The production of VFA should be controlled to ensure the stability of AD [2]. The methanogenic bacteria require a pH range of 6.5–7.2 for growth. Ref. [16] reported that the methanogens work best at an optimum pH of 7.0. According to these previous studies, the VFA inhibit microbial activity at concentrations greater than 2 g/L, and fermentation at more than 4 g/L. This is because the VFA diffuse rapidly through the hydrophilic layers of the bacterial cell wall and dissociate into anions which are unable to diffuse out of the cell. This leads to their accumulation within the cell leading to an intracellular pH drop.

8.2.4 Carbon-nitrogen ratio

Carbon (C) and nitrogen (N) are important elements in AD where carbon is a source of energy and nitrogen is used for microbial growth. The optimum C:N ratio was reported to be between 20 and 30 by Ref. [16]. Other studies indicate that the AD was efficient at C:N ratio of 15–20 and 19.6, respectively [2]. Ref. [4] showed that high C:N ratio of raw food waste (24.5) and liquid food waste (55.8) contributed to low biogas yield as compared to cooked food waste (17.9) which had the lowest C:N ratio. Low nitrogen concentrations will lead to low microbial growth hence inefficient digestion of carbon, whereas high concentrations will lead to increased ammonia production, which leads to a decrease in the pH and increased toxicity to the methanogens. The C:N ratio of the substrate and inoculum should be well balanced for an efficient AD of food waste.

8.2.5 Types of reactors

AD can be carried out in different reactors depending on the amount and type of waste to be treated. They can be either batch or continuous reactors. Batch involves a single step digestion where the substrate and inoculum are loaded at desired ratios, commonly called inoculum substrate (I/S) ratio, and incubated for a period of 25–60 days [22]. The I/S

Table 8.6: Summary of pH change during AD of different food waste.

Food waste	Operation mode	Temperature (°C)	Biochar	OLR	HRT (days)	Initial pH	Final pH	References
Cooked food waste*	Semicontinuous	35	No	16 g VS/L/d	12	7.3	6.8	[4]
Raw food waste*	Semicontinuous	35	No	16 g VS/L/d	11	7.4	4.5	[4]
Raw food waste*	Semicontinuous	35	No	6 g VS/L/d	28	7.5	7.3	[4]
Kitchen waste	Batch	35	Vermicompost (5% *w/w*)	50 g TS/kg	50	7.0	4.9	[9]
White bread	Batch	35	Pine saw dust (16.6 g/L)	0.82 g (*w/v*)	39	7.0	3.5	[20]
White bread	Batch	35	No	0.82 g (*w/v*)	39	7.0	6.7	[20]
Mixed food waste	Batch	37	*Ampelodesmos mauritanicus*	5 g (*w/w*)	15	7.0	7.5	[21]

g TS/kg, gram of total solids per kilogram; *g VS/L/d*, gram of volatile solids per liter per day; *HRT*, hydraulic retention time; *No*, no biochar added; *OLR*, organic loading rate; *w/v*, weight by volume basis; *w/w*, weight by weight basis; *, food waste from a restaurant.

ratio applied determines the amount of CH_4 produced. Ref. [22] performed several previous AD research works and reported that mesophilic batch operations at S/I ≤ 0.5 and 1.0 produced 417–529 L CH_4 kg^{-1} VS_{fed}, and was enough to prevent acidification of the reactors, which is a common problem in the single step reactors. Higher S/I ratios could lead to VFA accumulation and digester failure hence the batch reactors are not suitable for substrates with a high organic content like food wastes [22].

In continuous AD digestion, the substrate is loaded into the reactor and a seed sludge (inoculum) is added to provide the desired anaerobic bacterial cultures. The substrate is fed continuously or at intervals in a semicontinuous AD where some sludge is removed before feeding of a new substrate. An example of a continuous AD reactor is the upflow anaerobic sludge bed (UASB) reactor, which produces biogas with a high CH_4 content of about 70% [23]. Fig. 8.1 provides an illustration of a typical large-scale anaerobic digester.

8.3 Biochar properties and role in anaerobic digestion

8.3.1 Biochar production and characteristics

When heated in absence of oxygen, biomass, such as food waste, releases, some volatile compounds, including gas, such as carbon monoxide (CO) or methane (CH_4), and condensates, namely water and tars. It also gives rise to a carbon-rich, porous solid, called "biochar". This solid has properties closer to coal than those of raw waste. It is sometimes referred to as "pyrochar" when the thermal process is slow pyrolysis and "hydrochar" in the case of hydrothermal carbonization (HTC). The main characteristics of these processes are summarized in Table 8.7. It is important to note that pyrolysis is suitable with dry waste only since it requires heating between 350 and 600°C under inert atmosphere. Pyrolysis is generally carried out at atmospheric pressure, using slow heating rate in the range of a few °C/min and solid residence time of several hours. Such conditions enable to reach biochar yields up to 35 w% of dry biomass [24]. In the case of HTC, biochar yield

Figure 8.1
Anaerobic digesters (complete mix manure digester, www.regenis.net).

Table 8.7: Main characteristics of biochar production processes.

	Pyrolysis	HTC
Waste type	Dry	Wet
T (°C)	350–600	180–250
Gaseous atmosphere	1 bar inert	Water above saturated P
Heating rate (°C/min)	#10	#10
Residence time (hour)	1–5	1–10
Biochar yield (% dry weight biomass)	20–30	45–70

is even higher and lies in the range of 50–80 w% of dry biomass [24]. In contrast to pyrolysis, this process has the major advantage of being suitable with wet waste like food waste. Indeed, the reaction usually takes place between 180 and 250°C in water above saturated pressure. However, while pyrolysis is well-mastered on different scales, from very small and robust reactors to larger and more advanced ones, HTC is still at the demonstration stage.

Noteworthy is the major influence of process conditions, as well as feedstock used, on biochar physical and chemical properties [24]. For instance, surface area and porosity generally tend to be higher when pyrolysis temperature increases, typically from 500 to 700°C. However, this is not always the case: tests with one sample of wheat straw showed lower porosity of biochar at 700°C compared to 600°C [25]. The abundance of functional groups (carboxylic, hydroxyl and amino groups) decreases with an increase in temperature due to the highest carbonization degree of biomass [25]. The alkalinity of biochar strongly varies with feedstock and its initial inorganic composition. Biochar is usually alkaline, with pH ranging from 6.9 to 9.5, but oak wood showed acidic values between 4.8 and 4.9, and wastewater sludge between 4.8 and 6.1 [25].

8.3.2 Biochar sorption mechanisms

The physical and chemical properties of biochar influence its ability to sorb metals. Refs. [6,25] reported that the surface area and porosity, pH and surface charge, functional groups and mineral composition played a key role in the adsorption process of the metals.

Metal adsorption on the biochar surface can be achieved by any one of the mechanisms (Fig. 8.2) described by Refs. [6,25]; (1) physical adsorption where the metals settle on the surface, (2) complexation and surface precipitation where the metals form layers on the surface, (3) pore filling where the metals condense on the pores of the biochar, (4) hydrogen bonding and hydrophobic effect, (5) ion exchange, and (6) electrostatic attraction.

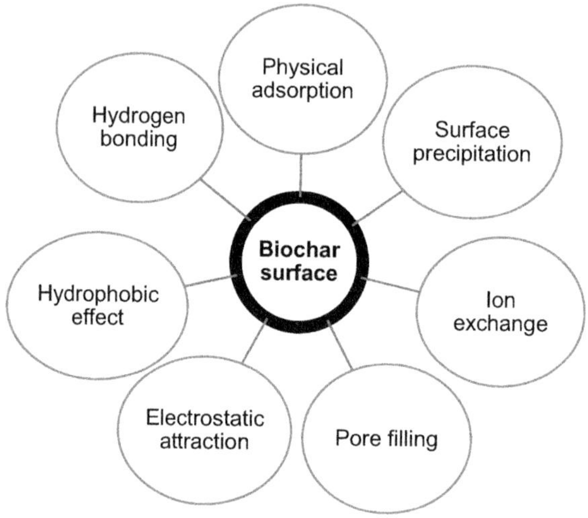

Figure 8.2
Mechanisms of metal sorption on the biochar surface. *Modified from Fagbohungbe MO, Herbert BMJ, Hurst L, Ibeto CN, Li H, Usmani SQ, Semple KT. The challenges of anaerobic digestion and the role of biochar in optimizing anaerobic digestion. Waste Management 2017;61:236–249.*

8.3.3 Role of biochar in AD

Recently, biochar has gained popularity in AD systems as it has been reported to reduce substrate-induced inhibition, increase the biogas yields and stability of the process (Table 8.8), which occurs in three ways [6]:

1. Adsorption of inhibitors
   Ammonia, long-chain fatty acids (LCFA) and heavy metals can be converted into toxic metabolites during AD which end up inhibiting the process. The addition of biochar as an adsorbent provides an alternative for the removal and reduction of the inhibitory effect of these compounds. Ref. [27] used three biochars for the AD of fruit waste and reported that the biochar alleviated the effect of limonene which led to high methane yields. Ref. [7] used fruitwood biochar in AD at elevated ammonium concentrations and reported that the biochar was able to alleviate ammonium inhibition by up to 7 g N/L.
2. Increasing the buffering capacity
   The buffering capacity of AD is achieved through digestion of the feedstocks, but it is also countered by the production of VFA which increase the acidity in the systems. The most common strategy to control the acidification is the addition of alkali compounds. Since most biochar are alkaline in nature, they can effectively be added in AD to increase the buffering capacity. Ref. [9] investigated the buffering capacity of pyrochar from vermicompost at 5%, 10%, 15% and 20% (*w/w*) biochar addition ratios on acetic, propionic, butyric and valeric acids. The authors reported that the buffering capacity

Table 8.8: Effects of biochar on the AD performance of different food waste.

Food waste (FW)	Biochar (B) source	Biochar type	Biochar addition (B:FW ratio)	Reactor type	Temperature (°C)	HRT days	Effect on AD	References
White bread from the supermarket	Pine sawdust	650°C	8.3, 16.6, 25.1, 33.3 g/L	Batch	35	8	- Increased the CH_4 production by 41.6% - Addition of 8.3 g/L biochar addition produced higher methane (55%–78%) while 33.3 g/L biochar addition showed the lowest yield	[20]
Municipal bio-waste	Holm oak residue	650°C	0%, 5%, 10% (by organic dry matter biochar/food waste)	Batch	40	30	- Increased CH_4 production by 5% (257–272 NL/kgODM) at 5% biochar ratio and 3% (252–267 NL/kgODM) at 10% biochar ratio	[26]

Continued

Table 8.8: Effects of biochar on the AD performance of different food waste.—cont'd

Food waste (FW)	Biochar (B) source	Biochar type	Biochar addition (B:FW ratio)	Reactor type	Temperature (°C)	HRT days	Effect on AD	References
Mixed kitchen waste	Vermicompost	500°C	5%, 15%, 20%: 50 g TS/kg	Batch	35	50	- Buffer to VFA impact, increased CH_4 production at 15% and 20% biochar ratio	[9]
Citrus peel waste	Wood, coconut shell, rice husks	450°C	3:1, 2:1, 1:1, 1:2	Batch	35	30	- Reduced the lag phase, increased the CH_4 yield and biofilm formation on the biochar surface	[27]
Mixed food waste	Ampelodesmos mauritanicus	450°C, 500°C, 550°C	0.9 (5 g of each biochar)	Batch	37	15	- pH increase from 7 to 7.5, cumulative CH_4 produced: 496.5 (450°C), 540.4 (500°C), 547.6 (550°C) ml CH4/g VS	[21]

HRT, hydraulic retention time; *NL/kgODM*, methane per kg of organic dry matter.

was higher with increased biochar addition ratio, and CH_4 content increased from 13.4% to 28.2% when the addition ratio was increased from 5% to 20%.

3. Immobilization of microbial cells
Immobilization is the process in which the microbial cells attach and grow on solid surfaces, and form biofilm layers. In AD, the immobilization of microbial cells allows the methanogens to have interspecies electron transfer thereby facilitating good CH_4 production [6]. The biofilms are important because they prevent wash out of microbial biomass and increase the acclimation rate of the microorganisms during substrate-induced inhibition. Ref. [27] carried out a comparative observation of the biochar before and after AD of sewage sludge and citrus fruit waste using scanning electron microscopy and observed a consortium of microorganisms attached to the pore spaces. This shows that biochar may also provide a surface area for the immobilization of microbial cells.

8.4 Conclusions and perspectives

The long-term AD of food waste is characterized by poor stability due to high accumulation of VFA which is mostly caused by the deficiency of essential trace elements such as manganese (Mn), cobalt (Co), nickel (Ni), zinc (Zn), and iron (Fe). TE supplementation is one of the solutions that can be used to stabilize the AD process depending on the inoculum/substrate ratio and the food waste characteristics. Biochar can also be used in the AD process due to its ability to inhibit the impact of metabolic byproducts such as ammonia on the methanogens, and it can also be a source of trace elements. The ability of biochar to enhance the AD process depends on its physical and chemical characteristics and hence the feedstock and operating conditions of the production process are very critical. Future studies should assess the suitability of the different types of biochar targeted for use in AD of food waste, which will ensure long-term stability of the digesters.

Acknowledgments

This work was sponsored by IHE Delft Institute for Water Education and the Dutch Government through the Netherlands Fellowship Program.

References

[1] Li L, Peng X, Wang X, Wu D. Anaerobic digestion of food waste: a review focusing on process stability. Bioresource Technology 2018;248(pt A):20–8.
[2] Zhang C, Su H, Baeyens J, Tan T. Reviewing the anaerobic digestion of food waste for biogas production. Renewable and Sustainable Energy Reviews 2014;38:383–92.

[3] Zhang L, Ouyang W, Lia A. Essential role of trace elements in continuous anaerobic digestion of food waste. Procedia Environmental Sciences 2012;16:102–11.
[4] Zhang C, Su H, Tan T. Batch and semi-continuous anaerobic digestion of food waste in a dual solid-liquid system. Bioresource Technology 2013;145:10–6.
[5] Chen Y, Cheng JJ, Creamer KS. Inhibition of anaerobic digestion process: a review. Bioresource Technology 2008;99:4044–64.
[6] Fagbohungbe MO, Herbert BMJ, Hurst L, Ibeto CN, Li H, Usmani SQ, Semple KT. The challenges of anaerobic digestion and the role of biochar in optimizing anaerobic digestion. Waste Management 2017;61:236–49.
[7] Lü F, Luo C, Shao L, He P. Biochar alleviates combined stress of ammonium and acids by firstly enriching Methanosaeta and then Methanosarcina. Water Research 2016;90:34–43.
[8] Qiang H, Niu Q, Chi Y, Li Y. Trace metals requirements for continuous thermophilic methane fermentation of high-solid food waste. Chemical Engineering Journal 2013;222:330–6.
[9] Wang D, Ai J, Shen F, Yang G, Zhang Y, Deng S, Zhang J, Zeng Y, Song C. Improving anaerobic digestion of easy-acidification substrates by promoting buffering capacity using biochar derived from vermicompost. Bioresource Technology 2017;227:286–96.
[10] Ortner M, Rameder M, Rachbauer L, Bochmann G, Fuchs W. Bioavailability of essential trace elements and their impact on anaerobic digestion of slaughterhouse waste. Biochemical Engineering Journal 2015;99:107–13.
[11] van Hullebusch ED, Guibaud G, Simon S, Lenz M, Yekta SS, Fermoso FG, Jain R, Duester L, Roussel J, Guillon E, Skyllberg U, Almeida CMR, Pechaud Y, Garuti M, Frunzo L, Esposito G, Carliell-Marquet C, Ortner M, Collins G. Methodological approaches for fractionation and speciation to estimate trace element bioavailability in engineered anaerobic digestion ecosystems: an overview. Critical Reviews in Environmental Science and Technology 2016;46(16):1324–66.
[12] Banks CJ, Zhang Y, Jiang Y, Heaven S. Trace element requirements for stable food waste digestion at elevated ammonia concentrations. Bioresource Technology 2012;104:127–35.
[13] Yazdanpanah A, Ghasimi DSM, Gu Kim M, Nakhla G, Hafez H, Keleman M. Impact of trace element supplementation on mesophilic anaerobic digestion of food waste using Fe-rich inoculum. Environmental Science and Pollution Research 2018.
[14] Hendriks ATWM, van Lier JB, de Kreuk MK. Growth media in anaerobic fermentative processes: the underestimated potential of thermophilic fermentation and anaerobic digestion. Biotechnology Advances 2018;36:1–13.
[15] Thanh PM, Ketheesan B, Yan Z, Stuckey D. Trace metal speciation and bioavailability in anaerobic digestion: a review. Biotechnology Advances 2016;34:122–36.
[16] Kondusamy D, Kalamdhad AS. Pre-treatment and anaerobic digestion of food waste for high rate methane production - a review. Journal of Environmental Chemical Engineering 2014;2:1821–30.
[17] de Lemos Chernicharo CA. Anaerobic reactors. In: Biological wastewater treatment in warm climate regions, vol. 1. London, UK: IWA Publishing; 2005. p. 659–827.
[18] Ariunbaatar J, Panico A, Esposito G, Pirozzi F, Lens PNL. Pretreatment methods to enhance anaerobic digestion of organic solid waste. Applied Energy 2014;123:143–56.
[19] El-Mashad HM, Zeeman G, van Loon WKP, Bot GPA, Lettinga G. Effect of temperature and temperature fluctuation on thermophilic anaerobic digestion of cattle manure. Bioresource Technology 2004;95:191–201.
[20] Sunyoto NMS, Zhu M, Zhang Z, Zhang D. Effect of biochar addition on hydrogen and methane production in two-phase anaerobic digestion of aqueous carbohydrates food waste. Bioresource Technology 2016;219:29–36.
[21] Luz FC, Cordiner S, Manni A, Mulone V, Rocco V, Braglia R, Canini A. *Ampelodesmos mauritanicus* pyrolysis biochar in anaerobic digestion process: evaluation of the biogas yield. Energy 2018;161:663–9.
[22] Braguglia CM, Gallipoli A, Gianico A, Pagliaccia P. Anaerobic bioconversion of food waste into energy: a critical review. Bioresource Technology 2018;248(pt A):37–56.

[23] Xu HL, Wang JY, Zhang H, Tay JH. A comparative study of anaerobic digestion of food waste in a single pass, a leachate recycle and coupled solid/liquid reactors. Water Science and Technology 2002;47(1):319–24.
[24] Libra JA, Ro KS, Kamman C, Funke A, Berge ND, Neubauer Y, Titirici MM, Führer C, Bens O, Kern J, Emmerich KH. Hydrothermal carbonization of biomass residuals: a comparative review of the chemistry, processes and applications of wet and dry pyrolysis. Biofuels 2011;2:71–106.
[25] Li H, Dong X, da Silva EB, de Oliveira LM, Chen Y, Ma LQ. Mechanisms of metal sorption by biochars: biochar characteristics and modifications. Chemosphere 2017;178:466–78.
[26] Meyer-Kohlstock D, Haupt T, Heldt E, Heldt N, Kraft E. Biochar as additive in biogas-production from bio-waste. Energies 2016;9(4):247.
[27] Fagbohungbe MO, Herbert BMJ, Hurst L, Li H, Usmani SQ, Semple KT. Impact of biochar on the anaerobic digestion of citrus peel waste. Bioresource Technology 2016;216:142–9.
[28] Zhang L, Lee Y-W, Jahng D. Anaerobic co-digestion of food waste and piggery wastewater: focusing on the role of trace elements. Bioresource Technology 2011;102:5048–59.
[29] Zhang R, El-Mashad HM, Hartman K, Wang F, Liu G, Choate C, Gamble P. Characterization of food waste as feedstock for anaerobic digestion. Bioresource Technology 2007;98:929–35.
[30] Zhang B, Zhang LL, Zhang SC, Shi HZ, Cai WM. The influence of pH hydrolysis and acidogenesis of kitchen wastes in two-phase anaerobic digestion. Environmental Technology 2005;26(3):329–39.

SECTION C

Food waste and chitin based biorefinery

CHAPTER 9

Theory of planned behavior on food waste recycling

Tiffany M.W. Mak[1], Iris K.M. Yu[1,2], Daniel C.W. Tsang[1]

[1]Department of Civil and Environmental Engineering, The Hong Kong Polytechnic University, Kowloon, Hong Kong, China; [2]Green Chemistry Centre of Excellence, Department of Chemistry, University of York, York, United Kingdom

9.1 Introduction

Extensive attention has been paid by policy makers and institutions at various administrative levels to the issue of sustainable food management [1–4]. A vast quantity of food available for individuals' consumption is wasted in various stages along the food supply chain. Globally, food losses and waste account for approximately 20% of supplied food for human consumption [5]. "Food losses" refer to those losses in the processes of preparation, postharvest and processing, and "food waste" refers to wastage during distribution and consumption stages [1,5]. In countries with higher gross domestic product per capita nominal such as Switzerland (USD$85,157) and Singapore (USD$62,984), food distribution and consumption account for the biggest wastage in household food waste. In contrast, countries with lower gross domestic product per capita nominal, such as the Central African Republic (USD$490), have the highest food loss in the agricultural and postharvest stages [5,6]. Annually around the globe, one-third of food, which is equivalent to 1.32 billion tonnes of food produced for human consumption, is lost or wasted [1]. In 2007, over 3 giga tonnes of carbon dioxide were released from food wastage that includes agricultural production, postharvest at handling and storage, processing, distribution, and consumption [7]. Organizations, such as the European Commission, are taking preventive measures to resolve the problem, reduce economical costs, alleviate the environmental impacts of food wastage, and prevent social impacts related to this phenomenon [8–10]. To address such a global issue, a well-established theory is indispensable to illustrate determinants in bettering sustainable food management to assist in related decision-making processes. The literature review of this paper would systematically introduce the development of the theory of planned behavior (TPB), its current implementation to

Waste Biorefinery. https://doi.org/10.1016/B978-0-12-818228-4.00009-5

predict food consumption pattern and to promote safe food handling, and food waste recycling behavior in household and commercial sectors.

9.2 Development of the theory of planned behavior

Important predictors of recycling behavior have been identified as environmental attitudes and situational and psychological variables. However, a theoretical framework for systematically identifying the determinants of recycling behavior is required to explore the further implications of these factors. The TPB as noted by Ajzen [11], provides such a theoretical framework. TPB has been applied successfully in many areas such as investigating dishonest actions [12]. It is extended from the earlier theory of reasoned action (TRA) as suggested by Ajzen and Fishbein [13]. From the original TRA, the major factor in TPB is the individual's intention leading to a behavior. Intentions involving motivational factors influence behavior, which indicate the extent that individuals are willing to attempt or plan to take an action. Generally, the stronger the intention, the more likely one would turn the intention into an action. Both TRA and TPB can be applied to situations involving choices of behavior and reasons could be provided for justifying such actions [14].

As shown in Fig. 9.1, TRA hypothesizes two factors that influence intentions, including attitude and subjective norm. "Attitude" refers to an individual's favorable or unfavorable evaluation leading to a behavior while "subjective norm" refers to the perception of social pressure leading to a behavior [13]. For example, if one believes that recreational drug use (the behavior) is acceptable within one's social group, one will more likely be willing to engage in the activity. Alternatively, if one's friends groups perceive that the behavior is

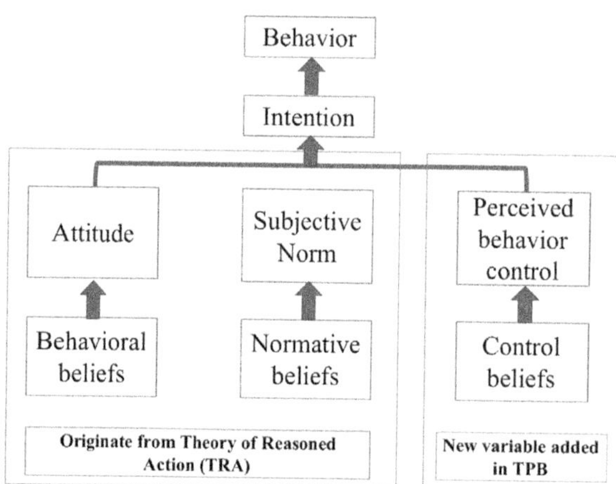

Figure 9.1
Framework of TPB and TRA.

bad, one will be less likely to engage in recreational drug use. TRA assumes that most behaviors are under volitional control and individuals can decide on their own whether or not to take an action. Liska [15] suggests that such behavior would be restricted by the lack of resources. In view of this, TPB could be extended to TRA by proposing the third variable – that is perceived behavioral control. It measures an individual's perception of the ease or difficulty in having a certain behavior. The concept is most compatible with Bandura's concept of perceived self-efficacy, which concerns "judgments of how well one can execute courses of action required to deal with prospective situations" [16]. Previous research conducted by Adams and Beyer [17] suggested that individuals' actions are greatly influenced by their confidence in their ability to perform them (i.e., by perceived behavioral control). Therefore, perceived behavioral control can be used directly in TPB to predict behavioral achievement. Overall, the relative importance of attitude, subjective norm, and perceived behavioral control in the prediction of intention vary across different conditions. TPB is currently one of the most popular and well-established social-psychological models to understand and predict human behaviors. Generally, the more favorable the attitude and subjective norm with respect to engaging in the behavior, and the greater the perceived behavioral control, the more likely an individual would come up with an intention, which may turn into a behavior. Beyond the factors that constitute the theory itself as discussed above, the potential importance of other variables in TPB, such as demographic characteristics, personality traits and emotions, were recognized. Previous literature demonstrates that demographic characteristics such as age would affect food waste generation. Food waste generation rate decreased dramatically when age increased in Australia [18], and the United Kingdom [19]. These variables are considered as background factors in TPB, which are expected to indirectly influence intentions and behavior [20].

Empirical validation of TPB is well-justified, with researches indicating that it reliably explains 40%–50% of the variance in intention, with intention subsequently explaining 20%–40% of the variance in behavior [21]. Although TPB is well accepted as an important framework for predicting behavior and health behavior specifically, it could not capture all the determinants of a more complex behavior. This may be explained by other related researches which do not incorporate exploratory studies in investigating the nature of the behavioral beliefs. Aizen argued that measuring underlying beliefs is of utmost importance as attitudes, intentions, and behaviors are most successfully altered when such beliefs are fully understood [22].

9.2.1 Current implementation of TPB on food management study

TPB models have been widely adopted in food management from pre- to post-consumer processes. They assist in predicting the behavior of individuals to provide conceptual order that allows decision makers to identify the behavior-driving substantive elements and to design effective interventions. A systematic literature review was conducted to identify relevant studies on two aspects, including studies predicting food consumption patterns and food handling, and food waste recycling in household and commercial sectors.

9.2.1.1 Application of TPB to predict food consumption and food handling behavior

Sustainability is defined as an integrated consideration of economic, environmental, and social aspects. Economically, there should be a balance between agricultural entrepreneurs and consumers. Environmentally, sufficient attention should be paid to the natural environment, including biotic and abiotic factors, the living environment and the quality of life for individuals. Socially, one should be concerned about how production processes match the priorities and needs of the society, thus implementing sustainability supporting policy. Consumers show increasing demands for convenience foods to reduce time and effort [23], and a growing concern on consumers consciously purchasing ethical or sustainable products [24]. Sustainable consumption is based on a decision-making process that considers not only individual needs such as taste, price, convenience, and health but also takes consumers' social responsibility into account [25]. Studies focus on investigating attitudes toward sustainability and sustainable consumption behavior. For instance, consumer attitudes are investigated with meat quality labels, which are introduced as part of the marketing response strategy by the meat industry. Results indicate that there are significant differences in consumer attitude toward meat quality labels across time and across age, gender, education level, buyer status and claimed television impact. Also, meat quality labels are a valuable and promising part of response strategies by the meat sector to negative media coverage [26]. On the other hand, consumption pattern of a green consumer is investigated in terms of variables directly related to purchase behavior, such as price consciousness and general care in shopping, interest in new products, and brand loyalty. Results suggest a green consumer to be an opinion leader and a careful shopper, who seeks comprehensive information on products and is rather skeptical of advertising [27,28].

Previous research suggested that perceived behavioral control reflected both inner factors such as attitude and external perceived factors such as perceived behavioral control [29], particularly significant in, perceived product availability [30] and perceived consumer effectiveness [31]. "Perceived availability" refers to a consumer feeling in terms of the ease to purchase or consume a product. Although consumers might be motivated to buy sustainable products, the intention to purchase sustainable products might be hampered if there was low perceived availability of the product [32]. "Perceived consumer effectiveness" refers to the extent that consumers thought that their personal efforts could contribute to ease the environmental problem, which stimulates consumers to express their positive attitudes toward sustainable products in actual consumption [31]. Previous literature also indicated that purchase intention of sustainable products could be independently predicted by attitudes, perceived behavioral control, and subjective norms [33]. Some studies argued that research on food consumption patterns should include self-related variables [30,33]. Contextual factors would prohibit positive attitudes from being expressed in action, which personal or situational factors predicted or translated the extent of attitudes that influence behavior intention [34]. For instance, researchers argued that youngsters have higher intention to buy sustainable products as they may be more interested about or aware of the potential impact of specific food production practices [35]. It was also significant that the more environmentally concerned an individual was, the more probably he/she would buy sustainable products [36].

In the context of food consumption, researches considered new factors such as perceived moral obligations into the TPB model [37–39]. However, studies casted doubts on the effect of the predictive ability of a moral measure on individual's behavior. Sparks et al. [38] believed that the inclusion of moral obligation only slightly assisted in the prediction of intentions while other researchers failed to find any significant improvement [40,41]. In a previous study of the moral issues in the context of organic foods, controlling attitudes, and subjective norms would influence consumers' choice between organic and conventional wine in the measure of personal norms [42]. The latter was also suggested as the most important determinant of consumers' ratings of their purchase frequency on various organic foods [43]. In addition, consumer's confidence in products and human values were considered as possible self-related factors on intention [32]. Generally, consumers were not confident in evaluating food quality [44], and to purchase sustainable products [33]. If an individual has relatively high confidence in the outcome of his/her behavior, he/she intends not to solely consider the behavior or opinion of others as a major source of information [43]. On the other hand, human values are considered as possible influencers of behavioral intentions toward sustainable food, which refer to food that is healthy for consumers and produced in an humane, ecologically benign, socially responsible and economically fair approach, such as grass-fed beef or lamb and organic eggs from local farmers [28]. Human values refer to personal or social desirability of behaviors and modes of existence [45,46]. Individuals living in a stable environment would result in cultivating stable values, which influenced both their sustainable attitudes and behavior in areas such as recycling [47] and green procurement [48]. Previous literature examines the impact of different cultural and psychological factors on green procurement behavior in China by constructing structural equation modeling. Results confirm the influence of consumers' degree of collectivism, ecological effect, and ecological knowledge on their attitude toward green procurement [48]. It was also indispensable in consumer decision-making processes, such as sustainable food choice [49]. Previous studies linked sustainable behavior to personal values [36,50–53]. Causal relation between certain values such as universalism, and a sustainable consumption pattern was confirmed, and it boosted these values through socialization and national institutions that could achieve sustainable consumption in the long run [53].

Apart from predicting food consumption pattern, food-handling behavior is also a popular application of TPB due to an increasing concern in food safety, which has been a global concern [54] that affects individual health and increase social expenses on medical welfare [55,56]. In particular, approximately one-quarter of the population in Australia and North America has suffered from illnesses caused by foodborne pathogens every year [57,58]. Recently, it has been evident that the number of foodborne diseases is increasing [57]. To reduce foodborne diseases, it was essential to handle food properly at all stages from preparation, storage, to disposal [59]. Previous research investigated people's knowledge about food safety behavior. For instance, consumer food safety information was compared and found that individuals had sufficient knowledge on cross-contamination [60]. Previous research explained and predicted safe food-handling behaviors by using various theoretical frameworks, including the health action process approach [61], and the health belief model [62,63]. However, TPB appeared to account for the most variance in behavior [64–69], applying to both overall safe food-handling behavior and specific behaviors such as hand hygiene [70].

In the context of safe food-handling behavior, TPB explained 34% of the variance in hand hygiene malpractices in the workplace [64], and the TPB model could successfully predict food safety practices in small-scaled food retailers [64]. The theory has also applied to predict consumer food-handling practices among Australian young adults, explaining over 60% of the variance in intention and over 20% of the variance in behavior. It was also revealed that only subjective norms and perceived behavior control were significant predictors, instead of attitudes [65]. Studies also show that TPB predicted 79% of intention and 97% of self-reported hygiene practice [71]. It is common to find that individuals are unable to translate their positive intention into behavior, which is often referred as "intention–behavior gap" [72]. For instance, intention was predicted to account for only about 20% of the variance in safe food-handling behavior, i.e., 80% remained unexplained [65]. Consequently, TPB was criticized as an incomplete model due to volitional characteristics [73]. Therefore, new variables were included to improve the predictive power of the model and to explain the phenomenon of "intention–behavior gap" that reflects the underlying psychological process that leads from intention to action [74–76]. Moral norm, which acted as both a preintentional predictor [77,78] and a direct predictor of behavior [79], has been added to the standard TPB to bridge the gap between intentions and behavior [80]. To investigate the influence of moral norms on actual behavior, a moderation analysis was conducted with data collected from five previous studies [79]. The study demonstrated that "morally aligned intentions," which formed on the basis of the perceived moral correctness, was a better predictor of behavior than "attitudinally aligned intentions" formed based on the likely outcomes [79]. Another commonly added variable to TPB was habit strength, which narrowed and explained the intention–behavior gap [81]. Habit strength refers to the degree to which a behavior becomes habitual or automatized [82]. It is particularly important in determining safe food-handling behavior in routine and regular food preparation. Researcher discovered that the conscious intention was unnecessary when behavior was constant in stable conditions and became habit eventually. Since safe food-handling behavior was typically regular, it was possible that the majority of people turn such behavior into habits [83]. In addition, habit and past experience were important determinants to engage in future safe food-handling behavior [84].

9.2.1.2 Application of TPB on household and commercial food waste recycling

Many resources such as energy, water, and land are required to produce food, and a significant portion of the greenhouse gases were emitted from households [85,86]. In the United Kingdom, it currently costs a family an estimated £680 a year to purchase and dispose of food without eating. Greenhouse gases which are equivalent to approximately 17 million tonnes of carbon dioxide are released [87,88]. In view of the serious environmental impacts, previous research has been performed to investigate food waste recycling in relation to consumers' perceptions and behaviors [19,89–93].

Some situational characteristics were identified in relation to the amount of household food waste. For example, the larger the household size was, the more food was wasted [94–97]. However, larger households generated less waste per capita than that of smaller households [19]. In particular, households with more children tended to generate more food waste by the means of calculating waste weights and conducting surveys with households in Oregon and Guelph [94,97], and parents reported challenges in predicting the quantity of food that children consume [90]. Moreover, as different family members preferred different types of foods, a large variety of foods was available in the market [98]. In the literature, a wide range of predictors were suggested that affected households' food waste recycling behavior such as awareness of food disposal [99], lifestyle [100], recycling attitude and behavior [101], impacts of recycling [99], and availability of packaging technologies and storage area [102]. Sociocultural drivers were quantified by previous studies, such as available knowledge of food waste [99], and the interaction of diverse factors along the globalized food chain [102]. For instance, the belief that awareness determines intention which in turn determines behavior has resulted in various campaigns seeking to educate consumers and provide guidelines to food waste reduction, such as Love Food Hate Waste, Feeding the 5000 and Think Eat Save in the United Kingdom. Multifaceted policy levers and public commitment are essential to improve performance in various major aspects of values, skills, and logistics [100,103].

Recognizing major factors to reduce household food waste and barriers is crucial in conducting qualitative research. Nevertheless, researchers argued that investigations should be theory-driven to discover the determinants of potentially modifiable behaviors. Theories can provide a framework to identify causal processes, which facilitate drawing up and implementing constructive, replicable, and parsimonious policies [104,105]. It is suitable to conceptualize the relationship between attitudes and behavior even when behavior is self-reported, according to the TPB principle. The TPB accounted for more than 11% of the variability in behavior, no matter the behavior is objective or observed [21,106]. It was also proved additional role of concepts can be easily and flexibly adopted upon the scope of the original model [107].

In the context of household food waste, the role of food-related practices and the core cognitive constructs specified by the TPB are explored. In particular, researchers investigated the impacts of attitude, subjective norm, and perceived behavioral control on individual's intention to reduce household food waste. Results revealed that only attitude is a significant factor to predict intention not to waste food, which comprised two constructs, i.e., moral attitude and lack of concern. Evidence showed that neither subjective norm nor perceived behavioral control drove intention. In addition, cross-sectional food waste behavior was not significantly related to intention of food waste reduction [91]. In the domestic sector, marketing and sales strategies of shops affect critical individuals' food-wasting habits [108]. The key predictors of domestic food waste behavior are mainly associated with attitude [106], followed by moral norms [109] and perceived behavioral control [110], while an indirect impact is caused by reuse/recycling habits [111]. Consumers' attitudes toward food waste are divided into two groups of measured variables, which are moral aspects and concern-based variables [91,112]. Results revealed that the moral aspects of attitudes significantly affected food waste in comparison to concern-based variables as consumers felt guilty when they wasted food [91,98,113]. TPB is further extended and applied to predict household waste collection behaviors among Iranians, which include attitudes, subjective norms, perceived behavioral control, moral obligations, self-identify, intention, action planning, and past behavior [114]. Researchers also applied TPB model to identify that culture, participation dimensions, and reputational concerns played important roles in influencing recycling behavior and shaping proenvironmental behaviors [115,116]. Previous literature explores the relationship between culture and waste recycling with data collected from the Italian Multipurpose Survey on Household Daily Life Aspects 2007. The survey was conducted by the National Institute of Statistics in 2007 on 19,170 Italian households for a total of 48,253 individuals [115]. Moreover, nonmonetary incentives such as gaining good or bad reputation toward household waste recycling are investigated with the modified public goods experiments in Costa Rica. The experiment can be characterized by three main features. First, four players can make contributions to a public account to ensure provision of a public good. Second, the public good is only provided in case of that contributions of a group of four players do reach a required preset threshold or target. The latter is a typical feature of threshold public goods game. Third, unlike traditional versions of the game, there is no direct redistribution of benefits to participants. Instead, the money of the public account is donated to a local NGO to finance services such as the set-up of recycling workshops in the community. Results indicate that individuals cooperate more if the situation is framed as avoiding shame (bad reputation) rather than as acquiring pride and gratitude (good reputation) [116]. The use of economic incentives, legislation, and public education are implemented in pilot recycling projects to motivate citizens. However, it is challenging to specify the exact effects of these factors through direct observation while previous studies suggest a volatile relationship between these factors and individual behaviors [117].

Previous studies suggested the significance of intention in food waste reduction [91,93]. However, planning and shopping routines were identified as additional determinants in the explanatory model and intention was not a significant determinant of food waste behavior anymore [91]. Both the intentions and behavior of consumers were affected by marketing strategies carried out by retailers. For instance, promotional offers in limited time drive consumers to purchase excessive quantities of food, emerging as one of the major factors of food waste generation [118,119].

Other daily activities are also considered to bring substantial behavioral changes in households. Firstly, understanding the food labeling information is essential and it has often been misunderstood by consumers. Recently in the European Union, a consumer market survey showed that only about 30% of consumers understood the meaning of the "best before" date. The meaning of the food labeling, including "best before," "expiry date," and "use by," are clarified to improve customer certainties and knowledge of food edibility. In particular, the "best before" date is related to the minimum durability while the "use by" date is related to safety, which can assist consumers to make informed decisions [120]. Secondly, consuming household leftovers is crucial as it can save money and reduce household food waste. For instance, household individuals can recook leftover from previous meals for next-day lunch at office. However, educational campaigns concerning the reuse of leftovers are inadequate. In Belgium, a series of cooking courses were arranged for citizens, targeted to assist households to reduce the food waste generation and increased their flexibility in meal planning. Lastly, a shopping list is advised. Planning routines can avoid unplanned purchases [91,121,122], and preparing a shopping list by checking food stocks prior to shopping can minimize food waste generation [119,123,124].

In view of the existing household solid waste separation and collection, scholars start to concern about the formation mechanism of household solid waste recycling behaviors. Such fundamental understanding helps promoting knowledge of the available recycling programs in community and thus encouraging the participation of individuals. Researchers believed that public relations could be used as an important platform to motivate involvement of individuals in recycling programs. Public relations activities such as separate collection of waste glass and environmental label must target at specific groups, with carefully designed projects, an analysis of the target group and choice of media. It was observed that there was limited public participation in recycling despite the strong governmental support and encouragement [125]. Concerns are thus raised to address such discrepancy. An extensive meta-analysis of over 60 empirical studies on the effect on recycling behavior by several variables was conducted. Incentives for social behavior and barriers to social behavior, which could either be internal or external to the individual were discovered [126]. An analysis of household recycling by apartment dwellers was conducted and two related strategies to motivate daily recycling were suggested. First, containers were placed at accessible locations for the convenience of residents nearby. Second, food was recycled during the preparation process [127]. Similarly, researchers proposed a different theoretical approach in recycling. An open-ended questionnaire was distributed to individuals to find out whether manageability contributes to desired recycling by organizing their activities with regard to effective self-regulation. It is indispensable for individuals to understand their environment in order to support desired behavior [128]. Social influence is defined as a concern over how one's recycling behavior be affected by friends and families. Social influence can significantly affect and sustain recycling behavior [129]. Moreover, motivation has a strong influence on recycling when individuals feel satisfied by contributing to the environment [130].

Recycling in which household waste must be sorted, prepared, and stored, is a behavior requiring substantial effort from individuals [131]. As a result, the recycling decision involves complex consideration of various factors, such as convenience. Researcher discovered that an effective motivator to drive recycling could make recycling with greater convenience [132]. Recently, researchers paid extensive attention to the motivational factors behind recycling attitude and behavior, aiming to isolate specific characteristics that contribute to recycling participation. The need to understand the influences of consumer environmental behavior is emphasized and predictors to such behavior are identified. To conclude this study, there was insignificant relationship between social norm and behavior [133]. It was contrary to the findings of a study conducted in California, in which peer pressure appears as a major predictor or motivational factor of recycling behavior. This implied that individuals intend to make more socially responsible decisions when their peers recycle [134]. In view of the determinants that promote sorting and collection, the formation of strong recycling habits across communities was investigated, and the social influences and altruistic and regulatory factors were considered important [129,135]. The TPB is combined with norm-activation theory to explain that recycling intentions are affected by perceived policy effectiveness in Hong Kong [136]. The importance of public understanding is emphasized in participation rates in recycling. Attention should be paid to the frequency and effectiveness of households instead of the number of householders participating in recycling [137]. The awareness of local authority and promotion campaigns are important as poorly designed and implemented campaigns lead to constantly low recycling participation rates [138].

Apart from investigating recycling behavior in households, the TPB has been applied in diverse areas such as the commercial sectors to understand food waste recycling. There is little discussion in the literature. An extended TPB was applied to investigate food waste recycling in commercial sector in Hong Kong. Researchers identified, prioritized, and quantified the relationships between key determinants that affect the food waste recycling behavior of hotel, food and beverages, and property management industries. Integrated qualitative content analysis and quantitative structural equation modeling were conducted to demonstrate that commercial food waste recycling behavior was influenced by three predictors, which were administrative incentives and corporate support, logistics and management incentives, and economic incentives. Predictors explained over 60% of the variance of recycling intention, which demonstrated substantial strength of the model. Indicative positive correlations among moral attitudes, logistics, and management incentives were observed. In particular, hotel, and food and beverages industries expressed great concern on administrative incentives and corporate support, whereas property management representatives paid greater attention to logistics and management incentives to promote recycling activities [139]. Moreover, a recent investigation discovers that insufficient resources can lead to difficulties in lowering food waste disposal and changing recycling behavior in restaurants [140].

9.2.2 National food waste policies and economies of food waste recycling

9.2.2.1 National food waste policies around the globe

In the United States, the proposed Food Recovery Act (2017) includes, among other things, federal grant and loan programs designed to improve food labeling and increase consumer awareness of food waste [141]. These principles are the centerpiece of the United States Department of Agriculture's effort to cut food waste in half by 2030 [142]. This is in line with the United Nation's Sustainable Development Goal of reducing global food waste at the retail and consumer level by 50% over this period [143]. Australia established a National Food Waste Baseline and methodology to measure progress to 2030, in which the policy goal is to achieve 50% reduction in food waste. In 2017, the Department of the Environment and Energy commissioned work to identify accessible food waste data sources for Australia, including amount of food waste entering or being diverted from landfills, food waste reduction by industry sectors or within supply and consumption chains, and the level of carbon emissions from Australian landfills [144]. The United Kingdom has implemented Waste and Resources Action Program (WRAP) to deliver consumer campaigns and voluntary industry commitments under their Courtauld Commitment 2025 [145].

9.2.2.2 Economies of food waste recycling

In order to quantify the costs and benefits of implementing food waste recycling, life-cycle cost (LCC) analysis is often employed. Cost-benefit of on-site food waste recycling system is evaluated and compared with large-scale treatment system in South Korea. From the local government point of view, results indicate that when large-scale centralized treatment system was replaced with on-site recycling system, there was significant cost reduction from the initial stage due to reduced investment, maintenance, and food wastewater treatment costs. As for the residents, the cost of using the on-site recycling system was higher than that of the large-scale system due to the facility installation cost at the initial stage. However, when taking continuous benefits into account (such as greenhouse gas emission reduction, compost utilization, and food wastewater reduction), cost reduction would result 6 years after the commission of the on-site recycling system [146]. Apart from LCC, cost-benefit analysis (CBA) is another common approach to examine the economic viability of projects. A life-cycle cost-benefit analysis (LC-CBA) framework, which integrates the life-cycle assessment (LCA) and CBA, guides decision-making in sustainable food waste management at the Hong Kong International Airport. Results indicate that on-site incineration scenario is the most sustainable option with the lowest life-cycle net costs of HKD 461.73/tonne (i.e., USD$59/tonne). The scenario achieves the highest energy recovery of 707 kWh/tonne, which leads to an economic savings of HKD 697.81/tonne (i.e., USD$89.2/tonne) and an environmental savings of HKD 470.96/tonne (i.e., USD$60.2/tonne) [147].

Another study investigates the overall sustainability of six scenarios for potato loss reduction in Switzerland [148]. Environmental, socio-economic and consumer dimensions are examined by conducting LCA, full-cost calculations, and an online consumer survey. Pearson correlation coefficients and linear regression analyses predict the influence of specific subjective items (e.g., intention) to avoid food loss, knowledge related to food loss, and consumers' price sensitivity on the assigned preference. Results show that perceived risks, perceived inconvenience, and the general acceptance of loss-reducing instruments influence consumers' preferences. Altogether, only three out of six studied scenarios seem realistic: selling unwashed potatoes in a lightproof box, selling unpacked potatoes, and improved quality of sorting at farms.

9.3 Conclusions and perspectives

In summary, this chapter provides comprehensive reviews on the current use of TPB in diverse areas of sustainable food management, including the prediction of food consumption pattern, safe food-handling behavior, and food waste recycling behavior in household and commercial sectors. In the context of food waste recycling, future research would be benefited from investigating the intermediate or moderating effects of determinants on recycling intention. Also, further studies may focus on exploring the price sensitivity of households and corporate toward their willingness to pay for food waste management, and the track of their actual separate sorting, collection, and recycling behaviors. Since the above studies are self-reported measures, there could exit bias that leads to under- or overestimation of the food consumption, food-handling, and recycling frequencies. Therefore, future researches should minimize such limitation by using objective measures.

References

[1] Gustavsson J, Cederberg C, Sonesson U, van Otterdijk R, Meybeck A. Global food losses and food waste: extent, causes and prevention. Rome: Food and Agriculture Organisation of the United Nations (FAO); 2011.

[2] Katajajuuri JM, Silvennoinen K, Hartikainen H, Heikkil€a L, Reinikainen A. Food waste in the Finnish food chain. Journal of Cleaner Production 2014;73:322–9.

[3] Miliute-Plepiene J, Plepys A. Does food sorting prevents and improves sorting of household waste? A case in Sweden. Journal of Cleaner Production 2015;101:182–92.

[4] Secondi L, Principato L, Laureti T. Household food waste behavior in EU-27 countries: a multilevel analysis. Food Policy 2015;56:25–40.

[5] Kummu M, de Moel H, Porkka M, Siebert S, Varis O, Ward PJ. Lost food wasted resources: global food supply chain losses and their impacts on freshwater, cropland, and fertiliser use. The Science of the Total Environment 2012;438:477–89.

[6] Parfitt J, Barthel M, Macnaughton S. Food waste within food supply chains: quantification and potential for change to 2050. Philosophical Transactions of the Royal Society B 2010;365:3065–81.

[7] FAO. Food wastage footprint impacts on natural resources: summary report. 2013 [Online], http://www.fao.org/docrep/018/i3347e/i3347e.pdf/.

[8] Graham-Rowe E, Jessop DC, Sparks P. Identifying motivations and barriers to minimising household food waste. Resources, Conservation and Recycling 2014;84:15–23.

[9] Garrone P, Melacini M, Perego A. Opening the black box of food waste reduction. Food Policy 2014;46:129–39.

[10] G€obel C, Langen N, Blumenthal A, Teitscheid P, Ritter G. Cutting food waste through cooperation along the food supply chain. Sustainability 2015;7:1429–45.

[11] Ajzen I. The theory of planned behavior. Organizational Behavior and Human Decision Processes 1991;50:179–211.

[12] Beck L, Ajzen I. Predicting dishonest actions using the theory of planned behavior. Journal of Research in Personality 1991;25:285–301.

[13] Ajzen I, Fishbein M. Understanding attitudes and predicting social behavior. Prentice-Hall; 1980.

[14] East R. Investment decisions and the theory of planned behavior. Journal of Economic Psychology 1993;14:337–75.

[15] Liska AE. A critical examination of the causal structure of the Fishbein/Ajzen attitude–behavior model. Social Psychology Quarterly 1984;47:61–74.

[16] Bandura A. Self-efficacy: toward a unifying theory of behavioral change. Psychological Review 1977;84:191–215.

[17] Bandura A, Adams NE, Beyer J. Cognitive processes mediating behavioral change. Journal of Personality and Social Psychology 1977;35:125–39.
[18] Hamilton C, Denniss R, Baker D. Wasteful consumption in Australia. Canberra, Australia: The Australia Institute; 2005.
[19] Quested TE, Marsh E, Stunell D, Parry AD. Spaghetti soup: the complex world of food waste behaviors. Resources, Conservation and Recycling 2013;79:43–51.
[20] Ajzen I. Consumer attitudes and behavior: the theory of planned behavior applied to food consumption decisions. Italian Review of Agricultural Economics 2015;2:121–38.
[21] Armitage CJ, Conner M. Efficacy of the theory of planned behavior. A meta-analytic review. British Journal of Social Psychology 2001;40:471–99.
[22] Aizen I. Icek Aizen: Theory of planned behavior. 2002 [Online], http://people.umass.edu/aizen/tpb.html.
[23] Faber M, Petersen T, Schiller J. Homo economicus and homo politicus in ecological economics. Ecological Economics 2002;40:323–33.
[24] Crane A. Unpacking the ethical product. Journal of Business Ethics 2001;30:361–73.
[25] Meulenberg M. Consument en burger, betekenis voor de markt van landbouwproducten en voedingsmiddelen. Tijdschrift voor Sociaal Wetenschappelijk Onderzoek van de Landbouw (TSL) 2003;18:43–56.
[26] Verbeke W, Viaene J. Consumer attitude to beef quality labels and associations with beef quality labels. Journal of International Food and Agribusiness Marketing 1999;10:45–65.
[27] Shrum LJ, McCarty JA, Lowrey TM. Buyer characteristics of the green consumer and their implications for advertising strategy. Journal of Advertising 1995;24:71–82.
[28] De Pelsmacker P, Driesen L, Rayp G. Are fair trade labels good business? Ethics and coffee buying intentions. Working Paper of Faculty of Economics and Business Administration, Ghent, Belgium. 2003.
[29] Sparks P, Guthrie CA, Shepherd R. The dimensional structure of the perceived behavioral construct. Journal of Applied Social Psychology 1997;27:418–38.
[30] Sparks P, Shepherd R. Self-identity and the theory of planned behavior: assessing the role of identification with green consumerism. Social Psychology Quarterly 1992;55:388–99.
[31] Roberts JA. Green consumers in the 1990s: profile and implications for advertising. Journal of Business Research 1996;36:217–31.
[32] Vermeir I, Verbeke W. Sustainable food consumption: exploring the consumer "attitude-behavioral intention" gap. Journal of Agricultural and Environmental Ethics 2006;19:169–94.
[33] Robinson R, Smith C. Psychosocial and demographic variables associated with consumer intention to purchase sustainable produced foods as defined by the Midwest Food Alliance. Journal of Nutrition Education and Behavior 2002;34:316–25.
[34] Tanner C, Kast SW. Promoting sustainable consumption: determinants of green purchases by Swiss consumers. Psychology and Marketing 2003;20:883–902.
[35] Bisonette MM, Contento IR. Adolescents' perspectives and food choice behaviors in terms of the environmental impacts of food production practices: application of a psychosocial model. Journal of Nutrition Education 2001;33:72–82.
[36] Grunert SC, Juhl HJ. Values, environmental attitudes, and buying of organic foods. Journal of Economic Psychology 1995;16:39–62.
[37] Raats M, Shepherd R, Sparks P. Including moral dimensions of choice within the structure of the theory of planned behavior. Journal of Applied Social Psychology 1995;25:484–94.
[38] Sparks P, Shepherd R, Frewer LJ. Assessing and structuring attitudes toward the use of gene technology in food production: the role of perceived ethical obligation. Basic and Applied Social Psychology 1995;16:267–85.
[39] Sparks P, Shepherd R. The role of moral judgments within expectancy-value-based attitude–behavior models. Ethics and Behavior 2002;12:299–321.
[40] Saba A, Rosati S, Vassallo M. Biotechnology in agriculture. British Food Journal 2000;102:114–21.

[41] Saba A, Rosati S. Understanding consumer perceptions of fermented yoghurt products using conjoint and generalised procrustes analysis. Italian Journal of Food Science 2002;4:339–50.
[42] Thøgersen J. Direct experience and the strength of the personal norm–behavior relationship. Psychology and Marketing 2002;19:881–93.
[43] Thøgersen J, Ölander F. The dynamic interaction of personal norms and environment-friendly buying behavior: a panel study. Journal of Applied Social Psychology 2006;36:1758–80.
[44] Verbeke W, Vermeir I, Brunsø K. Consumer evaluation of fish quality as basis for fish market segmentation. Food Quality and Preference 2007;18:651–61.
[45] Jager W, Vlek CAJ. Modelling consumer behavior. Faculty of Behavioral Social Sciences, and University of Groningen; 2000.
[46] Rokeach MJ. The nature of human values. New York: Free Press; 1973.
[47] McCarthy JA, Shrum LJ. The recycling of solid-wastes—personal values, value orientations and attitudes about recycling as antecedents of recycling behavior. Journal of Business Research 1994;30:53–62.
[48] Chan RYK. Determinants of Chinese consumers' green purchase behavior. Psychological and Marketing 2001;18:389–413.
[49] Burgess SM. Personal values and consumer research. Research in Marketing 1992;11:35–79.
[50] Finegan J. The impact of personal values on judgments of ethical behavior in the workplace. Journal of Business Ethics 1994;13:747–55.
[51] Sikula A, Costa AD. Are women more ethical then men. Journal of Business Ethics 1994;13:859–71.
[52] Fritzsche DJ. Personal values: potential keys to ethical decision making. Journal of Business Ethics 1995;14:909–22.
[53] Thogersen J. Consumer values, behavior and sustainable development. Asia Pacific Advances in Consumer Research 2001;4:207–9.
[54] Kuchenmüller T, Hird S, Stein C, Kramarz P, Nanda A, Havelaar A. Estimating the global burden of foodborne diseases. A collaborative effort. European Surveillance 2009;14:19201.
[55] Hall G, Kirk MD, Becker N, Gregory JE, Unicomb L, Millard G, Stafford R, Lalor K. Estimating foodborne gastroenteritis, Australia. Emerging Infectious Diseases 2005;11:1257–64.
[56] Mullan B. Using the theory of reasoned action to predict safe food handling. Knowledge, attitudes, and social norms of children and young adults regarding food hygiene behavior. Germany: VDM Publishing; 2009.
[57] McKercher CM. Monitoring the incidence and causes of diseases potentially transmitted by food in Australia. Annual report of the OzFoodNet network, 2010. 2012.
[58] Scallan E, Hoekstra R, Angulo F, Tauxe R, Widdowson M, Roy S, Jones JL, Griffin PM. Foodborne illness acquired in the U.S.-Major pathogens. Emerging Infectious Diseases 2011;17:7–15.
[59] National Health and Medical Research Council. NHMRC dietary guidelines for Australian adults. A guide to healthy eating. 2003.
[60] Redmond EC, Griffith CJ. Consumer food handling in the home: a review of food safety studies. Journal of Food Protection 2003;66:130–61.
[61] Chow S, Mullan B. Predicting food hygiene. An investigation of social factors and past behavior in an extended model of the Health Action Process Approach. Appetite 2010;54:126–33.
[62] Rimal RN. Closing the knowledge-behavior gap in health promotion. The mediating role of self-efficacy. Health Communication 2000;12:219–37.
[63] Bearth A, Cousin M-E, Siegrist M. Investigating novice cooks' behavior change. Avoiding cross-contamination. Food Control 2014;40:26–31.
[64] Seaman P, Eves A. Efficacy of the theory of planned behavior model in predicting safe food handling practices. Food Control 2010;21:983–7.
[65] Mullan B, Wong CL. Hygienic food handling behaviors. An application of the theory of planned behavior. Appetite 2009;52:757–61.
[66] Shapiro MA, Porticella N, Jiang L, Gravani RB. Predicting intentions to adopt safe home food handling practices. Applying the theory of planned behavior. Appetite 2011;56:96–103.

[67] Mari S, Tiozzo B, Capozza D, Ravarotto L. Are you cooking your meat enough? The efficacy of the theory of planned behavior in predicting a best practice to prevent salmonellosis. Food Research International 2012;45:1175–83.
[68] Mullan B, Wong C, Kothe EJ. Predicting adolescents' safe food handling using an extended theory of planned behavior. Food Control 2013;31:454–60.
[69] Arvola A, Vassallo M, Dean M, Lampila P, Saba A, Lahteenmaki L, Shepherd R. Predicting intentions to purchase organic food: The role of affective and moral attitudes in the Theory of Planned Behaviour. Appetite 2008;50:443–54.
[70] Clayton DA, Griffith C, Price P. An investigation of the factors underlying consumers' implementation of specific food safety practices. British Food Journal 2003;105:434–53.
[71] Jenner EA, Watson PWB, Miller L, Jones F, Scott GM. Explaining hand hygiene practice: an extended application of the Theory of Planned Behavior. Psychology, Health & Medicine 2002;7:311–26.
[72] Sheeran P. Intention-behavior relations. A conceptual and empirical review. European Review of Social Psychology 2002;12:1–36.
[73] Sniehotta FF, Presseau J, Araújo-Soares V. Time to retire the theory of planned behavior. Health Psychology Review 2014;8:1–7.
[74] Sniehotta FF, Scholz U, Schwarzer R. Bridging the intention–behavior gap. Planning, self-efficacy, and action control in the adoption and maintenance of physical exercise. Psychology and Health 2005;20:143–60.
[75] Reuter T, Ziegelmann JP, Wiedemann AU, Lippke S, Schüz B, Aiken LS. Planning bridges the intention-behavior gap. Age makes a difference and strategy use explains why. Psychology and Health 2010;25:873–87.
[76] Sainsbury K, Mullan B, Sharpe L. Gluten free diet adherence in coeliac disease. The role of psychological symptoms in bridging the intention-behavior gap. Appetite 2013;61:52–8.
[77] Conner M, Armitage CJ. Extending the theory of planned behavior. A review and avenues for further research. Journal of Applied Social Psychology 1998;28:1429–64.
[78] Manstead ASR. Attitudes, behavior, and social context: The role of norms and group membership. Applied social research. USA: Lawrence Erlbaum Associates Publishers; 2000. p. 11–30.
[79] Godin G, Gagnon H, Lambert L-D, Conner M. Determinants of condom use among a random sample of single heterosexual adults. British Journal of Health Psychology 2005;10:85–100.
[80] Godin G, Conner M, Sheeran P. Bridging the intention-behavior gap. The role of moral norm. British Journal of Social Psychology 2005;44:497–512.
[81] Gardner B, de Bruijn G-J, Lally P. A systematic review and meta-analysis of applications of the self-report habit index to nutrition and physical activity behaviors. Annals of Behavioral Medicine 2011;42:174–87.
[82] Verplanken B, Orbell S. Reflections on past behavior. A self-report index of habit strength. Journal of Applied Social Psychology 2003;33:1313–30.
[83] Ouellette JA, Wood W. Habit and intention in everyday life. The multiple processes by which past behavior predicts future behavior. Psychological Bulletin 1998;124:54–74.
[84] Brennan M, McCarthy M, Ritson C. Why do consumers deviate from best microbiological food safety advice? An examination of 'high-risk' consumers on the island of Ireland. Appetite 2007;49:405–18.
[85] Tukker A, Jansen B. Environmental impacts of products: a detailed review of studies. Journal of Industrial Ecology 2006;10:159–82.
[86] UNEP. Assessing the environmental impacts of consumption and production: priority products and materials. In: Hertwich E, van der Voet E, Suh S, Tukker A, Huijbregts M, Kazmierczyk P, Lenzen M, McNeely J, Moriguchi Y, editors. A report of the working group on the environmental impacts of products and materials to the international panel for sustainable resource management; 2010.
[87] FAO. Food wastage footprint: impacts on natural resources. Summary report. Rome: Food and Agriculture Organisation of the United Nations; 2013.

[88] WRAP. New estimates for household food and drink waste in the UK. 2011. Available from: http://www.wrap.org.uk/content/new-estimates-householdfood-and-drink-waste-uk/.

[89] WRAP. Household food and drink waste in the United Kingdom 2012: final report. 2013. Available from: http://www.wrap.org.uk/sites/files/wrap/hhfdw-2012-main.pdf/.

[90] Evans D. Blaming the consumer e once again: the social and material contexts of everyday food waste practices in some English households. Critical Public Health 2011;21:429–40.

[91] Stefan V, van Herpen E, Tudoran AA, Lähteenmäki L. Avoiding food waste by Romanian consumers: the importance of planning and shopping routines. Food Quality and Preference 2013;28:375–81.

[92] Abeliotis K, Lasaridi K, Chroni C. Attitudes and behavior of Greek households regarding food waste prevention. Waste Management and Research 2014;32:237–40.

[93] Graham-Rowe E, Jessop DC, Sparks P. Predicting household food waste reduction using an extended theory of planned behavior. Resources, Conservation and Recycling 2015;101:194–202.

[94] Van Garde S, Woodburn M. Food discard practices of householders. Journal of the American Dietetic Association 1987;87:322–9.

[95] Koivupuro H-K, Hartikainen H, Silvennoinen K, Katajajuuri J-M, Heikintalo N, Reinikainen A. Influence of socio-demographical, behavioral and attitudinal factors on the amount of avoidable food waste generated in Finnish households. International Journal of Consumer Studies 2012;36:183–91.

[96] Williams H, Wikstr€om F, Otterbring T, L€ofgren M, Gustafsson A. Reasons for household food waste with special attention to packaging. Journal of Cleaner Production 2012;24:141–8.

[97] Parizeau K, von Massow M, Martin R. Household-level dynamics of food waste production and related beliefs, attitudes, and behaviors in Guelph, Ontario. Waste Management 2015;35:207–17.

[98] Evans D. Beyond the throwaway society: ordinary domestic practice and a sociological approach to household food waste. Sociology 2012;46:41–56.

[99] Refsgaard K, Magnussen K. Household behavior and attitudes with respect to recycling food waste-experiences from focus groups. Journal of Environmental Management 2009;90:760–71.

[100] Thyberg KL, Tonjes DJ. Drivers of food waste and their implications for sustainable policy development. Resources, Conservation and Recycling 2015;106:110–23.

[101] Ko S, Poon CS. Domestic waste management and recovery in Hong Kong. Journal of Material Cycles and Waste Management 2009;11:104–9.

[102] Hebrok M, Boks C. Household food waste: drivers and potential intervention points for design: an extensive review. Journal of Cleaner Production 2017;151:380–92.

[103] Schmidt K. Explaining and promoting household food waste-prevention by an environmental psychological based intervention study. Resources, Conservation and Recycling 2016;111:53–66.

[104] Michie S, Abraham C. Interventions to change health behaviors: evidence-based or evidence-inspired. Psychology and Health 2004;19:29–49.

[105] Steg L, Vlek C. Encouraging pro-environmental behavior: an integrative review and research agenda. Journal of Environmental Psychology 2009;29:309–17.

[106] Ghani WAWAK, Rusli IF, Biak DRA, Idris A. An application of the theory of planned behavior to study the influencing factors of participation in source separation of food waste. Waste Management 2013;33:1276–81.

[107] Collins A, Mullan B. An extension of the theory of planned behavior to predict immediate hedonic behaviors and distal benefit behaviors. Food Quality and Preference 2011;22:638–46.

[108] Mondéjar-Jiménez JA, Ferrari G, Secondi L, Principato L. From the table to waste: an exploratory study on behavior towards food waste of Spanish and Italian youths. Journal of Cleaner Production 2016;138:8–18.

[109] Russell SV, Young CW, Unsworth KL, Robinson C. Bringing habits and emotions into food waste behavior. Resources, Conservation and Recycling 2017;125:107–14.

[110] Visschers VHM, Wickli N, Siegrist M. Sorting out food waste behavior: a survey on the motivators and barriers of self-reported amounts of food waste in households. Journal of Environmental Psychology 2016;45:66–78.

[111] Stancu V, Haugaard P, L€ahteenm€aki L. Determinants of consumer food waste behavior: two routes to food waste. Appetite 2016;96:7–17.
[112] Barr S. Factors influencing environmental attitudes and behaviors: a U.K. case study of household waste management. Environment and Behavior 2007;39:435–73.
[113] Bolton LE, Alba JW. When less is more: consumer aversion to unused utility. Journal of Consumer Psychology 2012;22:369–83.
[114] Pakpour A, Zeidi IM, Emamjomeh MM, Asefzadeh S, Pearson H. Household waste behaviors among a community sample in Iran: an application of the theory of planned behavior. Waste Management 2014;34:980–6.
[115] Alpízar F, Gsottbauer E. Reputation and household recycling practices: field experiments in Costa Rica. Ecological Economics 2015;120:366–75.
[116] Crociata A, Agovino M, Sacco PL. Recycling waste: does culture matter? Journal of Behavioral and Experimental Economics 2015;55:40–7.
[117] Valle P, Reis E, Menezes J, Rebelo E. Behavioral determinants of household recycling participation: the Portuguese case. Environment and Behavior 2004;36:505–40.
[118] Lyndhurst B. Food behavior consumer research – findings from the quantitative survey. Briefing Paper. United Kingdom: WRAP; 2007.
[119] WRAP. Understanding food waste – key findings of our recent research on the nature, scale and causes of household food waste. Banbury, United Kingdom. 2007.
[120] EU, European Union. 'Best before' date labels: protecting consumers and limiting food waste. EU Briefing. 2015.
[121] Bell D, Corsten D, Knox G. From point-of-purchase to path-to-purchase: how preshopping factors drive unplanned buying. Journal of Marketing 2011;75:31–45.
[122] Principato L, Secondi L, Pratesi CA. Reducing food waste: an investigation on the behavior of Italian youths. British Food Journal 2015;117(2).
[123] Chandon P, Wansink B. How biased household inventory estimates distort shopping and storage decisions. Journal of Marketing 2006;70:118–35.
[124] Exodus. Quantitative assessment of the nature, scale and origin of post-consumer food waste arising in great Britain. WRAP; 2006.
[125] Salhofer S, Issac NA. Importance of public relations in recycling strategies: principles and case studies. Journal of Environmental Management 2002;30:68–76.
[126] Hornik J, Cherian J, Michelle M, Narayana C. Determinants of recycling behavior: a synthesis of research results. The Journal of Socio-Economics 1995;4:105–27.
[127] Hormuth SE, Katzenstein H, Bruch B, Ringenberger B. Psychological studies on garbage avoidance and recycling. In: Paper presented at the international conference on healthy environments, Oklahoma City; 1993.
[128] Zimmerman BJ. A social cognitive view of self-regulated academic learning. Journal of Educational Psychology 1989;81:329–39.
[129] Vining J, Ebreo A. What makes a recycler? A compression of recyclers and non-recyclers. Environment and Behavior 1990;22:55–73.
[130] De Young R. Exploring the difference between recyclers and non-recyclers: the role of information. Journal of Environmental Systems 1988;18:341–51.
[131] Boldero J. The prediction of household recycling of newspapers; the role of attitudes, intentions and situational factors. Journal of Applied Social Psychology 1995;25:440–62.
[132] Nyamwange M. Public perception of strategies for increasing participation in recycling programs. The Journal of Environmental Education 1996;27:19–22.
[133] Bratt C. The impact of norms and assumed consequences on recycling behavior. Environment and Behavior 1999;31:630–56.
[134] Oskamp S, Harrington M, Edwards T, Sherwood D, Okuda S, Swanson D. Factors influencing household recycling behavior. Environment and Behavior 1991;23:494–519.

[135] Ewing G. Altruistic, egoistic and normative effects on curbside recycling. Environment and Behavior 2001;33:733–64.

[136] Wan C, Shen GQ, Yu A. The role of perceived effectiveness of policy measures in predicting recycling behavior in Hong Kong. Resources, Conservation and Recycling 2014;83:141–51.

[137] Thomas C. Public understanding and its effect on recycling performance in Hampshire and Milton Keynes. Resources, Conservation and Recycling 2001;32:259–74.

[138] Evison T, Read AD. Local authority recycling and waste awareness publicity/promotion in UK. Resources, Conservation and Recycling 2001;32:275–91.

[139] Mak MW, Yu KM, Tsang CW, Hsu SC, Poon CS. Promoting food waste recycling in the commercial and industrial sector by extending The Theory of Planned Behavior: a Hong Kong case study. Journal of Cleaner Production 2018;204:1034–43.

[140] Sakaguchi L, Pak N, Potts MD. Tackling the issue of food waste in restaurants: options for measurement method, reduction and behavioral change. Journal of Cleaner Production 2018;180:430–6.

[141] U.S. Congress House of Representatives. Food Recovery Act of 2016. 114th congress, 7th session. Washington, DC: U.S. House Resolution 4184; 2015.

[142] U.S. Department of Agriculture. USDA and EPA join with private sector, charitable organizations to set nation's first food waste reduction goals. 2015. Washington DC.

[143] United Nations. Global sustainable development report. 2015 [Online], https://sustainabledevelopment.un.org/globalsdreport/2015.

[144] Australian Government. National food waste strategy: halving Australia's food waste by 2030. 2017 [Online], https://www.environment.gov.au/system/files/resources/4683826b-5d9f-4e65-9344-a900060915b1/files/national-food-waste-strategy.pdf.

[145] Parry A, LeRoux S, Quested T, Parfitt J. UK food waste – historical changes and how amounts might be influenced in the future. London: Waste and Resources Action Programme; 2015.

[146] Lee KH, Oh JI, Chu KH, Kwon SH, Yoo SS. Comparison and evaluation of large-scale and on-site recycling systems for food waste via life cycle cost analysis. Sustainability 2017;9:2186.

[147] Lam CM, Yu IKM, Medel F, Tsang DCW, Hsu SC, Poon CS. Life-cycle cost-benefit analysis on sustainable food waste management: the case of Hong Kong International Airport. Journal of Cleaner Production 2018;187:751–62.

[148] Willersinn C, Mouron P, Mack G, Siegrist M. Food loss reduction from an environmental, socio-economic and consumer perspective – the case of the Swiss potato market. Waste Management 2017;59:451–64.

CHAPTER 10

Valorization of waste biomass for chitin and chitosan production

M. M. Tejas Namboodiri, Kannan Pakshirajan
Department of Biosciences and Bioengineering, Indian Institute Technology Guwahati, Guwahati, Assam, India

10.1 Introduction

Cellulose and chitin are the two most abundant naturally available biodegradable polymers. Despite its earlier discovery than cellulose, industrial application of chitin could not be realized mainly owing to the economics of its production [1].

Chitin is found in crustaceans, mollusca, fungi, roundworms, and insects, of which the cuticle and external are of major importance. It is a polysaccharide made up of β (1,4) linked 2-acetamido-2-deoxy-β-D-glucose, also known as N-acetylglucosamine monomer units (Fig. 10.1). Chitin occurs in three forms- α, β and γ chitin based on their degree of hydration, unit cell size, and chitin chains per cell (Fig. 10.2) [3]. Among these forms of chitin, α-chitin is the only extractable and most abundant form, occurring in crustaceans and cell wall of fungi. The removal of acetyl groups from chitin leads to the formation of chitosan, which is chemically 2-deoxy-β-D-glucopyranose. Till date, the major source of chitosan production is chitin, obtained from the shell of crustaceans, which is a waste from

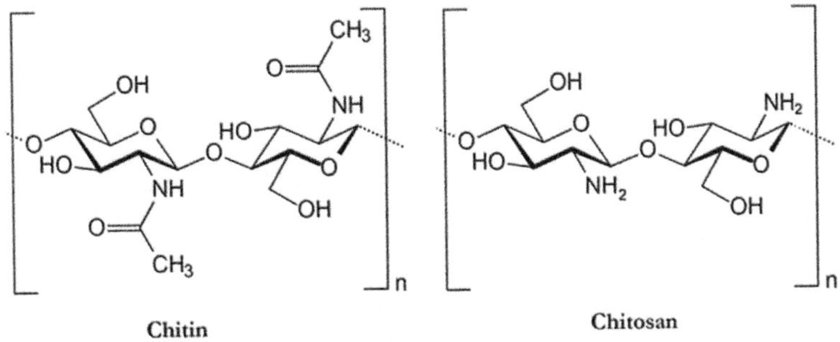

Figure 10.1
Structure of (A) chitin and (B) chitosan.

Waste Biorefinery. https://doi.org/10.1016/B978-0-12-818228-4.00010-1

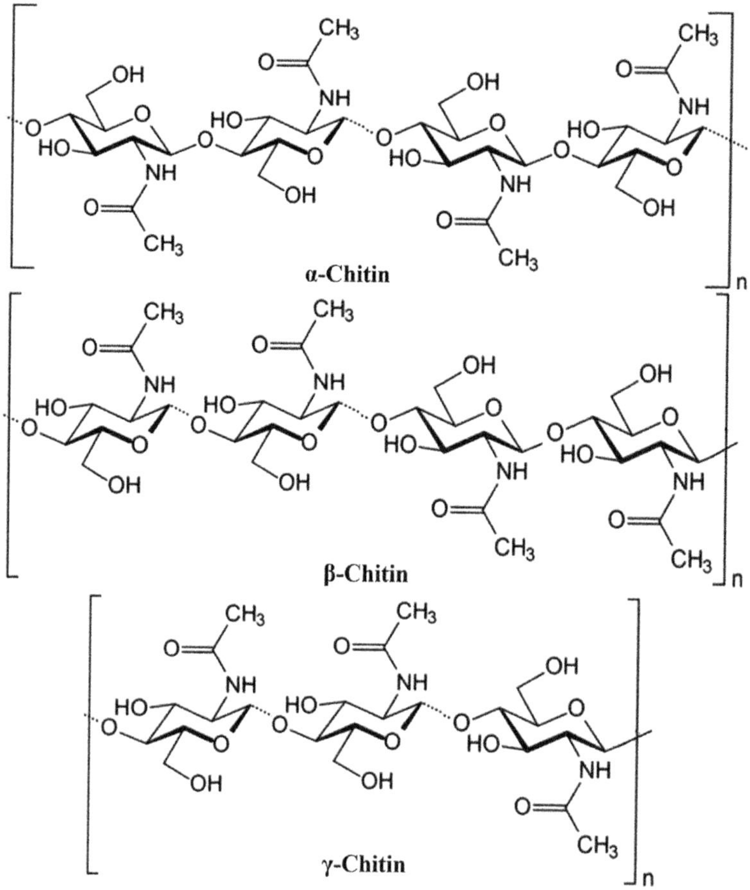

Figure 10.2
Structure of different forms of chitin [2].

the seafood industry. The annual chitin production is estimated to be around 10^{10}–10^{11} tons [4,5]. Extraction of chitin from crustacean wastes involves mechanical pretreatment for size reduction by grinding, followed by demineralization, deproteinization and removal of pigments (decoloration). Moreover, seasonal availability of the marine source and the highly variable property of the chitin are the major drawbacks of crustacean wastes as the feedstock for chitin and chitosan production and their application.

The cell wall of fungi contains chitin as one of its principal components, as it is highly essential for maintenance of the cell structure [6]. The zygomycetes class of fungi contain abundant chitosan than chitin in their cell walls. Along with chitin, glycoproteins and glucan make up the fungal cell wall. Chitin and glucans (both β-1,3 and β-1,6-glucans) form the structural component whereas the glycoproteins (glucuronoproteins, galactoproteins, mannoproteins, etc.) form the interstitial components. Thus, an intricately complex network is formed due to the cross-linking of chitin, glucan, and glucuronoproteins, thereby providing structural stability to the fungal cell wall [7].

Owing to its unique properties, chitosan is widely used in agriculture, environment, and pharmaceutical sectors as discussed in the next section. Therefore, there is an increased interest toward developing an efficient and economical process of chitosan production to meet the market demands which were estimated to reach 40,465 metric tons per year by the end of 2018 [8]. This chapter reviews the literature on various sources and routes of chitin and chitosan production along with alternative technologies which could be employed to achieve a sustainable way of chitin and chitosan extraction.

10.2 Chitosan-properties and application

Due to the diverse properties of chitosan, it has been studied extensively for various applications as listed in Fig. 10.3.

10.2.1 Physicochemical

Chitosan is a linear polyamine with reactive amino and hydroxyl groups. The presence of these groups is responsible for its solubility in acids and its availability in various forms such as flakes and gels. Degree of deacetylation and molecular weight are other two important parameters which determine the application of chitosan in different fields. Both these parameters are influenced during the conversion of chitin to chitosan, resulting in changes in charge distribution, which in turn affects the agglomeration. The average molecular weight of chitin and chitosan are $1 \times 10^6 - 2 \times 10^6$ Da and $1 \times 10^5 - 5 \times 10^5$ Da, respectively [9].

10.2.2 Bioactivity

The presence of reactive functional groups such as amino ($-NH_2$) and hydroxyl ($-OH$) groups renders it capable of being chemically or enzymatically modified to form several chitosan derivatives. Carboxymethylchitosan, chitosan 6-O-sulfate, alkylated chitosans, and carbohydrate branched chitosans are some of the derivatives of chitosan which can be produced by chemically or biologically mediated reactions (Fig. 10.4) [10].

10.2.3 Biodegradability

In vivo digestion of chitosan in mammals produces oligosaccharides of varying length, which being non-toxic can be taken up by the several metabolic pathways for further conversions or be converted to glucoproteins [12]. Chitosan biodegradability depends upon number of acetyl groups, their distribution and its molecular weight (chain length) [13].

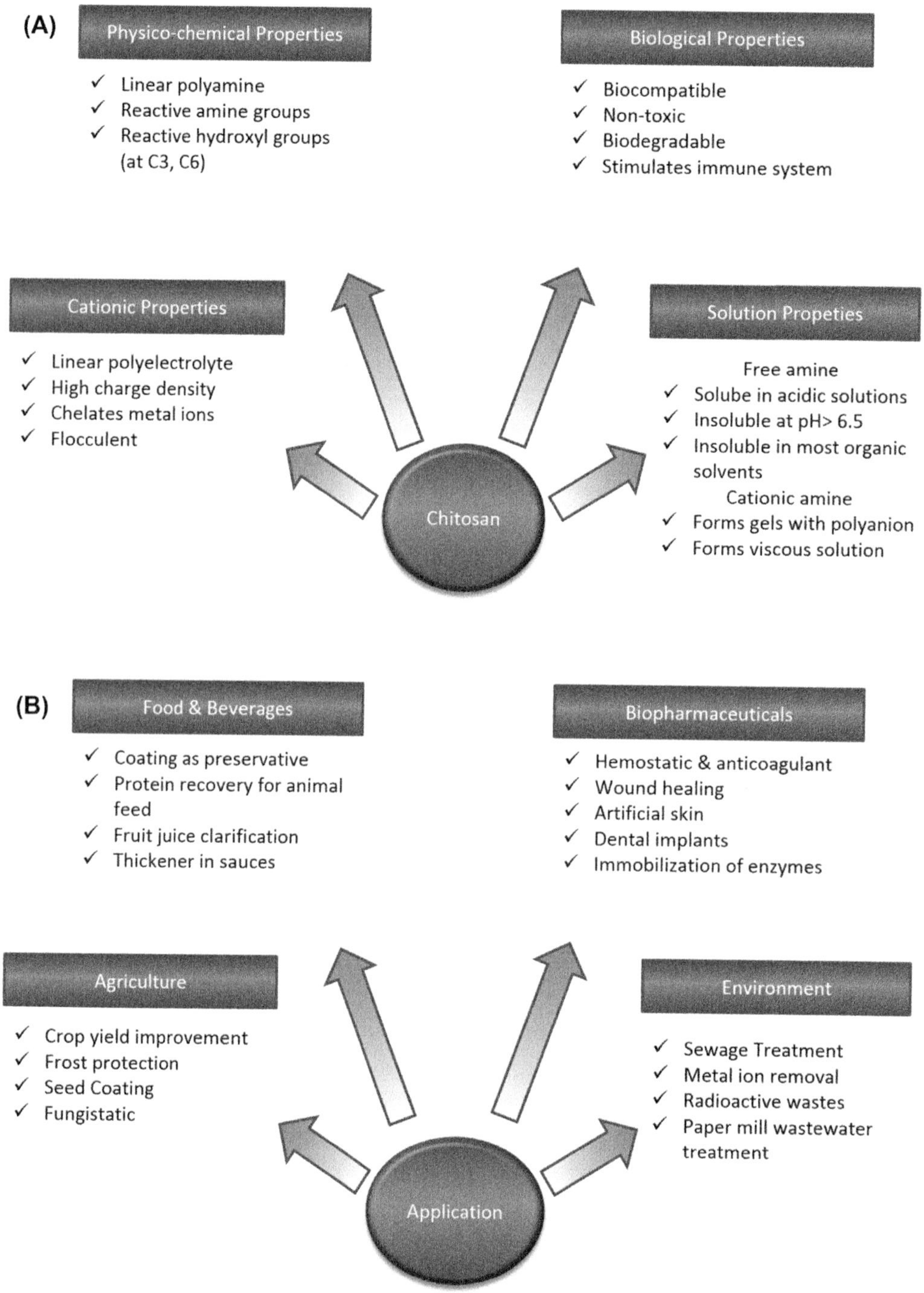

Figure 10.3
Chitosan- (A) properties and (B) applications [9,10].

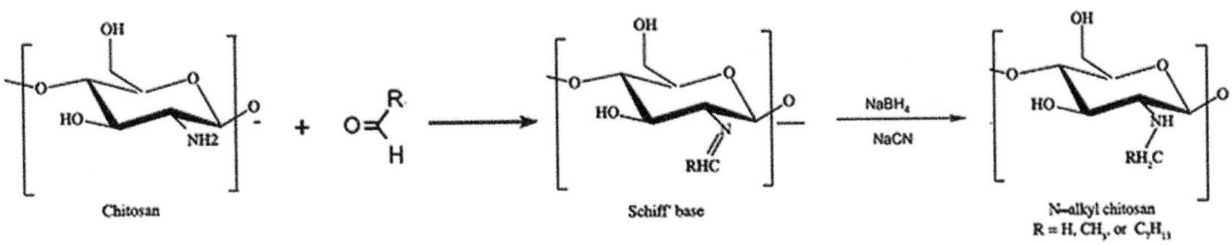

Figure 10.4
Alkyl chitosan derivatives obtained by modification reactions [11].

Chitosan biodegradability decreases with an increase in the degree of acetylation as well as length of the polymer [14–16].

10.2.4 Analgesic and anticholestrolemic

Chitosan has been reported to show analgesic properties. Chitosan being polycationic in nature, under low pH conditions the amine groups in chitosan protonate. The peptide, Bradykinin which is one of the reasons of pain was reported to be absorbed by the protonated chitosan, thus exhibiting the analgesic effect [17].

Chitosan has been found to have anticholestrolemic properties. It can bind to negatively charged fatty acids and lipids through their reactive amino and hydroxyl groups, and due to hydrophobic interactions [18,19].

10.2.5 Chelation and adsorption

Owing to its chelation and adsorption properties, chitosan has been reported to bind to metal ions, and adsorb phenolics. This property has been utilized for the removal of metal ions, phenols and other contaminants from water and wastewater for environmental applications [20,21].

10.2.6 Immobilization

Chitosan can be molded into different forms like membranes, beads, and gels. They are selectively permeable to various molecules and, therefore, can be used for various unit operations and enzyme immobilization [22–24].

10.3 Chitin and chitosan biosynthesis pathway

Chitin synthesis pathway is highly conserved and follows the same set of reactions for crustaceans, insects and fungi (Fig. 10.5). Chitin, a polymer of N-acetyl glucosamine (GlcNAc) is synthesized from different sugars or its storage compounds such as glycogen and trehalose [25,26]. The formation of glucosamine-6-phosphate from

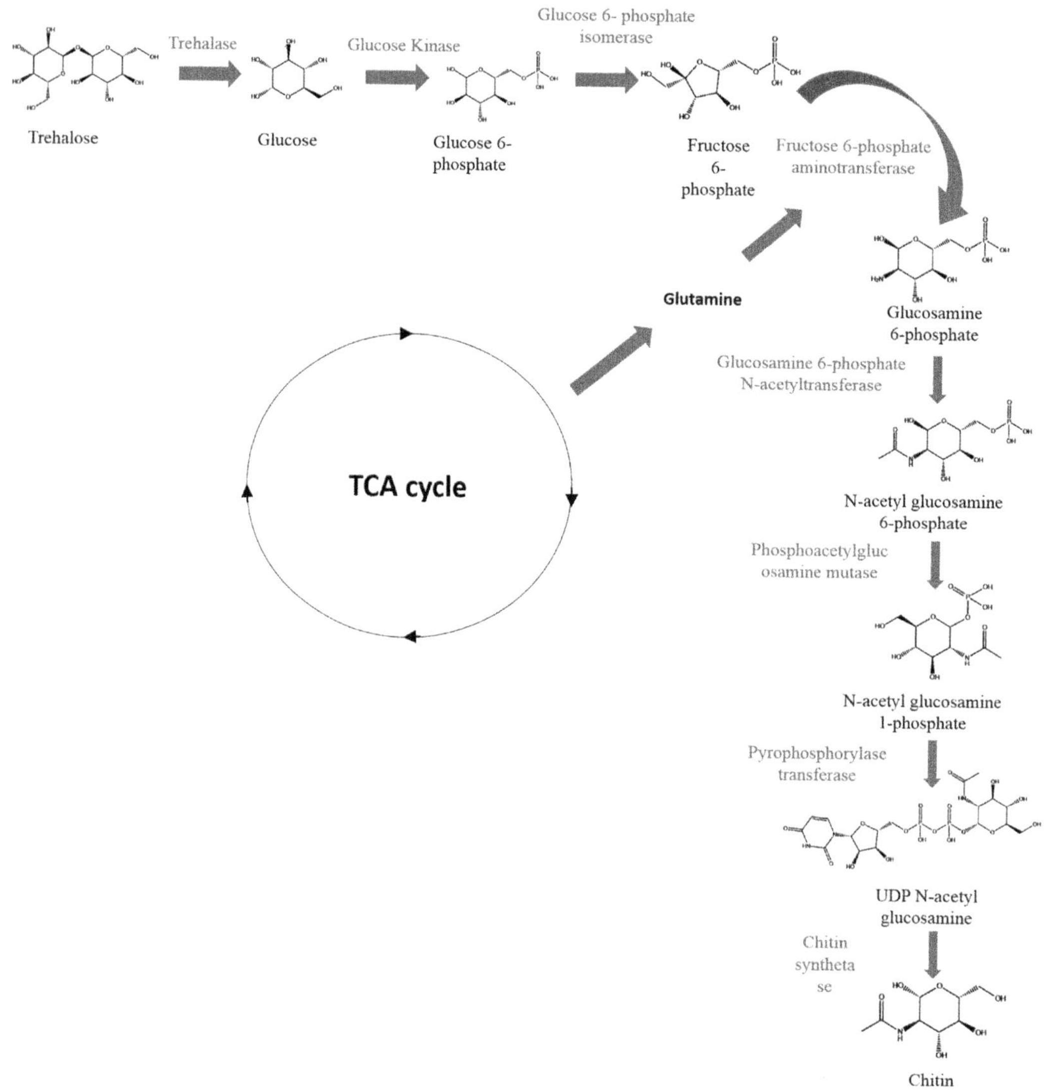

Figure 10.5
Chitin biosynthesis pathway.

fructose-6-phosphate is the first specific step toward chitin synthesis. Glucosamine-6-phosphate is subsequently converted to N-acetyl glucosamine-6-phosphate by the action of an acetylase enzyme, in the presence of Acetyl-CoA. A mutase enzyme is responsible for conversion of N-acetyl glucosamine-6-phosphate to N-acetyl glucosamine-1-phosphate, which subsequently reacts with UTP following a variant of the Leloir pathway yielding the activated amino sugar Uridine-diphospho-N-acetyl glucosamine (UDP-GlcNAc). Finally, chitin polymerization reaction is carried out by chitin synthetase using UDP-GlcNAc as the activated sugar donor [27].

10.4 Sources of chitin and chitosan

10.4.1 Crustaceans

Shrimp, lobster, crab, krill, oyster and squid are the most studied sources which have been applied for commercial extraction of chitin/chitosan. Table 10.1 provides information on

Table 10.1: Biological extraction of chitin from crustacean wastes.

Waste source	Microbial strains	Substrate/ incubation	Chitin content (%)	Process efficiency (%) (DP/DM)	References
Callinectes bellicosus	*Lactobacillus* sp. B2	Crab waste/ Sugarcane molasses 35°C/200 rpm/ 120 h	34	56/88	[89]
Metapeneaus monoceros	*Bacillus cereusSV1*,	Shrimp shell waste/5% glucose (w/v)	25.2	96/67	[90]
	Bacillus subtilis A26,			92/37	
	Bacillus mojavencis A21,			90/38	
	Bacillus pumilus A1,	37°C/200 rpm/5 d		88/37	
	Bacillus licheniformis RP1			94/59	
	Bacillus amyloliquefaciens An6			83/42	
Metapenaeus monoceros	*B. mojavensis A21*	Shell waste homogenate 60°C/6 h	18.5	88/–	[55]
	Bacillus subtilis A26			76/–	
	Bacillus licheniformis NH1			65/–	
	Bacillus licheniformis MP1			76/–	
	Vibrio metschnikovii J1			76/–	
	Aspergillus clavatusES1			59/–	
Metapeneaus monoceros	*B. pumilus A1*	70 g/L shrimp shells/50 g/L glucose 35°C/150 rpm/6 d	29	94/88	[91]
Cancer pagurus	*Exiguobacterium* spp. *Bacillus licheniformis* *Bacillus subtilis* + *Lactobacillus* spp. *Bacillus cereus* + *Pseudomonas* spp. *Pseudomonas* spp. *Pseudomonas migulae* *Enterococcus* sp.	Brown crab shell/ 10% glucose solution 30°C/175 rpm/7 d	14–16	–/99 (maximum)	[92]

chitin content in some crustaceans [28,29]. The shells of these crustaceans are abundant in chitin with a range of 13%–42% of shell mass. However, due to the intricate organization of these shells, the extraction of chitin from the other fractions of the shell requires very harsh chemical treatments (Fig. 10.6). These shells also contain proteins (30%–40%) and mineral salts (30%–50%) as their other major components [30].

The major sources of chitin and chitosan are crustaceans. According to the FAO report, 8666 thousand tons of crustaceans was produced and harvested in 2016 [31]. Different shellfish like crabs and shrimps are harvested for human consumption, but only 40%–50% of its weight is utilized for food processing. This production route leads to a large amount of waste generation, which is directly dumped into the seas and oceans. Low biodegradability of these wastes often results in major pollution of the coastal areas. Utilization of these wastes for the production of some value added compound is a promising way for addressing these environmental concerns [32,33]. Chitin, a major component of the waste shells, can be extracted along with other products such as protein and mineral salts. Crustacean shell wastes have been used as poultry feed owing to its high protein content. However, high fiber and mineral contents are some drawbacks of directly

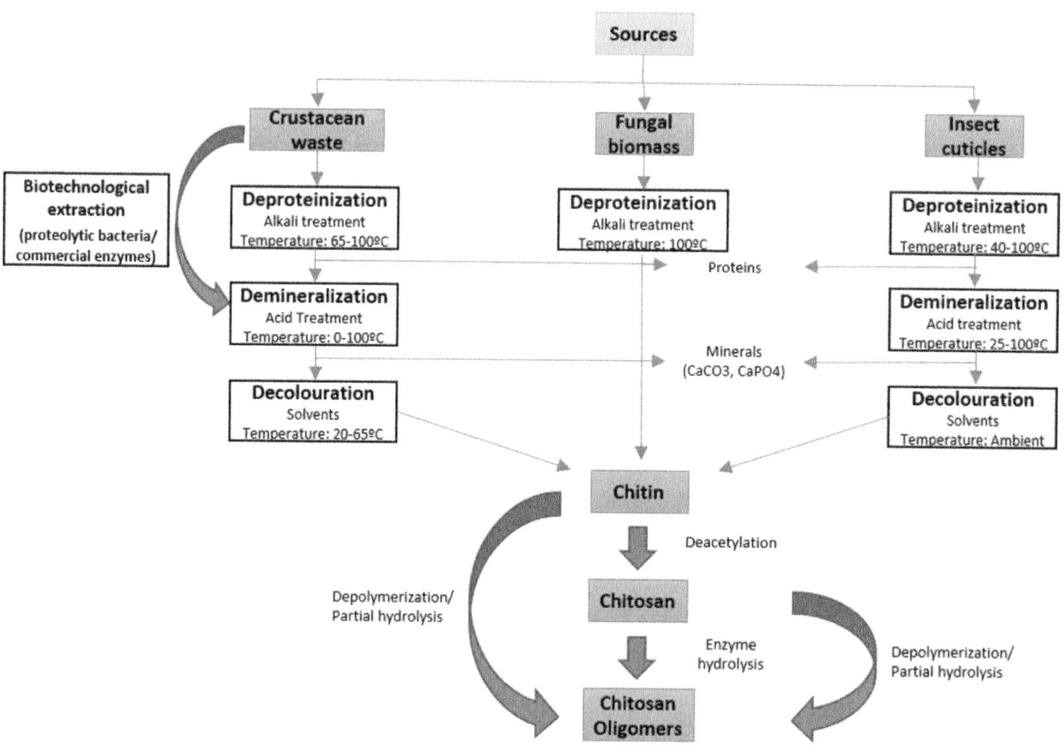

Figure 10.6
Methods of extraction of chitin and chitosan from crustaceans, fungi, and insects.

using the shell wastes as feed. The wastes are therefore subjected to demineralization before it could be utilized as poultry feed [34].

10.4.1.1 Chemical extraction

The industrial scale production of chitosan involves deproteinization, demineralization, and removal of pigments followed by deacetylation of chitin to form chitosan. Alkaline treatment by sodium hydroxide, potassium hydroxide, sodium carbonate, sodium bicarbonate, etc., or enzymatic treatment by proteases like trypsin, pepsin, etc., could be applied for deproteination of the ground shells. As the chitin polymer in cross-linked with the proteins inside the crustacean shells, depolymerization of the biopolymer is an important step to carry out the deproteinization. NaOH is the most widely used reagent for this purpose. NaOH at different concentrations and varying temperature and time regimes have been applied for efficient deproteination of the shells [35–37]. Demineralization of the calcified shells is carried out using strong acids such as hydrochloric acid, sulfuric acid, and nitric acid. HCl is the preferred acid for the removal of minerals, primarily calcium carbonate [30,38,39]. The reaction between the minerals and the acids results in the formation of soluble salts, for example, decomposition of calcium carbonate results in carbon dioxide and calcium chloride, which are soluble in water [40,41]. Melanin and carotenoid pigments can be removed by treating with potassium permanganate, hydrogen peroxide or sodium hypochlorite. The chitin obtained following these treatment steps is then converted to chitosan by deacetylation reactions by alkali and acid treatments. Alkali treatment using NaOH at mild to high temperatures, followed by acid treatment to remove impurities, leads to the formation of chitosan (Fig. 10.7). The use of harsh chemicals is extremely detrimental to the properties of chitosan, as they lead to depolymerization and, thus, affecting the molecular weight and viscosity. During this process, the waste effluent generated contains harmful chemicals, including unutilized mineral acid and other reagents [42,43]. Due to the threat of environmental pollution, the waste generated needs to be properly treated before it could be released into the nearby environment. Besides, the protein constituents of the shell are not available for application as animal feed [34,44].

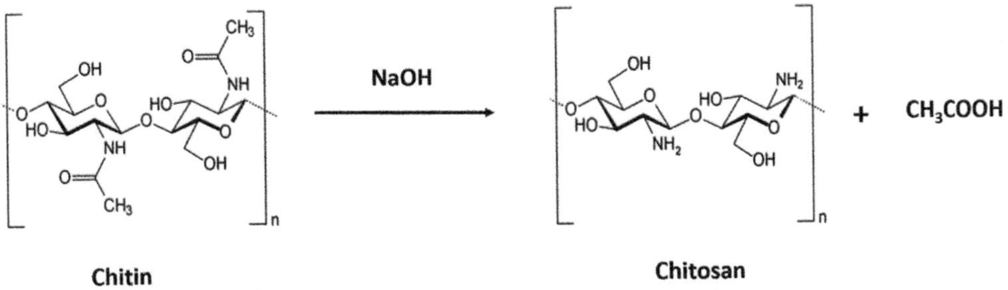

Figure 10.7
Deacetylation of chitin to form chitosan under alkali treatment.

The use of chemicals further results in high production costs. Due to these various limitations, milder extraction methods are of interest. Biological extraction of chitin using microorganisms and utilization of enzymes for extraction procedure are emerging to be a promising option [45–47].

10.4.1.2 Biological extraction

Bio-based products are widely accepted worldwide and are being favored against chemically synthesized ones. Drawbacks associated with the chemical extraction methods can be overcome by using enzymes or microorganisms (MOs). Biological extraction of chitin can be carried out by either using microbes directly for the extraction process which is known as fermentation mediated extraction or by using the enzymes obtained from the microbes for deproteinization and demineralization. Researchers have found that lactic acid fermentation when coupled with the traditional method of extraction, resulted in lower amounts of alkali and acid required along with high chitin yields [48].

Deproteinization of the waste shells could be carried out by the action of enzymes such as proteases. MOs could also be applied for this step after demineralization of the shells [49]. Proteolytic ability of the MOs leads to the formation of a protein and mineral rich liquid fraction which could be used as an animal feed and even for human consumption, along with a solid chitin fraction [50,51]. Khanafari et al. [52] carried out a comparative study between biological and chemical extraction methods for chitin. This study observed that the structure of chitin was preserved in case of the biological extraction process, which also yielded a high molecular weight of chitin, when the deproteinization was carried out by MOs with proteolytic activity [53]. Deproteinization by enzymes or demineralization and deproteinization by fermentation, which makes use of several MOs have been applied for biological extraction of chitin.

Proteases are used extensively for the deproteinization step, and crude proteases are preferred over purified enzymes, primarily because of the high cost of purified proteases. Bacteria and fish are the major sources of crude proteases, the most common among them being the bacterial proteases. Various bacterial and marine proteases have been reported for effective deproteinization of the shrimp waste for chitin recovery [29,42,54–56].

Fermentation, which utilizes MOs for production of various value-added products by the uptake of a wide range of substrates, has long played a significant role in the lives of human beings. Organic acid producing bacteria have found wide application in the extraction of chitin and chitosan. These bacteria utilize glucose and release various organic acids which lowers the pH of the media, thereby preventing the growth of other contaminating microbes. Calcium carbonate present in shrimp shell wastes is removed by reaction with lactic acid resulting in calcium lactate that is easily precipitated out.

Proteases produced by these bacteria act upon the proteins present in shrimp waste and mediates the deproteinization steps. Table 10.1 enlists the various studies that have been carried out on biological extraction of chitin and chitosan using different microorganisms.

10.4.2 Insects

Insects have been widely explored as a major source of medicines, food and pesticides across the world ranging from ancient traditional civilizations to the industrial age [57]. However, due to the high protein, fat and biopolymer content in their body, they have emerged as a major source of biomass in recent times. Moreover, the high dry matter percentage along with their ability to grow on organic wastes prove the potential of insects as the source of biomass [8]. One of the major components of the exoskeleton of insects is chitin. The extraction process of chitin from insects, slightly differs from the crustacean sources, where a harsher demineralization step is required in case of chitin extraction from the insect cuticle (Fig. 10.6). The chitin content of several insect species viz., *H. piceus, R. linearis, A. bipustulatus, A. imperator*, and *N. glauca* investigated were found to vary in the range 10%–20% w/w. Thermal stability and crystalline index of chitin were also found to vary in the different species. These insects could be a promising source of chitin, especially *Hydrophilus piceus* which showed highest chitin (20%) yield [58]. *Rhinolophus hipposideros* bat guano was estimated to have 28% chitin content. Fourier-transform infrared spectroscopy revealed that the chitin present in the bat guano was in α-chitin form. Around 79% chitosan was retrieved from the chitin. Scanning electron microscopy and X-ray diffraction experiments showed that the chitin and chitosan were present in the form of nanofibers [59]. Several studies have reported grasshoppers as a very good source of chitin. The structures of chitin obtained from seven different *Orthoptera* species were compared and it was observed that the chitin content from these various grasshopper species varied between 5.3% and 8.9%; the molecular weight of the chitin obtained was found to be low [60]. *Celes variabilis, Decticus verrucivorus, Melanogryllus desertus,* and *Paracyptera labiate* were investigated for the differences in chitin content in the male and female grasshoppers of the same species [61]. The chitin content was found to be higher in the males than the female and the highest chitin yield was found to be 11.84% of its dry weight in the case of *D. verrucivorus*. Marei et al. [62] estimated the chitin content from three different insects, viz. *Schistocerca gregaria* (locust), *Apis mellifera* (honey bee), and *Calosoma rugosa* (beetles) to be 12.2%, 5% and 2.5%, respectively. The chitin obtained was further deacetylated to obtain chitosan. The degree of deacetylation (DD) derived were 98%, 96% and 95%, respectively. These studies suggest that insects are a very promising source for chitin extraction with high yields (Table 10.2). Arthropods, including the insects, are so vast and biodiverse that their complete potential is yet to be realized. These are the most unexplored species with regard to their contribution to various areas of application and as a bioresource [8].

Table 10.2: Extraction of chitin from insects.

Insects	Chitin content (%)	Chitosan productivity (%)	Extraction	References
Agabus bipustulatus	14–15	71	1M HCl/90°C/1 h 1M NaOH/110°C/18 h Choloroform:methanol:water (1:2:4)	[58]
Anax imperator	11–12	67		
Hydrophilus piceus	19–20	74		
Notonecta glauca	10–11	69		
Ranatra linearis	15–16	70		
Rhinolophus hipposideros	28	79	4M HCl/50°C/24 h 4M NaOH/140°C/24 h Choloroform:methanol:water (1:2:4)/4 h	[59]
Aiolopus simulatrix	5.3	–	4M HCl/75°C/1 h 2M NaOH/175°C/18 h Choloroform:methanol:water (1:2:4)	[60]
Aiolopus strepens	7.4	–		
Duroniella fracta	5.7	–		
Duroniella laticornis	6.5	–		
Oedipoda caerulescens	8.9	–		
Oedipoda miniata	8.1	–		
Pyrgomorpha cognata	6.6	–		
Celes variabilis	9.93	–	4M HCl/75°C/2 h 2M NaOH/150°C/20 h water	[61]
Decticus verrucivorus	11.84	–		
Melanogryllus desertus	7.35	–		
Paracyptera labiate	7.60	–		
Apis mellifera	2.5	96	1M HCl/RT 1M NaOH/100°C/8 h Hot ethanol/Acetone	[62]
Calosoma rugosa	5	95		
Schistocerca gregaria	12.2	98		

10.4.3 Fungi

Chitin, a simple polysaccharide, is present in the cell wall of almost all fungi known till date. A high chitosan and chitin content is found to be present in fungi belonging to the zygomycetes family as with the other classes [63–65].

Chitin in the cell wall of fungi folds back on itself to form antiparallel chains and form intrachain hydrogen bonds that stiffen into immensely strong fibrous microfibrils structures tougher than any other molecule in nature. β(1,3)-glucan, which is another structural polysaccharide present in most fungal cell walls, is attached covalently to this chitin microfibril network. The composition of the cell wall is subject to change and may vary within a single fungal isolate depending upon the conditions and growth stage. The glycoprotein, glucan and chitin components are extensively cross-linked together to form a complex network, which forms the structural basis of the cell wall.

Zygomycetes are found to have a higher percentage of chitosan than chitin. Chitosan is synthesized by a deacetylase enzyme which can convert the chitin produced by the previously mentioned pathway to chitosan (Fig. 10.5). It is believed to be synthesized by the tandem action of chitin synthase and chitin deacetylase enzymes. Uridine-diphospho-N-acetyl glucosamine (UDP-GlcNAc) is added to the chitin chain by the action of the chitin synthases and directed to the cell wall simultaneously. These chitin synthases are localized in subcellular organelles called chitosomes. The chitosomes on reaching the cell surface may produce a tightly linked complex of long microfibrils or be present in a dispersed form. Chitin deacetylase present in the cell wall act upon this newly synthesized chitin to convert it into chitosan. However, only the chitin present in the dispersed form is susceptible to the action of this enzyme, and it is ineffective on crystalline chitin. As chitosan is reported to be more abundant than chitin in zygomycetes fungi, it could be inferred that the chitin synthase subunits are majorly present in a dissociate form in these fungi. The motive of this N-deacetylation of carbohydrate moieties is to offer resistance to lysozyme action, specifically from chitinases [66].

Zygomycetes group of fungi have nonseptate thalli and form dark thick-walled sexual spores which are called zygospores. Entomophthorales and mucorales make up most of the zygomyectes class of the fungi. Among these, Mucorales produce both chitin and chitosan in significant amounts in their cell wall. Mucorales are saprophytes which mainly target plants vertebrates as they are endoparasites [67]. Fungal chitosan production has received increased attention recently due to a number of significant advantages. Whereas the supply of crustacean wastes are limited by the sites of fishing industry as well as seasons, fungal fermentation processes are devoid of any geographical or seasonal limitation. Moreover, mycelia do not require a demineralization step due to its low inorganic material content, consistent properties of chitin and chitosan due to controlled fermentation conditions. Chitosan production in the cell wall of zygomycetes fungi eliminates the expensive chemical deacetylation step presently used in the conventional process of chitosan production from crustacean wastes. Additionally, zygomycetes can grow on a variety of waste materials, which makes it environmentally acceptable.

Absidia glauca, *Aspergillus niger*, *Aspergillus awamori*, *Mucor rouxii*, *Gongronella butleri*, *Absidia glauca*, *Cunninghaella elegans*, *Penicillium citrinum Rhizopus oryzae*, and *Lentinus elodes* have been investigated for chitin and chitosan production [39,68–72]. Among these species, *Mucor rouxii* is the most researched species as it contains chitosan up to 30%–35% of the dry weight of cell wall of the mycelia [29]. Chitosan production by various fungi along with their extraction procedure and DD are presented in detail in Table 10.3 [73].

Chitosan extraction from fungal mycelia involves a mild alkali treatment using 1M NaOH at 120°C for 15–20 min. Chitin and chitosan are insoluble in alkali whereas majority of

Table 10.3: Fungal chitosan from alternate carbon sources.

Fungi	Fermentation	Culture medium	Chitosan extraction	Chitosan yield	Degree of deacetylation	References
Aspergillus awamori	Submerged fermentation pH-3.7/150 rpm/ 30°C/96 h	Raw distillery thin stillage	0.5 N NaOH/ 121°C/20 min 1% (v/v)H_2SO_4/ 121°C/20 min	7%–9%	68.89%	[93]
Aspergillus niger	Solid-state fermentation pH-5.9–6.4/static/ 30°C/16 d	Soya bean, canola and corn seed residues	1N NaOH/120°C/ 20 min 2% (v/v) acetic acid/95°C/6 h	17.05 ± 0.95 g/kg ds soya bean 12.73 ± 1.22 g/kg canola residue	–	[80]
Cunninghamella bertholletiae	Submerged fermentation pH-4.5/150 rpm/ 28°C/7 d	Yeast peptone dextrose medium, sugarcane juice	NaOH/Acetic acid	55 mg/g dcw (YPD) 128 mg/g dcw (sugarcane)	88.20% 89.70%	[76]
Cunninghamella elegans	Submerged fermentation pH-5.6/150 rpm/ 28°C/72 h	Corn steep liquor and molasses	–	33.13 mg/g dcw		[94]
Cunninghamella elegans UCP-542	Submerged fermentation pH-7.0/150 rpm/ 28°C/96 h	Yam bean medium	2% (w/v) NaOH/ 90°C/2 h 10% (v/v)acetic acid/60°C/6 h	66 mg/g dcw	85%	[95]
Gongronella butleri	Submerged fermentation pH-4.5/150 rpm/ 30°C/110 h	Apple pomace	2% NaOH/90°C/2h (200 rpm) 10% (v/v) acetic acid/60°C/6 h (150 rpm)	1.19 g/L 20% dcw	–	[96]
Gongronella butleri CCT 4274	Submerged fermentation					
Lentinus edodes	Solid-state fermentation 28°C/ 19 d submerged fermentation 400 rpm/28°C/ 16 d	Wheat straw	1N NaOH/120°C/ 1.5 h 2% (v/v) acetic acid/95°C/14 h	6.18 g/kg biomass (SSF) 120 mg/L (SmF)	88%–90%	[97]

Mucor hiemalis	Submerged fermentation pH-5.5/150 rpm/32°C/72 h	Wheat hydrolysate	1% NaOH/120°C/20 min 2% (v/v) acetic acid/95°C/8 h	0.46 g/g AIM	–	[98]
Penicillium chrysogenum	–	Waste mycelia of pharmaceutical industry	1N NaOH/95°C/1.5 h 1M acetic acid/45°C/1.5 h	5.7%	86%	[81]
Penicillium citrinum	Submerged fermentation pH-4.5/200 rpm/28°C/72 h	Paper mill wastewater	1M NaOH/120°C/20 min 2% (v/v) acetic acid/95°C/8 h	13.8%	81%	[71]
Rhizomucor pusillus	Submerged fermentation pH-5.7–6.2/36–38°C/4 d	Xylose-rich wastewater from ethanol plant	1N NaOH/90°C/2 h 2% H_2SO_4/95–121°C/20 min	45.7% AIM	97.5%	[99]
Rhizopus arrhizus	Submerged fermentation pH-5.6/150 rpm/28°C/96 h	Corn steep liquor and honey	1N NaOH/121°C/15 min 2% (v/v) acetic acid/100°C/15 min	29.3 mg/g dcw	86%	[69]
Rhizopus oryzae	Submerged fermentation 200 rpm/35°C	Hemicellulose hydrolysate of corn straw after H_2SO_4 hydrolysis	1N NaOH/121°C/15 min 2% (v/v) acetic acid/95°C/24 h	0.58 g/L	89%–90%	[100]
	Solid-state fermentation	Rice straw	1N NaOH/121°C/20 min 2% (v/v) acetic acid/95°C/8 h	5.63 g/kg of fermented medium	73%–90%	[75]
	Solid-state fermentation 30°C/16 d	Soybean and mung bean residues	1N NaOH/121°C/20 min 2% (v/v) acetic acid/95°C/8 h	4.3 g/kg soybean residues 1.6 g/kg mung bean residues	–	[64]

the carbohydrates, lipids and proteins are dissolved in the supernatant. The alkali insoluble material (AIM) is washed thoroughly and subjected to an acid treatment using 2% v/v acetic acid at 90°C for 6–7 h. Chitosan can be recovered after this treatment by increasing the pH of the acid solution to 8.0–9.0 because chitosan precipitates out in alkaline environment. The precipitated chitosan is washed with water, ethanol and acetone and finally dried at 65°C for 12–24 h. A detailed schematic of the extraction procedure is shown in Fig. 10.8 [70].

10.4.3.1 Fungal chitosan production from waste resources

For an economical production of chitosan from fungi, it is imperative to use alternate cheap alternate carbon sources as substrate for the fungal growth. Several studies are being conducted with the focus on developing a viable process for chitosan production from fungi at a large scale. Fungal species belonging to *Aspergillus*, *Cunninghamella*, *Mucor*,

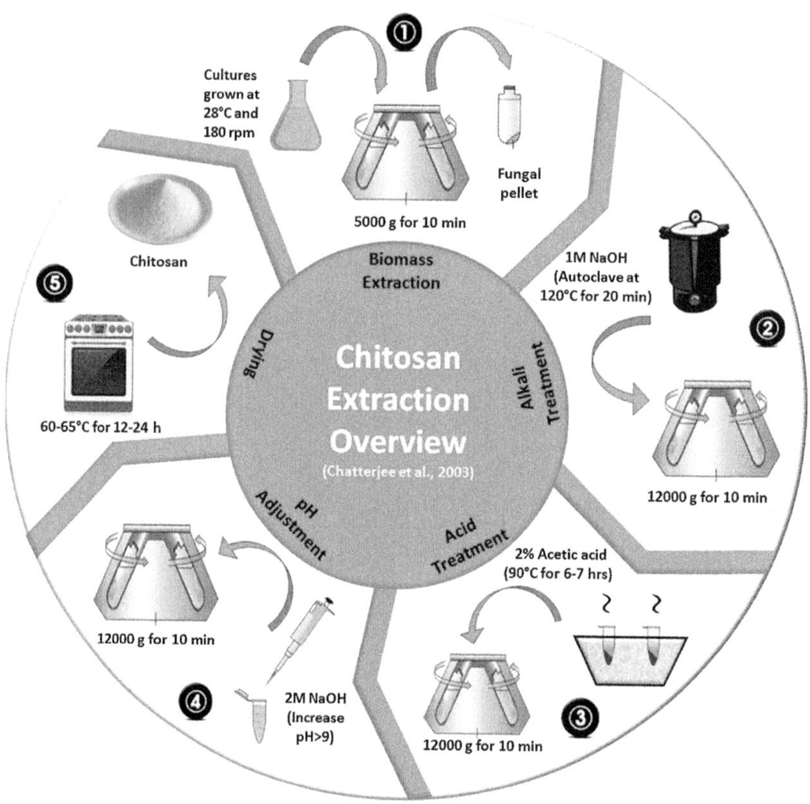

Figure 10.8
Schematic showing chitosan extraction from fungal biomass.

Penicillium, *Rhizopus*, etc. have been studied for chitosan production from inexpensive carbon sources (Table 10.3).

Rice and corn were used as substrates for growing *R. oryzae* by Hang [74] for chitosan production. 600 mg/L chitosan was produced after a cultivation period of 48 h at 30°C. Rice straw supplemented with minerals for fungal chitosan was investigated by Khalaf [75] using *R. oryzae, P. citrinium, A. niger*, and *F. oxysporium* under SSF conditions. *R. oryzae* gave the highest chitosan yield of 5.63 g/kg biomass followed by *A. niger* after 12 days of incubation. Soybean and mung bean residues were studied for chitosan production by four fungal strains and among these strains, *R. oryzae* gave the highest chitosan yield of 4.3 g/kg of substrate when grown on soybean in plastic bags with cotton plugs at 30°C after 16 days of cultivation [64].

Cunninghamella bertholletiae was reported to produce chitosan by Amorim et al. [76], from sugarcane juice and molasses as noncommercial substrates. A chitosan yield of 128 mg/g of dry mycelia was achieved under the optimal conditions of pH 4.5, 150 rpm agitation, 28°C for 7 days. The drawback which could be taken from this study is that the fungi was unable to efficiently utilize molasses for its growth and chitosan production when compared with sugarcane juice. A newly isolated strain of *Syncephalastrum racemosum* was used for chitosan production in media comprising molasses and sugarcane juice as carbon source. Highest chitosan yield of 74 mg/g dry mycelia was observed with the sugarcane juice based media when cultivated at 28°C, 400 rpm for 60 h [77].

Buckwheat/barley and sweet potato were used for chitosan production by *G. butleri* resulting in a yield of 730 mg/L chitosan using a sweet potato based medium obtained from shochu distillery wastewater at pH 5.0 and 30°C temperature after 5 days of fermentation [78]. Solid-state fermentation was applied for *G. butleri* using sweet potato as the substrate and the influence of different nitrogen sources was analyzed to obtain a chitosan yield of 11% dry cell weight [79]. *A. niger* produced highest amount of chitosan (17–18 g/kg DS) on soya bean residues, when compared with corn seed and canola residues, which were used as the naturally available noncommercial carbon sources [80]. *Penicillium* waste obtained from pharmaceutical industries served as a zero-cost source for chitosan extraction. A chitosan yield of 5.7% from *P. chrysogenum,* with an 86% DD was reported by Wang et al. [81]. Namboodiri and Pakshirajan [71] reported fungal chitosan production using paper mill wastewater by *Penicillium citrinum* biomass at 28°C temperature and pH 4.5. A maximum chitosan yield of 13.8% dry fungal biomass was achieved after acetic acid addition at low levels.

10.4.3.2 Bioreactor considerations

As discussed previously, fungal chitosan production has been the focus of research over the last two decades owing to the various advantages fungi possess as a source of chitin and chitosan. However, the recent focus is to scale-up this route of chitosan production to an industrial level. A number of researchers have successfully demonstrated strategies of fungal chitosan production by applying lab-scale bioreactors.

10.4.3.2.1 Solid-state fermentation

Solid-state fermentation has been widely used for cultivating fungi for the production of various metabolites, particularly enzymes. This mode of fermentation is highly suited to the growth characteristics as well as the metabolism of fungi. Low water demand, high product volume and absence of catabolite repression are the major advantages of employing solid-state fermentation. Various bioreactor designs have been developed for fungal fermentation of solid substrates depending upon the need of aeration (diffusion/forced), agitation (static/mixing), fungal morphology, substrate characteristics, and moisture content. Solid-state fermentation can be carried out using simple trays or sealed plastic bags with cotton plugs at small scale; Koji bioreactors, horizontal and rotating drum bioreactors are commonly used.

Perforated tray and rotating drum fermenters are the most widely applied bioreactor configurations for metabolite production by fungi (Fig. 10.9). The substrate is placed on

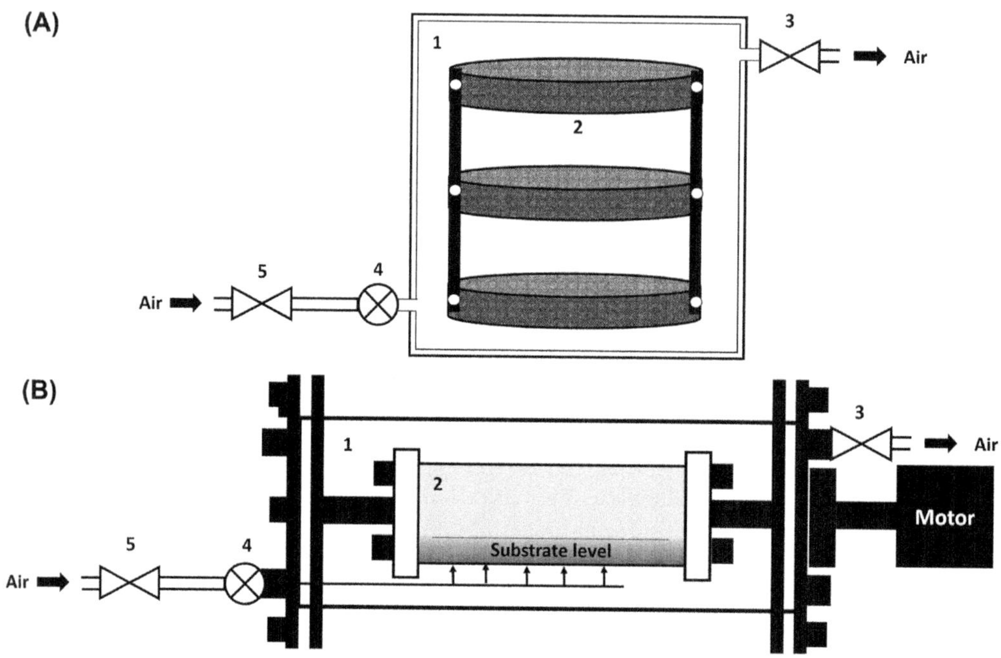

Figure 10.9
(A) Tray fermenter: (1) reactor vessel, (2) trays, (3,5) air filters, (4) humidifier. (B) Rotating drum filter: (1) reactor vessel, (2) rotating drum, (3,5) air filters, (4) humidifier.

trays that are perforated for better air transfer. These trays are placed in temperature and humidity-controlled environment such as inside a chamber or room. Although this set up is easy to scale-up but it suffers from a number of serious drawbacks. Substrate layer thickness along with the changes in porosity of the substrate due to the growth of fungi pose serious oxygen transfer limitations. Moreover, scaling up of this process is labor intensive and requires large operational area. Rotary drum fermenters make use of horizontal cylindrical vessel which is partially filled with the solid substrate. Aeration is done by passing the air through the headspace, which then diffuses into the solid phase. These cylinders or drums are rotated and therefore mixing is achieved due to the tumbling motion of the substrates inside. The mixing is carried out in continuous or intermittent modes. Continuous agitation is found to be detrimental for the fungal growth due to damage caused by the shear stress. The advantage of using a rotary drum fermenter over the static fermenters is that sufficient air transfer is achieved along with the dissipation of heat produced due to the fungal metabolism [82,83]. The major issues associated with scale-up of such solid-state reactors are the heterogeneous nature of the system along with the intense metabolic heat generation. Hence, several bioreactor designs with certain modifications have come up which attempt to mitigate these issues but only a few of these have been applied at a large scale. Several studies have reported to employ solid-state fermentation for fungal chitosan production which have been presented in Table 10.3.

10.4.3.2.2 Submerged fermentation

Filamentous fungi have been widely explored for primary and secondary metabolite production and biomass. Various factors such as oxygen demand and rheology of the culture medium govern the success of submerged fermentation using fungal biomass. However, they are highly susceptible to shear stress imparted due to agitation in the reactors. Moreover, viscosity of the broth increases with the growth of fungi and the fungal morphology is a crucial parameter which leads to highly viscous culture medium. The fungi grow as pellets or as free mycelia depending upon the agitation speed in the reactors. The formation of pellets is desirable and has been shown to enhance the overall efficiency of a fermentation process. Mycelial agglomeration can be overcome by maintaining various operational parameters such as aeration rate, agitation speed, etc. at their optimum levels as well as by a proper choice of baffle and reactor vessel design. Media components are also known to play a vital role in avoiding the formation of fungal pellets. Therefore, optimization of various process parameters along with modifications in bioreactor configurations is necessary.

The most commonly used reactor configurations for the production of fungal chitosan under submerged fermentation conditions are stirred tank and airlift fermenters (Fig. 10.10), with the former being the most widely used. Rushton impellers with radial flow are the default agitation system that is attached to the reactors. While impellers with

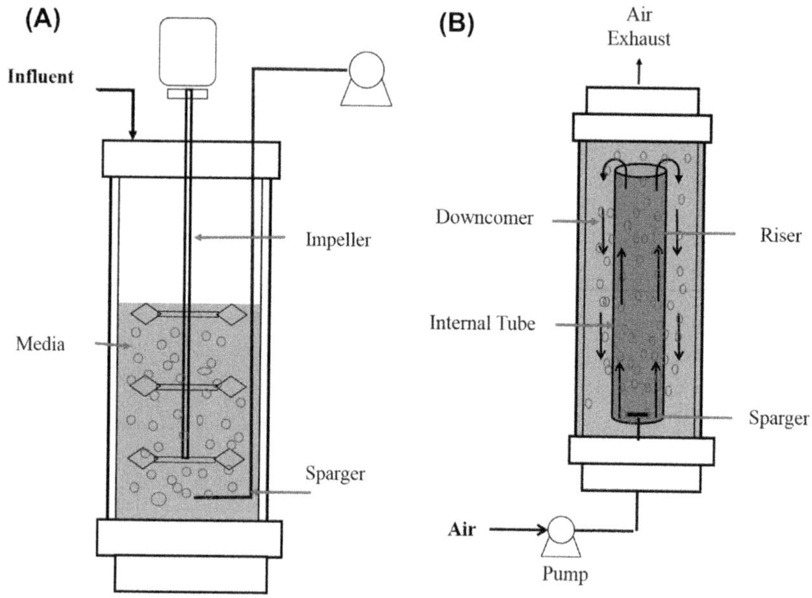

Figure 10.10
Schematic of (A) stirred tank reactor. (B) Airlift reactor.

alternate designs and radial flow have shown to be better than Rushton turbines for bulk mixing, they have been used in the case of exopolysaccharide production where high viscosity of the broth lead to acute mixing problems. Although, axial flow impellers provide better mixing than radial flow impellers, the shear stress imparted to the fungal mycelia is higher which is detrimental to the process. With the increase in scale however, the axial flow impellers provided superior agitation and the shear damage imparted to the mycelia decreased [84,85].

Four different configurations of the stirred tank reactor were studied to observe the difference in the morphology of the chitosan producing strain *Absidia coerula* at 25°C and pH 4.5. These configurations differed in the working volume (1–10 L) and arrangement of baffles and aeration setups. The changes in stirring speed and aeration led to varied sizes of pellets and accretion of mycelium on the impeller, baffles, probes, etc. This study showed the importance of impeller design as an important factor in the cultivation of fungi in bioreactors [86]. Rane and Hoover [87] carried out submerged fermentation using airlift and stirred tank reactors (STR) for chitosan production. The efficiency of the both these reactors were hampered due to mycelial agglomeration. Small pellets were obtained after the addition of an antifoam 289, which led to better results. Another important factor governing the design of the reactor and impellers is the medium composition.

The issues of shear damage and mixing associated with STRs have led to airlift fermenters being the choice of fungal fermentation involving fragile fungal mycelia and highly viscous media. Airlift reactors are specialized bubble column reactors that are divided into two sections: riser and downcomer. Circulation is achieved due to the difference in hydrostatic pressure and fluid density depending upon the point of sparging. Low shear damage, better mixing, reduced energy costs and low contamination risks due to the absence of impeller shaft have led to the use of airlift fermenters in industrial applications. Advances have been made where microbubbles are generated using fluidic oscillator which result in increased contact time between the gas and liquid, better mixing as well as less energy dissipation [88]. Different synthetic and alternate media have been employed for carrying out scale-up studies for fungal chitosan production.

10.5 Conclusions and perspectives

Chemical extraction methods to derive chitin and chitosan from marine sources is the widely used and accepted route of their production. However, harsh chemical treatments for demineralization and deproteinization result in hazardous wastes and undesired by-products. This in turn leads to high production costs of chitin and chitosan. Moreover, the variation in the properties of the derived chitosan is a serious drawback for applying the chemical mode of chitin extraction. The biological route of chitin and chitosan production is a very efficient and potent alternative. The use of microorganisms such as bacteria and fungi and utilization of their efficient enzyme systems would lead to a better and economically viable option to obtain chitin and chitosan from the marine wastes. Insects, alongside crustacean waste, are also an important source of chitin. High chitin content and the biodiversity of this group opens up a lot of options for a viable route of chitin and chitosan production. On the other hand, the economics of insect cultivation, time required along with the seasonal availability and inconsistent properties of the chitin and chitosan produced seem to be some of the serious drawbacks of using insects and crustaceans for chitosan production. Fungi are a promising option as a primary source of chitin and chitosan. Fungi contain both chitin and chitosan within their cell wall, which, therefore, negates the harsh demineralization step required in the case with marine sources. Fungi are known to utilize industrial and agricultural wastes as substrates for their growth, which can be applied to reduce the costs incurred for chitosan production. Moreover, properties of the chitosan are consistent and can be controlled by changing the different fermentation parameters. However, the technoeconomic feasibility of fungal-based chitosan production processes need to be evaluated for establishing the process as an economically viable and sustainable green technology for chitin and chitosan production.

Acknowledgments

The authors thank Department of Biosciences and Bioengineering, Indian Institute of Technology Guwahati, Assam, for supporting their knowledge dissemination and outreach activities related to this work.

References

[1] Roberts GAF. Thirty years of progress in chitin and chitosan. Progress on Chemistry and Application of Chitin 2008;13:7−15.

[2] Rufato K, Galdino J, Ody K, Pereira A, Hydrogels EC. Hydrogels based on chitosan and chitosan derivatives for biomedical application. intechopen.com; 2018. Available from: https://www.intechopen.com/download/pdf/64338.

[3] Carlström D. The crystal structure of α-chitin (poly-N-acetyl-D-glucosamine). The Journal of Cell Biology 1957;3(5):669−83.

[4] Gortari MC, Hours RA. Biotechnological processes for chitin recovery out of crustacean waste: a mini-review. Electronic Journal of Biotechnology 2013;16(3).

[5] Nair KG, Dufresne A. Crab shell chitin whisker reinforced natural rubber nanocomposites. 1. Processing and swelling behavior. Biomacromolecules 2003;4(3):657−65.

[6] Ruiz-Herrera J. Fungal cell wall: structure, synthesis, and assembly. CRC Press; 2016.

[7] Bowman SM, Free SJ. The structure and synthesis of the fungal cell wall. Bioessays 2006;28(8): 799−808.

[8] Philibert T, Lee BH, Fabien N. Current status and new perspectives on chitin and chitosan as functional biopolymers. Applied Biochemistry and Biotechnology 2017;181(4):1314−37.

[9] Kumar MNVR. A review of chitin and chitosan applications. Reactive and Functional Polymers 2000;46(1):1−27.

[10] Rinaudo M. Chitin and chitosan: properties and applications. Progress in Polymer Science 2006;31(7):603−32.

[11] Li J, Cai C, Li J, Li J, Li J, Sun T, et al. Chitosan-based nanomaterials for drug delivery. Molecules 2018;23(10):2661.

[12] Aranaz I, Mengíbar M, Harris R, Paños I, Miralles B, Acosta N, et al. Functional characterization of chitin and chitosan. Current Chemical Biology 2009;3(2):203−30.

[13] Huang M, Khor E, Lim LY. Uptake and cytotoxicity of chitosan molecules and nanoparticles: effects of molecular weight and degree of deacetylation. Pharmaceutical Research 2004;21(2):344−53.

[14] Hirano S. Bio-compatibility of chitosan by oral and intravenous administrations. Polymeric Materials Engineering and Science 1988;59:897−901.

[15] Tomihata K, Ikada Y. In vitro and in vivo degradation of films of chitin and its deacetylated derivatives. Biomaterials 1997;18(7):567−75.

[16] Zhang H, Neau SH. In vitro degradation of chitosan by a commercial enzyme preparation: effect of molecular weight and degree of deacetylation. Biomaterials 2001;22(12):1653−8.

[17] Okamoto Y, Kawakami K, Miyatake K, Morimoto M, Shigemasa Y, Minami S. Analgesic effects of chitin and chitosan. Carbohydrate Polymers 2002;49(3):249−52.

[18] Muzzarelli RAA, Orlandini F, Pacetti D, Boselli E, Frega NG, Tosi G, Muzzarelli C. Chitosan taurocholate capacity to bind lipids and to undergo enzymatic hydrolysis: an in vitro model. Carbohydrate Polymers 2006;66(3):363−71.

[19] Thongngam M, McClements DJ. Influence of pH, ionic strength, and temperature on self-association and interactions of sodium dodecyl sulfate in the absence and presence of chitosan. Langmuir 2005;21(1):79−86.

[20] Hadi AG. Dye removal from colored textile wastewater using synthesized chitosan. International Journal of Science and Technology 2013;2(4):359−64.

[21] Ngah WSW, Teong LC, Hanafiah MAKM. Adsorption of dyes and heavy metal ions by chitosan composites: a review. Carbohydrate Polymers 2011;83(4):1446−56.

[22] Ghanem A, Ghaly A. Immobilization of glucose oxidase in chitosan gel beads. Journal of Applied Polymer Science 2004;91(2):861−6.

[23] Muzzarelli RAA. Immobilization of enzymes on chitin and chitosan. Enzyme and Microbial Technology 1980;2(3):177−84.

[24] Taqieddin E, Amiji M. Enzyme immobilization in novel alginate-chitosan core-shell microcapsules. Biomaterials 2004;25(10):1937−45.
[25] Becker A, Schloder P, Steele JE, Wegener G. The regulation of trehalose metabolism in insects. Experientia 1996;52(5):433−9.
[26] Francois J, Parrou JL. Reserve carbohydrates metabolism in the yeast *Saccharomyces cerevisiae*. FEMS Microbiology Reviews 2001;25(1):125−45.
[27] Merzendorfer H. The cellular basis of chitin synthesis in fungi and insects: common principles and differences. European Journal of Cell Biology 2011;90(9):759−69.
[28] Cauchie HM, Murugan G, Thome JP, Dumont HJ. Intra- and interspecific variation in the chitin content of some anostracans. Hydrobiologia 1997;359:223−8.
[29] Synowiecki J, Al-Khateeb NA. Production, properties, and some new applications of chitin and its derivatives. Critical Reviews in Food Science and Nutrition 2003;43(2):145−71.
[30] Johnson EL, Peniston QP. Utilization of shellfish waste for chitin and chitosan production. Abstracts of Papers of the American Chemical Society 1979;(September):34.
[31] The state of world fisheries and aquaculture 2018. UN; 2018.
[32] Handayani AD, Sutrisno, Indraswati N, Ismadji S. Extraction of astaxanthin from giant tiger (*Panaeus monodon*) shrimp waste using palm oil: studies of extraction kinetics and thermodynamic. Bioresource Technology 2008;99(10):4414−9.
[33] Xuemei Z, Hawkins SJ. Interactions of aquaculture and waste disposal in the coastal zone. Journal of Ocean University of Qingdao 2002;1(1):8−12.
[34] Xu Y, Gallert C, Winter J. Chitin purification from shrimp wastes by microbial deproteination and decalcification. Applied Microbiology and Biotechnology 2008;79(4):687−97.
[35] Kurita K. Chitin and chitosan: functional biopolymers from marine crustaceans. Marine Biotechnology 2006;8(3):203−26.
[36] No HK, Meyers SP. Preparation and characterization of chitin and chitosan—a review. Journal of Aquatic Food Product Technology 1995;4(2):27−52.
[37] Percot A, Viton C, Domard A. Characterization of shrimp shell deproteinization. Biomacromolecules 2003;4(5):1380−5.
[38] No HK, Hur EY. Control of foam formation by antifoam during demineralization of crustacean shell in preparation of chitin. Journal of Agricultural and Food Chemistry 1998;46(9):3844−6.
[39] Teng WL, Khor E, Tan TK, Lim LY, Tan SC. Concurrent production of chitin from shrimp shells and fungi. Carbohydrate Research 2001;332(3):305−16.
[40] Palpandi C, Shanmugam V, Shanmugam A. Extraction of chitin and chitosan from shell and operculum of mangrove gastropod *Nerita* (*Dostia*) *crepidularia* Lamarck. International Journal of Medicine and Medical Sciences 2009;1(5):198−205.
[41] Tolaimate A, Desbrieres J, Rhazi M, Alagui A. Contribution to the preparation of chitins and chitosans with controlled physico-chemical properties. Polymer 2003;44(26):7939−52.
[42] Manni L, Ghorbel-Bellaaj O, Jellouli K, Younes I, Nasri M. Extraction and characterization of chitin, chitosan, and protein hydrolysates prepared from shrimp waste by treatment with crude protease from *Bacillus cereus* SV1. Applied Biochemistry and Biotechnology 2010;162(2):345−57.
[43] Pacheco N, Garnica-Gonzalez M, Gimeno M, Barzana E, Trombotto S, David L, Shirai K. Structural characterization of chitin and chitosan obtained by biological and chemical methods. Biomacromolecules 2011;12(9):3285−90.
[44] Al Sagheer FA, Al-Sughayer MA, Muslim S, Elsabee MZ. Extraction and characterization of chitin and chitosan from marine sources in Arabian Gulf. Carbohydrate Polymers 2009;77(2):410−9.
[45] Giyose NY, Mazomba NT, Mabinya LV. Evaluation of proteases produced by *Erwinia chrysanthemi* for the deproteinization of crustacean waste in a chitin production process. African Journal of Biotechnology 2010;9(5):707−10.
[46] Sini TK, Santhosh S, Mathew PT. Study on the production of chitin and chitosan from shrimp shell by using *Bacillus subtilis* fermentation. Carbohydrate Research 2007;342(16):2423−9.

[47] Sorokulova I, Krumnow A, Globa L, Vodyanoy V. Efficient decomposition of shrimp shell waste using *Bacillus cereus* and *Exiguobacterium acetylicum*. Journal of Industrial Microbiology and Biotechnology 2009;36(8):1123–6.
[48] Kaur S, Dhillon GS. Recent trends in biological extraction of chitin from marine shell wastes: a review. Critical Reviews in Biotechnology 2015;35(1):44–61.
[49] Jung WJ, Jo GH, Kuk JH, Kim YJ, Oh KT, Park RD. Production of chitin from red crab shell waste by successive fermentation with *Lactobacillus paracasei* KCTC-3074 and *Serratia marcescens* FS-3. Carbohydrate Polymers 2007;68(4):746–50.
[50] Arbia W, Arbia L, Adour L, Amrane A. Chitin extraction from crustacean shells using biological methods – a review. Food Technology and Biotechnology 2013;51(1):12–25.
[51] Rao SVS, Lekshmi R, Mani AGS, Sinha PK. Treatment of low level radioactive liquid wastes using composite ion-exchange resins based on polyurethane foam. Journal of Radioanalytical and Nuclear Chemistry 2010;283(2):379–84.
[52] Khanafari A, Marandi R, Sanatei S. Recovery of chitin and chitosan from shrimp waste by chemical and microbial methods. Journal of Environmental Health Science & Engineering 2008;5(1):1–24.
[53] Bustos RO, Healy MG. Microbial deproteinisation of waste prawn shell. In: Proceedings of the 2nd international symposium on environmental biotechnology. Institute of Chemical Engineers; 1994.
[54] Gildberg A, Stenberg E. A new process for advanced utilisation of shrimp waste. Process Biochemistry 2001;36(8–9):809–12.
[55] Younes I, Ghorbel-Bellaaj O, Nasri R, Chaabouni M, Rinaudo M, Nasri M. Chitin and chitosan preparation from shrimp shells using optimized enzymatic deproteinization. Process Biochemistry 2012;47(12):2032–9.
[56] Younes I, Hajji S, Frachet V, Rinaudo M, Jellouli K, Nasri M. Chitin extraction from shrimp shell using enzymatic treatment. Antitumor, antioxidant and antimicrobial activities of chitosan. International Journal of Biological Macromolecules 2014;69:489–98.
[57] Dossey AT. Insects and their chemical weaponry: new potential for drug discovery. Natural Product Reports 2010;27(12):1737–57.
[58] Kaya M, Baran T, Mentes A, Asaroglu M, Sezen G, Tozak KO. Extraction and characterization of alpha-chitin and chitosan from six different aquatic invertebrates. Food Biophysics 2014;9(2):145–57.
[59] Kaya M, Seyyar O, Baran T, Turkes T. Bat guano as new and attractive chitin and chitosan source. Frontiers in Zoology 2014;11.
[60] Kaya M, Erdogan S, Mol A, Baran T. Comparison of chitin structures isolated from seven *Orthoptera* species. International Journal of Biological Macromolecules 2015;72:797–805.
[61] Kaya M, Lelesius E, Nagrockaite R, Sargin I, Arslan G, Mol A, et al. Differentiations of chitin content and surface morphologies of chitins extracted from male and female grasshopper species. PLoS One 2015;10(1).
[62] Marei NH, Abd El-Samie E, Salah T, Saad GR, Elwahy AHM. Isolation and characterization of chitosan from different local insects in Egypt. International Journal of Biological Macromolecules 2016;82:871–7.
[63] Hu KJ, Hu JL, Ho KP, Yeung KW. Screening of fungi for chitosan producers, and copper adsorption capacity of fungal chitosan and chitosanaceous materials. Carbohydrate Polymers 2004;58(1):45–52.
[64] Pochanavanich P, Suntornsuk W. Fungal chitosan production and its characterization. Letters in Applied Microbiology 2002;35(1):17–21. Available from: http://www.ncbi.nlm.nih.gov/pubmed/12081543.
[65] Sousa Andrade V, de Barros Neto B, Fukushima K, Campos Takaki GM. Effect of medium components and time of cultivation on chitin production by *Mucor circinelloides* (*Mucor javanicus* IFO 4570) – a factorial study. Revista Iberoamericana de Micologia 2003;20(4):149–53. Available from: http://www.ncbi.nlm.nih.gov/pubmed/15456353.
[66] Muzzarelli RAA, Stanic V, Ramos V. Enzymatic depolymerization of chitins and chitosans. In: Carbohydrate biotechnology protocols. Springer; 1999. p. 197–211.

[67] Guarro J, Gené J, Stchigel AM. Developments in fungal taxonomy. Clinical Microbiology Reviews 1999;12(3):454–500. Available from: http://www.ncbi.nlm.nih.gov/pubmed/10398676.

[68] Arcidiacono S, Kaplan DL. Molecular-weight distribution of chitosan isolated from Mucor-rouxii under different culture and processing conditions. Biotechnology and Bioengineering 1992;39(3):281–6.

[69] Cardoso A, Lins CIM, dos Santos ER, Freitas Silva MC, Campos-Takaki GM, Silva MCF, Campos-Takaki GM. Microbial enhance of chitosan production by *Rhizopus arrhizus* using agroindustrial substrates. Molecules 2012;17(5):4904–14.

[70] Chatterjee S, Adhya M, Guha AK, Chatterjee BP. Chitosan from *Mucor rouxii*: production and physico-chemical characterization. Process Biochemistry 2005;40(1):395–400.

[71] Namboodiri MMT, Pakshirajan K. Sustainable and green approach of chitosan production from *Penicillium citrinum* biomass using industrial wastewater as a cheap substrate. Journal of Environmental Management 2019;240:431–40.

[72] Rane KD, Hoover DG. An evaluation of alkali and acid treatments for chitosan extraction from fungi. Process Biochemistry 1993;28(2):115–8.

[73] Dhillon GS, Kaur S, Brar SK, Verma M. Green synthesis approach: extraction of chitosan from fungus mycelia. Critical Reviews in Biotechnology 2013;33(4):379–403.

[74] Hang YD. Chitosan production from *Rhizopus oryzae* mycelia. Biotechnology Letters 1990;12(12):911–2.

[75] Khalaf SA. Production and characterization of fungal chitosan under solid-state fermentation conditions. International Journal of Agriculture and Biology 2004;6(6):1033–6.

[76] Amorim RVDS, Pedrosa RP, Fukushima K, Martínez CR, Ledingham WM, Campos-Takaki D, Maria G. Alternative carbon sources from sugar cane process for submerged cultivation of *Cunninghamella bertholletiae* to produce chitosan. Food Technology and Biotechnology 2006;44(4).

[77] Amorim RVS, Ledingham WM, Kennedy JF, Campos-Takaki GM. Chitosan from *Syncephalastrum racemosum* using sugar cane substrates as inexpensive carbon sources. Food Biotechnology 2006;20(1):43–53.

[78] Yokoi H, Aratake T, Nishio S, Hirose J, Hayashi S, Takasaki Y. Chitosan production from Shochu distillery wastewater by funguses. Journal of Fermentation and Bioengineering 1998;85(2):246–9.

[79] Nwe N, Stevens WF. Effect of urea on fungal chitosan production in solid substrate fermentation. Process Biochemistry 2004;39(11):1639–42.

[80] Maghsoodi V, Yaghmaei S, Beigi SM. Influence of different nitrogen sources on amount of chitosan production by *Aspergillus niger* in solid state fermentation. Iranian Journal of Chemistry and Chemical Engineering-International English Edition 2008;27(1):47–52.

[81] Wang TQ, Li HX, Wang MY, Tan TW. Integrative extraction of ergosterol, (1 → 3)-alpha-D-glucan and chitosan from *Penicillium chrysogenum* mycelia. Chinese Journal of Chemical Engineering 2007;15(5):725–9.

[82] Durand A. Bioreactor designs for solid state fermentation. Biochemical Engineering Journal 2003;13(2–3):113–25.

[83] Manan MA, Webb C. Design aspects of solid state fermentation as applied to microbial bioprocessing. Journal of Applied Biotechnology and Bioengineering 2017;4(1):91.

[84] de Carvalho CCCR. Fungi in fermentation and biotransformation systems. In: Biology of microfungi. Springer; 2016. p. 525–41.

[85] Gibbs PA, Seviour RJ, Schmid F. Growth of filamentous fungi in submerged culture: problems and possible solutions. Critical Reviews in Biotechnology 2000;20(1):17–48.

[86] Davoust N, Hansson G. Identifying the conditions for development of beneficial mycelium morphology for chitosan-producing *Absidia spp* in submersed cultures. Applied Microbiology and Biotechnology 1992;36(5):618–20.

[87] Rane KD, Hoover DG. Production of chitosan by fungi. Food Biotechnology 1993;7(1):11–33.

[88] AL-Mashhadani MKH, Wilkinson SJ, Zimmerman WB. Airlift bioreactor for biological applications with microbubble mediated transport processes. Chemical Engineering Science 2015;137:243–53.

[89] Flores-Albino B, Arias L, Gomez J, Castillo A, Gimeno M, Shirai K. Chitin and L(+)-lactic acid production from crab (*Callinectes bellicosus*) wastes by fermentation of *Lactobacillus sp* B2 using sugar cane molasses as carbon source. Bioprocess and Biosystems Engineering 2012;35(7):1193–200.
[90] Ghorbel-Bellaaj O, Younes I, Maalej H, Hajji S, Nasri M. Chitin extraction from shrimp shell waste using *Bacillus* bacteria. International Journal of Biological Macromolecules 2012;51(5):1196–201.
[91] Ghorbel-Bellaaj O, Hajji S, Younes I, Chaabouni M, Nasri M, Jellouli K. Optimization of chitin extraction from shrimp waste with *Bacillus pumilus* A1 using response surface methodology. International Journal of Biological Macromolecules 2013;61:243–50.
[92] Harkin C, Bruck WM, Lynch C. Isolation & identification of bacteria for the treatment of brown crab (*Cancer pagurus*) waste to produce chitinous material. Journal of Applied Microbiology 2015;118(4):954–65.
[93] Ray SG, Ghangrekar MM. Biodegradation kinetics of thin-stillage treatment by *Aspergillus awamori* and characterization of recovered chitosan. Applied Microbiology and Biotechnology 2016;100(4):1955–65.
[94] Berger LR, Stamford TC, Stamford-Arnaud TM, de Oliveira Franco L, do Nascimento AE, Cavalcante HM, et al. Effect of corn steep liquor (CSL) and cassava wastewater (CW) on chitin and chitosan production by *Cunninghamella elegans* and their physicochemical characteristics and cytotoxicity. Molecules 2014;19(3):2771–92.
[95] Stamford TCM, Stamford TLM, Stamford NP, Neto BD, de Campos-Takaki GM. Growth of *Cunninghamella elegans* UCP 542 and production of chitin and chitosan using yam bean medium. Electronic Journal of Biotechnology 2007;10(1):61–8.
[96] Streit F, Koch F, Laranjeira MCM, Ninow JL. Production of fungal chitosan in liquid cultivation using apple pomace as substrate. Brazilian Journal of Microbiology 2009;40(1):20–5.
[97] Crestini C, Kovac B, Giovannozzi-Sermanni G. Production and isolation of chitosan by submerged and solid-state fermentation from *Lentinus edodes*. Biotechnology and Bioengineering 1996;50(2):207–10.
[98] Vinche MH, Asachi R, Zamani A, Karimi K. Ethanol and chitosan production from wheat hydrolysate by *Mucor hiemalis*. Journal of Chemical Technology and Biotechnology 2013;88(2):255–60.
[99] Zamani A, Edebo L, Sjostrom B, Taherzadeh MJ. Extraction and precipitation of chitosan from cell wall of zygomycetes fungi by dilute sulfuric acid. Biomacromolecules 2007;8(12):3786–90.
[100] Tai C, Li S, Xu Q, Ying H, Huang H, Ouyang P. Chitosan production from hemicellulose hydrolysate of corn straw: impact of degradation products on *Rhizopus oryzae* growth and chitosan fermentation. Letters in Applied Microbiology 2010;51(3):278–84.

SECTION D

Non-edible oils based biorefinery and applications

CHAPTER 11

Potential of castor plant (Ricinus communis) for production of biofuels, chemicals, and value-added products

Ravneet Kaur[1,2], Thallada Bhaskar[2,3]

[1]Dr. B.R. Ambedkar National Institute of Technology, Jalandhar, Punjab, India; [2]Material Resource Efficiency Division, CSIR-Indian Institute of Petroleum, Dehradun, Uttarakhand, India; [3]Academy of Scientific and Innovative Research (AcSIR), New Delhi, India

11.1 Introduction

The dependence on fossil fuel to meet energy demand contributes to various environmental and human health issues. The rise in exploitation, technological progress, and potential reserves leads to increase of greenhouse gases (GHGs) emissions and change in the climate. The global atmospheric carbon dioxide level rises exponentially, which further leads to an increase in the global temperature. Energy conversion and proper utilization is the main challenge of our time. The selection of alternative energy sources depends on the (1) availability of the source, (2) environmental benefit, and (3) economic benefit. Biofuels are the good alternative to substitute petroleum-based fuels and can be utilized as transport fuels with little change in technologies as they also have the potential to improve environmental issues and reduce GHG emissions [1]. Biomass, a second-generation biofuel feedstock, has advantages such as a CO_2-neutral substitute of fossil fuel, abundant availability in all areas in the world, sustainable energy through renewable biomass, reduction in gases like NO_x/SO_x due to less amount present in biomass [2–4] to be utilized as an energy source. In the past three decades, global research on biomass contributes 56%, solar energy 26%, wind energy 11%, geothermal energy 5%, and hydropower 2% [5].

Castor plant (*Ricinus communis* L.) is a lignocellulosic biomass from Plantae kingdom, Tracheobinta Subkingdom, Magnoliopsida Class, and Euphorbiales Order. Castor plant is a nonedible oilseed plant belongs to the Euphorbiaceae family, includes 218 genera and 6745 species distributed globally as shown in Fig. 11.1. The genus "*Ricinus* L." is considered to be monotypic and comprised in the subfamily Acalyphoideae includes

Waste Biorefinery. https://doi.org/10.1016/B978-0-12-818228-4.00011-3

Figure 11.1
Castor (*Ricinus communis*) plant.

99 genera and 1865 species. This spurge family consists of about 300 genera and 7500 species [6]. It generally grows in moist, well-drained soils in wastelands, distributed area, stream banks, roadsides, river bottoms, and bottomlands. Cultivation of castor plant is done for seed oil as it has many industrial and medicinal applications. This chapter provides a brief introduction about castor plant origin and its cultivation process for the production of castor seed. It includes a discussion on the extraction of castor oil from its seed along with the physical and chemical properties of castor oil. The extraction process of oil from the seed is discussed along with its physical and chemical properties. This chapter explains the applications of each part of the castor plant in various fields along with an idea for its application toward environmental sustainability. The aim of the chapter is the appropriate utilization of castor plant residue for the production of value-added products makes the process more economical. With the same objective, this chapter gives a brief introduction to the potential value-added products formed from castor oil.

11.1.1 Castor plant: its origin

Theophrastus and Dioscorides in the 1st century gave a detailed description of the castor plant, including the extraction of oil from seeds and stated that the castor oil is not suitable for eating but has various medicinal applications. The origin of the castor plant is not very much evident, but it was said that the origin of the castor plant is in the tropical belt of India and Africa. The father of History Herodotus (c.484—c.425) stated that the Egyptians in the 4th century BC collected the oil from castor seeds and used it for burning purposes [7]. Egyptians used to name it Kiki, and until now the plantation in Greece is done by the same name.

The cultivation of castor plant was done until the 15th century, but the negative impacts of the plant make it unpopular by the 18th century. Later, in the 18th century, the castor has been extensively cultivated in Jamaica. During 1890, British India was the great castor-bean producing a country of the world. United States, England, France, Germany, Belgium, and Italy were imported castor oil from British India. As no records were there for the production of castor in British India, it was expected that the total exports of castor beans and oil to all countries depended on its output.

Castor plant was migrated to India and China during the Tang period (610–906 AD). After the discovery of Columbus, castor bean was naturalized in North America, and tropical and semitropical areas of the world [8]. It was believed that the castor plant was first originated in Ethiopian in the East Africa region. The four centers under these regions were: (1) Subcontinent of India, (2) Ethiopia, (3) Northwest and Southwest Asia and Arabian peninsula, and (4) China. It was endemic to the Ethiopian region of tropical East Africa and South Western United States around the world. It became a weed in many places, including the South Western U.S. In India, the castor oil is used in medicine and surgery and referred in *Susruta Sambita* written over 2000 years.

11.1.2 Nomenclature

The scientific name of castor plant is *Ricinus communis* has a more logical derivation. *Ricinus* is derived from the Latin word for "tick," as the seed has ticks and markings [9]. It is specific for the Mediterranean sheep tick (*Ixodes Ricinus*). *Communis* means "common" in Latin, and the castor plant is everywhere in the world in the 18th century so Swedish naturalist Carolus Linnaeus gave this name to castor plant. The residents Spaniards and Portuguese, confused it with a different plant, named the Vitex agnus castus, and then called it "agno casto". Due to this designation, the English person who traded in this oil coined the word castor, and thus gave rise to this name throughout the English-speaking world. Common names of castor plant are Wonder tree, Palma Christi, ricin, Eranda, Erando, Rendi, Vatari, Krapata, Djarak, Reer, Arandi, and Mexico seed. Bafureria, Baga, Mamona (in Brazil), Pi ma, Yuen Kin tse, Ta ma Tse (China), Kai Dudu Deu (Cochinchina), Palma Christi, Castor bean (Great Britain and USA), Catoputia major (Russia), and Wonderolie (Holland) are indigenous names of castor plant [10]. Castor plant, named as the Palm of Christ, is the oil derived from castor seed and can be used for medicinal purposes, which cures ailments and healing injuries. Castor bean also is known as "pfuta," "futa," and "fute" in Shona as it is the only source of hydroxylated fatty acid a feedstock of various industries ranging from pharmaceutical, cosmetics, fuel, personal care, lubrication, food, beverages, and many others. The other synonym of castor plant used in different languages is tabulated in Table 11.1.

Table 11.1: The synonym of castor plant in different languages.

Language	Name	Language	Name
Sanskrit	Gandharva- Hasta, panchi'gul, Vitiri	Malayalam	Ambanakka, Avanakku
Assam	Erri	Marathi	Erand, Erandee
Bengali	Bherenda, Rerira	Orrisa	Bheranda
English	Castor plant	Punjabi/Urdu	Erand
Gujrati	Erando, Divela	Tamil	Amanakku
Hindi	Erand, Rendee, Andu	Telegu	Amudanu, Amudmuchetu
Kannada	Harlu		

Source: Indian Agricultural Research Institute (2008).

11.1.3 Varieties of castor plant

Over the years, the Indian seed industry has evolved by adopting and innovating upon scientific improvements in variety development and quality seed production. The variation in castor varieties is based on the difference in the branching habits of the plant, the nature of the capsules, the color of the stem and branches, size of the seed, duration, and oil content in the seed. The high-yielding castor varieties in India are NPH-1 (Aruna), GAUCH-4, YRCH 1, TMV 5, TMV 6, CO-1, and TMVCH. Around 4307 accessions, of which 365 are exotic collections from 39 countries are present in The Germplasm Maintenance Unit in the Directorate of Oilseed Research (India) [11]. In India, Gujarat is the first state for the production of hybrid castor seeds in mid-sixties. First hybrid GCH (Gujarat castor hybrid) 3 was released in 1968. The second hybrid GCH 2 was released in the year 1985, while the Hybrid GCH 4 was reported first in 1986 and is in cultivation. To date, 15 hybrids and 18 varieties of castor are present in India [12,13]. Table 11.2 shows the different varieties of castor plant in India. GCH 6 is the commercial hybrid system based on NES type (a type of pistillate lines used for hybrid production) of sexual expression. The other GCH 3 commercial castor hybrid crop was also developed in India which had potential seed yield of 88% (higher than existing cultivars) with medium maturity of 140–210 days and oil content of 466 g/kg [14]. Some varieties which are also grown in the world for the production of castor oil are Hale, Brigham, BRS Nordestina, BRS Energia, Abaro, and Hiruy.

The ornamental types of castor plant are "Carmencita Rose," "Carmencita Bright Red," "Carmencita Pink," "Gibsonii," "Impala," "New Zealand purple," "Red Spire," "Sanguineus," and "Zanzibarensis" [15].

11.1.4 Production and protection of castor crop

11.1.4.1 Cultivation of castor crop

Cultivation of castor crop is done on a commercial scale for the production of seeds, and then extracting oil from it, in the countries between 400N and 400S latitudes. Castor is an

Table 11.2: Different varieties of castor plant in India.

State	Variety (V) and hybrid (H)
For all states	(V) Gujrat castor 2 (H) DCH-32 (Deepti), DCH-177 (Deepak), GCH-4, GCH-5, DCH 519
For entire India	(V) Sagar Shakti, DCS-107, Kohinoor, GCH-8 (H) M 574 x DCS 78, DCH 519
Gujrat	(V) GAUC-1, VI-9, S-20, J-1 and GCH 7 (H) CH-1, GCH-2, GCH-3, GCH-6, GCH-7, SBH-145, and SKP 84 x SKI 215, DSP-222
Andhra Pradesh	(V) Aruna, Bhagya, Jwala, Jyoti (DCS-9), Kranti (PCS-4), Kiran (PCS-136), Harintha (PCS-124), Sowbhagya
Rajasthan	(H) GCH-5, GCH-6, RHC-1
Tamil Nadu	(V) TMV-1, TMV-2, TMV-3, TMV-4, TMV-5, SA-1, SA-2, Jyothi, CO-1, DCS-9, TMV-6 (H) TMVCH-1, TMVCG
Telangana	Pragati
Maharashtra	(V) AKC-1, Girija (H) GCH-6
Haryana	(H) CH-1
Punjab	(V) Pb no. 1
Uttar Pradesh	(V) Kalpi-6, T-3, T-4, K-8501
Karnataka	(V) Jwala (48-1), RC-8, Jyothi, (H) HCH 6

annual crop, withstand on various types of soils but under consistent and appropriate rainfall. The cultivation of castor crop is done as a kharif crop and planted during July and August. Castor crop has a duration period of 4–5 months and harvested in December and January. Fig. 11.2 represents the time frame of the castor crop in Gujarat and Andhra Pradesh, India. The time of the planting varies from location to location as:

Australia: August to December
Brazil (north): January to March
Brazil (south): September to November
India: July
Illinois: early May
Venezuela: June or July
Morocco: March;
Taiwan: August or September to April or May.

The main factors that affect the castor growth are soil conditions, level of nutrients, and availability of moisture. Around 10–12 kg of seeds are required to cover 1 ha of land, whereas the variety of seeds and sowing method may affect the oil yield. Castor planting can be done on the land which is not suitable for commercial farming. For better germination of castor seed requires that soil should be soft and weed free and no clods should be there. Land preparation is also an essential task in castor cultivation. The land

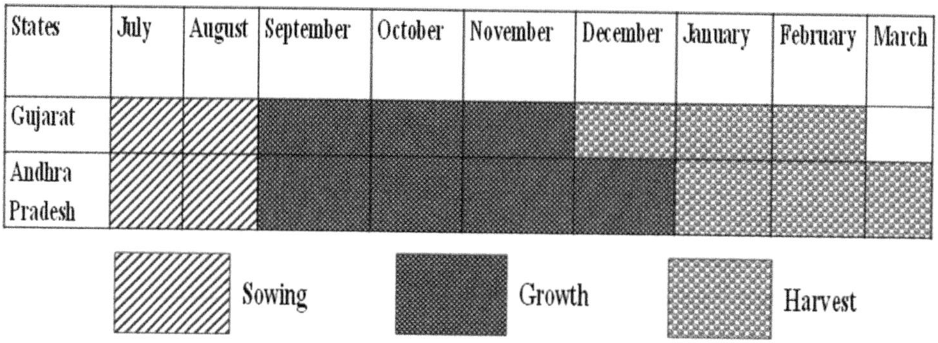

Figure 11.2
Time frame of castor crop in Gujarat and Andhra Pradesh, India.

preparation for effective growth of castor is made up with well-pulverized seeds, deep up to 45 cm with clay loam. An advantage of castor farming is that it can bear an annual temperature of 7.0–27.8°C and a pH of 4.5–6.3. The amount of rainfall required for castor crop is 38–50 cm. Castor crop requires sufficient moisture in the soil, for the same irrigation need to be done during its growth period. In some places, the cultivation of castor is done between rows. The required distance between rows is 90–120 cm, and the distance between plants is 40–60 cm in each row. Plow furrow, seed drill, and hand-dibbling techniques are used for sowing the castor seed. The mixing of the crop with others and sole crop are also varies depending on the country and regions as well. In India, intercropping of castor plant is done with Kharif crop (finger millet, groundnuts, cotton, dryland chilies, tobacco, black or green gram), i.e., sowing in July–August and arrivals from December onwards till March [16]. The intercropping of castor with peanuts or black gram is done with a ratio of 1:6, i.e., one row of castor seed and six rows of groundnut [17]. Intercropping of castor is beneficial to farmers, as they can earn extra income from it. Castor is intercropped with jatropha, as in the first 2–3 years the revenue generated from jatropha is low. During this period, castor becomes the source of the income as it can grow in a short period [18], and improving the economic viability on the commercial scale too. The required fertilizer dose of castor plant is 40 N- 40 P- 20 K kg/ha. In India, the highest yield of castor seeds requires 89 kg/ha of nitrogen while for the United States 45–135 kg/ha of nitrogen is needed in split applications. To obtain a high yield of castor crop, the other requirements of fertilizers per acre are organic fertilizer 6000–7000 kg, potassium sulfate 40–45 kg, diammonium phosphate 60–70 kg.

Harvesting of castor seed from the plant is done manually or mechanically. The appropriate period to start harvesting is when seeds are fully mature and leaves becomes dry. The period of harvesting varies with location. In India, it is done usually in November while in the United States, it starts in October. Harvesting of castor seed in India involves the collection of fruits, then sun drying while, in the U.S., drying of fruits is done by using frost or by defoliants. In Australia, chemical defoliants and modified wheat headers are used for seed drying. Mechanical harvesting is performed, when relative humidity is 45% or less. After collecting and drying, seeds are beaten out with sticks, winnowed and screened to remove hulls and trash. In Brazil, the U.S. and South Africa seeds are decorticated with a particular type of castor bean decorticators.

If the seeds are small in amount, it is decorticated just by rubbing on board. The castor seeds are then pass through the press for extraction of oil which includes preheating to reduce the viscosity of the oil. Castor bean can be stored for 3–4 years without deterioration. The normal storage temperature of seeds is 4°C, while for long term storage recommended temperature is −18°C. Premature harvesting of castor crop reduces seed weight, less oil content, and low germination.

11.1.4.2 Care from diseases and crop protection

Castor plant can withstand temperatures up to 38°C. High humidity can yield more plant diseases. The major diseases that affect the castor plant are leaf spot (*Cercospora reicinella*), leaf blight (*Alternaria*), and rust (*Melampsora oricini*), seedling blight, powdery mildew, and wilt. Weed control in the castor plant is done by simple harrowing. The best period to do harrowing is when the plant has two to five leaves. This process makes soil softer and removing all unwanted weed from castor crop. Herbicides such as Pendimethalin and Fluchloralin are applied after 3–4 days of sowing of castor seed [19]. The severe damage to the plant is fading of the seedling. To avoid the same, planting of the crop is not to be done in the low-lying and waterlogged areas. Pests that harms the castor crop can be controlled by a dusting of BHC 10% or 0.1% carbaryl during the early stages of cultivation. Several insects act as pests and harm the castor crop. In India, the castor semilooper (*Achoea janata*) are the worst pests while capsule borer (*Dichocrocis punctiferalis*) bores into young and ripening capsules. The other pests that effects the castor crop are Jassid (*Empoasca flavescene*) Fab, Whitefly (*Trialeurodes ricini*), Mishra Trerips (*Retithrips syriacus*), Serpentine leaf miner (*Liriomyza trifolii*), and Tobacco Caterpillar (*Spodoptera litura*); pests such as leaf-hoppers, leaf-miners, green stinkbugs, and grasshoppers feed on the leaves. Spraying of monocrotophos, a spray of neem seed kernel extract are the possible ways for prevention from pests.

11.1.5 Parts of plant and composition

The various part of the castor plant includes flower, seed or bean, leaves, and stem, as shown in Fig. 11.3. According to literature, per ton of castor plant produces 468 kg of seeds, 388 kg of stems, and 144 kg of leaves [20]. *Ricinus communis* is a herbaceous, evergreen plant and it became woody with age. The wood of the plant is soft, light with

Figure 11.3
Image of castor plant (A) flower, (B) seed, (C) leaf, (D) seed and oil, (E) stem.

thick central pith while the bark is smooth, light brown, display rings at the nodes and raised lenticels.

11.1.5.1 Flower

Flowers are formed in dense inflorescences 8–18 in tall and at the top of the stems, as shown in Fig. 11.3A. Each plant has male and female flowers. Male flowers are placed in under portion while the female flowers occupy the upper part of the spike [8]. Both male and female flowers have no corolla in it. Male flowers of castor plant have calyx, are green in color, and have three to five segments enclosing legion-branched yellow stamens. Female flowers have three narrow portions of the calyx, which are reddish. The ovary of the female flower is about the size of golf ball coronated from the center by profoundly divided carmine-red threads.

11.1.5.2 Seed and fruit

Castor bean, the fruits of castor plant, are produced mainly to extract oil from it. The seeds of castor plants have exquisite, shiny, and complicated designs, as shown in Fig. 11.3B. Fruit of castor plant is a globular spiny capsule less than 2.5 cm long with three multicolored (cream, brown, red, gray, yellow-brown, maroon and black). Each spherically shaped seed capsule is covered with soft, flexible spines and has three carpels in it. After maturity, the seed from each section is removed out by applying considerable force, breaking the carpel. Each seed has a small, spongy caruncle at one end which helps to absorb water for germination when planted. Hot pressing (>70°C) yields around 38%–48% of oil, while cold pressing yields around 30%–36% of oil. The wild variety of castor plant has small seed size and less oil in it. Improved varieties are producing more oil content, in a shorter period and easy to harvest. It contains 45.0%–50.6% oil, 12.0% –16.0% protein, 23.1%–27.2% crude fiber (CF), 5.1%–5.6% moisture, 3.1%–7.0% NFE, and 2.0%–2.2% ash. Among this, castor seed is rich in phosphorus. Castor seed also contains lipase, employed for commercial hydrolysis of fats, invertase, amylase, glycolic acid, maltase, endotrypsin, oxidase, ribonuclease, and a fat-soluble zymogen. Catalase, peroxidase, and reductase are the main components of sprouting seeds. The oil cake consists of 49.0% total carbohydrate, 20.5% protein, 9.0% moisture, 6.5% oil, and 15.0% ash (C.S.I.R., 1948–76) [16]. The oil from the castor meal has been extracted with hexane or carbon tetrachloride through a simple salting-out procedure [21]. The seed cake after oil extraction contains 6.6% N, 2.6% P_2O_5, and 1.2% K_2O (C.S.I.R., 1948–76) [16].

11.1.5.3 Leaves

Castor leaves are large over 10–75 cm, umbrella-like, and palmately arranged, as shown in Fig. 11.3C. Around five to eleven pointed finger-like lobes are there in each leaf, with serrated edges and bulging central veins. The leaf color can vary from dark green with a reddish tinge, green-bronze, black-purplish, maroon leaves or green with white leaves. The maximum amount of lignocellulosic material present in the castor plant is in its stems and leaves. Per 100 g of castor leaves contain 57.4 g total carbohydrate, 24.8 g protein, 10.3 g fiber, 5.4 g fat, 12.4 g ash, 2670 mg Ca, and 460 mg P. Leaves of the castor plant contains isoquercetin 2,5-dihydroxybenzoic acid and epicatechin [22].

11.1.5.4 Stem

The stem of castor plant is round, smooth, and red, with clear sap.

11.1.6 Production of castor seed and oil

11.1.6.1 Globally

According to the research, the global castor oil market is expected to grow at compound annual growth rate (CAGR) of around 4.3% during 2018–23 [23,24]. Globally, the cultivation of castor is done an area of $12.5 * 10^5$ ha with a production of $17.7 * 10^5$ tons of castor seed, and productivity of 1414 kg/ha [25]. Castor crop is cultivated in more than 30 countries lying in the tropical belt of the world. Fig. 11.4 shows the major castor oil producing countries worldwide. The world production of castor oil is approximately $5.5 * 10^5$ tons and India ranks first in the world for castor oil production from seeds. China, with a share of 23% and Brazil with a share of 7%, stood second and third position respectively for the production of castor worldwide. As per the news published in The

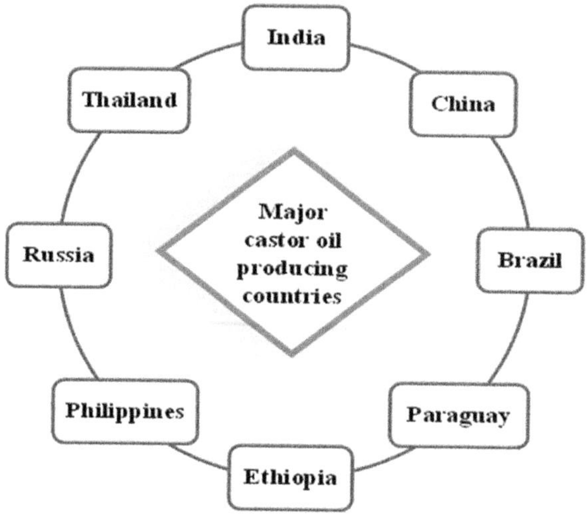

Figure 11.4
Major castor oil producing countries.

Sunday Mail [26] on June 24, 2018, Zimbabwe could soon join India as one of the top producers of castor bean. Isreal Isdory Kembo, Chief Operations Officer of Life Brand Agric Services, said that company has boarded on a hostile roll-out of castor bean production. They already propagate 800 ha under irrigation to produce seeds. The castor oil is the best fuel in the market to be used as a fuel additive. He furthers states that "We are encouraging youths and women in various farming districts throughout the country to grow castor beans under contract." The production of castor beans will create employment for over 500,000 peoples. Mr. Kembo said that "Castor is a more valuable, easier to grow a crop which needs less labor and provides more by-products as compared to jatropha in less maturity period of 120-day cycle".

Fig. 11.5 represents the global consumer countries of castor oil with consumption (in brackets). China market holds higher share with the rapid advancement of the downstream industry, for example, production of sebacic acid, a raw material used to produce plastics, cosmetics, lubricants, candles, and painting material. Asia–Pacific is the largest consumer of the castor oil, and it was observed that the maximum use of castor oil is done for the production of pharmaceuticals and cosmetics industry and it is expected that the consumption will rise in the next few years. The application of castor oil in Europe was for the production of biobased cosmetics, while in North America castor oil it is used for the production of biodiesel. California is ready to use the castor bean oil for the production of biofuels [27]. In 1949, castor was introduced in an article in the University California's *California Agriculture* magazine as a potential new crop for California agriculture.

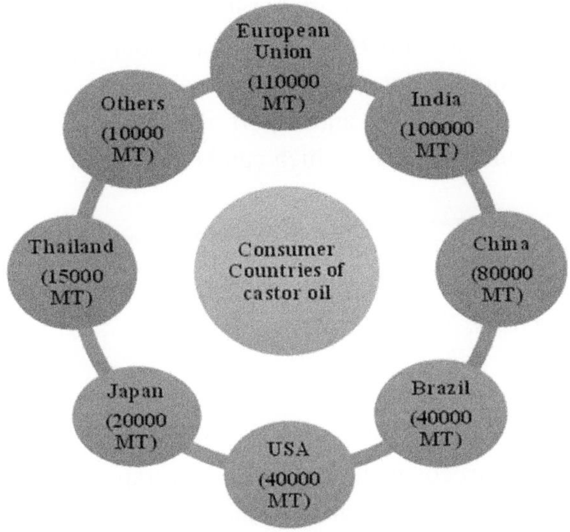

Figure 11.5
Global consumer countries of castor oil with consumption in brackets.

The import of castor oil within leading countries are shown in Fig. 11.6. India, China, and Brazil are the leading exporters of castor oil. India meets more than 90% of castor oil demand and leads to fulfilling the domestic and international market demands. India exports castor oil to China, Japan, Europe, and the U.S. The global castor oil market was $1180 million in 2018, and it is expected to touch $1470 million by the end of 2025.

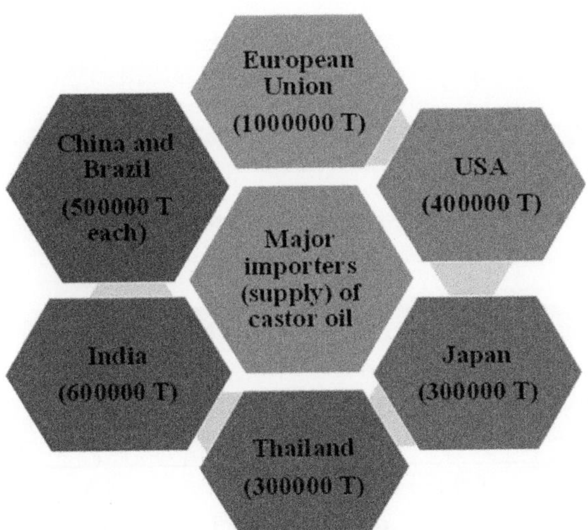

Figure 11.6
Major importers of castor oil.

The estimated export of India from castor oil and its derivatives are over US$ 1.1 billion annually while the global estimation is US$ 3.0 billion. India exports castor oil in two types, namely, Castor Oil Commercial, and First Special grade. Although India ranks first in the castor market both for production and export, it is just a price taker and not a price settler, all because of poor infrastructure, which can be improved by exporting the castor derivatives.

The current price of castor as on January 01, 2019 was Rs. 5080.00 for 100 kg [28]. The great fluctuations on the prices of castor seed over time was observed as in late 2015, and in early 2016 it was around Rs. 31,000/ton.

11.1.6.2 India

Oils and oilseeds have played a vital role in the Indian economy for a long time. India produces a vast diversity of oilseeds including groundnut, rapeseed, sunflower, etc. that make the extensive country share of foreign exchange. Fig. 11.7 shows the production of castor seeds across India from 2013 to 2018. The significant castor producing states of India during 2018 are shown in Fig. 11.8. Gujrat ranked second for the production of top 10 most abundant oilseeds crops and first for castor crop production in India. Agriwatch, also known as Indian Agribusiness Systems Ltd, surveyed field crop of Gujarat, Rajasthan, Andhra Prades, and Telangana. The maximum output, i.e., 90% of castor seed, is done in Gujrat. Mehsana, Banaskantha, Sabarkantha, Gandhinagar, Ahmedabad, and Kutch are the

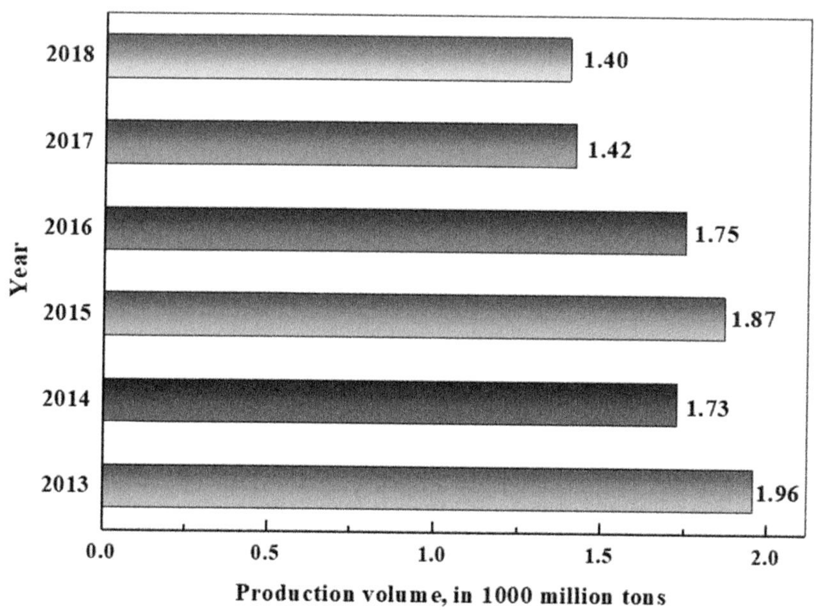

Figure 11.7
Production of castor seeds in India from 2013 to 2018.

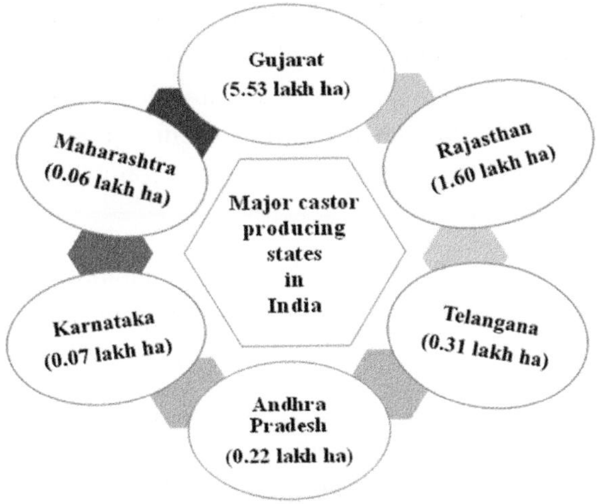

Figure 11.8
Major castor producing states of India during 2018.

districts of Gujrat which are indulged in the production of castor. The output of Andhra Pradesh and Rajasthan are 0.14 and 0.10 million tons of castor respectively. The districts of Andhra Pradesh namely Nalgonda, Mehboobnagar, Prakasam, Guntur, and Ranga Reddy, produces castor seed.

In the year 2017, the total area used for the production of castor seed in India was around $8.23 * 10^5$ ha, which further produces $14.21 * 10^5$ tons of castor seed and productivity of 1713 kg/ha. The estimated area, estimated production, an estimated yield for the year 2017 are shown in Fig. 11.9. Gujarat recorded 5.3% rise in sowing area at 595,600 ha, compared to 564,400 ha last year. As per estimation, the production in Gujarat was 1.22 million tons, up by 42% rise over last year's estimate of 861,000 tons. The decrease in the castor area in Rajasthan, Andhra Pradesh and Telangana were observed leads to decrease

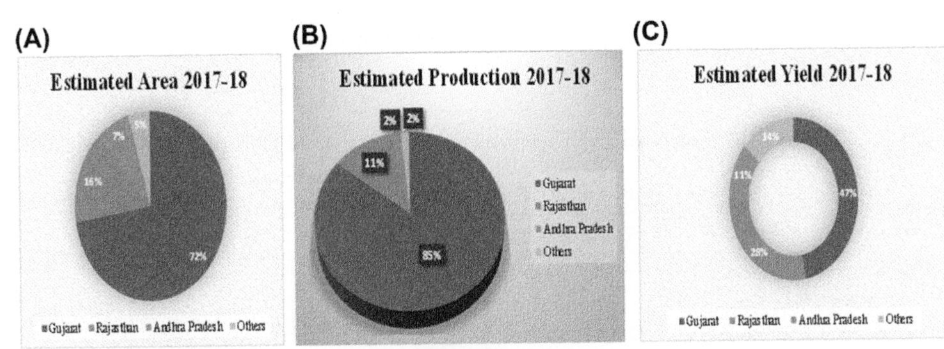

Figure 11.9
Estimated (A) area, (B) production and (C) yield for the year 2017 in India.

in the yield of seeds. Around 5000 ha of land is used for the production of Arandi in Chhattisgarh. The production of castor seed is also done in West Bengal [29]. The main parameters that affect the castor seed production are climate conditions, soil, irrigation, and cultural practices. Abhay Udeshi, chairman of Solvent Extractors' Association (SEA) castor seed and oil council, projects that by 2025, castor crop will be 2.9 mt equivalents to 1.3 mt of oil [30].

Table 11.3 shows the estimates of 2017–18 for cotton, groundnut, and castor crops in India, a sharp decrease in yield was observed. The main reason for this decline was because of unusual rains in the key growing regions of Gujarat [31].

The other factor that affects the castor crop production are: monsoon and level of rainfall, production and acreage variations, prices of other competitive oils like hydrogenated oil, dehydrated oil, sulfonated oil and sulfated oil, yield of other countries, demands of different countries and domestic, development of new applications of the oil, seasonal price variations, carryover stocks.

11.2 Castor oil

11.2.1 Extraction and purification of castor oil

The oil from castor seeds was extracted using:

1. Mechanical extraction using screw and hydraulic press
2. Biochemical and enzyme-assisted aqueous extortion process
3. Solvent or chemical extraction of oil using n-hexane or n-heptane
4. Reactive extraction

The first part of the extraction process involves prepressing of seeds in oil expeller. Oil recovery of 42%–46% was obtained at low temperatures by applying mechanical pressing, and up to 80% of oil yield was obtained using high temperature hydraulic pressing. Solvent extraction is the preferred process used to extract oil from castor cake in the presence of solvents such as hexane, heptane, or petroleum ether.

Table 11.3: Estimation of various seed crops in 2017 and 2018.

	Kharif 2018	Kharif 2017	
Crops	First adv est	Final est	% Change
Castor	1,173,000	1,484,000	−20.96
Soybean	90,000	115,000	−21.74
Groundnut	2,695,000	3,843,000	−29.87
Cotton	8,828,000	10,187,000	−13.34
Guar seed	123,000	136,000	−9.56

The oil extracted from castor seeds needs purification, and the process involved in the same are shown in Fig. 11.10. Initially, filtration of oil is performed to remove impurities such as dissolved gases, particulates, water, and acids. Then the crude oil goes through degumming of the oil, neutralization, bleaching, deodorization, and winterization.

1. Degumming of oil: involves removal of insoluble matter such as gums. It is the initial stage of refining of oil. The stirring of the mixture is done with hot water, and then phase separation is performed in separating funnel. The aqueous layer is the insoluble matter present in the oil, and this process is repeated several times to extract insoluble matter from the oil (Fig. 11.10A).

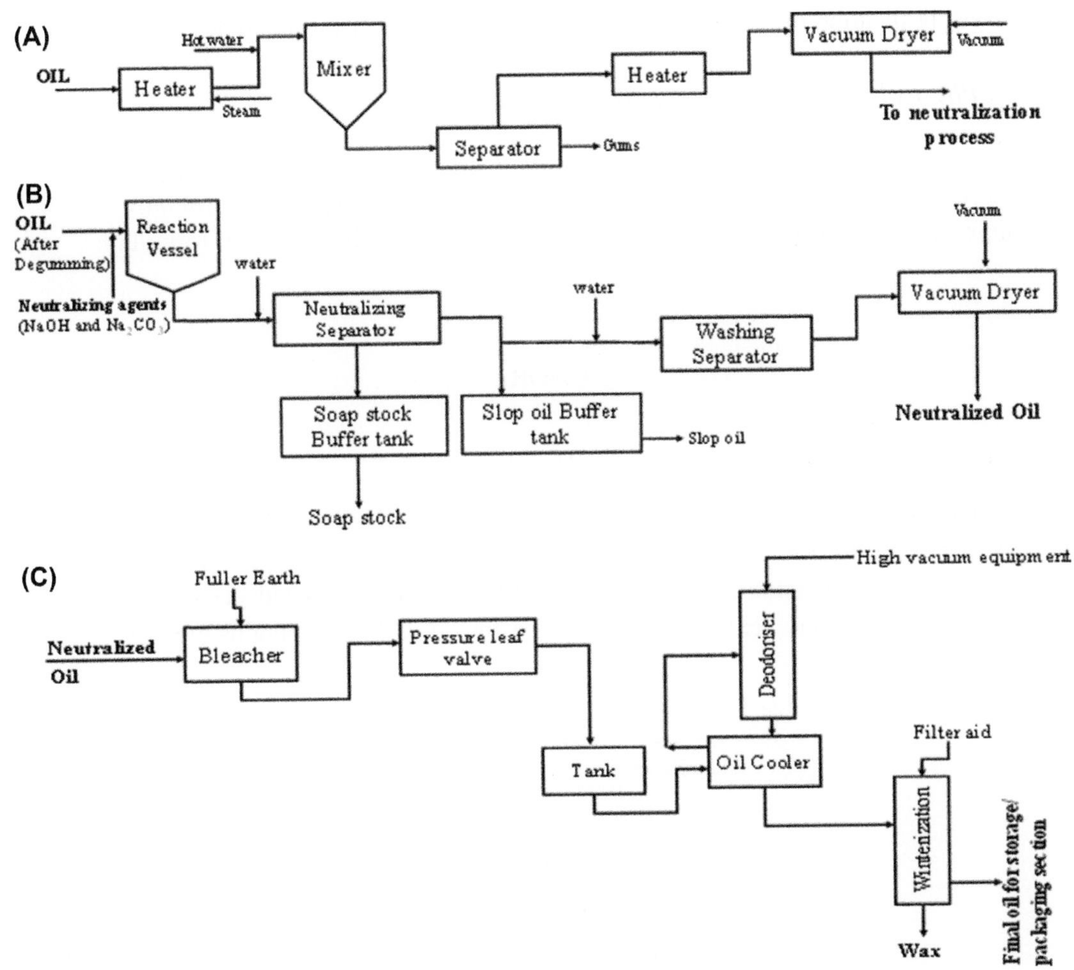

Figure 11.10
Purification process of oil.

2. Neutralization: is the second stage and involves the elimination of fatty acids present in the oil. This process involves heating of oil at 80°C and the addition of 40 mL of sodium hydroxide and later further addition of sodium chloride. The purpose of the acquisition of sodium hydroxide is to neutralize the fatty acid, and sodium chloride helps to remove fatty acids that appear in the form of soap (Fig. 11.10B).
3. Bleaching: it involves the removal of coloring material, oxidation products, and phospholipids from the oil. In this process, oil is heated at 90°C, and then fuller earth bleaching agent is added and stirred at 30 min. The mixture is then filtered at 70°C; the obtained oil is clear oil (Fig. 11.10C).
4. Deodorization: is the last stage process that includes removal of odors from the oil.
5. Winterization: is the process where a filtering process removed waxes present in the oil. The filter aid used is Kieselguhr. The process is almost similar to "dewaxing," and involves clarification of oil so that they no cloud formation will take place (Fig. 11.10C).

11.2.2 Physical and chemical properties of castor oil

According to the *International Cosmetic Ingredient Dictionary and Handbook* [32], the castor seed oil is defined as the fixed oil that is obtained from the seeds of *Ricinus communis*. As given by TNO BIBRA International Ltd. [33], the structural formula is:

$$\begin{array}{c} CH_2OR \\ | \\ CHOR \\ | \\ CH_2OR \end{array}$$

where R represents a fatty acyl group [$CH_3(CH_2)_5CH(OH)CH_2CH{=}CH(CH_2)_7COOH$] naturally derived from Ricinoleic acid. Castor oil is triglycerides, an approximate 90% of fatty acid chains are of ricinoleate or ricinoleic (12 hydroxy-*cis* 9-octadecaenoic) acid. Table 11.4 shows the fatty acid composition of castor oil. Oil also contains globulin,

Table 11.4: Fatty acid composition of castor oil.

Fatty acid	Composition, %
Ricinoleic acid	87 to 90
Oleic acid	2 to 7
Linoleic acid	1 to 5
α-Linoleic acid	0.5 to 1
Palmitic acid	1 to 2
Stearic acid	0.5 to 1
Dihydrostearic acid	0.3 to 1
Others	0.2–0.5

cholesterol, lipase, vitamin E, and β-sitosterol [10,34]. Ricinoleic acid (RA), a monounsaturated, 18 carbon atoms and has a hydroxyl functional group at C12. Table 11.5 represents the various physical and chemical properties of castor oil. Castor oil is a natural polyol, provides oxidative stability to the oil, and high shelf life as compared to other oils by preventing peroxide formation. The presence of the hydroxyl group in RA and its derivatives helps to perform various reactions such as dehydration, halogenation, alkoxylation, esterification, and sulfonation. Due to its functional properties, it is more polar, having high dielectric constant and used as a dielectric fluid in high voltage capacitors. The massive concentration of RA in castor oil consents the manufacture of high purity derivatives.

Table 11.5: Physical and chemical properties of castor oil.

Property	Value	References
Color	Colorless to pale-yellow viscous liquid	[35]
Taste	Slightly acrid	[36]
Density	0.953–0.965 g/mL at 20°C	[36]
Viscosity	6 to 8 P at 25°C	[36]
	283 cP at 37°C	[37]
Solubility	<1 mg/mL at 20°C in water	[36]
	≥100 mg/mL at 20°C in DMSO	[35]
	≥100 mg/mL at 20°C in 95% ethanol	
	Miscible in methanol	
	≥100 mg/mL at 20°C in acetone, glacial acetic acid, chloroform and ether	
Flash point	229°C	[36]
Melting point	−12°C	[36]
Boiling point	313°C	[36]
Pour point	2.7°C	[36]
Autoignition temperature	448°C	[36]
Freezing point	−10°C	[36]
Acid value	<4	[36]
Saponification value	178	[36]
Iodine value	85	[36]
Acetyl value	144 to 150	[36]
Hydroxyl value	161 to 169	[36]
Polenske value	<0.5	[36]
Reichert-Meissl value	<0.5	[36]
Surface tension	39.0 dyn/cm at 20°C; 35.2 dyn/cm at 80°C	[36]
Refractive index	1.4784 at 20°C; 1.473 to 1.477 at 25°C; 1.466 to 1.473 at 40°C	[36]
Optical rotation	Not less than + 3.5°	[36]
Thermal conductivity	4.727 W/m °C	[36]
Specific heat	0.089 kJ/kg/K	[36]

11.2.3 Ricin: a poison

Castor seed is poisonous due to the presence of ricinine ($C_8H_8O_2N_2$), a water-soluble toxin in it. In the year 1888, Stillmark named ricin after testing the castor bean on red blood cells [38,39]. Ricin is cytotoxic and suppresses protein synthesis in eukaryotic cells. Castor seed consists of two components (1) Alkaloid ricinine and (2) toxalbumine ricin. Ricinolein, a triglyceride is present in 60% in the seed [40]. The high amount of ricin is present in its seed as compared to the whole plant and are as follows: 0.43−7.0 g/kg in seeds, 3 g/kg in roots, 10.7 g/kg in flowers, 0.16 g/kg in shoots, 2.3−32.9 g/kg on leaves, and 2.4 g/kg in stems [41−49].

Ricin is considered as a biological and chemical weapon and is prohibited by the Biological and Toxin Weapons Convention (BTWC). Castor acts as slow-poison, with death occurring after one to three days. The other toxic present in the castor seed *Ricinus communis agglutinin* (RCA), pastes the red blood cells together. A small dose of RCA into the bloodstream causes a person's blood to clot internally. One single ricin molecule when it enters the cytosol of a cell can inactivate over 1,500 ribosomes and damage the cell. The effect of seed decreases if it is just swallowed without chewing, and it does not affect the digestive tract. It is estimated that ricin is 12,000 times more poisonous than rattlesnake venom and 6,000 times more poisonous than cyanide [50]. Castor seed has been stated as seven times more toxic than cobra venom [27]. Consumption of four ingested seeds can cause the death of an adult, and the lesser amount may lead to side effects of poisoning, such as vomiting, severe abdominal pain, diarrhea, etc. The degree of intoxication depends upon the amount of intake, age, and the general health of the individual. As per the *Morck Index: An Encyclopedia of Chemicals, Drugs, and Biologicals* (1997), a dose of only 70 μg or two-millionths of ricin is enough to kill a 160-pound person. Ricin is considered as the most poisonous in the world by 2007 edition of *Guinness Book of World Records*. The amount of ricin is not expressed in oil as it remains in press cake because of the poisonous nature of cake it is not used as an animal feed. Poultry and livestock can also be affected if they consume seeds or meal left from the seeds.

The function of ricin in castor plants is not evident, but it is believed that it acts as a toxin or insect repellent. Ricin is used as an insecticide and fungicide [51]. The other dual role of ricin is for plant defense, nitrogen (N) storing, and translocation. The young leaves have high ricin and have more N stored in it. In senescing leaves, N becomes translocated and has low ricin content as well. A study concludes that the ricin is not toxic against microorganism as the stimulatory effect of ricinine was done on lactic acid fermentation by *Lactobacillus leichmannii* Henneberg [52]. The other study by Xu et al. [43], reported that certain fungi could degrade the ricin of ca. 94% after 15 days fermentation process.

Considering the point of toxicity of castor seed, Dick Auld, oilseed crop specialist and Research Scientist at Texas Tech University, explains that "[t]he long term solution is needed to develop castor varieties that greatly reduce toxicity and we are well on the road to achieving this goal." The concern over ricin levels in castor is not that much; that it's been grown in the past in the United States and nobody died from it. Calvin Trostle, Associate Professor and research scientist at Texas A&M Agrilife in Lubbock, confirmed that no study has been reported yet in literature which proves that handling castor results in the person sick or dead [27]. Research has been conducted to reduce the effect of the ricin in the USDA Research Center in Lubbock and a new variety known as the Brigham reported by R.D. Brigham, which has potential to minimize ricin toxicity by 70%–90%.

Further studies are mandatory to understand the importance of ricinine and used as a potential agronomic tool which has been largely neglected by the scientific community. It includes the effect of environmental factors and agronomic practices on ricinine concentration, assesses the variability on ricinine content among genotypes, importance of the compound in conferring resistance to pests and diseases of castor crop, and the risks of human and animal intoxication from ricinine.

11.3 Castor oil derivatives

Castor oil is the only oil which is rich in RA and has the potential to be used in various applications. The biodegradable and eco-friendly nature of the castor oil also enhance its applications by which it becomes an essential commodity to the chemical industry. The production of various petrochemicals from castor oil was due to its unique properties over the other oils. Even the castor oil itself has multiple advantages and applications, but numerous chemical derivatives are produced from castor oils, which have many industrial and domestic applications as well.

11.3.1 Classifications of derivatives

The classifications of the castor oil derivatives are divided into four types, as shown in Table 11.6. In 2016, the market size of the Basic grades and Generation I derivatives were \$ 1.3 billion, Generation II derivatives \$ 600 million, and Generation III derivatives \$ 400 million. Basic grade category is the commodity products which have shallow margins, i.e., (<5%). Generation I derivatives and Generation II derivatives are value-added chemicals and have margins of low (5%–10%) and medium (10%–20%) respectively. Specialty Chemicals are categorized under Generation III derivatives and have medium to high margins [53].

11.3.2 Key derivatives of castor oil

The brief introduction of key derivatives of castor oil is based on their categories as Main grades, Generation I derivatives, Generation II derivatives, and Generation III derivatives.

Table 11.6: Classification of castor derivatives.

Classification of castor oil derivatives	
Main grades	
• commercial, FSG, BSS • first pressed degummed grade castor oil • refined castor oil extra pale grade • castor oil pharmaceutical grade • blown castor oil • urethane grade • pale pressed grade	
Generation I derivatives	**Structure**
Dehydrated castor oil	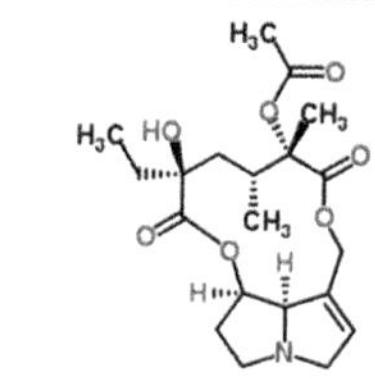
Hydrogenated castor oil	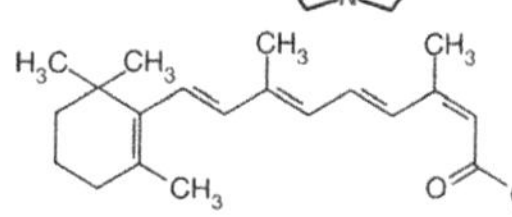
Ethoxylated castor oil	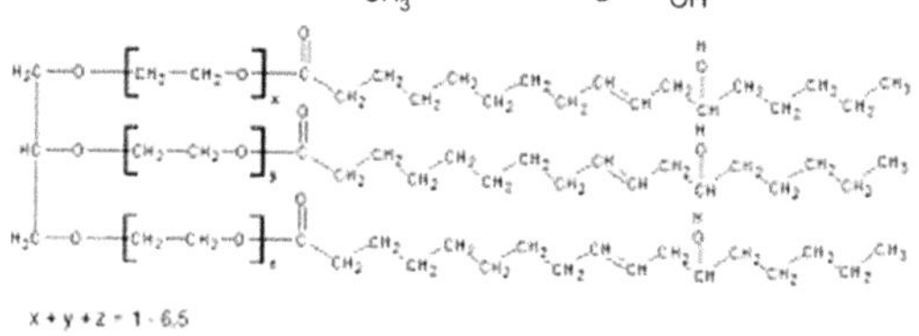
Sulfonated castor or Turkey red oil	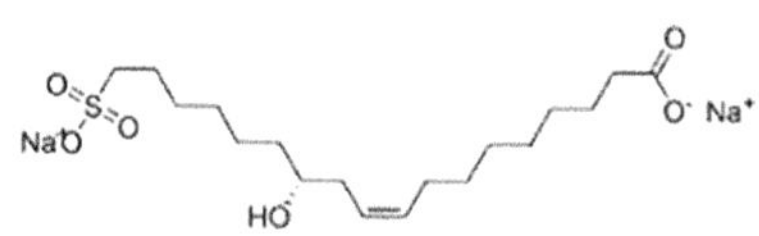
12-Hydroxy stearic acid	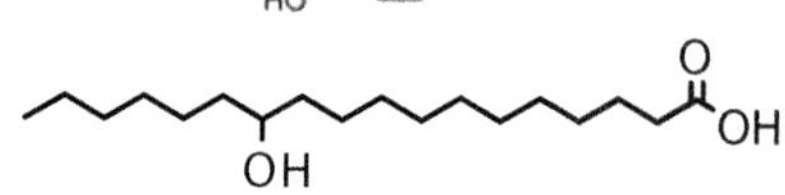
Sebacic acid	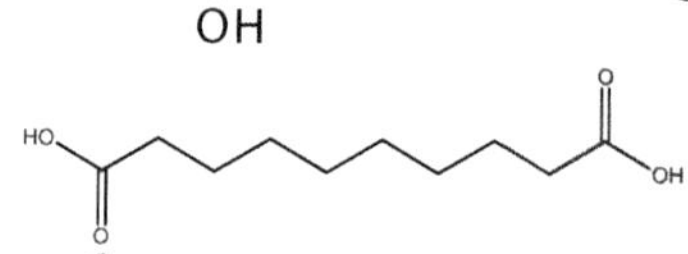
Ricinoleic acid	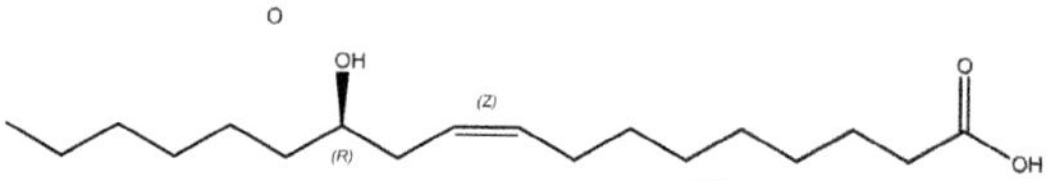
Heptaldehyde	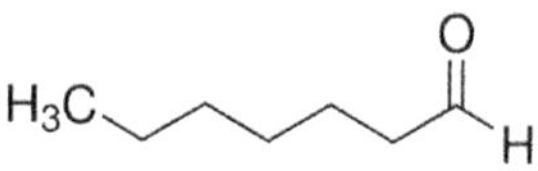
Polyols	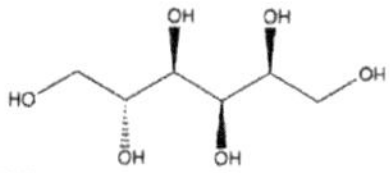

Table 11.6: Classification of castor derivatives.—cont'd

Generation I derivatives	Structure
Undecylenic acid	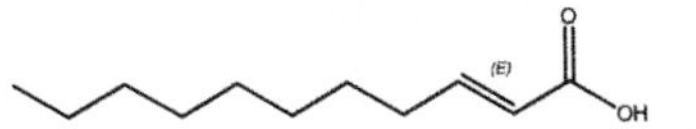
Undecylenic aldehydes	
2-Heptanol, 2-octanol	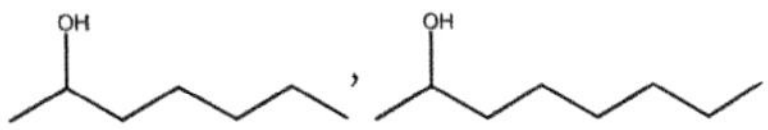
Other dimer acids	
Methyl 12-HAS	
Methyl ricinoleate	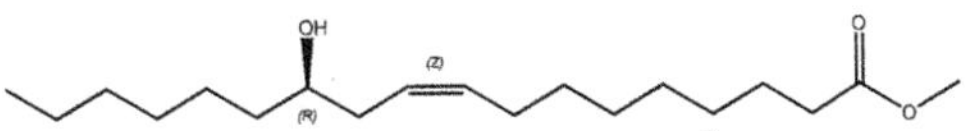
Methyl undecylenate	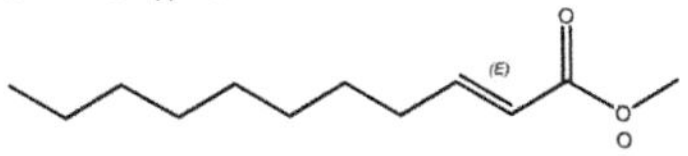
Zinc ricinoleate	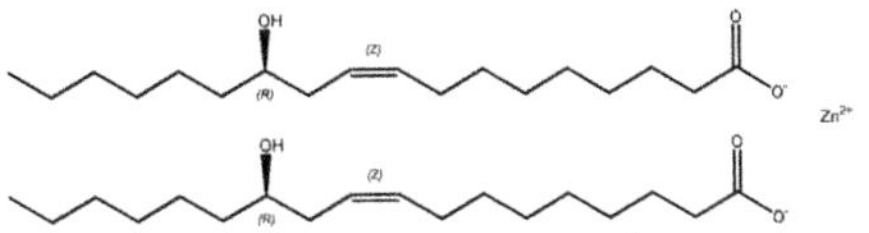
Zinc undecylenate	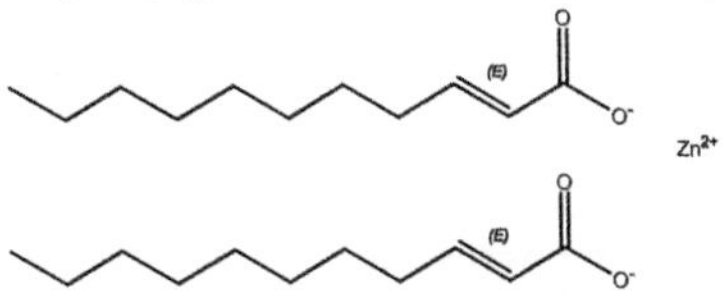
Calcium undecylenate	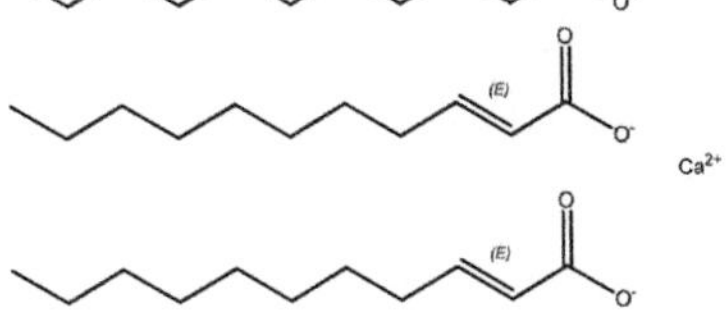
Others	

Commercial, FSG, BSS: Commercial Grade castor oil is obtained from a mixture of first pressing of castor seed as well as the second phase of solvent extraction of seeds. The other commercial grade FSG was obtained by refining of oil using bleaching and filtering process.

First pressed degummed grade castor oil: This product is the initial stage product, obtained by the first pressing of castor seeds and then using the degumming process. This product is an improved version of castor oil by a change in texture, color, etc.

Refined castor oil extra pale grade: This product was produced by just pressing the castor bean.

Castor oil pharmaceutical grade: BP grade castor oil and European Pharmacopoeia grade castor oil is the commercial grade product of castor oil is obtained by neutralization process and are highly refined. The applications of these products are in pharmacy and medicinal purposes.

Blown castor oil: The blown oil is prepared by blowing oxygen or air at a temperature of 80–130°C, in the presence/absence of a catalyst. This type of castor oil is also known as oxidized castor oil.

Urethane grade castor oil: It is the oil obtained in the first pressing of castor seeds without losing its any medicinal value of oil.

Pale pressed grade: It is the premium product of castor oil obtained from the first pressing of the castor seeds.

Dehydrated castor oil: Dehydrated castor oil is obtained when oil is treated at a temperature of about 250°C in the presence of a catalyst such as activated earth and concentrated sulfuric acid under inert atmosphere or vacuum. In the dehydration process, the hydroxyl group and the adjacent hydrogen atom from the C-11 or C-13 position of the RA present in the oil are removed as water. Two acids were formed containing two double bonds, and one of them is conjugated. The acid having conjugated double bonds has similar properties as tung oil.

Hydrogenated castor oil: Hydrogenation is the conversion process of unsaturated radicals of fatty glycerides in the presence of hydrogen and nickel as a catalyst to form completely or highly saturated glycerides. Hydrogenated castor oil (HCO) is obtained by reacting the RA in the presence of hydrogen and nickel, which becomes fully saturated, and a waxy product, which has a melting point of 80°C. HCO is also known as catalyst wax.

Ethoxylated castor oil: When 1 mol of castor oil reacts with 35 mol of ethylene oxide, then ethoxylated castor oil will form and has the capability of reacting as a nonionic detergent in solutions.

Sulfonated castor or Turkey red oil: The production of sulfonated castor oil is done by adding sulfuric acid to castor oil. The reaction of sulfuric acid and castor oil is done at 25–30°C for several hours, followed by washing and neutralizing with sodium hydroxide solution. The sulfonated castor oil is also considered as the first detergent.

12-Hydroxystearic acid: 12-Hydroxylstearic acid (12-HSA) is an off-white solid fatty acid. When 12-HSA reacts with an ester, forms a hard finish which is used in automotive and small appliance industries. 12 HAS is also used to produce lithium and calcium-based lubricating greases.

Sebacic acid: It is also known as 10-carbon dicarboxylic acid, synthesized by heating of castor oil at a temperature of 250°C or more in the presence of alkali. It is a white flake or powdered crystal, easily dissolved in ethanol, ether and slightly soluble in water.

The formation of sebacic acid will also be done from phenols and cresols, but oxidation of castor oil is considered as a greener process.

Ricinoleic acid: It is an unsaturated omega-9 fatty acid that is naturally present in castor seeds. The castor oil has its content of around 90%.

Heptaldehyde: It is castor oil C-7 derivatives and produced from pyrolysis of castor oil at a temperature of 700°C in the presence of benzoyl peroxide (0.5%). The colorless liquid obtained is known as heptaldehyde along with undecylenic acid.

Polyols: The formation of polyols are done by reacting the castor oil with mercaptoethanol. The reaction involved in this process is known as thiolene reaction. Polyols are organic compounds that have multiple hydroxyl groups.

Undecylenic acid: It is castor C-11 derivatives. The synthesis of undecylenic acid was done by two routes:

(i) Pyrolysis of castor oil at 700°C along with heptaldehyde.
(ii) Hydrolysis of methyl undecylenate.

2-Octanol: The production of 2-octanol is done along with sebacic acid in the caustic fusion of castor oil under high temperature. It is known as capryl alcohol and monohydric.

Methyl ricinoleate: It is the castor oil C-18 derivatives and obtained by the chemical reaction between the castor oil and methyl alcohol. It is tested lubricity additive for petrol and diesel.

Zinc ricinolate: Zinc ricinolate is zinc salt of RA and is produced from castor oil. The main applications of zinc ricinolate are in deodorants and categorized as castor oil C-18 derivatives.

Zinc undecylenate: is the zinc salt of undecylinic acid. The applications of zinc undecylenate acid are to treat and to prevent superficial fungus infections of the skin. The other applications are to reduce itching, burning, and irritation.

Calcium undecylenate: It is the calcium salt of undecylenic acid and produced by vacuum distillation of castor oil and has antifungal properties of high chain fatty acids.

The various importers of castor oil and its derivatives in India are Jayant Agro-Organics, Gujarat Ambuja Exports, NK Proteins, Gokul Overseas, RPK Agrotech, Adya Oil and Chemicals Ltd., Sree Rayalaseema Alkalies and Chemicals Ltd., N.K. Industries, Amee Castor, and derivatives.

11.3.3 Application of castor products

The products derived from castor oil has many applications both in terms of medicinal as well as industrial purposes. Fig. 11.11 shows the various uses of the castor plant and its

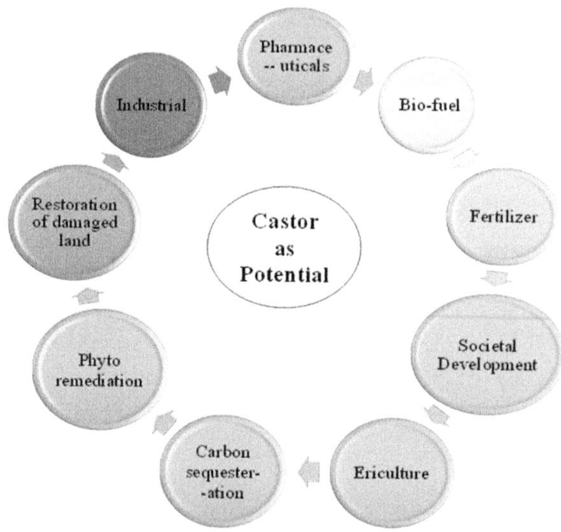

Figure 11.11
Applications of castor plant and its products.

products. Castor oil is feedstock of a wide range of products used in transportation, cosmetics, manufacturing, and pharmaceuticals.

11.3.3.1 Medicinal applications

Castor is considered as a reputed remedy for all kinds of rheumatic affections. Overall, it is useful in gulma, constipation, amadosa, inflammations, fever, cough, ascites, bronchitis, leprosy, skin diseases, and vitiated situations of vata, colic, and lumbago. The root of the castor plant is useful in pains, asicites, asthma, eructations, glands, rectum diseases, and head. Leaves of castor plant are helpful in "kapha" and "vata," night blindness, intestinal worms, earache, increases biliousness, burns, nyctalopia, strangury, and for bathing and fermentation and vitiated conditions of vata, especially in rheumatoid arthritis and arthralgia. Leaves are applied externally to the head to relieve headache. Fresh leaves are used by nursing mothers as an external application to increase the flow of milk [54]. Flowers of castor plant are used as an appetizer in tumors and pain, "vata," piles, diseases of the liver and spleen. Seeds are beneficial in dyspepsia and for preparing a poultice to treat arthralgia. The seed of the castor plant is aphrodisiac and cathartic. The oil is helpful in tumors, heart diseases, fever, asicites, inflammations, back pain, typhoid, lumbago, leprosy, elephantiasis, convulsions; increases "kalpa" causes biliousness. Castor oil can strengthen the immune system, and it is considered an excellent remedy to treat the major illnesses and ailments like Cerebral Palsy, Multiple Sclerosis, Parkinson's Disease Pain from Arthritis and Rheumatism, Hair loss. Problems like Acne, Athlete's Foot, Yeast Infections, Inflammation, Menstrual Disorders, Sunburn, Ringworm, Gastrointestinal

Problems. Migraines are also cured by castor oil. Castor oil is used to induce labor, but pregnant women should always consult a doctor before using it. Seed oil is good for hair growth. Castor oil is used as a solvent in Ophthalmic surgery. Sulfonated castor oil is used for ointments [55]. RA is used in contraceptive jellies. Castor oil is applied externally for eye ailments and dermatitis [56]. The other medicinal use of castor oil as abscess, antifertility, anticancer, antioxidant, antimicrobial, antidiabetic, antiulcer, antifungal, bone regeneration, boils, dog bite, antiasthmatic, tuberculosis, tumors, uteritis, venereal diseases, wound healing, etc. [57,58]. Regular use of castor oil packs helps to improve your lymphatic drainage, detoxify the body, improve bowel movements, reduce food sensitivities, strengthen the immune system, prevent diseases, decrease inflammation, and boosts overall health. Castor oil packs can be used in cases of fibroids, menstrual irregularities, and ovarian cysts. The castor oil is considered as safe if it is used in moderation. However, it creates a blockage in intestinal to pregnant and lactating women, and it is recommended that not be used before a doctor's consult. A nominal dosage of not more than one-half to one full teaspoon per day is recommended. Overdosage may cause nausea, vomiting, diarrhea, abdominal pain, and cramping. Castor oil is used in many veterinary medicines. It is used as a soothing medium for animals to remove foreign bodies.

11.3.3.2 Industrial applications

In World War I, castor oil was used as a motor lubricant in internal combustion engines, racing cars, and airplanes. Fig. 11.12 shows the various industrial applications of castor oil. Dehydrated castor oil is used in paints, enamels, and varnishes, whereas hydrogenated oil is used for the manufacture of polishes, waxes, carbon paper, crayons, and decorated candles [59]. The oil itself does not contain ricin and can be used as a potential ingredient for the manufacture of cosmetics for skin and hair. Industrially castor oil is used in production of adhesives, brake fluids, caulks, cox, dyes, electrical liquid medium, flypapers, moisturizers, organic fertilizers, food packaging, nylon 11 plastics, hydraulic oil, printing ink, leather, light oil, grease, machine oil, paints, pigments, polyurethane adhesives, a polyamide nylon type fiber named Rilson, surfactants, soaps, polishes, frozen lubrication agent, plastics and rubber, sealants, textiles, detergent and wax. Castor oil is used as an additive for flavoring and used as a mold inhibitor in the food industry. Castor oil has its applications in the aerospace and precision instruments advanced lubricants defense, and the electronics and telecommunications industry [60]. One of the significant applications of castor oil is to use as a modifier in the manufacture of chocolate bars.

11.3.3.3 Other applications

In the 1940s, the US military experimented ricin as a warfare agent and also by Iraq and terrorist organizations in Afghanistan. In ancient Egyptian tombs back to 4000 BC, castor oil is used in lamps for lighting and body anointments. A large amount of castor seed was

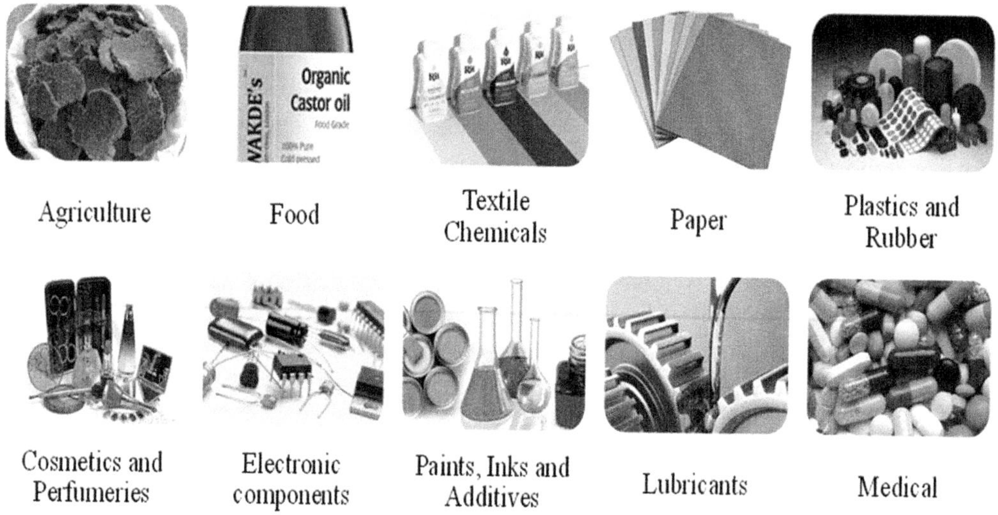

Figure 11.12
Industrial applications of castor oil.

used as a force by the paramilitary as it contains toxic protein Ricin [61]. During World War II, British and the United States collaborated to test castor bean and found that it includes Ricin, and they developed a Ricin Bomb named "compound W," but they never implemented this in battle [57].

It was mentioned by Watt [62] that a variety of castor can be used as an edible purpose as no poison is there. The Nigerian peoples grow castor plant, collect seeds, and after fermentation of 15–30 days, they eat the same as food [63]. Fermented castor meal is used in various dishes as a spice and food condiment. The leaves, stems, and branches of castor plant often burned to prepare the soil for next planting [64,65]. Castor plant is also grown as a decorative plant in public places. It is food for many insects like Lepidoptera, along with *Hypercompe hambletoni* and nutmeg (*Discestra trifolii*). Castor seeds are also used for making girls ornaments like necklaces and bracelets. Children use castor seeds for slingshot balls as they have right hardness, weight, and size. The export of castor cake is done due to the presence of the high content of protein and other minerals in it. After detoxification, it can be used as organic fertilizer. Castor cake manure is suitable for sugarcane, paddy, tobacco, etc. France uses this fertilizer for grape plantations, which further used to produce wine. Castor leaves with relish are eaten by cattle, goats, and sheep, and in cows and buffaloes, it increases the flow of milk. The leaves of castor are rich in protein and used for ericulture. Leaves in powder form are used for repelling aphids, mosquitoes, whiteflies and rust mites. Ricin can be said the body is a treasure as castor leaves are nutrient-rich and used as feed to silkworm in Assam and China. The plant stalks of castor are used as fuel in household building or as thatching material or preparing pulp for paper manufacture, ropes, and building material clapboard. Fibers from the castor plant can be used for paper and wallboard. The advantage of using castor oil as a feedstock for biodiesel production is it is soluble in acetone and does not require heat and energy to convert it into fuel.

Considering the vast applications of the castor plant and its oil, it was stated by Mutlu and Meier [66] that "castor oil is one of the most promising raw material to be used in chemical and polymer industries due to its diverse uses and well-established industrial procedures that yields a variability of renewable platform chemicals".

11.4 Way to sustainability: potential of value addition in castor and research reported

Different researchers have worked on different parts of the castor plant. The main area on which research has been carried is hybrid castor seed production to obtain maximum oil, reduction of toxic content from seed and cake, production of biodiesel from castor oil, utilization of castor stem as paperboards, bioremediation, medicinal applications of the castor plant, the formation of flavonoids and tannins, etc. Model castor farm project is a good idea with the motto to enhance the production of castor with minimal cost.

11.4.1 Model castor farm project

An article by Shekhar Ghosh in *Agriculture* on May 1, 2018, reports that an experiment in Gujarat on castor seed was conducted and stated that the yield gets double without additional cost. The tests have been conducted in 160 ha land of six districts using the GCH-7 variety, which is developed by Sardarkrushinagar Dantiwada Agricultural University (SDAU), Palanpur. The castor farming during this was done under the guidance and continuous monitoring of SEA of India, results the output per hectare from most of the areas is an average of four tons without any additional input cost. GCH-7 developed by SDAU has given superior results, in terms of better growth, resistance to disease, and higher productivity. The organic inputs, better farm practices like watering techniques, precise spacing, the intercropping, wind, and sun use were involved under model castor farm project. The most important aspect of the Model Castor Farm Project is to provide continuous technical guidance and inputs which are free of cost. The President of SEA suggests the farmers do intercrop castor with other leguminous plants to maintain the fertility of the soil. Intercropping of crops helps to keep nitrogen fixation, healthy soil as well as a source of additional income to the farmers. Haresh Vyas, Head of Castor Model Farms projects at SEA briefs that it can be easily replicated at all castor growing areas in India like Gujarat, Telangana, Rajasthan to get a higher return. The main reason for model farming is to encourage farmers to produce castor seed [67]. Abhay Udeshi, Chairman,

SEA Castorseed and Oil Promotion Council states that demand of castor oil and derivatives is rising year on year, and this type of projects helps to encourage farmers, to grow castor crop which increases the income of farmers and boost the exports as well. A help from Government agencies is required to increase productivity and production in India, to fulfill the world demand make India be a world leader [68].

11.4.2 Seed, oil and cake

McKeon et al. [69] performed the experiments to know the toxic content of commercial castor cultivars. The work aims to reduce ricin content or no ricin in castor so that it can be used as a potential energy crop. The detection of ricin content and homologous RCA in seeds was done by validating a sandwich enzyme-linked immunosorbent assay. It was observed that this method supports the castor breeding effort in developing low or no ricin—containing cultivars of castor which can be further used as a principal energy crop for the production of fuels and products so to replace the petroleum-derived products. The field experiment to test the impact of hydrogel application on rainfed castor was performed by Ramanjaneyulu et al. [70]. The study failed to enhance castor production using this technology, but the improvement in the soil was observed. For the case of castor crop, the addition of hydrogel reduced the benefit per rupee as well as net returns. Lavanya et al. [71], conducted a study in India to determine high yielding castor genotypes ideal for Biodiesel production. They concluded that female line Genotype DPC-9 could be a likely parent in a breeding program for the development of castor hybrids with the economic level of biodiesel traits in the future.

The work on production, chemistry, and commercial applications of various chemicals from castor oil was done by Naughton. In the year 1973, the same work was presented in the symposium entitled "Novel uses of Agricultural oils" at AOCS Spring Meeting, New Orleans, Louisiana [72]. Onunniyi [73] reviewed a paper on castor oil: A vital industrial raw material. In this paper, the extraction of oil from seeds, the refining process of castor oil, properties, and industrial applications of castor oil were well described. Most of the research is done using castor oil as raw material for the production of biodiesel [74—87] as it has technical and ecological benefits. The challenge for using raw castor oil as biodiesel is due to its high viscosity. Blending of oil with other diesel may help to overcome this problem. Catalysts were used to reduce the viscosity, improve the quality and performance of the biodiesel so that it can be used as transportation fuel [88—90]. Alkaline transesterification is the process where alcohol (methanol) is mixed with alkali catalyst (NaOH and KOH), and then it is mixed with fatty acid so that transesterification takes place. The complete reaction is triglyceride is reacting with 3 mol of methanol in the presence of catalyst (NaOH) to yield a mixture of methyl esters and glycerin as a by-product. The mixture of methyl esters is known as biodiesel. A side reaction of triglycerides with water forms free fatty acids which further used to produce soap. Alkaline transesterification of castor oil was done by Panwar et al. [91], for engine applications. The work aims to reduce the viscosity of castor oil from 226.82 to 8.50 cS at 38°C so that it can be used as a transportation fuel. The process is environmentally friendly, as using 10% of the total production of castor seed for the production of biodiesel saves around 79,862 tons of CO_2 emissions with biodiesel yield of 95.8% was obtained in the presence of alkali catalyst. Marin and Garcia [92] investigated the characteristics of biodiesel produced by castor oil and sunflower oil. It was observed that sunflower oil is best in quality, but in performance biodiesel produced from castor oil is better. AL-Harbawy and AL-Mallah [93] used hot and cold extraction methods for the production of oil from castor seeds and then obtained oil is used for biodiesel production. It was concluded that Methanolysis was the suitable method for the production of biodiesel from oils extracted by the hot and cold method. Optimization of processing parameters and kinetic study for biodiesel production using castor oil were also reported [81,89,94,95]. It was concluded that catalyst concentration and methanol to oil molar ratio are the most significant parameters that affect the yield of fatty acid methyl esters. A positive interaction effect of catalyst concentration with both methanol to oil molar ratio and reaction temperature were observed [81]. The other study by Ramezani et al. [89], concluded that among the various catalysts ($NaOCH_3$, NaOH, $KOCH_3$, KOH), potassium methylate gave the highest yield, and the optimization was performed by Taguchi method for the same. The reaction temperature and mixing intensity were the main parameters that can be optimized. A 2^4 full factorial central composite design (CCD) was developed by Goswami et al. [94], for the maximum bioconversion of castor oil. The optimum conditions were temperature 40°C, pH = 7.72, enzyme concentration 5.28 mg/g oil and buffer concentration 1 g/g oil with the conversion of 65.5% in 6 h. Dias et al. [95], conclude that the product yield varies from 43.3 to 74.1 (w/w) depending on reaction conditions. The best temperature and reaction time for biodiesel production of 73.62% were 65°C and 8 h with the purity of 83.41%. For the continuous production of biodiesel in Brazil, a plant was designed by Santana et al. [90], where castor oil was used as feedstock and alkali metal as a catalyst. The simulation part of this work is done by using HYSYS. Total capital investment, manufacturing cost, and annual equivalent cost were calculated. It was observed that the use of biodiesel in Brazil would not only provide financial, environmental benefits, but it also helps to boost the agricultural sector as well. Estolide 2-ethylhexyl ester was synthesized from castor oil, lauric acid and 2-ethylhexyl alcohol. The catalyst used in this work was perchloric acid. Data obtained from FTIR and NMR were used to know the characteristics of product and the obtained product has been used a biolubricant material [96]. Wang et al. [97], worked on novel ambient curable coating made up of aceroactylated castor oil and a multifunctional acrylate. The effect on various properties like tensile strength, cross-linking density, thermal properties, were observed, and it was concluded that Michael cross-linking technology provides a highly biobased coating system. The mercaptenized castor oil (MCO) and diethyl allyl phosphonate (DEAP) polyol were used for the formation of polyurethane foams. The foams showed closed cell content, which is greater than 95%; the addition of fire retardant does not show any hostile effect on the cellular structure of foams. The pyrolysis of castor bean cake was carried out and highest bio-oil yield of 63% w/w was obtained at 400°C for 60 min [98]. The pyrolysis kinetics of castor seed deoiled cake (*Ricinus communis*) using thermogravimetric technique was done by Sokoto and Bhaskar [99]. The kinetic parameters were calculated by using Friedman (FD), Kissinger—Akahira—Sunose (KAS), and Flynn—Wall—Ozawa (FWO). The average apparent activation energy values calculated were 124.61, 126.95, and 129.80 kJ/mol respectively. The calculated data can further utilize in designing a model and develop a suitable thermochemical system for the conversion of deoiled cake to energy carrier. The other kinetic studies reported on castor bean presscake was reported by different researchers [100,101] and concludes that DTG curves showed that heating rate affects the maximum decomposition temperature. Model-free methods such as FWO, KAS, and Kissinger showed the obtained values of kinetic parameters are in good agreement and contribute a better understanding of the process of biomass pyrolysis. The activation energy for *Jatropha Curcas* (Jatropha), *Pongamia pinnata* (Pongamia), and *Ricinus communis* (Castor) were found to be 108.73, 94.74, and 126.91 kJ/mol respectively. The characterization studies (physical, chemical, and thermal) properties of babassu, canola, castor seed and sunflower residual cake was evaluated by de Castro et al., [102]. The properties like bulk density, elemental analysis, the compositional analysis were performed, and the obtained results can be beneficial for wide range of applications of the biomass. Zapata et al. [103], investigated the yield and quality of oil produced by *E. lathyris*, *B. napus*, and *R. communis*. The press cake after extraction of oil from all the three plants can be used for the production of biofertilizers or thermal energy. Studies on detoxification of castor meal were also reported. The simple way for the detoxification of castor meal is just by boiling for 2 h. The other methods include boiling, autoclaving, steam heating, fermentation, ionizing radiation, the addition of sodium hypochlorite, alkali, or acid substance, the addition of calcium hydroxide followed by extrusion [104—107]. Addition of lime is considered as simplest and effective way for the detoxification of castor meal.

11.4.3 Castor plant (leaves, stem, root)

R. communis is the natural plant of India. Research on castor increases tremendously to improve productivity and to enhance the new castor varieties using biotechnology [108]. The root system and leaves of castor enhance the refill quality of the soil. As per the article published in Express News Service on July 17, 2017 entitled castor oil to reduce soil pollution, claims that the castor seed plant grows in areas where the soil is highly polluted. Castor plant absorbs toxic heavy metals from soil. A study conducted by

University of Hyderabad's Plant Sciences Department highlights that the castor bean plant proves a boon in the remediation of areas in and around Hyderabad where the soil is highly polluted, i.e., in industrial areas. The other study done by the same department reports the application of castor plant in the industrial areas [109]. The roots, leaves, and stem of the plant were tested, and the presence of lead was observed. The high amount of lead 19.53 mg was observed in the roots, while in the leaves and stem the content of lead is less. It was observed that the presence of some chemicals in the plant known as chelators increases the capability of castor bean plant to accumulate heavy metals in it. It was also stated that the Government needs to take some action for remediation of polluted areas [110]. The various phytochemical constituents and pharmacological activities of castor plant were discussed by Kumar [111]. This review confirms that *R. communis* is very effective for therapeutic value and a new, safe, cheap, and effective drug in the future. The pharmacological investigation, clinical trials, more exploration, and public awareness proves that *R. communis* is the best utilization of its medicinal properties. There is a need for industrial entrepreneurs to come forward and to take new steps for the better utilization of *R. communis* for medicinal purposes. The flavonoids and tannins present in each part of the castor plant were examined by Alugah and Ibraheem [112]. Leaves, stems, seeds, and roots were rich in flavonoids and tannins while the capsules contain flavonoids. The antioxidant, antihemolytic, and antibacterial effects were also elucidated. Further studies should be done on cytotoxicological test, anticoagulant activity, and structural characterization of various parts of the plant. The cytotoxicological test is used to obtain information about the level of drugs and another chemical present in a body. The effect of retarding or inhibiting the coagulation of the blood is observed by anticoagulant activity. Heparin is a commonly used anticoagulant activity. Structural characterization of protein provides variation between various types of proteins, based on the structure and amino acids sequences.The fibrous residues of castor bean culture include leaves, branches, and stem was tested to know their chemical and physical properties, and its application in pulp and paper production. It was observed from physicochemical properties of castor residue contains 80.10% of holocellulose (50.56% of cellulose, and 29.64% of hemicellulose), 17.34% of lignin, 1.48% of ash, and 1.88% of extractives soluble in organic solvents and 1.24% of extractives soluble in hot water. It was concluded that due to high cellulose content, the fibrous crop residues of castor, it can be used as an alternate for pulp and paper manufacturing [113]. Shah et al. [114], evaluate the cellulose and lignin contents present in castor stems residue. The castor stems used in this study was of GCH-7 hybrid. The work is done for the utilization of castor stem for pulp and paper industries. The obtained results show that castor stems have 45.7% and 17.2% of cellulose and lignin. As per the study reported by Grigoriou et al. [115], castor stalks are widely available and can be used as particleboards as lignocellulosic material is present in it. The research work aims to produce a value-added product that would enhance the overall success of castor as a viable agricultural crop. This study concludes that the castor stalks can be used only as a supplement to wood for particleboard production. The work on alkaloids and flavonoids from *Ricinus communis* were reported by Kang et al. [116]. Two alkaloids, ricinine, and N-demethylricinine, and six flavonol glycosides: kaempferol-3-O-β-D-glycopyranoside, kaempferol-3-O-β-rutinoside, kaempferol-3-O-β-D-xylopyranoside, quercetin-3-O-P-rutinoside, quercetin-3-O-β-D-glycopyranoside, quercetin-3-O-β-D-xylopyranoside were present in dried leaves of the castor plant. Bateni et al. [88], discussed that the whole castor plant was used for biofuel production. Extracted castor oil was used as feedstock for biodiesel production while the residue (leaves, seedcake, and stems) were employed for ethanol and biogas generation. The processes involved in the conversion of castor residue to biofuels are (1) pretreatment and (2) anaerobic digestion for the production of methane. For the production of ethanol the processes involved are (1) pretreatment and (2) simultaneous saccharification and fermentation (SSF). It was observed that the pretreatment helps to increase the biogas production from the castor stem while for leaves and seedcake imparts negative effects on product yield. Another study reported by Bateni et al. [20], for the production of biofuels from castor plant concludes that there is a production of 149.6 gm of biodiesel and 30.1 g ethanol from 1 kg of the castor plant. Pretreatment process was employed at high temperature to open up the lignocellulosic structure of residue, and it helps to improve ethanol production. Brazilian Government encourages for the production of castor crop and use it as biofuel production. Ladda and Kamthane [22] explained the various application of *Ricinus communis* L., and it was concluded that among the numerous studies had been carried out in different part of the plant. The literature mentioned in this research supports the potential of castor plant as a medicinal tree. Further, more research has been carried out for unexplored benefits of this plant. The pyrolysis kinetics and thermodynamic parameters of castor residue (stems and leaves) using thermogravimetric analysis were reported by Kaur et al. [117]. The kinetic parameters were calculated out using iso-conversional methods like FWO, KAS, and Kissinger method. It was reported that the obtained data could be useful for model prediction and plant designing. Hilioti et al. [118], worked on the characterization of biochar produced by slow pyrolysis at 550°C from castor stalks and deoiled castor cake. Although both biochars are highly alkaline in nature, the difference of biochars was observed in terms of their surface area, nutrients content, C:N, H:C ratio as well as the morphology of the biochars. The biochars can be used to improve seed germination, achieving 90% success rate earlier when compared to control. Addition of biochars increased the development of castor, affects the soil exchange capacity, PO_4, total N, K, P, Ca, Na and Al levels. Patel and Virdia [119] conducted the experiments to know the effect of weed on castor yield in South Gujrat. It was observed that to obtain a high yield; the castor crop should be weed free. It is advised that pendimethalin @ 1.0 kg/h can be used as a preemergence supplement either with one interculturing + one hand weeding at 30 days after sowing. The postemergence application of fenoxaprop-p-ethyl @ 50 g a. i. ha^{-1} at 25 DAS under south Gujarat conditions.

The life cycle assessment of castor-based biorefinery was done by Khoshnevisan et al. [120]. The main products formation was the biodiesel produced from castor seed oil and biogas from castor plant residuals. The additional formed products include bioethanol and biomethane. The aim of this research work is the production of castor biodiesel and formation of B35 as a transportation fuel. Three schemes were chosen: Scheme 1: Biodiesel production from castor oil by transesterification with ethanol and coproduction of bioethanol and biogas from castor plant residuals. Scheme 2: biodiesel production from castor oil by transesterification with methanol and coproduction of bioethanol and biogas from castor plant residuals. Scheme 3 involves the biodiesel production from castor oil by transesterification with methanol and coproduction of biogas from castor plant residuals. Pretreatment of castor residuals was done in the presence of NaOH, then pretreated material was sent to SSF unit and then anaerobic conditions were maintained. Fifteen midpoints were considered in this method, which include: carcinogens, respiratory inorganics, global warming, ozone layer depletion, nonrenewable energy, etc. It was observed that Scheme 3 had a positive net energy gain (NEG) by which it may be replaced in place of fossil fuel. Considering the parameter GHG emissions, castor biorefinery could help to reduce the emissions and compensate for the depletion of fossil fuels. In terms of sustainability, its effects on human health and ecosystem so, further research need to be done as to minimize such impacts.

Research has been carried out using various types of oils for the production of biodiesel and make it as a transportation fuel commercially. Table 11.7 shows the comparison of castor oil with other oils for economic viability [121].

Castor is the only raw material, rich in RA among many of vegetable oils. For every 2.2–2.5 kg of castor seed extract, 1 kg of castor oil. The life of castor growth is up to 5–10 years.

11.5 Residue generation and utilization

The castor residual includes castor leaves, stems, and seed cake. More than 50% of the residuals were generated from castor leaves and stems but the utilization is not that much

Table 11.7: Comparison of castor oil with other oils.

	Castor oil	Jatropha oil	Palm oil
Assumption with same planting cost per acre	2300 USD	2300 USD	2300 USD
Nursery period	Not required	2 months	9 months
Flower and Fruiting	3–4 months	6–9 months	3 years
Annual yield per acre	8–10 MT	4 MT	14 MT
Supply seed price	USD 373 per MT	USD 200 per MT	USD 150 per MT
Oil extraction rate	50%	32%	20%
Crude oil price (year 2010)	1700 USD per MT	800 USD per MT	750 USD per MT

so considered as waste. The various applications of these residuals were discussed above in the application section. The castor cake is rich in protein, starch, and lignocellulosic material, and approximately 0.55 MT of deoiled cake is produced per ton of castor seed. As all the parts of the castor residuals contain lignocellulosic content can be converted into value-added products and helps to enhance the economy of the process. On the other hand, this massive amount of waste creates environmental severe and disposal problems. As the availability of the castor plant is more in tropical regions, by self-pollination and by cultivation (for oil from seeds), the residual generation is also increasing globally. The availability of residue is more in India, as the maximum amount of castor seed is produced in India as well.

In the literature, various studies were reported where lignocellulosic biomass is used, and conversion is done into fermented sugars by saccharification. In this method, pretreatment of biomass was involved where a change in chemical, physical and morphological characteristics takes place. Pretreatment method is unlocking the cell wall and formation of a porous structure and further affect the hydrolysis/fermentation process [122]. Pretreatment process is an energy-intensive process which also enhances the cost of the process. Different types of pretreatment (acids, alkali) were used, but no single technology has proved ideal or superior for all biomass feedstocks. The various parameters that affect the pretreatment process are that it should be a green process, meaning no use of harmful chemicals, maximizing the conversion of substrate, minimum loss of sugars, less energy requirement, low cost and have the potential to be used in large scale. Combination of more than two technologies was used for the conversion of castor residue into valuable bioproducts. In the complete process, first pretreatment of the biomass is done to increase the surface area; then enzyme accessibility takes place. The detoxication of castor cake using the same technology was also reported. The production of ethanol from castor leaves and stems were done using alkali-treated pretreatment and enzymatic hydrolysis. Pretreatment process involves the removal of lignin and hemicellulose, which further enhances the rate of hydrolysis. The enzymes used in this study was cellulase and β-glucosidase followed by fermentation, which yields 82.2%, 77.6% and 33.7% ethanol from castor stem, castor leaves and castor seed cake respectively [88]. The other technique used for production of ethanol from castor seed cake was SSF. The significant advantages of this process are (1) elimination of one fermentor and other equipment; (2) presence of ethanol in the fermentation broth which reduces the contamination; (3) reduction in end-product inhibition of cellulase which is produced from sugars [123]. This technology is used for the production of value-added enzymes and biochemicals. In this process, the growth of microorganisms is done on the feedstock (agricultural, etc.) having components like carbohydrates, proteins, and minerals and then enzymes and metabolites were generated on it. Several studies are stated in the literature on SSF, out of which one was done by Brazilian researchers where castor cake was used as feedstock. In this work, *Penicillium fungus* was used, and the products enzyme and fermented castor cake was separated using solid-liquid phase differential extraction. This study proves SSF as a promising process for catalytic transformation of castor cake [124].

The conversion of lignocellulosic biomass into value-added products is also performed by using various thermochemical conversion routes. Combustion, gasification, pyrolysis, and hydrothermal liquefaction (HTL) are the main routes, out of these HTL gained attention due to its numerous advantages over other processes. The conversion of lignocellulosic biomass using the HTL process will produce various value-added products which may be used in the petrochemical industry or used as medicinal or industrial applications. The HTL of castor residue was also performed by Kaur et al. [125] under different operating conditions. A total bio-oil (TBO) yield of 15.8 wt.% was obtained at 300°C and 60 min. The obtained bio-oil is a mixture of various chemical compounds, out of which maximum area% is of phenolic compounds. The obtained phenolic compounds can be a useful feedstock for the production of biophenolics and chemical industry [125]. The optimization study of castor residue was reported, and it was observed that temperature is a significant parameter that influences the TBO and conversion yield [126]. The utilization of these residuals in an effective manner is required, and the suggested way is an option for making the process cost-effective. Further research on the different parameters (optimum reaction conditions, etc.) will help to use the process on a large scale.

11.6 Challenges and opportunities

Castor is an ancient and important nonedible oilseed crop, playing an essential role in the Indian agriculture economy. Per the definition presented by the International Energy Agency Bioenergy task 42, the objective of biorefining is the conversion of biomass to final valuable/marketable products and bioenergy is to maximize the profitability of the system [127]. The whole castor plant is a potential candidate to be used as a biorefinery feedstock. The main advantage of using castor as feedstock is its easy availability. The renewable energy sources are the unique energy sources, as they are carbon sources, stores solar energy, and can be converted into solid, liquid, and gaseous biofuels. The estimated World castor plantation is more than 300 million hectares, with an annual output of castor seed of about 1.5 million tons but still, the requirements did not meet the basic needs of the market. The potential for the development of planting castor industry chain provides an excellent opportunity both as market and employment. Castor model farming initiated by SEA will help farmers to increase the yield of castor seed without paying extra input cost. This project provides hybrid castor seeds that are free of charge, education to farmers on modern farming practices, reduction in costs by reducing water, pesticides, fertilizers, enhances the yield of seeds to fulfill the target or global demand.

The various advantages of using castor oil as feedstock are noncompetition with food, biodegradability, low cost, renewable nature, and eco-friendliness. Castor oil is a recognized competitor in the industrial chemical market, as many industrial products/derivatives are manufactures from it. The properties of castor oil as low cost, high boiling point, and viscosity makes it suitable to be used as a biobased chemical resource. The sustainability for the production of petrochemicals from castor oil includes better climate, reduction in oil spills which damaging marine environments, reduction in pollution and smog, reduction to damage to ecosystems as we dig for coal, oil and the like, etc. Strong export demand for castor oil for the manufacture of various products lifted castor future internationally and made it as one of the highly demanded cash crops. Research has been carried out to overcome the multiple problems related to its application as biodiesel in an appropriate blend on a commercial scale. The seed oil has numerous applications; the remaining portion of the castor plant consists of 90% lignocellulosic biomass and can be converted into value-added products. The suggested techniques and studies reported may help to convert castor residue into biobased products on a large scale.

There were several workshops and seminars to be conducted out by various organizations in India where technology transfer related to castor was done. Out of this, one is at Technology Information, Forecasting, and Assessment Council (TIFAC), Department of Science and Technology, Government of India started with an aim on brainstorming workshop cum discussion meetings considering the opportunities and potential of value addition in castor. Different organizations, Industry personals, and Institutes researchers were part of the meeting. The status, opportunities, prospects of castor oil along its derivatives were discussed. The research and development (R&D) issues, technology availability, and challenges of hybrid castor products were also part of the meeting. For the higher production of castor seed and oil, the future strategy was planned. Overall, the recommendations and suggestions are that there is a need to develop new high yielding hybrids and varieties of castor seeds that are diseases resistant and have inert and multiple cropping systems. Development of ricin-free castor seeds using conventional and biotechnological approaches. Facilities, including R&D, should be provided on a pilot or full plant scale for the production of castor oil derivatives. These facilities may be based on the Public Private Partnership (PPP). The appropriate and straightforward process needs to be developed for the detoxification of castor meal so that it can be used for various aspects. There should be a center of excellence to enhance the value addition of castor farm residues. Complete utilization of castor plant needs to be done, which demands a range of unit operations, separation techniques, reactions as well as product distribution for castor biorefinery. Castor crop biorefinery was suggested for the production of bioethanol, biosurfactants, biochar, and biolubricants with the development based on green chemical processes. The development of biopolyol from castor oil was done by IICT, having 90% biobased content and replaces petroleum-based polyol. Further research should be carried out on each part of the plant so that it can be used as a potential cash crop and produces more biobased products.

11.7 Conclusions and perspectives

Ricinus communis (castor) is one of the most versatile plants, abundantly available in nature. In 1983, Brazil was the leading castor producing country in the world, but now, India has played a significant role in the production of castor seed and oil and by fulfilling global demand. Gujrat (India) produces 80% of the total castor seed production and has the potential to double the yield under suitable conditions. The growth of castor-based industries demands more castor oil which can be fulfilled by intercropping of crop or by cultivating it with Rabi/semi-Rabi conditions. The significant applications of castor oil are for the production of derivatives such as sebacic acid, RA, etc. The other part of the plant (stems and leaves) can be a potential feedstock for the production of biofuels (bioethanol, biomethanol) and biochemicals (biophenolics). The whole part of the plant has numerous industrial and medicinal advantages and applications, but only a limited number of industries used this. Castor industry needs to concentrate more on value as compared to volume, by installing more industries and producing more value-added products from castor oil. For the effective utilization of whole castor plant, further research using advanced technology is required. Some initiatives from government agencies and tie-ups will also help to make castor, a potential cash crop.

References

[1] Carere CR, Sparling R, Cicek N, Levin DB. Third generation biofuels via direct cellulose fermentation. International Journal of Molecular Sciences 2008;9:1342–60.

[2] Demirbas A. Biofuels sources, biofuel policy, biofuel economy and global biofuel projections. Energy Conversion and Management 2008;49:2106–16.

[3] Kucuk MM, Demirbas A. Biomass conversion processes. Energy Conversion and Management 1997;38:151–65.

[4] Mckendry P. Energy production from biomass (part 1): overview of biomass. Bioresource Technology 2002;83:37–46.

[5] Manzano-Agugliaro F, Alcayde A, Montoya FG, Zapata-Sierra A, Gil C. Scientific production of renewable energies worldwide: an overview. Renewable and Sustainable Energy Reviews 2013;18:134–43.

[6] Notes on taxonomy and nomenclature. Online available at: https://www.cabi.org/isc/datasheet/47618.

[7] Castor and its derivatives. Online available at: http://www.crnindia.com/commodity/castor.html.

[8] Purseglove JW. In: Purseglove JW, editor. Leguminosae. Tropical crops: dicotyledons. Essex, UK: Longman Group Ltd.; 1981. p. 250–4.

[9] Tyler EV, Brady LR, Robbers JE. Pharmacognosy. Philadelphia, USA: Lea & Febiger; 1976.

[10] Scarpa A, Guerci A. Various uses of the castor oil plant (*Ricinus communis* L.): a review. Journal of Ethnopharmacology 1982;5:117–37.

[11] Anjani K. Castor genetic resources: a primary gene pool for exploitation. Industrial Crops and Products 2012;35:1–14.

[12] High yield castor varieties in India. Online available at: http://castoroil.in/b/2015/01/high-yield-castor-varieties-in-india/.
[13] Lavanya C, Mukta N. Varieties and hybrids of castor. Oilseeds Research, Hyderabad 2008:88.
[14] Severino LS, Auld DL, Baldanzi M, Cândido MJD, Chen G, Crosby W, Tan D, He X, Lakshmamma P, Lavanya C, Machado OLT, Mielke T, Milani M, Miller TD, Morris JB, Morse SA, Navas AA, Soares DJ, Sofiatti V, Wang ML, Zanotto MD, Zieler H. A review on the challenges for increased production of castor. Agronomy Journal 2012;104:853–80.
[15] Cultivars. Online available at: https://en.wikipedia.org/wiki/Ricinus.
[16] Castor crop information. Castor Oil World. Online available at: https://www.castoroilworld.com/castor-crop-information/.
[17] Inter-crop in castor cultivation. Online available at: http://www.agrifarming.in/castor-cultivation-information-guide.
[18] Lele S. The cultivation of jatropha cursas ratan jyot. Online available at: http://www.svlele.com/jatropha plant.html.
[19] Weed management in castor cultivation. Online available at: http://www.agrifarming.in/castor-cultivation-information-guide.
[20] Bateni H, Karimi K. Biodiesel production from castor plant integrating ethanol production via a biorefinery approach. Chemical Engineering Research and Design 2016;107:4–12.
[21] Franz DR, Jaax NK. Ricin toxin. Medical Aspects of Chemical and Biological Warfare 1997:631–42.
[22] Ladda PL, Kamthane RB. *Ricinus communis* (Castor): an overview. International Journal of Research in Pharmacology and Pharmacotherapy 2014;3:136–44.
[23] Castor oil market: global industry trends, share, size, growth, opportunity and forecast 2018-2023. Online available at: https://www.researchandmarkets.com/reports/4514597/castor-oil-market-global-industry-trends-share.
[24] Global castor oil market 2018-2023. Online available at: https://www.prnewswire.com/news-releases/global-castor-oil-market-2018-2023-300631295.html.
[25] FAO online. Online available at: http://faostat.fao.org.
[26] Castor beans to oil the future. Online available at: https://www.sundaymail.co.zw/castor-beans-to-oil-the-future.
[27] Castor an oilseed crop that can cure, kill you. Online available at: https://www.farmprogress.com/management/castor-oilseed-crop-can-cure-kill-you.
[28] Castor seed rate, today's castor seed price in India: castor seed news on Economic Times. Online available at: https://economictimes.indiatimes.com/commoditysummary/symbol-Castorseed.cms.
[29] Top 10 largest oilseeds producing states in India. Online available at: http://www.thedailyrecords.com/2018-2019-2020-2021/world-famous-top-10-list/india/largest-oilseeds-producing-states-india/18363/.
[30] Rise of about 30% expected in castor oil export this year. Online available at: https://www.business-standard.com/article/markets/castor-oil-exports-to-rise-33-this-year-115032600142_1.html.
[31] Gujarat estimates 2018-19 cotton, groundnut, castor crops sharply lower. Online available at: http://www.commoditiescontrol.com/eagritrader/login/login.php?referer=/eagritrader/common/newsdetail.php?type=MKN&varid=0&cid1=&no=1.
[32] Gottschalck TE, McEwen Jr GN, editors. International cosmetic ingredient dictionary and handbook. 10th ed. Washington, DC: CTFA; 2004.
[33] Toxicity profile. Castor oil. Surrey, United Kingdom: TNO BIBRA International Ltd.; 1999.
[34] Castor plant chemistry. Castor Oil World. Online available at: https://www.castoroilworld.com/castor-crop-information/.
[35] Lewis RJ. Sax's dangerous properties of industrial materials. New York: John Wiley & Sons; 1996.
[36] NTP Chemical Repository. Castor oil. Research Triangle Park, NC: NTP; 2003.
[37] Fredholt K, Larsen DH, Larsen C. Modification of in vitro drug release rate from oily parenteral depots using a formulation approach. European Journal of Pharmaceutical Sciences 2000;11:231–7.
[38] Dangerous/poisonous. Online available at: http://ntbg.org/plants/plantsdetails.php.

[39] Cooperative Extension Service, Purdue University. Castor bean, castor oil plant; September 2007.Online available at: http://www.vet.purdue.edu/depts/addl/toxic/plant11.html.
[40] Castor seeds. Online available at: http://www.agriculturalproductsindia.com/seeds/seeds-castor-seeds.html.
[41] de Melo Cazal C, Batalhao JR, de Cassia Domingues V, Bueno OC, Rodrigues Filho E, Forim MR, da Silva MFGF, Vieira PC, Fernandes JB. High-speed counter-current chromatographic isolation of ricinine, an insecticide from *Ricinus communis*. Journal of Chromatography. A 2009;1216:4290–4.
[42] Wen Y, Feng Y, Zheng M. Extraction and content of the ricinine for different parts of *Ricinus communis*. Agrochemicals 2008;8.
[43] Xu ZH, Shao GL, Nie JP. Studies on the detoxification of castor bean meal by microorganisms. Journal of Hubei Institute for Nationalities (Natural Science Edition) 2007;25:451–4.
[44] Leite AC, Cabral EC, dos Santos DAP, Fernandes JB, Vieira PC, Silva M. Isolation of the alkaloid ricinine from the leaves of *Ricinus communis* (Euphorbiaceae) through counter-current chromatography. Quimica Nova 2005;28:983–5.
[45] Ali RM. Effect of nicotinic acid and nicotinamide adenine dinucleotide on growth and content of oil, glycerol and ricinine alkaloids of salinity stressed *Ricinus communis* L. Phyton 2002;42: 269–77.
[46] Holfelder MGAH, Steck M, Komor E, Seifert K. Ricinine in phloem sap of *Ricinus communis*. Phytochemistry 1998;47:1461–3.
[47] Moshkin VA, editor. Castor. New Delhi: Amerind; 1986.
[48] Waller GR, Skursky L. Translocation and metabolism of ricinine in the castor bean plant, *Ricinus communis* L. Plant Physiology 1972;50:622–6.
[49] Waller GR, Tang MS, Scott MR, Goldberg FJ, Mayes JS, Auda H. Metabolism of ricinine in the castor plant. Plant Physiology 1965;40:803.
[50] Ricin: a deadly protein. Online available at: waynesword.palomar.edu/plmar99.html.
[51] Toxicity. Online available at: https://en.wikipedia.org/wiki/Ricinus.
[52] Singh SP, Sandani G, Dubey JK, Kumar A, Kumar B, Shamim MD. Lactic acid fermentation by *L. leichmannii* exposed to some alkaloids. Asian Journal of Chemistry 1996;8:571.
[53] Castor oil derivatives. Online available at: storage.unitedwebnetwork.com/files/23/cbf660358ff1916355283bae5d6aa6e0.pdf.
[54] Bentley R, Trimen H. Medicinal plants. In: Medical Plants. J & A Churchill; 1981.
[55] Nielsen F, Hill B, de Jongh J. Castor (*Ricinus communis*). Potential of castor for bio-fuel production. FACT-Arrakis. FACT Foundation. FACT Project No. 146 2011.
[56] Medicinal uses. Online available at: http://ntbg.org/plants/plant_details.php.
[57] Uses. Online available at: www.iucngisd.org./gisd/species.php?sc=1000.
[58] Bhakta S, Das SK. In praise of the medicinal plant *Ricinus communis* L.: a review. Global Journal of Research on Medicinal Plants and Indigenous Medicine 2015;4:95–105.
[59] Ogunniyi DS, Njikang GN. Preparation and evaluation of alkyd resin from castor oil. Pakistan Journal of Scientific and Industrial Research 2000.
[60] Indigenous practices. Online available at: http://ntbg.org/plants/plant_details.php.
[61] Castor oil. Online available at: http://www.castoroil.in.html.
[62] Watt G. The commercial product of India. New Delhi, India: Today and Tomorrow Printers and Publishers; 1986.
[63] Coursey DG, Nigerian Federal Institute of Industrial Research. In: Technical Monograph, vol. 22; 1964.
[64] Azzini A, de Barros Salgado AL, Savy Filho A, Banzatto NV. Restos vegetais da cultura da mamona como materia-prima para celulose. Bragantia, Campinas 1981;40:115–24.
[65] Azzini A, Savy Filho A, de Barros Salgado AL, Arnaldi FZ. Deslignificacao dos residuos agricolas da cultura da mamona para producao de celulose e papel. Bragantia, Campinas 1984;43:519–30.
[66] Mutlu H, Meier MAR. Castor oil as a renewable resource for the chemical industry. European Journal of Lipid Science and Technology 2010;112:10–30.

[67] Gujarat experiment in new seed might double castor output without additional cost – report. Online available at: https://www.indoasiancommodities.com/2018/05/01/gujarat-experiment-new-seed-might-double-castor-output-without-additional-cost-report/.
[68] Hindmata Prime News: doubling farm income in castor, SEA shows the way: the model castor farms project gets farmers excited across India. Online available at: http://hindmatanews.blogspot.com/2018/04/doubling-farm-income-in-castor-sea_25.html.
[69] McKeon TA, Auld D, Brandon DL, Leviatov S, He X. Toxin content of commercial castor cultivars. Journal of American Oil Chemist's Society 2014;91:1515–9.
[70] Ramanjaneyulu AV, Madhavi A, Anuradha G, Ramana MV, Suresh G, Naik BB, Seshu G. Agronomic and economic evaluation of hydrogel application in rainfed castor grown on Alfisols. International Journal of Current Microbiology and Applied Sciences 2018;7:3206–17.
[71] Lavanya C, Murthy IYLN, Nagaraj G, Mukta N. Prospects of castor (*Ricinus communis* L.) genotypes for biodiesel production in India. Biomass and Bioenergy 2012;39:204–9.
[72] Naughton FC. Production, chemistry, and commercial applications of various chemicals from castor oil. Journal of American Oil Chemist's Society 1974;51:65–71.
[73] Ogunniyi DS. Castor oil: a vital industrial raw material. Bioresource Technology 2006;97:1086–91.
[74] Castor research. Online available at: http://castoroil.in/b/castor-research/.
[75] Barbosa DDC, Serra TM, Meneghetti SMP, Meneghetti MR. Biodiesel production by ethanolysis of mixed castor and soybean oils. Fuel 2010;89:3791–4.
[76] Canoira L, García Galeán J, Alcántara R, Lapuerta M, García-Contreras R. Fatty acid methyl esters (FAMEs) from castor oil: production process assessment and synergistic effects in its properties. Renewable Energy 2010;35:208–17.
[77] Berman P, Nizri S, Wiesman Z. Castor oil biodiesel and its blends as alternative fuel. Biomass and Bioenergy 2011;35:2861–6.
[78] Deligiannis A, Anastopoulos G, Karavalakis G, Mattheou D, Karonis D, Zannikos F, Stournas E, Lois E. Castor (*Ricinus communis* L.) seed oil as an alternative feedstock for the production of biodiesel. In: Proceedings of 11th International Conference on Environmental Science and Technology, Chania, Crete, Greece; 2009.
[79] Albuquerque MCG, Machado YL, Torres AEB, Azevedo DCS, Cavalcante CL, Firmiano LR, Parente EJS. Properties of biodiesel oils formulated using different biomass sources and their blends. Renewable Energy 2009;34:857–9.
[80] Deep A, Sandhu SS, Chander S. Experimental investigations on the influence of fuel injection timing and pressure on single cylinder C.I. engine fueled with 20% blend of castor biodiesel in diesel. Fuel 2017;210:15–22.
[81] Pradhan S, Madankar CS, Mohanty P, Naik SN. Optimization of reactive extraction of castor seed to produce biodiesel using response surface methodology. Fuel 2012;97:848–55.
[82] Amouri M, Mohellebi F, Zaid TA, Aziza M. Sustainability assessment of *Ricinus communis* biodiesel using LCA approach. Clean Technologies and Environmental Policy 2017;19:749–60.
[83] Liang S, Xu M, Zhang T. Life cycle assessment of biodiesel production in China. Bioresource Technology 2013;129:72–7.
[84] Salmani A, Adl M, Pazoki M. Comparative life cycle assessment of biodiesel production from castor *Ricinus cummunis* and arugula *Eruca sativa*. Tehran, Iran: 11th International Energy Conference; 2016 (in Persian).
[85] Sreenivas P, Mamilla VR, Sekhar KC. Development of biodiesel from castor oil. International Journal of Energy Science 2011;1:192–7.
[86] Hincapie G, Mondragon F, Lopez D. Conventional and in situ transesterification of castor seed oil for biodiesel production. Fuel 2011;90:1618–23.
[87] Meneghetti SMP, Meneghetti MR, Wolf CR, Silva EC, Lima GES, de Lira Silva L, Serra TM, Cauduro F, de Oliveira LG. Biodiesel from castor oil: a comparison of ethanolysis versus methanolysis. Energy & Fuels 2006;20:2262–5.

[88] Bateni H, Karimi K, Zamani A, Benakashani F. Castor plant for biodiesel, biogas, and ethanol production with a biorefinery processing perspective. Applied Energy 2014;136:14–22.
[89] Ramezani K, Rowshanzamir S, Eikani MH. Castor oil transesterification reaction: a kinetic study and optimization of parameters. Energy 2010;35:4142–8.
[90] Santana GCS, Martins PF, de Lima da Silva N, Batistella CB, Maciel Filho R, Wolf Maciel MR. Simulation and cost estimate for biodiesel production using castor oil. Chemical Engineering Research and Design 2010;88:626–32.
[91] Panwar NL, Shrirame HY, Bamniya BR. CO_2 mitigation potential from biodiesel of castor seed oil in Indian context. Clean Technologies and Environmental Policy 2010;12:579–82.
[92] García MJ, Prieto F. Obtaining and characterization of biodiesel from castor oil (*Ricinus communis*) and sunflower (*Helianthus annuus*) grown in Tabasco, Mexico. International Journal of Applied Science and Technology 2012;2:58–74.
[93] AL-Harbawy AW, AL-Mallah MK. Production and characterization of biodiesel from seed oil of castor (*Ricinus communis* L.) plants. International Journal of Science and Technology 2014;3:508–13.
[94] Goswami D, Sen R, Basu JK, De S. Maximization of bioconversion of castor oil into ricinoleic acid by response surface methodology. Bioresource Technology 2009;100:4067–73.
[95] Dias JM, Araujo JM, Costa JF, Alvim-Ferraz MCM, Almeida MF. Biodiesel production from raw castor oil. Energy 2013;53:58–66.
[96] Tajuddin NA, Rosli NH, Salimon J, Yusoff MFM, Abdullah N. Synthesis of estolide 2-ethylhexyl ester from *Ricinus communis*. Malaysian Journal of Analytical Science 2013;17:185–92.
[97] Wang T, Wang J, He X, Cao Z, Xu D, Gao F, Zhong J, Shen L. An ambient curable coating material based on the Michael addition reaction of acetoacetylated castor oil and multifunctional acrylate. Coatings 2019;9:1–10.
[98] Aldobouni IA, Fadhil AB, Saied IK. Conversion of de-oiled castor seed cake into bio-oil and carbon adsorbents. Energy Sources, Part A: Recovery Utilization and Environmental Effects 2015;37: 2617–24.
[99] Sokoto AM, Bhaskar T. Pyrolysis of waste castor seed cake: a thermo-kinetics study. European Journal of Sustainable Develepment Research 2018;2:1–12.
[100] Thiagarajan J, Srividya PK, Balasubramanian P. Thermogravimetric and decomposition analysis of Jatropha, Castor and Pongamia deoiled seed cakes. International Journal of Innovations in Engineering and Technology 2016;7:417–25.
[101] V Santos NA, Magriotis ZM, Saczk AA, Fa ssio GTA, Vieira SS. Kinetic study of pyrolysis of castor beans (*Ricinus communis* L.) presscake: an alternative use for solid waste arising from the biodiesel production. Energy & Fuels 2015;29:2351–7.
[102] de Castro AM, dos Reis Castilho L, Freire DMG. Characterization of babassu, canola, castor seed and sunflower residual cakes for use as raw materials for fermentation processes. Industrial Crops and Products 2016;83:140–8.
[103] Zapata N, Vargas M, Reyes JF, Belmar G. Quality of biodiesel and press cake obtained from *Euphorbia lathyris*, *Brassica napus* and *Ricinus communis*. Industrial Crops and Products 2012;38:1–5.
[104] Gardner Jr HK, D'aquin EL, Koltun SP, McCourtney EJ, Vix HLE, Gastrock EA. Detoxification and deallergenization of castor beans. Journal of American Oil Chemist's Society 1960;37:142–8.
[105] Kling SH. Estudo da solubilidade das proteinas da mamona. Rio de Janeiro: M.S. diss. Univ. Federal do Rio de Janeiro; 1974. in Portuguese.
[106] Freitas J. Efeito da radiacao ionizante sobre as proteinas da torta de mamona. Rio de Janeiro: M.S. diss. Univ. Federal do Rio de Janeiro; 1974. in Portuguese.
[107] Mackinnon PJ, Alderton MR. An investigation of the degradation of the plant toxin, ricin, by sodium hypochlorite. Toxicon 2000;38:287–91.
[108] Wen-bin ZHUBLI. Progress in castor biotechnology. China Biotechnology 2007;10.
[109] Kiran BR, Prasad MNV. *Ricinus communis* L.(Castor bean), a potential multi-purpose environmental crop for improved and integrated phytoremediation. The EuroBiotech Journal 2017;1:101–16.

[110] Castor oil can reduce soil pollution. Online available at: http://www.newindianexpress.com/cities/hyderabad/2017/jul/17/castor-oil-can-reduce-soil-pollution-1629688.html.

[111] Kumar M. A review on phytochemical constituents and pharmacological activities of *Ricinus communis* L. plant. International Journal of Pharmacognosy and Phytochemical Research 2017;9:466–72.

[112] Alugah CI, Ibraheem O. Whole plant screenings for flavonoids and tannins contents in castor plant (*Ricinus communis* L.) and evaluation of their biological activities. International Journal of Herbal Medicine 2014;2:68–76.

[113] Vasconcelos JS, de Almeida DH, de Almeida TH, Vasconcelos JCS, Christoforo AL, Caraschi JC, Lahr FAR. Physical and chemical properties of fibrous residues of castor bean culture. International Journal of Materials Engineering 2014;4:75–8.

[114] Shah SK, Patel DK, Patel RM, Patel PS. Seed and oil quality characteristics of some castor (*Ricinus communis* L.) inbred lines. International Journal of Agriculture, Environment and Biotechnology 2015;8:177–81.

[115] Grigoriou AH, Ntalos GA. The potential use of *Ricinus communis* L.(Castor) stalks as a lignocellulosic resource for particleboards. Industrial Crops Products 2001;13:209–18.

[116] Kang SS, Cordell GA, Soejarto DD, Fong HHS. Alkaloids and flavonoids from *Ricinus communis*. Journal of Natural Products 1985;48:155–6.

[117] Kaur R, Gera P, Jha MK, Bhaskar T. Pyrolysis kinetics and thermodynamic parameters of castor (*Ricinus communis*) residue using thermogravimetric analysis. Bioresource Technology 2018;250:422–8.

[118] Hilioti Z, Michailof CM, Valasiadis D, Iliopoulou EF, Koidou V, Lappas AA. Characterization of castor plant-derived biochars and their effects as soil amendments on seedlings. Biomass and Bioenergy 2017;105:96–106.

[119] Patel AJ, Virdia HM. Effect of weed management on weeds and nutrient uptake in Rabi castor (*Ricinus communis* L.) under south Gujarat conditions. Advance Research Journal of Crop Improvement 2011;2:147–50.

[120] Khoshnevisan B, Rafiee S, Tabatabaei M, Ghanavati H, Mohtasebi SS, Rahimi V, Shafiei M, Angelidaki I, Karimi K. Life cycle assessment of castor-based biorefinery: a well to wheel LCA. The International Journal of Life Cycle Assessment 2018;23:1788–805.

[121] Comparison between castor and other seed oil. Online available at: https://www.castoroilworld.com/comparison-between-castor-and-other-seed-oil/.

[122] Cara C, Moya M, Ballesteros I, Negro MJ, Gonzalez A, Ruiz E. Influence of solid loading on enzymatic hydrolysis of steam exploded or liquid hot water pretreated olive tree biomass. Process Biochemistry 2007;42:1003–9.

[123] Gauss WF, Suzuki S, Takagi M. Manufacture of alcohol from cellulosic material using plural ferments. 1976. US patent 3990944.

[124] Polyanna NH, Keila AM, Raquel PB, Tatiana SP, Cristina MdS, Ana LFP. Potential application of waste from castor bean (*Ricinus communis* L.) for production for xylanase of interest in the industry. Biotech 2016;6(2):144.

[125] Kaur R, Gera P, Kumar M, Bhaskar T. Reaction parameters effect on hydrothermal liquefaction of castor (*Ricinus communis*) residue for energy and valuable hydrocarbons recovery. Renewable Energy 2019;141:1026–41.

[126] Kaur R, Gera P, Jha MK, Bhaskar T. Optimization of process parameters for hydrothermal conversion of castor residue. The Science of the Total Environment 2019;686:641–7.

[127] Cherubini F, Jungmeier G. LCA of a biorefinery concept producing bioethanol, bioenergy, and chemicals from switchgrass. International Journal of Life Cycle Assessment 2010;15:53–66.

CHAPTER 12

Utilization of nonedible oilseeds in a biorefinery approach with special emphasis on rubber seeds

Sutapa Das[1], Ali S. Reshad[1], Nilutpal Bhuyan[2], Debashis Sut[2], Pankaj Tiwari[1], Vaibhav V. Goud[1], Rupam Kataki[2]
[1]*Department of Chemical Engineering, Indian Institute of Technology Guwahati, Guwahati, Assam, India;* [2]*Department of Energy, Tezpur University, Tezpur, Assam, India*

12.1 Introduction

Energy is available in innumerable forms, however, energy dependency on conventional fuels such as coal, petroleum, methane, natural gas, etc. needs to be reduced for the betterment of the environment [1]. In the early 20th century, with rapid growth in population and industrialization, the subsequent energy consumption also increased, posing an enhanced burden on fossil fuels. They serve as the major energy sources in today's world however, studies report that with the current consumption rate, this feedstock will last only for 70–80 years [2]. Thus, prolonged use of these resources may leave us with the scarcity of fuels in the future [3]. The exploitation of these conventional energy sources also presents the looming danger of emission of harmful greenhouse gases like carbon dioxide, sulfur dioxide, nitrogen dioxide, and carbon monoxide, among others [4]. Hence, in recent times the main focus has shifted toward utilization of renewable energy resources including solar, wind, biomass, hydro, marine etc. [5], which in turn will also reduce the emission of greenhouse gases to the surroundings. So new techniques need to be developed to overcome the challenges posed by fossil fuels and the use of biomass can play a major role in mitigating this obvious danger [2,6]. The term, biomass, refers to all organic materials that are obtained from plants (including woody material, grass, algae, trees, and crops) through natural processes or through by-product of human activities. It is a primary source of energy in which the energy of sun light is stored in the form of chemical bonds. Currently, the biomass source of energy contributes 10%–14% of the world energy supply, but theoretically, it has the potential to supply 100% of world energy supply [2,6]. According to the report presented by the European Environment Agency

Waste Biorefinery. https://doi.org/10.1016/B978-0-12-818228-4.00012-5

(EEA, 2013), 20% of the final energy consumption has to be provided using renewable resources (i.e., by biofuel) by 2020.

The energy scenario in India is quite promising for biomass energy plants. As per the statistics report by the Central Statistics Office, Govt. of India the estimated energy consumption was 1,066,268 GWh during 2016–17. Out of this, the renewable sector contributed nearly 1,001,132 MW wherein the contribution from biomass sources was around 18,601 MW which increased from previous year's contribution of 17,538 MW [7]. India has ~5 GW capacity biomass plants, out of which 83% are connected to the grid while rest are off-grid plants. The leading states in terms of biomass energy production are Karnataka, Uttar Pradesh, and Maharashtra, each with more than 1 GW of biomass grid-connected power [8].

India generally produces ~450–500 million tons of biomass/year, which provides 32% of the primary energy use in the country at present. Estimates of energy production from biomass sources may increase from 18 to 50 GW with the incorporation of state of the art technologies and better implementation of policies [8]. Presently, the contribution of biofuels is limited to the formation of blends with ethanol (5%) in gasoline. However, the country plans to meet 20% of the overall diesel requirements by employing biodiesel until the year 2020. In this regard, plants like *Jatropha curcas*, *Neem*, and *Mahua* are primarily identified as the chief sources of this green fuel [9].

The utilization of biomass would not only reduce the dependence on fossil energy sources but also improve energy in a sustainable and carbon-neutral manner. However, it should be noted that the biomass utilization in the traditional way has many drawbacks like the quality of products thus obtained are not of commercial grade and require further upgrading. Also, the energy efficiency of the biodiesel or bio-oil derived from biomass is less as compared to fossil fuels. Most present-day production and use of biomass for energy are carried out in a very unsustainable manner with a great many negative environmental consequences [6,10].

There are mainly four generations of biofuels based on the feedstocks from which they are derived. First-generation feedstocks consist of food crops like sugarcane, rice, wheat etc. [11]. They are converted by biochemical methods like hydrolysis, fermentation but the main issue with the first generation is that it raises the food versus fuel debate [12]. Presently about 95% of the biodiesel is being produced from the first generation i.e., edible feedstocks. However, recent studies reported that utilization of edible vegetable oils and fats has become a great concern as they compete with the food materials. As the demands of vegetable oils are increasing day by day, so its use for biodiesel production cannot be justified. Developing countries like India are not self-sufficient in terms of demand for vegetable oils and hence cannot afford to use edible oil for biodiesel production. The second-generation feedstocks overcome this issue as it utilizes the nonfood crops, i.e., lignocellulosic biomass like wood, forestry residues, agricultural residues, organic wastes, etc. which are converted by biochemical and thermochemical processes. The third generation utilizes algae and other microbes for the production of biofuels by biochemical and thermochemical processes. Finally, the fourth generation is an extension of the third wherein the algae are modified via genetic engineering to alter its properties and are then used for the production of biodiesel. In 2012, India's vegetable oil import touched 10.3 million tons and 46% of it was for edible purpose [13]. So to partially close the issue of food versus fuel crisis and also considering the high cost of edible feedstock, researchers focused on the second-generation biofuels encompassing nonedible feedstocks [14].

12.2 Diversity of nonedible oil seed bearing tree species of northeastern India

The northeast region of India is known for high ethnic and biological biodiversity. The northeast region embraces two biogeography zones viz. Eastern Himalayas comprising Arunachal Pradesh, and northeast India comprising the states of Assam, Manipur, Meghalaya, Mizoram, Nagaland, Tripura, and Sikkim. The Eastern Himalayas biogeography zone is accorded as mega diversity in plant wealth. Assam (89°50/E to 96°10/E and 24°30/N to 28°10/N) is one of the richest biodiversity zones in northeast India. The total area of Assam is 78,438 km^2 out of which 26,781.91 km^2 is demarcated as forest area. A number of tropical rain forests are available in Assam. Moreover, there are revering grasslands, bamboo, orchards, and numerous wetland ecosystems. Many of these areas have been protected by developing National Parks, Wildlife Sanctuary and Reserve Forest. The state is enriched with extensive forest areas and famous for floras and faunas. Similarly, Arunachal Pradesh (28.22°N, 94.73°E) possesses a total forest area of about 51,540 km^2. Revenues from forests form the lion's share of the total. There are approximately 700 woody plants that grow in these hills. The forests in Arunachal have been mainly exploited for timber. Other minor forest producers are normally given secondary importance. Due to the alarming rate of forest destruction for various purposes like agriculture, urbanization, over exploitation, etc., forest land is being depleted and timber becomes scarce. As a result, the minor forest produces are gaining importance. The different minor forest produces like cane, bamboos, resins, gums, medicinal plants, spices have been well recognized. Another important forest product having vast potentials is the nonedible tree borne oil seeds [15].

It is well known that many of our forest trees produce seeds in excess quantities; more than what is required for its own regeneration. However, bulks of these are destined to perish in nature. These excess seeds if properly exploited can yield sustained returns for the population of the region. The seeds could be employed for generating oil. In India over

100 species of tree seeds are known to yield an oil which may find utility as edible oils or as raw materials for manufacturing industries like cosmetics, paints, lubricants, candles, vanaspathi, varnish, etc. At present at the national level, only a part of the total potential of seeds being tapped due to various constraints associated with accessibility, collection, transportation, extraction, etc. Unfortunately, there is a total lack of knowledge about these tree species in this region, especially in Assam and Arunachal Pradesh. Some of these trees if properly assessed may prove to be better than the already known ones. However, many of the indigenous tree species are still underutilized and need to be explored as a source of energy (Table 12.1).

From Table 12.1, it is observed that there is a fundamental need for judicious exploitation of forest wealth available in northeast India to assess the resources and potentiality. There remains a vast scope for tapping this particular resource of tree borne oil seeds for sustainable energy supply in a decentralized manner.

Some of the nonedible oilseed bearing plants available in northeast India along with the rest of the country are *Jatropha curcas*, *Pongamia pinnata* L. (Karanja), *Cascabela thevetia* (Yellow oleander), *Mesua ferrea* (Nahor), *Ricinus communis* (Castor), *Hevea brasiliensis* (Rubber tree), *Moringa oleifera*, *Calophyllum inophyllum* L. (Polanga), etc. These tree species have been investigated well for their biodiesel production potential [16]. Apart from this, the recent report on few other nonedible seeds from the northeastern region is also reported [17]. These are mainly *Citrus maxima*, *Cucurbita moschata*, *Anisomeles indica*, *Luffa acutangula*, *Cucumis sativus*, *Parkia timoriana*, *Meyna spinose*, *Abelmoschus moschatus* to study the influence of key parameters [18]. A comparative compositional analysis of the above-mentioned feedstocks is given in Table 12.2 which shows their potential as a feedstock for biodiesel production.

There are various literature on the production of biodiesel from the various nonedible seeds. Ameen et al. [27] used rubber seeds to produce green diesel oil. They used catalytic dehydrogenation of rubber seed oil over synthesized Ni–Mo/γ-Al_2O_3 catalyst and obtained a yield of 80.87 wt% of diesel range hydrocarbons. Naik et al. [28], studied the Karanja oils for the production of biodiesel. They discussed the mechanism involving two-process step for the production of biodiesel from Karanja oil. The first step includes acid-catalyzed esterification using 0.5% H_2SO_4, alcohol 6:1 M ratio with respect to the high free fatty acid (FFA) Karanja oil to produce methyl ester by lowering the acid value, and the second step is alkali-catalyzed transesterification. The yield of biodiesel by this process is reported to be 96.6%–97% [28]. Another study on Moringa by Salaheldeen et al. [29], revealed that moringa seeds can be used as a potential feedstock for the production of biodiesel. High cetane number i.e., 60.2 and low free fatty acid i.e., 0.35% makes it a potent for biodiesel.

Table 12.1: Potential tree borne oil seeds of northeast India (Assam and Arunachal Pradesh) [15,16].

Sl no.	Feedstock	Family	Vernacular name	Kernel (%)	Oil content (%)	Place of availability
1	*Cinnamomum glaucescens*	Lauraceae	Gonsorai	74.00	34.48	Assam
2	*Ostodes paniculata*	Euphorbiaceae	Tasichangne	76.00	36.25	Arunachal Pradesh
3	*Styrax serrulatum*	Styracaceae	Phulkat	54.40	20.25	Arunachal Pradesh
4	*Aphanamixis polystachya*	Meliaceae	Bogamari	82.30	30.95	Assam
5	*Magnolia griffithii*	Magnoliaceae	Gahorisopa	83.90	21.23	Arunachal Pradesh
6	*Creteva nurvala*	Capparaceae	Barun	82.60	30.65	Arunachal Pradesh
7	*Croton oblongifolius*	Euphorbiaceae	Mahunda	79.80	26.25	Arunachal Pradesh
8	*Croton triglium*	Euphorbiaceae	Jamalgota	64.20	25.98	Arunachal Pradesh
9	*Gironniera cuspidata*	Ulmaceae	-	73.13	27.64	Arunachal Pradesh
10	*Neolitsea cuipala*	Lauraceae	Phulsopa	87.80	55.27	Assam
11	*Knema linifolia*	Myristiceaceae	Amool	74.00	7.32	Arunachal Pradesh
12	*Mallotus albus*	Euphorbiaceae	Morali	-	30.20	Arunachal Pradesh
13	*Talauma hodgsonii*	Magnoliaceae	Boranthuri	-	39.26	Arunachal Pradesh
14	*Canarium strictum*	Burseraceae	Dhuna	-	43.20	Arunachal Pradesh
15	*Gynocardia odarata*	Flacourtiaceae	Chaulmugra	8.80	17.78	Arunachal Pradesh
16	*Putranjiva roxburghii*	Putranjivaceae	Putranjiva	38.6	10.34	Arunachal Pradesh
17	*Sterculia villosa*	Malvaceae	Udal	66.7	17.21	Arunachal Pradesh
18	*Ailanthus grandis*	Simaroubaceae	Borpat, botpat	74.2	33.33	Arunachal Pradesh
19	*Cedrus deodara*	Pinaceae	Deodaru	85.0	21.83	Arunachal Pradesh
20	*Claoxylon khasianum*	Euphorbiaceae	-	-	49.34	Arunachal Pradesh
21	*Knema angustifolia*	Myristicaceae	Tezranga	85.1	18.94	Arunachal Pradesh
22	*Olia dioca*	Oleaceae	-	61.6	20.12	Arunachal Pradesh

Continued

Table 12.1: Potential tree borne oil seeds of northeast India (Assam and Arunachal Pradesh) [15,16].—cont'd

Sl no.	Feedstock	Family	Vernacular name	Kernel (%)	Oil content (%)	Place of availability
23	*Terygota alata*	Sterculiaceae	Karibadam	55.2	26.61	Arunachal Pradesh
24	*Artocarpus chaplasha*	Moraceae	Sam	94	9.21	Arunachal Pradesh
25	*Artocarpus heterophyllus*	Moraceae	Kathal	95	2.71	Arunachal Pradesh
26	*Artocarpus lakoocha*	Moraceae	Dewa-Sali	56	1.69	Arunachal Pradesh
27	*Baccaurea sapida*	Euphorbiaceae	Leteku	82	-	Arunachal Pradesh
28	*Bauhinia variegata*	Caesalpinaceae	Boga katra	92	14.90	Arunachal Pradesh
29	*Bombax ceiba*	Malvaceae	Semul	-	24.30	Arunachal Pradesh
30	*Bischofia javanica*	Euphorbiaceae	Urium	-	17.60	Arunachal Pradesh
31	*Canarium strictum*	Burseraceae	Dhuna	5	2.16	Arunachal Pradesh
32	*Croton caudatus*	Euphorbiaceae	Ghahe-lewa	22	5.21	Arunachal Pradesh
33	*Endospermum chinense*	Euphorbiaceae	Phulgamari	-	35.30	Arunachal Pradesh
34	*Garcinia stipulate*	Glusiaceae	-	-	17.52	Arunachal Pradesh
35	*Sapium baccatum*	Euphorbiaceae	Saleng	-	38.60	Arunachal Pradesh
36	*Vangueria spinosa*	Rubiaceae	Ketkora	31	34.30	Arunachal Pradesh
37	*Vaticalancae folia*	Dipterocarpaceae	Morhal	78	-	Arunachal Pradesh
38	*Mangium chinensis*	Fabaceae or Leguminosae	Bogamarulia	-	26.20	Assam, Arunachal Pradesh
39	*Chisocheton panniculatus*	Meliaceae	Bandordima	-	32.20	Assam, Arunachal Pradesh
40	*Elaeocarpus aristatus*	Elaeocarpaceae	Gaharisopa		36.40	Assam, Arunachal Pradesh
41	*Prunus jenkinsii*	Rosaceae	Thereju	-	14.10	Assam
42	*Gmelina arborea*	Lamiaceae	Gamari		57.40	Assam, Arunachal Pradesh
43	*Litsea angustifolia*	Lauraceae	-	-	24.00	Assam, Arunachal Pradesh
44	*Litsea confertiflora*	Lauraceae	-	-	58.00	Arunachal Pradesh

Continued

Table 12.1: Potential tree borne oil seeds of northeast India (Assam and Arunachal Pradesh) [15,16].—cont'd

Sl no.	Feedstock	Family	Vernacular name	Kernel (%)	Oil content (%)	Place of availability
45	*Litsea cubeba*	Lauraceae	Mejangkori	-	19.10	Assam, Arunachal Pradesh
46	*Litsea glutinosa*	Lauraceae	Baghnola	-	34.00	Assam, Arunachal Pradesh
47	*Litsea laeta*	Lauraceae	Bon sualu	-	52.50	Assam
48	*Litsea lanuginosa*	Lauraceae	-	-	58.60	Assam, Arunachal Pradesh

12.3 Rubber seeds: a by-product of booming rubber industry of northeast India

Favorable agro-climate condition in India has led to an increase in rubber plantation which economically benefits the local society [30]. In India, Tripura and Assam are the main rubber tree producing states next to Kerala. Tripura itself has dedicated 50,070 ha of land

Table 12.2: Comparative compositional analysis of nonedible seeds.

Non-edible seed	Species family	Oil yield (wt%)	Fatty acid composition (wt%)
Rubber [19]	*Hevea brasiliensis*	40–50	Linoleic acid, oleic acid, palmitic acid
Karanja [19]	*Pongamia pinnata*	27–39	Oleic acid, linoleic acid, eicosenoic acid, behenic acid
Yellow oleander [20]	*Cascabela thevetia*	62	Palmitic acid, stearic acid, oleic acid, linoleic acid, arachidic acid
Moringa [21]	*Moringa oleifera*	37	Oleic acid, stearic acid, palmitic acid, palmitoleic acid
Nahor [22]	*Mesua ferrea*	69	Oleic acid, linoleic acid, stearic acid, palmitic acid
Pomelo [23]	*Citrus maxima*	0.01	Limonene, linalool, nerolidol, farnesol
Catmint [24]	*Anisomeles indica*	0.04	Farnesyl acetone, nootkatone, jasmatone
Butternut squash [25]	*Cucurbita moschata*	33.2	245-Ethyl 5ot-cholesta-7,25-dien-31β-ol, 24S-ethyl 5αt-cholesta-7,22E-dien-3β-ol, 24S-ethyl 5ot-cholesta-7,22E,25-trien-3β-ol
Physic nut [26]	*Jatropha curcas*	34.4	Linoleic acid, oleic acid, palmitic acid

for rubber plantation followed by Assam with 23,705 ha. After producing rubber from the latex of the rubber tree, the seeds left can be used for various purposes. Basically, rubber seed is a by-product of a rubber tree. It is used for producing rubber tree oil but commercially is not of much use since it is not edible. But, in case of high food sector demand, it can replace the edible raw material such as soybean oil, sunflower oil, palm oil, etc. for biodiesel production. Nonedible feedstocks like rubber seeds are reported to be potential producers of energy [31,32] that pave way for utilizing these resources for harvesting energy.

Rubber seed gives 40%−50% (by weight) of kernel which contain 42%−50% (% wt.) nonedible oil comprising of palmitic, stearic, oleic, linoleic, linolenic fatty acid in 12.2 wt %, 8.7 wt%, 21.8 wt%, 39.6 wt% and 16.3 wt% respectively, promising for the production of biodiesel [30,33,34]. Rubber seeds are abundant in the northeastern State of India and are generally considered as waste. Also, India is currently the sixth largest producer and the second largest consumer of natural rubber (NR) in 2017 according to the Statistical and Planning Department Rubber Board, Kerala. The International Rubber Study Group (IRSG), reports that there has been an increase in NR production worldwide by 7.4%. The global NR consumption increased from 12.67 million tons in 2016 to 13.23 million tons in 2017 registering a growth of 4.4%. The production rate of the seed and promising content of oil (1553.19 kg/ha/yr) [35] makes it a potential candidate for biodiesel production. Along with this, other value-added products from rubber seed like animal feedstock, activated carbon, bio-oil, metal soap, gases produced can also be used in industrial processes etc. establishes it as a promising candidate for biorefinery [36]. Balkose et al. [37], reported that rubber seed oil having 2.2% myristic acid, 7.6% palmitic acid, 10.7% stearic acid, 20.61% oleic acid, 36.62% linoleic acid, 22.5% linolenic acid was used to making barium, calcium, cadmium and zinc soaps. The thermal behavior of soaps (Ba, Ca, Cd and Zn) of rubber seed oil were used as additives in the processing of poly(vinyl chloride). The soaps were found to be thermally stable with values of apparent activation energy for decomposition varying from 52 to 96 kJ/mol. The rubber seed cake can be further pyrolyzed to produce bio-oil, gases, and biochar which can be subsequently be used for producing activated carbon which is discussed in Section 12.6.

12.4 Renewable energy scenario

Renewable energy is basically the energy derived from natural resources that get replenishes with time and are nonexhaustive in nature like solar, wind, hydropower, and biomass. Solar, wind, and hydropower has been extensively used and is well established whereas, converting biomass into useful energy has still not been exploited to its full potential thereby demands intensive research in this direction [38,39]. Compared to the hydrocarbon molecules obtained from fossils, biorefinery products like ethanol, butanol, and lignin are more oxidized. A biorefinery plant targets to completely exploit the biomass. Biomass has several conventional and nonconventional uses, though they are mostly used in producing biodiesel. For example, wood in pulp or paper mill is used to obtain biodiesel [40]. These biomasses mostly become the feedstock in a biorefinery where they end up becoming useful oils in an efficient process releasing zero wastes into the environment, thereby helping in reducing greenhouse gas (GHG) emissions. Other sources of biomass are the products of sugar-based industry, for example, sugar beet, and sugarcane), starch crops (e.g., cereals, grains, such as corn, cassava, or wheat) or oleaginous crops (e.g., rape, soy). They are called the "first-generation" raw material or feedstock. The biorefineries that use these products as feedstock are already established and the biofuels that they produce are commonly called as "first-generation" biofuels like bioethanol and biodiesel [41] (Tables 12.3 and 12.4). Besides sugar mills, paper mills also produce feedstock which generates similar first-generation biofuels. Oleaginous crops like rapeseed, sunflower, palm oil, canola, jatropha, etc. are the basic raw materials for first-generation biofuel production.

12.5 Biofuel/biodiesel production from oil seeds

Biofuels can be derived from edible and nonedible oil seeds. Most common oilseed crops are soybean, rapeseed and canola, mustard, castor bean, safflower, sunflower, jatropha etc. [42]. In the United States, soybean-based biorefineries produce 1.5 gallons of oil per bushel. This is equivalent to an oil yield of 66 gallons per acre. Though not so popular, the yield of oil from rapeseed and canola are much higher and have a yield of 75–240 gallons of oil per acre [43]. The yield of oil from such crops is also dependent on weather conditions. Rapeseed, canola, jatropha, soybean are rotational crops that are commonly cultivated for oil production having application in various areas.

Basically, biodiesel is produced from edible oil seeds, this tremendous use of food sources for automated use may bring additional problems to society. Arable lands that would otherwise have been used to grow food are now directed toward the production of fuel-

Table 12.3: First-generation biodiesel producers in EU in 2014 [42].

Company	Plants	Total production capacity (t/year)
Avril	13	2,700,000
Neste oil	3	1,180,000
ADM biodiesel	3	975,000
Infinita	2	600,000
Marseglia group	2	560,000
Verbio AG	2	450,000

Table 12.4: First-generation bioethanol producers in the USA [42,43].

Company	No. of plants	Total Capacity (million liters/year)
Poet biorefining	26	6315
Green plains	14	4600
Valero renewable fuels	11	4960
Archer Daniels Midland Co.	8	>5970
Pacific Ethanol	8	1950
Flint River Resources, LLC	7	3050

based crops. Thus, research to produce biodiesel from nonedible oils has gained momentum. The nonedible vegetable oils such as *Madhuca indica*, *Jatropha curcas*, and *Pongamia pinnata* are found to be suitable for biodiesel production under the experimental conditions [44].

The energy scenario depicts that commercially viable biorefineries are already established in developed countries [45,46]. In developing countries like India, it is still in its nascent stage. The government, however, has laid down new policies to establish the principles and modalities of their functioning and development. Currently, according to the "National Policy on Biofuel 2009" [47], a minimum of 10% blending of biofuel with petrol and diesel should be undertaken and this level has to be increased to 20% in the coming years. The surplus would be exported to the Asian markets for catering their need for fuel consumption.

India's vast energy requirements can be broadly classified into three sectors, wherein, domestic and transport sectors constitute nearly 40% each, and the rest 20% of the energy requirements are for the industrial sectors. The conventional sources of energy such as crude oil and gases provide energy to nearly 90% of the primary and transport sectors and the remaining 10% is required for the production of industrial chemicals [48]. This tremendous strain on the fossil reserves is only increasing exponentially every year. This, in turn, leads to escalating prices of crude oil in the international market to never-seen-before levels. Coping with such prices will not be economically feasible for many nations and along with the danger of this resource in terms of energy security, the developing nations have been forced to explore alternative and inexpensive energy sources of energy to meet the growing energy demand. In this aspect, the renewable sources of energy are the go-to sources for the future and in it, one of the promising candidates for providing support to many countries, and in particular, to agrarian economies, is biomass whose effective utilization in biorefineries are to be carefully explored for its tremendous potential [49].

The drive for establishing biorefineries in India comes from its enormous reliance on the consumption of liquid fuels [48]. According to reports, India uses an estimated 50 million metric tons (MMT) of liquid fuels per year. However, with the vast reserves of biomass it possesses, with effective utilization, India is touted to produce almost double this amount in a single year [48]. The biomass source estimates provided here constitutes only the available crop residues and primarily the second-generation fuels as the usage of first-generation crop-based fuels have been embroiled in debates as it is basically the source of nutrients and livelihood in a much food-starved nation.

Hence, the focus has to be on relying on agricultural wastes and nonedible resources which have the potential to generate biofuel. One promising candidate which ticks many boxes has been rubber seeds owing to its high oil content and rubber seed cake produced thereafter. Rubber seeds can be effectively utilized for the production of biofuels and other value-added products satisfying a no-waste approach in compliance with the biorefinery concept.

12.6 Biorefinery concept

The basic concept of a biorefinery is to comprehensively utilize a single feedstock i.e., biomass to produce different types of end products, wherein each component from the process is converted in a way to add value. Hence sustainability to the whole plant, but it releases no waste during the process operation i.e., biorefineries would generate zero waste [50]. A biorefinery could, for example, produce one or several low-volume, but high-value chemical or nutraceutical products and a low-value, but high-volume liquid transportation fuel such as biodiesel or bioethanol, while also generating electricity and process heat through combined heat and power (CHP) technology for its own use, and perhaps enough for sale of electricity to the local utility [51]. By producing multiple products, a biorefinery takes advantage of the various components in biomass and their intermediates, thereby maximizing the value derived from the biomass feedstock [52]. In this scenario, the high-value products increase profitability, the high-volume fuel helps meet energy needs, and the power production helps to lower energy costs and reduce greenhouse gas emissions, as compared to traditional power plant facilities. Although there are few biorefineries present, there exists a need to expand more of them across wider geographic regions. Future, biorefineries may also play a major role in assisting the production of chemicals and byproducts which are traditionally produced from petroleum [53–55].

The subsequent diagram or flowchart below shows the concept of the proposed biorefinery [56].

The nonedible seed consists of seed kernel and seed cover. The seed kernel can be exploited to obtain vegetable oil by various techniques like soxhlet extraction, ultrasonic extraction, and supercritical fluid extraction. Among these, soxhlet is the simplest to use. It is also the cheapest, does not require filtration, also the solvent recovery is high, the contact time of solvent with the solid matrix is high but have few disadvantages associated with it. The main shortcoming of this process is that it takes very long hours of extraction time and requires a large amount of solvent. The ultrasonic extraction is also similar to soxhlet in terms of cost and simplicity. The main disadvantage is that filtration is required, also a large amount of solvent is needed. Supercritical is an automated fast extraction process but has a very high operational cost. After the extraction of vegetable oil, the deoiled cake is further used for the pyrolysis oil to obtain bio-oil, biochar, and gas.

The ground seed kernels are kept inside a filter paper bag and then placed inside the cylindrical soxhlet extractor, which is then placed onto the round bottom flask containing the organic solvent. The top of the soxhlet is connected to a condenser. The solvent is heated by using heating mantle to reflux. The solvent is recycled by heating it and condensing simultaneously. The basic mechanism is that the solvent vapors move up in the soxhlet extractor through the provided bypass tube arm. Once the vapor cools down it condenses and falls drip-wise into the soxhlet chamber. The chamber gets flooded with the solvent, i.e., the ground kernel seeds in the filter bag get flooded. When the chamber is full, it is automatically emptied through a siphoning action along a siphon tube and the solvent runs down to the flask. This cycle is repeated for around 2–12 h, depending on the time required for extraction of oil. The extraction solvents used for this are mainly hexane, ethyl acetate, and methanol, etc. After the extraction, the sample is further purified i.e., separation of solvent from oil using rotary vacuum evaporator. To produce biodiesel, the oil sample is further esterified or transesterified based on its acid value.

The obtained vegetable oil after extraction is converted to biodiesel by esterification and transesterification processes. If the acid value of the vegetable oil is less than three then directly esterifications is done. But if the acid value is higher than three then a two-step process is required, i.e., esterification followed by transesterification process. After this process biodiesel is obtained with glycerol as the by-product. Glycerol is in a crude form containing different impurities such as oily, alkali, and soap components, as salt or diols, depending on the type of feedstocks. Thompson and He [57] reported that crude glycerol percentage obtained in the production of biodiesel from different feedstocks is generally between 60 wt% and 70 wt%. This crude glycerol can be converted into purified glycerol through different processes such as distillation, filtration, chemical treatment, adsorption (using activated carbon), ion-exchange (using resin), extraction, and crystallization etc. [58]. Purified glycerol has numerous industrial uses especially in the chemical industry, pharmaceuticals, food, and cosmetic industries [58,59].

Pure glycerol has been used in various chemical reactions e.g., in production of dihydroxyacetone and glyceraldehydes through catalytic oxidation reaction (Figs. 12.1 and 12.2), as a raw material in the production of 1,3-propanediol, dendrimers, hyper-branched polyester, etc. [63–65]. Moreover, the dehydration of glycerol in the presence of catalyst

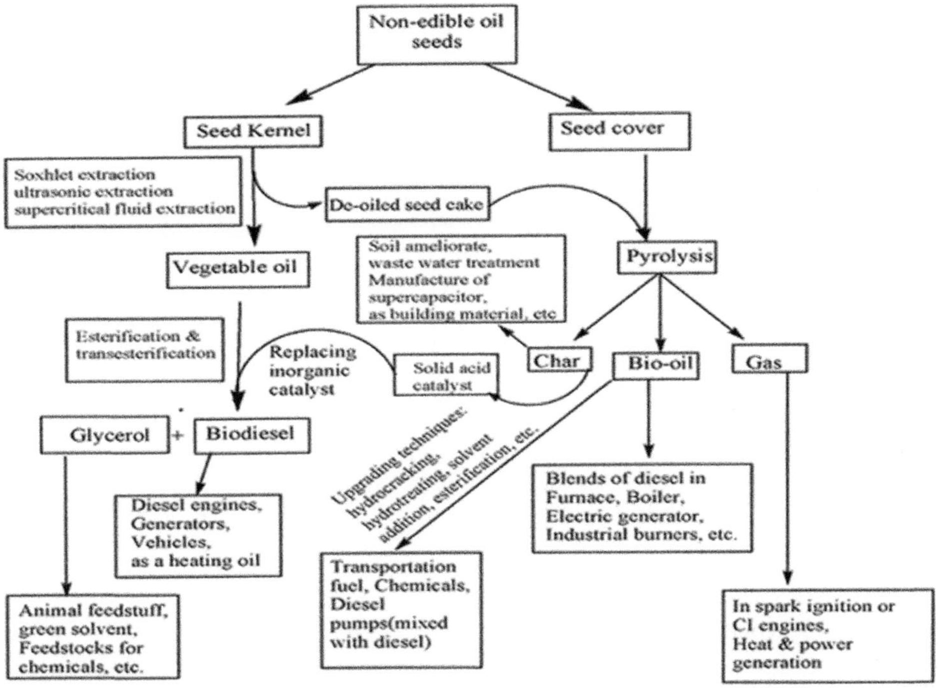

Figure 12.1

A schematic diagram of biorefinery approach for biofuel production from nonedible oil seeds [56].

leads to the production of acrolein [66]. It is also a good source for the production of hydrogen through steam reforming, partial oxidation, autothermal reforming process [65]. Glycerol is used in the production of various medicine like cough syrup, ear infection medicines, as a carrier for antibiotics and antiseptics and plasticizers for medicine capsules [67]. In the cosmetic industry, it is used primarily for providing lubrication, enhancing smoothness, and as moistener [68]. In the food industry, glycerol acts as a solvent, sweetener and preservative agent. It is normally ingested in manufacturing extracts of tea, coffee, ginger, and other vegetable substances. It is also used as a softening agent in some foodstuff like bread, cakes, etc. [68].

The deoiled cake used for the pyrolysis process further produces bio-oil, biochar, and gas. The obtained bio-oil can be used with blends of diesel in furnace and boilers. To enhance the quality of bio-oil, so that it can be used in the transportation sector, upgradation is done. There are various upgrading techniques like hydrocracking, hydrotreating, solvent addition, esterification, fuel blending or microemulsion, supercritical treatment.

The seed cover is further exploited to obtain valuable products like bio-oil, biochar, and gas. Thermochemical conversion techniques like pyrolysis, gasification, etc. are widely

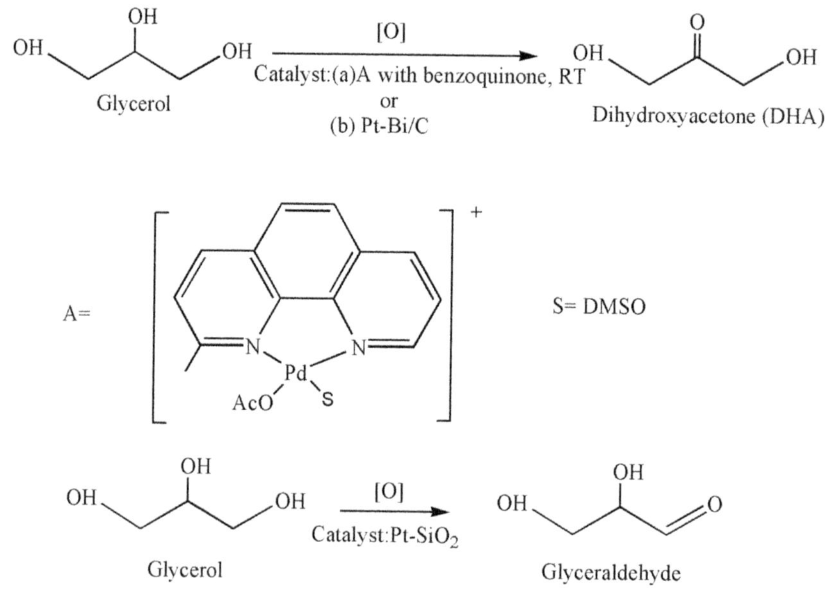

Figure 12.2
Schematic diagram of catalytic oxidation of glycerol [60–62].

used. Out of this pyrolysis process is convenient, cost-effective and well established is mostly used. Pyrolysis has long been established as a viable technology for the conversion of biomass into charred material; however, pyrolysis can be further segregated into several types including slow, fast, vacuum, flash and microwave pyrolysis. Each process utilizes different equipment and production conditions to maximize individual products and properties. Slow pyrolysis or conventional carbonization utilize lower heating rates to moderate temperatures (<700°C), and long vapor residence times to generate higher char yields than other variations of pyrolysis [69–71]. Slow pyrolysis is therefore regarded as the more favorable technology to maximize biochar yield for soil application while also generating valuable coproducts for heat and power generation. Fast pyrolysis is designed to rapidly heat biomass (>200 K/min) to peak temperature in a very short time scale resulting in short vapor residence times (<2s). These conditions are designed to favor the formation of bio-oil while also inhibiting char formation [70,71]. The reaction conditions do not just affect the yield of products but also influence other properties such as composition, viscosity, heating value, etc. Although "fast" and "slow" pyrolysis are the leading routes for bio-oil and biochar production respectively, alternative methods for biomass conversion such as gasification, intermediate pyrolysis, flash pyrolysis, and microwave pyrolysis also exist, with their own specific advantages and disadvantages.

In pyrolysis C, H, and O containing large molecules of biomass were broken down into smaller molecular fragments in the form of condensable vapors or bio-oils, noncondensable gases, and biochar. The probable reaction of pyrolysis process can be represented as follows [72]:

$$\text{Biomass} \xrightarrow[\text{Heat.}]{\text{Inert gas.}} \text{Solid char} + \text{Liquid (Organic phase} + \text{Aqueous phase)} + \text{Gas } (CO_2,\ CO,\ H_2,\ CH_4)$$

Although biomass pyrolysis reaction pathway is complex process it can be shown in three main steps:

$$\text{Biomass} \rightarrow \text{Water} + \text{Nonreacted residue} \quad (A)$$

$$\text{Nonreacted residue} \rightarrow (\text{Volatile} + \text{Gases})_1 + (\text{Char})_1 \quad (B)$$

$$(\text{Char})_1 \rightarrow (\text{Volatile} + \text{Gases})_2 + (\text{Char})_2 \quad (C)$$

(Subscript 1 & 2 indicates − primary and secondary decomposition respectively.)

In the first step (A), decomposition of biomass occurs and involves dehydration, bond rupture, free radical formation, formation of carbonyl, carboxyl groups or some rearrangement reactions. The second step (B) can be called as the major pyrolysis step, involved the primary decomposition of nonreacted residue to the pyrolysis products at a high rate. In the third step (C), secondary decomposition of char occurs at a very slow rate.

Lignocellulosic biomass consists of three major constituents, viz. hemicellulose, cellulose, and lignin. The order in which thermal degradation of these constituents occurs, can be summarized as follows [73]:

$$\text{Hemicelluloses} > \text{cellulose} > \text{lignin}$$

Among these components, the decomposition of cellulose has been most widely analyzed and best comprehend [74]. Various schemes such as Broido–Shafizadeh model [75], Waterloo model [76], and Varhegyi–Antal model [77] were reported in the literature to explain the cellulose pyrolysis reactions. All of these models support the formation of anhydrosugars through depolymerization, which competes with the formation of char. The mechanism of products formation as proposed by Kawomoto et al. [78] is shown in Figs. 12.3 and 12.4. They found carbonization reactions gradually transformed levoglucosan to methanol and water-soluble residue, and eventually into a fraction totally insoluble to water and methanol.

12.6.1 Bio-oil

Pyrolysis liquids are created via rapid and simultaneous depolymerization and fragmentation of cellulose, hemicelluloses and lignin following intense heating and are

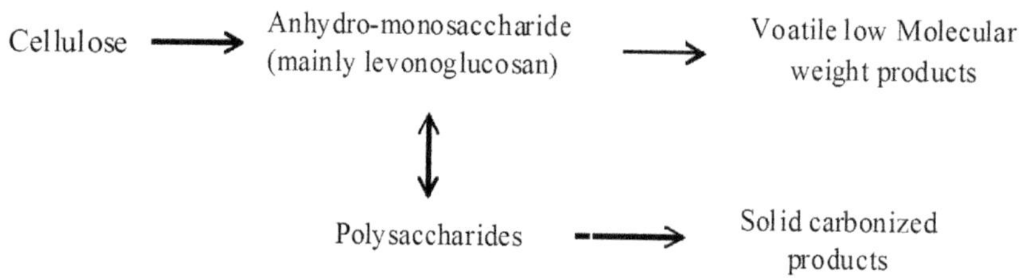

Figure 12.3
Scheme of cellulose pyrolysis mechanism.

separated from the gas stream through rapid cooling [80,81]. The properties of bio-oil are largely dependent on its chemical composition which is closer in origin to the elemental composition of its parent biomass than that of petroleum oils. The main reason for the differences experienced between pyrolysis oils and hydrocarbon fuels are due to the presence of oxygen in the majority of the 300+ compounds that have been identified in bio-oil [80,81]. These compounds consist of very complex oxygenated hydrocarbons and species such as carboxylic acids, alcohols, ketones, phenols, alkenes, syringols, sugars, etc. [82]. The high oxygen content has a direct effect on the energy density of bio-oil (50% of that of conventional fuel oils) while also causing instability and immiscibility with hydrocarbon fuels. However, continuing to increase the severity (HTT, heating rate, residence time, etc.) of pyrolysis can lead to the cracking of vapors and formation of gases resulting in a reduced organic liquid yield with less oxygen [80]. Similarly to petroleum

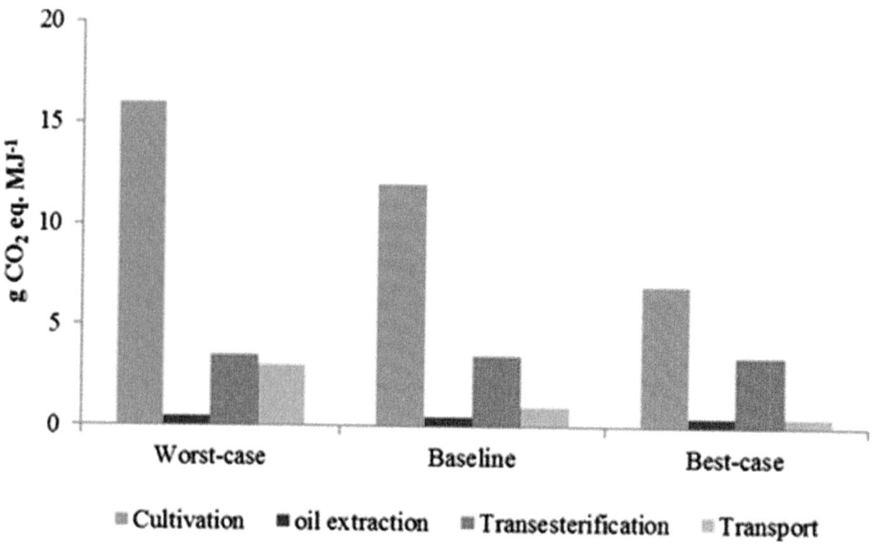

Figure 12.4
GHG emissions from rubber seed–based biodiesel production (worst case, base case and best case scenarios) [79].

feedstock, bio-oil can also be used for the synthesis of chemicals with high and comparable revenue to energy and fuel products [83,84].

12.6.2 Gaseous product

Following the removal of the condensable liquids from the vapor stream the remaining noncondensable gases can also be burned directly for heat and power generation [85]. Pyrolysis gas consists of varying amounts of carbon monoxide (CO) (16%–51%) and hydrogen (H_2) (2%–43%) as well as CO_2 (9%–55%), methane (CH_4) (4%–11%) and small amounts of C_2 hydrocarbon gases such as ethane (C_2H_6) [70]. Depending on the production parameters chosen for pyrolysis, the yield, and composition of pyrolysis gas can change greatly, with the greatest volume and energy content produced by flash pyrolysis [85].

12.6.3 Biochar

The heating value of chars are comparable with lignite and coke while the heating values of bio-oil and pyrolysis gases are much lower than that of petroleum fuels and natural gas respectively [86,87]. Other than heating value, the most essential indicator of biochar quality is its pH, high adsorption and cation exchange capacities, and high aromatic carbon content which is generally depend on the feedstock type [88]. Due to these properties, biochar has much beneficial application such as incorporation into the soil to increase the long term storage of carbon while reducing GHG and providing soil amendment benefits [89,90]. Biochar amelioration into the soil provides a good habitat for soil microorganisms, can increase the total organic carbon, improve the cation exchange capacity (CEC), improving air and water circulation in the soil and may stimulate root growth [91]. Also, the water holding capacity as well as nutrients retention capability of biochar makes it favorable for plants growth. Biochar can also be used to provide additional heat to the pyrolysis system through combustion [82,92]. However as mentioned in other literature the mitigation impact of biochar is about 25% larger, on an average, than the impact obtained if the same biomass was fully combusted for energy [93]. If additional benefits of biochar, such as increased plant growth, reduced N_2O emissions, etc., are considered then biochar production could be a favorable option compared to the combustion of biomass or production of bio-oils [94].

The bio-oil obtained after pyrolysis contains around 30 wt% of water which leads to its low calorific value. To make it viable for industrial applications, it needs to be upgraded. There are various types of upgrading techniques like hydrodeoxygenation, catalytic hydrogenation, steam reforming, hot-gas vapor filtration, etc. which reduce the oxygen content and hence enhance the property of bio-oil [95–97]. The biochar produced can further be used as the catalyst in esterification and transesterification reactions to produce biofuels from vegetable oils [98,99]. Also, biochar is demonstrated as promising electrode materials for supercapacitor applications due to their specific porosity, low-cost, and electrochemical stability. Various researchers reported on using biochar obtained from different biomass feedstock for electrode fabrication of supercapacitor [100,101].

Biochar could also be used as a cost-effective sorbent for wastewater remediation [102]. Biochar can be used for absorption of hydrocarbons, some inorganic metals such as lead (Pb), cadmium (Cd), mercury (Hg), Chromium (Cr), Iron (Fe), etc. and it has been found that biochar is comparable to some commercial activated carbon as well as some low-cost sorbent such as coal, coconut shell etc. [103,104]. Komkiene and Baltrenaite [105], studied the absorption of heavy metals (cadmium (II), lead (II), copper (II) and zinc (II)) on the Scots pine and silver birch biochar. They varied the metal concentration (one to four folds), biochar dosage (1.6–140 g) at constant pH of leaching solution, temperature and contact time. The adsorption capacity was assessed by Freundlich isotherm. The maximum adsorption capacity of copper (II) was 128.7 μg/g on silver birch and the maximum adsorption capacity of zinc (II) was 107 μg/g on scotus pine biochar. The adsorption of lead (II) varied in between 1.3 and 3.8 μg/g on silver birch and 2.4–4.5 μg/g on scotus pine biochar.

It has also been found that biochar has the potential to be used as a carbon sequestering admixture in concrete constructions and as a building material which offers the possibility of carbon negative construction providing a way to waste recycling [106,107]. Two basic properties of biochar, i.e., low thermal conductivity and high water absorbability indicates its importance as the right material for insulating buildings and regulating humidity [107].

12.7 Current challenges in the use of rubber seed for energy generation

Use of rubber seed for energy generation has some challenges, starting from the synthesis of bio-oil derived from rubber seed. The biorefinery process to derive bio-oil from rubber seed oil involves transesterification reaction which can be carried out by using either homogeneous (acid or base) or heterogeneous (acid or base, or enzymes) catalyst [108–110]. Usually, homogeneous catalysts help to have a faster reaction than that of heterogeneous catalysts, but it is considerably costly, and also it is very difficult to separate the homogeneous catalyst from the reaction mixture [111,112]. Choice of acid or alkali catalyst depends on the presence of FFA content in the raw oil. If the FFA content is beyond 3%, acid esterification followed by alkaline transesterification process is carried out whereas if the FFA content is below 3%, only alkaline transesterification process is carried out [113]. It is reported that the variables influencing the transesterification process have to be optimized to enhance product efficiency [30,114].

The yield of the bio-oil from rubber seed is also dependent on the type of rubber seed based on geographical region, altitude, etc., which makes its availability dependable on the superiority of plantation. The properties of biodiesel derived from rubber seed oil are generally compared with the standard specification for biodiesel (ASTM D 6751) [115]. Some studies report that sodium metal and sodium hydroxide catalyst are more suitable for transesterification of refined rubber seed oil, while sulfuric acid and phosphoric acid catalyst are more suitable for crude rubber seed oil [116].

Also, unlike other biomass-derived bio-oil, bio-oil generated from rubber seed is not zero-waste process. Sensitivity analysis assessment was done by Wagner et al. [19] to study the influence of key parameters (e.g., rubber seed yield) on the GHG mitigation potential. They reported that the largest share of total GHG emissions is caused by the cultivation of the rubber tree stands.

Thus, some of the challenges in the use of rubber seed generation are in its synthesis due to lack of cost-effectiveness in the development of heterogeneous catalyst required for the biorefinery process to derive bio-oil. Besides that, there could be geographical limitations for the production of bio-oil from rubber seed. The process is yet to reach zero-waste generation.

12.8 Scope for production of variable products using oil seeds

Various nonedible seeds have been reported to produce significant valuable products *Jatropha curcas* L. is very a versatile and a high generator of numerous value-added products. Activated carbon produced from *Jatropha curcas* seeds demonstrated excellent thermal, surface, and regeneration properties [117–120]. Derived activated carbon can be further used in applications like purification, water treatment, and dye removals. Jatropha seed comprises a high quantity of oil which are mostly similar to fatty acids composition [121,122]. Fatty acids are biosurfactant and numerous researchers have derived surfactant and polymeric surfactants from jatropha [123,124]. Surendra et al. [125] very recently synthesized polymer from Jatropha seeds. Similar to Jatropha, Castor seeds have been used for the synthesis of surfactant and polymeric surfactant [126]. Adsorbent derived from Karanja and *Moringaoleifera* seeds exhibited outstanding surface area [127] (i.e., 343 m^2/g) and are capable of removal of 95% of dye [128]. Although very limited studies have been reported, but *Hevea brasiliensis* and *Calophyllum inophylum* L. have the ability in the synthesis of cosmetics, soap, detergent, and paint [129–131]. The above reports suggest that nonedible feedstocks produce a wide range of other value-added products along with the biofuel.

12.9 Conclusions and perspectives

In this chapter, the biorefinery approach using nonedible and biowaste feedstocks for conversion to biofuel along with the extraction of valuable products is discussed. This approach could be a better and economically feasible approach over the conventional techniques used. From the biomass processing point-of-view, many techniques have been tried and different approaches have been employed however, no absolute formula has yet been devised which can comprehensively convert the biomass feedstock into valuable products. This chapter attempts to shed light on a promising avenue for biomass processing, which includes, incorporating the biorefineries system of creating wealth from waste which is economically attractive. Although many industries around the world have established biorefineries, with promising outcomes especially in the financial front, nonetheless, In India, its journey has started some time back. It has yet to attain a fully functional and operational status but in time, the development of biorefineries in India will play a key role in shaping the energy sector of this country as it possesses huge amounts of feedstock material which is only increasing every year.

India being an agrobased economy, relies largely on its expansive pastures and vast fertile lands for sustaining the economic growth. It is blessed with enormous quantities of biomass source which for a long period of time has, in terms of energy generation, has largely remained untapped. Most of the biomass resources are diverted toward inefficient processes, however, by adapting a prudent approach this renewable source of energy can effectively contribute toward and fulfill the growing energy demands. Effective utilization of this resource, which also includes the conversion and development of biomass to biofuels, can substantially reduce the burden on conventional sources and contribute toward a lower carbon footprint.

References

[1] Wüstenhagen R, Wolsink M, Bürer MJ. Social acceptance of renewable energy innovation: an introduction to the concept. Energy Policy 2007;35(5):2683–91.
[2] REN 21, renewables 2017 global status report. Paris: REN21 Secretariat.
[3] Capellán-Pérez I, Mediavilla M, de Castro C, Carpintero Ó, Miguel LJ. Fossil fuel depletion and socio-economic scenarios: an integrated approach. Energy 2014;77:641–66.
[4] Höök M, Tang X. Depletion of fossil fuels and anthropogenic climate change—a review. Energy Policy 2013;52:797–809.
[5] Droege P. Renewable energy and the city: urban life in an age of fossil fuel depletion and climate change. Bulletin of Science, Technology & Society 2002;22(2):87–99.
[6] https://www.ecotricity.co.uk/our-green-energy/energy-independence/the-end-of-fossil-fuels.
[7] Industry, G. o. I. R. B. M. o. C.; 2018. http://www.mospi.gov.in/central-statistics-office-cso-1.
[8] http://indien.um.dk/en/innovation/sector-updates/renewable-energy/biomass-energy-in-india/.
[9] https://mnre.gov.in/.

[10] Di Blasi C. Modeling chemical and physical processes of wood and biomass pyrolysis. Progress in Energy and Combustion Science 2008;34(1):47–90.

[11] Naqvi M, Yan J. First-generation biofuels. Handbook of clean energy systems. 2015. p. 1–18.

[12] Martin MA. First generation biofuels compete. New Biotechnology 2010;27(5):596–608.

[13] Banković-Ilić IB, Stamenković OS, Veljković VB. Biodiesel production from non-edible plant oils. Renewable and Sustainable Energy Reviews 2012;16(6):3621–47.

[14] Carriquiry MA, Du X, Timilsina GR. Second generation biofuels: economics and policies. Energy Policy 2011;39(7):4222–34.

[15] Basumatary S. Non-edible oils of Assam as potential feedstocks for biodiesel production: a review. Journal of Chemical, Biological and Physical Sciences 2012–13;3(1):551–8.

[16] Kotoky R, Pathak MG, Kanjilal PB. Physio-chemical characteristics of seed oils of some *Litsea* species found in North-East India. Natural Product Radiance 2007;6(4):297–300.

[17] Mazumdar P, Borugadda VB, Goud VV, Sahoo L. Physico-chemical characteristics of *Jatropha curcas L.* of North East India for exploration of biodiesel. Biomass and Bioenergy 2012;46:546–54.

[18] Barua P, Dutta K, Basumatary S, Deka DC, Deka DC. Seed oils from non-conventional sources in North-East India: potential feedstock for production of biodiesel. Natural Product Research 2014;28(8):577–80.

[19] Wagner M, Lippe M, Lewandowski I, Salzer M, Cadisch G. CO_2 footprint of the seeds of rubber (*Hevea brasiliensis*) as a biodiesel feedstock. Source 2018;9:548–65.

[20] Sut D, Bordoloi NJ, Narzari R, Chutia RS, Kataki R. Complete utilization of non-edible oil seeds of *Cascabela thevetia* through a cascade of approaches for biofuel and byproducts. Bioresource Technology 2016;213:111–20.

[21] Leone A, Spada A, Battezzati A, Schiraldi A, Aristil J, Bertoli S. *Moringa oleifera* seeds and oil: characteristics and uses for human health. International Journal of Molecular Sciences 2016;17(12):2141–55.

[22] Chakraborty T, Das MK. Oil of *Mesuaferrea* L. seed as a promising pharmaceutical excipient in lipid based nanoformulation. Journal of Applied Pharmaceutical Science 2017;7:133–41.

[23] Duñg NX, Pha NM, Lô VN, An NTK, Leclercq PA. The essential oil from the flowers of *Citrus maxima* (J. Burman) Merrill from Vietnam. Journal of Essential Oil Research 1991;3(5):359–60.

[24] Basappa G, Kumar V, Sarojini BK, Poornima DV, Gajula H, Sannabommaji TK, et al. Chemical composition, biological properties of *Anisomeles indica Kuntze* essential oil. Industrial Crops and Products 2015;77:89–96.

[25] Schinas P, Karavalakis G, Davaris C, Anastopoulos G, Karonis D, Zannikos F, et al. Pumpkin (*Cucurbita pepo* L.) seed oil as an alternative feedstock for the production of biodiesel in Greece. Biomass and Bioenergy 2009;33:44–9.

[26] Ahad B, Reshi ZA, Rasool H, Shahri W, Yousuf AR. *Jatropha curcas*: a prospective energy crop. In: Biomass and bioenergy. Cham: Springer; 2014. p. 289–306.

[27] Ameen M, Azizan MT, Ramli A, Yusup S, Alnarabiji MS. Catalytic hydrodeoxygenation of rubber seed oil over sonochemically synthesized Ni–Mo/γ-Al2O3 catalyst for green diesel production. Ultrasonics Sonochemistry 2019;51:90–102.

[28] Naik M, Meher L, Naik S, Das L. Production of biodiesel from high free fatty acid Karanja (*Pongamia pinnata*) oil. Biomass and Bioenergy 2008;32(4):354–7.

[29] Salaheldeen M, Aroua M, Mariod A, Cheng SF, Abdelrahman MA. An evaluation of *Moringa peregrina* seeds as a source for bio-fuel. Industrial Crops and Products 2014;61:49–61.

[30] Ramadhas AS, Jayaraj S, Muraleedharan C. Biodiesel production from high FFA rubber seed oil. Fuel 2005;84(4):335–40.

[31] Balat M. Potential alternatives to edible oils for biodiesel production—a review of current work. Energy Conversion and Management 2011;52:1479–92.

[32] Kole C, Joshi CP, Shonnard DR. Handbook of bioenergy crop plants. CRC Press; 2012.

[33] Aigbodion A, Pillai C. Preparation, analysis and applications of rubber seed oil and its derivatives in surface coatings. Progress in Organic Coatings 2000;38(3–4):187–92.
[34] Ikwuagwu O, Ononogbu I, Njoku O. Production of biodiesel using rubber [*Hevea brasiliensis* (Kunth. Muell.)] seed oil. Industrial Crops and Products 2000;12(1):57–62.
[35] Zhu Y, Xu J, Li Q, Mortimer PE. Investigation of rubber seed yield in Xishuangbanna and estimation of rubber seed oil based biodiesel potential in Southeast Asia. Energy 2014;69:837–42.
[36] Ng WPQ, Lim MT, Lam HL, Yusup S. Overview on economics and technology development of rubber seed utilisation in Southeast Asia. Clean Technologies and Environmental Policy 2014;16(3):439–53.
[37] Balköse D, Egbuchunam TO, Okieimen FE. Thermal behaviour of metal soaps from biodegradable rubber seed oil. Journal of Thermal Analysis and Calorimetry 2010;101(2):795–9.
[38] Akella A, Saini R, Sharma MP. Social, economical and environmental impacts of renewable energy systems. Renewable Energy 2009;34(2):390–6.
[39] Klass DL. Biomass for renewable energy, fuels, and chemicals. Elsevier; 1998.
[40] Ragauskas AJ, Nagy M, Kim DH, Eckert CA, Hallett JP, Liotta CL. From wood to fuels: integrating biofuels and pulp production. Industrial Biotechnology 2006;2(1):55–65.
[41] Rødsrud G, Lersch M, Sjöde A. History and future of world's most advanced biorefinery in operation. Biomass and Bioenergy 2012;46:46–59.
[42] Murugesan A, Umarani C, Chinnusamy T, Krishnan M, Subramanian R, Neduzchezhain N. Production and analysis of bio-diesel from non-edible oils—a review. Renewable and Sustainable Energy Reviews 2009;13(4):825–34.
[43] https://farm-energy.extension.org/oilseed-crops-for-biodiesel-production/.
[44] Mushtaq A, Mir AK, Muhammad Z, Shazia S. Biodiesel from non edible oil seeds: a renewable source of bioenergy. In: dos Santos Bernardes MA, editor. Economic effects of biofuel production; 2011.
[45] Kamm B. Biorefineries–their scenarios and challenges. Pure and Applied Chemistry 2014;86(5):821–31.
[46] http://sorona.com/.
[47] Energy, G. o. I. M. o. N. R. National Policy on Biofuels. https://mnre.gov.in/file-manager/UserFiles/biofuel_policy.pdf.
[48] https://www.bioenergyconsult.com/biorefinery-india/.
[49] Purohit P, Dhar S. Biofuel roadmap for India. UNEP DTU Partnership; 2015.
[50] Cherubini F. The biorefinery concept: using biomass instead of oil for producing energy and chemicals. Energy Conversion and Management 2010;51(7):1412–21.
[51] Chowdhury R, Ghosh S, Debnath B, Manna D. Indian agro-wastes for 2G biorefineries: strategic decision on conversion processes. In: De S, Bandyopadhyay S, Assadi M, Mukherjee D, editors. Sustainable energy technology and policies. Springer; 2018. p. 353–73.
[52] de Jong E, Jungmeier G. Biorefinery concepts in comparison to petrochemical refineries. In: Industrial biorefineries & white biotechnology. Elsevier; 2015. p. 3–33.
[53] Demirbas A. Biorefineries: current activities and future developments. Energy Conversion and Management 2009;50(11):2782–801.
[54] Fernando S, Adhikari S, Chandrapal C, Murali N. Biorefineries: current status, challenges, and future direction. Energy & Fuels 2006;20(4):1727–37.
[55] Clark J, EI Deswarte FJ, Farmer T. The integration of green chemistry into future biorefineries. Biofuels, Bioproducts and Biorefining 2009;3(1):72–90.
[56] Bhuyan N, Sut D, Gogoi L, Kataki R, Goud VV. Rural biorefinery: a viable solution for production of fuel and chemicals in rural India. In: Rai M, Ingle AP, editors. Sustainable Bioenergy. Elsevier; 2019. p. 21–47.
[57] Thompson JC, He BB. Characterization of crude glycerol from biodiesel production from multiple feedstocks. Applied Engineering in Agriculture 2006;22:261–5.
[58] Stamatelatou K. Advanced oil crop biorefineries. Cambridge: RSC Publishing; 2011.
[59] Brandner A, Lehnert K, Bienhonz A, Lucas M, Claus P. Production of biomass-derived chemicals and energy: chemocatalytic conversion of glycerol. Topics in Catalysis 2009;52:278–87.

[60] El Roz A, Fongarland P, Dumeignil F, Capron M. Glycerol to glyceraldehyde oxidation reaction over Pt-based catalysts under base-free conditions. Frontiers in Chemistry 2019;7:156.
[61] Painter RM, Pearson DM, Waymouth RM. Selective catalytic oxidation of glycerol to dihydroxyacetone. Angewandte Chemie International Edition 2010;49(49):9456–9.
[62] Hu W, Lowry B, Varma A. Catalytic Oxidation of glycerol to dihydroxyacetone. In: 21st international symposium on chemical reaction engineering; 2016.
[63] Demirel GS, Lucas M, Claus P. Liquid phase oxidation of glycerol over carbon supported gold catalysts. Catalysis Today 2005;102–103:166–72.
[64] Haag R, Sunder A, Stumbe J. An approach to glycerol dendrimers and pseudo-dendritic polyglycerols. Journal of the American Chemical Society 2000;122:2954–5.
[65] Fan XH, Burton R, Zhou YC. Glycerol (byproduct of biodiesel production) as a source for fuels and chemicals—mini review. The Open Fuels & Energy Science Journal 2010;3:17–22.
[66] Alhanash A, Kozhevnikova EF, Kozhenikov IV. Gas phase dehydration of glycerol to acroleincatalyzed by caesium heteropoly salt. Applied Catalysis 2010;378:11–8.
[67] Tan HW, Aziz ARA, Aroua MK. Glycerol production and its applications as a raw material: a review. Renewable and Sustainable Energy Reviews 2013;27:118–27.
[68] Ampaitepin S, Tetsuo T. A perspective on incorporation of glycerin purification process in biodiesel plants using waste cooking oil as feedstock. Energy 2010;35:2493–504.
[69] Jeffery S, Verheijen FG, Van Der Velde M, Bastos AC. A quantitative review of the effects of biochar application to soils on crop productivity using meta-analysis. Agriculture, Ecosystems & Environment 2011;144(1):175–87.
[70] Brownsort PA. Biomass pyrolysis processes: performance parameters and their influence on biochar system benefits. 2009.
[71] Manyà JJ. Pyrolysis for biochar purposes: a review to establish current knowledge gaps and research needs. Environmental Science & Technology 2012;46(15):7939–54.
[72] Balat M, Kırtay E, Balat H. Main routes for the thermo-conversion of biomass into fuels and chemicals. Part 1: pyrolysis systems. Energy Conversion and Management 2009;50(12):3147–57.
[73] Balat M. Mechanisms of thermochemical biomass conversion processes. Part 1: reactions of pyrolysis. Energy Sources, Part A 2008;30(7):620–35.
[74] Kan T, Strezov V, Evans TJ. Lignocellulosic biomass pyrolysis: a review of product properties and effects of pyrolysis parameters. Renewable and Sustainable Energy Reviews 2016;57:1126–40.
[75] Bradbury AG, Sakai Y, Shafizadeh FA. Kinetic model for pyrolysis of cellulose. Journal of Applied Polymer Science 1979;23(11):3271–80.
[76] Piskorz J, Radlein DSA, Scott DS, Czernik S. Pretreatment of wood and cellulose for production of sugars by fast pyrolysis. Journal of Analytical and Applied Pyrolysis 1989;16(2):127–42.
[77] Varhegyi G, Jakab E, Antal Jr MJ. Is the Broido-Shafizadeh model for cellulose pyrolysis true? Energy & Fuels 1994;8(6):1345–52.
[78] Kawamoto H, Murayama M, Saka S. Pyrolysis behavior of levoglucosan as an intermediate in cellulose pyrolysis: polymerization into polysaccharide as a key reaction to carbonized product formation. Journal of Wood Science 2003;49(5):469–73.
[79] No S-Y. Inedible vegetable oils and their derivatives for alternative diesel fuels in CI engines: a review. Renewable and Sustainable Energy Reviews 2011;15:131–49.
[80] Czernik S, Bridgwater A. Overview of applications of biomass fast pyrolysis oil. Energy & Fuels 2004;18(2):590–8.
[81] Mohan D, Pittman CU, Steele PH. Pyrolysis of wood/biomass for bio-oil: a critical review. Energy & Fuels 2006;20(3):848–89.
[82] Bridgwater AV. Review of fast pyrolysis of biomass and product upgrading. Biomass and Bioenergy 2012;38:68–94.
[83] Bridgwater T. Biomass for energy. Journal of the Science of Food and Agriculture 2006;86(12):1755–68.

[84] Isahak WNRW, Hisham MW, Yarmo MA, Hin T-yY. A review on bio-oil production from biomass by using pyrolysis method. Renewable and Sustainable Energy Reviews 2012;16(8):5910–23.
[85] Becidan M, Skreiberg Ø, Hustad JE. Products distribution and gas release in pyrolysis of thermally thick biomass residues samples. Journal of Analytical and Applied Pyrolysis 2007;78(1):207–13.
[86] Laird DA, Brown RC, Amonette JE, Lehmann J. Review of the pyrolysis platform for coproducing bio-oil and biochar. Biofuels, Bioproducts and Biorefining 2009;3:547–62.
[87] Yaman S. Pyrolysis of biomass to produce fuels and chemical feedstocks. Energy Conversion and Management 2004;45:651–71.
[88] Glaser B, Lehmann J. Ameliorating physical and chemical properties of highly weathered soils in the tropics with charcoal a review. Biology and Fertility of Soils 2002;35:219–30.
[89] Shackley S, Sohi S, Brownsort P, Carter S, Cook J, Cunningham C, Gaunt J, Hammond J, Ibarrola R, Mašek O. An assessment of the benefits and issues associated with the application of biochar to soil. London: Department for Environment, Food and Rural Affairs, UK Government; 2010.
[90] Sohi S, Krull E, Lopez-Capel E, Bol R. A review of biochar and its use and function in soil. Advances in Agronomy 2010;105:47–82.
[91] Hutapea S. Waste management with the technology of biopore hole absorption (LRB) based on biochar in Medan, Indonesia. IOSR Journal of Environmental Science, Toxicology and Food Technology 2018;12(2):77–82.
[92] Yoder J, Galinato S, Granatstein D, Garcia-Pérez M. Economic tradeoff between biochar and bio-oil production via pyrolysis. Biomass and Bioenergy 2011;35:1851–62.
[93] Woolf D, Amonette JE, Street-Perrott FA, Lehmann J, Joseph S. Sustainable biochar to mitigate global climate change. Nature Communications 2010;1:1.
[94] Hammond J, Shackley S, Sohi S, Brownsort P. Prospective life cycle carbon abatement for pyrolysis biochar systems in the UK. Energy Policy 2011;39:2646–55.
[95] Gollakota AR, Subramanyam MD, Kishore N, Gu S. Upgradation of bio-oil derived from lignocellulose biomass—a numerical approach. In: Proceedings of the first international conference on recent advances in bioenergy research. Springer; 2016. p. 197–212.
[96] Yang H, Yao J, Chen G, Maa W, Yan B, Qi Y. Overview of upgrading of pyrolysis oil of biomass. Energy Procedia 2014;61:1306–9.
[97] Hoekstra E, Hogendoorn KJA, Wang X, Westerhof RJM, Kersten SRA, van Swaaij WPM, et al. Fast pyrolysis of biomass in a fluidized bed reactor: in situ filtering of the vapors. Industrial & Engineering Chemistry Research 2009;48(10):4744–56.
[98] Chellappan S, Nair V, Sajith V, Aparna K. Synthesis, optimization and characterization of biochar based catalyst from sawdust for simultaneous esterification and transesterification. Chinese Journal of Chemical Engineering 2018;26(12):2654–63.
[99] Dehkhoda AM, West AH, Ellis N. Biochar based solid acid catalyst for biodiesel production. Applied Catalysis A: General 2010;382:197–204.
[100] Qu WH, Xu YY, Lu AH, Zhang XQ, Li WC. Converting biowaste corncob residue into high value added porous carbon for supercapacitor electrodes. Bioresource Technology 2015;189:285–91.
[101] Ali GAM, Divyashree A, Supriya S, Chong KF, Ethiraj AS, Reddy MV, et al. Carbon nanospheres derived from *Lablab purpureus* as a high performance supercapacitor electrode: green approach. Dalton Transactions 2017;46:14034–44.
[102] Rasheed A, Sana S, Kashif S, Umer Z, Khatoon M. To evaluate the efficiency of char and biochar for waste water treatment. Journal of Waste Recycling 2017;2(2):7.
[103] Mohan D, Sarswat A, Ok YS, Pittman Jr CU. Organic and inorganic contaminants removal from water with biochar, a renewable, low cost and sustainable adsorbent – a critical review. Bioresource Technology 2014;160:191–202.
[104] Inyang MI, Gao B, Yao Y, Xue Y, Zimmerman A, Ahmed M, et al. A review of biochar as a low-cost adsorbent for aqueous heavy metal removal. Critical Reviews in Environmental Science and Technology 2016;46(4):406–33.

[105] Komkiene J, Baltrenaite E. Biochar as adsorbent for removal of heavy metal ions [cadmium (II), copper (II), lead (II), zinc (II)] from aqueous phase. International Journal of Environmental Science and Technology 2016;13(2):471–82.

[106] Gupta S, Kua HW, Low CY. Use of biochar as carbon sequestering additive in cement mortar. Cement and Concrete Composites 2018;87:110–29.

[107] https://www.ithaka-institut.org/en/ct/97.

[108] Lopez DE, Goodwin Jr JG, Bruce DA, Lotero E. Transesterification of triacetin with methanol on solid acid and base catalysts. Applied Catalysis A: General 2005;295:97–105.

[109] Dossin TF, Reyniers MF, Marin GB. Kinetics of heterogeneously MgO-catalyzed transesterification. Applied Catalysis B: Environmental 2006;62(1–2):35–45.

[110] Van Geem KM, Reyniers M-F, Marin G. Taking optimal advantage of feedstock flexibility with COILSIM1D. In: AIChE spring meeting: ethylene producers conference; 2008.

[111] Ma F, Hanna MA. Biodiesel production: a review. Bioresource Technology 1999;70:1–15.

[112] Du W, Liu D, Zeng J. Comparative study on lipase-catalyzed transformation of soybean oil for biodiesel production with different acyl acceptors. Journal of Molecular Catalysis B: Enzymatic 2004;30:125–9.

[113] Meena Devi R, Subadevi R, Paul Raj S, Sivakumar M. Comparative studies on biodiesel from rubber seed oil using homogeneous and heterogeneous catalysts. International Journal of Green Energy 2015;12(12):1215–21.

[114] Sharma Y, Singh B, Upadhyay S. Advancements in development and characterization of biodiesel: a review. Fuel 2008;87:2355–73.

[115] Kannan T, Marappan R. Comparative study of performance and emission characteristics of a diesel engine fueled by emulsified biodiesel/diethyl ether blended biodiesel. Journal of Applied Sciences 2011;11:2961–7.

[116] Omorogbe S, Ikhuoria E, Aigbodion A, Obazee E, Momodu V. Production of rubber seed oil based biodiesel using different catalysts. Current Research in Chemistry 2013;5:11–8.

[117] Hsu S-H, Huang C-S, Chung T-W, Gao S. Adsorption of chlorinated volatile organic compounds using activated carbon made from *Jatropha curcas* seeds. Journal of the Taiwan Institute of Chemical Engineers 2014;45(5):2526–30.

[118] Karthikeyan S, Sivakumar P, Palanisamy P. Novel activated carbons from agricultural wastes and their characterization. Journal of Chemistry 2008;5(2):409–26.

[119] Xin-Hui D, Srinivasakannan C, Jin-Hui P, Li-Bo Z, Zheng-Yong Z. Comparison of activated carbon prepared from *Jatropha* hull by conventional heating and microwave heating. Biomass and Bioenergy 2011;35(9):3920–6.

[120] Kumar R, Chinnaiya N. Development and characteristics of activated carbons from *Jatropha* husk, an agro industrial solid waste, by chemical activation methods. Journal of Environmental Engineering and Management 2009;19(3):173–8.

[121] Jingura RM, Musademba D, Matengaifa R. An evaluation of utility of *Jatropha curcas* L. as a source of multiple energy carriers. International Journal of Engineering, Science and Technology 2010;2(7).

[122] Achten WM, Mathijs E, Verchot L, Singh VP, Aerts R, Muys B. *Jatropha* biodiesel fueling sustainability? Biofuels, Bioproducts and Biorefining: Innovation for a Sustainable Economy 2007;1(4):283–91.

[123] Kumar S, Saxena N, Mandal A. Synthesis and evaluation of physicochemical properties of anionic polymeric surfactant derived from Jatropha oil for application in enhanced oil recovery. Journal of Industrial and Engineering Chemistry 2016;43:106–16.

[124] Kumar S, Kumar A, Mandal A. Characterizations of surfactant synthesized from Jatropha oil and its application in enhanced oil recovery. AIChE Journal 2017;63(7):2731–41.

[125] Surendra B, Veerabhadraswamy M. Microwave assisted synthesis of polymer via bioplatform chemical intermediate derived from *Jatropha* deoiled seed cake. Journal of Science: Advanced Materials and Devices 2017;2(3):340–6.

[126] Babu K, Pal N, Bera A, Saxena V, Mandal A. Studies on interfacial tension and contact angle of synthesized surfactant and polymeric from castor oil for enhanced oil recovery. Applied Surface Science 2015;353:1126–36.

[127] Ganvir VN, Meshram ML, Dwivedi AP. Preparation of adsorbent from karanja oil seed cake and its characterization. In: IJCA proceedings on international conference on emerging frontiers in technology for rural area (EFITRA-2012),. IJCA Journal EFITRA (No. 3); 2012.

[128] Reck IM, Paixão RM, Bergamasco R, Vieira MF, Vieira AMS. Removal of tartrazine from aqueous solutions using adsorbents based on activated carbon and *Moringa oleifera* seeds. Journal of Cleaner Production 2018;171:85–97.

[129] Leksono B, Windyarini E, Hasnah T, Rahman S, Baral H. *Calophyllum inophyllum* for green energy and landscape restoration: plant growth, biofuel content, associate waste utilization and agroforestry prospect. In: International conference and utility exhibition on green energy for sustainable development (ICUE). IEEE; October, 2018. p. 1–7.

[130] Iyayi A, Akpaka P, Ukpeoyibo U. Rubber seed processing for value-added latex production in Nigeria. African Journal of Agricultural Research 2008;3(7):505–9.

[131] Widyaningsih S, Chasani M, Diastuti H. Formulation of antibacterial liquid soap from Nyamplung seed oil (*Calophyllum inophyllum* L.) with addition of *Curcuma heyneana* and its activity test on *Staphylococcus aureus*. In: IOP conference series: materials science and engineering. IOP Publishing; 2018. p. 012062.

CHAPTER 13

Waste biorefinery based on waste carbon sources: case study of biodiesel production using carbon based catalysts and mixed feedstocks of nonedible and waste oils

Ritesh S. Malani[1], Hanif A. Choudhury[2], Vijayanand S. Moholkar[1,3]

[1]*Centre for Energy, Indian Institute of Technology, Guwahati, Guwahati, Assam, India;* [2]*Department of Chemical Engineering, Texas A&M University at Qatar, Doha, Qatar;* [3]*Department of Chemical Engineering, Indian Institute of Technology, Guwahati, Guwahati, Assam, India*

13.1 General introduction on waste biorefinery

Energy derived from coal and crude oil in terms of electricity and transportation fuels constitute the primary needs of human civilization and development of society in present scenario [1]. Limited reservoirs of fossil fuels and their uneven distribution make energy utilization through–out the world inappropriate. Organization of the Petroleum Exporting Countries (OPEC) countries have ~75% of world's ascertained oil reservoirs with only 6% of world population. OPEC currently contributes to ~50% of global oil production [2]. Population in countries with high oil reserves enjoy far greater energy per capita as compared to rest of the world.

As per the recent reports of International Energy Agency (IEA), the global energy demand in 2017 grew by 2.1%, which was more than twice as compared to 2016, and has resulted in enhancement of 1.4% or ~450 MT (metric tons) global CO_2 emissions [3].

It is estimated that global energy demand will reach 680 quadrillion British Thermal Units (qBTU) in 2040, which is ~25% rise from 575 qBTU in 2015 [4]. The energy consumption statistics published by IEA, gives the breakdown of delivered energy consumption by different sectors for 2015 as depicted in Fig. 13.1.

The energy consumption by the different transportation sector in 2018 is depicted by Fig. 13.2. It is interesting to note here that ~48% of global energy consumption is from

Waste Biorefinery. https://doi.org/10.1016/B978-0-12-818228-4.00013-7

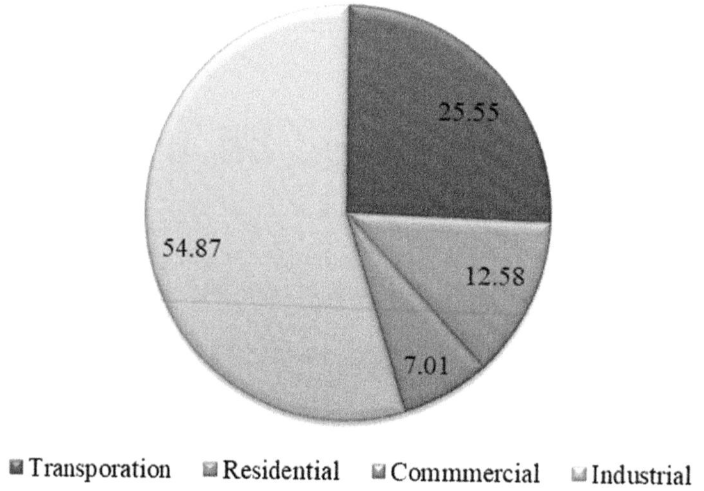

Figure 13.1
Total world delivered energy consumption sector-wise. *Modified from A report on "2018 Outlook for Energy: A view to 2040" from "ExxonMobil". Available from https://corporate.exxonmobil.com/en/energy-and-environment/energy-resources/outlook-for-energy/2018-outlook-for-energy-a-view-to-2040#aViewTo2040.pdf.*

the light-duty vehicles alone, whereas only 1.1% was contributed by two-to-three-wheel vehicles.

Today, 81% of global energy demand has been fulfilled by fossil fuels over last three decades, despite strong growth in the renewable energy sector [5]. The average per capita energy consumption of the globe in 2017 was nearly 1.77 tons of oil equivalent (toe), which is varied significantly from region to region. For example, the average per capita energy consumption in North America is 7.6 toe; Middle-East, 3.3 toe; Europe, 2.65 toe; Latin America, 1.06 toe; and Asia, 1.03 toe [3,4,5]. Asia is the major contributor in global energy demand followed by Africa. China and India together contributed 40% of the increase in Asia's increase in global energy demand, although per capita energy consumption was much below in these region as compared to global average [5].

Average growth of global oil demand was 1.6% higher in 2017, resulted in 1.5 mb/d (million barrels per day) of oil production. This was much higher than average annual growth rate of 1% over a decade [3]. The major contributors for this growth in oil demand were transport sector followed by petrochemical sector. Modern efficient engines showed reduction in average fuel consumption per vehicle but rapid increase in the share of sport utility vehicles (SUVs) and other large vehicle sales goes counter to the engine energy efficiency [3].

On the other hand, rapid increase in plastic and other petrochemical products promoted petrochemical sector as fastest-growing sector in last few years. Asia becomes the major

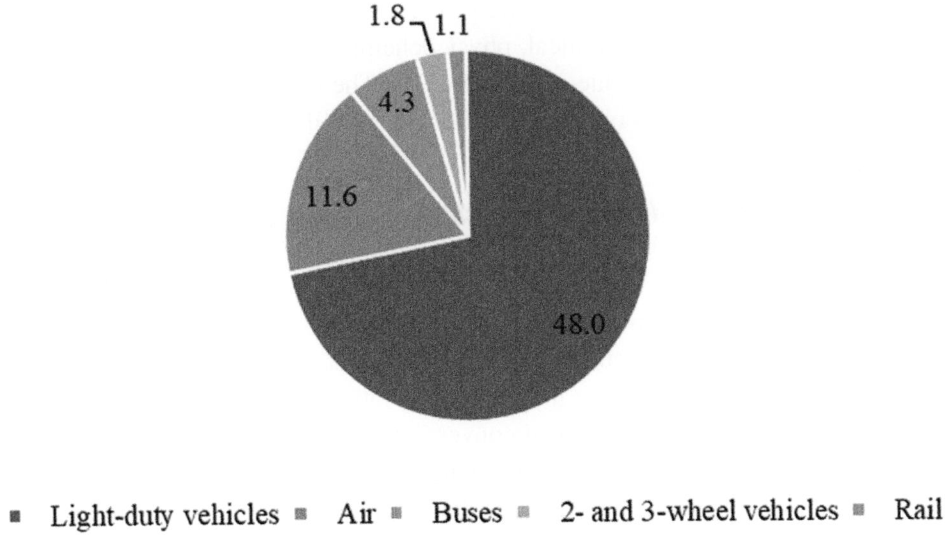

Figure 13.2

Global energy consumption by different transportation sector. *Modified from A report on "2018 Outlook for Energy: A view to 2040" from "ExxonMobil". Available from https://corporate.exxonmobil.com/en/energy-and-environment/energy-resources/outlook-for-energy/2018-outlook-for-energy-a-view-to-2040#aViewTo2040.pdf.*

contributor to global oil demand with ~60% of total oil growth [3]. China has emerging economy but also the main contributor in oil demand after India in Asia-pacific region. India has surpassed the oil demand of China and became rising contributor to incremental oil demand [6,7].

Rise in energy related CO_2 emissions in 2017 is a strong indication of climate change and illustrates that present attempts are insufficient to meet goals of 2015 Paris agreement. Fast depletion of oil reservoirs, constantly increasing energy (fuel and electricity) demand, fluctuating prices of crude oil and detrimental exhaust emissions from excess consumption of fossil fuels leading to environmental pollution have intensified the search for alternative energy sources globally. The environmental pollution has proven detrimental to human health and has also poised threat of climate change.

Use of biomass-derived fuels and chemicals is a potential solution to lower the dependency of fossil sources. The biorefinery concept can be considered similar to petroleum refinery, where biomass can be treated to obtain various value-added products with clean processes, which has a constructive impact on ecology [8]. Several technologies and processes were developed over last decade which utilize the wastes from agriculture residue, nonedible feedstock and food waste to yield range of biofuels, biochemicals as well as biomaterials including marketable biopolymers [9]. Waste biorefinery will assist in sustainable waste management with generation of renewable energy [8]. Commonly, the waste can be treated with thermochemical, physicochemical or biochemical techniques in biorefinery to convert them into value-added products. The most common method to convert this waste material to energy is thermal process [8,9].

13.2 Alternative methods for conversion of waste carbon source to energy/fuel

Conventionally, waste carbon can be originated from agricultural/forest waste or municipal solid waste. The most common method to convert this waste material to energy is thermal process. The common thermal process includes incineration, pyrolysis or gasification. Alternative to thermal processes biological conversion processes such as anaerobic digestion or fermentation were also used to produce biofuels. The basic principles of some common routes to convert waste to energy are discussed as follows.

Combustion: The oldest and frequently practice thermal process uses a wide variety of carbon source to produce energy in terms of heat. Recent advances in complete combustion of waste leads to various designs of incinerators which improve the efficiency of the process and lowers the energy consumption. The heat generated in this process commonly used to produce steam from water which can be converted into electricity. The modern design of incinerator will also incorporate the waste heat recovery units for further improvement of efficiency. The only drawback of this process is the ash produced after combustion needs proper processing and disposal, otherwise leads to pollution of the environment.

Gasification: Similar to incineration gasification can also use a wide range of waste material. The major difference between incineration and gasification is that, in incineration complete combustion occurs in presence of excess oxygen, whereas in gasification combustion of waste material occurs in limited presence of oxygen. Basically, gasification is a thermochemical process, which can convert the solid waste into mixture of numerous combustible gases. The most common use of gasification is production of ‘syngas’, which can be further used to produce various chemicals or electricity. The major hurdle in commercialization of gasification processes is obtaining same composition of producer gas from different kinds of feedstocks.

Pyrolysis: Pyrolysis is an another thermal process that can convert the waste carbon material into oil (called as bio-oil) and combustible gases in absence of oxygen. Recently pyrolysis technology has gained significant attention over the combustion and gasification techniques due to higher efficiency and more value-added products and byproducts. The pyrolysis process produces the fine carbon bio-char, which not only retains most of the micronutrients from waste biomass, but is also rich in carbon content. This bio-char can be used to enrich the fertility of soil.

Anaerobic digestion: Landfilling is a common process for handling the municipal solid waste. When the waste remain undisturbed for a long time generates numerous gases (such as CO_2, methane, etc.) through decomposition of organic matter in absence of oxygen known as anaerobic digestion. It can occur naturally in landfills or it can be targeted to occur in digester under control process conditions. Major advantage of anaerobic digestion is utilization of semi-solid or wet waste material, which cannot be processed through thermal processes. The product gas (biogas) can be burned in an engine to generate electricity or in a boiler to produce steam.

Fermentation: In a fermentation process numerous bacteria and yeast were used to convert waste carbon material in fuels such as bioethanol and biobutanol. Fermentation process needs a series of chemical reactions and processes to convert the waste biomass or agricultural waste to bioalcohols. The most common difficulties in fermentation route is the efficiency of the process and time required to convert the waste into energy.

The increasing global population will generate more and more waste and also increases the demand of energy. Thus, efficient technologies are needed which can solve the issue of waste generation as well as energy demand. As stated earlier, biodiesel is one of the most promising alternative to mineral diesel and can be used directly or in blending with biodiesel. Thus, the subsequent section deals with the identification of various organic carbon wastes and their conversion to value-added products (such as catalyst) and biodiesel using biorefinery concept.

13.3 Prospects of biodiesel production in waste biorefinery

As per the data published by Central Statistics Organization (Ministry of Statistics and Program Implementation, Government of India), the production of crude oil within India has hardly shown any growth in last 5 years (41 MMT or million metric tons from 2010 to 11 to 43 MMT in 2015–16) with continuously declining production of natural gas (44.3 Mtoe or million ton of oil equivalent in 2010–11 to 26.3 Mtoe in 2015–16) [10]. India's crude oil import rose to 202.85 MMT in 2015–16 with an annual rise of 7.08%, which not only has created an enormous burden on country's economy but also adversely affected the climate and environment, and has led to global challenges associated with sudden increment in greenhouse gas emissions [11]. Some of the important features of India's energy economics and security policy are as follows.

- To meet the continuous increasing demand for consumption of petroleum products and electricity without contributing excess CO_2 emissions, the energy economics and security policy of Government of India, has focused toward generation and utilization of alternate fuels, which are technically efficient, economically viable and environmentally sustainable.

- Renewable energy from nonconventional sources such as wind, solar (both photovoltaic and thermal), geothermal, hydropower and biomass have a great potential for generation of clean energy. Central Statistics Organization (Ministry of Statistics and Program Implementation, Government of India) estimates the total power generated through renewable energy sources as 1,198,856 MW (detail distribution is as depicted in Table 13.1 [10].
- Research and development in new technologies for efficient synthesis of biofuels from non–food feed stocks, should be promoted. Policy proposes to undertake intensive R&D in advanced conversion technologies for first-generation biofuels and emerging technologies for second-generation biofuels including conversion of lignocellulosic and nonfood grade feed stocks. The policy also emphasizes development of technologies for end-use applications including modification and development of engines for the transport sector based on large-scale centralized approach, and for nonstationary applications such as electricity production through decentralized approach.
- The policy identifies minimum support price (MSP) for oil seeds as major instrument to promote plantation of oil seeds by farmers. This will, of course, be decided after consultation with Government agencies, states, local bodies such as Gram Panchayat or Gram Sabha and other stakeholders. Policy also purposes extension of statutory minimum price (SMP) mechanism (similar to that for sugarcane procurement) for oil seeds to be purchased by biodiesel processing units.

Transport sector has the highest share among all major energy consumption sectors. The projected demand for diesel and petrol will rise to 110 MMT and 31.1 MMT, respectively by the year 2021–22 from the current demand of 80.4 MMT and 26.1 MMT in 2017–18 [11]. Thus, biofuels provide a higher degree of national energy security by reducing dependence on imported fossil fuels and help to meet the energy needs of India's massive

Table 13.1: Category-wise estimated potential of renewable power in India 2017–18.

Renewable energy source	Estimated power (in MW)	% Distribution
Wind power (both @ 80 m and @100 m)	405,023	33.78
Small-hydro power (SHP)	19,749	1.65
Biomass power	17,538	1.46
Bagasse-based cogeneration	5000	0.42
Waste to energy	2556	0.21
Solar power	748,990	62.48
Total	1,198,856	100.00

Based on A report on "Energy Statics" from "Central Statistics Office Ministry of Statistics and Programme Implementation, Government of India". Available from http://www.mospi.nic.in/sites/default/files/publication_reports/Energy_Statistics_2017r.pdf.pdf.

population. Planning Commission, Government of India, has proposed 20% biofuel blending target for both bioethanol and biodiesel to be achieved in 12th five-year plan by the end of 2022 [10].

Biofuels are envisaged as carbon neutral fuels that do not contribute net carbon to atmosphere and ultimately improve air quality [12]. Bioethanol and biodiesel have been studied extensively among other biofuels such as biobutanol, biohydrogen, biomethane, etc. due to ease of blending with conventional fuel and compatible with existing engine configuration [13]. World's biofuel production has increased from 27.9 Mtoe in 2006 to 82.3 Mtoe in 2016 with an average 14% annual growth rate [3].

Biodiesel basically is a mixture of $C_{12}-C_{22}$ fatty acid monoalkyl esters (FAMEs), and it has several advantages over conventional diesel such as, biodegradable, sulfur-free, sustainable with higher lubricity, flash point etc. [14,15]. It is an alternative to petroleum diesel for reducing emissions of gaseous pollutants such as CO_2, SO_x, particulate matter and organic compounds. It can be produced from wide range of available feedstocks (more than 350 oil-bearing crops, waste cooking oil, animal fat, algal oil, etc.) making it one of the popular liquid biofuels worldwide [16].

13.4 Waste carbon sources for biodiesel production

The availability and type of the feedstock is the major contributor to the overall cost of the biodiesel production [17]. Due to food security and higher cost of edible oils, countries like India, edible crops cannot be diverted for biodiesel production. Nonedible oil seeds contain 10%–40% oil which can be extracted using different techniques such as solvent extraction, mechanical expeller, etc. These nonedible oil crops have not grown by comprising food crops on fertile land, their availability is a major issue. Alternatively, the waste cooking oil, animal fat (waste tallow and lard) and waste grease can be used as feedstock for biodiesel production. Thus, the feedstock for biodiesel production has been classified under first-, second-, and third-generation feedstock.

First-generation feedstock for biodiesel production: Transesterification in earliest years (~1930) were carried out using edible oils such as soybean or palm oil, and thus, edible oils are considered as first-generation feedstock for biodiesel production. Edible oils (rapeseed and soybean) are commonly used for biodiesel production in Europe and Latin America. Whereas, Philippines and Malaysia are also utilizing edible oils such as coconut oil and palm oil, respectively, for biodiesel production. The edible oils such as soybean, rapeseed, palm, coconut, sunflower and linseed oils, etc. are exploited commercially by some developed countries as raw material for biodiesel production [18].

Second-generation feedstock for biodiesel production: Countries like India, where food security is on priority, edible oils cannot be diverted for fuel production. Also, the higher cost of first-generation feedstock will lead to uneconomical production of biodiesel, thus second-generation feedstock is a feasible alternate for large-scale biodiesel production. Nonedible oils have toxic substances, which restrict their use for human and animal consumption. Thus, nonedible oils with higher free fatty acids are used as feedstock for biodiesel production and considered as second-generation feedstock. Nonedible oils such as *Jatropha curcas* (Jatropha), *Ficus elastica* (rubber), *Camelina sativa* L. Crantz (Camelina), *Madhuca indica* (mahua), *Pongamia pinnata* (karanja), *Nicotina tabacum* (tobacco), *Calophyllum inophyllum* (polanga) etc., are explored as potential raw material for biodiesel production [18,19].

Third-generation feedstock for biodiesel production: In order to reduce the cost of biodiesel production, micro- and macroalgae oils have also been used by researchers as a source of feedstock for the production of biodiesel. Therefore, micro- and macroalgae oils are categorized as third-generation of feedstock for biodiesel production. Due to higher biomass production, faster growth and photosynthetic efficiency of microalgae as compared to other energy crops could be a potential source of low cost feedstock for large-scale production of biodiesel [17,19]. However, Knothe and Razon [13] have cautioned on source of microalgae (or cyanobacteria) as feedstock for biodiesel production. The microalgal lipids have high content of polyunsaturated fatty acids (PUFAs) which make their oxidative stability weak. The life-cycle assessment (LCA) studies of microalgae have revealed greater nitrogen fertilizers and energy consumption as compared to terrestrial plants. The postharvest treatments such as lipid extraction, purification and partial hydrogenation prior to transesterification also raises the production cost. All of these factors adversely affect the economic feasibility of microalgae–based biodiesel production. Thus many technological challenges need to be overcome before commercialization of this technology.

The percentage of raw materials utilized for commercial biodiesel production in the world are rapeseed oil (84%), sunflower oil (13%), palm oil (1%), soybean oil, and others (2%), which include second- and third-generation feedstock [18]. Looking at the data, the nonedible oil sources and waste triglycerides (waste cooking oil, grease, tallow, lard, etc.) has not explored as per their capability. The probable cause underlying for this negligence may be either consistent availability of these sources for biodiesel production at large scale or lack of research and development in converting these processed into commercial technologies. In following sections, we have attempted to explore the various possibilities and opportunities for biodiesel production using nonedible oils as well as waste carbon sources either in terms of feedstock or in terms of catalyst preparation.

Knothe and Razon [13] have compiled a comprehensive collection of alternate terrestrial plant sources for biodiesel, which has been reproduced in Table 13.2.

Table 13.2: Engine tested alternative terrestrial plant sources for biodiesel production.

Common name/seed name or plant part	*Scientific name*	Family
Apricot	*Prunus armeniaca*	Rosaceae
Beach almond	*Terminalia belerica*	Combretaceae
Camelina	*Camelina sativa*	Brassicaceae
Caper Spurge	*Euphorbia lathyris*	Euphorbiaceae
Castor	*Ricinus communis*	Ricinus
Chinese parasol	*Firmiana platanifolia*	Malvaceae
Chinese pistache	*Pistacia chinensis*	Anacardiaceae
Desert Date	*Balanites aegyptica*	Zygophyllaceae
Ethiopian mustard	*Brassica carinata*	Brassicaceae
Hazelnut	*Corylus avellana*	Betulaceae
Jatropha	*Jatropha curcas*	Euphorbiaceae
Jojoba	*Simmondsia chinensis*	Simmonsiaceae
Karanja	*Pongamia pinnata*	Fabaceae
Kuntze tea	*Camelia sinensis*	Theaceae
Kusum	*Schleichera oleosa*	Sapindaceae
Linseed	*Linum usitatissimum*	Linaceae
Mahua	*Madhuca indica*	Sapotaceae
Manchurian apricot	*Prunus mandshurica*	Rosaceae
Moringa	*Moringa oleifera*	Moringaceae
Mukinduri	*Croton megalocarpus*	Euphorbiaceae
Neem	*Azadirachta indica*	Meliaceae
Niger	*Guizotia abyssinica*	Asteraceae
Olive pomace	*Olea europaea*	Oleaceae
Paradise tree	*Simarouba glauca*	Simaroubaceae
Pilu	*Salvadora oleoides*	Salvadoraceae
Polanga	*Calophyllum inophyllum*	Clusiaceae
Poon	*Sterculia foetida*	Malvaceae
Radish, Turnip	*Raphanus sativus*	Brassicaceae
Ribbed melon	*Hodgsonia macrocarpa*	Cucurbitaceae
Rice bran	*Oryza sativa*	Poaceae
Rocket	*Eruca sativa*	Brassicaceae
Rubber	*Hevea brasiliensis*	Euphorbiaceae
Safflower	*Carthamus tinctorius*	Asteraceae
Sesame	*Sesamum indicum*	Pedaliaceae
Siberian apricot	*Prunus sibirica*	Rosaceae
Terebinth	*Pistacia terebinthus*	Anacardiaceae
Tobacco	*Nicotiana tabacum*	Solanaceae
Weeping forsythia	*Forsythia suspense*	Oleaceae
Yellow horn	*Xanthoceras sorbifolia*	Sapindaceae

Adopted from Knothe G, Razon LF. Biodiesel fuels. Progress in Energy and Combustion Science 2017;58:36–59 with permission of copyright @ 2017 Elsevier BV.

13.5 *Waste carbon-based catalysts for biodiesel production*

Conventionally, biodiesel was produced using edible oil reacting with short chain primary or secondary monohydric aliphatic alcohols having one to eight carbon atoms in presence of homogenous base catalyst such as NaOH/KOH [20]. The process occurs in three consecutive steps, where in each step one molecule of alcohol get consumed and one molecule of fatty acid ester is formed as product. The detail step wise reaction scheme is as shown in Fig. 13.3. The nonedible oil/waste cooking oil or animal fat has higher free fatty acid (FFA) content, which in presence of NaOH/KOH react with methanol to form soap instead of FAME. Thus, to lower the FFA content of the feedstock, homogeneous acid catalyst (e.g. conc. H_2SO_4) was used [15]. The chemical reaction involved in reducing FFA content of the feedstock is called as esterification reaction. The general reaction scheme for esterification process is as shown in Fig. 13.3. Besides H_2SO_4 other acids were applied for esterification reactions are HF, HCl, H_3PO_4, p-toluensulfonic acid, etc. These acids have shown high effectiveness in reduction of FFA content of nonedible and waste oils, but suffers with severe concern of corrosion of reactor as well as required excess water to remove them at the end of reaction. In recent years, endeavors were made to overcome these issues with commercialization of large-scale biodiesel production processes from nonedible oils using heterogeneous catalysts. Some of these heterogeneous acid catalysts are as follows: sulfated metal oxides, mesoporous silica, modified zeolites, metal organic framework (MOF) structures, ion exchange resins, polymer supported sulfonic groups, carbon-based supports with functionalized acid groups, etc. [21,22,23,24,25,26,27,28,29].

Among these various catalysts, carbon-based heterogeneous acid catalysts have shown a great potential as a sustainable solid acid catalyst as replacement for homogeneous acid catalysts as well as costlier and complex heterogeneous acid catalysts. Carbon-based (acid/base) catalysts have advantages of being environmentally friendly and help in waste utilization with economical production of biodiesel. These catalysts are basically divided in two broad categories based on the carbon source selection: (1) organic matter or carbohydrate, and (2) agriculture waste.

Organic matter: These catalysts were prepared using carbohydrates materials such as sucrose, glucose, starch, cellulose, etc. The incomplete carbonization in presence of strong acids will result in polycyclic aromatic sheets material which has strong thermal stability and highly stable acidic sites.

Agricultural waste: Catalyst prepared by sulfonation and/or calcination at high temperature of agricultural waste or residue such as rice husk, de-oiled seed cake, fruit peels etc. as starting material. These catalysts have utilized the waste generated in biorefinery and solve the problem of waste management to some extent. Table 13.3 presents the summary of published literature in the area of carbon-based catalyst and their application in biodiesel industry.

It can be seen from Table 13.3, that very few researchers have explored the area of catalyst preparation using waste carbon as a starting/supporting material. Most of these

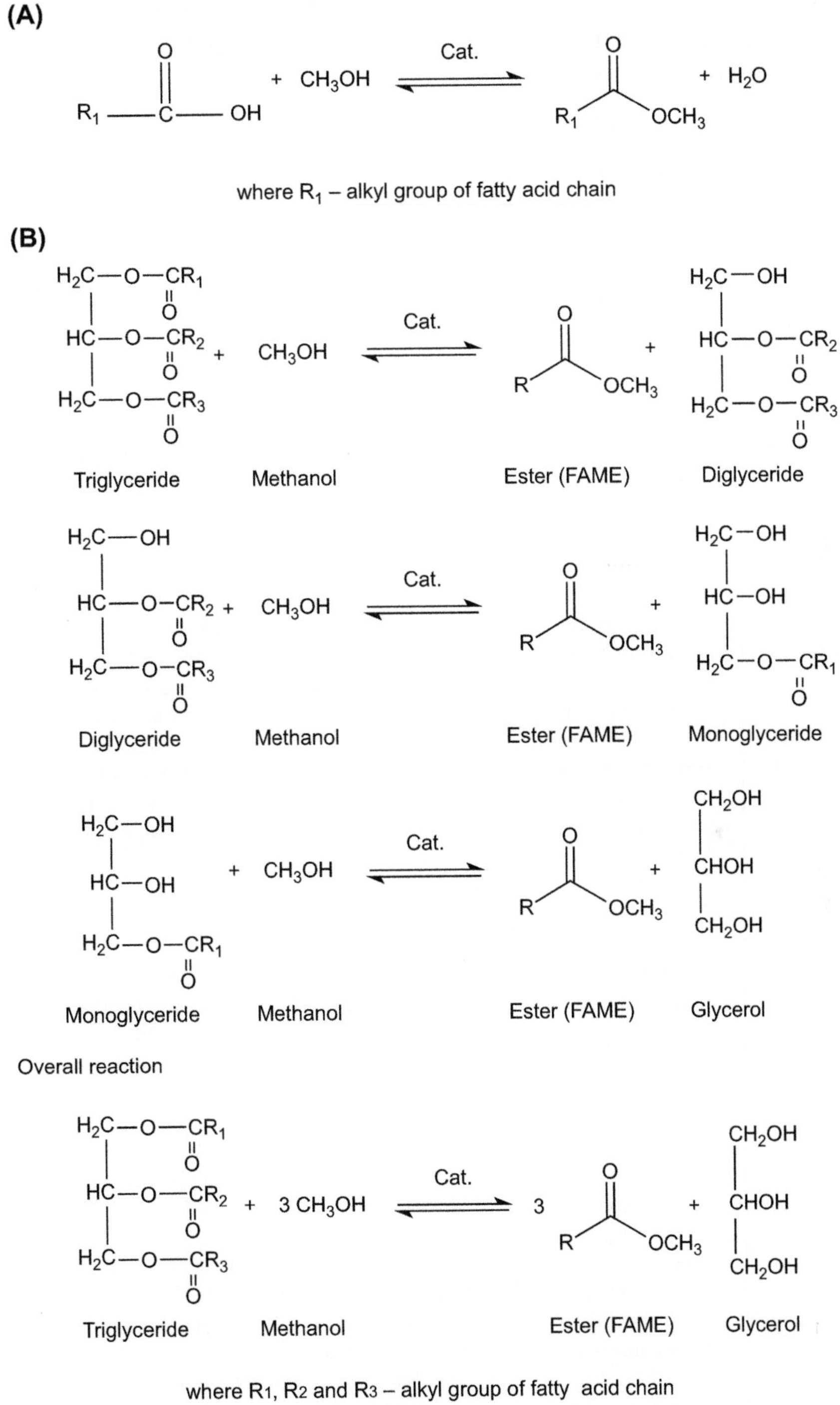

Figure 13.3
(A): Reaction scheme of esterification of FFA using methanol. (B) Reaction Scheme of transesterification reaction of triglyceride using methanol.

catalysts were prepared by incomplete carbonization followed by sulfonation of carbonized material. Few authors have used hydrothermal or impregnation method for preparation of catalyst.

Table 13.3: Literature summary of carbon-based catalyst for biodiesel production.

Support/starting material	Preparation method	Application	Yield	References
D-glucose	Incomplete carbonization followed by sulfonation	Esterification of oleic acid and transesterification of waste oils	95% for esterification; 90% for transesterification	[30]
Vegetable oil and petroleum asphalt	Sulfonation of carbonized vegetable oil/petroleum asphalt	Esterification of oleic acid and transesterification of waste oils	>80% for both esterification and transesterification	[31]
Microcrystalline cellulose powder	Incompletely carbonization followed be sulfonation	Esterification of oleic acid and transesterification of triolein	100% for esterification and 98% for transesterification	[32]
Rise husk	Carbonization followed by sulfonation	Esterification of glycerol with acetic acid	90% glycerol conversion into triglycerides	[33]
De-oiled seed waste cake of *J. curcas, P. pinnata and M. ferrea* L.	Radical sulfonation/ direct sulfonation/ hydrothermal sulfonation	Esterification of oleic acid	>97% oleic acid conversion with all three types of catalyst	[34]
Activated carbon	Wet impregnation with Tungstophosphoric acid	Transesterification of crude Jatropha oil	87.33% biodiesel yield	[35]
Cyclodextrin	One-step hydrothermal carbonization	Transesterification of waste cooking oil	90.8% transesterification yield	[36]
Musa acuminata peel	Completely burning to prepared activated ash	Transesterification of soybean oil	98.95% biodiesel yield	[37]
Glucose or *C. inophyllum* seed cake	Incompletely carbonization followed be sulfonation	Transesterification of nonedible seed oil (*Calophyllum inophyllum*)	99% conversion of high FFA content oil	[38]
Activated carbon	Wet impregnation with potassium acetate	Transesterification of bitter almond oil	93.21% biodiesel yield from bitter almond oil	[39]
Rubber de-oiled cake	Direct sulfonation followed by calcination	Transesterification of blended nonedible oil	Biodiesel yield of 91.2% in single-step process and 93.7% in two-step process	[40]

Zong et al. [30] first reported the preparation and use of sugar catalyst for biodiesel production. Authors used D-Glucose as starting material, heated the powder glucose for 15 h at 400°C under inert atmosphere to completely carbonized the glucose. The obtained material was ground into fine powder and treated with conc. sulfuric acid at 150°C for 15 h under inert atmosphere of nitrogen. The final black product was washed several times with distilled water to remove unreacted acid. The final catalyst was dried at 60°C under vacuum and tested for its catalytic activity in esterification as well as transesterification. The esterification of oleic acid showed that 95% conversion was achieved whereas in transesterification of waste oil ~90% biodiesel yield was obtained at the end of 5 h reaction. Authors found that the prepared catalyst has higher acidic ($-SO_3H$) active sites than sulfonated zirconia as well as amberlyst-15.

Shu et al. [31] have used the vegetable oil or petroleum asphalt for preparation of carbon-based heterogeneous acid catalyst. Authors obtained the vegetable oil asphalt from commercial biodiesel plant and treat it to remove any water and ester present in it. Further, the treated asphalt was oxidized and then sulfonation was carried out using conc. sulfuric acid at 210°C for 10 h under reflux conditions. The prepared catalyst was washed with de-ionized water and the filtrate was dried at 120°C under vacuum to remove the moisture. The catalyst was characterized using various techniques to determine surface properties. The activity of catalyst was tested in transesterification of cotton seed oil and oleic acid mixture (50% each by volume). The results showed that the catalyst can convert over 80% cotton seed oil and oleic acid into products.

Nakajima and Hara [32] have synthesized the carbon based heterogeneous acid catalyst using microcrystalline cellulose powder. Authors used incomplete carbonization method followed by sulfonation to prepare the catalyst. The structure of active sites of synthesized catalyst is as shown in Fig. 13.4. Authors studied the effect of acid/support ratio on surface properties. Authors have also attempted to correlate the results of transesterification of triolein with the structural properties of various prepared catalyst. Authors found that maximum 98% triolein conversion was achieved at the end of 4 h reaction and complete conversion of oleic acid can be obtained with same operating conditions. Authors also tested the catalyst for reusability study and found that catalyst can be recycled up to five cycled without losing significant activity. The performance of catalyst was found to be superior than commercial catalyst such as Amberlyst-15, Nafion NR50, and Nafion SAC-13.

Konwar et al. [34] prepared the three different carbon-based catalyst using three different methods. Authors used de-oiled waste seed cake of *J. curcas, P. pinnata*, and *M. ferrea* L as a starting material. The catalysts were prepared using radical sulfonation/direct sulfonation and hydrothermal sulfonation method and compared for

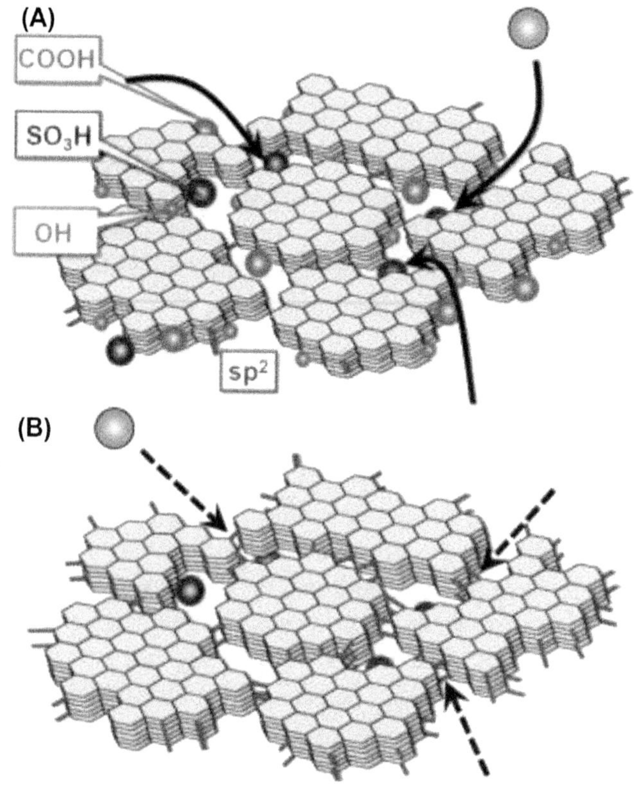

Figure 13.4

Schematic structures of proposed SO3H-bearing CCSA materials carbonized at different temperatures: (A) carbonized below 450°C and (B) carbonized above 550°C. *Adapted from Nakajima K, Hara M. Amorphous carbon with SO_3H groups as a solid Brønsted acid catalyst. ACS Catalysis 2012;2(7):1296–1304 with permission of copyright @ 2012 American Chemical Society.*

their surface properties. The chemical reactions involved in preparation of catalyst are shown in Fig. 13.5. Authors tested these catalysts for esterification of oleic acid and found that catalyst prepared using radical sulfonation method has slightly higher conversion with superior surface properties such as BET surface area, $-SO_3H$ density, etc. Authors also compared their results with published literature of heterogeneous acid catalysts from noncarbon supports as well as commercial catalysts and found that catalyst prepared by them has excellent properties than the many previously reported catalysts.

Phatak et al. [37] developed the waste-derived catalyst from banana peels. Authors collected the banana peels, washed it to remove dirt and other impurities from it. The washed peels were dried to remove moisture and further cut into small pieces. Authors completely burned these peels to get activated carbon as catalyst. Authors analyzed the

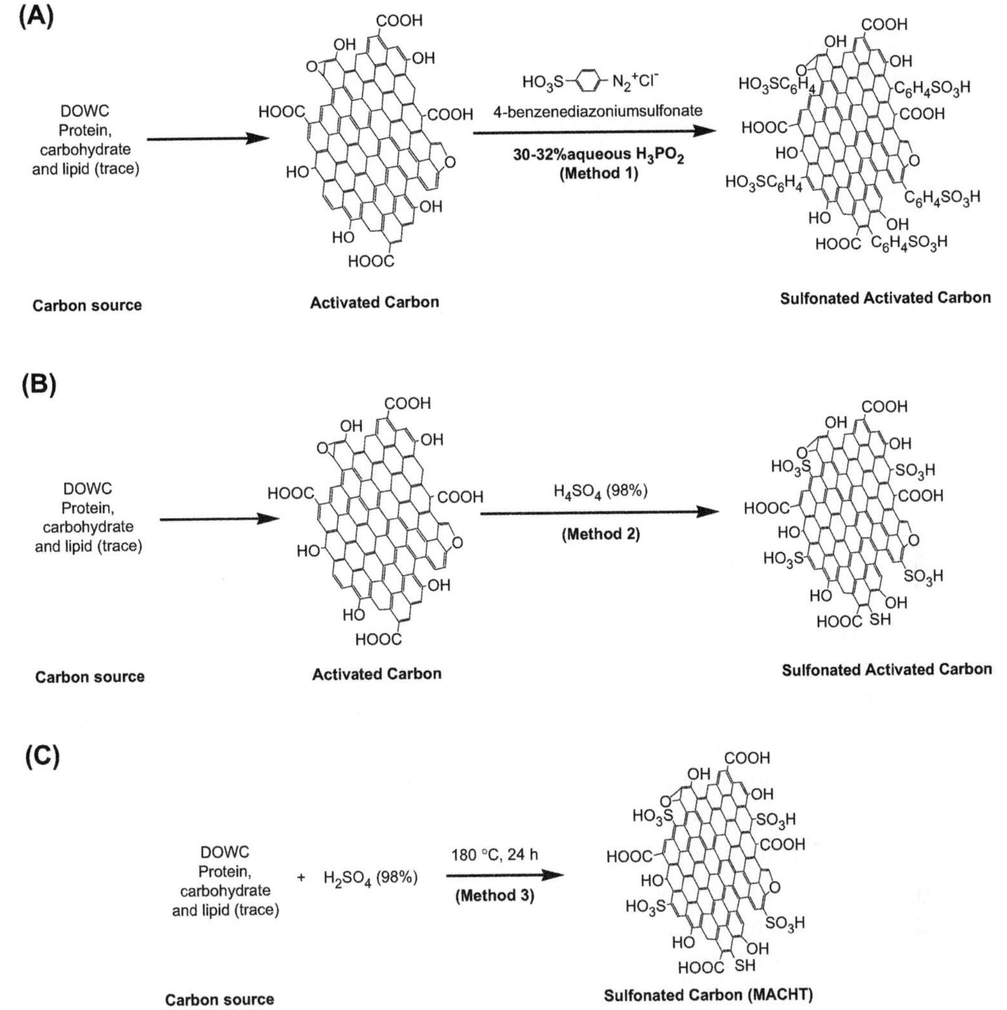

Figure 13.5

Reaction schemes in preparation of carbon catalyst (A) radical route; (B) direct sulfonation route and (C) one-step hydrothermal route. *Adopted from Konwar LJ, Mäki-Arvela P, Salminen E, Kumar N, Thakur AJ, Mikkola JP, Deka D. Towards carbon efficient biorefining: multifunctional mesoporous solid acids obtained from biodiesel production wastes for biomass conversion. Applied Catalysis B: Environmental 2015;176:20–35 with permission of copyright @ 2015 Elsevier BV.*

activated carbon obtained from burning and found that various metals were present which can catalyze transesterification reaction. Authors tested this catalyst for transesterification of soybean oil using methanol and found that >98% biodiesel yield can be obtained at the end of 4 h reaction. Authors also found that the catalyst can be recycled up to four cycles. Authors conclude that the waste generated can be converted in environment friendly catalyst with simple preparation techniques.

Recently, Malani et al. [40] have reported synthesis of heterogeneous acid catalyst from rubber de-oiled seed cake using two different acids. Authors prepared the heterogeneous carbon catalyst using conc. sulfuric acid and conc. chlorosulfonic acid through direct sulfonation followed by calcination process. The catalyst was characterized with different analytical techniques. The (FE-SEM) micrographs of the synthesized catalysts are shown in Fig. 13.6. The de-oiled cake treated with chlorosulfonic acid had superior surface properties and active sites as compared to sulfuric acid treated de-oiled cake.

Authors used chlorosulfonated catalyst for transesterification of mixed nonedible oil in single as well as two-step process. The results showed that 91.2% and 93.7% triglyceride conversion was achieved in single and two-step process at the end of 3 and 1 h reaction, respectively. Authors also studied the reusability study and found that catalyst can be recycled up to three successive cycles without significant changes in activity. Authors also found that two-step process was more beneficial over single-step process and minimum denaturing of catalyst was observed in two-step process.

In summary, the waste-derived catalysts have shown promising results as reported by several authors. These catalysts help in utilization of waste generated in biorefinery. The cost of catalysts is usually a significant component of overall production cost of the process, as catalysts are expensive components of process. Moreover, frequent replacement of the catalysts due to deactivation during the process is also not feasible. Waste carbon-based catalysts offer a solution to these issues. These catalysts are not only much cheaper than conventional catalysts – but also can be manufactured on–site (within biorefinery itself) instead of importing it from outside. Thus, further research and development on this matter is crucial to commercialization of biodiesel processes and making the economy of such processes attractive to entrepreneurs.

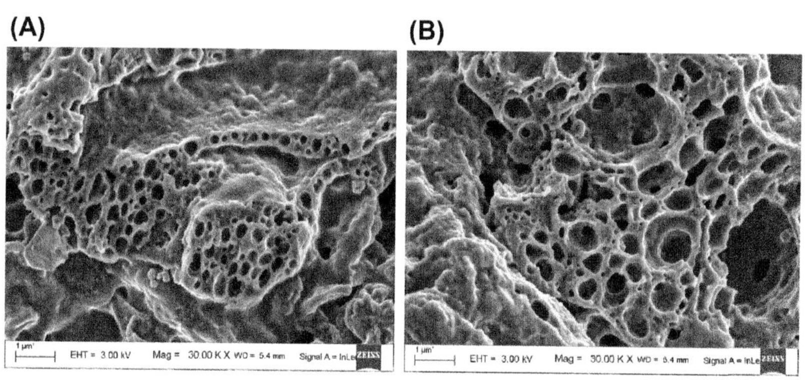

Figure 13.6
FE-SEM micrograph images of synthesized carbon catalyst (A) sulfuric acid treated and (B) chlorosulfonic acid treated.

13.6 Opportunities/advantages of using mixed feedstocks for biodiesel and case studies

As stated earlier, the potential of nonedible as well as waste oils are not discovered fully. The major problem underlying the utilization of these feedstocks is their continuous availability and supply throughout the year. Extensive research has been carried out by researchers to explore the various sources of nonedible oil as well as fatty acid or triglyceride from waste sources. The short summary of these studies in recent years has been presented here in tabulated form as in Table 13.4.

Form literature presented in Table 13.4, it can be seen that most previous studies have used jatropha, karanja, rubber, and waste cooking oil, as compared to other nonedible and waste sources. Even India's National biodiesel mission has focused on biodiesel derived from *Jatropha curcus* oil, showing the potential of Jatropha at large-scale production [10,11]. But at the same time, Jatropha is herb and due to its toxicity and other demerits it cannot be grown by comprising the other edible/food crops, thus continuous availability of single seed oil is tuned-up as major issue in many Jatropha based biodiesel plants in India.

The feasible solution to improve the economy of biodiesel production and utilization of waste and nonedible oil sources is to utilize mixed or blended oil feedstock and modify the process accordingly. This will help in continuous production of biodiesel at large scale through short supply of single feedstock. This will also help in generation of employment locally, who can collect and supply these feedstocks to biodiesel production units. Commonly, blending of two or more nonedible oils or waste oil sources solve the problem to great extent and also helps to treat the waste generated from branded restaurants, poultry farms, meat shops, etc. Thus, blended feedstock for biodiesel production through waste biorefinery model is an attractive alternative to conventional process. These routes are not explored fully and need major research and development before converting these processes in commercial techniques.

13.7 Case studies for biodiesel production using mixed nonedible and waste oils

Mixed feedstocks using waste resources for biodiesel production have been used by few researchers in recent years. Use of mixed feedstock makes process flexible and adds to its viability, as compared to processes employing a single specific feedstock. Given below is a summary of few studies that report biodiesel production and characterization using blends of nonedible and waste oils.

(1) Yogish et al. [55] have used a blend of nonedible Jatropha and Karanja oils as feedstock for transesterification. Transesterification process was carried out in 2 L lab scale

Table 13.4: Literature summary of biodiesel production using various nonedible and waste oils.

Oil (source)	Catalyst	Molar ratio (methanol/oil)	Catalyst loading (w/w)	Reaction temperature (K)	Time (min)	Mode of mixing	% FAME (yield)	References
Mustard oil	KOH	6:1	0.75%	333	45	Mechanical agitation 600 rpm	95.54	[41]
Rubber seed oil	$Ba(OH)_2 \cdot 8H_2O$	7.5:1	4%	338	10	Ultrasound 30 kHz/100 W	94.5	[42]
Karanja oil	KOH	6:1	1%	328	30	Mechanical agitation 600 rpm	95.5	[43]
Pongamia oil	CaO	15:1	12%	338	180	Mechanical agitation 600 rpm	97.28	[44]
Papaya seed oil	KOH	10:1	1%	318	60	Mechanical agitation 600 rpm	96.48	[45]
Salvadora alii oil	CaO	10:1	3%	338	30	Ultrasound 20 kHz/100 W	92	[46]
Thespesia populneoides oil	CaO	6:1	3.5%	338	30	Ultrasound 20 kHz/100 W	88.6	[46]
Calophyllum inophyllum oil	Cellulose based solid acid catalyst	15:1	5%	453	240	Mechanical agitation 600 rpm	99	[47]
Nahor oil	Li doped egg shell derived CaO	10:1	5%	338	240	Mechanical agitation 900 rpm	94	[48]
Distaff thistle oil	NaOH	5:1	0.64%	333	120	Mechanical agitation 500 rpm	97	[49]
Waste cooking oil	Montmorillonite clay K-30	12:1	3%	363	180	Mechanical agitation 200 rpm	78.4	[50]
Mesua ferrea oil	Co doped ZnO	9:1	2.5%	333	180	Mechanical agitation 600 rpm	98.03	[51]
Waste cooking oil	Sulfonated carbon catalyst from cyclodextrin	16:1	11.5%	390	8.8	Ultrasound 25 kHz	90.8	[36,52]
Waste cooking oil	Hydrotalcite	15:1	0.08 g/g	330	60	Ultrasound 20 kHz/11 W	76.45	[53]
Waste cooking oil	Coal fly ash	10.71:1	4.97%	333	1.41	Ultrasound 20 kHz/100 W	95.57	[54]

Kusum oil	$Ba(OH)_2$	9:1	3%	323	80	Ultrasound 20 kHz/ 250 W	96.8	[61]
Karanja oil	$Ba(OH)_2$	9:1	5%	303	60	Ultrasound 30 kHz/ 100 W	83.87	[61]
Palm oil	SrO/Al_2O_3	9.2:1	1.6%	333	30.2	Ultrasound 20 kHz/ 200 W	80.2	[62]
Pistacia khinjuk seed oil	Sulfated tin oxide impregnated with silicon dioxide	13:1	3%	338	50	Ultrasound 20 kHz/ 100 W	88	[63]
Waste cooking oil	Calcium diglyceroxide	9:1	1%	333	30	Ultrasound 22 kHz/ 120 W	93.5	[64]
Crude palm oil	Fly ash on CaO	12:1	4%	318	30	Ultrasound 20 kHz/ 700 W	97.04	[65]
Xanthium sibiricum Patr oil	KOH	10:1	0.8%	338	60	Mechanical agitation 500 rpm	98.7	[66]
Jatropha oil	CaO	8:1	2%	338	78	Mechanical agitation 600 rpm	90	[67]
Calophyllum inophyllum oil	Sulphonated carbon catalyst	30:1	10%	453	300	Mechanical agitation 600 rpm	99	[38]
Yellow oleander oil	MgO	5:1	0.2%	363	120	Mechanical agitation 600 rpm	93.1	[68]
Crude Jatropha oil	$Na_2SiO_3@Fe_3O_4/C$	7:1	5%	328	80	Ultrasound 25 kHz/36 W	94.7	[69]
Sesame oil	$Ba(OH)_2$	6.69:1	1.79%	305	40.3	Ultrasound 20 kHz/ 1200 W	98.6	[70]
Waste cooking oil	$Ba(OH)_2$	6:1	0.75%	333	2	Ultrasound 25 kHz/ 200 W	83.5	[71]
Palm oil	Ostrich eggshell-derived CaO	9:1	8%	333	60	Ultrasound 20 kHz/ 120 W	92.7	[72]
Milk thistle oil	TiO_2 doped with $C_4H_4O_6HK$	16:1	5%	333	30	Ultrasound 40 kHz/ 250 W	90.1	[73]
Crude Jatropha oil	Boiler scale deposits	12:1	8%	338	180	Mechanical agitation 600 rpm	82.85	[74]
Bitter Almond oil	Potassium acetate supported on activated carbon	9:1	2.5%	338	150	Mechanical agitation 600 rpm	93.21	[39]
Silybum marianum L. seed oil	KOH	6:1	0.9%	333	100	Mechanical agitation 600 rpm	95	[75]

Continued

Table 13.4: Literature summary of biodiesel production using various nonedible and waste oils.—cont'd

Oil (source)	Catalyst	Molar ratio (methanol/oil)	Catalyst loading (w/w)	Reaction temperature (K)	Time (min)	Mode of mixing	% FAME (yield)	References
Bitter almond oil	NaOH/KOH	6:1	0.75%	323	60	Mechanical agitation 700 rpm	97.75	[76]
Bitter almond oil	Potassium acetate impregnated CaO	9:1	2%	333	120	Mechanical agitation 600 rpm	91.22	[77]
Waste cooking oil	Smoke deposited nano MgO	5:1	1.5%	328	45	Ultrasound 24 kHz/200 W	98.7	[78]
Crude Jatropha oil	CaO	11:1	5.5%	337	60	Ultrasound 35 kHz/35 W	96	[79]
Palm fatty acid distillate	Sulfonated cellulose	6:1	3%	333	180	Ultrasound 20 kHz/120 W	81.2	[80]
Waste cooking oil	Tripotassium phosphate	6:1	3%	323	90	Ultrasound 22 kHz/375 W	92	[81]
Crude Jatropha oil	Carbon-supported heteropoly acid	20:1	4%	323	60	Ultrasound 20 kHz/400 W	87.33	[35]
Crude Jatropha oil	Cesium doped heteropoly acid	25:1	3%	327	34	Ultrasound 20 kHz/400 W	90.5	[82]
Castor oil/Pongamia oil/Neem oil	Cs-tungstosilicic acid/Zr-KIT-6	27:1	7%	393	300	Mechanical agitation	98	[83]
Brucea javanica seed oil	NaOH	6:1	1%	338	300	Mechanical agitation	94.34	[84]
Ailanthus altissima seed oil	KOH	8.5:1	1.01%	323	4.71	Ultrasound 24 kHz/400 W	92.26	[85]
Karanja oil	KOH	9:1	1.13%	-	6.4	Microwave irradiation 180 W	90.14	[86]
Terebinth oil	H_2SO_4	20:1	2%	323	720	Mechanical agitation	77	[87]
Chlorella protothecoides micro-algal oil	KOH	8:1	0.5%	333	60	Mechanical agitation 400 rpm	98	[88]
Karanja oil	KOH	6:1	1.5%	333	90	Ultrasound 20 kHz/120 W	92	[89]
Sea mango oil	Sulfated zirconia	12:1	1%	423	180	Mechanical agitation	94.1	[90]
Alexandrian Laurel kernel oil	NaOH	9:1	0.8%	333	75	Mechanical agitation 1000 rpm	97.14	[91]

Waste cooking oil/Jatropha oil	Activated carbon	6:1	1%	333	120	Mechanical agitation 400 rpm	93.95/ 93.27	[92]
Crude Jatropha oil	Ca–Mg mixed oxide	25:1	3%	393	180	Mechanical agitation	90	[93]
Babassu oil	KOH	6:1	1%	303	10	Ultrasound 20 kHz/ 600 W	97%	[94]
Crude Jatropha oil	Bi_2O_3–La_2O_3	15:1	2%	423	240	Mechanical agitation 400 rpm	93	[95]
Castor oil	Mussel shell	6:1	2%	333	180	Mechanical agitation	91.17	[96]
Crude Jatropha oil	CaO–MgO mixed oxide	38.67:1	3.7%	389	200	Mechanical agitation 500 rpm	93.55	[97]
Aleurites trisperma oil	NaOH	6:1	1%	333	60	Mechanical agitation	96.62	[98]
Crude Jatropha oil	KOH	5:1	0.75%	318	7	Ultrasound 24 kHz/ 200 W	97.63	[99]
Chlorella protothecoides micro-algal oil	KOH	8:1	0.92%	333	90	Mechanical agitation 100 rpm	97.25	[100]
Annona diversifolia seed oil	KOH	12:1	0.5%	338	120	Mechanical agitation	91.9	[101]
Waste cooking oil	Hydrocalumite	6:1	7%	338	300	Mechanical agitation	97	[102]
Aamla oil	F–Ca (magnetic oxide)	6:1	2.5	333	120	Mechanical agitation 600 rpm	90.31	[103]
Crude Jatropha oil	Nano-architectured $Ca(OCH_3)_2$/ Activated carbon	12:1	3%	333	60	Mechanical agitation 600 rpm	98.65	[104]
Ceiba Pentandra oil	KOH	6:1	1%	338	45	Mechanical agitation 600 rpm	99.5	[105]
Crude Jatropha oil	CaO–La_2O_3	25:1	3%	433	180	Mechanical agitation	98.76	[106]
Nagchampa oil	KOH	6:1	1%	313	40	Ultrasound 20 kHz/ 120 W	92.6	[107]
Palm oil	CaO	15:1	6%	338	240	Mechanical agitation 600 rpm	93.5	[108]

Continued

Table 13.4: Literature summary of biodiesel production using various nonedible and waste oils.—cont'd

Oil (source)	Catalyst	Molar ratio (methanol/oil)	Catalyst loading (w/w)	Reaction temperature (K)	Time (min)	Mode of mixing	% FAME (yield)	References
Crude Jatropha oil	$Ca(OCH_3)_2$	15:1	2%	338	90	Mechanical agitation 900 rpm	95	[109]
Silybum marianum oil	TiO_2 doped with $C_4H_4O_6HK$	16:1	5%	333	30	Ultrasound 40 kHz/250 W	90.1	[110]
Crude Jatropha oil	CaO	5.15:1	2%	333	133	Mechanical agitation 500 rpm	98.54	[111]
Karanja oil	Cement waste	30:1	4%	333	180	Mechanical agitation 1000 rpm	80	[112]
Kusum oil	NaOH	9:1	0.9%	332	58.5	Mechanical agitation	97.37	[113]
Crude Jatropha oil	Al-SBA-15	12:1	3%	453	1440	Mechanical agitation 600 rpm	99.8	[114]
Madhuca indica oil	KOH	9:1	1.5%	333	90	Mechanical agitation	88.71	[115]
Algal oil	CaO/Al_2O_3	3.2:1	1.56%	323	125	Mechanical agitation	88.89	[116]
Silybum eburneum oil	MgO, Al_2O_3−CaO, TiO_2	3:1	0.1%	343	180	Mechanical agitation 600 rpm	91	[117]
Wild radish seed oil	KOH	9:1	1%	323	30	Mechanical agitation 500 rpm	94.58	[118]
Schleichera oleosa L oil	NaOH/KOH	8:1	1%	328	90	Mechanical agitation 1000 rpm	96	[119]
Kusum oil	$Ba(OH)_2$	9:1	3%	323	80	Ultrasound 20 kHz/250 W	96.8	[61]
Babasssu oil	CaO/SnO_2	10:1	6%	327	120	Mechanical agitation 250 rpm	89.58	[120]
Annona squamosa seed oil	Phosphoric acid	18:1	5%	338	180	Mechanical agitation	85	[121]
Jatropha oil	H_2SO_4	7:1	6%	343	60	Ultrasound 35 kHz/35 W	80.12	[122]
Milk thistle oil	NaOH	6:1	0.75%	333	120	Mechanical agitation 600 rpm	89.51	[123]

reactor using NaOH as a catalyst. Jatropha and Karanja oils were mixed in different proportions to make five blends, although the exact composition of blends (on volume or weight basis) has not been mentioned by authors. For each blend biodiesel was synthesized using two-step acid/base catalyzed esterification–transesterification process. Authors have evaluated the effects of various operating parameters, viz. methanol quantity, catalyst concentration, reaction temperature, time of reaction, etc. on the yield and quality of biodiesel. The results showed that minimum of 10.7 mol% methanol was required for the transesterification process. The optimum quantity of methanol was 35% for which maximum 94% yield of biodiesel was obtained. Ester yield (or biodiesel yield) was relatively insensitive to methanol concentration for minor variations in methanol content of reaction mixture: slightly above or below than optimum value of 35%. Nonetheless, the glycerol formation was influenced by methanol concentration. Moreover, increase in reaction temperature from 40 to 65°C increased biodiesel yield from 80% to 93%. Biodiesel yield increased with catalyst concentration from 82% yield for 0.5 wt% catalyst to 93% yield for 1.5 wt% catalyst. However, with further increase of catalyst concentration to 2 wt%, biodiesel yield reduced to 76%. The optimum reaction time for maximum biodiesel yield was 90 min, and further reaction did not yield any additional biodiesel.

Yogish et al. [55] have evaluated the fuel properties, viz. density, viscosity, flash and fire point, calorific value, iodine value, acid value, sulfur content, water content, glycerol content as well as sulfated ash content, of biodiesel obtained from all five mixed feedstock. The average values of these properties for the five blends are as follows: density, 844 ± 16.96 kg/m^3; viscosity, 3.22 ± 0.73 cSt at 40°C; flash point, 134 ± 14.5°C; fire point, 161 ± 13.5°C; Calorific value, 46.62 ± 0.78 MJ/kg; iodine value, 83.8 ± 16.2; acid value, 0.4 ± 0.072 mg KOH/g; sulfur content, 40.8 ± 7.36 mg/kg; water content, 440.6 ± 43.05 mg/kg; glycerin content, 0.13 ± 0.11% and sulfated ash, 0.0018 ± 0.0008%. The results of analysis confirmed that all the biodiesel samples have the fuel properties within the specified limits.

(2) Fadhil et al. [56] investigated the biodiesel production using mixture of nonedible oil and waste fish oil. Authors have studied different blends of castor oil and waste fish oil by varying the weight of waste fish oil from 10% to 50% in mixture. The blends were homogeneously mixed using mechanical stirrer. Various basic physical properties of raw oils and different blends were analyzed, and based on the results of initial screening, the blended feedstock with equal quantities of waste fish oil and castor oil (50 wt% each) was selected for biodiesel production. Transesterification of blended oil was done using three types of alkali catalysts, viz. NaOH/KOH/CH_3ONa, in a mechanically agitated three-neck round bottom flask. Using one parameter at a one-time approach, authors have optimized the process and also the type of alkali catalyst. The optimum conditions for biodiesel production were determined as

follows: catalyst = KOH; catalyst concentration = 0.5 wt%; molar ratio = 8:1; temperature = 32°C and reaction time = 30 min with optimum agitation speed of 600 rpm. At these conditions maximum 95.2 wt% yield of transesterification reaction from mixed oil feedstock was achieved with biodiesel purity of 97.66%.

Authors have also analyzed the various fuel properties of biodiesel prepared from individual nonedible oil and blended oil feedstock. The results showed that biodiesel obtained from blended feedstock and waste fish oil are in the specified limits of ASTM D 6751 and EN 14214, whereas biodiesel obtained from castor oil had higher density, viscosity and pour point than the specified limit due to higher density and viscosity of raw castor oil.

Fadhil et al. [56] had also carried out the cost analysis of transesterification process using blended oil feedstock. Authors have reported that biodiesel production from waste (*Cyprinus carpio*) fish oil had an attractive economy as compared to biodiesel production from waste (*Salmon*) fish oil and also biodiesel production using castor oil. Biodiesel production using blended feedstock helped in reduction of production cost of biodiesel from individual castor oil and also improves the fuel properties of biodiesel, which makes it merchantable.

(3) Malani et al. [57] has extended the concept of using blended feedstock for biodiesel production by using mixture of four nonedible oils and waste cooking oil. Authors have investigated the ultrasound-assisted biodiesel production using mixture of five different oils (viz. jatropha, castor, cotton seed, rubber seed and waste cooking oil) in packed bed as well as batch slurry reactor with heterogeneous base catalyst. Powder Cu_2O was used as heterogeneous base catalyst for transesterification reaction in two-step process. The study had three parts: esterification of blended feedstock; optimization of operating parameters using packed bed reactor and kinetic modeling of the process using Eley–Rideal kinetic model. In first part acid value of blended feedstock was reduced using homogeneous acid catalyzed esterification process. The results of esterification experiments showed that acid value of feedstock reduced by 88% and the final acid value of blended feedstock was 1.03 mg KOH/g. This esterified oil was further used for transesterification in packed bed reactor. Reaction temperature, residence time, catalyst packing height and molar ratio were optimized using Box–Behnken statistical design. The schematic experimental of packed bed set-up used by authors is shown in Fig. 13.7. The analysis of statistical design resulted in optimum conditions for maximum biodiesel yield of 90% as follows: catalyst packing height = 35.6 mm; molar ratio = 10.62:1; temperature = 62.5°C; and residence time = 33.5 min.

The authors have attempted to understand the mechanistic features of heterogeneously catalyzed transesterification reaction using kinetic modeling of the process. For this purpose, packed bed process parameters were converted into batch reaction parameters

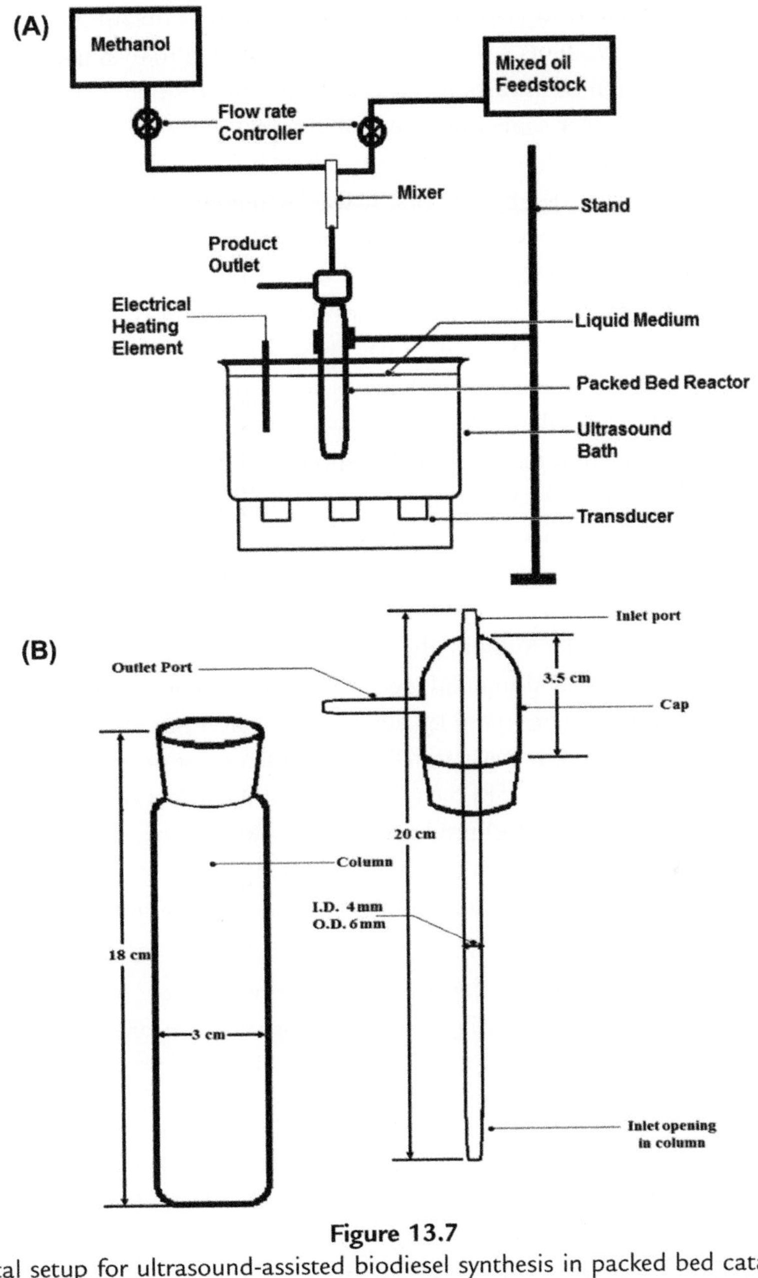

Figure 13.7

Experimental setup for ultrasound-assisted biodiesel synthesis in packed bed catalytic reactor. (A) Schematic of the complete assembly comprising reactor, ultrasound bath and feed/outlet system for reaction mixture. (B) Schematic (with dimensions) of the glass column and cap used for making the packed catalyst bed. *Adopted from Malani RS, Patil S, Roy K, Chakma S, Goyal A, Moholkar VS. Mechanistic analysis of ultrasound-assisted biodiesel synthesis with Cu_2O catalyst and mixed oil feedstock using continuous (packed bed) and batch (slurry) reactors. Chemical Engineering Science 2017;170:743–755 with permission of copyright @ 2017 Elsevier BV.*

and experiments in batch slurry reaction were performed to obtain time profile of various reactants and products. Authors had also attempted to evaluate the intensification done by application of ultrasound by comparing the results of mechanically agitated experiments under same operating conditions. Eley–Rideal mechanism in which one reactant get adsorbed on catalyst active site and react with another reactant from the bulk phase was used by authors. It was assumed that methanol is the adsorbing reactant due to hydrophilic surface of catalyst (and also the fact that reaction mixture contained excess methanol) and react with triglyceride in three steps to form biodiesel. The final kinetic equations were developed starting from elementary reaction steps (as tabulated in Tables 13.5 and 13.6). These equations were solved in MATLAB using Genetic Algorithm coupled with Runge–Kutta fourth-order method.

The results showed that methanol adsorption was the slowest step in heterogeneously catalyzed transesterification reaction. Application of ultrasound reduces the mass transfer between the reactant and catalyst molecule and boosts the kinetics of elementary reaction steps (i.e., triglyceride to diglyceride; diglyceride to monoglyceride; and monoglyceride to glycerol formation). On the other hand, application of ultrasound had adverse impact on methanol adsorption step due to the shock waves generated through transient cavitation. Authors also found that successive conversion of triglyceride to diglyceride and monoglyceride had higher kinetics and lower mass transfer barrier. It was also observed by authors that heterogeneously catalyzed transesterification process was mass transfer controlled even in presence of ultrasound.

(4) In another study, Malani et al. [58] have investigated the effect of various oil fractions in mixture feedstock for biodiesel production using lab synthesized KI impregnated ZnO as catalyst. A blend of four nonedible oils, viz. jatropha, castor, rubber, and palm oils along with waste cooking oil was employed. The effect of individual component

Table 13.5: The steps and corresponding kinetic expressions in Eley–Rideal mechanism of transesterification using a solid heterogeneous catalyst.

Step in the mechanism	Chemical equation	Rate expression
1. Methanol adsorption	$* + CH_3OH \rightleftarrows CH_3OH^*$	$r_1 = -k_1[*]_f[CH_3OH]$
2. Transesterification reactions	$CH_3OH^* + T \rightleftarrows D^* + F$	$r_2 = -k_2[T][CH_3OH^*]$
	$CH_3OH^* + D \rightleftarrows M^* + F$	$r_3 = -k_3[D][CH_3OH^*]$
	$CH_3OH^* + M \rightleftarrows G^* + F$	$r_4 = -k_4[M][CH_3OH^*]$
3. Desorption of adsorbed species	$D^* \rightleftarrows D + *$	$r_5 = -k_5[D^*]$
	$M^* \rightleftarrows M + *$	$r_6 = -k_6[M^*]$
	$G^* \rightleftarrows G + *$	$r_7 = -k_7[G^*]$

Symbols: *D*, diglyceride; *F*, fatty acid methyl ester (biodiesel); *G*, glycerol; *M*, monoglyceride; *T*, triglyceride; *, free catalyst active site.

Adopted from Malani RS, Patil S, Roy K, Chakma S, Goyal A, Moholkar VS. Mechanistic analysis of ultrasound-assisted biodiesel synthesis with Cu_2O catalyst and mixed oil feedstock using continuous (packed bed) and batch (slurry) reactors. Chemical Engineering Science 2017;170:743–755 with permission of copyright @ 2017 Elsevier BV.

Table 13.6: Final kinetic rate expressions in transesterification process based on Eley–Rideal mechanism.

Rate of consumption of triglyceride	$r_T = \frac{d[T]}{dt} = -\frac{k_2[T]}{1+\frac{k_2[T]}{k_5}+\frac{k_3[D]}{k_6}+\frac{k_4[M]}{k_7}+\frac{k_2[T]+k_3[D]+k_4[M]}{k_1[CH_3OH]}}$
Rate of consumption of diglyceride	$r_D = \frac{d[D]}{dt} = -\frac{k_3[D]-k_2[T]}{1+\frac{k_2[T]}{k_5}+\frac{k_3[D]}{k_6}+\frac{k_4[M]}{k_7}+\frac{k_2[T]+k_3[D]+k_4[M]}{k_1[CH_3OH]}}$
Rate of consumption of monoglyceride	$r_M = \frac{d[M]}{dt} = -\frac{k_4[M]-k_3[D]}{1+\frac{k_2[T]}{k_5}+\frac{k_3[D]}{k_6}+\frac{k_4[M]}{k_7}+\frac{k_2[T]+k_3[D]+k_4[M]}{k_1[CH_3OH]}}$
Rate of consumption of methanol	$r_{CH_3OH} = \frac{d[CH_3OH]}{dt} = -\frac{k_2[T]+k_3[D]+k_4[M]}{1+\frac{k_2[T]}{k_5}+\frac{k_3[D]}{k_6}+\frac{k_4[M]}{k_7}+\frac{k_2[T]+k_3[D]+k_4[M]}{k_1[CH_3OH]}}$
Rate of formation of glycerol	$r_G = \frac{d[G]}{dt} = \frac{k_4[M]}{1+\frac{k_2[T]}{k_5}+\frac{k_3[D]}{k_6}+\frac{k_4[M]}{k_7}+\frac{k_2[T]+k_3[D]+k_4[M]}{k_1[CH_3OH]}}$
Rate of formation of FAME or biodiesel	$r_F = \frac{d[F]}{dt} = \frac{(k_2[T]+k_3[D]+k_4[M])}{1+\frac{k_2[T]}{k_5}+\frac{k_3[D]}{k_6}+\frac{k_4[M]}{k_7}+\frac{k_2[T]+k_3[D]+k_4[M]}{k_1[CH_3OH]}}$

Adopted from Malani RS, Shinde V, Ayachit S, Goyal A, Moholkar VS. Ultrasound-assisted biodiesel production using heterogeneous base catalyst and mixed non-edible oils. Ultrasonics Sonochemistry 2018;52:232–243 with permission of copyright @ 2018 Elsevier BV.

of the oil blend on the physical properties of the blend was evaluated using pseudo-component mixture design. The process was carried out using two-step trans-esterification process, where individual oil was esterified using acid catalyst in first step and their free fatty acid (FFA) content lowered. In second step total 36 experiments were performed with varying volume fractions of different oils and keeping other reaction conditions, such as catalyst loading, reaction volume, temperature, reaction time and molar ratio constant. Authors observed that raw oils having high viscosity and high FFA content lowered the yield of biodiesel from blended oil feedstock when they have significantly high-volume fractions. Thus, castor, palm, and jatropha oils should be blended in volume fraction less than 20% to achieve higher yield and improved properties of biodiesel. The composition of oil blend (in volume fraction) used for optimization and kinetic analysis of the process was as follows: castor = 0.05; jatropha = 0.05; rubber = 0.5; palm = 0.2; and waste cooking oil = 0.2. Optimization of process parameters was done using Box–Behnken statistical design. The optimum conditions of transesterification reaction yield 92.35% triglyceride conversion at catalyst loading = 7 wt%; molar ratio = 11.68:1 and reaction temperature = 59°C in 1 h. Effect of temperature, molar ratio and catalyst loading on biodiesel yield is studied through surface plots as shown in Fig. 13.8.

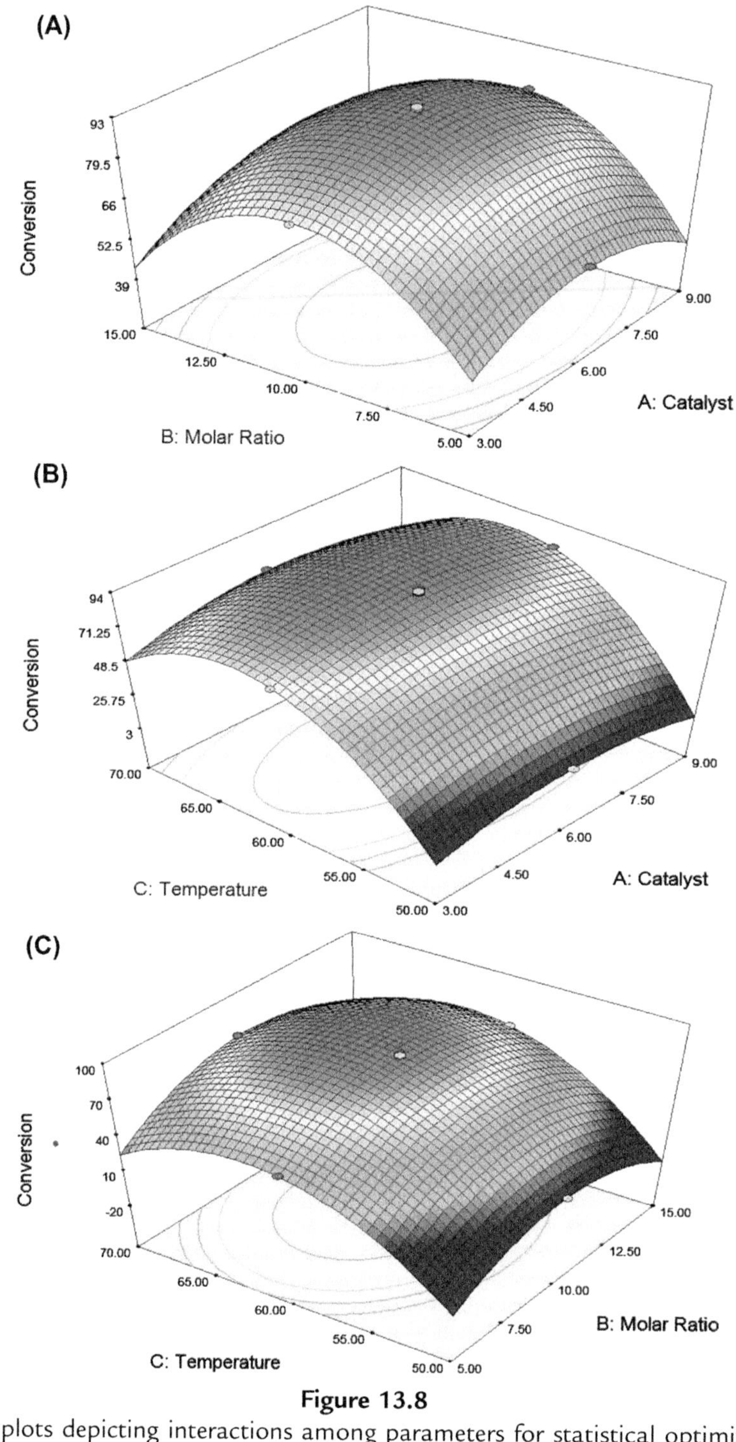

Figure 13.8

Contour plots depicting interactions among parameters for statistical optimization of transesterification process (A) molar ratio v/s catalyst loading; (B) temperature v/s catalyst loading and (C) temperature v/s molar ratio.

Similar to their previous study, Malani et al. [58] have investigated the mechanistic features of ultrasound-assisted KI/ZnO catalyzed biodiesel synthesis from mixed oil feedstock using Eley–Rideal kinetic model. The intensification of transesterification process was compared by performing the experiments with mechanical agitation at same operating conditions. The profile fitting of experimental results and model predicted data at various operating conditions is as shown in Fig. 13.9. The results of kinetic modeling revealed that all the rate constants except k_1 (methanol adsorption rate constant) showed 25%–50% enhancement with application of ultrasound. Application of ultrasound lowers the adsorption of methanol on active sites of catalyst. The overall process was mass transfer controlled and more than half of overall activation energy is required to overcome mass transfer limitations. Application of ultrasound lowers the overall activation energy of transesterification process from 135.04 kJ/mol to 123.65 kJ/mol. At the same time, application of ultrasound rises the fraction of activation energy utilize to overcome mass transfer barrier from 52% in mechanically agitated system to 63%. As demonstrated in earlier study, authors observed that successive conversion of triglyceride to diglyceride and monoglyceride lowers the requirement of activation energy through lowering the interfacial surface tension between the phases.

The biodiesel resulting from the blended feedstock was tested for various properties using standard norms. The properties of the biodiesel are listed in Table 13.7. It could be seen that this biodiesel met most of the required properties. This study has essentially confirmed feasibility of use of blend of nonedible oils to produce good quality biodiesel.

(5) Recently, in another study Malani et al. [40] utilized the blend of five different oils (four nonedible oils, viz. jatropha, castor, rubber, palm oil, and waste cooking oil) to produce waste-derived heterogeneous acid catalyst. The catalyst was prepared by treating the de-oiled seed cake of rubber seeds with sulfuric and chlorosulfonic acids. The characterization shows that chlorosulfonic acid treated catalyst have higher active sites as well as other surface properties as compared to sulfuric acid treated catalyst. Authors carried out the biodiesel synthesis in single-step and two-step process using ultrasound as a mode of mixing and intensification. The different oils were blended in following volumetric ratios to prepare the feedstock for biodiesel production: jatropha = 15%; castor = 10%; rubber = 30%; palm = 20%; and waste cooking oil = 25%. The blended feedstock was characterized for their basic properties and these properties were found as density = 0.921 g/mL; viscosity = 28.95 mPa s (at 40°C) with acid value = 12.42 mg KOH/g and average molecular weight of 883.4 g/mol.

The study was carried out in two parts, viz. (1) single-step biodiesel production and process optimization using statistical design; (2) two-step biodiesel production process

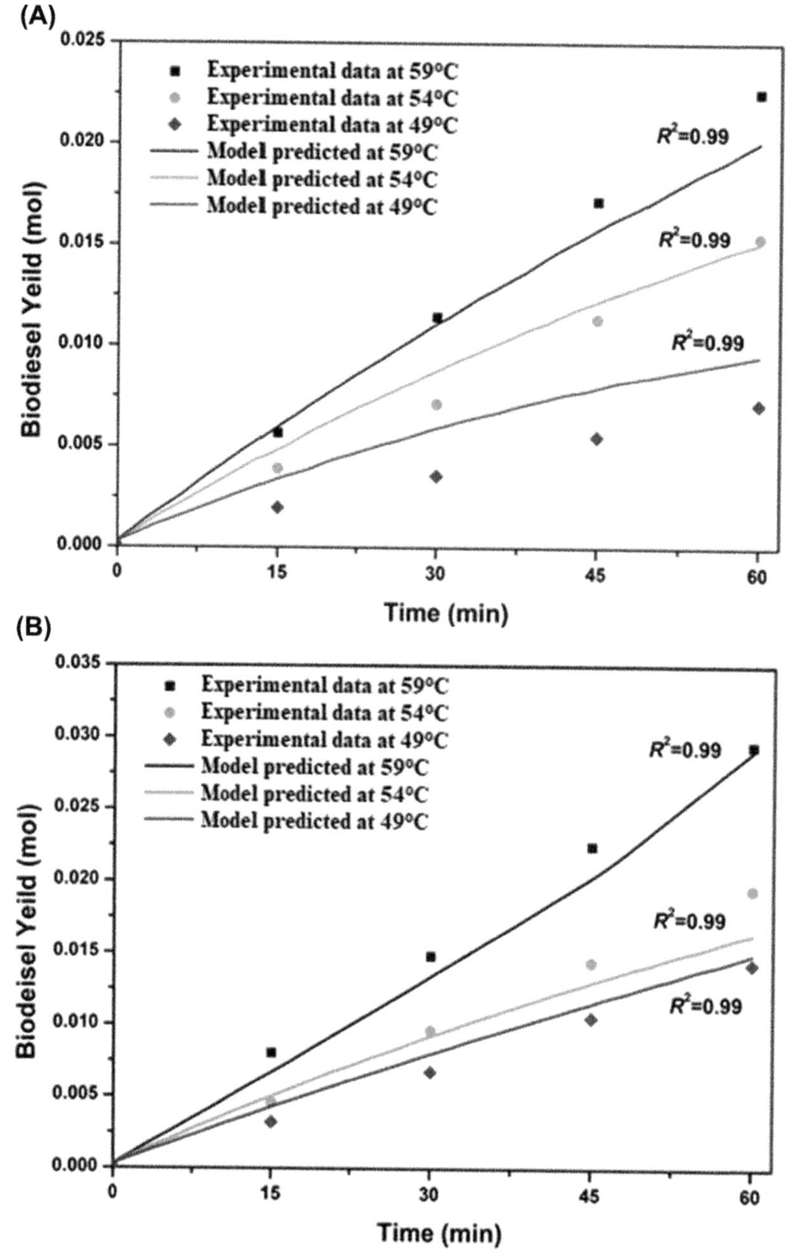

Figure 13.9

Fitting of experimental and model predicted data for biodiesel yield (A) with mechanical agitation and (B) with ultrasound treatment. *Adopted from Malani RS, Shinde V, Ayachit S, Goyal A, Moholkar VS. Ultrasound-assisted biodiesel production using heterogeneous base catalyst and mixed non-edible oils. Ultrasonics Sonochemistry 2018;52:232–243 with permission of copyright @ 2018 Elsevier BV.*

Table 13.7: Properties of biodiesel synthesized using mixed nonedible oil feedstock.

Property	Present study	ASTM D 6751 standard	EN 14214 standard
Density at 15°C (kg/m^3)	887	860–900	860–900
Flash point (°C)	128	Min. 93	Min. 101
Kinematic viscosity at 40°C (mm^2/s)	4.08	1.9–6.0	3.5–5.0
Cloud point (°C)	2	–	–
Oxidation stability at 110°C (h)	5.58	Min. 3	Min. 6
Calorific value (MJ/kg)	36.23	–	–

Adopted from Malani RS, Shinde V, Ayachit S, Goyal A, Moholkar VS. Ultrasound-assisted biodiesel production using heterogeneous base catalyst and mixed non-edible oils. Ultrasonics Sonochemistry 2018;52:232–243 with permission of copyright @ 2018 Elsevier BV.

for analyzing the role of process intensification induced by application of ultrasound. Moreover, authors have also modified the Eley–Rideal model used in their previous studies and used it to investigate the mechanistic points in heterogeneously acid catalyzed biodiesel production in single- and two-step process. Authors assume that instead of intermediates the final product biodiesel or FAME would be adsorbed on catalyst active sites during formation and derived the modified final kinetic expressions. The modified reaction steps and final kinetic expressions are tabulated in Tables 13.8 and 13.9 as shown below.

Result of single-step process optimization study showed that the optimum parameters will yield 91.2% biodiesel yield at catalyst loading = 8.18 wt%; reaction temperature = 63°C and alcohol molar ratio = 12.8:1 at the end of 3 h reaction. Fig. 13.10 represents the effect of process variables on the conversion of triglyceride to biodiesel.

These optimum conditions were used in two-step process, where first step of esterification was carried out using H_2SO_4 catalyst. The esterification experiment of blended oil feedstock resulted in 82.29% reduction of acid value, with final acid value of 2.2 mg KOH/g and viscosity of 14.7 mPa s (at 40°C). This reduced FFA content blended oil

Table 13.8: Elementary reaction steps with corresponding forward kinetic rate expressions based on modified Eley–Rideal mechanism.

Step in the mechanism	Chemical equation	Rate expression
1. Methanol adsorption	$* + CH_3OH \rightleftarrows CH_3OH^*$	$r_1 = -k_1[*]_f[CH_3OH]$
2. Transesterification reactions	$CH_3OH^* + T \rightleftarrows F^* + D$	$r_2 = -k_2[T][CH_3OH^*]$
	$CH_3OH^* + D \rightleftarrows F^* + M$	$r_3 = -k_3[D][CH_3OH^*]$
	$CH_3OH^* + M \rightleftarrows F^* + G$	$r_4 = -k_4[M][CH_3OH^*]$
3. Desorption of adsorbed species	$F^* \rightleftarrows F + *$	$r_5 = -k_5[F^*]$

Adopted from Malani RS, Sardar H, Malviya Y, Goyal A, Moholkar VS. Ultrasound–intensified biodiesel production from mixed nonedible oil feedstock using heterogeneous acid catalyst supported on rubber de-oiled cake. Industrial & Engineering Chemistry Research 2018;57(44):14926–14938 with permission of copyright @ 2018 American Chemical Society.

Table 13.9: Final kinetic rate expressions in transesterification process based on modified Eley–Rideal mechanism.

Rate of consumption of triglyceride	$r_T = \frac{d[T]}{dt} = -\frac{k_2[T]}{1+\frac{k_2[T]+k_3[D]+k_4[M]}{k_5}+\frac{k_2[T]+k_3[D]+k_4[M]}{k_1[CH_3OH]}}$
Rate of consumption of diglyceride	$r_D = \frac{d[D]}{dt} = -\frac{k_3[D]-k_2[T]}{1+\frac{k_2[T]+k_3[D]+k_4[M]}{k_5}+\frac{k_2[T]+k_3[D]+k_4[M]}{k_1[CH_3OH]}}$
Rate of consumption of monoglyceride	$r_M = \frac{d[M]}{dt} = -\frac{k_4[M]-k_3[D]}{1+\frac{k_2[T]+k_3[D]+k_4[M]}{k_5}+\frac{k_2[T]+k_3[D]+k_4[M]}{k_1[CH_3OH]}}$
Rate of consumption of methanol	$r_{CH_3OH} = \frac{d[CH_3OH]}{dt} = -\frac{k_2[T]+k_3[D]+k_4[M]}{1+\frac{k_2[T]+k_3[D]+k_4[M]}{k_5}+\frac{k_2[T]+k_3[D]+k_4[M]}{k_1[CH_3OH]}}$
Rate of formation of glycerol	$r_G = \frac{d[G]}{dt} = \frac{k_4[M]}{1+\frac{k_2[T]+k_3[D]+k_4[M]}{k_5}+\frac{k_2[T]+k_3[D]+k_4[M]}{k_1[CH_3OH]}}$
Rate of formation of FAME or biodiesel	$r_F = \frac{d[F]}{dt} = \frac{k_2[T]+k_3[D]+k_4[M]}{1+\frac{k_2[T]+k_3[D]+k_4[M]}{k_5}+\frac{k_2[T]+k_3[D]+k_4[M]}{k_1[CH_3OH]}}$

feedstock was used in transesterification reaction with the same optimum conditions resulted in 93.47% triglyceride yield at the end of 1 h reaction. The separate esterification–transesterification process resulted in superior biodiesel yield with much lower reaction time. To investigate the enhancement in biodiesel yield authors analyzed the process using modified Eley–Rideal kinetic model. To understand the underlying role of process intensification by sonication authors carried out the experiments using mechanical agitation and analyzed the results using kinetic modeling approach. The kinetic analyses (Arrhenius plots as shown in Fig. 13.11) have revealed interesting facets of the role of sonication as well as the limitations in single-step process as compared to two-step process.

Authors have found that application of ultrasound in both (single as well as two-step process) enhances the rate constants and thus triglyceride conversion. Moreover, the enhancement in single-step process was lower as compared with two-step process. The activation energies of single-step process were found to be 75.98 kJ/mol and 69.62 kJ/mol for mechanically agitated system and ultrasound-assisted system, respectively. Whereas, the activation energies of two-step process were found to be 73.28 kJ/mol and 55.98 kJ/mol for mechanically agitated system and ultrasound-assisted system, respectively. The difference in overall activation energies of single-step process was ~9% and it escalated to ~23% in a two-step process. At the same time, there was negligible change in overall activation was observed when

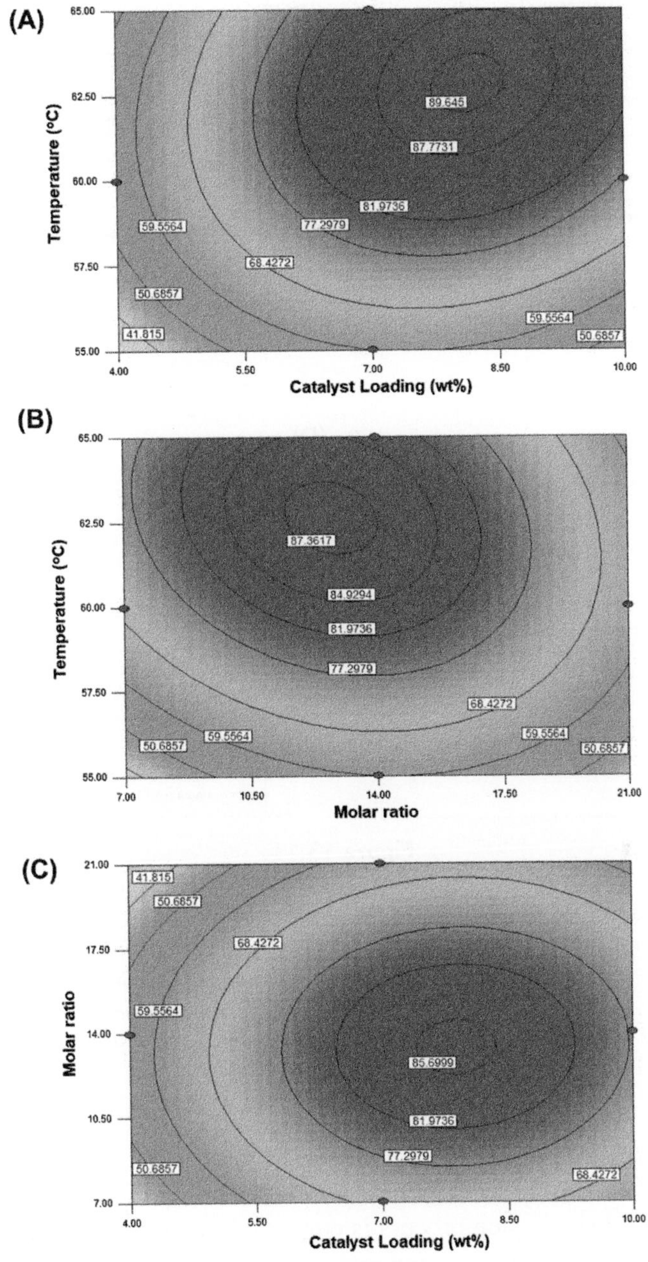

Figure 13.10

Contour plots depicting interaction between the process parameters in single-step trans-esterification process (A) catalyst loading v/s temperature; (B) molar ratio v/s temperature and (C) catalyst loading v/s molar ratio. *Adopted from Malani RS, Sardar H, Malviya Y, Goyal A, Moholkar VS. Ultrasound–intensified biodiesel production from mixed nonedible oil feedstock using heterogeneous acid catalyst supported on rubber de-oiled cake. Industrial & Engineering Chemistry Research 2018;57(44):14926–14938 with permission of copyright @ 2018 American Chemical Society.*

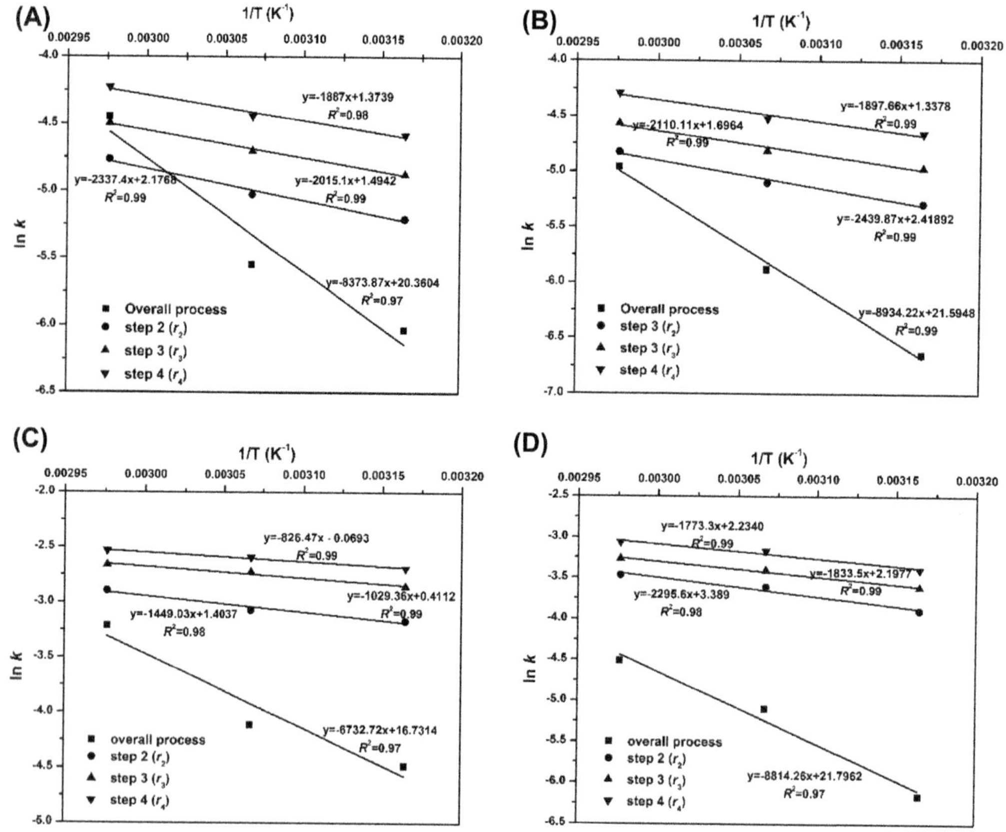

Figure 13.11

Arrhenius plot for overall transesterification process and three reaction steps (A) single-step process with ultrasound; (B) single-step process with mechanical agitation; (C) second step (in two-step) process with ultrasound and (D) second step (in two-step) process with mechanical agitation. *Adopted from Malani RS, Sardar H, Malviya Y, Goyal A, Moholkar VS. Ultrasound–intensified biodiesel production from mixed nonedible oil feedstock using heterogeneous acid catalyst supported on rubber de-oiled cake. Industrial & Engineering Chemistry Research 2018;57(44):14926–14938 with permission of copyright @ 2018 American Chemical Society.*

mechanically agitated single-step process was converted in two-step process, though significant improvement in rate constant was observed. On the other hand, when ultrasound-assisted single-step process is converted in two-step process resulted in ~20% reduction in overall activation energy.

Eley–Rideal kinetic analysis showed that all the systems were strongly mass transfer influenced, the sum of activation energies of three reaction steps was much lower than the overall activation energies. Analysis showed that sum of activation energies of three steps of the transesterification process (27.47 kJ/mol) was 47% smaller than the gross activation energy of 51.87 kJ/mol. Authors have linked this reduction of energy to removal of water molecule which formed as a byproduct during the esterification reaction and leads to reversible nature of transesterification reaction. Separate esterification step eliminates these water molecules and thus required lower activation energy as well as time to achieve the same (or superior) conversion of triglyceride. Moreover, the similar effect in mechanically agitated system has no (negligible) impact on activation energy, but shown boosting of rate constants. This probably due the macromixing done by mechanical agitation which could not formed the fine emulsion between oil and methanol molecules as achieved in ultrasound-assisted system.

Authors also studied the reusability study of catalyst in ultrasound-assisted single as well as two-step process and found that simultaneous esterification and transesterification (single-step) process results in reduction in more active sites as compared to separate esterification and transesterification (two-step) process. The catalyst was found to retain ~90% active sites after three cycles of reusability in two-step process.

(6) Vinayaka et al. [59] have also investigated biodiesel production using mixture of nonedible Karanja and Neem oil. Authors selected two-step process for production of biodiesel. The oils were blended in the ratio of 70:30 v/v (i.e., 70 mL Karanja oil and 30 mL Neem oil) to prepare feedstock. In first step esterification was carried out using conc. sulfuric acid. Central composite design was used to optimize the NaOH catalyzed transesterification process by varying the molar ratio, reaction time and catalyst loading. Maximum 86.3% yield was achieved under optimum process conditions of reaction time = 77 min, molar ratio = 6:1 and catalyst loading = 0.67 wt% at reaction temperature of 65°C. The biodiesel was characterized for basic fuel properties. The results showed that biodiesel synthesized using mixture of Karanja and Neem oil had density = 0.82 g/mL; kinetic viscosity = 4.5 mm^2/s at 40°C; cetane number = 66.26; flash point = 178°C; cloud and pour point of 8°C and −2°C, respectively. These properties of synthesized biodiesel were under the specified limits of ASTM D 6751 and EN 14214 specifications and can be used for commercial application.

(7) Malani et al. [60] have recently investigated the enzyme catalyzed biodiesel synthesis from mixed nonedible oil and methanol using commercial immobilized lipase from *Thermomyces lanuginosus.* The optimum conditions for the process were determined using Box–Behnken statistical design as alcohol/oil molar ratio = 7.64:1, enzyme loading = 3.55% (w/w) and temperature = 36°C, which gives ~ 90% conversion at the end of 2 h reaction. Authors also found that addition of water to reaction mixture (10% v/v) not only boosted biodiesel yield from 90% to 94% but also lowered activation energy to 78.7 kJ/mol. Analysis of reaction kinetics using ordered Bi–Bi model revealed drastic rise (~1.6×) in reaction velocity with concurrent changes in Michaelis–Menten and dissociation constant of both substrate and products.

Additionally, authors found that sonication of reaction mixture resulted in lowering of activation energy to 100.4 kJ/mol (from 124.4 kJ/mol under mechanical agitation). The reusability study carried out by authors revealed that, the recycled enzyme showed good retention of activity up to six cycles in presence of ultrasound, which further enhanced with water addition to reaction mixture.

13.8 Conclusions and perspectives

The literature and case studies reviewed in this chapter clearly demonstrate the potential of mixed oil feedstock-based production of biodiesel using waste (or nonedible) oils. Use of mixed feedstocks or nonedible oil blends adds both viability and economic feasibility to the process. Especially, the waste carbon based acid/base catalysts are potential alternative to the conventional catalyst as they have smaller production cost than inorganic catalysts. Although, laboratory scale studies have reported yields with blended feedstocks at par with single feedstocks, further experiments on bench/pilot scale are necessary, as the physical properties of the oil blends vary significantly with composition. Application of heterogeneous catalysts simplifies the downstream processing and also avoids contamination of the glycerol produced during transesterification, which could fetch additional revenue for industry. In summary, processes with feedstock flexibility outlines, as basis of sustainable biorefinery in near future, and biodiesel production from waste oil blends and with application of waste carbon derived catalyst is a vivid example of the same.

References

[1] Moholkar VS, Choudhary HA, Singh S, Khanna S, Ranjan A, Chakma S, Bhasarkar S. Physical and chemical mechanisms of ultrasound in biofuel synthesis. In: Fang Z, Smith Jr RL, Qi X, editors. Biofuels and biorefineries. Production of biofuels and chemicals with ultrasound, vol. 4. Dordrecht: Springer; 2015. p. 35–86.

[2] A report on "World Energy Resources Oil 2016", from World Energy Resources. Available from https://www.worldenergy.org/wp–content/uploads/2017/03/WEResources_Oil _2016.pdf.

[3] A report on "Global Energy & CO2 Status Report 2017" from International Energy Agency. Available from https://www.iea.org/publications/freepublications/publication/GECO2017.pdf.

[4] A report on "2018 Outlook for Energy: A view to 2040" from "ExxonMobil". Available from https://corporate.exxonmobil.com/en/energy-and-environment/energy-resources/outlook-for-energy/2018-outlook-for-energy-a-view-to-2040#aViewTo2040.pdf.

[5] A report on "Statistical Review of World Energy June 2017" from "BP Statistical Review". Available from https://www.bp.com/content/dam/bp/en/corporate/pdf/energy–economics/statistical–review–2017/bp–statistical–review–of–world–energy–2017–full–report.pdf.

[6] A report on "Global Economic Prospects 2018" from "A World Bank Group". Available from https://openknowledge.worldbank.org/bitstream/handle/10986/.../9781464812576.pdf.

[7] A report on "IEA: OIL 2017 – Analysis and Forecasts to 2022". Available from https://www.iea.org/publications/freepublications/publication/Market_Report_Series_Oil 2017.pdf.

[8] Mohan SV, Nikhil GN, Chiranjeevi P, Reddy CN, Rohit MV, Kumar AN, Sarkar O. Waste biorefinery models towards sustainable circular bioeconomy: critical review and future perspectives. Bioresource Technology 2016;215:2–12.
[9] Ruiz HA, Rodríguez-Jasso RM, Fernandes BD, Vicente AA, Teixeira JA. Hydrothermal processing, as an alternative for upgrading agriculture residues and marine biomass according to the biorefinery concept: a review. Renewable and Sustainable Energy Reviews 2013;21:35–51.
[10] A report on "Energy Statics" from "Central Statistics Office Ministry of Statistics and Programme Implementation, Government of India". Available from http://www.mospi.nic.in/sites/default/files/publication_reports/Energy_Statistics_2017r.pdf.pdf.
[11] An "Annual Report 2017–18" from "Ministry of Petroleum and Natural Gas, Government of India". Available from http://petroleum.nic.in/sites/default/files/APR_E_1718.pdf.
[12] Knothe G. Historical perspectives on vegetable oil-based diesel fuels. Inform 2001;12(11):1103–7.
[13] Knothe G, Razon LF. Biodiesel fuels. Progress in Energy and Combustion Science 2017;58:36–59.
[14] Su F, Guo Y. Advancements in solid acid catalysts for biodiesel production. Green Chemistry 2014;16(6):2934–57.
[15] Mittelbach M, Remschmidt C. Biodiesel, the comprehensive handbook. 2nd ed.. Vienna, Graz, Austria: Boersedruck Ges.m.b.H; 2004.
[16] Giakoumis EG, Rakopoulos CD, Dimaratos AM, Rakopoulos DC. Exhaust emissions of diesel engines operating under transient conditions with biodiesel fuel blends. Progress in Energy and Combustion Science 2012;38(5):691–715.
[17] Nurfitri I, Maniam GP, Hindryawati N, Yusoff MM, Ganesan S. Potential of feedstock and catalysts from waste in biodiesel preparation: a review. Energy Conversion and Management 2013;74:395–402.
[18] Ramachandran K, Suganya T, Gandhi NN, Renganathan S. Recent developments for biodiesel production by ultrasonic assist transesterification using different heterogeneous catalyst: a review. Renewable and Sustainable Energy Reviews 2013;22:410–8.
[19] Ng JH, Ng HK, Gan S. Recent trends in policies, socioeconomy and future directions of the biodiesel industry. Clean Technologies and Environmental Policy 2010;12(3):213–38.
[20] Demirbas A. Biodiesel production from vegetable oils via catalytic and non-catalytic supercritical methanol transesterification methods. Progress in Energy and Combustion Science 2005;31(5–6):466–87.
[21] Zabeti M, Daud WMAW, Aroua MK. Activity of solid catalysts for biodiesel production: a review. Fuel Processing Technology 2009;90(6):770–7.
[22] Lam E, Luong JH. Carbon materials as catalyst supports and catalysts in the transformation of biomass to fuels and chemicals. ACS Catalysis 2014;4(10):3393–410.
[23] Toda M, Takagaki A, Okamura M, Kondo JN, Hayashi S, Domen K, Hara M. Green chemistry: biodiesel made with sugar catalyst. Nature 2005;438(7065):178.
[24] Stellwagen DR, van der Klis F, van Es DS, de Jong KP, Bitter JH. Functionalized carbon nanofibers as solid-acid catalysts for transesterification. ChemSusChem 2013;6(9):1668–72.
[25] Hara M. Environmentally benign production of biodiesel using heterogeneous catalysts. ChemSusChem 2009;2(2):129–35.
[26] Wang H, Covarrubias J, Prock H, Wu X, Wang D, Bossmann SH. Acid-functionalized magnetic nanoparticle as heterogeneous catalysts for biodiesel synthesis. Journal of Physical Chemistry C 2015;119(46):26020–8.
[27] Melero JA, Bautista LF, Morales G, Iglesias J, Briones D. Biodiesel production with heterogeneous sulfonic acid-functionalized mesostructured catalysts. Energy & Fuels 2008;23(1):539–47.
[28] Xie W, Wang H, Li H. Silica-supported tin oxides as heterogeneous acid catalysts for transesterification of soybean oil with methanol. Industrial & Engineering Chemistry Research 2011;51(1):225–31.
[29] Liu H, Chen J, Chen L, Xu Y, Guo X, Fang D. Carbon nanotube-based solid sulfonic acids as catalysts for production of fatty acid methyl ester via transesterification and esterification. ACS Sustainable Chemistry & Engineering 2016;4(6):3140–50.

[30] Zong MH, Duan ZQ, Lou WY, Smith TJ, Wu H. Preparation of a sugar catalyst and its use for highly efficient production of biodiesel. Green Chemistry 2007;9(5):434–7.
[31] Shu Q, Nawaz Z, Gao J, Liao Y, Zhang Q, Wang D, Wang J. Synthesis of biodiesel from a model waste oil feedstock using a carbon-based solid acid catalyst: reaction and separation. Bioresource Technology 2010;101(14):5374–84.
[32] Nakajima K, Hara M. Amorphous carbon with SO_3H groups as a solid Brønsted acid catalyst. ACS Catalysis 2012;2(7):1296–304.
[33] Galhardo TS, Simone N, Goncalves M, Figueiredo FC, Mandelli D, Carvalho WA. Preparation of sulfonated carbons from rice husk and their application in catalytic conversion of glycerol. ACS Sustainable Chemistry & Engineering 2013;1(11):1381–9.
[34] Konwar LJ, Mäki-Arvela P, Salminen E, Kumar N, Thakur AJ, Mikkola JP, Deka D. Towards carbon efficient biorefining: multifunctional mesoporous solid acids obtained from biodiesel production wastes for biomass conversion. Applied Catalysis B: Environmental 2015;176:20–35.
[35] Badday AS, Abdullah AZ, Lee KT. Transesterification of crude Jatropha oil by activated carbon-supported heteropolyacid catalyst in an ultrasound-assisted reactor system. Renewable Energy 2014;62:10–7.
[36] Maneechakr P, Samerjit J, Uppakarnrod S, Karnjanakom S. Experimental design and kinetic study of ultrasonic assisted transesterification of waste cooking oil over sulfonated carbon catalyst derived from cyclodextrin. Journal of Industrial and Engineering Chemistry 2015;32:128–36.
[37] Pathak G, Das D, Rajkumari K, Rokhum L. Exploiting waste: towards a sustainable production of biodiesel using Musa acuminata peel ash as a heterogeneous catalyst. Green Chemistry 2018;20(10):2365–73.
[38] Dawodu FA, Ayodele O, Xin J, Zhang S, Yan D. Effective conversion of non-edible oil with high free fatty acid into biodiesel by sulphonated carbon catalyst. Applied Energy 2014;114:819–26.
[39] Fadhil AB, Aziz AM, Altamer MH. Potassium acetate supported on activated carbon for transesterification of new non-edible oil, bitter almond oil. Fuel 2016;170:130–40.
[40] Malani RS, Sardar H, Malviya Y, Goyal A, Moholkar VS. Ultrasound-intensified biodiesel production from mixed nonedible oil feedstock using heterogeneous acid catalyst supported on rubber de-oiled cake. Industrial & Engineering Chemistry Research 2018;57(44):14926–38.
[41] Al-dobouni IA, Fadhil AB, Saeed IK. Optimized alkali-catalyzed transesterification of wild mustard (*Brassica juncea* L.) seed oil. Energy Sources, Part A: Recovery, Utilization, and Environmental Effects 2016;38(15):2319–25.
[42] Reshad AS, Barman P, Chaudhari AJ, Tiwari P, Kulkarni V, Goud VV, Sahoo N. Rubber seed oil methyl ester synthesis, engine performance, and emission characteristics of blends. Energy & Fuels 2015;29(8):5136–44.
[43] Aniya VK, Muktham RK, Alka K, Satyavathi B. Modeling and simulation of batch kinetics of non-edible karanja oil for biodiesel production: a mass transfer study. Fuel 2015;161:137–45.
[44] Anjana PA, Niju S, Meera Sheriffa Begum KM, Anantharaman N. Utilization of limestone derived calcium oxide for biodiesel production from non-edible pongamia oil. Environmental Progress & Sustainable Energy 2016;35(6):1758–64.
[45] Anwar M, Rasul MG, Ashwath N. Production optimization and quality assessment of papaya (*Carica papaya*) biodiesel with response surface methodology. Energy Conversion and Management 2018;156:103–12.
[46] Asif S, Chuah LF, Klemeš JJ, Ahmad M, Akbar MM, Lee KT, Fatima A. Cleaner production of methyl ester from non-edible feedstock by ultrasonic-assisted cavitation system. Journal of Cleaner Production 2017;161:1360–73.
[47] Ayodele OO, Dawodu FA. Production of biodiesel from *Calophyllum inophyllum* oil using a cellulose-derived catalyst. Biomass and Bioenergy 2014;70:239–48.
[48] Boro J, Konwar LJ, Deka D. Transesterification of non-edible feedstock with lithium incorporated egg shell derived CaO for biodiesel production. Fuel Processing Technology 2014;122:72–8.

[49] Ahmad M, Sultana S, Keat Teong L, Abdullah AZ, Sadia H, Zafar M, Hadda TB, Ashraf MA, Tareen RB. Distaff Thistle Oil: a possible new non-edible feedstock for bioenergy. International Journal of Green Energy 2015;12(11):1066–75.
[50] Ayoub M, Ullah S, Inayat A, Bhat AH, Hailegiorgis SM. Process optimization for biodiesel production from waste frying oil over montmorillonite clay k-30. Procedia Engineering 2016;148:742–9.
[51] Borah MJ, Devi A, Borah R, Deka D. Synthesis and application of Co doped ZnO as heterogeneous nanocatalyst for biodiesel production from non-edible oil. Renewable Energy 2019;133:512–9.
[52] Maneechakr P, Samerjit J, Karnjanakom S. Ultrasonic-assisted biodiesel production from waste cooking oil over novel sulfonic functionalized carbon spheres derived from cyclodextrin via one-step: a way to produce biodiesel at short reaction time. RSC Advances 2015;5(68):55252–61.
[53] Anuar MR, Abdullah AZ. Ultrasound-assisted biodiesel production from waste cooking oil using hydrotalcite prepared by combustion method as catalyst. Applied Catalysis A: General 2016;514:214–23.
[54] Xiang Y, Wang L, Jiao Y. Ultrasound strengthened biodiesel production from waste cooking oil using modified coal fly ash as catalyst. Journal of Environmental Chemical Engineering 2016;4(1):818–24.
[55] Yogish H, Chandrashekara K, Kumar MP. Optimization of experimental conditions for composite biodiesel production from transesterification of mixed oils of *Jatropha* and *Pongamia*. Heat and Mass Transfer 2012;48(11):1955–60.
[56] Fadhil AB, Al-Tikrity ET, Albadree MA. Biodiesel production from mixed non-edible oils, castor seed oil and waste fish oil. Fuel 2017;210:721–8.
[57] Malani RS, Patil S, Roy K, Chakma S, Goyal A, Moholkar VS. Mechanistic analysis of ultrasound-assisted biodiesel synthesis with Cu_2O catalyst and mixed oil feedstock using continuous (packed bed) and batch (slurry) reactors. Chemical Engineering Science 2017;170:743–55.
[58] Malani RS, Shinde V, Ayachit S, Goyal A, Moholkar VS. Ultrasound-assisted biodiesel production using heterogeneous base catalyst and mixed non-edible oils. Ultrasonics Sonochemistry 2018;52:232–43.
[59] Vinayaka AS, Mahanty B, Rene ER, Behera SK. Biodiesel production by transesterification of a mixture of *pongamia* and neem oils. Biofuels 2018:1–9.
[60] Malani RS, Umriwad SB, Kumar K, Goyal A, Moholkar VS. Ultrasound-assisted enzymatic biodiesel production using blended feedstock of non-edible oils: kinetic analysis. Energy Conversion and Management 2019;188:142–50.
[61] Sarve AN, Varma MN, Sonawane SS. Ultrasound assisted two-stage biodiesel synthesis from non-edible *Schleichera triguga* oil using heterogeneous catalyst: kinetics and thermodynamic analysis. Ultrasonics Sonochemistry 2016;29:288–98.
[62] Saha R, Goud VV. Ultrasound assisted transesterification of high free fatty acids karanja oil using heterogeneous base catalysts. Biomass Conversion and Biorefinery 2015;5(2):195–207.
[63] Asif S, Ahmad M, Bokhari A, Chuah LF, Klemeš JJ, Akbar MM, Sultana S, Yusup S. Methyl ester synthesis of *Pistacia khinjuk* seed oil by ultrasonic-assisted cavitation system. Industrial Crops and Products 2017;108:336–47.
[64] Gupta AR, Yadav SV, Rathod VK. Enhancement in biodiesel production using waste cooking oil and calcium diglyceroxide as a heterogeneous catalyst in presence of ultrasound. Fuel 2015;158:800–6.
[65] Ho WWS, Ng HK, Gan S, Chan WL. Ultrasound-assisted transesterification of refined and crude palm oils using heterogeneous palm oil mill flyash supported calcium oxide catalyst. Energy Science & Engineering 2015;3(3):257–69.
[66] Chang F, Hanna MA, Zhang DJ, Li H, Zhou Q, Song BA, Yang S. Production of biodiesel from non-edible herbaceous vegetable oil: *Xanthium sibiricum* Patr. Bioresource Technology 2013;140:435–8.
[67] Chavan SB, Kumbhar RR, Madhu D, Singh B, Sharma YC. Synthesis of biodiesel from *Jatropha curcas* oil using waste eggshell and study of its fuel properties. RSC Advances 2015;5(78):63596–604.
[68] Dawood S, Ahmad M, Ullah K, Zafar M, Khan K. Synthesis and characterization of methyl esters from non-edible plant species yellow oleander oil, using magnesium oxide (MgO) nano–catalyst. Materials Research Bulletin 2018;101:371–9.

[69] Zhang F, Fang Z, Wang YT. Biodiesel production directly from oils with high acid value by magnetic Na_2SiO_3@ Fe_3O_4/C catalyst and ultrasound. Fuel 2015;150:370–7.

[70] Sarve A, Sonawane SS, Varma MN. Ultrasound assisted biodiesel production from sesame (*Sesamum indicum* L.) oil using barium hydroxide as a heterogeneous catalyst: comparative assessment of prediction abilities between response surface methodology (RSM) and artificial neural network (ANN). Ultrasonics Sonochemistry 2015;26:218–28.

[71] Martinez-Guerra E, Gude VG. Transesterification of used vegetable oil catalyzed by barium oxide under simultaneous microwave and ultrasound irradiations. Energy Conversion and Management 2014;88:633–40.

[72] Chen G, Shan R, Shi J, Yan B. Ultrasonic-assisted production of biodiesel from transesterification of palm oil over ostrich eggshell-derived CaO catalysts. Bioresource Technology 2014;171:428–32.

[73] Takase M, Zhang M, Feng W, Chen Y, Zhao T, Cobbina SJ, Yang L, Wu X. Application of zirconia modified with KOH as heterogeneous solid base catalyst to new non-edible oil for biodiesel. Energy Conversion and Management 2014;80:117–25.

[74] Eswararao Y, Niju S, Begum KMS, Anantharaman N. Investigation of boiler scale deposits as heterogeneous base catalyst for biodiesel production from jatropha oil. Biofuels 2017:1–6.

[75] Fadhil AB, Ahmed KM, Dheyab MM. *Silybum marianum* L. seed oil: a novel feedstock for biodiesel production. Arabian Journal of Chemistry 2017;10:S683–90.

[76] Fadhil AB, Al-Tikrity ET, Ibraheem KK. Transesterification of bitter almond oil as a new non-edible feedstock with mixed alcohols system: parameter optimization and analysis of biodiesel. Waste and Biomass Valorization 2017:1–12.

[77] Fadhil AB, Al-Tikrity ET, Khalaf AM. Transesterification of non-edible oils over potassium acetate impregnated CaO solid base catalyst. Fuel 2018;234:81–93.

[78] Sivakumar P, Sankaranarayanan S, Renganathan S. Studies on sono-chemical biodiesel production using smoke deposited nano MgO catalyst. Bulletin of Chemical Reaction Engineering and Catalysis 2013;8(2):89–96.

[79] Choudhury HA, Goswami PP, Malani RS, Moholkar VS. Ultrasonic biodiesel synthesis from crude Jatropha curcas oil with heterogeneous base catalyst: mechanistic insight and statistical optimization. Ultrasonics Sonochemistry 2014;21(3):1050–64.

[80] Gaikwad ND, Gogate PR. Synthesis and application of carbon based heterogeneous catalysts for ultrasound assisted biodiesel production. Green Processing and Synthesis 2015;4(1):17–30.

[81] Pukale DD, Maddikeri GL, Gogate PR, Pandit AB, Pratap AP. Ultrasound assisted transesterification of waste cooking oil using heterogeneous solid catalyst. Ultrasonics Sonochemistry 2015;22:278–86.

[82] Badday AS, Abdullah AZ, Lee KT. Ultrasound-assisted transesterification of crude Jatropha oil using cesium doped heteropolyacid catalyst: interactions between process variables. Energy 2013;60: 283–91.

[83] Gopinath S, Kumar PV, Kumar PSM, Arafath KY, Sivanesan S, Baskaralingam P. Cs–tungstosilicic acid/ Zr-KIT-6 for esterification of oleic acid and transesterification of non-edible oils for green diesel production. Fuel 2018;234:824–35.

[84] Hasni K, Ilham Z, Dharma S, Varman M. Optimization of biodiesel production from *Brucea javanica* seeds oil as novel non-edible feedstock using response surface methodology. Energy Conversion and Management 2017;149:392–400.

[85] Hoseini SS, Najafi G, Ghobadian B, Mamat R, Ebadi MT, Yusaf T. *Ailanthus altissima* (tree of heaven) seed oil: characterisation and optimisation of ultrasonication-assisted biodiesel production. Fuel 2018;220:621–30.

[86] Iyyaswami R, Halladi VK, Yarramreddy SR, Bharathaiyengar SM. Microwave-assisted batch and continuous transesterification of karanja oil: process variables optimization and effectiveness of irradiation. Biomass Conversion and Biorefinery 2013;3(4):305–17.

[87] İlkılıç C, Çılğın E, Aydın H. Terebinth oil for biodiesel production and its diesel engine application. Journal of the Energy Institute 2015;88(3):292–303.

[88] Kumar M, Sharma MP. Kinetics of transesterification of Chlorella protothecoides microalgal oil to biodiesel. Waste and Biomass Valorization 2016;7(5):1123–30.

[89] Joshi SM, Gogate PR, Kumar SS. Intensification of esterification of karanja oil for production of biodiesel using ultrasound assisted approach with optimization using response surface methodology. Chemical Engineering and Processing–Process Intensification 2018;124:186–98.

[90] Kansedo J, Lee KT. Process optimization and kinetic study for biodiesel production from non-edible sea mango (*Cerbera odollam*) oil using response surface methodology. Chemical Engineering Journal 2013;214:157–64.

[91] Kshirsagar CM, Anand R. Homogeneous catalysed biodiesel synthesis from Alexandrian Laurel (*Calophyllum inophyllum* L.) kernel oil using ortho-phosphoric acid as a pretreatment catalyst. International Journal of Green Energy 2017;14(9):754–64.

[92] Kamel DA, Farag HA, Amin NK, Fouad YO. Biodiesel synthesis from non-edible oils by transesterification using the activated carbon as heterogeneous catalyst. International Journal of Environmental Science and Technology 2017;14(4):785–94.

[93] Taufiq-Yap YH, Lee HV, Yunus R, Juan JC. Transesterification of non-edible *Jatropha curcas* oil to biodiesel using binary Ca–Mg mixed oxide catalyst: effect of stoichiometric composition. Chemical Engineering Journal 2011;178:342–7.

[94] Paiva EJ, da Silva MLC, Barboza JC, de Oliveira PC, de Castro HF, Giordani DS. Non-edible *babassu* oil as a new source for energy production–a feasibility transesterification survey assisted by ultrasound. Ultrasonics Sonochemistry 2013;20(3):833–8.

[95] Nizah MR, Taufiq-Yap YH, Rashid U, Teo SH, Nur ZS, Islam A. Production of biodiesel from non-edible Jatropha curcas oil via transesterification using Bi_2O_3–La_2O_3 catalyst. Energy Conversion and Management 2014;88:1257–62.

[96] Nurdin S, Rosnan NA, Ghazali NS, Gimbun J, Nour AH, Haron SF. Economical biodiesel fuel synthesis from castor oil using mussel shell-base catalyst (MS–BC). Energy Procedia 2015;79:576–83.

[97] Lee HV, Yunus R, Juan JC, Taufiq-Yap YH. Process optimization design for jatropha-based biodiesel production using response surface methodology. Fuel Processing Technology 2011;92(12):2420–8.

[98] Kumar KR, Chandrika K, Prasanna KT, Gowda B. Biodiesel production and characterization from non-edible oil tree species *Aleurites trisperma Blanco*. Biomass Conversion and Biorefinery 2015;5(3):287–94.

[99] Kumar G, Kumar D, Johari R. Time reducing process for biofuel production from non-edible oil assisted by ultrasonication. Ultrasonics Sonochemistry 2014;21(5):1618–23.

[100] Kumar M, Sharma MP. Optimization of transesterification of *Chlorella protothecoides* oil to biodiesel using Box–Behnken design method. Waste and Biomass Valorization 2016;7(5):1105–14.

[101] Reyes-Trejo B, Guerra-Ramírez D, Zuleta-Prada H, Cuevas-Sánchez JA, Reyes L, Reyes-Chumacero A, Rodríguez-Salazar JA. *Annona diversifolia* seed oil as a promising non-edible feedstock for biodiesel production. Industrial Crops and Products 2014;52:400–4.

[102] Sankaranarayanan S, Antonyraj CA, Kannan S. Transesterification of edible, non-edible and used cooking oils for biodiesel production using calcined layered double hydroxides as reusable base catalysts. Bioresource Technology 2012;109:57–62.

[103] Ullah K, Ahmad M, Qureshi FA, Qamar R, Sharma VK, Sultana S, Zafar M. Synthesis and characterization of biodiesel from Aamla oil: a promoting non-edible oil source for bioenergy industry. Fuel Processing Technology 2015;133:173–82.

[104] Teo SH, Islam A, Ng CH, Mansir N, Ma T, Choong ST, Taufiq-Yap YH. Methoxy-functionalized mesostructured stable carbon catalysts for effective biodiesel production from non-edible feedstock. Chemical Engineering Journal 2018;334:1851–68.

[105] Sivakumar P, Sindhanaiselvan S, Gandhi NN, Devi SS, Renganathan S. Optimization and kinetic studies on biodiesel production from underutilized *Ceiba Pentandra* oil. Fuel 2013;103:693–8.

[106] Lee HV, Juan JC, Taufiq-Yap YH. Preparation and application of binary acid–base CaO–La_2O_3 catalyst for biodiesel production. Renewable Energy 2015;74:124–32.

[107] Gole VL, Gogate PR. Intensification of synthesis of biodiesel from nonedible oils using sonochemical reactors. Industrial & Engineering Chemistry Research 2012;51(37):11866–74.
[108] Latchubugata CS, Kondapaneni RV, Patluri KK, Virendra U, Vedantam S. Kinetics and optimization studies using response surface methodology in biodiesel production using heterogeneous catalyst. Chemical Engineering Research and Design 2018;135:129–39.
[109] Teo SH, Rashid U, Taufiq-Yap YH. Green nano-catalyst for methanolysis of non-edible Jatropha oil. Energy Conversion and Management 2014;87:618–27.
[110] Takase M, Chen Y, Liu H, Zhao T, Yang L, Wu X. Biodiesel production from non-edible *Silybum marianum* oil using heterogeneous solid base catalyst under ultrasonication. Ultrasonics Sonochemistry 2014;21(5):1752–62.
[111] Reddy ANR, Saleh AA, Islam MS, Hamdan S, Maleque MA. Biodiesel production from crude Jatropha oil using a highly active heterogeneous nanocatalyst by optimizing transesterification reaction parameters. Energy & Fuels 2015;30(1):334–43.
[112] Kumar D, Singh B, Banerjee A, Chatterjee S. Cement wastes as transesterification catalysts for the production of biodiesel from Karanja oil. Journal of Cleaner Production 2018;183:26–34.
[113] Sarve A, Varma MN, Sonawane SS. Optimization and kinetic studies on biodiesel production from Kusum (*Schleichera triguga*) oil using response surface methodology. Journal of Oleo Science 2015;64(9):987–97.
[114] Meloni D, Perra D, Monaci R, Cutrufello MG, Rombi E, Ferino I. Transesterification of *Jatropha curcas* oil and soybean oil on Al-SBA-15 catalysts. Applied Catalysis B: Environmental 2016;184:163–73.
[115] Muthukumaran C, Praniesh R, Navamani P, Swathi R, Sharmila G, Kumar NM. Process optimization and kinetic modeling of biodiesel production using non-edible *Madhuca indica* oil. Fuel 2017;195:217–25.
[116] Narula V, Khan MF, Negi A, Kalra S, Thakur A, Jain S. Low temperature optimization of biodiesel production from algal oil using CaO and CaO/Al_2O_3 as catalyst by the application of response surface methodology. Energy 2017;140:879–84.
[117] Shaheen A, Sultana S, Lu H, Ahmad M, Asma M, Mahmood T. Assessing the potential of different nano-composite (MgO, Al_2O_3-CaO and TiO_2) for efficient conversion of *Silybum eburneum* seed oil to liquid biodiesel. Journal of Molecular Liquids 2018;249:511–21.
[118] Senthilkumar C, Krishnaraj C, Sivakumar P, Sircar A. Statistical optimization and kinetic study on biodiesel production from a potential non-edible bio-oil of wild radish. Chemical Engineering Communications 2018:1–10.
[119] Silitonga AS, Masjuki HH, Mahlia TMI, Ong HC, Kusumo F, Aditiya HB, Ghazali NNN. *Schleichera oleosa* L oil as feedstock for biodiesel production. Fuel 2015;156:63–70.
[120] Solis JL, Alejo L, Kiros Y. Calcium and tin oxides for heterogeneous transesterification of Babasssu oil (*Attalea speciosa*). Journal of Environmental Chemical Engineering 2016;4(4):4870–7.
[121] Parthiban KS, Perumalsamy M. Kinetic studies on oil extraction and biodiesel production from underutilized *Annona squamosa* seeds. Fuel 2016;180:211–7.
[122] Choudhury HA, Malani RS, Moholkar VS. Acid catalyzed biodiesel synthesis from Jatropha oil: mechanistic aspects of ultrasonic intensification. Chemical Engineering Journal 2013;231:262–72.
[123] Ullah K, Ahmad M, Qiu F. Assessing the experimental investigation of *milk thistle* oil for biodiesel production using base catalyzed transesterification. Energy 2015;89:887–95.

CHAPTER 14

Production of biodiesel and its application in engines

Shailendra Kumar Shukla, Pushpendra Kumar Singh Rathore
Centre for Energy and Resources Development, Department of Mechanical Engineering, Indian Institute of Technology (BHU), Varanasi, Uttar Pradesh, India

14.1 Introduction

Energy is one of the most important building blocks in human development, and as such acts as a key factor in determining the economic development of all the countries. In an effort to meet the demands of a developing nation, the energy sector has witnessed rapid growth. This increase in the energy demand has been met using fossil resources, which leads to fossil fuel depletion, increase in the price and serious environmental impacts. A major portion of the global energy demand is used by the transportation sector and consequently responsible for the highest greenhouse gas (GHG) emissions. In India, the transport sector is the second-largest consumer of fossil-based fuels after the industry sector. The transport sector consumes 66% and 99% of diesel and petrol respectively. Currently, diesel alone meets an estimated 46% of transportation fuel demand, followed by petrol at 24%. Combustion of the petroleum-based product leads to the generation of carbon monoxide (produced when combustion is incomplete), nitrogen oxides (produced when combustion occurs at very high temperature), sulfur oxides (produced when elemental sulfur is present in the fuel), and particulates. In India, the transport sector is the third largest GHG producer after electricity and industry sector.

Limited reserves of fossil fuel have caused to search for the new and alternative sources of the energy that would be economically efficient, socially equitable, environmentally sound, and meets the increased demand for the energy. Biofuels can act as a potential alternative of fossil-based fuels and have all the above-mentioned features. Several types of biofuels such as biogas, syngas, biodiesel, and ethanol can be a potential replacement of fossil-based petroleum products. All these fuels are renewable in nature as they are produced from different types of biomass such as vegetable oils, sugar cane, wheat, agricultural waste, municipal waste, etc. Among these fuels, biodiesel appears to have exceptional

Waste Biorefinery. https://doi.org/10.1016/B978-0-12-818228-4.00014-9

importance as they are renewable and widely available, biodegradable, nontoxic, and environmentally friendly. Biodiesel is a clean-burning alternative fuel, produced from renewable resources like pure or used vegetable oil (edible and nonedible). It can be stored just like petroleum diesel fuel and hence does not require any separate infrastructure for storage. Biodiesel can be used as a potential fuel for compression ignition (diesel) engines without any significant technical modification in the engines. Additionally, it has very similar physicochemical properties like diesel. Table 14.1 shows the comparison of physicochemical properties between the biodiesel and diesel. Biodiesel can reduce the emissions of unburned hydrocarbons (HC), carbon monoxide (CO), and oxides of nitrogen (NO_X) [1].

Use of biodiesel in conventional diesel engines results in a substantial reduction of unburned hydrocarbon, CO, and PM. Its high cetane number improves the ignition quality even when blended with conventional diesel. Seeds from the plants such as jatropha (*Atrophy curcas*), mahua (*Madhuca Indica*), karanja (*Pongamia pinnata*), neem (*Azadirachta indica*), and castor (*Ricinus communis*) could be widely used for its production [2]. The nonedible oilseeds of trees such as Karanja, Jatropha, Mahua, and neem are being considered as sources of straight vegetable oil and biodiesel [3].

Table 14.1: Comparative chart of physicochemical properties of diesel and biodiesel.

Parameter	Diesel	Disadvantage	Biodiesel	Benefits
Cetane number	45	High emissions and noise	60	Avoid knocking
Carbon residue	0.35%	Carbon deposits in the engine and in fuel pump	0.002%	Optimized maintenance
Flash point (°C)	60	Flammable and explosive	145	Safe
Sulfur	0.50	Contributes to PM emissions and wear of the system	Nil	Environmental friendly
Biodegradability	No	Contributes to Green House effect	Yes	Safe for environmentally sensitive areas and aquatic life
Oxygen	Nil	Poor combustion/ Incomplete combustion	10%	Increase in net fuel efficiency
Aromatic	5%	Increased engine deposits and tailpipe emissions	Nil	Minimized soot emissions
Lubricity	2000–5000	High wear and tear	>7000	Protection against wearing, scuffing

14.2 Biodiesel production

Biodiesel is mainly produced from vegetable oil and animal fat via transesterification reaction with alcohols, usually methanol, to form fatty acid methyl esters (FAME). The end products of the transesterification process are raw biodiesel and raw glycerol. In a further process, these raw products undergo a cleaning step. In the case of using methanol as alcohol FAME biodiesel is produced. The purified glycerol can be used in the food and cosmetic industries, as well as in the chemical industry. The glycerol can also be used as a substrate for the anaerobic digestion. Several such methods have been reported for quantitative or semiquantitative determination of FAMEs in biodiesel and for common metric classification of biodiesel feedstock. Those include optical spectroscopic methods, various chromatographic methods, commonly with mass spectrometry (MS) detectors [4—7]. Vegetable oil can be converted into diesel by using the following methods:

14.2.1 Direct blending

In this method, the raw vegetable oil is directly mixed with the diesel. This blending is done on the basis of certain proportion. However, this method has certain drawbacks like high viscosity, high acid value and gum formation which makes its use difficult in engines.

14.2.2 Microemulsions

This process involves the mixing of vegetable oil with some suitable emulsifying agents like methanol, ethanol or butanol. This process faces drawbacks of incomplete combustion and carbon deposits.

14.2.3 Catalytic cracking

This process involves the catalytic transformation of the nonedible oil to the liquid products having similar properties to that of diesel in the absence of air or oxygen. The pyrolyzed material contains a sufficient amount of sulfur, moisture, and sediments.

14.2.4 Transesterification

The transesterification process is a reversible reaction and carried out by mixing the reactants—fatty acids, alcohol, and catalyst as shown in Fig. 14.1. A strong base or a strong acid can be used as a catalyst. At the industrial scale, mostly sodium or potassium methanol is used. Fig. 14.2 depicts a transesterification unit located at IIT (BHU) and is used for the production of biodiesel. The percentage of biodiesel yield can be evaluated by using the following equation:

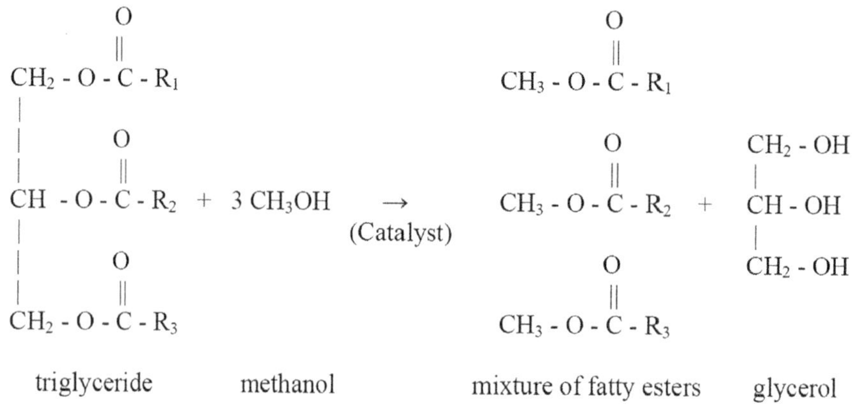

Figure 14.1
Transesterification reaction.

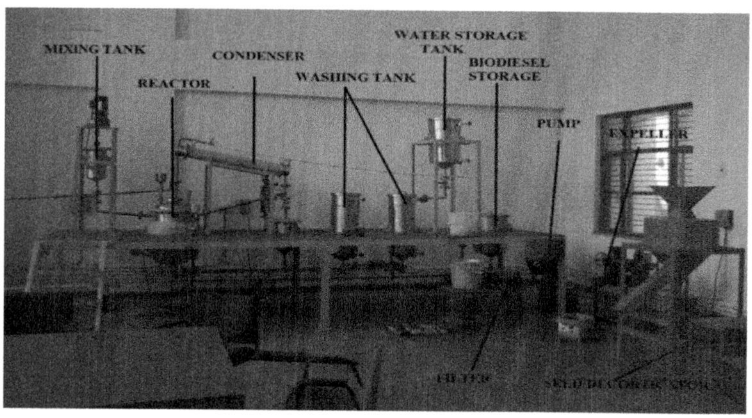

Figure 14.2
Biodiesel transesterification unit.

$$\text{Process yield } (\%) = \frac{\text{Pure biodiesel (g)}}{\text{Oil used (g)}} \times 100$$

14.3 Policy considerations

Policies and regulations play a very vital role in the growth and development of biodiesel production. Stable and long-term policy frameworks can facilitate the expansion of the advanced biofuels industry and enable capital and production cost reduction potential. Several countries have implemented policy measures for the increased production and use of biodiesel. In Brazil, the National Biodiesel Production and Use program was launched in 2004 to increase the uptake of the biodiesel in the country. In 2005, Brazilian government mandated a blend of minimum 2% biodiesel by 2008 which was to increase to 5% by 2013. Brazil's current standard states a 27% blend of the ethanol in the gasoline and for the biodiesel, it is 10% [8]. China has become the fourth-largest producer of biofuels in the world after the United States, Brazil, and the European Union, because of several policy measures that were involved in last four five-year plans for the development of renewable biofuels. Chinese implementation plan concerning the Expansion of Ethanol Production and Promotion for transportation fuel was jointly announced by the National Development and Reform Commission (NDRC), National Energy Administration (NEA), Ministry of Finance, and 12 other Ministries in 2017. The plan calls for China to achieve nationwide use of 10% ethanol (E10) by 2020. The United States provides several types of incentives on the production and usage of the biodiesel. Tax credits for blending provide $1.00 per gallon of the biodiesel, agribiodiesel, or renewable diesel that is blended with the petroleum diesel to produce a mixture that includes at least 0.1% diesel fuel [9]. Related tax credits exist for the delivery of 100% biodiesel as an on-road fuel. In India on May 2018, the Indian government has approved the National Policy on Biofuels. The Policy categorizes biofuels as "basic biofuels" viz. first-generation (1G) bioethanol and biodiesel and "advanced biofuels" second-generation (2G) ethanol, municipal solid waste (MSW) to drop-in fuels, third-generation (3G) biofuels, bio-CNG, etc. to enable the extension of appropriate financial and fiscal incentives under each category. The Policy expands the scope of raw material for ethanol production by allowing the use of sugarcane juice, sugar-containing materials such as sugar beet, sweet sorghum, and starch-containing materials such as corn, cassava, damaged food grains (wheat, broken rice, rotten potatoes) unfit for human consumption are widely used for ethanol production. In India, with an emphasis on advanced biofuels, the policy indicates a viability gap funding scheme for 2G ethanol, biorefineries of Rs. 5000 crore in six years in addition to additional tax incentives and higher purchase price as compared to 1G biofuels. The policy also encourages setting up of supply chain mechanisms for the biodiesel production from nonedible oilseeds, used cooking oil and short gestation crops [10].

14.4 Life-cycle and economic analysis

A Life-Cycle Assessment (LCA) is a tool that is used to determine or assess the impact of a product or a process on the environment. It evaluates to use the energy and raw material consumption, wastes and emissions of a product's life cycle. LCA on waste cooking oils and virgin oil showed that the production of the biodiesel from the waste cooking oil was more eco-friendly in comparison to the virgin oil [11]. The processes using waste cooking oil have the greatest impact on the resources, whereas the processes using virgin oil have the greatest impact on human health. The economic analysis suggests that the production of the biodiesel from waste cooking oil has a reasonable low payback period. LCA of biodiesel in the Greek transport sector revealed that from the environmental point of view the biodiesel appeared attractive because its use resulted in significant reductions of

GHG emissions in comparison to gasoline and diesel [12]. It also has lower well-to-wheel emissions of methane. However, the use of biodiesel as transportation fuel increases emissions of PM10, nitrous oxide, nitrogen oxides (NOx) as well as nutrients such as nitrogen and phosphorous; the latter is the main agents for eutrophication. LCA of the biofuel production process from sunflower oil, rapeseed oil, and soybean oil suggested that the processes involved have a greater environmental impact. This impact was reflected in general in all impact categories, highlighting the category of carcinogens, respiratory inorganics, a decline of fossil fuels and land use. Thus, it was advisable to use rapeseed and soybean for biofuel production because their impact was less than those from the sunflower seed. A study on life-cycle energy and GHG emission analysis of using biodiesel in the United States showed that soy biodiesel could achieve an 80% reduction in fossil energy consumption and 66%–72% reduction in overall GHG emissions, relative to its petroleum counterpart [13]. LCA of soybean biodiesel produced in the Pampean region of Argentina and its analysis on the influence of different tillage systems on the Energy Return on Investment (EROI) showed that agricultural practices that increased the crop yields did not always lead to energy improvement, as additional flows of the materials and energy was needed to optimize the productivity [14]. LCA of biodiesel production from soybean, jatropha, and microalgae in China showed that the choice of allocation method, transport distance, uncertainty in jatropha and microalgae yield and oil content, and recycling rate of harvest water of microalgae have a significant influence on the life cycle environmental performance of biodiesel [15]. An economic analysis on the production of biodiesel from waste cooking oil (dedicated to the production of 39,208 metric tons of biodiesel per year) was done by Avinash and Murugesan [16]. The total capital investment on this project was estimated to be Rs. 1615.133 millions. At a value of Rs. 15.00 per kg for feedstock waste cooking oil, biodiesel cost of Rs. 55.00 per kg was estimated. Along with the main revenue from biodiesel, additional revenue of around Rs. 22.00 per kg could be generated from the sale of glycerol. On the whole, the profitability analysis of the study disclosed that the modeled biodiesel production plant has the ability to produce biodiesel with a unit production cost of Rs. 51.00 per kg and unit production revenue of Rs. 58.00 per kg.

14.5 Case studies

To mitigate the climate change scenario, several studies have been conducted around the globe for finding out the compatibility of biodiesel in internal combustion (IC) engines. Since the major application of the biodiesel is in transport sector therefore, it is important to evaluate the performance characteristics of the biodiesel in terms of emissions, efficiency, specific fuel consumption and output power. Several studies have been conducted to optimize the performance of the biodiesel in the engines. Carraretto et al. [17] experimentally examined the combustion characteristics of methyl esters of neem, waste cooking oil and their diesel blends in a diesel engine. Results suggested that up to

30% of methyl esters did not affect the performance, combustion, and emission characteristics and above B30, led to the reduction in performance [18]. It was evident in the study that for all test fuels the brake thermal efficiency increased with an increase in brake power and blends up to B30 had a maximum brake thermal efficiency. With an increase in the biodiesel blends the value of brake-specific fuel consumption (BSFC) also increased.

BSFC is the parameter, which reflects the rate of fuel consumption to produce output power and depends upon the RPM, torque, and fuel consumption rate of the engine. The diesel engine produces a lesser amount of CO and HC emissions than spark-ignition engines. Moreover, in case of biodiesel fueled engines, presence of airborne oxygen as well as its presence in the molecules of biodiesel aids nearly complete combustion of fuel. The emission of diesel at maximum load was noted to be 960 ppm, whereas for B100 it was 890 ppm. This reduced NO_x emission for B100 biodiesel, when compared to diesel, maybe due to the reduced premixed combustion rate leading to lower NO_x emissions for B100 biodiesel operation. The experimental results proved that up to B30 blend of biodiesel blends, the performance and emission characteristics were not much affected. When the blend ratio increases, incomplete combustion takes place because of the less time available for mixture formation, which leads to a reduction in the brake thermal efficiency of the engine as well as an increase in the emission level. The combustion analysis revealed that the overall combustion characteristics of B30 biodiesel blends were closer to the diesel than pure biodiesel.

Liaquat et al. [19] studied the effect of storage on the physio-chemical properties of biodiesel produced from *Ricinus communis* (Castor), *Hevea brasiliensis*(Rubber), *Gossypium hirsutum* (Cotton), *Azadirachta indica* (Neem), *Glycinmax* (Soyabean), and *Jatropha curcas* (Jatropha oils) stored in an open-air environment for a period of 10 months. The peroxide value increased with storage. Muralidharan and Vasudevan [20] determined the performance and emission characteristics of CNG and neem blends in CI engine. The maximum achievable neem biodiesel replacement by natural gas varied with engine loads. From the Comparison of results, CNG1 (4% CNG + 96% neem oil), CNG3 (8%CNG + 92% neem oil), CNG5 (12% CNG + 88% neem oil) were optimum. The CNG fueled engine produced more power and pressure cylinder than that of neem oil operation.

However, biodiesel as an engine fuel seemingly is very cost-competitive overall compared with methanol and CNG alternative fuels [21]. For example, converting a vehicle to run on CNG or methanol requires huge infrastructure investment, whereas minor modifications are required to run the vehicle on biodiesel The increase in the brake thermal efficiency is attributed mainly due to the higher ignition delay, which allows more time for preflame reaction in CNG air mixture, which on initiation of combustion results in a higher rate of heat release. The neem-CNG operation produced lower emission of CO, CO_2 for all operating conditions. Overall, the neem-CNG dual-fueled engines have a great possibility to be comparable to that of neem oil. Nevertheless, to reach the optimum performance, the CNG fueled engine required some modification that may be studied further. A study with different blends (B10, B20) of neem oil and diesel at various loads showed that the brake thermal efficiency of diesel was slightly higher at all the loads, followed by the blends of the neem oil and diesel [22]. It has been established that 20% of neem oil biodiesel can be used as a substitute for diesel without any engine modification, thus neem oil as nonedible oil could be a good renewable raw material for biodiesel production. Studies have shown that the blends of neem oil and diesel could be successfully used with acceptable performance up to a certain extent [23]. However, due to some properties of neem oil, it cannot be used directly as IC engine fuel due to higher viscosity and density, which will result in low volatility and poor atomization of oil during oil injection in the combustion chamber, causing incomplete combustion and carbon deposits. The biodiesel blends produce lower brake thermal efficiency and higher brake specific fuel consumption than diesel because of low calorific value. The properties results of all blends show that blends up to 20% of neem oil have a value of viscosity and density equivalent to the specified range for IC engine fuel, therefore it can be concluded that up to 20% blends can be used to run the IC engine at short term basis [23]. Performance, and emissions of different blends (B10, B20, and B40) of PME, JME, and NME in comparison to diesel has shown that the B20 has a closer performance to the diesel and the B100 had lower brake thermal efficiency mainly due to its high viscosity compared to the diesel. However, its diesel blends showed reasonable efficiencies, lower smoke, CO and HC. Jatropha and neem-based methyl esters (biodiesel) can be directly used in the diesel engines without any engine modifications. Brake thermal efficiency of B10, B20, and B40 blends are better than B100 but still inferior to diesel. Properties of different blends of biodiesel were very close to the diesel and B20 gave best results. It is not advisable to use B100 in IC engines unless its properties are comparable with diesel fuel. Smoke, HC, CO emissions at different loads were higher for diesel, compared to B10, B20, B40 blends [24]. The brake thermal efficiency was reduced by about 5% for castor oil ester when compared to diesel. The brake specific fuel consumption was increased for neem oil ester by about 11%–13% when compared to diesel fuel. The brake power was reduced by about 12% for neem oil ester when compared to that of diesel. The carbon mono oxide was reduced for neem oil ester by about 16% when compared to that of diesel. It was concluded that the carbon monoxide emission for vegetable oil ester was less when compared to diesel fuel. The concentration of hydrocarbon decreased 15% for neem oil ester when compared to diesel fuel. The formation of nitric oxides was decreased by about 3% for neem oil ester when compared to that of diesel fuel. The smoke level was decreased by about 12% for neem oil ester when compared to diesel fuel. Thus, the multi-zone combustion model could be an efficient tool to calculate the effect of design and operating parameter. Hence, in general it could be concluded that in terms of the performance characteristics and emissions, vegetable oil esters could be regarded as a potential substitute for diesel fuel. The neem-based methyl esters (biodiesel) can be directly used in the diesel engines without any major engine modifications.

De and Panua [24] analyzed the effects on diesel engine performance when fueled with the blends of biodiesel and diesel in various proportions on a volume basis [25]. It was concluded that the biodiesel blends were satisfactory in the diesel engine without any major modifications in the hardware of the system. The fuel consumption of the engine was somewhat higher at low loads and speeds due to lower gross heat of combustion, and the mass of the fuel consumed increased with the increase in the injection pressure. The BTE of neem blends were lower than the diesel throughout the entire range, showing poor combustion characteristics of methyl ester due to the high viscosity and poor volatility. The emissions of the hydrocarbons and carbon monoxide were considerably reduced for all biodiesel and additive blends as the injection pressure increased, the emissions go on decreasing due to the complete combustion of fuels. The knocking was not observed for biodiesel blend at all the operating conditions [26].

14.6 Conclusions and perspectives

During the last century, the consumption of energy has increased a lot due to the change in the lifestyle and significant growth of the population. A major portion of this energy is utilized by the transportation sector, which makes it a major source of GHG emissions. Biodiesel can play a crucial role in reducing the dependency on diesel and petrol fuels for transportation. Because of large similarity in physical and chemical properties of the diesel and biodiesel, the latter one can be used as an alternative fuel for diesel engines. The biodiesel with different volume proportion in its different blendings, such as B10, B20, and B30 could be used in diesel engines. Engine performance properties such as brake power, mechanical efficiency, and brake thermal efficiency generally show comparable performance with biodiesel blends.

Biodiesel is technically and economically competitive to diesel for its application in engines. However, the lack of policies and regulations seemingly create a bottleneck for its growth and development. The governments must focus on the infrastructure side of these alternative fuels and must provide every possible support in terms of research and development, technologically, economically. As far process development for the production of biodiesel, improvement, and modification of continuous transesterification and recovery of high-quality glycerol are needed. Also, use of various fuels additive for improving properties, better engine performance, and emission control should be looked into.

References

[1] Demirbas A. Biodiesel. Springer London; 2008. p. 111–9.
[2] Gupta NK, Rathore PS, Sinha S. Biodiesel production from waste cooking oil using ultrasonic cavitation & its characteristics. In: 2017 International conference on advances in mechanical, industrial, automation and management systems (AMIAMS). IEEE; 2017, February. p. 139–43.
[3] Radha VK, Manikandan G. Novel production of biofuels from neem oil. Chennai, India: Department of chemical Engineering, Anna University; 2011.
[4] Van Gerpen JH, Peterson CL, Goering CE. Biodiesel: an alternative fuel for compression ignition engines. In: 2007 Agricultural equipment technology conference Louisville, Kentucky, USA; 2007.
[5] Galib AA, Roknuzzaman M. Biodiesel from Jatrofa oil as alternative fuel for diesel engine. Khulna, Bangladesh: KUET; 2009.
[6] Azam MM, Waris A, Nahar NM. Prospects and potential of fatty acid methyl esters of some non-traditional seed oils for use as biodiesel in India. Biomass and Bioenergy 2005;29(4):293–302.
[7] Kumar A, Shukla SK, Tirkey JV. A review of research and policy on using different biodiesel oils as fuel for C.I. Engine. In: 5th International conference on advances in energy research. Mumbai, India: ICAER 2015; 2015.
[8] U.S. Department of Agriculture (USDA). Brazil biofuels annual 2016; GAIN report number BR 16009; Brazilian law 13.263/2016. Washington, DC, USA: USDA; 2016.
[9] U.S. Public Law 114-113. Available from: https://www.gpo.gov/fdsys/pkg/PLAW-114publ113/html/PLAW-114publ113.htm.
[10] Press Information Bureau. Government of India. 2018. Available from: http://pib.nic.in/newsite/PrintRelease.aspx?relid=179313.
[11] Varanda MG, Pinto G, Martins F. Life cycle analysis of biodiesel production. Fuel Processing Technology 2011;92(5):1087–94.
[12] Nanaki EA, Koroneos CJ. Comparative LCA of the use of biodiesel, diesel and gasoline for transportation. Journal of Cleaner Production 2012;20(1):14–9.
[13] Chen R, Qin Z, Han J, Wang M, Taheripour F, Tyner W, O'Connor D, Duffield J. Life cycle energy and greenhouse gas emission effects of biodiesel in the United States with induced land use change impacts. Bioresource Technology 2018;251:249–58.
[14] Piastrellini R, Arena AP, Civit B. Energy life-cycle analysis of soybean biodiesel: effects of tillage and water management. Energy 2017;126:13–20.
[15] Hou J, Zhang P, Yuan X, Zheng Y. Life cycle assessment of biodiesel from soybean, jatropha and microalgae in China conditions. Renewable and Sustainable Energy Reviews 2011;15(9):5081–91.
[16] Avinash A, Murugesan A. Economic analysis of biodiesel production from waste cooking oil. Energy Sources, Part B: Economics, Planning and Policy 2017;12(10):890–4.
[17] Carraretto C, Macor A, Mirandola A, Stoppato A, Tonon S. Biodiesel as alternative fuel: experimental analysis and energetic evaluations. Energy 2004;29(12–15):2195–211.
[18] Subramaniam D, Murugesan A, Avinash A. A comparative estimation of CI engine fuelled with methyl esters of punnai, neem and waste cooking oil. International Journal of Energy and Environment 2013;4(5):859–70.
[19] Liaquat AM, Masjuki HH, Kalam MA, Fattah IR, Hazrat MA, Varman M, Shahabuddin M. Effect of coconut biodiesel blended fuels on engine performance and emission characteristics. Procedia Engineering 2013;56:583–90.
[20] Muralidharan K, Vasudevan D. Performance, emission and combustion characteristics of a variable compression ratio engine using methyl esters of waste cooking oil and diesel blends. Applied Energy 2011;88(11):3959–68.
[21] Ahouissoussi NB, Wetzstein ME. A comparative cost analysis of biodiesel, compressed natural gas, methanol, and diesel for transit bus systems. Resource and Energy Economics 1998;20(1):1–15.

[22] Kumar V, Anuprasad SG, Mahesh BG. Production of bio-diesel to neem oil and its performance and emission analysis in two stroke diesel engine. International Journal of Engineering Science and Technology 2013;5:391–5.
[23] Tamboli Y, Selokar GR, Paul A, Zala J. Feasibility Testing of VCR Engine using various blend of neem oil. International Journal of Innovations in Engineering and Technology 2013;2:170–82.
[24] De B, Panua RS. An experimental study on performance and emission characteristics of vegetable oil blends with diesel in a direct injection variable compression ignition engine. Procedia Engineering 2014;90:431–8.
[25] Demirbaş A. Biodiesel fuels from vegetable oils via catalytic and non-catalytic supercritical alcohol transesterifications and other methods: a survey. Energy Conversion and Management 2003;44(13):2093–109.
[26] Raheman H, Ghadge SV. Performance of compression ignition engine with mahua (*Madhuca indica*) biodiesel. Fuel 2007;86(16):2568–73.

Further reading

[1] Requena JS, Guimaraes AC, Alpera SQ, Gangas ER, Hernandez-Navarro S, Gracia LN, Martin-Gil J, Cuesta HF. Life Cycle Assessment (LCA) of the biofuel production process from sunflower oil, rapeseed oil and soybean oil. Fuel Processing Technology 2011;92(2):190–9.

SECTION E

Sewage sludge biorefinery

CHAPTER 15

A biorefinery approach for sewage sludge

Ayan Banerjee[1,2], Thallada Bhaskar[1,2], Debashish Ghosh[1,2]
[1]Material Resource Efficiency Division, CSIR-Indian Institute of Petroleum, Dehradun, Uttarakhand, India; [2]Academy of Scientific and Innovative Research (AcSIR), New Delhi, India

15.1 Introduction

The temporal shifts in civilization with the advent of technology and consequent affluence have since long contributed to emerging challenges with undesirable outcomes. The rapid industrial development and urban sprawl led to uneven settlement patterns posing a serious challenge to waste management. In the early stages of development, the land was utilized for waste disposal regardless of the waste's characteristics [1], with growing population and demand for natural resources, new technologies for consumption were discovered, which further increased the amount of waste. As a result, new methods of handling, treating, and disposing of the waste were traced out. Sewage is a major concern for human settlements which was directed to sewage treatment plants (STPs) in the developed economies [2], the underdeveloped regions of the world struggled with limited resources to treat sewage. Sewage treatment facilities improved but the sludge stream from the treatment plants became a concern as it required further treatment. When the treatment load on the plants increase, so increases the quantity of sludge which must be handled to avoid any public nuisance. The population explosion in the developing world has left the urban areas with limited space to expand the existing STP facilities which are already overburdened with million liters per day (MLD) of sewage inflow. Sewage sludge contains a large portion of organic pollutants like polyaromatic hydrocarbons (PAHs), polychlorinated biphenyls (PCBs), heavy metals and pathogens which on direct possible contact is detrimental to public health. Removal of these pollutants from the waste stream before it finds a way to the natural ecosystem is a challenge for public welfare [3]. Treating sewage sludge is a challenge but it is also a source of organic carbon, nitrogen, phosphorous, as well as inorganic compounds such as silicates, aluminates, etc., which if recycled can be utilized in industries and agriculture [4,5]. Looking into the perspective of the environment, sewage sludge holds a position of both asset and liability depending on

Waste Biorefinery. https://doi.org/10.1016/B978-0-12-818228-4.00015-0

its fate after the wastewater treatment process. It is estimated that sewage sludge treatment and management accounts for 50% of the cost for wastewater treatment and contributes to 40% of greenhouse gas emission related to wastewater [6]. The potential challenge today is the recovery of material from sewage sludge. Sludge is either directed to fertilizer or is incinerated for energy recovery but in both cases, there is either loss of energy or material. Sewage sludge when dried and used as fertilizer, the compounds adds up in the food chain affecting the consumers and when incinerated the recoverable materials left out in the ash gets disposed of in the landfills [7,8], to avoid such problems the pollutants must be removed and materials must be recovered. Achieving resource recovery within waste treatment facilities is a viable option for combating resource depletion. There are different conversion processes for sludge which can be broadly grouped into two classes thermochemical and biochemical. Pyrolysis, liquefaction, and gasification form the basis of thermochemical treatment while biochemical treatment is achieved by aerobic or anaerobic digestion. Utilizing a single process cannot lead to resource recovery as all of them pose limitation with respect to the final product. The quality of sludge varies with places and consumption pattern of the local population which mandates designing a process that can operate accordingly with material available in the sludge and product desired after treatment. Thus, there is a need for integrating different conversion processes that can be utilized for energy and material recovery [9]. Since loss of nutrient and recoverable compounds as heavy metals, salts and nutrients leads to further depletion of available resource stock, the approach should be of recirculating materials into market which means the compounds which are generally made unavailable due to a change in their physical form or chemical structure in which they are utilizable needs to be processed so that they can be reutilized in different activities. The heavy metals which form a complex with organic compound if recovered in elemental form can be reintroduced in the market. The organic matter which is majorly the product of photosynthesis somewhere at the beginning of the food chain can be used up in producing char to sequester carbon or can be used in methane production for energy generation. An understanding of multiple concepts and their application is necessary for designing a sewage sludge biorefinery. An integrated sewage sludge biorefinery will open the scope of obtaining multiple products within a single treatment plant and will lead to resource efficiency and waste minimization which are the current hurdles in sustainable development.

15.1.1 Sewage sludge: present status

The International Water Association in its wastewater report of 2018 stated that the developed countries have the capacity to treat 70% of sewage they generate while for the developing world the capacity stands at only 8% [10]. In India, about only 30% of wastewater is processed and the rest is discharged as untreated [11]. The unavailability of data for the generation, treatment and use of wastewater further exacerbates the challenge for the management of sewage sludge. Sato et al. reported after examining the wastewater data from 181 countries that only 30.4% of the assessed countries had data available for production, treatment and use of wastewater [12]. The available data in most cases are not recent and they may not be useful in estimating current trend and contemporary issues related to wastewater and sludge. China as a developing nation with 70% of its population expected to be urbanized by 2050 has 3513 wastewater treatment plants (WWTPs) which consume 17.5 billion kilowatts/hour of electricity which rather is mostly produced in thermal power plants [13,14]. The WWTPs contributed to 72% of methane and 26% of nitric oxide emission as of 2013 which are potent greenhouse gases and a major challenge for developing countries like China and India [15]. Even about 60% of the wasted sludge is sent for dewatering and thickening which utilize energy and renders valuable nitrogen and sulfur that can be utilized as chemical materials untapped [16].

In India, almost 80% of the water in domestic use is returned back as wastewater. It led to a generation of about 1.7 million tons of fecal waste per day where the treatment potential stood at 51% as of 2013, while it was only 8% for smaller cities. The central pollution control board reported a total 522 operational STPs with a total treatment capacity of 18,883.2 MLD and estimated 215 STPs to be installed which would make the total number of STPs both operational and not-operational to 816 with a treatment capacity of 23,277.36 MLD as of 2015 [17]. The estimated sewage generation was 29,129 MLD in 2017. The increasing rate of sewage generation will surpass the treatment capacity and it is a concern for future development and welfare. Public concern and a potential threat to public health have made the wastewater and sewage sludge a priority. Treatment advancement and resource recovery plays an important role in the developing world and will prove to be a new age approach to sustainability.

15.1.2 Wastewater treatment background: potential sources of sewage sludge

Most of the countries have a basic method of treating the municipal sewage sludge in the STP: In primary treatment, the sewage passes through the screen that removes bulk material and grit to eliminate any hindrance in further stages. The primary treatment covers the grit and fine particle removal in the screens and settling chambers better known as sedimentation tanks which exhibits a variety of design depending on the rapidity of treatment required. After settling out heavy and fine particles in the sedimentation tanks, the sewage is directed to secondary treatment facility which majorly takes place in the aeration chambers and clarifiers but some places favor anaerobic digestion as per STP design.

The target is reduction in the biochemical oxygen demand (BOD) which is achieved by the action of microbes using different treatment techniques viz. activated sludge process,

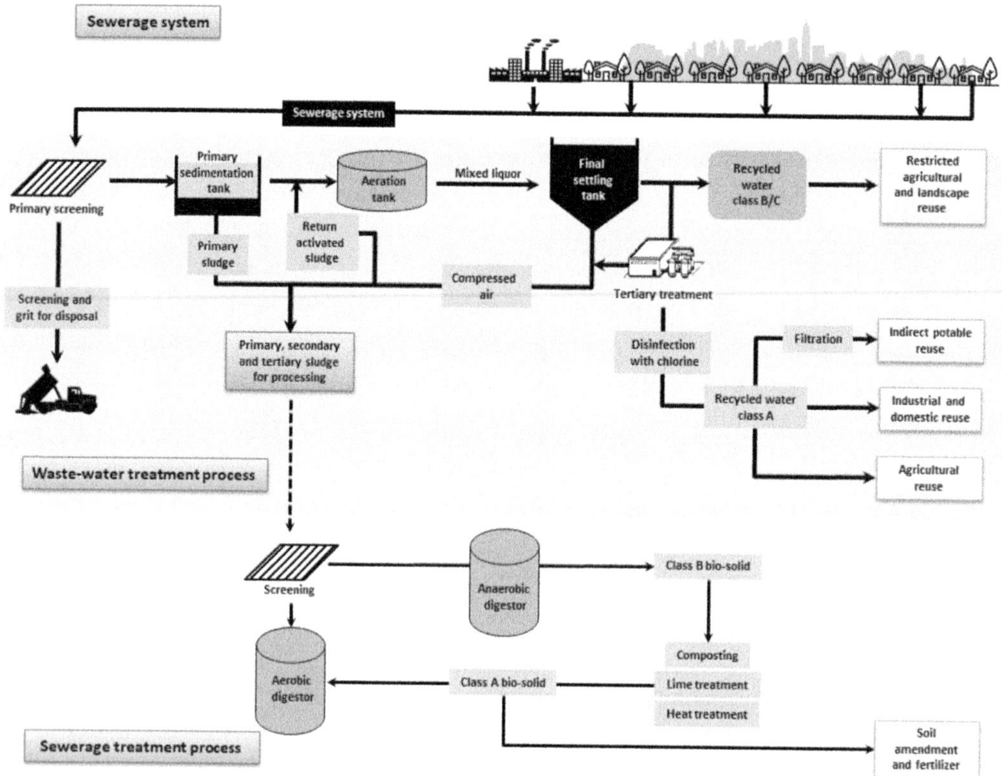

Figure 15.1
Schematic of the wastewater and sludge treatment process.

moving bed biofilm reactor (MBBR), sequencing batch reactor, upflow anaerobic sludge blanket followed by activated sludge process (UASB-ASP), membrane bioreactor, waste stabilization pond, constructed wetland. A most common and effective method is activated sludge process (ASP) (Fig. 15.1) in which a portion of sludge from the clarifier is recirculated to the aeration chamber which adds up the population of aerobic microbes to aid in the digestion of fresh sewage. MBBR incorporates the use of plastic media in circular drum surface in which the biological mass grows and the organic matter is oxidized reducing the BOD. In the sequencing batch reactor, the sewage is processed in a number of the circular process each of them having a combination of anoxic filling tank, aeration, sedimentation/clarification and decantation, sludge withdrawal. UASB-ASP uses hydraulic force to uplift the sewage sludge in the UASB reactor similar to ASP. Membrane bioreactors separate the suspended matter from sewage stream using microfiltration membrane with pore sizes ranging from 0.1 to 0.4 μm without clarification. In waste stabilization pond the aerobic microbes oxidize the organic matter in the upper layer of the pond while anoxic condition prevails which reduces the BOD. Constructed wetlands are shallow artificial units which utilize the use of the aquatic plant, algae or aquatic microbes to degrade the organic matter and accumulate pollutants from the sewage stream. For any of the process, the transformation of degradable organic matter in sewage to carbon dioxide, water and microbial biomass is the aim of secondary treatment. The biological treatment is the most time consuming and major treatment process in STPs. A comparison between different biological treatment techniques and their demerits is given in Table 15.1 [25].

The advanced stage of treatment, which is not usually applicable in developing countries, is the tertiary treatment which focuses on the removal of dissolved nutrients and inorganics in the treated sewage. These processes are costly and rarely used in wastewater treatment. They include membrane filtration, infiltration/percolation, and disinfection by chlorine or ozone etc. However, disinfection by chlorination is employed in some of the developing countries at present but in most cases, it is done occasionally when the STP is unable to reduce the BOD to the required standard.

Depending on the treatment the sludge generated in each stage of treatment is named as primary, secondary and tertiary sludge respectively. They usually vary in their composition being subjected to specifically targeted removal processes. According to the process mechanism during primary treatment, settling being the principal phenomenon, a major portion of organic matter is remained unused while during secondary or biological treatment the microorganisms utilize the organic matter and produce a flock of microbial biomass which leads to a change in the sludge composition. Hence establishing a standard database for the composition of sludge depending on their source and type of treatment is vital for further sludge processing.

Primary and secondary sludge, being the byproducts are usually subjected to further treatment. Mainly the raw sludge is screened for grit removal and dewatered either thermally in drying bed or by filtration. Dewatering the sludge concentrate its solids content to 20%–40%. The process of conditioning sometimes aids in separating the solids from the liquid phase which is achieved by the action of inorganic salts like alum, lime, ferrous or ferric salts, or polyelectrolytes. After a reduction in water content, the dried sludge is directed to various stabilization technologies, mainly to produce type A biosolid (directly applicable as fertilizer) and type B biosolid (for upgrading to type A biosolid) as categorized by US-EPA [26,27]. But all these steps lead to loss of energy or material. To manage the material resource present in the sludge an interdisciplinary approach is necessary. Material flow balance, lifecycle assessment with an understanding of the present needs of the market is essential in achieving such targets. There is at present limited information about such integration of different processes which can handle feedstock with variable composition and property, and it is where the idea of this chapter focuses. An

Table 15.1: Merits and demerits of the different biological wastewater treatment process.

Process	Merit	Demerits	Typical loading	References
Activated sludge process (ASP)	• Good process flexibility • Reliable operation • Proven track record in all plant sizes • Fewer land requirements • Low odor emission • Energy production • Ability to withstand nominal changes in water characteristics	• High energy consumption • Skilled operators needed • An uninterrupted power supply is required • Requires sludge digestion and drying • Less nutrient removal	20,000 m^3/d	[18]
Moving bed biofilm reactor (MBBR)	• Moving bed biofilm reactor needs less space since there is no primary clarifier and detention period in the reactor is generally 4–5 h. • Ability to withstand shock load with equalization tank option • High operator oversight is not required	• High operating cost due to large power requirements • Not much experience available with larger capacity plants (>1.5 MLD) • Skilled operators needed • No energy production • Effluent quality not up to the mark in India • Much less nutrient removal • Designed criteria not well established	4217 g SS/m^3	[19]
Sequencing batch reactor (SBR)	• Excellent effluent quality • Smaller footprint because of the absence of primary, secondary clarifiers and digester • Recent track record available in large applications in India also • Biological nutrient (N&P) removal	• Comparatively high energy consumption • To achieve high efficiency, complete automation is required • Highly skilled operators needed • No energy production • Uninterrupted power supply required	≤5 million gallon/d	[20]

Table 15.1: Merits and demerits of the different biological wastewater treatment process.—cont'd

Process	Merit	Demerits	Typical loading	References
Upflow anaerobic sludge blanket followed by activated sludge process (UASB-ASP)	• A high degree of coliform removal • Less chlorine dosing required for post disinfection • Ability to withstand hydraulic and organic shock loads • Relatively simple operation and maintenance • No external energy requirement and hence less vulnerable to power cuts • No primary treatment required • Energy production possible but generally not achieved • Low sludge production • No special care or seeding required after interrupted operations • Can absorb hydraulic and organic shock loading	• Post-treatment required to meet the effluent standard • Anoxic effluent exerts a high oxygen demand • Large land area requirement • More man-power require for O&M • Effluent quality is not up to the mark and poor fecal and total coliform • Removal • Foul smell and corrosion problems around STP area • High chlorine dosing required for disinfection. • Less nutrient removal	1.0–2.0 kg COD/m^3 d	[21]
Membrane bioreactor (MB)	• Low hydraulic retention time and hence low footprint (area) requirement • Less sludge production • High-quality effluent in terms of low turbidity, TSS, BOD and bacteria • Stabilized sludge • Ability to absorb shock loads	• High construction cost • Very high operation cost periodic cleaning and replacement of membranes • High membrane cost • High automation • Fouling of membrane • No energy production	0.5–3.0 kg COD/m^3 d	[22]

Table 15.1: Merits and demerits of the different biological wastewater treatment process.—cont'd

Process	Merit	Demerits	Typical loading	References
Waste stabilization pond (WSP)	• Simple to construct and operate and maintain • Low operating and maintenance cost • Self-sufficiency, ecological balance, and economic viability is greater • Possible recovery of the complete resources • Good ability to withstand hydraulic and organic load fluctuations	• Requires extremely large areas • Large evaporation loss of water • If the liner is breached, groundwater is impacted • Effluent quality may vary with seasons • No energy production • Comparatively inferior quality of effluent • Less nutrient removal • High chlorine dosing for disinfection • Odor and vector nuisance • Loss of valuable greenhouse gases to the atmosphere	100–350 kg BOD_5/m^3 d	[23]
Constructed wetlands (CW)	• Simple to construct and operate and maintain • Low operating and maintenance cost • Self-sufficiency, ecological balance, and economic viability is greater • Possibility of complete resource recovery • Good ability to withstand	• Requires large area • Large evaporation loss of water • Not easy to recover from a massive upset • If the liner is breached, groundwater is impacted • Effluent quality may vary with seasons • No energy production • No nutrient removal	8.2 cm/d	[24]

Table 15.1: Merits and demerits of the different biological wastewater treatment process.—cont'd

Process	Merit	Demerits	Typical loading	References
	hydraulic and organic load fluctuations	• Odor and vector nuisance • Loss of valuable greenhouse gases to the atmosphere		

integrated sewage sludge biorefinery may provide a basis for waste management solution in the current future when the capacity of the conventional process will fall insufficient.

15.2 Characterization of sewage sludge

Sewage sludge is a multiphase floc medium with various components which could be parted into organic and inorganic particles, aggregates of microbial biomass, extracellular polymeric substances (EPSs), suspended in a large volume of water. However, the composition and type of the raw wastewater govern the characteristics of sewage sludge at the end of the treatment stage with different treatment process leading to a variable proportion of organic and inorganics (Table 15.2 [28]; Fig. 15.2).

15.2.1 Organic fraction

The organic fraction of sewage sludge, in large proportion, is composed of the EPSs which are a complex high molecular weight polymer originated from secretion and lysis of microbial cells and/or adsorption on the organic matter already present in the raw wastewater. EPSs add up the proteins, nucleic acid, humic substances, polysaccharides and lipids in the organic sludge fraction [29].

The presence of EPSs influences the structure, flocculation, settling properties, dewatering properties, surface charge and adsorption ability of the microbial aggregate. The EPSs

Table 15.2: Typical sludge characteristics and effect of treatment methods on sludge characteristics.

	Primary sludge	Secondary sludge	Mixed sludge
Dry matter (DM), g/L	12	7	10
Volatile matter (VM), %DM	65	77	72
C, %VM	51.5	53	51
H, %VM	7	6	7.4
O, %VM	35.5	33	33

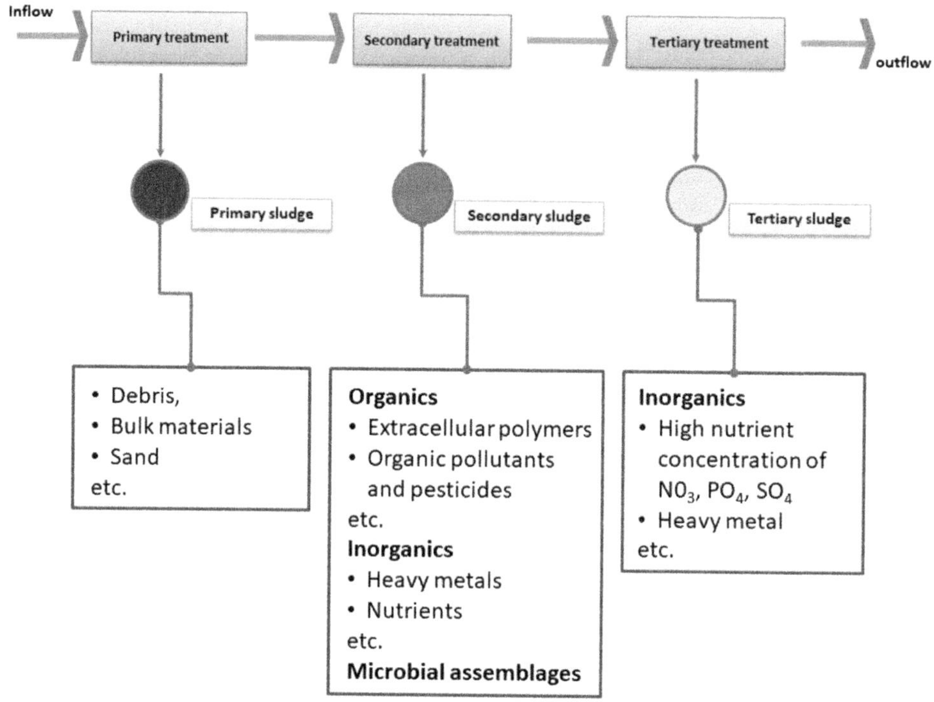

Figure 15.2
Major constituents of sewage sludge.

forms a complex net-like structure after binding with the microbial cell interlocking plenty of water which protects the cell against dewatering [29] and from toxic substances [30]. The EPSs are of two forms, first being bound to the microbial cells and not easily soluble in the aqueous phase of wastewater called bound EPSs. These are composed of the sheaths, capsular polymers, condensed gels and loosely bound polymers. The second form is unbound EPSs which are the biopolymers freely suspended in the aqueous phase like soluble macromolecules, colloids, and slimes [31,32].

There is variability in the composition of EPSs, but significant sludge components of EPSs are carbohydrates and proteins with humic substances contributing to about 20% of total EPSs mass [33,34]. The type of culture, growth phase, bioreactor type, extraction method and analytical tool used leads to variability in reporting for the composition of extracted EPSs [31].

15.2.1.1 Adsorption characteristics of ESPs

The proteins and carbohydrates of EPSs have aromatics and apathetic, and hydrophobic regions, leading to the availability of sites for adsorption of metal ions and organic matter [35]. This accounts for heavy metal adsorption which may be transported to the environment under insufficient treatment conditions. There are functional moieties reported forming a complex with heavy metals, as carboxyl, phosphoric, sulfhydryl, phenolic and hydroxyl group [36,37]. The high strength and bonding between EPS and heavy metal species lead to follow Langmuir and Freundlich equations for adsorption [38,39,40]. Due to the presence of hydrophobic regions in EPSs, the organic pollutants in the sludge are adsorbed. In comparison to the microbial cells, the EPSs adsorb more benzene, toluene and m-xylene due to the abundance of the negative functional group or the net negative surface charge [41,42].

15.2.1.2 Biodegradability of EPSs

The bacteria can degrade the protein and carbohydrate in the EPSs as a source of carbon and energy. Also, the EPSs, produced by the aerobic bacteria in the ASP can be of similar source for metabolic activity of other bacteria present in the sludge [43]. However, the ESPs in the inner layer of aerobic granular sludge are readily biodegradable than those that are present in the outer layers due to their hydrophilic nature [44]. The presence of both hydrophilic and hydrophobic group in EPSs makes it amphoteric, with higher hydrophilic fraction than hydrophobic one [45]. The hydrophilic group is dominated by proteins whereas the carbohydrate is found in the hydrophobic fraction, which is readily degradable. The nondegradable EPSs fraction adds up an organic load in the effluent stream after biological treatment, affecting the process cost, due to the requirement of additional treatment steps.

15.2.1.3 Importance of EPSs

The EPSs in sewage sludge has a direct influence on the microbial aggregates affecting its functioning. The EPSs forms a boundary layer surrounding the microbial aggregates which reduces the permeability of substrate [46]. The decrease in the permeability reduces the mass transfer coefficient of the substrate while the increase in adsorption by EPSs favors availability of organic substances to microbial cells [47]. The EPSs consisting of the proteins having amino group neutralizes the carboxyl and phosphate groups present in the sewage sludge which rather aids in flocculation. But high content of EPSs specially rich in protein reduces the settleability of microbial aggregated by increasing the negative charge leading to repulsive action between the aggregates [48,49,50]. The sludge volume index (SVI) is a good indicator of ESPs and settleability of the aggregates. A lower SVI indicates good settleability as the EPSs content is generally low.

Sludge volume index (mL/g) = ([Settled sludge volume, mL/L] × [1000, mg/g])/MLSS, mg/L.

The EPSs plays major role in the stability of microbial aggregates by giving it mechanical and hydrodynamic stability due to its layer formation around the cell aggregates. These factors leads to the consideration of EPSs while planning for sewage treatment or the utilization of the sewage sludge. The content of EPSs in the sewage sludge thus depends on the proportion of the net negative charged groups i.e., proteins as an increase in the negative charge effects floc formation and settleability negatively.

15.2.1.4 Other organic chemicals

Sewage sludge is rich in organic chemicals which finds it source from the available organic food, pharmaceuticals, pesticides in urban areas to the organic dyes and solvents used in household and institutional purposes. Harrison et al. [51] reviewed the presence of organic chemicals in sewage sludge to list 152 organic chemicals. Out of the 111 organic priority pollutants and 143 targeted compounds as per US-EPA, 90 and 101 were present in sewage sludge respectively [52]. Polychlorinated biphenyl was the most abundant followed by pesticides. The pesticides present in sewage sludge are the priority pollutants having a direct impact on the health of people as well as livestock when discharged in streams even in lower concentrations.

15.2.2 Inorganic fraction

Other than organic matters, sewage sludge contains inorganic fractions, which can potentially enter the food chain affecting consumer health. Those include nitrogen, potassium and phosphorous as major ingredients with a variable amount of heavy metals.

15.2.2.1 Heavy metals in sewage sludge

The inorganic metals present in the sewage sludge are Zn, Pb, Cu, Cr, Ni, Cd, Hg, and As with concentration ranging between below 1 ppm and beyond 1000 ppm [53]. Their presence in respective concentration in the sludge is the decider for its downstream processing in energy and materials recovery since they find a route in the produced liquid, char and gas after thermal treatment [54]. The number of heavy metals leached to the final product after treatment depends on the boiling point of the respective metals at atmospheric pressure with metals having lower boiling point escaping to gaseous product and vice versa. The heavy metals originate mainly from the anthropogenic sources as chemical, glass and petroleum industries, electronics, dentistry and jewelry production [55]. Apart from the conventional sources of metals, the recent advances in nanotechnology have introduced the application of metallic nanoparticles in food (Ag, TiO_2, Si, Pt), in medicine (Au, Si), in textiles (Ag) etc. [56,57,58]. These metallic nanoparticles are effectively removed in wastewater treatment facilities and hence get added up in the sewage sludge [59]. This also opens up an option for precious metals recovery from sludge although present in lower concentrations.

A survey in the United States showed that all the heavy metals present in the sewage sludge had an enrichment factor of more than one, indicating their sources were of anthropogenic origin rather than from soil particle and dust [60]. Post-incineration ash of sewage sludge has variable metal concentration as reported in different regions of the world. The presence of Si, Ca, Fe and Al were abundant in Germany while rare earth metals like Sm, Eu, Tb, Sc, Gd, La, and Ce were found in Japan as readily used in electronic gadgets [61,62]. These signified the lifestyle and technological advancement in the different world economy had a contribution toward types and quantity of metals present in sewage sludge.

15.2.2.2 Macronutrients in sewage sludge

Nitrogen is a major part of people's diet which finds its way to the sewage sludge via domestic wastewater stream. Another major source of nitrogen is ammonium which comes from the urea used in garden and household premises. Major forms of nitrogen present in raw sewage sludge are ammonia (40%), organic nitrogen (60%) and nitrate nitrogen (<1%). The influent nitrogen concentration varies between 20 and 85 ppm in the wastewater stream. After biological nitrification, the ammonia is converted to nitrate nitrogen by nitrifying bacteria which settles down with the sludge. The nitrogen content in activated sludge dry mass is reported to be 24–67 g N/kg dry solids while the remaining N is denitrified to N_2 gas [63,64]. The nitrate present in sewage sludge is the basis of utilizing it as a fertilizer as the available N source for plant metabolism.

The phosphorous content in wastewaters varies between 4 and 16 ppm depending on the source [65]. The cellular metabolism of microbes in the wastewater treatment utilizes about half of the phosphorous in the wastewater while rest is discharged with the effluent stream. Chemical precipitation and biological phosphorous removal in tertiary treatment lead to accumulation of phosphorous in tertiary sludge with different bioavailability. The precipitated phosphorous due to binding with iron is not readily bioavailable while biologically sequestered sludge has phosphorous which is readily available to plants for uptake and thus preferred in tertiary treatment. Phosphorous is both an essential nutrient for microbial function in sewage treatment and an initiator of eutrophication if discharged into streams without removal from effluent.

Potassium is present in the sewage sludge as potassium dioxide (K_2O). There is a concentration of 10–30 ppm of potassium in the treated wastewater stream while the potassium content in the sewage sludge varies between 50 and 70 g K_2O/kg of dry solids. Potassium present in sludge is bioavailable and can be utilized by plants and microbes in their metabolic activity. The application of sewage sludge as soil amendment thus increases the NPK content and bioavailability of nutrients. But the leaching potential of nutrients in excess and presence of heavy metals and priority pollutants in sewage sludge indicates the scope for risk assessment and environmental impact analysis before the direct application of sewage sludge.

15.2.3 Microbial assemblages and pathogens

Microbial diversity is a vital transforming unit of materials in molecular and ionic level in biological systems. Due to the open and dynamic nature of WWTP system, there is temporal and spatial variability in microbial community members especially on the basis of liming and redox conditions [66]. In anaerobic conditions, iron and sulfur take an active part in redox reactions [67]. Ferric ion is stabilized to ferrous by reducing bacteria such as *Ferrimicrobium*, *Nitrospira*, and *Acidithiobacillus*, which are also the mediators of C and N transformations [31,66,67]. The major pathways for N transformation in sewage sludge by the microorganism are as:

$$NH_4^+ + 2O_2 \rightarrow NO^{3-} + 2H^+ + H_2O \tag{15.1}$$

$$NO_3^- + 0.1C_{10}H_{19}O_3N \rightarrow 0.5N_2 + CO_2 + 0.3H_2O + 0.1NH_3 + OH^{-4} \tag{15.2}$$

$$NO_3^- + 0.83CH_3OH \rightarrow 0.5N_2 + 0.83CO_2 + H_2O + OH^- \tag{15.3}$$

In aerobic condition the nitrifying bacteria as *Nitrobacter* sp. oxidize ammonia to nitrate (Eq. 15.1). In anaerobic condition the organic carbon source (BOD) represented as $C_{10}H_{19}O_3N$ is utilized by the microorganisms to reduce the nitrate to gaseous nitrogen. Since the process of denitrification (Eq. 15.2) utilize the carbon source, the efficiency of nitrogen removal in downstream digestion process is reduced during sewage treatment. To facilitate BOD and nitrogen removal biologically the recirculation of activated sludge is adopted for upstream denitrification prior to oxidation in wastewater treatment or alcohol or acetate is added (Eq. 15.3) [68].

Sulfur is the source of energy in anaerobic processes. The toxic heavy metals present in sewage sludge have not been seen having an impact on the microbial community structure. Activated sludge is a very heterogeneous assembly of microorganisms dominated by certain bacterial taxa. It consists of a complex interactive cluster of bacteria, archaea, fungi and protists which degrade organic compounds including petroleum products, toluene and benzopyrene [69]. The removal of nitrogen, phosphorous and aromatic compounds including organic pollutants are undertaken by proteobacteria. Nitrogen is removed by aerobic nitrifying and anaerobic denitrifying bacterial genus while *Nitrosomonas* sps. and *Nitrospira* sps. are ammonia oxidizer in sewage sludge. *Proteobacteria* which are involved in nutrient cycling in the environment are abundant in aerobic systems while *Bacteroidetes* are abundant in anaerobic systems which are involved in degrading proteins to volatile acid and ammonia [70]. There were 77 bacterial genera reported in sewage sludge despite differences in source and treatment which were dominated by *Clostridium*, *Treponema*, *Syntrophus*, and *Comamonas*. Also, a total of 913 operational units have been reported including 48 classes, 172 families and 474 genera of microorganisms in the wastewater stream. *Acinetobacter* was the dominating genus in incoming sewage. It is suggested that microorganisms such as *E.coli*, *Enterococcus* spp.

and *C. perfringens* which have sanitary significance may be allowed to function before disposal of the final dehydrated sludge to minimize pathogens [71]. Although, *Escherichia* and *Enterococcaceae* are present in the initial waste stream but they are found almost absent after treatment [72].

15.3 Concept of integrated sewage sludge biorefinery

Sewage sludge is a potential resource for the recovery of different materials and proper treatment and process integration can contribute to multiple product procurement with resource efficiency. Integration of process is necessary because in the case of sewage sludge the desired products can be organic matter based or inorganic metal based or chemical or nutrient based. For a different type of products, a single process is insufficient. For example, desired protein cannot be obtained by thermal conversion and the valuable metals in elemental form cannot be obtained by biochemical conversion. A biorefinery which has a design of integrating different conversion processes with the flexibility of optimizing the capacity of each process with changing sewage sludge composition is a promising option for waste minimization and resource optimization.

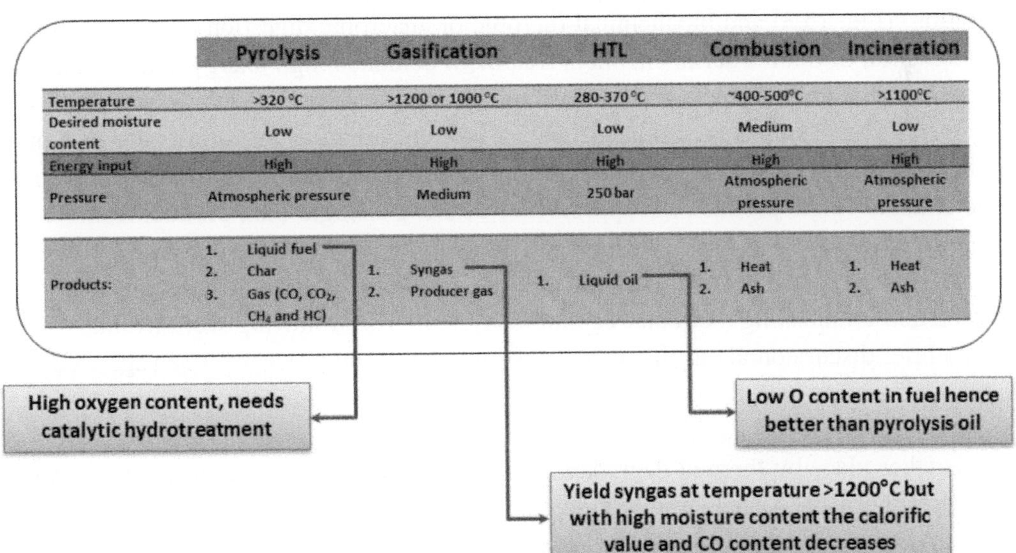

Figure 15.3
Thermochemical conversion platform of sewage sludge.

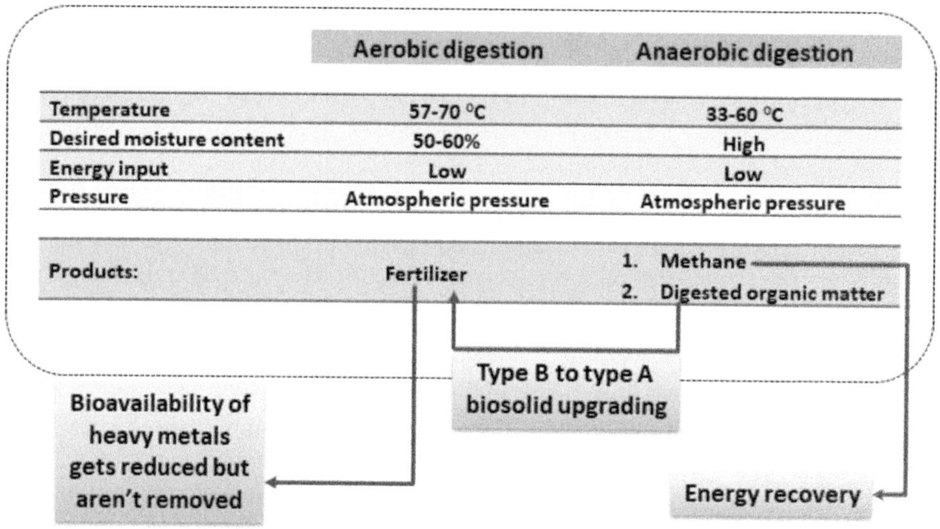

Figure 15.4
Biochemical conversion platform of sewage sludge.

15.3.1 Thermochemical and biochemical platforms for sewage sludge

The concept of such a biorefinery cannot be framed without the basic understanding of existing conversion platforms for sewage sludge. The conversion platform for sewage sludge is either thermochemical (pyrolysis, gasification, liquefaction, combustion and incineration) (Fig. 15.3) and biochemical (aerobic or anaerobic digestion) (Fig. 15.4) or the combination of both. Thermochemical treatment demands high energy and capital/ running cost where biochemical treatment demands time and specific reaction environment. Thermochemical treatment yields different products for different ranges of operating parameters while the biochemical treatment itself limits to narrow range of the parameter and a single product. An approach of desiring a single bulk product from biochemical treatment and then a range of products by thermochemical treatment of the byproducts from the biochemical treatment can solve the problems of achieving waste utilization and establishing a circular economic system. The following sections thus contain a brief discussion about the major existing conversion process for sewage sludge.

15.3.1.1 Pyrolysis

Pyrolysis takes place by thermal degradation of molecules in an inert atmosphere. Sewage sludge when thermally degraded at a temperature between 300 and 900°C yields liquid bio-oil/tar, combustible gases, fixed carbon, ash and water vapor [73]. Pyrolysis of biomass-based feedstock and municipal solid waste has been studied in details but the consideration of sewage sludge as feedstock has recently gained attention. The temperature, gas atmosphere, heating rate, type of reactor, reaction and residence time are the affecting parameters for quality and composition of yield. When the heating rate is higher it favors the vapor formation while at lower heating rates the charring is more due to secondary polymerization of liquid/tar. Organic compounds and water is the major constituent of the liquid yield from sewage sludge. This liquid yield is inferior compared to other biomass feedstock but the lower ash content in sewage sludge makes it superior when the bio-oil yield is expressed in terms of dry ash free basis. The solid residue left after pyrolysis has a high concentration of the heavy metals except for Cd and Hg [74]. The concentrating of heavy metals in the final residue from sewage sludge pyrolysis is suitable for metal recovery. The liquid and tar produced in sewage sludge pyrolysis are rich in 1-alkenes, monoaromatic hydrocarbons and its alkyl derivatives as PAHs [5]. The tar also contains heavy oxygenated hydrocarbons as carboxylic acids, ketones and esters, halogenated aromatics, alkyl aromatic hydrocarbons and cycloalkenes [75]. The oil requires upgrading before utilization or simply hydrotreatment in the presence of catalysts for reducing oxygen content. The water content of up to 30% is produced due to degradation and dehydration of the heavy oxygenated compounds at a temperature above 200°C [76]. Extraction of N and S from the liquid yield is also an added step in directing bio-oil for application as fuel [77]. High N and S content in the fuel will add up to the concentration of oxides of nitrogen and sulfur after its combustion. The gas fraction from pyrolysis contains majorly H_2, CO, CO_2, CH_4 and light hydrocarbons (C_1-C_4). Primary pyrolysis of sewage sludge can yield a high quantity of gas fraction of up to 45 wt%. In a study of sewage sludge pyrolysis in pilot plant Kasakura and Hiraoka reported of H_2 (5.5 vol%), CO (3.65 vol%), methane (1.48 vol%) as the major constituents with the other light hydrocarbons in lower concentrations, while the other identified products were HCN, nitrogen oxides, hydrogen chloride, and sulfur oxides [77]. The pyrolytic char which is the solid residue consisting mainly of fixed carbon, ash and heavy metals. The lixiviation in pyrolytic char is less compared to ash from sewage sludge combustion and incineration making it a better residue for disposal purpose [16,78]. The high concentration of heavy metals renders it unsuitable for use as a soil amendment in the fields. Alternative use of the char as an absorbent for pollutants of the acid group like SO_2, SH_2, phenols, and its derivatives is new but better than the earlier disposal option [79].

15.3.1.2 Gasification

Gasification is an accepted process for producing syngas which is further converted into transport fuel and fine chemicals. Waste biomass has been tested for its suitability in the gasification process, but a few studies have been undertaken with sewage sludge as a feedstock. Main products from gasification are producer gas, synthesis gas and residual char. The reaction temperature defines the end product desired from gasification, as the temperature above 1200°C, H_2 and CO are the major component producing syngas, and below 1000°C, the products formed are H_2 and CO (50% of total product), CH_4, C_xH_y

aliphatic hydrocarbons, benzene, toluene, tars, CO_2, and H_2O, known as producer gas [80]. The elevated temperature in gasification leads to high conversion rates in comparison with most of the available alternative treatment processes for carbon-rich feedstock [81]. Gasification is a reductive process and thus eliminates the problem escaping nitrogen oxides, sulfur oxides, chlorinated dibenzodioxins and dibenzofurans to the atmosphere when compared to combustion and incineration of sewage sludge. Gasification of sewage sludge has 99.99% destruction removal efficiency (DRE) for refractory organic compounds [82]. The hydrocarbon-rich combustible gas is suitable for burning and energy generation. Hydrogen production from gasification is a better option but demands high temperature (900°C), high air flow rate and catalyst raising the cost of the process. Gasification yield a single product easy to handle than the products obtained from pyrolysis. For conversion of sewage sludge to fuel or fine chemicals the process must be suitable for feedstock with a high moisture content which affects the desired product. The moisture in the feedstock affects the composition and calorific value of syngas due to the conversion of CO into CO_2 and H_2 rich gas which is referred to as water-gas-shift [83]. When compared to pyrolysis high moisture content in sewage sludge favors gasification since the former process is suitable for dry feedstock. Syngas also contains trace compounds as alkali metals, particulates and hydrocarbons of which the later two are primary air pollutants and makes gas cleaning necessary. The cleaning of gas from gasification is the major challenge yet to overcome.

15.3.1.3 Hydrothermal liquefaction

Hydrothermal treatment is the processing of the feedstock in the presence of water. When the reaction conditions are less severe, this process is called hydrothermal carbonization as it leads to the formation of solid carbon-rich residue similar to coal, which is either combusted for energy production or used for soil amendment. When the process parameters are severe with temperature between 280 and 350°C and pressure up to 250 bar this process is termed as hydrothermal liquefaction leading to the production of liquid residue.

The major advantage of the hydrothermal carbonization process is that it can easily operate with the high moisture content of sewage sludge. The reaction conditions are moderate with a temperature range between 180 and 220°C and pressure range between 20 and 25 bars with a reaction time of 1–75 h [84]. Additional catalyst is seldom used in this process and hence it reduces the material cost of the treatment. The chemical processes involved in hydrothermal liquefaction are decarboxylation, dehydration, and polymerization (aromatization). The removal of carboxyl and −OH group increased energy density and high heating value (HHV) depending on feedstock [85]. This process is suitable for feedstock with low initial energy density and hence is preferable with sewage sludge. The hydrothermal liquefaction of sewage sludge powder in both water and ethanol with or without catalyst was shown to increase the heating value of heavy oil to about 30 from 15 MJ/kg of dry sludge powder [86]. Hydrothermal liquefaction is efficient in increasing the energy density of fuel produced from sewage sludge but the process limited to lab scale at present.

15.3.1.4 Combustion

Combustion is the oxidation of sewage sludge at high temperature. The required ignition temperature for sewage sludge is 344–432°C and the self-ignition temperature is 439–481°C which varies accordingly with the moisture content as a large amount of energy is lost as latent heat. Organic matter in the sewage sludge is converted to CO_2, H_2O, and heat while the metals, fixed carbon, and ash remains after combustion. The combustion of sewage sludge destroys organic contaminants and pathogens. The problem with the combustion of sewage sludge is that the flue gas has a high concentration of particulate matter, oxides of nitrogen and sulfur, and toxicants like dioxin and furans, which mandates the installation of emission control technologies [87]. Sewage sludge combustion is different than the combustion of other carbon-rich solid fuels as it has high moisture, volatile matter and ash content with low fixed carbon content. The heavy metals in the sewage sludge are retained in the ash. The metal from the sewage sludge ash does not leach readily into the ground after disposal but dumping should be the second option after metal and phosphorus recovery. The metals can be recovered after acid digestion and ion-exchange precipitation/filtration. In the process of metal recovery, the remaining slag and ash after combustion is dissolved in sulfuric acid. It is then passed through a set of ion exchangers where the first ion exchanger collects iron ions which is followed by collection of sulfate ions in the form of potassium sulfate in an anionic exchanger. Lastly by the addition of HCl the phosphate ions are collected as phosphoric acid [88]. The residual solution is rich in heavy metals which can be then recovered instead of conventionally way of disposal. This also helps in recovering phosphorous in the form of phosphoric acid. When the combustion efficiency of the sewage sludge is reduced due to high moisture and ash content co-combustion with other carbon-rich and less polluting solid fuel as wood and municipal solid waste is viable.

15.3.1.5 Incineration

The use of incinerators is for energy recovery through heat generation or for the volume reduction of sewage sludge. The initial volume of sewage sludge is reduced by 90%–95% since all of the organic matter is destroyed by the high temperature of incineration. The ash after incineration has a composition similar to clinker. Incinerated ash is rich in calcium dioxide, silicon dioxide, iron (III) oxide, and alumina. The sewage sludge ash is suitable for a raw material substitute in cement industries and sewage sludge has been incinerated in cement kilns for material recovery and heat generation [89]. The heat recovered from the kilns can be used for drying of raw sewage sludge at the preparatory stage for other treatment processes. The phosphorous concentration of sewage sludge ash is a difficulty in utilizing it with clinker. The desired phosphorous content for a raw material to be used in clinker is less than 2% [90,91]. The heavy metals in the sewage sludge are concentrated in the incinerated ash in a similar manner like in pyrolysis and combustion but the metal lixiviation in this process is comparatively more which limits application in fields for soil amendment. Thermal incineration is the most practiced conventional process for volume reduction and treatment but has a lot of negative impacts on the environment and has less resource recovery options.

15.3.1.6 Aerobic digestion

Aerobic digestion or composting is a widely accepted way of sludge treatment. The sewage sludge is acted upon by microbes in the presence of oxygen to break the biodegradable organic matter into CO_2, H_2O and humic-like materials. The nitrogen in the sludge is converted to ammonium nitrogen which is further acted by ammonifying bacteria to produce NH_3 or is nitrified to NO_2 and NO_3 by aerobic nitrifying bacteria. Composting or aerobic digestion has a thermophilic range of 57–70°C when the microbial digestion rate is maximum. Due to this high temperature, the pathogens are killed and biogenic toxicity of the produced manure is reduced. The thermophilic stage is followed by conditioning mesophilic stage when the product is stabilized and the remaining organic matter is digested. The carbon to nitrogen ratio, moisture content and pH are vital parameters in this process. The optimum C/N ratio for composting is 30 with 50%–60% moisture content and pH range from 6 to 8 [92].

Toxic organic compounds like PAHs [93] and PCBs [94] in sewage sludge are reported to be removed by aerobic digestion through various mechanisms. The heavy metals cannot be eliminated from the product but digestion reduces their mobility and bioavailability which has a positive environmental impact and reducing waste toxicity [95]. Aerobic digestion is an economically viable process for sludge conversion but the fate of heavy metals and capturing CO_2 released (being the most contributing GHG) are major challenges in including it in the biorefinery process.

15.3.1.7 Anaerobic digestion

Anaerobic digestion (AD) utilizes microbes to convert biodegradable organic matter into methane-rich biogas and digests the material in the absence of oxygen. The methane is recovered for energy generation while the digested material being rich in nutrient is utilized as fertilizer after drying [96]. The process of AD has sequential steps with different microbial consortia producing specific products in every step. The process of AD is completed in five steps namely hydrolysis, acidogenesis, acetogenesis and methanogenesis. Hydrolysis breaks complex organic compounds like fat, carbohydrates and proteins into soluble compounds. These soluble organic compounds are fermented to acetic acid by acideogens which produce acid, H_2 and CO_2, further used to produce acetate by acetogens. The methanogens which are majorly archaebacteria finally convert the acid and acetates to methane and CO_2. Biogas is an energy-rich product of AD, by this process 0.35 L of biogas can be obtained out of 1 g of organic matter. Biogas can generate 21.5 MJ/m^3 of energy and it can be processed to increased energy content similar to natural gas by purification [97]. AD is temperature sensitive and a combination of mesophilic as well as thermophilic groups of microbes dominate in every step. High temperature decreases the quality of digested material as well as creates odor problems. Pretreating the feedstock can increase yield quality, e.g., alkali pretreatment aids in breaking the cross-linkage of the lignocellulosic material while thermal pretreatment increases solubility and biodegradability of the organic matter present in feedstock [98]. One promising process for sewage sludge is co-digestion. If a suitable substrate is added with sewage sludge in digesters, the process economics can be improved by producing a multi-feed single product unit. Many substrates are used as a substrate in co-digestion of sewage sludge as waste paper, food waste and animal waste from meat processing industries [98]. It must be noted that the temperature, water content, addition of substrates in co-digestion and roles of the microorganisms are key players in AD. The integration of these factors is necessary for improving the process economics. With high moisture content and organic fraction, sewage sludge is a suitable feedstock for AD and production of biogas in biorefineries.

15.3.2 Biorefinery approach

A biorefinery can be established depending on the existing processes and sludge stream in the treatment facilities or by introducing changes in the facilities being planned to be installed. For both the cases, it is necessary that basic information of the sludge composition be figured to resolute the treatment processes and the range of expected products. Sludge from the primary, secondary and tertiary stream can be directed to a mixed stream or can be treated individually yielding different products. Since the elemental content of the sludge varies with stream i.e., it differs for the mixed stream, so the products will also vary as per the input stream in the biorefinery.

Fig. 15.5 gives a typical composition of mixed sludge stream which contains dry matter (DM), volatile matter (VM) which has C, H, O, N and S. The net protein content is expressed as % DM. With an assumed 1 MLD input in a sewage treatment plant, there is typically 10^4 kg of dry matter of which 72% i.e., 7200 kg is volatile matter. There is a scope of recovering 3672 kg of C, 532.8 kg of H, 2376 kg of O, 511.2 kg of N and 108 kg of S in a day. The net protein content from the sludge that can be recovered is 3000 kg in a day with 1 MLD input.

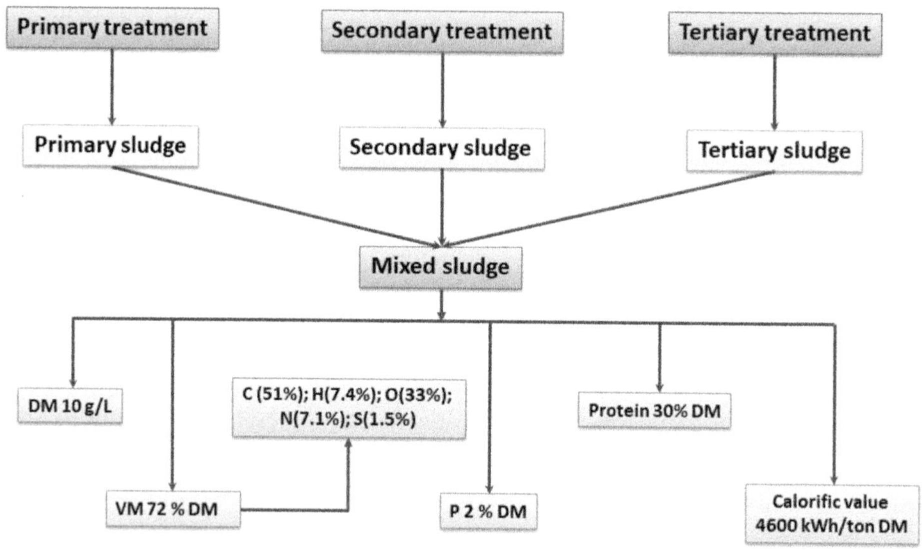

Figure 15.5
Typical composition of mixed sewage sludge stream.

For a treatment facility generating sludge in different streams, the composition and content of C, H, N, O, and S vary. The carbon content in the secondary sludge increases due to the formation of nondegradable EPSs by the acting microbes. The polymeric carbon in the form of organic macromolecules is high in secondary sludge. Thus, the calorific value of secondary sludge is higher than the primary sludge. Biological treatments utilize the oxygen and hence the oxygen content in the secondary sludge reduces. The ammonium nitrogen is converted to nitrate and rest is lost as dinitrogen gas to the atmosphere. The wastewater treatment and technologies used have a major influence on the constituents of the sludge generated. Thus, it is necessary to draw baseline information about the treatment applied and desired outcomes only then a viable setup for a biorefinery will have economic returns.

The treatments which are majorly classed as thermochemical and biochemical produce varied products. The thermochemical treatment yields oil, gas and char but the carbohydrate and protein content is lost due to the high reaction temperature. The quantity of sludge generated is in tons, so the setup and amount of energy required for thermochemical treatment will be huge. This necessitates the cost and energy consideration for the recovery process. The possibility of recovering organic macromolecule or polymers thus favor biochemical pathway. Biochemical treatment can be applied as either a cell-free or a whole cell system. If there is a cell-free system, then the process cost rises which questions the economic viability of the refinery when compared to the product generated. The whole cell system depends on the desired product.

For obtaining a specific protein or lipid the system must favor intracellular metabolism while for obtaining antibiotics or enzyme-like protease the system must favor extracellular metabolism. It is better to consider thermochemical treatment after biochemical treatment in sludge biorefinery since the biochemical treatment leads to volume reduction of up to 90% which can then be directed to the thermochemical pathway. The high moisture content in the sludge renders it unsuitable for direct application in thermochemical treatments thus favoring biochemical conversion. In every treatment process, the volatile components escape as gas in some form. To achieve resource efficiency the escaping gases must be trapped and condensed or recirculated. But the conversion of an open system of sludge treatment to a semiopen or closed system will hinder the capacity and flow of treatment. The capacity of treatment when compromised will need an additional place for storage of influent sewage thus raising demand for land. Thus, the system must be designed as a combination of an open, semiopen and closed system to optimize resource efficiency. The recovery of metals from the sludge streams can be done using ion-exchange resins in filters. The backwashed solution will contain the exchanged metals that can be recovered by chemical separation in the metal recovery unit. From the tertiary sludge stream, the nutrients like nitrate and phosphate can be precipitated out adding it to the fertilizer for quality enhancement.

The approach in the biorefinery must focus on the simplification of the process rather than installing multiple process units and creating a complex. The principle of industrial ecology is helpful in framing a simple multiprocess system. The byproducts of one unit can be used as feed in other unit or the intermediate product from one unit can be sent as a resource for other industry as depicted in Fig. 15.6. It is not necessary to reutilize all the products generated in the biorefinery itself. Canalizing the intermediates as resources for other industries and units outside the biorefinery will also bring resource efficiency in the macro level though it may seem as a small contribution.

15.3.3 Economic benefits

Installing a biorefinery needs long-term planning. The products desired from sewage sludge have the potential of giving good economic returns. The products may not be final but intermediate products for other industries. Channelizing the recovered material in the market will open a prospect of waste to wealth from sewage sludge. The reduction in the pollution load on the stream will be a direct effect which will lead to improved environmental health in the region as a secondary benefit. This will benefit the people in the form of reduced vulnerability to pollutants. The organizations and companies may be included in establishing the infrastructure as a part of their Social Corporate Responsibilities (SCRs). The biorefinery as a whole will have a net positive impact on the economy and employment generation.

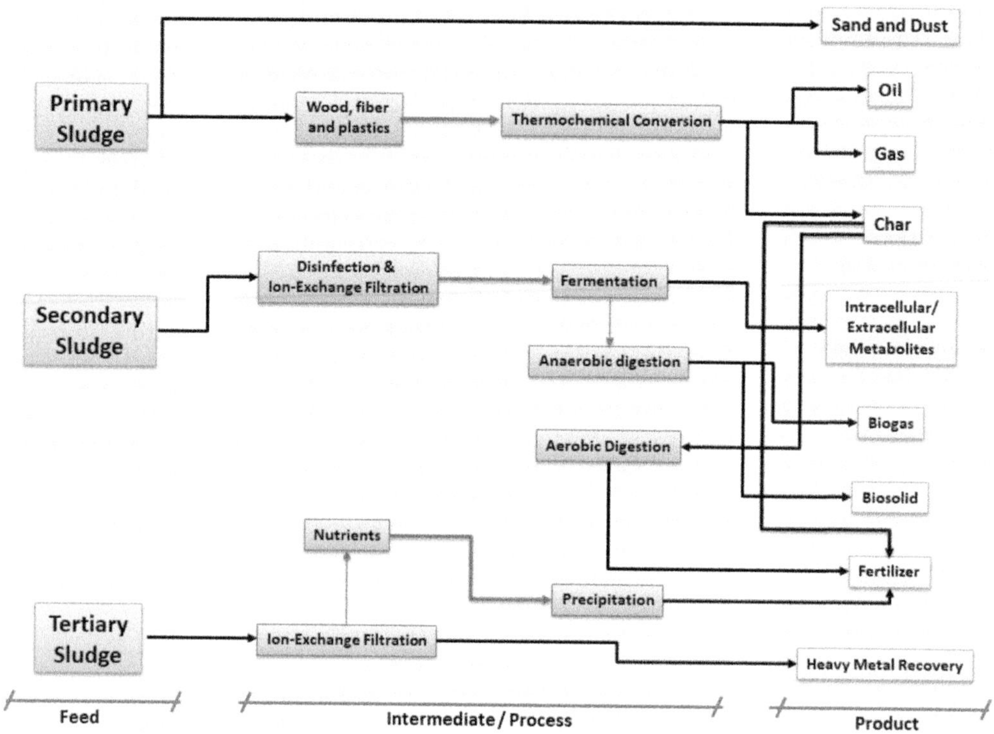

Figure 15.6
Proposed integrated sewage sludge biorefinery.

15.3.4 Environmental benefits

The direct environmental benefits are that the priority pollutants and heavy metals from the sewage sludge won't reach the ecosystem. Once part of the ecosystem, it travels along the food chain reaching people and affecting their health. Improved resource efficiency will aid in combating resource depletion. The land area required in future for landfilling activity related to sludge will be utilized for development purposes. There will be a direct positive influence on land, water, energy and material resource conservation.

15.4 Conclusions and perspectives

The sewage sludge biorefinery is a new way approach of integrating the existing processes to optimize resource recovery and reduce the environmental impacts of sludge disposal. Integrating the processes needs quantification of the process scale and its suitability based on the geographic location. A complete biorefinery design must undergo an assessment of the environmental impacts and net carbon footprint reduction. The viability of the sludge

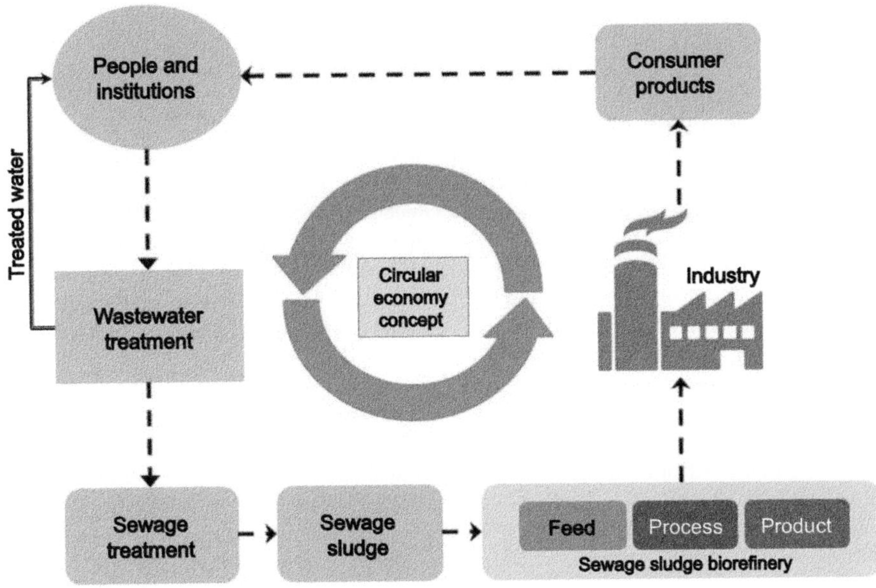

Figure 15.7
Sewage sludge based biorefinery and circular economy.

biorefinery depends on the efficiency of the designed system to function with the changing sewage load and characteristics. The biorefinery must be flexible in design to sustain both short term and long-term changes. The economy of the region and benefits that the biorefinery can provide in socio-economic development either in terms of improved environmental health or measurable enhancement in living standards is the basis for deciding on the presence of the biorefinery in that region.

References

[1] Maier RM, Pepper IL, Gerba CP. Introduction to environmental microbiology. In: Maier RM, Pepper IL, Gerba CP, editors. Environmental microbiology, vol. II. San Diego, USA: Academic Press; 2009. p. 3–7.
[2] Campbell HW. Sludge management – future issues and trends, vol. 41(8). San Diego, USA: Academic Press; 2000. p. 1–8.
[3] Remy C, Lesjean B, Waschnewski J. Identifying energy and carbon footprint optimization potentials of a sludge treatment line with life cycle assessment. Water Science and Technology 2013;67(1):63–73.
[4] Dentel SK. Contaminants in sludge: implications for management policies and land application. Water Science and Technology 2004;49(10):21–9.
[5] Fytili D, Zabaniotou A. Utilization of sewage sludge in EU application of old and new methods – a review. Renewable and Sustainable Energy Reviews 2008;12(1):116–40.
[6] Liu B, Wei Q, Zhang B. Life cycle GHG emissions of sewage sludge treatment and disposal options in Tai Lake watershed, China. The Science of the Total Environment 2013;447:361–9.
[7] Inguanzo M, Dominguez A, Menendez JA. On the pyrolysis of sewage sludge: the influence of pyrolysis conditions on solid, liquid, and gas fractions. Journal of Analytical and Applied Pyrolysis 2002;63(1):209–22.

[8] Lu Q, He ZL, Stoffella PJ. Land application of biosolids in the USA: a review. Applied and Environmental Soil Science 2012;62:12–4.
[9] Rulkens WH, Bien JD. Recovery of energy from sludge – comparison of the various options. Water Science and Technology 2004;50(9):213–21.
[10] Shan Y. The reuse oppurtunity, wastewater report 2018. 2018.
[11] Cieślik BM, Namieśnik J, Konieczka P. Review of sewage sludge management: standards, regulations and analytical methods. Journal of Cleaner Production 2015;90:1–15.
[12] Sato T, Qadir M, Yamamoto S, Endo T, Zahoor A. Global, regional, and country level need for data on wastewater generation, treatment, and use. Agricultural Water Management 2013;130:1–13.
[13] OECD economic surveys: China. Paris, France: OECD Publishing; 2013.
[14] Technical specification for management of municipal wastewater treatment plant operation (HJ 2038–2014). Beijing, China: Ministry of Environmental Protection of the People's Republic of China, China Environmental Science Press; 2014.
[15] Second national communication on climate change of the People's Republic of China. Beijing, China: Climate Change Division of National Development and Reform Commission of the People's Republic of China, China Economic Press; 2013.
[16] Tian Y, Zuo W, Ren Z, Chen D. Estimation of a novel method to produce biooil from sewage sludge by microwave pyrolysis with the consideration of efficiency and safety. Bioresource Technology 2011;102:53–61.
[17] CPCB. Status of STPs; 2017. Available at: https://www.cpcb.nic.in/status-of-stps/?&page_id=status-of-stps.
[18] Islam MA, Amin MSA, Hoinkis J. Optimal design of an activated sludge plant: theoretical analysis. Applied Water Science 2013;3(2):375–86.
[19] Mcquarrie JP, Boltz JP. Moving bed biofilm reactor technology: process applications, design, and performance. Water Environment Research 2011;83(6):560–75.
[20] Wastewater technology fact sheet sequencing batch reactors. 1999 [Washington DC, US tech].
[21] Daud MK, Rizvi H, Akram MF, Ali S, Rizwan M, Nafees M, Jin ZS. Review of upflow anaerobic sludge blanket reactor technology: effect of different parameters and developments for domestic wastewater treatment. Journal of Chemistry 2018:1–13.
[22] Iorhemen O, Hamza R, Tay J. Membrane bioreactor (MBR) technology for wastewater treatment and reclamation: membrane fouling. Membranes 2016;6(2):33.
[23] von Sperling M. Waste stabilization ponds. In: von Sperling M, editor. Biological wastewater treatment series, vol. III. London: IWA Publication; 2007. p. 8–40.
[24] Vymazal J. Constructed wetlands for wastewater treatment. Water 2010;2(3):530–49.
[25] Ministry of Environment, Forest and Climate Change (MoEFCC). Sewage treatment in class I towns. 2010. p. 1–23. Available at: http://www.moef.nic.in/sites/default/files/ngrba/003_EQP_S&R_02.pdf.
[26] Gerba CP, Pepper IL. Municipal wastewater treatment. In: Maier RM, Pepper IL, Gerba CP, editors. Environmental microbiology, vol. II. San Diego: Academic Press; 2014. p. 583–605.
[27] European Commission, DG environment. Disposal and recycling routes; 2001. Available at: http://ec.europa.eu/environment/waste/sludge/sludge disposal.htm.
[28] Guangyin Z, Youcai Z. Sewage sludge generation and characteristics. Pollution Control and Resource Recovery; 2016. p. 1–11.
[29] Wingender J, Neu TR, Flemming HC. What are bacterial extracellular polymeric substances? In: Wingender J, Neu TR, Flemming HC, editors. Microbial extracellular polymeric substances: characterization, structures and function. Berlin Heidelberg: Springer-Verlag; 1999. p. 1–18.
[30] Sutherland IW. Biofilm exopolysaccharides: a strong and sticky framework. Microbiology 2001;147:3–9.
[31] Nielsen PH, Jahn A. Extraction of EPS. In: Wingender J, Neu TR, Flemming HC, editors. Microbial extracellular polymeric substances: characterization, structure and function. Berlin, Heidelberg: Springer-Verlag; 1999. p. 49–72.

[32] Laspidou CS, Rittmann BE. A unified theory for extracellular polymeric substances, soluble microbial products, and active and inert biomass. Water Resource 2002;36:2711–20.

[33] Frolund B, Griebe T, Nielsen PH. Enzymatic activity in the activated-sludge floc matrix. Applied Microbiology and Biotechnology 1995;43:755–61.

[34] Frolund B, Palmgren R, Keiding K, Nielsen PH. Extraction of extracellular polymers from activated sludge using a cation exchange resin. Water Resource 1996;30:1749–58.

[35] Flemming HC, Leis A. Sorption properties of biofilms. In: Flemming HC, Bitton G, editors. Encyclopedia of environmental microbiology, vol. V; 2002. p. 2958–67.

[36] Liu H, Fang HHP. Characterization of electrostatic binging sites of extracellular polymers by linear programming analysis of titration data. Biotechnology and Bioengineering 2002;80:806–11.

[37] Joshi RM, Juwarkar AA. In vivo studies to elucidate the role of extracellular polymeric substances from Azotobacter in immobilization of heavy metals. Environmental Science & Technology 2009;43:5884–9.

[38] Ha J, Gelabert A, Spormann AM, Brown GE. Role of extracellular polymeric substances in metal ion complexation on *Shewannella oneidensis*: batch uptake, thermodynamic modeling, ATR-FTIR, and EXAFS study. Geochimica et Cosmochimica Acta 2010;74:1–15.

[39] Bhaskar PV, Bhosle NB. Bacterial extracellular polymeric substances (EPS): a carrier of heavy metals in the marine food-chain. Environment International 2006;32:191–8.

[40] Moon SH, Park CS, Kim YJ, Park YI. Biosorption isotherms of Pb (II) and Zn (II) on Pestan, an extracellular polysaccharide, of *Pestalotiopsis* sp. KCTC 8637P. Process Biochemistry 2006;41:312–6.

[41] Zhang DY, Wang JL, Pan XL. Cadmium sorption by EPSs produced by anaerobic sludge under sulfate-reducing conditions. Journal of Hazardous Materials 2006;138:589–93.

[42] Spath R, Flemming HC, Wuertz S. Sorption properties of biofilms. Water Science and Technology 1998;37(4–5):207–10.

[43] Sheng GP, Yu HQ, Li XY. Extracellular polymeric substances (EPS) of microbial aggregates in biological wastewater treatment systems: a review. Biotechnology Advances 2010;28(6):882–94.

[44] Boyd A, Chakrabarty AM. Role of alginate lyase in cell detachment of *Pseudomonas aeruginosa*. Applied and Environmental Microbiology 1994;60:2355–9.

[45] Zhang XQ, Bishop PL. Biodegradability of biofilm extracellular polymeric substances. Chemosphere 2003;50:63–9.

[46] Zheng YM, Yu HQ. Determination of the pore size distribution and porosity of aerobic granules using size-exclusion chromatograph. Water Research 2007;41:39–46.

[47] Hinson RK, Kocher WM. Model for effective diffusivities in aerobic biofilms. Journal of Environmental Engineering 1996;122:1023–30.

[48] Morgan, J.W., Forster, C.F., and Evison, L. "A comparative study of the nature of biopolymers extracted from anaerobic and activated sludges. Water Research, 24: 743–753.

[49] Wang ZW, Liu Y, Tay JH. Distribution of EPS and cell surface hydrophobicity in aerobic granules. Applied Microbiology and Biotechnology 2005;69:469–73.

[50] Jorand F, Boue-Bigne F, Block JC, Urbain V. Hydrophobic/hydrophilic properties of activated sludge exopolymeric substances. Water Science and Technology 1998;37(4–5):307–15.

[51] Harrison EZ. Organic chemicals in sewage sludges. The Science of the Total Environment 2006;367:481–97.

[52] U.S. Environmental Protection Agency. Superfund analytical services, contract laboratory program (CLP). 2004. Available at: http://www.epa.gov/superfund/programs/clp/target.htm.

[53] Chancerel P, Meskers CEM, Hageluken C, Rotter VS. Assessment of precious metal flows during preprocessing of waste electrical and electronic equipment. The Journal of Industrial Ecology 2009;13:791–810.

[54] Jamali, M.K., Kazi, T.G., Arain, M.B., Afridi, H.I., Jalbani, N., and Memon, A.R. "Heavy metal contents of vegetables grown in soil, irrigated with mixtures of wastewater and sewage sludge in Pakistan, using ultrasonic-assisted pseudo-digestion. Journal of Agronomy and Crop Science, 193: 218–228.

[55] Saveyn H, Ferrasse JH, Hernandez AB, Rose J, Meeren PV, Roche N. The distribution of heavy metals following sewage sludge gasification. In: 2nd European conference on sludge management ECSM; 2010.
[56] Chaudhry Q, Castle L. Food applications of nanotechnologies: an overview of opportunities and challenges for developing countries. Trends in Food Science & Technology 2011;22:595–603.
[57] Gao Y, Cranston R. Recent advances in antimicrobial treatments of textiles. Textile Research Journal 2008;78:60–72.
[58] Marchesan S, Prato M. Nanomaterials for (nano)medicine. ACS Medicinal Chemistry Letters 2013;4:147–9.
[59] Westerhoff PK, Kiser A, Hristovski K. Nanomaterial removal and transformation during biological wastewater treatment. Environmental Engineering Science 2013;30:109–17.
[60] Westerhoff P, Lee S, Yang Y, Gordon GW, Hristovski K, Halden RU, Herckes P. Characterization, recovery opportunities, and valuation of metals in municipal sludges from U.S. wastewater treatment plants nationwide. Environmental Science & Technology 2015;49(16):9479–88.
[61] Kruger O, Grabner A, Adam C. Complete survey of German sewage sludge ash. Environmental Science & Technology 2014;48:11811–8.
[62] Zhang F, Yamasaki S, Kimura K. Rare earth element content in various waste ashes and the potential risk to Japanese soils. Environment International 2001;27(5):393–8.
[63] Sedlak R. Phosphorus and nitrogen removal from municipal wastewater. 2018.
[64] Shammas NK, Wang LK. Characteristics and quantity of biosolids. In: Wang LK, Shammas NK, Hung YT, editors. Biosolids treatment processes. New York: The Humana Press Inc.; 2007.
[65] Rittmann BE, Mayer B, Westerhoff P, Edwards M. Capturing the lost phosphorus. Chemosphere 2011;84:846–53.
[66] Poortinga AT, Bos R, Norde W, Busscher HJ. Electric double layer interactions in bacterial adhesion to surfaces. Surface Science Reports 2002;47:1–32.
[67] Farazaki M, Gikas P. Nitrification-denitrification of municipal wastewater without recirculation, using encapsulated microorganisms. Journal of Environmental Management 2019;242:258–65.
[68] Popova T, Kaleva MD. Decontamination of sewage sludge by treatment with calcium oxide. International Journal of Current Microbiology and Applied Sciences 2004;3(9):184–92.
[69] Durmaz B, Sanin FD. Effect of carbon to nitrogen ratio on the composition of microbial extracellular polymers in activated sludge. Water Science and Technology 2001;44(10):221–9.
[70] Mulchandani A, Westerhoff P. Bioresource Technology Recovery opportunities for metals and energy from sewage sludges. Bioresource Technology 2016;215:215–26.
[71] Hu M, Wang X, Wen X, Xia Y. Microbial community structures in different wastewater treatment plants as revealed by 454-pyrosequencing analysis. Bioresource Technology 2012;117:72–9.
[72] Nascimento AL. Sewage sludge microbial structures and relations to their sources, treatments, and chemical attributes. Frontiers in Microbiology 2018;9:1–11.
[73] Park ES, Kang BS, Kim JS. Recovery of oils with high calorific value and low contaminant content by pyrolysis of digested and dried sewage sludge containing polymer flocculants. Energy & Fuels 2008;22:1335–40.
[74] Kaminsky W, Kummer AB. Fluidized bed pyrolysis of digested sewage sludge. Journal of Applied Pyrolysis 1989;16:27–35.
[75] Zhang B, Xiong S, Xiao B, Yu D, Jia X. Mechanism of a wet sewage sludge pyrolysis in a tubular furnace. International Journal of Hydrogen Energy 2011;36:355–63.
[76] Karayildirim T, Yanik J, Yuksel M, Bockhorn H. Characterisation of products from pyrolysis of waste sludges. Fuel 2006;85:1498–508.
[77] Kasakura T, Hiraoka M. Pilot plant study on sewage sludge pyrolysis-I. Water Research 1982;16:1335–48.
[78] Hwang IH, Ouchi Y, Matsuto T. Characteristics of leachate from pyrolysis residue of sewage sludge. Chemosphere 2007;68:1913–9.

[79] Radovic LR, Silva IF, Ume JI, Menendez JA, Leon Y, Leon CA, Scaroni AW. An experimental and theoretical study of the absorption of aromatics possessing electron-withdrawing and electron-donating functional groups by chemically modified activated carbons. Carbon 1997;35:1339–49.
[80] Boerrigter H, Rauch R. Review of applications of gases from biomass gasification. ECN Biomassa, Kolen en Milieuonderzoek; 2006. p. 20.
[81] Psomopoulos CS, Bourka A, Themelis NJ. Waste-to-energy: a review of the status and benefits in USA. Waste Management 2009;29(5):1718–24.
[82] Marrero T, McAuley B, Sutterlin W, Morris S, Manahan S. Fate of heavy metals and radioactive metals in gasification of sewage sludge. Waste Management 2003;24:193–8.
[83] Couto ND, Silva VB, Monteiro E, Rouboa A. Assessment of municipal solid wastes gasification in a semi-industrial gasifier using syngas quality indices. Energy 2015;93:864–73.
[84] Behrendt F, Neubauer Y, Oevermann M, Wilmes B, Zobel N. Direct liquefaction of biomass. Chemical Engineering & Technology 2008;31:667–77.
[85] Kirtania K. Thermochemical conversion processes for waste biorefinery. In: Bhaskar T, Pandey A, Mohan SV, Lee D-J, Khanal SK, editors. Waste biorefinery, vol. I. Elsevier B.V.; 2018. p. 129–51.
[86] Li H, Yuan X, Zeng G. The formation of bio-oil from sludge by deoxy-liquefaction in supercritical ethanol. Bioresource Technology 2010;101(8):2860–6.
[87] Werther J, Ogaba T. Sewage sludge combustion. Progress in Energy and Combustion Science 1999;25(1):55–116.
[88] Lundin M, Olofsson M, Pettersson GJ. Environmental and economic assessment of sewage sludge handling options. Resources, Conservation and Recycling 2004;41(4):255–78.
[89] Zabaniotou A, Theofilou C. Green energy at cement kiln in Cyprus—use of sewage sludge as a conventional fuel substitute. Renewable and Sustainable Energy Reviews 2008;12(2):531–41.
[90] Ariyaratne WKH, Melaaen MC, Eine K. Meat and bone meal as renewable energy source in cement kilns: investigation of optimum feeding rate. International conference on renewable energies and power quality (ICREPQ'11), Las Palmas de Gran Canaria, Spain, 13–15 April 2010. 2010. Available at: http://icrepq.com/icrepq'11/609-ariyaratne.pdf.
[91] Schneider M, Romer M, Tschudin M. Sustainable cement production – present and future. Cement and Concrete Research 2011;41(7):642–50.
[92] United States Environmental Protection Agency, editor. Decision maker's guide to solid waste management, vol. II; 1995 [chapter 7: Composting]. Available at: http://www.epa.gov/osw/nonhaz/municipal/dmg2/chapter7.pdf.
[93] Amir S, Hafidi M, Merlina G. Fate of polycyclic aromatic hydrocarbons during composting of lagooning sewage sludge. Chemosphere 2005;58(4):449–58.
[94] Michel Jr FC, Quensen J, Reddy CA. Bioremediation of a PCB contaminated soil via composting. Compost Science & Utilization 2001;9(4):274–84.
[95] He MM, Tian GM, Liang XQ. Phytotoxicity and speciation of copper, zinc and lead during the aerobic composting of sewage sludge. Journal of Hazardous Materials 2009;163(2–3):671–7.
[96] Martins das Neves LC, Converti A, Vessoni Penna TC. Biogas production: new trends for alternative energy sources in rural and urban zones. Chemical Engineering & Technology 2009;32(8):1147–53.
[97] Rittmann BE, McCarty PL. Environmental biotechnology: principles and applications. New York, USA: McGraw Hill; 2001.
[98] Nijaguna BT. Biogas technology. New Delhi, India: New Age International (P) Limited; 2002.

SECTION F

Modelling and LCA studies

CHAPTER 16

Multiscale modeling approaches for waste biorefinery

A.K.M. Kazi Aurnob, Ahaduzzaman Nahid, Kazi Bayzid Kabir, Kawnish Kirtania
Department of Chemical Engineering, Bangladesh University of Engineering and Technology, Dhaka, Bangladesh

16.1 Introduction

Multiscale modeling is a generic term for modeling approaches ranging from "nanoscale" to "system scale." Irrespective of the processes, modeling has the benefit of being nonspecific. The same modeling approach can be applied to a variety of processes or, methods. By avoiding numerous experimental investigations for optimized operation of a waste biorefinery, modeling provides the liberty to become creative and efficient. With multiscale modeling, the understanding and optimization of processes in different scales become viable. With computational power at our disposal, it is possible to study each reaction in a process in detail [1].

As biorefinery is a combination of various pathways for conversion [2], it is natural to go through a selection process to obtain an optimized state for technical and economic feasibility. This is typically achieved by using system-scale modeling approach. The application of system-scale modeling has a tremendous impact on waste biorefinery design, for example, it could be used for process hierarchy optimization or, simple process optimization through parametric variation [3]. Each process in the biorefinery can be a simple input-output model or, little more complicated thermodynamic model [4]. Either way, the goal is to reach a point with a range of products with optimized yields for a better economics and process performance.

It is intriguing to note that a single reaction might play as the limiting factor for some biorefineries. In such cases, modeling in smaller scale is relevant. Typically, reactions are modeled using thermodynamic models for simplicity in system scale. A more detail approach is to conduct kinetic modeling to obtain realistic data. When other factors are affecting the reaction comes into play, i.e., heat transfer, mass transfer and fluid dynamics in a reactor, complex modeling approaches are adopted [5]. In case of systems with

Waste Biorefinery. https://doi.org/10.1016/B978-0-12-818228-4.00016-2

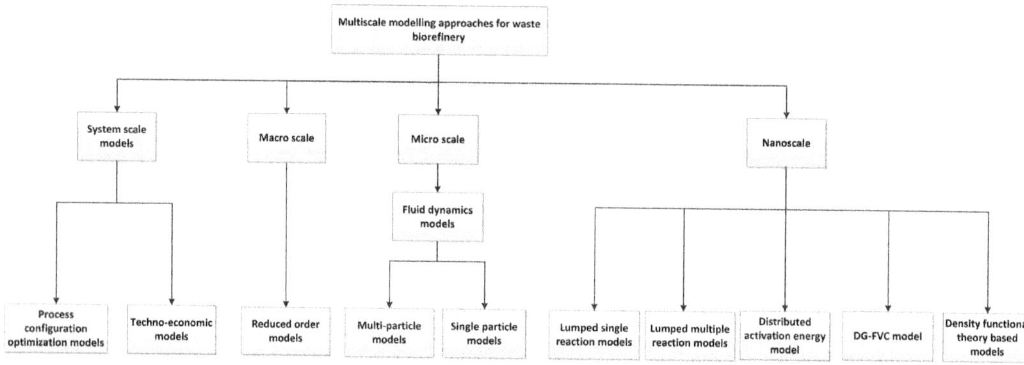

Figure 16.1
Modeling approaches available for waste biorefineries.

dominating reactive systems, intricate reaction kinetics algorithms are used for a better description of the process [6]. This type of modeling is considered as the molecular level or, nanoscale approach. To summarize, an overview of the relevant modeling methodologies for waste biorefinery is provided in Fig. 16.1.

In the figure, it could be seen that the ramification of the nanoscale models is greater than the rest. This happened due to the recent advances in molecular-level research enhancing our understanding. In the middle, there is reduced order model (macro scale) which is a hybrid of system and microscale model [7]. In this book chapter, the modeling approaches are discussed gradually from nanoscale to system-scale model developments and applications in relation to waste biorefinery following a "bottom-up" approach.

16.2 Modeling strategies for biorefineries

Both biochemical and thermochemical pathways for waste conversion are equally important for waste-based biorefineries. Of the two, biochemical reactions take place at low temperature close to atmospheric pressure. Therefore, the reaction rates in biochemical processes are relatively slower than the thermochemical pathways which take place at an elevated temperature and pressure. A typical biochemical conversion of biomass to ethanol in a batch reactor would require 2 days for the completion of the reaction. On the other hand, a thermochemical route using continuous reactor for the same reaction would need only 7 min [8]. Due to this advantage, biochemical reactive processes are easier to model in molecular level without considering the impact of fluid dynamics, heat and mass transfer. The determining factors are related to the reactants and, the enzymes which catalyze the reactions. In this way, biochemical processes are typically considered for molecular scale (reaction level) and system-scale modeling (considers the whole biorefinery).

In the early stage of the development of waste-based biorefineries, the typical modeling approach for biochemical pathways was basically empirical, supported with system-scale optimization of product yields to improve overall economics. For example, methanol and vegetable oil are the raw materials for biodiesel production. The amount of methanol that is lost during the process poses a heavy penalty on biodiesel production. Methanol recovery by separating from the final product and recycling in the biodiesel production loop can improve process performance and economics significantly [9]. The empirical approach would be by optimization of parameters using factorial design [10]. For the case of empirical modeling, typical factors for biochemical processes are temperature, concentration of reactants and time [11]. However, experiments are expensive to perform provided that the process modeling approach offers a cheaper way out through assessing the viable cases. Now with the advancement of life sciences, our understanding of molecular-level biological reactions has increased significantly. Currently, it is possible to dive into more detail modeling (at molecular level) complementing the system-scale modeling.

In contrast, modeling approaches on thermochemical processes for waste biorefinery has evolved over the last century. The understanding has become greater due to the consistent development of thermochemical processes through industrialization to meet the energy demand. Now it is possible to generate a number of products other than primary energy from these processes and thus, converging to biorefineries whenever using biogenic feedstocks. Typically, thermochemical processes use high pressure and moderate to high temperature for waste processing. This makes the modeling processes more complicated due to the involvement of several phenomena like fluid dynamics, heat and mass transfer along with reaction kinetics. Any of these processes can become dominating depending on the process condition (e.g., heating rate).

As mentioned earlier, the reaction rates for biochemical pathways are slower than those of thermochemical pathways. A simple comparison on reaction rates can be made from the production of ethanol from biomass.

Whether it is a biochemical or thermochemical pathway-dominated biorefinery, appropriate modeling approach can serve as a tool from preliminary design stage to procurement phase. The following sections of this chapter are devoted to the discussion on different modeling strategies available for adoption based on the nature of the process.

16.3 Nanoscale modeling

Nanoscale modeling (also known as molecular modeling) can be divided into two major categories. One considers the statistical possibility of interaction of the molecules with each other to react. The calculation for the interaction is typically based on quantum chemistry while some other models use a preloaded database [12]. These models can take into account several factors such as bond length and affinity of the molecules to react with each other. The other approach considers the reaction among different reactant molecules and addresses the reaction propagation using the famous Arrhenius correlation, i.e., activation energy. This is also known as kinetic modeling of bio and thermochemical processes. The benefit of this method relies in the fact that, it is applicable to any reactive process and can be used to predict the product composition at any stage of the reaction. Kinetic modeling is especially effective because in many cases for waste biorefinery, the reactive process does not reach equilibrium [13].

Among several approaches available for quantum chemistry-based modeling, density functional theory (DFT) is the most popular due to its ease of use and acceptable accuracy. For DFT, all the pertinent factors are related to the electron density of the atoms. This is a probabilistic approach, however, highly successful in many cases by producing accurate results. Only demerit of this kind of modeling is the computational cost with respect to the output. Creating a possible new molecule under reaction conditions using DFT is subject to serious computational load. So, it is not always practical to simulate a whole biorefinery which might involve hundreds or, thousands of reactions. Nevertheless, this type of modeling has been already proven useful in describing the mechanism for tar formation during pyrolysis and gasification process [14]. It holds the potential to become one of the major modeling approaches for simulating larger molecules and more complicated reactive processes with the development of newer algorithms and increase in computational capabilities. Another method of molecular simulation is by adopting the "functional group-devolatilization, vaporization, and crosslinking" (FG-DVC) model. In this model, some pathways for molecule formation are already identified and fed as the input, then based on the process conditions; typically, a Monte-Carlo simulation is performed to derive the final array of products [15]. This model was specifically successful in modeling pyrolysis process by addressing the complexity of the process. Although, some researchers found that distributed activation energy model (a type of Arrhenius model) was more accurate in several cases even after inclusion of all the complexity in the pyrolysis process by FG-DVC [16].

Thermochemical conversion of biogenic material involves a series of temperature dependent decomposition reactions. Kinetics plays a major role in determining the rate of the decomposition process at lower temperatures. As the decomposition temperature increases, the control gradually shifts from the chemical process to transport processes.

The number of reactions occurring during devolatilization is numerous and so is the number of products. It is impossible to generalize the products due to the heterogeneous nature of the starting material. Hence, researchers took an overall approach: to lump the products into three groups—gas, liquid, and solid.

The lumped kinetic models can have a single reaction or multiple parallel reactions leading to formation of these three groups of products. Some of the models consider the reactions to be of first-order, while the others use *n*th-order reaction models. Some researchers also considered the activation energy as a nonunique value. Conveniently, a distribution is used to describe the variation in activation energy with the level of conversion.

16.3.1 Density functional theory approach

DFT provides an approximate solution to the intractable Schrödinger equation of many-particle system. While using DFT, the wave function of the Schrödinger equation is replaced by the ground state electron density as the fundamental variable [17]. To simplify, DFT takes the electron density, which is a function of the position coordinate, as an input. The output is the energy of the system. Due to better accuracy and low computational cost compared to other similar techniques, DFT has become one the leading tool in molecular modeling [18].

Waste biogenic materials are composed of complex molecules. Their conversion usually involves destruction and generation of thousands of molecules following numerous reaction pathways. Therefore, it is sometimes necessary to study the formation of value-added products from the conversion of biogenic materials. Researchers are using DFT for noncatalytic as well as catalytic mechanistic study leading to formation of targeted products in biorefineries. Some of these studies involved conversion of major components of biomass (e.g., lignin, cellulose) or processing of the intermediates produced from the conversion of biomass.

Elder and Beste studied several concerted mechanisms of pyrolysis of substituted lignin models [19]. Photo-oxidation of lignin by using $P_{25}TiO_2$ leading to production of dehydrodiisoeugenol (DHDIE) was studied by Zielinski et al. [20]. Based on the DFT study, they proposed a stage-wise mechanism of oxidation of isoeugenol to DHDIE.

Cellulose pyrolysis leading to formation of levoglucosan (LG) was studied by Zhang et al. [21]. Out of the three studied pathways, their calculation showed that levoglucosan chain-end mechanism is the most preferred pathway for LG formation. All three calculations mentioned here were performed using Gaussian 09 (simulation software).

However, it should be noted that the use of DFT in modeling biorefineries is relatively new and used only to a limited extent. The biorefinery systems are highly complex, involving thousands of molecules and reactions. The cost of computing will be high, and it is unlikely that a single DFT model will include all the major compounds and reaction pathways of a biorefinery. This means the use of DFT models will be selectively applicable whenever a detail molecular modeling is of absolute necessity.

16.3.2 FG-DVC modeling approach

FG-DVC model is a combination of two devolatilization structural models: functional group (FG) model and depolymerization-vaporization crosslinking (DVC) model [22,23]. FG component of the model is based on the concept that the weaker aliphatic and ether bridges between the aromatic clusters breakdown during thermal decomposition resulting the release of light gases and tar precursors. The step-wise release is due to the variation in strength of these bridges. The DVC component includes depolymerization, crosslinking and transport of tar molecules. The model provides a molecular-weight distribution of char and tar by using Monte-Carlo simulation [15] or, percolation theory [24]. These two components of the FG-DVC model are well supplemented by each other: FG component provides the number of light gaseous molecules which is then used for calculating the probability of crosslinking utilizing the DVC component.

Though, the model was initially developed for coal pyrolysis, it has also been successfully implemented for biomass [25]. The combustion of wheat straw waste was studied in a drop-tube reactor and the model using FG-DVC method satisfactorily predicted the experimental results. Chen et al. [26] developed a "streamlined" version of FG-DVC, also known as FG-biomass, for predicting pyrolysis behavior of several biomass samples. They concluded that the model needed further improvement for predicting pyrolysis at heating rates in the range of 10^4-10^5 K/s. A similar conclusion was obtained by Jong et al. [27] for the FG-biomass model while studying pyrolysis of waste biomass using TG-FTIR. Overall, FG-DVC model showed prospect in describing pyrolysis of biogenic wastes in a more mechanistic manner. Though significant development is made, further research and development focusing on waste pyrolysis would facilitate its wider application in the field of biorefinery.

16.3.3 Lumped models based on single and multiple reactions

Thermochemical conversion of biogenic materials always starts with pyrolysis. In biorefineries, pyrolysis is sometimes followed by combustion or gasification. During pyrolysis, biomass starts to lose weight as the heat increases. First, the moisture in removed. As the temperature is increased further, the components of biomass start to break up. Hemicellulose is the least stable of all components and starts breaking down at 498 K. Lignin decomposes between 523 and 773 K, while the decomposition temperatures for cellulose are between 598 and 648 K [28]. As the heating rate is higher, the decomposition reactions of biomass components tend to merge together. This has been the basis of single-reaction models. The generic form of reaction mechanism for this group of models was proposed by Shafizadeh and Chin [29], comprised of three parallel reactions leading to formation of gas, liquid (tar) and solid (char), as shown in Fig. 16.2. Reaction initiation

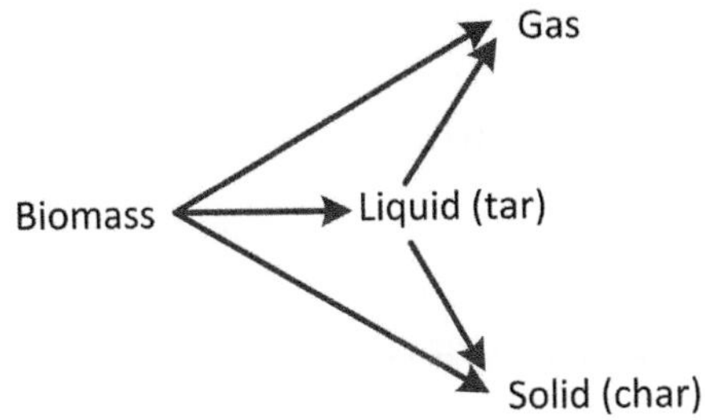

Figure 16.2
Devolatilization of biomass: single-reaction model.

temperature depends on the type of biomass; however, in most cases the reaction is observed from temperatures as low as 500 K.

Several equipment (e.g., thermogravimetric analyzer, fixed- and fluidized-bed reactors, wire-mesh reactors, drop-tube, and entrained-flow reactors [30]) have been used for conducting the studies leading to single-reaction kinetic models. The mathematical analysis of the experimental data led to the determination of activation energies between 56 and 174 kJ/mol. Such a wide range indicates the effect of the equipment of choice as well as the heating conditions on the pyrolysis reaction.

Di Blasi [31] analyzed the kinetic data and categorized them into several groups based on the severity of heat treatment and activation energies obtained. They are:

1. High temperature pyrolysis with activation energies between 69 and 91 kJ/mol
2. Low temperature pyrolysis
   a. With activation energies between 56 and 106 kJ/mol
   b. With activation energies between 125 and 174 kJ/mol

Pyrolysis being a heterogeneous process is affected by both chemical and transport processes. At lower temperatures, the chemical process is slower than the transportation of pyrolysis products from the surface to the bulk fluid phase. The interpretations of the data obtained from the low temperature treatments are therefore more likely to provide the true kinetics (hence true activation energies) of the process. With increased temperature, the reaction becomes faster and the transport processes become rate limiting. Hence, the treatment of data provides apparent activation energies, which is much lower than the actual values.

From Di Blasi's classification of single-reaction kinetic models [31], it can be noted that low temperature data yield both a low-activation energy group and a high activation energy group. Some of the low temperature studies were conducted in tube furnaces, fluidized-beds, entrained-flow reactors, etc. Carrying out pyrolysis experiments in such reactors are usually influenced by significant heat and mass transfer limitations and, therefore, can result in providing apparent kinetics. Also, use of large sample size, or using large particles can lead to heat or mass transfer-controlled operations. This is the reason behind the kinetics observed in the category 2(a).

Multiple reaction models assume biomass as a heterogeneous mixture of different components. Therefore, the mechanism of multicomponent devolatilization assumes reactions, usually parallel to each other for each of the pseudocomponents. The reaction scheme is [31]:

$$Y_i \rightarrow V_i; \quad k_i = A_i \exp(-E/RT_i); \quad i = 1, n \tag{16.1}$$

where, Y_i is the volatile fraction of the ith pseudocomponent, V_i is the corresponding devolatilization products and k_i is the corresponding reaction rate constant. For three-reaction models of biomass devolatilization, these pseudocomponents are usually hemicellulose, cellulose and lignin.

Thermogravimetric studies leading to estimation of the kinetic parameters for multiple reaction models usually performed at very slow heating rates. Therefore, the studies are usually kinetically controlled. Generally, heating rate below 10 K/min were used [31]. It must be mentioned that the kinetic parameters obtained by mathematical treatment of experimental data obtained at a single heating rate cannot ensure that the true kinetics is obtained. Experimental data fitting at a single heating rate can produce several sets of preexponential factor and activation energy, all of which can predict well the same mass loss curve. It is debatable whether this observation is due to a true or false compensation effect. Chornet and Roy [32] used published data to conclude the presence of true compensation effect during pyrolysis of cellulosic materials. However, Agarwal's critical analysis [33,34] on their work showed that the compensation effect is rather an computational artifact and originated from the computational and experimental errors. Presence of significant thermal lag due to endothermicity of the pyrolysis reaction changes the actual heating rate the sample experiences and causes the compensation effect [35]. In order to obtain the actual kinetics, one needs to fit experimental data from different heating rates with the same kinetic parameters [36,37]. Only then it can be assured that the reported model parameters correspond to the true kinetics.

Simultaneous treatment of data for different heating rates has also used for estimation of the kinetic parameter [38–41]. These treatments were performed from heating rate as low as 0.5 K/min to 108 K/min [41]. Activation energies for cellulose decomposition from the single heating rate experiments are similar to that obtained from multiple heating rate experiments, 195–286 kJ/mol and 192–250 kJ/mol, respectively [31]. However, for hemicellulose, the reported activation energies are higher for the multiple heating rate treatments, 154–200 kJ/mol compared to 80–116 kJ/mol for single heating rate experiments [31]. Activation energies obtained from the single heating rate experiments were reported to vary between 18 and 65 kJ/mol for lignin. Lignin decomposition activation energies from the multiple heating rate experiments have been reported to be in 54–61 kJ/mol and 188–219 kJ/mol ranges, with a strong dependence of the preexponential factor on the heating rates.

In order to improve accuracy of the model prediction, modification of the three pseudocomponent mechanism have been proposed. Grønli et al. [42] proposed a five reaction model for several hardwood and softwood species. Three reactions for hemicellulose, cellulose and lignin (activation energies 100, 236 and 46 kJ/mol) were able to predict the devolatilization at temperature above 553 K. They proposed an inclusion of additional two reactions for low temperature decomposition of extractive contents with activation energies of 105 and 127 kJ/mol, respectively. Müller-Hagedorn et al. [43] considered power-law dependence on the volatile mass fraction for five parallel reactions, two for both hemicellulose and cellulose, and one for lignin. They reported reaction orders between 0.8 and 1.8. Mészáros et al. [44] studied pyrolysis of three biomass species between 10 K/min and 40 K/min. They concluded that at least six parallel reactions are needed for first-order kinetics and at least four parallel reactions are required for *n*th order kinetics to describe the devolatilization behavior.

Composition-wise, algal biomass differs from the lignocellulosic counterpart. Microalgae are typically composed of 40%–60% proteins, 5%–60% lipids, 8%–30% carbohydrates and 5%–10% other components such as pigments, antioxidants, fatty acid etc. [45]. Like biomass, considering algae as a heterogeneous mixture of its components, multiple reaction models have been proposed for the pyrolysis of algal biomass. Sharara et al. [46] used a four pseudocomponent model for pyrolysis of two freshwater algae. The low temperature decomposition corresponds to moisture and light hydrocarbon loss. Decomposition of protein and starch were observed between 573 and 673 K, while the fourth pseudocomponent peaked around 873 K. Bui et al. [47] considered oil-extracted microalgae residues to be composed of lipid, protein and lignocellulosic cell wall (e.g., cellulose, hemicellulose and lignin). The five pseudocomponent model predicted the devolatilization process with first-order and *n*th-order parallel reactions. For *n*th order model, the reaction order varied between 1 and 1.5. Bach and Chen [48,49] used five models to fit the pyrolysis of Chlorella vulgaris ESP-31 at 10 K/min hearing rate. The single-reaction model did not fit the experimental data well. However, the multiple reaction models estimated the pyrolysis behavior better than single-reaction model. Seven reaction model were more accurate in predicting the experimental data than four-, three- or two-reaction models. In this case, the proposed seven pseudocomponents were low thermal resistant carbohydrate, high thermal resistant carbohydrate, low thermal resistant protein, high thermal resistant protein, lipids, and other intermediates. It must be mentioned that increasing number of pseudocomponents are more likely to increases the data fitting quality of a model. However, one must consider whether increasing the number of pseudo-components, and hence the reactions, are physically meaningful or not. The selection of number of reactions in a model must be well-supported by the composition or components present in a material.

16.3.4 Distributed activation energy model (DAEM)

The distributed activation energy model (DAEM) method is widely accepted as a relatively accurate method for estimating activation energy (E). While other single heating methods assume that a single lumped reaction is taking place, DAEM is a leap forward. A brief description on DAEM is provided by Bhaskar and Dhyani in Chapter 2 of *Waste Biorefinery*, vol. 1 [50], while a more detail approach is taken in this chapter to elucidate further. DAEM takes into account that reaction mechanisms, models are dynamic and utilizes the value of apparent activation energy. A continuous distribution model is employed to express the energy distribution. The most common distribution function used is Gaussian. However, researchers had worked on several other types of distributions based on the feedstock characteristics. The most popular continuous distributions used are: Gaussian, Weibull, and Logistic.

The DAEM can be represented by the following equation

$$\frac{w}{w^*} = \alpha = \int_0^\infty \left\{ 1 + (1-n)\int_0^t k_0 \exp\left(-\frac{E}{RT}\right)dt \right\}^{\left(\frac{1}{1-n}\right)} f(E)dE; \quad \text{for } n \neq 1 \tag{16.2}$$

$$\frac{w}{w^*} = \propto = \int_0^\infty \left[1 - \exp\left(- \int_0^t k_0 e^{-\frac{E}{RT}} dt \right) \right] f(E)dE; \quad \text{for } n = 1 \tag{16.3}$$

$\frac{w}{w^*}$ is the ratio of the mass of volatiles released at time t to the total mass of volatiles. This is directly proportional to fraction conversion, $\alpha . f(E)$ stands for the distribution function. Different types of distributions are expressed below

Gaussian distribution

$$f(E) = \left\{ \frac{1}{\sqrt{2\pi\sigma^2}} \exp\left(-\frac{(E - mE)^2}{2\sigma^2} \right) \right\} dE \tag{16.4}$$

The preexponential factor (k_0), mean activation energy (mE), and the standard deviation of distribution (σ) are estimated.

Weibull distribution

$$f(E) = \left(\frac{\beta}{\eta}\right)\left(\frac{E-\gamma}{\eta}\right)^{\beta-1}\exp\left[-\left(\frac{E-\gamma}{\eta}\right)^{\beta}\right]; \quad \text{for } E \geq \gamma, \eta > 0, \text{ and } \beta > 0 \qquad (16.5)$$

where E is the activation energy expressed in kcal/mol; η is the scale parameter; β is the shape parameter and determines the shape of the Weibull distribution, for $\beta = 4$ is almost indistinguishable from a normal distribution; γ is the threshold parameter for the activation energy, it is assumed that reactions with activation energy less than the threshold parameter will not take place.

Logistic distribution

$$f(E) = \frac{b}{\sigma}\frac{e^{-b(E-\mu)/\sigma}}{\left[1+e^{-b(E-\mu)/\sigma}\right]^2} \qquad (16.6)$$

where $b = \pi/\sqrt{3}$, μ and σ are the mean and standard deviation of the distribution, respectively.

For complex reactions such as pyrolysis, it is extremely difficult to approximate the k_0 accurately, and the Gaussian distribution is not very accurate especially at the initial and final stages of the pyrolysis process. On the other hand, the Weibull distribution is a three-dimensional probability density function [51]. To model complex biogenic waste pyrolysis, Cai et al. [52,53] introduced logistic distribution in DAEM. The disparity between the Gaussian and logistic distribution is the slightly longer tails of the logistic distribution. Due to this characteristic of the logistic distribution, it is claimed to be suitable to be used to describe the activation energy distribution for the pyrolysis of complex solid fuels i.e., waste biomass.

16.3.4.1 General modeling approach with DAEM

The general approach to address the DAEM equation can be divided into two types: integral approach and differential approach. Some of the popular integral approaches applied on waste conversion were developed by a number of researchers including Doyle [54], Coats and Redfern [55], Senum and Yang [56], Agarwal and Sivasubramanian [57], Miura and Maki [58], and Kirtania and Bhattacharya [59]. On the other hand, the same approaches can be applied in differential form for thermochemical conversion of biomass. Comparatively, differential approaches have been more erroneous than the integral method. In some cases, differential method can lead to significant error in estimating the value of preexponential factor. Of the above approaches, one of the recent and robust integral approaches was proposed by Kirtania and Bhattacharya which was successfully applied on algae pyrolysis [59], blends of fuels [1] and coupled with single particle model (See Section 4.1) for a number of biomass wastes [6]. For this approach, the DAEM model Eq. (16.2) is represented as

$$\frac{w}{w^*} = \alpha = \int_0^\infty Z(E,t)f(E)dE; \quad \text{for } n \neq 1 \tag{16.7}$$

where *Z(E, t)* contains the preexponential factor, k_0, and the order of the reaction, *n*. *f(E)* represents the continuous distribution model. If the heating rate is τ K/min, then *Z(E, t)* can be written as *Z(E, T)* and expressed as follows:

$$Z(E,T) = \left\{1 + (1-n)\int_0^T \frac{k_0}{\tau}\exp\left(-\frac{E}{RT}\right)dT\right\}^{\left(\frac{1}{1-n}\right)}; \quad \text{where } T = T_0 + \beta t \tag{16.8}$$

The limits of this integration work as long as the value of T_0 selected is low enough where no reaction is taking place [60]. Eq. (16.8) is evaluated using asymptotic approximation when $E/R \rightarrow \infty$ [61]:

$$Z(E,T) = \left\{1 + (1-n)\left[-\frac{k_0 RT^2}{aE}\exp\left(-\frac{E}{RT}\right)\right]\right\}^{\left(\frac{1}{1-n}\right)} \tag{16.9}$$

Kirtania and Bhattacharya [59] assumed that k_0 is independent of temperature while studying the kinetics for the pyrolysis of algae. Eq. (16.9) can then be written as:

$$Z(E,T) = \{1 + (1-n)k_0\xi(E,T)\}^{\left(\frac{1}{1-n}\right)}; \quad \text{where } \xi(E,T) = -\frac{RT^2}{aE}\exp\left(-\frac{E}{RT}\right) \tag{16.10}$$

The temperature integral is evaluated at each value of decomposition and a matrix of $\xi(E_i, T)$ is formulated for each value of activation energy. Eq. (16.7) can then be written as follows:

$$\frac{1}{w^*}\begin{bmatrix} w(T_1) \\ w(T_2) \\ w(T_3) \\ \cdots \\ w(T_F) \end{bmatrix} = \left\{1 + (1-n)k_0 \begin{bmatrix} \xi(E_1, T_1) & \xi(E_2, T_1) & \cdots & \xi(E_N, T_1) \\ \xi(E_1, T_1) & \xi(E_2, T_1) & \cdots & \xi(E_N, T_2) \\ \xi(E_1, T_1) & \xi(E_2, T_1) & \cdots & \xi(E_N, T_3) \\ \cdots & \cdots & \cdots & \cdots \\ \xi(E_1, T_F) & \xi(E_2, T_F) & \cdots & \xi(E_N, T_F) \end{bmatrix}\right\}^{\left(\frac{1}{1-n}\right)} \begin{bmatrix} w(E_1) \\ w(E_2) \\ w(E_3) \\ \cdots \\ w(E_N) \end{bmatrix} \tag{16.11}$$

$$\frac{\mathbf{w}}{w^*} = \{1 + (1-n)k_0\xi\}^{\left(\frac{1}{1-n}\right)}\mathbf{f} \tag{16.12}$$

T_1 and T_F are the minimum temperature and the maximum temperature, respectively. The typical estimation accuracy by applying this approach can be seen from Fig. 16.3. The regression analysis yielded a R^2 value of 0.9996 for the pyrolysis curve at 10 K/min heating rate [59].

16.4 Fluid dynamics modeling

During waste conversion process in a biorefinery, fluid dynamics may play an important role along with heat and mass transfer. Furthermore, reaction progress is controlled by kinetics or, molecular-level interaction. To account for all these phenomena accurately, it is often necessary to consider the fluid dynamics in modeling approaches. This modeling approach has been popular in the field of aerodynamics of flying objects and also, for cars. Eventually, airplane and car manufacturing companies became some of the major consumers of computational fluid dynamics (CFD) simulations.

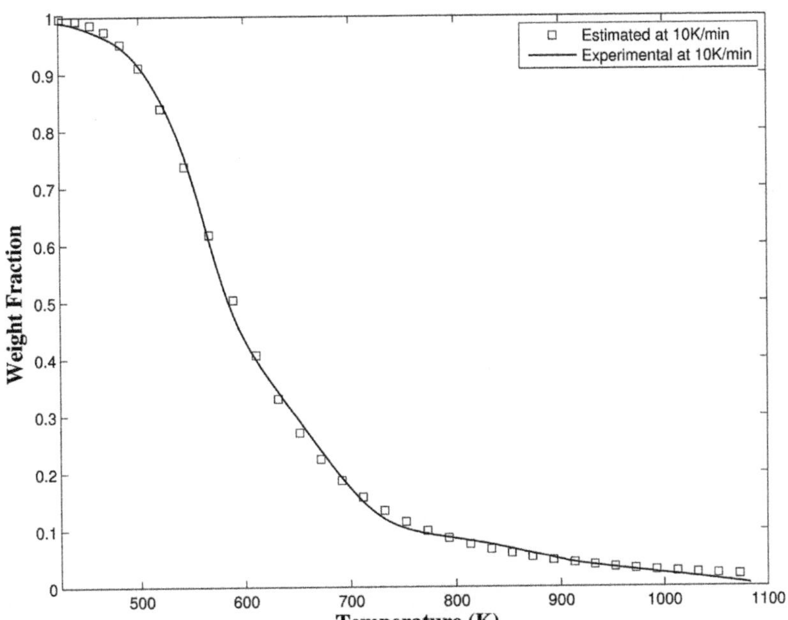

Figure 16.3
Comparison of estimated and experimental data for algae using DAEM. *Adapted from Kirtania K, Bhattacharya S. Application of the distributed activation energy model to the kinetic study of pyrolysis of the fresh water algae* Chlorococcum humicola. *Bioresource Technology 2012;107:476–81, Copyright 2011, with permission from Elsevier.*

In the field of waste biorefinery, application of fluid dynamics has been visible only recently. One reason was the limited computational capability of the computers while the other was with defining the inherent complexity of simulating a process inside a reactor (i.e., hydrothermal treatment, pyrolysis, gasification, etc.). Thanks to the effort over last 20 years, it is now possible to model the reactors with sufficient details including the turbulence involved along with the heat and mass transfer. As each additional step in a process makes the overall simulation more cumbersome, it takes more time to simulate. To reduce the time for simulation, researchers often make several assumptions to simplify the calculations. As for example, a model involving a single particle in motion, the fluid flow is typically considered to be laminar [62] while turbulence models are more commonly applied for multi-particle models involving particle interaction [63]. In many cases, researchers simplify the reaction kinetics using lumped first-order reaction to reduce the calculation time [64]. This definitely reduces the computational cost but induces some error in simulation. Another strategy is to simplify the fluid dynamic calculation using a single particle model and introducing more complicated reaction kinetics [65]. Both strategies have trade-offs with their performance and accuracy level while it is to be remembered that the most accurate model might not be the most useful one in practical cases. With careful consideration, it is possible to simplify such complicated models rendering them to be more useful.

16.4.1 Single particle modeling approach

Single particle model is the idealistic approach available for CFD simulation. This is simplified significantly considering the flow as laminar in most cases. Although it is highly idealistic, it is particularly useful for simulating a stagnant or, dropping particle in a flowing fluid. Single particle model was very popular in describing pyrolysis of large biomass particles stagnant in a fixed bed reactor [66]. The objective of this type of modeling is to determine the differential temperature distribution inside a particle with the conversion profile along the characteristic length. This method has the potential to generate ample information to know about both physical and chemical transformations inside a particle including heating rate, overall reaction rate, mass loss etc. For a moving particle, the model can track the particle along the reactor. The actual benefit lies in the low computational cost for this model compared to multi-particle simulation while generating a lot of valuable information. Fig. 16.4 shows a black liquor (a waste/byproduct from pulp mill) particle moving down an entrained flow reactor under pyrolysis conditions stated in the study by Bach et al. [67]. Black liquor has a tendency to inflate during pyrolysis which is also simulated during the process. The heat transfer limitation imposed by the inflation of the particle is also considered in this simulation. This kind of detail is difficult to obtain in more complex approaches i.e., multiparticle models.

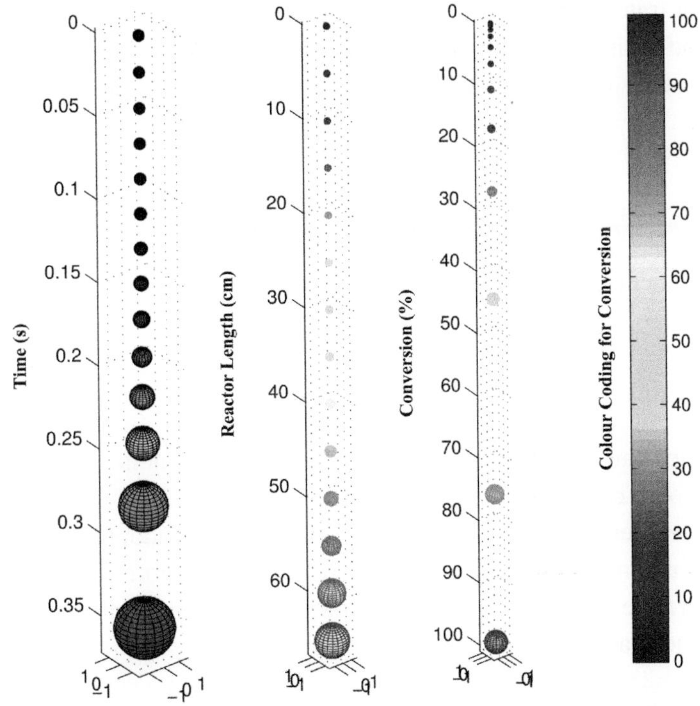

Figure 16.4
Swelling, conversion and motion of a black liquor particle during pyrolysis.

Energy balance at position, r and time, t inside a biogenic waste particle can be written as (considering local equilibrium)

$$\rho C_p \frac{\partial T_p}{\partial t} = \lambda \left[\frac{\partial^2 T_p}{\partial r^2} + \frac{b-1}{r} \frac{\partial T_p}{\partial r} \right] + q_{\text{reac}} \left(-\frac{d\rho}{dt} \right) \tag{16.13}$$

In case of apparent kinetics, q_{reac} (heat of reaction) is typically neglected whereas it is important to include for the intrinsic kinetics. Here, b is the shape factor that varies from 1 to 3. Mass conservation could be considered as a nondominating phenomenon for pyrolysis. Typical initial conditions $(0 \leq r \leq R)$ are: $T_p = 300$ K; $\rho_b = \rho_{b0}$; $\rho_c = 0$. The boundary condition of the particle center $(t \geq 0)$ is expressed as

$$\lambda \frac{\partial T_p}{\partial r} = 0 \tag{16.14}$$

where λ is thermal conductivity of the biomass particle and T_p is the particle temperature. The boundary condition on the particle surface $(t \geq 0)$ is given by

$$\lambda \frac{\partial T_s}{\partial r} = h(T_g - T_s) + \sigma\varepsilon \left(T_g^4 - T_s^4 \right) \tag{16.15}$$

where h is the heat transfer coefficient, T_g is the gas temperature, T_s is the temperature at particle surface and σ_b is the Stefan–Boltzmann constant. The density (ρ), specific heat (C_p), and conductivities (λ) were calculated by Eqs. (16.16)–(16.18).

$$\rho = \rho_b + \rho_c \tag{16.16}$$

$$C_p = \left(\rho_b C_{p,b} + \rho_c C_{p,c}\right)/\rho \tag{16.17}$$

$$\lambda = \left(\rho_b \lambda_b + \rho_c \lambda_c\right)/\rho \tag{16.18}$$

Heat transfer coefficient is calculated by Nusselt number ($Nu = hD/\lambda$)

$$Nu = 2 + 0.6\mathrm{Re}_D^{0.5}\mathrm{Pr}_D^{1/3} \tag{16.19}$$

where Re_D and Pr_D are the Reynolds and Prandtl number respectively.

The reaction kinetics can be modeled with simple first-order kinetics [68] or, complex DAEM [6]. The model can be used to calculate the required reactor length for optimum conversion of waste. Despite its elegance and usefulness, this modeling approach cannot be used in generic manner for different types of reactors. For more realistic industrial scale reactors where the particle interaction becomes prominent, multi-particle modeling approach will be more appropriate.

16.4.2 Multiparticle modeling approach

Multiparticle model works on the same principle as the single particle model with significantly more complicated fluid flow considerations with particle interactions. This type of modeling is typically carried out using commercial fluid dynamics simulation software like Ansys-Fluent, COMSOL etc. Currently, open source software is also available for this sort of simulation (OpenFOAM), however, requires more customization than the commercial counterparts. In this case, Navier–Stokes equation is embedded in the software to define the fluid flow pattern along with the capability of modifying the geometry of the reactor. Turbulence models are typically used to describe the eddy formation and heat/mass transfer coefficient determination. The multi-particle model provides detail information regarding the particle movement, particle interaction, eddy formation and, heat and mass transfer. Theoretical development of CFD models are covered in detail in several books [69–71], therefore, considered as beyond the scope of this chapter. However, it is interesting to note that application of CFD calculations is rather in the early stage for biorefinery applications considering the availability of accurate data and computational resources.

For this genre of models, reaction kinetics is often simplified to lumped first-order to reduce the computational cost. While it is possible to include the detail reaction modeling, that can easily increase the computational cost by an order of magnitude. Due to this

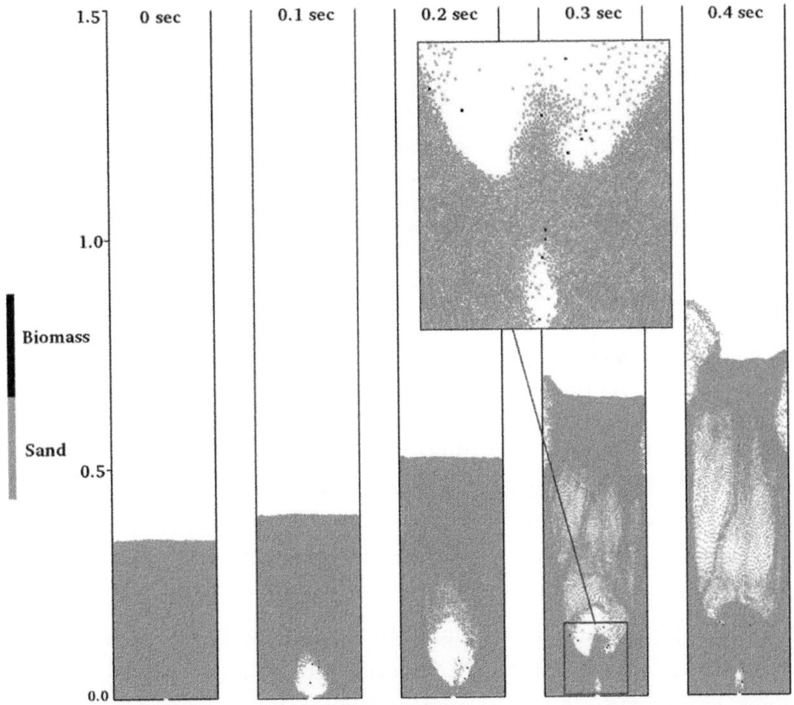

Figure 16.5
Flow pattern in a biomass gasifier (fluidized) with respect to time. *Adapted from Ku X, Li T, Løvås T. CFD–DEM simulation of biomass gasification with steam in a fluidized bed reactor. Chemical Engineering Science 2015;122:270–83. Copyright 2015, with permission from Elsevier.*

reason, it is more common to include detail reaction model in single particle models while multi-particle models focus on the complex physical phenomena influencing the process. For example, to simulate a fluidized bed gasifier, multi-particle model will consider the interaction of each moving particle with other particles and the reactor wall. The flow pattern can be observed in Fig. 16.5 where multi-particle biomass gasification is simulated in a fluidized bed with sand as the bed material [72].

This means large industrial scale gasification process can benefit from multi-particle CFD simulations. It is to be noted that fluidized bed gasifier simulation using multi-particle CFD started during the last decade of the 20th century and has progressed now to the development of more realistic and complicated models [63]. On the other hand, entrained-flow biomass gasification (EBG) models are only developed at the beginning of the 21st century [73] using Ansys-CFX following the development of pressurized entrained-flow biomass gasification (PEBG) using another commercial simulation software, Ansys-Fluent in 2007 [74]. After 7 years of the aforementioned study, a biomass entrained-flow gasifier

Table 16.1: Selection criteria for single and multi-particle models.

Reactor type for waste conversion	Single particle model	Multi-particle model
Fixed bed	Preferred	Not preferred over single particle
Fluidized bed	Not preferred over multi-particle	Preferred
Entrained-flow	Useful	Preferred

model is developed using open source simulation software, OpenFOAM at Norwegian University of Science and Technology [75].

It is noticeable that the development of multi-particle CFD models has been slower for biomass and waste processing compared to other industries. This development has gained pace recently due to more pilot scale studies performed to obtain reliable data for comparison with the simulation results. With the tremendous potential of this modeling approach to simulate almost every conceivable physical and chemical phenomenon inside the reactor, it is gaining more attention among the waste and biomass processing industries and researchers. Nonetheless, as multiparticle model computationally expensive, it is possible to choose single particle model over multiparticle simulation in many cases as described in Table 16.1.

16.5 Reduced order modeling

Reduced order models (ROM) for waste biorefinery are also known as "Equivalent Reactor Network Models." This modeling strategy is adopted when the multi-particle simulation models become computationally expensive and impractical for large scale biorefinery simulations. In this approach, a single reactor can be divided into several reactors (typically, a combination of CSTRs and PFRs) or, processes to simplify its functions. Then these segmented processes or, reactors could be joined together to simulate the whole process inside a single reactor [76,77]. The benefit is that, it bypasses simulating the whole reactor altogether and realistic simplifications are applied in each segment of the reactor. Fig. 16.6 shows a demonstration of the breakdown of a waste gasifier using blocks as separate processes. The gasifier is divided in different zones to identify the type of phenomenon that could take place and allocated accordingly. For example, the reactants will mix in the recirculation zones to further react and therefore, a "well stirred reactor" or, a "continuous stirred tank reactor" is an appropriate choice.

Each of these blocks in the figure is simulated sequentially to generate the result for a complete gasifier. Any recirculation or, interconnection among the sections is simulated

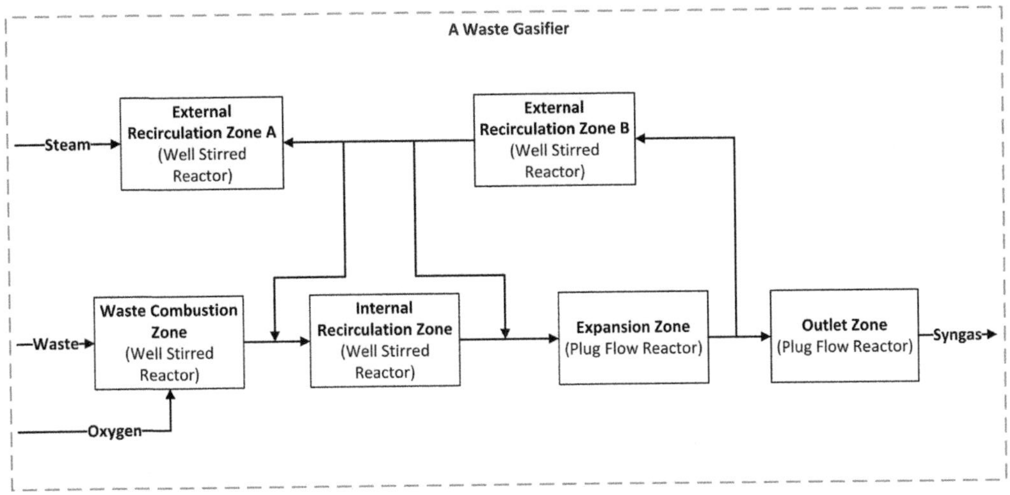

Figure 16.6
Block representation of reduced order model for a waste gasifier.

using recycle streams [7]. These models are fast to simulate and, capable of generating large amount of practical data for the process and the overall biorefinery. Typically, a large biorefinery with several processes is addressed by combining thermodynamics, reaction kinetics and simplified fluid dynamics whenever appropriate [7]. This means practical experience in industrial scale is required for applying this approach to simulate a biorefinery with acceptable accuracy.

16.6 System-scale modeling

16.6.1 Process configuration optimization

In case of process configuration optimization (i.e., automated targeting), a number of conversion processes for waste processing are considered for energy and biofuel production. The individual conversion processes are combined to improve the overall process efficiency and economics. An appropriate combination of technologies and conversion processes play the key role for the development of a possible biorefinery [2]. Based on laboratory scale experiments and literature data, it is possible to provide the input parameters for each process. Considering, the waste stream can be passed through n number of conversion processes, there are $(m \times n)$ number of choices available for formation of biorefineries. Fig. 16.7 shows all the probable combinations. Here, m is the number of exclusive parallel process pathways for conversion.

Each conversion process receives data from the previous level to generate output suitable for the next stage. The output from the final stage is a function of yield (y)

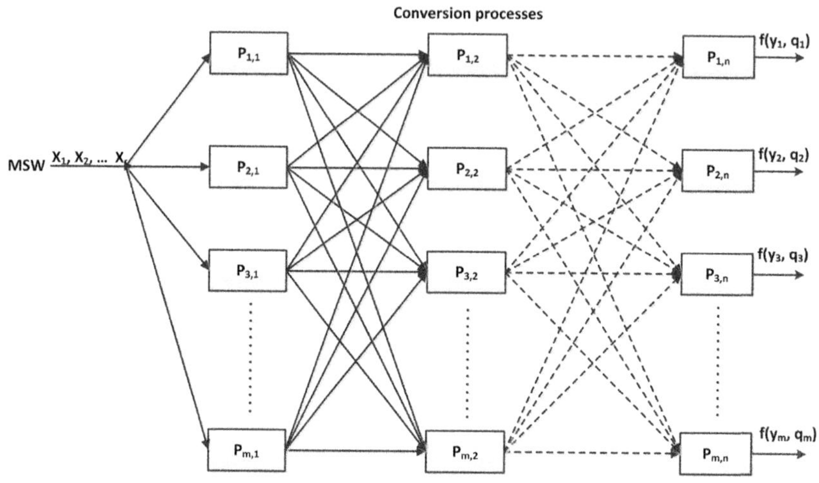

Figure 16.7
Possible process configurations for a municipal solid waste (MSW) based biorefinery (here, *p* denotes any waste conversion process, m and n mean the combination of choices available, *y* denotes product yield and *q* means product quality respectively).

and quality (q) of product. Quality of product is directly connected to the economics of the process. Therefore, maximizing the function of yield and quality provides the best possible configuration for the biorefinery.

The probable configurations can be mathematically expressed as

$$C_1 = \prod_{j=1}^{n} P_{1,j}; \ldots\ldots\ldots\ldots C_{10} = \prod_{j=1}^{n} P_{10,j}; \ldots\ldots\ldots\ldots C_m = \prod_{j=1}^{n} P_{m,j} \tag{16.20}$$

In the first stage of a biorefinery, b (i.e., $X_1, X_2 \ldots \ldots X_r$) is input as a vector containing the input parameters. Each configuration (C_1, C_2 etc.) can process X as input and provide an output function—$f(y, q)$. The configurations can also be denoted as a vector (C) as well. Element wise multiplication of these two vectors would generate an array of output functions

$$C_1X + C_2X + C_3X + \ldots\ldots\ldots\ldots\ldots\ldots\ldots\ldots\ldots\ldots + C_mX = \sum_{k=1}^{m} f(y_k, q_k) \tag{16.21}$$

The output functions can be compared for minimization with respect to the variation in the input parameters ($X_1, X_2 \ldots \ldots X_r$). The best possible configurations with maximum yield and product quality (hence, product price) will be determined by maximizing the objective function

$$OF = \max \sum_{k=1}^{m} f(y_k, q_k) \tag{16.22}$$

Nonlinear multivariable optimization is employed for the best possible accuracy; however, linear optimization with necessary simplification is also used in many cases [3].

16.6.2 Technoeconomic assessment

Technoeconomic assessment (TEA) is used to evaluate technical and economic viability of a process. TEA involves performing material and energy balances (MEB), estimating CAPEX (Capital Expenditure), OPEX (Operating Expenditure) and revenue. This technique is used at all levels of process design, starting from the process creation to detailed engineering [4,78]. However, for biorefineries the studies conducted so far are limited to either in the process creation [79–82] or concept demonstration stages [83–86] of design, mostly due to the level of maturity of the technologies in consideration.

Biochemical pathways of biorefineries mostly use input-output models for MEBs, while input-output, thermodynamic and kinetic models, or a combination of these are frequently used for thermochemical pathways. Vlysidis et al. [87] examined four schemes of biodiesel biorefineries for co-production of biodiesel and succinic acid. They performed a single-objective optimization to maximize net present value (NPV) and a multi-objective optimization to optimize a trade-off between profitability and CO_2 emission. Several scenarios for production of bio-jet-fuel from jatropha oil were studied by Zech et al. [88]. A combination of power-to-gas unit to provide the hydrogen for the hydrotreatment of oil they found that the cost of jet fuel is three to four times higher than the fossil jet fuel.

Fornell et al. [89] performed a TEA for kraft pulp mill-based biorefinery producing both ethanol and DME. A synergistic effect of combining two production processes was observed. However, the feasibility of the combination found to be highly dependent on the prices of ethanol and DME, and the investment cost.

Gasification, pyrolysis and HTL pathways leading to five different fuels (ethanol, DME, methanol, gasoline, and diesel) were analyzed by Zhu et al. [90]. Methanol and DME were found to have better energy efficiency than the production of FT diesel or ethanol. The capital requirements were found to be substantial. To make the products economically attractive, cost reduction is essential for both liquefaction and gasification-based pathways. A summary of these TEA studies is provided in Table 16.2.

Assessment of environmental impacts of biorefineries is not a part of TEA. However, environmental impacts (i.e., process effluent treatment) are often associated with cost and hence needs to be considered in TEA. Life-cycle analysis (LCA) is a technique for

Table 16.2: Scale and CAPEX requirements for biorefinery TEA.

Biorefinery	Products	Scale	Capital requirement	Lifetime
Biodiesel Biorefineries [87]	Biodiesel Succinic acid Glycerol Rapeseed meal	7.92 kt/year biodiesel	5.3–10.7 million euro	20 years
Hydrotreatment of vegetable oil with power-to-gas [88]	Naphtha Jet fuel Diesel	Professing of 500 kt of vegetable oil	214–378 million euro	30 years
Kraft pulp-mill-based [89]	Ethanol Dimethyl ether	2065 t dry wood/day	492 million euro	15 years
Biomass-to-liquid [90]	Gasoline Diesel	2000 t dry wood chip/day	332 million USD	20 years

assessing the environmental impacts of a product during its production. Brown et al. [78] mentioned TEA and LCA to have a symbiotic relationship as LCA quantifies the environmental effects associated with the TEA assumptions, while TEA quantifies the corresponding cost associated with the environmental effects. Biorefinery products are usually set to replace the fossil fuels leading to GHG and pollution emission offset. These offsets, quantified by LCA, can be considered as economic incentives and used in the comparative TEA and can identify policy-level support for implementation of such processes.

Alongside the technical and economic barriers, LCA evaluates the environmental barriers for the implementation of a process. Hence, combining TEA with LCA can play a key role in decision making during process development and design. This has been a reason for the increasing trend for coupling TEA and LCA [91–94]. It can be noted that the system boundaries for TEA and LCA are not necessarily the same for the study of the same system.

Due to the assumptions made during the estimation of CAPEX, OPEX, and environmental impacts, models for TEA or LCA contains some inherent uncertainty. TEA studies, therefore, also involve uncertainty analysis. Two methods are consecutively used for uncertainty analysis: sensitivity analysis and Monte-Carlo simulation [4]. Sensitivity analysis considers the impact of uncertainties on the profitability (i.e., NPV, etc.). On the other hand, Monte-Carlo simulation allows studying the combined effect of uncertainties based on the sensitivity analysis. Therefore, researchers are combining TEA and LCA with the Monte-Carlo simulation to study the effect of parameters on economic and environmental performance of biorefineries [95–98].

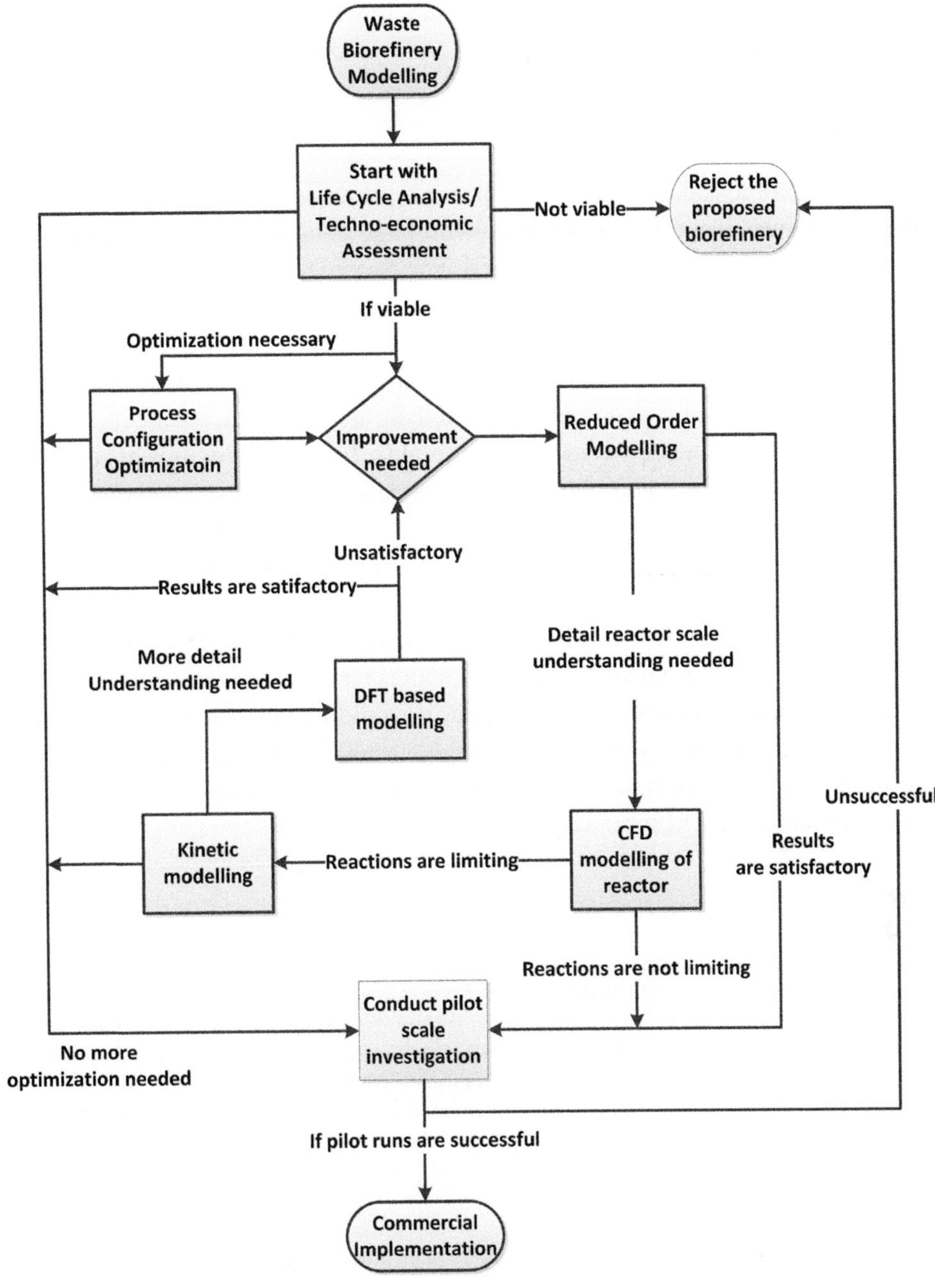

Figure 16.8
A guideline to approach waste biorefinery modeling.

Performing TEA does not essentially require use of software. However, use of software provides reduces the volume of work to be performed as the software are equipped with necessary databases for property calculations and cost analysis. Process simulations are widely used for TEA. For TEA of biorefineries, ASEPN Plus and ASPEN Economic Analyzer are widely used for process modeling and economic analysis, respectively [99].

16.7 Conclusions and perspectives

The models in different scales have made it possible for us to visualize waste biorefineries from a bird's eye view. Commercial software with large databases is developed to serve multiple purposes to meet the application need. For researchers, it is still preferable to develop their own in-house models to retain the freedom to customize the model accordingly. In all modes of application, it is undeniable that modeling had a profound impact on the waste biorefinery design through making the process more efficient and economically feasible. There are occasions when there is a need for practical considerations to make a choice for models, preferably, based on previous experiences. The most common factors are complexity of the models and the associated computational cost with it. Even with the available computational power at our disposal, it is almost always preferable to model a process with plausible simplifications without losing its integrity. For this reason, waste biorefinery modeling is to be approached from a high-level modeling (system scale) and low-level (micro/nanoscale) models are to be used only if necessary. Fig. 16.8 shows a generic strategy to address the waste biorefinery modeling in a more structured manner. Based on this method, it would be possible to identify the feasibility of a waste biorefinery in the early stage of model development. In future, this modeling strategy will evolve based on the development of newer and improved modeling approaches relevant to waste biorefinery.

References

[1] Kirtania K, Bhattacharya S. Pyrolysis kinetics and reactivity of algae-coal blends. Biomass and Bioenergy 2013;55:291–8. https://doi.org/10.1016/j.biombioe.2013.02.019.

[2] Kirtania K. Thermochemical conversion processes for waste biorefinery. In: Bhaskar T, Pandey A, Mohan SV, Lee D-J, Khanal SK, editors. Waste biorefinery: potential and perspectives. 1st ed. Elsevier Inc.; 2018. p. 129–56. https://www.sciencedirect.com/science/article/pii/B9780444639929000045.

[3] Stuart PR, El-Halwagi MM, editors. Integrated biorefineries – design, analysis and optimization. CRC Press, Taylor & Francis Group; 2013. https://doi.org/10.1016/B978-0-12-409548-9.10138-1.

[4] Hytönen E, Stuart PR. Technoeconomic assessment and risk analysis of biorefinery processes. In: Stuart PR, El-Halwagi MM, editors. Integrated biorefineries. CRC Press; 2012. p. 1376–80. https://doi.org/10.1201/b13048-8.

[5] Anca-Couce A. Reaction mechanisms and multi-scale modelling of lignocellulosic biomass pyrolysis. Progress in Energy and Combustion Science 2016;53:41–79. https://doi.org/10.1016/j.pecs.2015.10.002.

[6] Kirtania K, Bhattacharya S. Coupling of a distributed activation energy model with particle simulation for entrained flow pyrolysis of biomass. Fuel Processing Technology 2015. https://doi.org/10.1016/j.fuproc.2015.04.014.

[7] Andersson J, Umeki K, Furusjö E, Kirtania K, Weiland F. Multi-scale reactor network simulation of an entrained flow biomass gasifier – model description and validation. Energy Technology 2017;5:1484–94. https://doi.org/10.1002/ente.201600760.

[8] Basu P. Introduction. In: Biomass gasification, pyrolysis and torrefaction; 2010. p. 1–25. https://doi.org/10.1016/B978-0-12-374988-8.00001-5.

[9] Dhar BR, Kirtania K. Excess methanol recovery in biodiesel production process using a distillation column: a simulation study. Chemical Engineering Research Bulletin 2009;13:55–60. https://doi.org/10.3329/cerb.v13i2.3538.

[10] Vicente G, Coteron A, Martinez M, Aracil J. Application of the factorial design of experiments and response surface methodology to optimize biodiesel production. Industrial Crops and Products 1998;8:29–35. https://doi.org/10.1016/S0926-6690(97)10003-6.

[11] Bouaid A, Martinez M, Aracil J. A comparative study of the production of ethyl esters from vegetable oils as a biodiesel fuel optimization by factorial design. Chemical Engineering Journal 2007;134:93–9. https://doi.org/10.1016/j.cej.2007.03.077.

[12] Lissianski V, Zamansky V, Rizeq G. Integration of direct combustion with gasification for reduction of NOx emissions. Proceedings of the Combustion Institute 2002;29:2251–8. https://doi.org/10.1016/S1540-7489(02)80274-7.

[13] Defoort F, Dupont C, Durruty J, Guillaudeau J, Bedel L, Ravel S, Campargue M, Labalette F, Da Silva Perez D. Thermodynamic study of the alkali release behavior during steam gasification of several biomasses. Energy & Fuels 2015;29:7242–53. https://doi.org/10.1021/acs.energyfuels.5b01755.

[14] Zhang X, Yang W, Blasiak W. Kinetics of levoglucosan and formaldehyde formation during cellulose pyrolysis process. Fuel 2012;96:383–91. https://doi.org/10.1016/j.fuel.2012.01.006.

[15] Solomon PR, Serio MA, Hamblen DG, Yu ZZ, Charpenay S. Advances in the FG-DVC model of coal devolatilization. In: ACS division of fuel chemistry preprints; 1989. p. 479–88.

[16] Arenillas A, Rubiera F, Pevida C, Pis JJ. A comparison of different methods for predicting coal devolatilisation kinetics. Journal of Analytical and Applied Pyrolysis 2001;58–59:685–701.

[17] Baseden KA, Tye JW. Introduction to density functional theory: calculations by hand on the helium atom. Journal of Chemical Education 2014;91:2116–23. https://doi.org/10.1021/ed5004788.

[18] Lejaeghere K, Bihlmayer G, Björkman T, Blaha P, Blügel S, Blum V, Caliste D, Castelli IE, Clark SJ, Dal Corso A, De Gironcoli S, Deutsch T, Dewhurst JK, Di Marco I, Draxl C, Dułak M, Eriksson O, Flores-Livas JA, Garrity KF, Genovese L, Giannozzi P, Giantomassi M, Goedecker S, Gonze X, Grånäs O, Gross EKU, Gulans A, Gygi F, Hamann DR, Hasnip PJ, Holzwarth NAW, Iuşan D, Jochym DB, Jollet F, Jones D, Kresse G, Koepernik K, Küçükbenli E, Kvashnin YO, Locht ILM, Lubeck S, Marsman M, Marzari N, Nitzsche U, Nordström L, Ozaki T, Paulatto L, Pickard CJ, Poelmans W, Probert MIJ, Refson K, Richter M, Rignanese GM, Saha S, Scheffler M, Schlipf M, Schwarz K, Sharma S, Tavazza F, Thunström P, Tkatchenko A, Torrent M, Vanderbilt D, Van Setten MJ, Van Speybroeck V, Wills JM, Yates JR, Zhang GX, Cottenier S. Reproducibility in density functional theory calculations of solids. Science 2016;351:1415–22. https://doi.org/10.1126/science.aad3000.

[19] Elder T, Beste A. Density functional theory study of the concerted pyrolysis mechanism for lignin models. Energy & Fuels 2014;28:5229–35. https://doi.org/10.1021/ef5013648.

[20] Zielinski M, Burke LA, Samokhvalov A. Selective activation of C=C bond in sustainable phenolic compounds from lignin via photooxidation: experiment and density functional theory calculations. Photochemistry and Photobiology 2015;91:1332–9. https://doi.org/10.1111/php.12509.

[21] Zhang X, Yang W, Dong C. Levoglucosan formation mechanisms during cellulose pyrolysis. Journal of Analytical and Applied Pyrolysis 2013;104:19–27. https://doi.org/10.1016/j.jaap.2013.09.015.

[22] Whelan JK, Solomon PR, Deshpande GV, Carangelo RM. Thermogravimetric Fourier transform infrared spectroscopy (TG-FTIR) of petroleum source rocks. Initial results. Energy & Fuels 1988;2:65–73. https://doi.org/10.1021/ef00007a010.

[23] Carangelo RM, Solomon PR, Gerson DJ. Application of TG-FT-i.r. to study hydrocarbon structure and kinetics. Fuel 1987;66:960–7. https://doi.org/10.1016/0016-2361(87)90336-X.

[24] Solomon PR, Hamblen DG, Yu Z-Z, Serio MA. Network models of coal thermal decomposition. Fuel 1990;69:754–63. https://doi.org/10.1016/0016-2361(90)90042-O.

[25] Jones JM, Pourkashanian M, Williams A, Hainsworth D. A comprehensive biomass combustion model. Renewable Energy 2000;19:229–34. https://doi.org/10.1016/S0960-1481(99)00036-1.
[26] Chen Y, Charpenay S, Jensen A, Wójtowicz MA, Serio MA. Modeling of biomass pyrolysis kinetics. Symposium on Combustion 1998;27:1327–34. https://doi.org/10.1016/S0082-0784(98)80537-7.
[27] de Jong W, Di Nola G, Venneker B, Spliethoff H, Wojtowicz M. TG-FTIR pyrolysis of coal and secondary biomass fuels: determination of pyrolysis kinetic parameters for main species and NOx precursors. Fuel 2007;86:2367–76. https://doi.org/10.1016/j.fuel.2007.01.032.
[28] Sharma A, Pareek V, Zhang D. Biomass pyrolysis—a review of modelling, process parameters and catalytic studies. Renewable and Sustainable Energy Reviews 2015;50:1081–96. https://doi.org/10.1016/j.rser.2015.04.193.
[29] Shafizadeh F, Chin PPS. Thermal deterioration of wood. In: Wood technology: chemical aspects; 1977. p. 57–81. https://doi.org/10.1021/bk-1977-0043.ch005.
[30] Lehto J. Development and characterization of test reactor with results of its application to pyrolysis kinetics of peat and biomass fuels. Tampere University of Technology; 2007.
[31] Di Blasi C. Modeling chemical and physical processes of wood and biomass pyrolysis. Progress in Energy and Combustion Science 2008;34:47–90. https://doi.org/10.1016/j.pecs.2006.12.001.
[32] Chornet E, Roy C. Compensation effect in the thermal decomposition of cellulosic materials. Thermochimica Acta 1980;35:389–93. https://doi.org/10.1016/0040-6031(80)87140-1.
[33] Agrawal RK. On the compensation effect. Journal of Thermal Analysis 1986;31:73–86. https://doi.org/10.1007/BF01913888.
[34] Agrawal RK. Compensation effect in the pyrolysis of cellulosic materials. Thermochimica Acta 1985;90:347–51. https://doi.org/10.1016/0040-6031(85)87114-8.
[35] Narayan R, Antal MJ. Thermal lag, fusion, and the compensation effect during biomass pyrolysis. Industrial & Engineering Chemistry Research 1996;35:1711–21. https://doi.org/10.1021/ie950368i.
[36] Anca-couce A, Berger A, Zobel N. How to determine consistent biomass pyrolysis kinetics in a parallel reaction scheme. Fuel 2014;123:230–40. https://doi.org/10.1016/j.fuel.2014.01.014.
[37] Conesa JA, Marcilla A, Caballero JA, Font R. Comments on the validity and utility of the different methods for kinetic analysis of thermogravimetric data. Journal of Analytical and Applied Pyrolysis 2001;58–59:617–33.
[38] Völker S, Rieckmann T. The potential of multivariate regression in determining formal kinetics of biomass pyrolysis. In: Bridgwater AV, editor. Progress in thermochemical biomass conversion. Oxford, UK: Blackwell Science Ltd; 2008. p. 1076–90. https://doi.org/10.1002/9780470694954.ch87.
[39] Caballero JA, Conesa JA, Font R, Marcilla A. Pyrolysis kinetics of almond shells and olive stones considering their organic fractions. Journal of Analytical and Applied Pyrolysis 1997;42:159–75. https://doi.org/10.1016/S0165-2370(97)00015-6.
[40] Teng H, Lin H-C, Ho J-A. Thermogravimetric analysis on global mass loss kinetics of rice hull pyrolysis. Industrial & Engineering Chemistry Research 1997;36:3974–7. https://doi.org/10.1021/ie970017z.
[41] Manyà JJ, Velo E, Puigjaner L. Kinetics of biomass pyrolysis: a reformulated three-parallel-reactions model. Industrial & Engineering Chemistry Research 2003;42:434–41. https://doi.org/10.1021/ie020218p.
[42] Grønli MG, Várhegyi G, Di Blasi C. Thermogravimetric analysis and devolatilization kinetics of wood. Industrial & Engineering Chemistry Research 2002;41:4201–8. https://doi.org/10.1021/ie0201157.
[43] Müller-Hagedorn M, Bockhorn H, Krebs L, Müller U. A comparative kinetic study on the pyrolysis of three different wood species. Journal of Analytical and Applied Pyrolysis 2003;68–69:231–49. https://doi.org/10.1016/S0165-2370(03)00065-2.
[44] Mészáros E, Várhegyi G, Jakab E. Thermogravimetric and reaction kinetic analysis of biomass samples from an energy plantation. Energy & Fuels 2004;18:497–507.
[45] Uggetti E, Sialve B, Trably E, Steyer J-P. Integrating microalgae production with anaerobic digestion: a biorefinery approach. Biofuels, Bioproducts and Biorefining 2014;8:516–29. https://doi.org/10.1002/bbb.1469.

[46] Sharara MA, Holeman N, Sadaka SS, Costello TA. Pyrolysis kinetics of algal consortia grown using swine manure wastewater. Bioresource Technology 2014;169:658–66. https://doi.org/10.1016/j.biortech.2014.06.111.

[47] Bui H-H, Tran K-Q, Chen W-H. Pyrolysis of microalgae residues – a kinetic study. Bioresource Technology 2016;199:362–6. https://doi.org/10.1016/j.biortech.2015.08.069.

[48] Bach Q-V, Chen W-H. A comprehensive study on pyrolysis kinetics of microalgal biomass. Energy Conversion and Management 2017;131:109–16. https://doi.org/10.1016/j.enconman.2016.10.077.

[49] Bach Q-V, Chen W-H. Pyrolysis characteristics and kinetics of microalgae via thermogravimetric analysis (TGA): a state-of-the-art review. Bioresource Technology 2017;246:88–100. https://doi.org/10.1016/j.biortech.2017.06.087.

[50] Dhyani V, Bhaskar T. Kinetic analysis of biomass pyrolysis. In: Bhaskar T, Pandey A, Mohan SV, Lee D-J, Khanal SK, editors. Waste biorefinery: potential and perspectives. 1st ed. 2018. p. 39–80.

[51] Lakshmanan CC, Bennett ML, White N. Implications of multiplicity in kinetic parameters to petroleum exploration: distributed activation energy models. Energy & Fuels 1991;5:110–7.

[52] Cai J, Jin C, Yang S, Chen Y. Logistic distributed activation energy model – part 1: derivation and numerical parametric study. Bioresource Technology 2011;102:1556–61. https://doi.org/10.1016/j.biortech.2010.08.079.

[53] Cai J, Yang S, Li T. Logistic distributed activation energy model – part 2: application to cellulose pyrolysis. Bioresource Technology 2011;102:3642–4. https://doi.org/10.1016/j.biortech.2010.11.073.

[54] Doyle CD. Estimating isothermal life from thermogravimetric data. Journal of Applied Polymer Science 1962;6:639–42. https://doi.org/10.1002/app.1962.070062406.

[55] Coats AW, Redfern JP. Kinetic parameters from thermogravimetric data. Nature 1964;201:68–9.

[56] Senum GI, Yang RT. Rational approximations of the integral of the Arrhenius function. Journal of Thermal Analysis 1977;11:445–7.

[57] Agrawal RK, Sivasubramanian MS. Integral approximations for nonisothermal kinetics. AIChE Journal 1987;33:1212–4. https://doi.org/10.1002/aic.690330716.

[58] Miura K, Maki T. A simple method for estimating f(E) and k0(E)in the distributed activation energy model. Energy & Fuels 1998;12:864–9. https://doi.org/10.1021/ef970212q.

[59] Kirtania K, Bhattacharya S. Application of the distributed activation energy model to the kinetic study of pyrolysis of the fresh water algae *Chlorococcum humicola*. Bioresource Technology 2012;107:476–81. https://doi.org/10.1016/j.biortech.2011.12.094.

[60] Miura K. A new and simple method to estimate f(E) and $k_0(E)$ in the distributed activation energy model from three sets of experimental data. Energy & Fuels 1995;9:302–7.

[61] Burnham AK, Braun RL. Global kinetic analysis of complex materials. Energy & Fuels 1999;13:1–22.

[62] Farazi S, Sadr M, Kang S, Schiemann M, Vorobiev N, Scherer V, Pitsch H. Resolved simulations of single char particle combustion in a laminar flow field. Fuel 2017;201:15–28. https://doi.org/10.1016/j.fuel.2016.11.011.

[63] Kraft S, Kirnbauer F, Hofbauer H. Model development for fluid mechanics and chemical reactions in a dual fluidized bed steam gasification system by means of CPFD. In: 24th European biomass conference and exhibition, Amsterdam, Netherlands; 2016. p. 494–500.

[64] Mehrabian R, Zahirovic S, Scharler R, Obernberger I, Kleditzsch S, Wirtz S, Scherer V, Lu H, Baxter LL. A CFD model for thermal conversion of thermally thick biomass particles. Fuel Processing Technology 2012;95:96–108. https://doi.org/10.1016/j.fuproc.2011.11.021.

[65] Fuentes-Cano D, Go A, Nilsson S, Ollero P, Gómez-Barea A, Nilsson S, Ollero P. Kinetic modeling of tar and light hydrocarbons during the thermal conversion of biomass. Energy & Fuels 2015;30:377–85. https://doi.org/10.1021/acs.energyfuels.5b02131.

[66] Babu BV, Chaurasia AS. Modeling for pyrolysis of solid particle: kinetics and heat transfer effects. Energy Conversion and Management 2003;44:2251–75. https://doi.org/10.1016/S0196-8904(02)00252-2.

[67] Bach-oller A, Kirtania K, Furusjö E, Umeki K. Co-gasification of black liquor and pyrolysis oil at high temperature: Part 1. Fate of alkali elements. Fuel 2017;202:46–55. https://doi.org/10.1016/j.fuel.2017.04.013.

[68] Umeki K, Kirtania K, Chen L, Bhattacharya S. Fuel particle conversion of pulverized biomass during pyrolysis in an entrained flow reactor. Industrial & Engineering Chemistry Research 2012;51:13973–9. https://doi.org/10.1021/ie301530j.

[69] Anderson J. Computational fluid dynamics. 1st ed. McGraw-Hill Education; 1995.

[70] Lomax H, Pulliam TH, Zingg DW. Fundamental of computational fluid dynamics. 1st ed. Springer-Verlag Berlin Heidelberg; 2001. https://doi.org/10.1007/978-3-662-04654-8.

[71] Ferziger JH, Peric M. Computational methods for fluid dynamics. 2nd ed. Springer Berlin Heidelberg; 2003.

[72] Ku X, Li T, Løvås T. CFD-DEM simulation of biomass gasification with steam in a fluidized bed reactor. Chemical Engineering Science 2015;122:270–83. https://doi.org/10.1016/j.ces.2014.08.045.

[73] Fletcher DF, Haynes BS, Christo FC, Joseph SD. A CFD based combustion model of an entrained flow biomass gasifier. Applied Mathematical Modelling 2000;24:165–82.

[74] Marklund M, Tegman R, Gebart R. CFD modelling of black liquor gasification: identification of important model parameters. Fuel 2007;86:1918–26. https://doi.org/10.1016/j.fuel.2006.12.015.

[75] Ku X, Li T, Løvås T. Eulerian–Lagrangian simulation of biomass gasification behavior in a high-temperature entrained-flow reactor. Energy & Fuels 2014;28:5184–96. https://doi.org/10.1021/ef5010557.

[76] Sahraei MH, Duchesne MA, Yandon R, Majeski A, Hughes RW, Ricardez-sandoval LA. Reduced order modeling of a short-residence time gasifier. Fuel 2015;161:222–32. https://doi.org/10.1016/j.fuel.2015.07.096.

[77] Sahraei MH, Yandon R, Duchesne MA, Hughes RW, Ricardez-Sandoval LA. Parametric analysis using a reactor network model for petroleum coke gasification. Energy & Fuels 2015;29:7681–8. https://doi.org/10.1021/acs.energyfuels.5b01731.

[78] Brown TR, Wright MM, Román-Leshkov Y, Brown RC. Techno-economic assessment (TEA) of advanced biochemical and thermochemical biorefineries. In: Advances in biorefineries. Elsevier; 2014. p. 34–66. https://doi.org/10.1533/9780857097385.1.34.

[79] Papalexandrou MA, Pilavachi PA, Chatzimouratidis AI. Evaluation of liquid bio-fuels using the analytic hierarchy process. Process Safety and Environmental Protection 2008;86:360–74. https://doi.org/10.1016/j.psep.2008.03.003.

[80] Sammons NE, Yuan W, Eden MR, Aksoy B, Cullinan HT. Optimal biorefinery product allocation by combining process and economic modeling. Chemical Engineering Research and Design 2008;86:800–8. https://doi.org/10.1016/j.cherd.2008.03.004.

[81] Cohen BJ, Janssen M, Chambost V, Stuart P. Critical analysis of emerging forest biorefinery (FBR) technologies for ethanol production. Pulp and Paper Canada 2010;111:24–30.

[82] Kabir KB, Kirtania K. An assessment of gasification of lipid-extracted algae by thermodynamic simulation. Journal of Chemical Engineering 2017;30:59–63.

[83] Frederick WJ, Lien SJ, Courchene CE, DeMartini NA, Ragauskas AJ, Iisa K. Co-production of ethanol and cellulose fiber from Southern Pine: a technical and economic assessment. Biomass and Bioenergy 2008;32:1293–302. https://doi.org/10.1016/j.biombioe.2008.03.010.

[84] Laser M, Jin H, Jayawardhana K, Lynd LR. Coproduction of ethanol and power from switchgrass. Biofuels, Bioproducts and Biorefining 2009;3:195–218. https://doi.org/10.1002/bbb.133.

[85] Laser M, Jin H, Jayawardhana K, Dale BE, Lynd LR. Projected mature technology scenarios for conversion of cellulosic biomass to ethanol with coproduction thermochemical fuels, power, and/or animal feed protein. Biofuels, Bioproducts and Biorefining 2009;3:231–46. https://doi.org/10.1002/bbb.131.

[86] Carvalho L, Furusjö E, Kirtania K, Lundgren J, Anheden M, Wolf J, Wetterlund E, Lundgren J, Anheden M, Wolf J. Techno-economic assessment of catalytic gasification of biomass powders for methanol production. Bioresource Technology 2017;237:167–77. https://doi.org/10.1016/j.biortech.2017.02.019.

[87] Vlysidis A, Binns M, Webb C, Theodoropoulos C. A techno-economic analysis of biodiesel biorefineries: assessment of integrated designs for the co-production of fuels and chemicals. Energy 2011;36:4671–83. https://doi.org/10.1016/j.energy.2011.04.046.

[88] Zech KM, Dietrich S, Reichmuth M, Weindorf W, Müller-Langer F. Techno-economic assessment of a renewable bio-jet-fuel production using power-to-gas. Applied Energy 2018;231:997–1006. https://doi.org/10.1016/j.apenergy.2018.09.169.

[89] Fornell R, Berntsson T, Åsblad A. Techno-economic analysis of a kraft pulp-mill-based biorefinery producing both ethanol and dimethyl ether. Energy 2013;50:83–92. https://doi.org/10.1016/j.energy.2012.11.041.

[90] Zhu Y, Tjokro Rahardjo SA, Valkenburt C, Snowden-Swan LJ, Jones SB, Machinal MA, Valkenburg C, Snowden-Swan LJ, Jones SB, Machinal MA. Techno-economic analysis for the thermochemical conversion of biomass to liquid fuels. PNNL report – 19009. Washington, DC, United States: United States Department of Energy; 2011. https://doi.org/10.2172/1128665.

[91] You F, Tao L, Graziano DJ, Snyder SW. Optimal design of sustainable cellulosic biofuel supply chains: multiobjective optimization coupled with life cycle assessment and input-output analysis. AIChE Journal 2012;58:1157–80. https://doi.org/10.1002/aic.12637.

[92] Quinn JC, Davis R. The potentials and challenges of algae based biofuels: a review of the techno-economic, life cycle, and resource assessment modeling. Bioresource Technology 2015;184:444–52. https://doi.org/10.1016/j.biortech.2014.10.075.

[93] Beal CM, Gerber LN, Sills DL, Huntley ME, Machesky SC, Walsh MJ, Tester JW, Archibald I, Granados J, Greene CH. Algal biofuel production for fuels and feed in a 100-ha facility: a comprehensive techno-economic analysis and life cycle assessment. Algal Research 2015;10:266–79. https://doi.org/10.1016/j.algal.2015.04.017.

[94] Tao L, Tan ECD, McCormick R, Zhang M, Aden A, He X, Zigler BT. Techno-economic analysis and life-cycle assessment of cellulosic isobutanol and comparison with cellulosic ethanol and n-butanol. Biofuels, Bioproducts and Biorefining 2014;8:30–48. https://doi.org/10.1002/bbb.1431.

[95] Wang K, Ou L, Brown T, Brown RC. Beyond ethanol: a techno-economic analysis of an integrated corn biorefinery for the production of hydrocarbon fuels and chemicals. Biofuels, Bioproducts and Biorefining 2015;9:190–200. https://doi.org/10.1002/bbb.1529.

[96] Pérez-López P, Montazeri M, Feijoo G, Moreira MT, Eckelman MJ. Integrating uncertainties to the combined environmental and economic assessment of algal biorefineries: a Monte Carlo approach. The Science of the Total Environment 2018;626:762–75. https://doi.org/10.1016/j.scitotenv.2017.12.339.

[97] Kelloway A, Daoutidis P. Process synthesis of biorefineries: optimization of biomass conversion to fuels and chemicals. Industrial & Engineering Chemistry Research 2014;53:5261–73. https://doi.org/10.1021/ie4018572.

[98] Souza SP, Gopal AR, Seabra JEA. Life cycle assessment of biofuels from an integrated Brazilian algae-sugarcane biorefinery. Energy 2015;81:373–81. https://doi.org/10.1016/j.energy.2014.12.050.

[99] Cristóbal J, Caldeira C, Corrado S, Sala S. Techno-economic and profitability analysis of food waste biorefineries at European level. Bioresource Technology 2018;259:244–52. https://doi.org/10.1016/j.biortech.2018.03.016.

CHAPTER 17

Application of life-cycle assessment in biorefineries

Stella Bezergianni, Loukia P. Chrysikou
Chemical Process & Energy Resources Institute - CPERI, Centre for Research & Technology Hellas CERTH, Thessaloniki, Greece

17.1 Introduction

Considering the depletion of petroleum sources and the increase of greenhouse gas (GHG) emissions the exploration of renewable energy systems is imperative. Biomass utilization for biofuels and high added-value products is being systematically explored as a promising and innovative pathway toward the investigation of renewable energy sources, the climate change mitigation and the decrement of fossil resources dependence. As a main premise of the biorefinery concept, biomass conversion processes are integrated leading to the coproduction of biofuels, energy, and high added-value products. Biorefining is defined as the "sustainable synergetic processing of biomass into a spectrum of marketable products (chemicals, materials) and energy (fuels, power, heat)" [1]. In this essence, the biorefinery is depicted as a process, a facility or a cluster of facilities integrally encompassing the upstream, midstream and downstream processing of bio-based feedstocks. The upstream processes involve the biomass production, transportation and pretreatment, the midstream process involves the biomass conversion to the targeted products, while the downstream process entails the products distribution [2].

Biorefineries are usually distinguished according to the employed biomass feedstock. Specifically, first-generation biorefineries use food/feed crop resources (e.g., sugar, corn, vegetable oils, etc.), whereas the second-generation biorefineries process nonfood/feed feedstocks (e.g., energy crops, organic residues, agro-industrial/forestry wastes, etc.) [3]. Agricultural and forest residues along with herbaceous materials and municipal wastes are potential lignocellulosic-based materials leading to plant-derived sustainable biofuel sources via biotechnological conversion. Recently, specific focus is directed toward waste-based biorefineries valorizing animal waste streams and industrial waste residues, as well. A combination of different novel technologies is commonly applied involving mechanical pretreatment, chemical processes (hydrolysis, transesterification,

Waste Biorefinery. https://doi.org/10.1016/B978-0-12-818228-4.00017-4

hydrogenation, and oxidation to change the biomass chemical structure), thermochemical (pyrolysis) and biochemical processes (enzymatic conversion, anaerobic digestion, and fermentation) for the production of the targeted products [4]. Recently, algae-based biomass is being utilized in the third generation biorefineries that are presently under investigation utilizing microalgal biomass for the coproduction of biofuels and high value-added products [5,6].

Particularly, the biorefinery exploits all the biomass fractions maximizing the products yields per biomass input, as conventional refineries have optimized the coproduction of a cascade of products from crude oil. The combined energy and dedicated platform chemicals coproduction characterizes the specific biorefinery system, as well as the variant processes (biochemical, thermocatalytical) encompassed. More specifically, under the biorefinery concept technologies from various fields (including agriculture, engineering, chemistry, microbiology) are applied to an integrated process to separate the biomass into its building blocks that are further converted to biofuels and biomaterials (Fig. 17.1) [4]. It should be highlighted that the biorefineries exploiting biobased feedstocks and wastes residues to cogenerate fuels, chemicals, and energy are the pillar of the circular economy [7].

A typical biorefinery chain is structured in the next stages involving bio-based feedstock production, transportation, conversion through processes to the envisioned products followed by their distribution and end-use [8]. As biorefineries technological schemes continuously emerge exploiting bio-based feedstocks for value-added products (biofuels, energy, chemicals), there is an increasing necessity to evaluate their overall environmental performance. An approach to investigate the biorefinery's sustainability is prominent, validating the processing of the selected biomass through conversion processes into bioenergy (biofuels, energy) and bio-based valuable

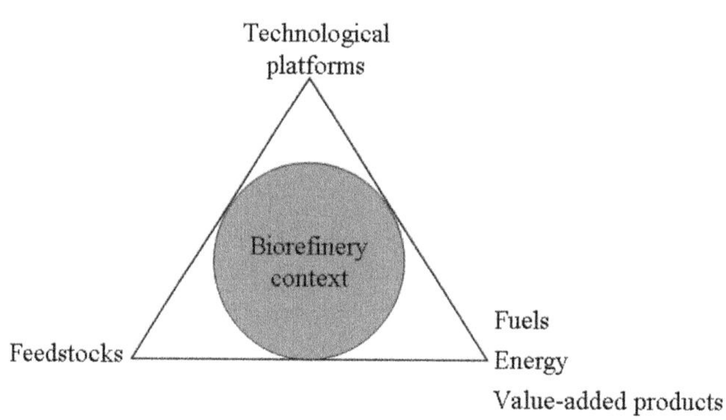

Figure 17.1
Biorefinery's system depiction.

marketable products (chemicals, materials, etc.) from an environmental perspective. Furthermore, the systematic approach of the biorefineries environmental dimension would verify their validity in terms of greenhouse gas (GHG) emissions, energy performance and other environmental impacts [9]. The life cycle assessment (LCA) constitutes a valuable and expedient tool quantifying the environmental impacts of production technologies leading to new products, providing conclusions based on a multitude of individual assessment results. LCA is commonly applied for the biorefinery's environmental profile assessment providing data for its environmental performance, optimal bio-based feedstock, processes configuration, etc. Nevertheless, the sustainability incorporation during the design phase of a biorefinery is central for its bio-based economy development, improving further the energy supply security [10].

17.2 What is LCA?

LCA provides a quantitative estimation of the potential environmental problems of an examined system in terms of environmental indicators, proposing concurrently ways to overcome the environmental burdens, thus addressing thoroughly the issue of sustainability. These results obtained could potentially contribute catalytically to the sustainability-targeted decision support on innovative technological systems. In particular, the LCA results could provide the basis for decision support establishing new technologies, processes or products, for industrial applications and policymaking for mitigation of climate change or fossil resource dependency. Based on the biorefinery system the assessment of parameters related to its implementation potentials (e.g., feedstock availability), feasibility (e.g., technical), and stability (e.g., durability, yield stability) add valuable aspects of the new products and production technologies. Moreover, these results constitute the cornerstone of robust conclusions and future-oriented recommendations for the industry [10].

The life cycle of a product is defined as "consecutive and inter-linked stages of a product system, from raw material acquisition or generation from natural resources to final disposal" [11]. A study system encloses the production path of a product and enables the quantification of its environmental impacts following either the "cradle-to-gate" or the "cradle-to-cradle" approach. Specifically, a "cradle-to-gate" life cycle, indicates a system boundary until the production stage of a product, while "cradle-to-cradle" refers to a life cycle that includes reuse, recovery or recycle of a product or its coproducts, if any [8]. Thus, in both approaches the associated energy and resource consumption, together with the emissions for the whole system are quantified, as well [12].

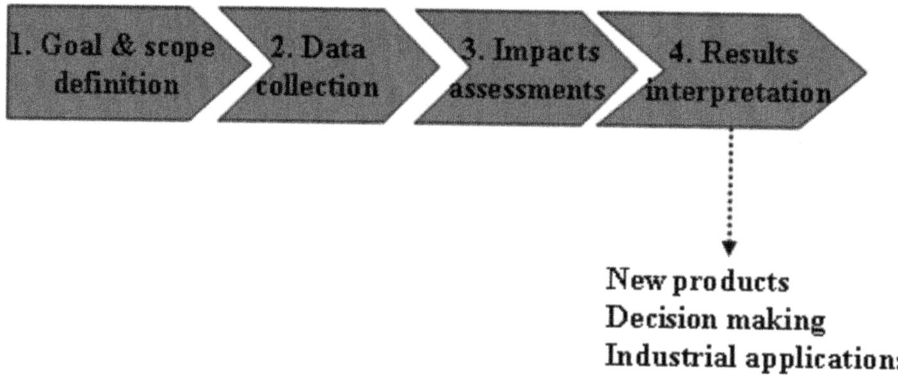

Figure 17.2
LCA methodology steps.

To systemically carry out an LCA, four steps (Fig. 17.2) have to be followed according to ISO 14040 standard [5,11] that are listed below:

(1) the goal and scope definition orienting the objective of the study and basic elements such as the system boundary and the functional unit;
(2) the data collection and inventory analysis, presenting the life-cycle inventory data utilized;
(3) the life-cycle impact assessment, selecting the impact categories for the emissions generated, the energy, and resources utilization;
(4) the results from interpretation and presentation, analyzing thoroughly the results providing also suggestions on impact reductions, if applicable.

Moreover, the cornerstones of the LCA study are defined during the goal and scope definition, forming the basis for the results interpretation and further considerations during the study, if required. The study's scope also involves the description of the investigated process in terms of the functional unit assigning accordingly the inputs and outputs (e.g., materials, energy etc.) [12]. In detail, the systematic record of the mass and energy balances provides the fundamental data for the accurate and concise quantification of the environmental impacts in the following phases of an LCA study. These key points determine the fundamental methodology of the study.

The lack of available inventory data is a major hurdle for LCA practice therefore commercial software tools (for instance, Global Emissions Model for Integrated Systems [GEMIS], SimaPro, GABI, etc.) could simplify the process evaluating the environmental impacts of biorefinery systems. Frequently assumptions have to be considered in order the used data to be representative for an industrial-scale technology. Other software can be utilized for the whole process simulation and for mathematical calculations, if necessary (for instance Aspen Plus, MATLAB, etc.).

Considering that the products from a biorefinery might have different applications and physical attributes (biofuels, chemicals, energy etc.) their management complicates the LCA study. In addition, in multifunctional technological schemes usually an activity fulfills more than one function, for instance the waste management process handles the wastes and also generates energy. The multifunctional schemes involve the feedstock cultivation and the subsequent biorefinery processes, whereas in the LCA study such schemes are divided in subprocesses connected to specific products. For instance, in the case of a pulp mill generating pulp and heat, the use and recovery of energy and chemicals are often integrated to connect each subprocess to either pulp or heat. It is then necessary to find a rational basis for allocating the environmental burdens between the processes [13,14].

Hence, in multifunctional systems that involve the coproduction of valuable products as well as recycled materials, products allocation is commonly applied. However, the allocation of environmental burdens from biorefineries involving feedstock provision and logistics, and its conversion via multi-steps processes (with coproducts utilization) is a rather challenging and complicated task. Nonetheless, the LCA allocation constitutes a topic of great debate and the difficulties are intensified in cases that the multidimensional processes chains cannot be divided into subprocesses associated with specific products [14]. Allocation can be applied in the biorefineries systems via two ways:

(1) by system expansion, if the functional unit can be redefined to include the functions of all coproducts, or in the case that the main product is selected and is given credit for the avoided environmental burdens from products assumed to be substituted by its coproducts.
(2) by partitioning, if the environmental burdens are allocated to coproducts based on their mass, volume or energy content, or an economic attribute such as production cost or market value [12,14,15].

In LCA of biorefinery products, particularly in studies applying system expansion and in cases focusing on nondominant coproducts, the allocation method should be warily selected [14]. The choice of the allocation method is recommended to be implemented systematically and methodically, especially in the biorefineries focusing on coproducts with low flows, even though the results can respectably vary between different applied methods. Core for consistent allocating is the assessment of multidisciplinary and updated databases that could allow the comparison of the allocation methods that is essential especially for the biomass cultivation stages [14].

These aforementioned allocation methods are commonly applied in LCA studies and are acknowledged by ISO 14044 [16], while their selection can also be based in other standards and directives, as well such as the international reference life-cycle data system (ILCD) handbook [17] or the Fuel Quality Directive [18]. Shortly, ISO 14044 specifies requirements and provides guidelines for the LCA studies including the goal and scope definition, the life-cycle inventory analysis, assessment and interpretation phases as well as the study's limitations and critical review [16]. The ILCD handbook provides technical guidance for detailed LCA studies and the technical basis to derive product-specific criteria, guides, and simplified tools, based on and conformed to the ISO 14040 and 14044 standards. Particularly, this handbook provides also assistance for life-cycle emission and resource consumption inventory data [17]. The Fuel Quality Directive introduces a mechanism to monitor and reduce GHG emissions from road transportation fuels [18].

The life-cycle impact assessment incorporates the assignment of specific environmental effects to the targeted products, as defined in the previous steps of the LCA study. The selection of the impact categories is based on the defined goal and scope of each study and each category indicator quantitatively represents a defined impact category [12]. The LCA impact categories can be categorized in the following characteristic groups: GHG emissions, resource depletion, land use ecological impacts, regional environmental impacts, human health impacts, untreated hazardous environmental impacts. Specifically, for the LCA studies of the biorefineries projects the impacts are related notably with the GHG emissions that are commonly accounted with the Intergovernmental Panel on Climate Change (IPCC) global warming potential (GWP) as an indicator of greenhouse effect, whereas other impact categories commonly involve the acidification potential (AP) (indicator of acid rain phenomenon), the eutrophication potential (EP) (indicator of over fertilization of water and soil), the ozone layer depletion potential (ODP) (indicator of ozone layer degradation), the photochemical ozone creation potential (POCP) (indicator of photo-smog creation), the fossil depletion (related to the use of fossil fuels) etc. The impact category of climate change is frequently emphasized in the biorefineries LCA studies, however for decision-making regarding design activities of biorefineries, a comprehensive set of impact categories is widely applied [14]. The appropriate selection of the impact categories provides a complete and comprehensive environmental profile of the examined biorefinery [8,19,20].

Additionally, from the LCA viewpoint, it is significant to consider a relevant reference system for comparison purposes. Commonly in the case of biorefineries producing biofuels, a reference system is selected for instance fossil-derived diesel or biodiesel from vegetable oils, including also their life-cycle impacts results in the LCA study as a comparison basis. The consideration of a relevant reference system against the system under study assesses the sustainability of the transition from fossil resources to renewable biomass. In several biorefineries the reference system is a fossil-based fuel pathway regarding GHG balance, although in cases that the biorefinery coproducts could substitute an existing product, a reference substituted product is also defined. In particular, a representative substitute product has to be selected in the reference system to include significant GHG emissions savings from coproducts and land use if possible that could drastically influence the impact results [21,22].

LCA studies are categorized into attributional and consequential, accounting the processes to be included in the examined system boundary, deriving from a concrete definition of the study's goal. The attributional LCA study quantifies the environmental impacts of a product through current well established processes identifying the hot spots of the process, while the consequential study produces data describing the consequences of future strategic decisions involving also the system expansion in multi-products systems [11].

17.3 Basics of LCA in biorefineries

LCA has been extensively applied for the biorefinery's sustainability assessment that is often considered inherently sustainable due to the renewability of the biomass. By contrast, the biorefineries system's environmental effects related to different types of bioenergy and bioproducts coproduction vary primarily according to the feedstock, agricultural practices and the bioconversion processes employed [5].

In general, in a biorefinery LCA study, the system boundaries commonly define the biomass production chain and its subsequent conversion to the targeted products as well as byproducts valorization if possible and energy supply systems. Ultimately, the LCA system boundaries concrete delimitation ensures that the relevant processes are embraced, diminishing the obstacle of burden shifting from one stage of the life-cycle to another, especially in multifunctional processes (Fig. 17.3). The LCA data are collected from all the stages encountered in the system boundaries. The functional unit provides a quantitative illustration of the basic function examined and should be related to the targeted products of the biorefinery. Thereupon the functional units of the biorefineries LCA studies commonly refer to mass (kg) or energy (MJ) basis, accounting high added-value chemicals and biofuels, respectively as final products. In biorefineries producing bioethanol, frequently the impacts assessment are analyzed considering the 1 km as the functional unit [23].

The inventory data applied in LCA studies of biorefineries originate mainly from site-specific data in nexus with literature sources and databases, adopting modeling approaches, as well. For the evolving biorefineries technologies the up-scaling data from pilot-plant or even laboratory scale into industrial-extent is rather uncertain and exigent

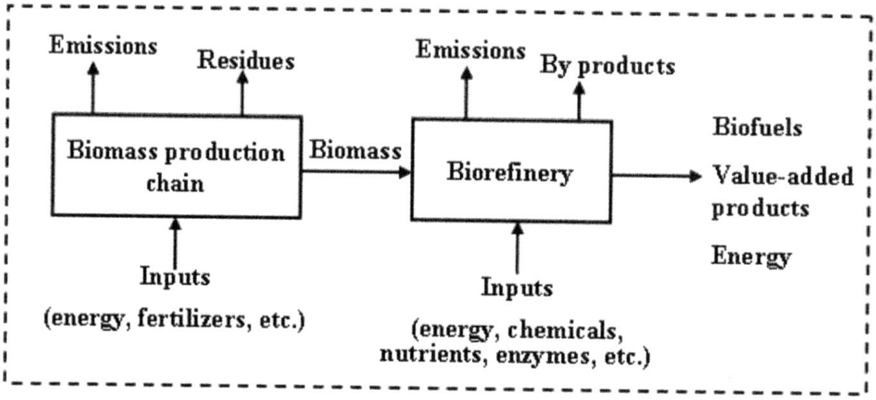

Figure 17.3
Generalized system boundaries of a biorefinery LCA approach.

and not always applicable, hence should be performed scrupulously [23]. The sustainability impacts of the biorefineries are selected in order to be consistent with the study's goal and the scope, reflecting a comprehensive set of environmental issues related to the product system being studied. A substantial number of studies focus on assessing climate change impacts (e.g., GHG emissions) including also other environmental impacts (acidification, photo-oxidant formation, fossil depletion, etc.) [16].

Furthermore, parameters affecting the accuracy and reliability of the LCA results involve low data quality/accuracy, local/regional conditions, and assumptions due to lack of required data and variations/fluctuations in data with time. Other challenges and limitations of the biorefineries LCA studies comprise rigid system boundaries, deficient data, products selectivity, local environmental conditions etc. [24,25]. Core points arise apart from the direct environmental impacts assessment of a biomass supply chain that influence the biorefinery's overall sustainability, including the fossil resources depletion and the competition with food and feed production, if applicable. Therefore, the sustainable use of resources and the energy efficiency of the biorefinery technologies are the major parameters that should be extensively appraised by an LCA study based upon a sustainable production route [20].

17.3.1 Nonfood/feed-based biorefineries

Considering that the biomass resources should not directly compete with food crops cultivations, the way toward the exploitation of residual biomass (forest and agricultural residues), waste streams (industrial and urban) and algae biomass is explored [20]. Particularly, the lignocellulosic feedstocks constitute the world's largest bioethanol renewable resource originating from cellulose containing biomass, forestry woods, agricultural residues and municipal solid wastes. Lately, marine algae biomass is explored as a potential biorefinery feedstock leading to biodiesel and to biojet production. The environmental performance of biofuels production from lignocellulosic-based biorefinery is substantially influenced by methodological aspects, such as the system boundaries, the functional units, the inventory data and the allocation methods. The LCA applications of such biorefineries are focused mainly on the GHG intensity and the integrated management of the coproducts (electricity, heat) presenting their favorable environmental profile compared to a reference fossil-based system [9,22]. The LCA studies of the technological pathways leading to biofuels production follow the standardized methodology as previously analyzed, on a well-to-wheel perspective (incorporating the biofuel's production and combustion's emissions) quantifying mainly the associated GHG emissions and several other pollutants (e.g., SO_2, NOx, CO, particulates, etc.)

However, uncertainties of the quantified environmental impacts are frequently arisen, evoking the necessity for sensitivity analysis. Particularly, the highest uncertainty is assigned to the cultivation stage in case of crops biomass due to the N_2O emissions from the fertilization practices and to the carbon stock changes. The GHG balances of energy crops depict wide variability, while the inclusion of land use changes further modifies the results. Thereby, the land use changes and the indirect land use changes with the N_2O emissions are occasionally not considered in the system boundaries of the biorefineries [20]. At the same time, some environmental concerns are raised due to the soil quality deterioration from the perpetual removal of the lignocellulosic-based feedstocks that is considerably influenced by regional parameters (soil type, location, season, tillage practices, etc.) [24].

Sustainability concerns of microalgae-based biorefineries are particularly related to the appropriate microalgae species for targeted products and the thermochemical processes applied in order to minimize the environmental burdens. LCA studies have identified potential environment benefits of microalgae biofuels compared to fossil-derived fuels, focusing on the recovery of the lipids with integration on anaerobic digestion utilizing the produced methane for heat and power cogeneration and recycling the nutrient rich effluent. Nevertheless, the obtained results are influenced by the system boundaries and especially the lipid content of the biomass that can be extracted during the cultivation process.

17.3.2 Waste-based biorefineries

As aforementioned, recent systematic research efforts are targeted toward wastes-based biorefineries that valorize wastes as renewable feedstocks to obtain bioenergy and a gamut of bio-based chemicals through sustainable biotechnology routes. These integrated approaches incorporate multi-step bioprocesses exploiting low-value wastes streams for biofuels production and in order to maximize the productivity of the intermediate products. In addition to lignocellulosic feedstocks, municipal and industrial solid wastes are also a potential raw material for biofuel production. Moreover, wastes are accounted as an economic feedstock that substitutes fossil-based energy resources generating a cascade of bio-based products (food, feed, chemicals) and bioenergy (biofuels, power, and heat) by integrated and sustainable technologies [7].

The research interest on the valorization of biomass-based wastes is continuously enhanced since they could be used in large-scale applications, while no-competition with food is guaranteed. The bioprocesses enabling the waste biorefinery adopt technological processes including for instance acidogenesis, bioelectrogenesis, photosynthesis, photofermentation, etc. [25]. The biological conversion technologies encompassed in the biorefineries are considered an environmentally friendly substitute to the thermochemical processes, improving the utilization of wastes as feedstocks [7]. In this sense currently extensive research is continuing to deploy targeted efforts to the conversion of wastes to high-added products attempting to become economically feasible and therefore commercially viable [25].

Shortly, the LCA methodology steps applied for a waste-based biorefinery begin with the goal definition scoping the study's boundaries, guiding the selection of the functional unit and the environmental impacts, as well (a generic scheme is presented in Fig. 17.4). The inventory data (obtained from literature, databases, software, and/or experimental data adjusted to a mature industrial-scale technology) are applied for the quantification of the chosen environmental impacts indicators generating the study's results. Afterward, the interpretation of the LCA findings compares the results with previous studies and reference systems, drawing conclusions for the examined production system and making pointed recommendations for the further impact reductions in the future, if possible.

The LCA of the waste-based biorefineries is a rather demanding task necessitating several factors elucidation e.g., waste feedstocks types, amounts and characterization, energy inputs, biorefinery products selections, etc. Particularly, the LCA approaches primarily focus on the estimation of direct impacts of waste biorefineries and on the comparison of their performance with conventional fossil-based systems to verify their validity in terms of GHG emissions, energy utilization etc. The LCA results commonly provide the environmental benefits of such systems, identifying the hotspots of the processes that induce the highest impacts that could be potentially improved. Therefore, an LCA study of any waste biorefinery should be performed on a basis of a detailed design study, clarifying parameters about its configuration and operation, even though uncertainty aspects (yields, byproducts exploitation) commonly affect the biorefinery sustainability. More specifically, the uncertainty issues are related with the selected waste-based biomass and the conversion pathways to the envisioned bioenergy and high added-value products. The uncertainty affects the accuracy of the bioprocesses simulation models that are often

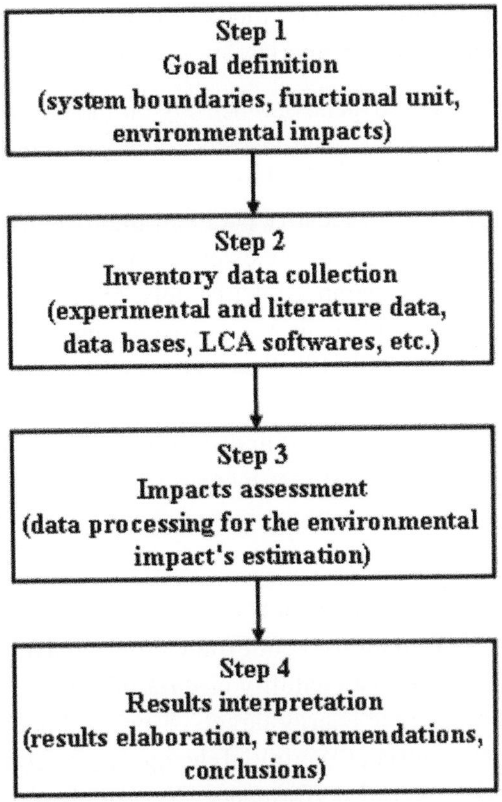

Figure 17.4
Schematic representation of the LCA methodology steps for a waste-based biorefinery.

employed in LCA studies and the optimal biochemical routes for maximizing the economical profit. In that setting, in order to augment the robustness and reliability of the LCA results, a sensitivity analysis is frequently designed elaborating the specific influence of an input. A sensitivity analysis inquires how the results are influenced by the input data, assessing their fluctuation to variations in input data and to modeling options, if applied [23].

17.3.3 Impact of LCA

The biorefinery concept utilizing a variety of biomass feedstocks and through technologies (for instance including fermentation, gasification, pyrolysis, hydrothermal liquefaction, hydrogenation, hydrotreatment, hydrothermolysis, oxidation, and hydrodeoxygenation) produce a palette of products, biofuels and chemicals. The sustainability assessment of a biorefinery and the quantification of its environmental impacts is an overarching aspect during the designing of a waste-based technological scheme. Impact categories such as

GHG intensity, energy consumption, etc. are thoroughly accounted prior to the implementation of industrial-scale biorefineries.

The GHG balance of a biorefinery depends immensely on the life-cycle stages encompassed based upon the waste-derived feedstock and the technology implemented. Uncertainties of these emissions are frequently arisen, evoking the necessity for sensitivity analysis. The studies usually can also identify potential improvement actions for further reduction of the environmental burdens of the examined biorefinery (for instance, byproducts utilization for energy generation). Ultimately, research studies for the design of biorefinery projects present their performance assessment connected with the conversion processes and the biomass supply chain [8]. Especially, the second-generation biorefineries producing biofuels from wastes display favorable environmental performance, while byproducts subsequent valorization further diminishes their environmental burdens. Additionally, the processes integration further increases the environmental benefits of a biorefinery via the complete conversion of the biomass constituents to biofuels and high value-added products [5].

Controversies about the sustainability impacts of biorefineries are being continuously raised, accounting that the sustainability is not founded exclusively on renewable bio-based biomass and the bioenergy sustainability criteria of the European Commission in order to count in national renewable energy targets. Toward this direction the application of sustainability criteria at the initial design stages could potential enhance substantially the biorefineries overall performance. Nevertheless, it is acknowledged that for the development of sustainable biorefinery projects sustainability's dimensions should be carefully considered, for instance, byproduct valorization, land use, etc. [8,26].

17.4 Representative case studies

Numerous LCA studies have assessed the environmental sustainability of biorefineries considering mainly global warming impacts and energy balances as impacts categories, whereas other environmental burdens such as acidification, eutrophication are also quantified, though characterized as site-oriented. Particularly, the biorefineries can potentially save up to 60% of GHG emissions compared to fossil-based refineries, producing bioproducts with substantial environmental benefits. Nonetheless, the LCA approach of a waste-based biorefinery system depends on the life-cycle stages and the inputs and outputs considered generating discrepancies between results in several cases, however a generic schematic overview can be presented (Fig. 17.5). This subsection describes selectively LCA case studies of biorefineries distinguished on the basis of the biomass feedstock considered.

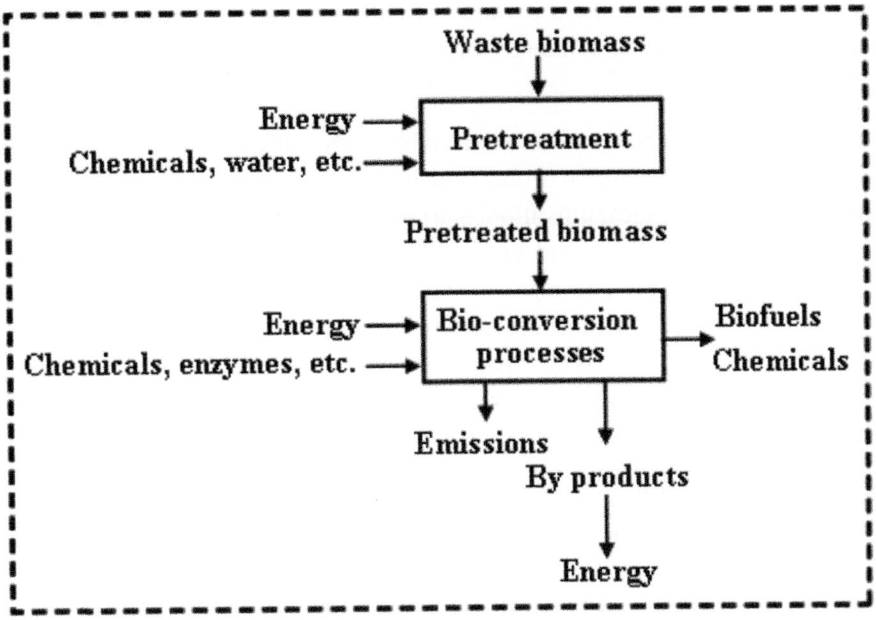

Figure 17.5
Generic overview of the main inputs and outputs for a waste-based biorefinery LCA study.

17.4.1 Energy crops derived feedstock

The environmental profile of a biorefinery concept coproducing bioethanol, phenols and energy (electricity, heat) from switchgrass, combining several conversion stages was accessed. The biochemical route followed after the pretreatment of the lignocellulosic-based biomass (hemicellulose depolymerization and lignin separation) includes enzymatic hydrolysis of the cellulose to glucose monomers, fermentation and distillation of sugars to bioethanol, anaerobic digestion of wastewaters, fast pyrolysis of lignin, residues combustion (heat and power joint production). The biorefinery system was compared with fossil-derived references system, and in particular gasoline, heat and electricity (from natural gas) and conventional phenols. The biorefinery system releases lower GHG emissions (60.5 kt CO_2 eq) than the fossil reference system (281 kt CO_2 eq), reducing the GHG emissions by ~78%. The highest contribution to the total GHG balance of the biorefinery was associated with the lignocellulosic biomass production chain, whereas the eutrophication (2.82 kt PO_4 eq) and acidification (1.23 kt SO_2 eq) impacts were also prominent, pinpointing future actions for sustainable cultivation practices. The energy requirements of the biorefinery's were higher than the fossil reference system, although it is basically based on renewable energy, rendering 80% savings of nonrenewable energy. Therefore, the examined lignocellulosic-based biorefinery presented great potential toward the coproduction of bioenergy and value-added chemicals, toward climate change mitigation and reduction the dependence on fossil-energy sources [22].

The environmental sustainability of a "sugar-power-ethanol" biorefinery system has been assessed in order to improve its environmental profile. The investigated biorefinery system involved the sugarcane cultivation, harvesting, milling and the byproduct utilization i.e., bagasse for steam and electricity, molasses for ethanol and vinasse for fertilizer and soil conditioner. The results revealed that the mechanized farming combined with integrated utilization of biomass residues (e.g., cane trash, vinasse for fuels and fertilizers) could potentially reduce the climate change, acidification and photo-oxidant formation and particulate matter formation by ~40%, 60%, 90%, and 63%, respectively [27].

An LCA approach assessed the GHG emissions and energy efficiency of a lignocellulosic-based biorefinery coproducing bioethanol and high added-value products. The study considered the environmental impacts of a perennial gross plant production chain (*Phalaris aquatica* L.) and its conversion steps for bioethanol and succinic acid production. Three different scenarios were developed covering variant biomass plantations examining regionally different biomass yields, inquiring two biochemical routes for the joint production of the targeted products. The quantified GHG emissions of the bioethanol production process ranged from 31.67 to 38.12 g CO_2 eq/MJ lower compared to other bioethanol systems (15–123 g CO_2 eq/MJ). The GHG emissions for the succinic acid ranged from 58.74 to 193.2 g CO_2 eq/kg in the examined scenarios. The high energy ratios values quantified are indicative of the biorefinery's energy performance, due to the use of lignin and residues for energy cogeneration, covering part of its energy requirements. On view of the above, the examined biorefinery presented a favorable environmental profile identifying the potential of a well-established plantation for GHG balance optimization [28].

Three potential variant feedstocks (rice straw, napier grass, *Eucalyptus* spp.) were examined for bioethanol production via biochemical avenues, exploiting in parallel the produced wastes (pellet fuel, paper mold) for energy and secondary product generation. The system boundaries involved the biomasses supply chain followed by their biochemical conversion in nexus with the wastes utilization. Napier grass an energy crop, appeared to be the most suitable feedstock for bioethanol production, presenting 47% lower negative impacts (global warming, acidification, eutrophication, etc.) compared to the other plants. The environmental performance of the specific crops presented differences associated with the agricultural practices and the bioethanol conversion yields. In addition, the fermentation waste utilization routes could potentially improve the efficiency in electricity generation and reduce the GHG emissions. Therefore, the comparative analysis of crops in a study can lead to expedient and practical results paving the way for sustainable biorefineries development [29].

The environmental consequences of a rapeseed-based biorefinery for joint production of biofuels and value-added products were studied utilizing inventory data and mass and energy balances. The environmental impact categories assessed were global warming, acidification and terrestrial eutrophication. The LCA results identified that the studied process the enzymatic transesterification and advanced oil extraction processes showed improved environmental benefits than conventional processes, whereas the exploitation of rapeseed straws for energy generation further improved the GHG footprint (9%–29%). Industrial-scale data are required to validate the favorable environmental profile of the biorefinery [25].

The environmental profile of an oil palm-based biorefinery cogenerating cellulosic ethanol and phytochemicals was evaluated via a cradle-to-gate approach. The processes considered involve the biomass cultivation and transportation to the biorefinery, its pretreatment and biochemical processing (via simultaneous saccharification, fermentation and distillation). GWP (2265.69 kg CO_2 eq), acidification potential (355.34 kg SO_2 eq) and human toxicity potential (142.79 kg DCB eq) for 1 ton of bioethanol were the most substantial environmental impact categories attributed to the use of fossil fuels, pesticides etc. in the examined system. The simultaneous saccharification and fermentation processes emerged as the most thermodynamically sustainable and environmentally friendly production system, indicative of the future optimization steps toward sustainable biorefineries systems [30].

Different crops feedstocks (grass-clover, ryegrass, alfalfa, and festulolium) were converted to energy and livestock feed in a biorefinery applying LCA to quantify the environmental impacts provoked. The approach followed includes the biomass production chain followed by its bioconversion under a green biorefinery umbrella, via a five-stage sequence (pretreatment, fractionation, coagulation, separation, drying). The results showed alfalfa as the most suitable crop in terms of the quantified environmental impacts (GWP, EP, nonrenewable energy, potential freshwater ecotoxicity), except of agricultural land occupation. This crop presented also the highest yields and the lowest fertilization demands. Respectable variation was observed for the corresponding environmental impacts of the examined biomass types, ascribed to the agricultural practices. Indeed, the agricultural phase affected significantly the environmental impacts, while the biorefinery emissions contribute immensely to the GWP and nonrenewable energy. The hotspots processes identified (coagulation, drying) could be enhanced by system optimization or by integration of alternative heat sources. Relevant holistic LCA approaches of different biomass are instrumental in the phase of decision-making introducing a specific feedstock in a biorefinery concept with environmental benefits [31].

The environmental impacts of ethanol production from corn crop applying LCA methodology have been studied, encompassing the agricultural (fertilizing, sowing, harvesting, drying) and the biorefinery (milling, liquefaction, saccharification, distillation, dehydration, and stillage treatment) subsystems. The results demonstrated that the fertilizers application, the seeds production, the harvesting etc. had the most significant impacts of the agricultural system in the environmental categories quantified (acidification, eutrophication and climate change). In the biochemical treatment of biomass, the supplied heat and burned natural gas contributed significantly to the determined impacts (e.g., acidification/eutrophication, climate change etc.). Thereupon, a cogeneration system could presumably lead to more efficient environmental performance [32].

17.4.2 Waste-derived feedstock

An LCA was performed via different scenarios reflecting the management, treatment and handling of municipal solid waste (plastics) by alternative thermochemical processes. In particular, the technologies investigated were low temperature pyrolysis recovering platform chemicals (e.g., gases, naphtha, waxes, heat) and hydrogenation producing syncrude and e-gas (comparable to natural gas). GWP, AP, POCP, and EP were the considered impact categories. The processing of solid wastes via hydrogenation resulted in the highest savings in terms of EP, due to the avoided impacts related to the naphtha production, while in terms of GWP the solid waste management practices appear to be an environmentally friendly option. Thereby these results highlight that the waste-based biorefineries can provide insight into renewable energy generation systems [33].

The integration of waste lipid feedstocks (waste cooking oil) in a refinery via coprocessing with petroleum fractions has been examined toward the production of market diesel with 10% v/v biocontent. The LCA study involved the cohydrotreatment of waste cooking oil with petroleum fraction producing a bio-based fuel with 10% v/v biocontent. Catalytic hydroprocessing constitutes a thermocatalytic process that has been applied for the production of renewable diesel from residual lipids. The targeted bio-based fuel abides by the target appointed for 2020 in order to raise the proportion of biofuels and other renewable fuels to 10% of the total transportation fuel needs [34]. The life-cycle stages encompassed the crude oil refining process, the coprocessing of the waste lipids with the petroleum fraction and the biodiesel production (from energy crops) that is added also in the final product to reach the 10% biocontent. The GHG emissions of the new bio-based fuel (243 kg CO_2 eq-/m^3) are significantly lower compared to fossil diesel production (315 kg CO_2 eq-/m^3). Based on these LCA results the coprocessing approach of waste lipids with petroleum fractions presents an environmentally friendly profile promoting the valorization of this specific residual biomass contributing toward climate change mitigation [35].

Forest residues and corn stover were employed for the production of bioethanol via two competing processes, thermochemical (alcohols coproduction) and biochemical (electricity coproduction), respectively. The life-cycle stages included were extraction and pretreatment of the waste-based biomass, transportation, waste feedstocks conversion, bioethanol use and waste management. The LCA results highlighted four major impacts of both biorefineries contexts including land use change, GWP, nonrenewable energy use and respiratory inorganics. Thermochemical processes using forest waste were found to be more environmentally favorable than the biochemical processes of using corn stover as feedstock. However, uncertainty parameters such as low-quality data, regional differences in the employed data, etc. were pointed out [36].

The GHG performance and energy balance of a biorefinery using straw and forest residues for coproduction of ethanol, biogas, and electricity was analyzed. The examined bioprocesses involved a pretreatment stage followed by simultaneous hydrolysis, fermentation, and anaerobic digestion. The ethanol derived from forest residue-based generally showed lower GHG emissions compared with the straw-based, while the GHG savings for both examined feedstocks were 51%−84% relative to fossil fuel. The study highlighted that influential inputs of the LCA study are the enzymes and the changes in soil organic carbon content due to removal of residues. Therefore, careful consideration of the enzyme dose is crucial decreasing accordingly the impacts from enzyme production in the environmental profile of the process [37].

The valorization of refinery side-stream products was analyzed via two different valorizing routes of sugar beet pulp, a byproduct of the sugar industry, with the aim of obtaining pectin-derived oligosaccharides, a product with prebiotic properties. Two different scenarios at pilot scale were considered under thermal (conventional autohydrolysis at high temperature) and enzymatic treatment. The environmental effects of the case studies were highly dependent on the production yield of the targeted products and the valorization routes followed. In fact, the pectin-derived oligosaccharides yield of the autohydrolysis approach is around 20% higher than in the enzymatic one however, this route was related with worst results considering a functional unit based on the amount of valorized material. The profile entirely differentiates accounting a unit based on the economic revenue (1 €). Therefore, the valorizing sequences of sugar beet pulp investigated appear to be promising and attractive options to produce high added-value products with multiple applications in a waste-based biorefinery [19].

The valorization of food waste biomass to hydroxymethylfurfural - a versatile platform chemical has been studied via eight scenarios with different combinations of solvents, catalysts, and experimental conditions. The specific study aimed to evaluate the environmental burdens and benefits originated from the raw material acquisition, the material processing, the production and the yield of the final products. The scenarios included the catalytic conversion of food waste-based biomass to the value-added chemical, and in particular the use of solvent and cosolvents, the addition of catalysts, heating, and the bioproduct yield. The inventory data originated from experimental data, while the environmental impacts associated with the use of water solvent, organic cosolvents, metal catalysts, the reaction temperature and time were quantified. The optimal food waste valorization route was identified applying the conversion of bread waste (using water-acetone medium with the catalyst aluminum chloride), due to the utilization of less polluting cosolvent (acetone) and catalyst (aluminum chloride) leading to the relatively high yield of the platform chemical (27.9 mol%). Metal depletion impacts attributed mainly to the production of metal chlorides catalyst displayed the highest among the categories, followed by the toxicity impacts (marine ecotoxicity, freshwater toxicity and human toxicity) associated mostly with the production of organic cosolvents [26].

The environmental consequences of a brewery waste-based biorefinery system for joint production of bioethanol and xylooligosaccharides were determined following the LCA methodology. The examined biorefinery system encompasses the following processes: (1) reconditioning and storage, (2) autohydrolysis pretreatment, (3) xylooligosaccharides purification, (4) fermentation, and (5) bioethanol purification. The GWP quantified were 7.39 kg CO_2 eq/kg bioethanol (including enzymes production) with the autohydrolysis stage displaying the highest contribution in this impact. The identified environmental hotspots of the biorefinery were the steam generation (from natural gas) required for the autohydrolysis (contributing >50% to GWP and to acidification) and the enzymes production for the simultaneous saccharification and fermentation (contributing > 95% to terrestrial and marine aquatic ecotoxicity potentials). Toward the biorefinery's sustainability improvement actions, steam generation via renewable sources (e.g., wood chips) renders significant environmental reductions, whereas the enzymes specific activity may also directly affect the environmental burdens [12].

Sugarcane residues were employed as biorefineries feedstock coproducing bioethanol, methanol and lactic acid, with electricity surplus. The LCA applied in order to define the optimal biochemical route among the biomass candidates in conjunction with the coproducts and the energy supply systems. The lactic acid biorefinery pathway presented the highest energy demands with the highest chemical consumptions as well as the highest conversion of biomass carbon input to products. The examined biorefineries approaches producing ethanol or ethanol-lactic comprising biomass residues with coal cocombustion were environmental favorable compared to the biorefinery leading to methanol production.

The supply of the process energy by the cocombustion of coal with biomass residues augmented the available biomass for valorization [38].

The analysis of the aforementioned LCA studies has showed that the valorization of residual and nonfood biomass for the production of a palette of bio-based products achieves a quantifiable reduction of GHG emissions integrating energy-efficient technologies and exploiting byproducts. Specifically, the improvement of the environmental benefits of the produced biofuels relative to fossil fuels is attributed to the favorable energy balance by optimal process integration and exploitation of the excess energy of the processes.

Accounting that the majority of the studies focus on the GHG emissions Table 17.1 attempts to summarize them and juxtapose the corresponding reference system values, when available. Nonetheless, the comparison of the GHG emissions of the biorefineries is rather sinuous due to variations in methodological aspects of the LCA applications and to the examination of scenarios widening the range of the results. However, the environmental benefits of the bioproducts produced from variant biorefineries contexts are clearly illustrated regarding the CO_2 eq emissions mitigation.

17.4.3 Algae-biomass derived feedstock

Apart from the waste-based biorefineries, microalgae-based biomass for biofuel production under a biorefinery platform could also establish a sustainable system, supposing concrete improvement actions. Accounting that the research in the field of microalgae-based biorefineries is intense leading to product portfolio, some characteristic case studies will be described.

Table 17.1: Greenhouse gas (GHG) emissions of variant biorefinery bioproducts.

Feedstock	Bio-product	GHG emissions	Reference system	References
Switchgrass	Bioethanol	60.5 kt CO_2 eq	281 kt CO_2 eq	21[a]
Sugarcane	Bioethanol	309 kg CO_2 eq	509 kg CO_2 eq	28[b]
Phalaris aquatica L.	Bioethanol	31.67–38.12 g CO_2 eq/MJ	15–123 g CO_2 eq/MJ	29
Oil palm fronds	Bioethanol	2.26 kg CO_2 eq/kg	-	31[c]
Residual lipids	New bio-based fuel	243 kg CO_2 eq/m^3	315 kg CO_2 eq/m^3	35
Straw	Bioethanol	15.46 g CO_2 eq/MJ	83.8 g CO_2 eq/MJ	37
Forest residues	Bioethanol	14.86 g CO_2 eq/MJ	83.8 g CO_2 eq/MJ	37

[a]The functional unit was appointed based on the amount of the treated biomass per year.
[b]The functional unit was set based on the final targeted products.
[c]Reference system was not defined.

LCA has been applied as foundation tool in evaluating the environmental impacts of microalgae-based feedstock conversion to biofuels and coproducts production. In particular, microalgal biodiesel production from *Chlorella vulgaris* was studied based on a hybrid cultivation system that couples airlift tubular photobioreactors with raceway ponds with relatively the same production capacity, in a two-stage process for high lipid accumulation. The downstream processes for the microalgal biodiesel production include harvesting, centrifugation, drying, cell disruption, extraction and transesterification. The results depicted that the microalgal biodiesel production would have 42% and 38% savings in GWP and fossil-energy requirements compared to fossil-derived diesel, respectively. However, bottlenecks identified include the energy requirement for microalgal cultivation and drying as well as burdens regarding the nutrient supply and the construction materials [39].

An LCA study of an algae biorefinery considering multiproducts (biodiesel, protein, and succinic acid) was carried out to estimate the environmental impacts compared to a fossil-based reference system. Seawater algae strain (*C. vulgaris*) was cultivated in raceway ponds and the extracted deoiled algae biomass was converted to biodiesel by transesterification and a biochemical route was followed for the succinic production. The remaining algal biomass of the biorefinery was assumed to be utilized for biogas production (in anaerobic digester) that is further used in the combined heat and power plant covering partially its energy requirements. For biodiesel, protein and succinic acid joint production system the CO_2 emissions reduction was 23%–31% based on the composition of the algae oil compared to the fossil-based reference system of biodiesel and protein. In biodiesel and protein coproduction system, the CO_2 emissions were 18%–24% lower than the reference fossil system, validating the environmental benefits of the specific algae biorefinery. Thereby, an algae-based biorefinery coproducing biodiesel, protein, and succinic acid could be a potential renewable production scheme, mitigating the climate change [21].

17.5 Future research directions of LCA in biorefineries

The biorefinery has emerged as an alternative to conventional fossil-based refineries utilizing bio-based feedstocks and valorizing waste streams through biotechnological technologies. Commonly the biorefinery integrates thermochemical and biochemical processes for the conversion of biomass to energy and to a wide array of value-added products evolving sustainable production technologies. The recent perspectives concepts that enable biorefineries to convert biogenic wastes to high-value products are being systematically promoted. Waste streams as renewable resources are being systematically exploited, ensuring independency from fossil sources. Moreover, waste-targeted biorefineries venture to develop environmentally friendly technological schemes enforcing the circular economy.

The future research directions should be immensely oriented toward sustainable biorefineries valorizing wastes and upgrading byproducts, depleting the dependence on fossil-energy source. The research around waste-based processes paves the way for sustainable technological avenues that will contribute to climate change mitigation [40].

In order to verify the sustainability of the biorefineries the LCA studies should subtend comparisons of potential future systems via different scenarios, toward the implementation of sustainable technologies. Indeed, the integration of biotechnological processes expanding to wastes valorization under the biorefinery structure is a challenging and complicated task. However, this is the optimal solution for bioenergy generation in nexus with value-added products in sustainable management of biomass waste in long term presenting a favorable environmental profile. In particular, the role of waste-to-energy processes in a biorefinery could be profoundly optimized, by highlighting proven energy-efficient technologies providing further innovation incentives. Thereupon, the sustainability approach of such systems will elucidate issues regarding the feedstock upgrading process affecting also the impending design activities decisions. Accounting that biorefineries have high energy demands, the maximization of its energy self-sufficiency could further ensure the sustainability aspects related to circular economy. Nevertheless, future LCA studies should also emphasize on the biomass supply chain optimization regarding its availability and yields [8].

Presently several research programs aim to valorize residual feedstocks streams in new technological systems aimed at GHG emissions reduction validating and optimizing design processes at pilot scale. Toward this direction, an innovative technological scheme is investigated via the conversion of residues and nonfood/feed plants (straw and miscanthus) with thermocatalytic processes (ablative fast pyrolysis and hydroprocessing), into high-quality bio-based intermediates. The biobased intermediates could be directly integrated to a refinery via coprocessing with petroleum fractions achieving significant environmental advantages relative to fossil fuels and to conventional biofuels supporting the development of sustainable energy technologies [41].

The sustainable conversion of waste-based biomass into energy, chemicals and biobased products is the aim of several research programs, validating and optimizing design processes at pilot scale. In addition other research efforts are focused on the sustainable conversion of renewable biomass into bio-based products, chemicals and energy under the umbrella of a second-generation cellulosic ethanol biorefinery [42]. Furthermore, the establishment of an advanced circular economy is endorsed via the valorization of the lignin-rich industrial waste stream from second generation biofuel plants into higher value products, for instance marine fuels, fuel additives, and chemical building blocks [43]. Such research designated orientations aspire to contribute to the valorization of waste streams and to the drastic increase of nonfood/feed biomass utilization for the production of greener transportation fuels and high-added products. These sustainable and effective production pathways have the potential to mitigate GHG emissions compared to fossil reference systems.

It should also be highlighted that the inclusion of social and economic aspects along with the sustainability consideration is anticipated accounting that the wastes valorization processes are maturely and fully integrated in a biorefinery. Such approaches are still in their infancy steps, even though the LCA study could be accounted as a milestone for the optimal waste valorization systems selection. This multidimensional characterization of waste-based biorefineries could also contribute toward the development of a comprehensive decision-supporting tool guiding and assisting future designs activities. It is also envisioned that when the development of large-scale valorization systems becomes more mature the decision-supporting tool could be expanded for the pilot and industrial-scale systems evaluation [26]. In this sequel, it is suggested that relative decisions should be preferably stem from context-specific LCA studies utilizing generalized results.

17.6 Conclusions and perspectives

This chapter presented an overview of LCA applications in biorefineries exploring recent developments and advances in their environmental performance. Recently, efforts have been focused on the progression and elevation of the waste-based feedstock biorefineries, as these materials present considerable potential for biofuels and bioproducts coproduction. Variant conversion technologies are embedded in the biorefineries either biochemical or thermochemical, whereas in some systems are both integrated converting the waste streams into a palette of higher added-value products. The key factors influencing the conversion process selection are the quantity and type of biomass feedstock, the biorefinery configuration and the end form of the targeted products. The system boundaries of a biorefinery commonly encompass the biomass cultivation, harvesting, transportation and intermediate storage, pretreatment, conversion, and use of products with potential exploitation of byproducts. Apparently, in the waste-based biorefinery the stage of the biomass provision is omitted.

LCA methodologies are applied in order to quantify the environmental impacts of biorefineries producing a cascade of valuable marketable products. Particular attention has been drawn to multifunctional systems boundaries, inventory data collection, while the allocation cannot be omitted in a biorefinery with high complexity valorizing several waste streams. In general, as previously denoted the environmental burdens of the biorefineries depend on the boundary conditions determined and are centered on the GHG balance, as being energy-intensive and to land use changes, regarding the bio-based feedstocks supply chain.

As consecutively biorefineries configurations emerge, their parallel comparison with conventional fossil-based refineries is conceptual for their environmental profile verification. Compared to the fossil reference systems, the biorefineries environmental profile is favorable, even though the results present considerable variation, attributed principally to the biomass supply chain and to the sequence of the conversion technologies, generating conflicting outcomes. Indeed, the biorefineries incorporating bio-based feedstocks or waste streams can result to environmental savings in comparison with conventional fossil refineries. However, these savings are correlated with the biorefinery's configuration and complexity and specifically with the bioprocesses involved to the conversion stages and with the complete utilization of the byproducts. Nonetheless, it is acknowledged that such direct comparisons are rather arduous and time-demanding, accounting parameters of biomass production chain, biorefineries configuration and energy supply etc. and thus frequently avoided.

Several studies have been conducted on the LCA applications of the lignocellulosic-based biorefineries, incorporating the biomass provision, pretreatment and conversion to biofuels and to a range of value-added chemicals. However, the energy consumption-related impacts (including climate change, photochemical oxidant formation, acidification, fossil fuel depletion), of such systems is significantly high, hence the maximization of a biorefinery's energy self-efficiency should be addressed. For instance, the energy balance of a biorefinery system could be optimized via the internal utilization of residues and byproducts for energy production.

Furthermore, intense research has also been devoted to algae-based biorefineries for the production of high value products. These results have identified the potential environment benefits of microalgae biofuels over petroleum-derived fuels, analyzing primarily the GHG emissions and the energy efficiency. However, the LCA outcome is dependent on the system boundaries depicting variant microalgae strains cultivations and the conversion technologies utilized. By contrast, concerns regarding the cultivation of the appropriate microalgae strains and the multi-stage systems are rising to alleviate the environmental burdens via process integration actions.

Toward this direction, future research efforts will explore waste valorization technologies to enhance the biorefinery sustainability via comprehensive LCA studies broadening industrial-scale implementation. Waste residues are subjected to advanced biological treatment with special attention to the sustainability of the biorefinery and to the achievable added-value bioproducts. The LCA practitioners are encouraged to orientate their efforts toward systematic collection and evaluation of the inventory data ensuring the reliability of the calculated results in conjunction with the study's goal. The reliability and representativeness of the outcomes could be validated with sensitivity analyses on identified hotspots of the examined biorefinery. In particular, for the waste-based biorefineries comprehensive and well-appointed modeling studies are required, based upon distinct frameworks depicting the system under investigation. Such studies could presumable enhance the completeness and transparency of the LCA outcome.

References

[1] van Ree R, van Zeeland A. Sustainable and synergetic processing of biomass into marketable food & feed ingredients, products (chemicals, materials) and energy (fuels, power, heat). In: Task 42 biorefining. International Energy Agency: Bioenergy, Wageningen UR Food and Biobased Research; 2014.

[2] Pandey A, Höfer R, Taherzadeh M, Nampoothiri M, Larroche C. Industrial biorefineries & white biotechnology. Amsterdam: Elsevier; 2015 [chapter 1].

[3] Guo M, Song W, Buhain J. Bioenergy and biofuels: history, status, and perspective. Renewable and Sustainable Energy Reviews 2015;42:712–25.

[4] Jin Q, Yang L, Poe N, Huang H. Integrated processing of plant-derived waste to produce value-added products based on the biorefinery concept. Trends in Food Science & Technology 2018;74:119–31.

[5] Parajuli R, Knudsen MT, Birkved M, Djomo SN, Corona A, Dalgaard T. Environmental impacts of producing bioethanol and biobased lactic acid from standalone and integrated biorefineries using a consequential and an attributional life cycle assessment approach. The Science of the Total Environment 2017;598:497–512.

[6] Chew KW, Yap JY, Show PL, Suan NH, Juan JC, Ling TC, Lee DJ, Chang JC. Microalgae biorefinery: high value products perspectives. Bioresource Technology 2017;229:53–62.

[7] Liguori R, Faraco V. Biological processes for advancing lignocellulosic waste biorefinery by advocating circular economy. Bioresource Technology 2016;215:13–20.

[8] Parada MP, Osseweijer P, Duque JAP. Sustainable biorefineries, an analysis of practices for incorporating sustainability in biorefinery design. Industrial Crops and Products 2017;106:105–23.

[9] Limayem A, Ricke SC. Lignocellulosic biomass for bioethanol production: current perspectives, potential issues and future prospects. Progress in Energy and Combustion Science 2012;38(4):449–67.

[10] Keller H, Rettenmaier N, Reinhardt GA. Integrated life cycle sustainability assessment – a practical approach. Applied Energy 2015;154:1072–81.

[11] International Standardization Organization (ISO 14040). Environmental management - life cycle assessment - principles and framework. 2006.

[12] González-García S, Morales BC, Gullón B. Estimating the environmental impacts of a brewery waste–based biorefinery: bio-ethanol and xylooligosaccharides joint production case study. Industrial Crops and Products 2018;123:331–40.

[13] Kralisch D, Ott D, Gericke D. Rules and benefits of life cycle assessment in green chemical process and synthesis design: a tutorial review. Green Chemistry 2015;17:123–45.

[14] Sandin G, Røyne F, Berlin J, Peters GM, Svanström M. Allocation in LCAs of biorefinery products: implications for results and decision-making. Journal of Cleaner Production 2015;93:213–21.

[15] Martin M, Svensson N, Fonseca J, Eklund M. Quantifying the environmental performance of integrated bioethanol and biogas production. Renewable Energy 2014;61:109–16.

[16] International Standardization Organization (ISO 14044). Environmental management — Life cycle assessment — Requirements and guidelines. 2006.

[17] European Commission - Joint Research Centre - Institute for Environment and Sustainability. International reference life cycle data system (ILCD) handbook - general guide for life cycle assessment - detailed guidance. 1st ed. Luxembourg: Publications Office of the European Union; March 2010. EUR 24708 EN.

[18] European Commission. Directive 2009/30/EC of the European Parliament and of the Council of 23 April 2009. 2009.

[19] González-García S, Gullon B, Moreira MT. Environmental assessment of biorefinery processes for the valorization of lignocellulosic wastes into oligosaccharides. Journal of Cleaner Production 2018;172:4066–73.

[20] Kajaste R. Chemicals from biomass - managing greenhouse gas emissions in biorefinery production chains - a review. Journal of Cleaner Production 2014;75:1–10.

[21] Gnansounou E, Raman JK. Life cycle assessment of algae biodiesel and its co-products. Applied Energy 2016;161:300–8.

[22] Cherubini F, Jungmeier G. LCA of a biorefinery concept producing bioethanol, bioenergy and chemicals from switchgrass. International Journal of Life Cycle Assessment 2010;15(1):53–66.

[23] Laurent A, Clavreul J, Bernstad A, Bakas I, Niero M, Gentil E, Christensen TH, Hauschild MZ. Review of LCA studies of solid waste management systems – part II: methodological guidance for a better practice. Waste Management 2014;34(3):589–606.

[24] Gabrielle B, Gagnaire N. Life-cycle assessment of straw use in bio-ethanol production: a case-study based on deterministic modelling. Biomass and Bioenergy 2018;32:431–41.

[25] Nizami AS, Rehan M, Waqas M, Naqvi M, Ouda OKM, Shahzad K, Miandad R, Khan MZ, Syamsiro M, Ismail IMI, Pant D. Waste biorefineries: enabling circular economies in developing countries. Bioresource Technology 2017;241:1101–17.

[26] Lam C-M, Yu IKM, Hsu S-C, Tsang DCW. Life-cycle assessment on food waste valorisation to value-added products. Journal of Cleaner Production 2018;199:840–8.

[27] Silalertruksa T, Pongpat P, Gheewala SH. Life cycle assessment for enhancing environmental sustainability of sugarcane biorefinery in Thailand. Journal of Cleaner Production 2017;140:906–13.

[28] Chrysikou LP, Bezergianni S, Kiparissides C. Environmental analysis of a lignocellulosic-based biorefinery producing bioethanol and high-added value chemicals. Sustainable Energy Technologies and Assessments 2018;28:103–9.

[29] Chang F-C, Lin L-D, Ko C-H, Hsieh H-C, Yang B-Y, Chen W-H, Hwang W-S. Life cycle assessment of bioethanol production from three feedstocks and two fermentation waste reutilization schemes. Journal of Cleaner Production 2017;143:973–9.

[30] Ofori-Boateng C, Lee KT. An oil palm-based biorefinery concept for cellulosic ethanol and phytochemicals production: sustainability evaluation using exergetic life cycle assessment. Applied Thermal Engineering 2014;62(1):90–104.

[31] Corona A, Parajuli R, Ambye-Jensen M, Hauschild MZ, Birkved M. Environmental screening of potential biomass for green biorefinery conversion. Journal of Cleaner Production 2018;189:344–57.

[32] Pieragostini C, Aguirre P, Mussati MC. Life cycle assessment of corn-based ethanol production in Argentina. The Science of the Total Environment 2014;472:212–25.

[33] Al-Salem SM, Evangelisti S, Lettieri P. Life cycle assessment of alternative technologies for municipal solid waste and plastic solid waste management in the Greater London area. Chemical Engineering Journal 2014;244:391–402.

[34] Directive 2009/28/EC of European Parliament and of the Council of 23 April 2009 on the promotion of the use of energy from renewable sources and amending and subsequently repealing Directives 2001/77/EC and 2003/30/EC.

[35] P Chrysikou L, Dagonikou V, Dimitriadis A, Bezergianni S. Waste cooking oils exploitation targeting EU 2020 diesel fuel production: environmental and economic benefits. Journal of Cleaner Production 2019;219:566–75.

[36] Stuart PR, Mahmoud M, El-Halwagi. Integrated biorefineries: design, analysis and optimization. CRC Press; 2012 [chapter 27].

[37] Karlsson H, Börjesson P, Hansson PA, Ahlgren S. Ethanol production in biorefineries using lignocellulosic feedstock – GHG performance, energy balance and implications of life cycle calculation methodology. Journal of Cleaner Production 2014;83:420–7.

[38] Mandegari M, Farzad S, Görgens JF. A new insight into sugarcane biorefineries with fossil fuel co-combustion: techno-economic analysis and life cycle assessment. Energy Conversion and Management 2018;165:76–91.

[39] Adesanya VO, Cadena E, Scott SA, Smith AG. Life cycle assessment on microalgal biodiesel production using a hybrid cultivation system. Bioresource Technology 2014;163:343–55.
[40] Venkata Mohan S, Nikhil GN, Chiranjeevi P, Nagendranatha Reddy C, Rohit MV, Naresh Kumar A, Sarkar O. Waste biorefinery models towards sustainable circular bioeconomy: critical review and future perspectives. Bioresource Technology 2016;215:2–12.
[41] http://www.BioMates.eu.
[42] https://bioskoh.eu/.
[43] http://www.falcon-biorefinery.eu/.

CHAPTER 18

Life-cycle assessment of food waste recycling

Chor-Man Lam, Iris K.M. Yu, Shu-Chien Hsu, Daniel C.W. Tsang
Department of Civil and Environmental Engineering, The Hong Kong Polytechnic University, Kowloon, Hong Kong, China

18.1 Introduction

Globally, one-third of the food produced for human consumption—1.3 billion tonnes—is lost or wasted every year, which costs US$680 billion and US$310 billion in industrialized countries and in developing countries, respectively [1]. The notable amount of food waste embeds a substantial carbon footprint of 3.3 billion tonnes of CO_2-equivalent [2]. Proper food waste management measures are crucial to environmentally friendly and sustainable development. Numerous techniques, such as composting, anaerobic digestion (AD) and incineration, have been well-developed and adopted by municipalities to handle food waste. Other newly emerged technologies valorizing food waste into valued-added products have also been widely studied and tested in lab-scale food waste treatment. To opt for the most environmentally beneficial food waste management strategies, a comprehensive decision-guiding tool is important to evaluate and compare the performance of the handling techniques.

Life-cycle assessment (LCA) is a widely recognized decision-guiding tool which could systematically evaluate the environmental sustainability of products and processes. LCA is a “cradle-to-grave” approach that investigates the environmental impacts of the whole life-cycle of the products, including the life-cycle stages of raw material acquisition, material processing, production manufacturing, transportation, consumption, disposal and recycling [3]. The associated energy and resource consumption, together with the emissions for the whole system are being assessed [4]. As LCA comprehensively covers a wide range of environmental issues and life-cycle stages [5], it avoids the shifting of environmental burdens between life-cycle stages and impact categories [6–8]. The LCA approach has been standardized by the International Organization for Standardization (ISO), and its method and procedures are given in the international standards ISO 14040 and 14044 [9,10].

Waste Biorefinery. https://doi.org/10.1016/B978-0-12-818228-4.00018-6

A standard LCA includes four main phases: goal and scope definition, life-cycle inventory (LCI) analysis, life-cycle impact assessment (LCIA), and data interpretation. In the first phase, the aim and objective of the study, the intended users, practitioners, and stakeholders of LCA results and the expected application of the findings are stated. The scope of LCA, which is represented by the system boundary, defines the life-cycle stages, the flow of material and energy, as well as the associated environmental emissions, to be covered in the study. The system boundary also determines the temporal and geographic coverage of the LCA. The functional unit (FU) in an LCA defines the primary function of the product or the system being analyzed, and it is quantitatively defined in this phase for the convenient comparison between entities with equivalent function [11,12]. The LCI analysis phase involves data collection and analysis on all the inputs and outputs, per FU, of the processes that are included in the scope of the LCA study. This phase deals with differentiation between the product system and environment system, cut-off rule and allocation rule, in order to construct a reliable inventory of material and energy flow for further analysis in the next phase. LCIA is the phase that processes the outcome of the LCI analysis, which is the inventory table and interprets it in terms of environmental impacts. This phase includes the steps of selection of impact category, characterization, normalization, grouping and weighting. Numerous impact categories have been commonly included in LCA studies, such as climate change, human toxicity, acidification, and eutrophication. Selection of impact categories should be consistent with the goal and scope of the LCA study. In the characterization step, the category indicators, which represent the environmental impacts in different categories, are derived from individual pollutants using characterization factors. For example, greenhouse gases, such as carbon dioxide and methane, are converted into climate change factor based on their global warming potentials and are reported in terms of kg CO_2-equivalent. Normalization is the step to compute the environmental impacts represented by category indicators relative to the reference information, which could be the impacts generated from region or country, a community or a system. Grouping involves the aggregation of normalized results of various impact categories into one or more sets, such as impacts of human health, ecosystem, and resources. Weighting is the optional step in which normalized results are assigned weightings to reflect their relative importance based on government policies, scientific supports or expert advice. Data interpretation, the last phase of LCA, includes the evaluation and documentation of the assumptions, choice of models and results so that sound conclusions and recommendations could be made based on the findings of the LCA.

18.2 Life-cycle assessment of food waste management

As an inevitable municipal solid waste (MSW) with high organic contents, food waste has been recognized as an essential environmental issue globally. With growing interests from the government authorities, practitioners and researchers on the environmental sustainability of food waste management, numerous research studies which evaluated the environmental performance of different handling techniques using LCA approach have been conducted. Previous studies have demonstrated LCA as a suitable tool to guide decisions on selecting environmentally favorable food waste handling techniques that fits the actual situations of the countries or regions. LCA studies on food waste management have been reviewed and summarized in this chapter. This chapter is structured as follows: Firstly, early LCAs on mixed solid wastes, including food waste, are reviewed. Then, LCAs on conventional food waste treatment technologies and more recently developed techniques, such as bioconversion and valorization to valued-added chemicals, are presented. To provide more comprehensive information on the procedures of conducting LCA on food waste management options, the details of two of the published studies, which serve as the examples of conventional macroscale management strategies and laboratory-scale valorization techniques, are then summarized and presented.

18.2.1 Early LCA studies on solid wastes

Early research studies, instead of specifically evaluating food waste treatment options, compared treatment options of mixed MSW. For example, Morris compared recycling and waste-to-energy incineration options for 25 types of MSW including food waste [13]. The author estimated the conserved energy from food waste via incineration and anaerobic digestion to be 2744 kJ/kg and 4215 kJ/kg, respectively, thus implied that food waste recycling via AD is a more favorable option. In the research study conducted by Dalemo et al., simulation model for urban organic waste handling was developed to simulate the emissions of different treatment and disposal scenarios for organic wastes including food waste [14]. Some studies systematically evaluated the environmental performance of different solid waste management options, such as combustion, using the LCA approach [15,16]. It was not until the 21st century that LCA on food waste management has started to develop. Instead of evaluating the food waste management approaches, Ohlsson stated the importance of considering the environmental performance of the production stages [17]. The LCA on food production conducted by Ohlsson focuses on the energy consumption and eutrophication of different processes within the food production chain [17].

Early studies rarely investigated the treatment alternatives specifically for food waste but evaluated the environmental performance of management options for mixed solid wastes. The few reviewed studies that evaluated food waste separately from other solid wastes focuses on energy consumption and the number of environmental impact categories included is limited. However, in the past 15 years, LCA has been demonstrated in numerous research studies to be an appropriate and practical tool to guide decisions on food waste management strategies. The flexibility and comprehensiveness of LCA enhance its applicability to evaluate different food waste recycling alternatives.

18.2.2 LCA on conventional food waste management technologies

The first LCA study applied to food waste management was conducted by Lundie and Peters in 2005 [18]. The LCA study evaluated four food waste treatment alternatives, including household in-sink food waste processor, home composting, landfilling and centralized composting. The FU of this study was the management of food waste produced by a Sydney household in 1 year. Covering eight environmental impact categories, the LCA concluded that aerobic home composting outperformed the other alternatives in all impact categories. The study also highlighted the importance of assembly stage, market penetration rate and transportation distance in the overall environmental performance of the options.

Ogino et al. conducted an LCA to evaluate the environmental performance of food waste recycling to animal feeds [19]. Three food waste recycling or disposal options, including wet feeds production by sterilization with heat, dry feeds production by dehydration and incineration, were evaluated in the case of Japan. This study focused on environmental impact in the global warming category and estimated the CO_2, CH_4 and N_2O emissions. Based on the GHG emissions of the options, wet feeds production by sterilization was identified to be the most environmentally friendly choice. The energy consumption of the three options showed similar pattern as the GHG emissions; water consumption of the animal feed production options was significantly lower than that of food waste incineration. Such findings further supported the superiority of recycling food residues into wet feeds.

Recognizing the pollution caused by food waste landfilling, Korea has banned direct disposal of food waste into landfills in 2005, which provided an incentive for examining the environmental impacts of other handling alternatives. An LCA case study for Seoul, Korea, assessed the environmental burdens of food waste landfill, incineration, composting, and feed manufacturing [20]. The study evaluated the environmental impacts of individual treatment and disposal methods, and then compared the changes of impacts between year 1997 and 2005 caused by the change of food waste management systems, that is shifting from landfilling to recycling. The findings revealed a substantial shift of environmental impacts from global warming and human toxicity to acidification, eutrophication and ecotoxicity.

Since then LCA has been more commonly applied to inform decision-making in selection of different conventional food waste management strategies. Most of the reviewed study in this section applied LCA approach and models to assess different food waste treatment and recycling methods.

Khoo et al. [21] conducted an LCA on food waste recycling in Singapore, using the Environmental Development of Industrial Products (EDIP), [22]. The study investigated the environmental performance of four food waste recycling options:

(1) recycling of food waste through AD and composting (with a capacity of 300 tons of food waste per day), with the rest incinerated,

(2) recycling of food waste through AD and composting (with a capacity of 500 tons of food waste per day), with the rest incinerated,
(3) recycling of food waste through AD and composting (with a total capacity of 800 tons of food waste per day), with the rest incinerated, and
(4) recycling of food waste through AD and composting (with a total capacity of 800 tons of food waste per day), with 50% of the rest incinerated and 50% treated by aerobic composting. Five impact categories—namely global warming, acidification, eutrophication, photochemical oxidation, and energy use—have been evaluated.

The findings suggested scenario 4 to be the most environmentally favorable food waste recycling option.

Kim and Kim [23] evaluated a single environmental impact indicator, GWP, using LCA to compare four food waste handling options, including dry feeding, wet feeding, composting and landfilling in Korea. The environmental impacts of the collection, transportation, treatment, and disposal stages have been covered. The study also evaluated the risk of redisposal of treatment byproducts (animal feed and compost). Wet feeding was revealed to be the most favorable option if the byproducts were used properly, while dry feeding was the most favorable option if the byproducts were incinerated or landfilled.

The LCA conducted by Bernstad and Jansen evaluated four food waste management options: (A) Incineration of food waste and organic waste with energy recovery, (B) Composting of food waste, (C1) AD of food waste, from which biogas was yielded and upgraded to substitute petrol in light vehicles, and digestate was used to substitute commercial fertilizers, and (C2) AD of food waste, from which biogas is yielded for electrical and thermal energy recovery, and digestate was used to substitute commercial fertilizers [24]. The LCA model EASEWASTE has been adopted to evaluate the environmental impacts in five impact categories [25]. Considering the aggregated net environmental impacts, scenario C1, with the use of digestate on sandy soils, was revealed to be the most favorable option. The second-best option was revealed to be scenario C2. The findings of the study also revealed the importance of waste recycling for energy and material recovery in the overall environmental performance of the different waste handling strategies.

The impacts of treating food waste together with other organic waste, such as sewage sludge, have also been investigated. In the LCA study conducted by Nakakubo et al., the combinations of two food waste management options, namely incineration and AD, with six sewage sludge treatment technologies were evaluated [26]. Three environmental performance indicators, including greenhouse gas reduction, phosphorus recovery, and health impacts, were assessed for the FU of the capacity to provide food waste and wastewater treatment services for 100,000 people. The results showed that the combination of each of the six sludge treatment technologies with codigestion was superior to their counterparts integrated with food waste incineration. The study revealed that considering the management of both food waste and sewage sludge, codigestion is more environmentally favorable compared to incineration.

Kim et al. assessed the environmental impact of three scenarios for treating food waste in Korea, including (1) AD, (2) codigestion with sewage sludge, and (3) drying followed by incineration with energy recovery [27]. Although six environmental impact categories have been covered in this LCA, the study focused on the global warming potential (GWP) indicator as it was revealed to be the most significant environmental impact when comparing the normalized impacts of the six categories. Considering the electricity balance, thermal energy recovered, and primary materials avoided, the dryer-incineration option presented the highest net environmental benefits, thus was considered to be the most favorable option for food waste recycling in urban Korea.

The LCA conducted by Saer et al. investigated the overall environmental impacts, as well as process contribution, of food waste composting [28]. Nine impact categories were assessed using the Tool for the Reduction and Assessment of Chemical and other environmental Impacts (TRACI) developed by the US Environmental Protection Agency (US EPA) [29]. The study concluded that when the environmental offsets of using the food-waste-derived compost were accounted for, food waste composting could achieve net environmental benefits in all the impact categories. The results also indicated that the decomposition emissions from the compost processing stage contributed the most to GWP, acidification, and eutrophication.

Zhao and Deng [30] conducted an LCA using the EASEWASTE model to compare the environmental performance of three food waste management options in Hong Kong, including landfilling, composing and combined AD with composting. The FU used in this study was treatment of 3584 tonne of food waste per day. The processes of food waste treatment covered in the LCA included collection, transportation, pretreatment, biotreatment, energy recovery, byproduct consumption, and final disposal. The influence of using different fuel mixes for electricity generation was also investigated. The LCA findings revealed that energy recovery achieved in combined AD and composting brought higher environmental benefits than other options in most of the impact categories, such as acidification, nutrient enrichment and global warming. Change in electricity fuel mix presented higher impacts on landfilling and combined AD with composting than on composting, as it affected the environmental offsets achieved by energy recovery.

Eriksson et al. [31] conducted an LCA to evaluate six food waste management scenarios, namely landfilling, incineration, composting, AD, animal feeds and donations, for handling the food waste generated from supermarkets. The study investigated the carbon footprint, which was presented as the environmental impact in the GWP category in an LCA, of handling five types of food waste, including banana, chicken, lettuce, beef, and bread. Generally considering all the five types of food waste, AD and donation were revealed to be the options with the lowest level of greenhouse gas emissions. Incineration with energy recovery was also a favorable option for treating bread and chicken. The potential of greenhouse gas reduction varied among different types of food waste. Bread had the highest potential for greenhouse gas reduction if proper management scenarios could be applied to treat the waste.

An LCA has been conducted to evaluate the environmental performance of three food waste management options in China which were capable of biogas generation, namely codigestion with sewage sludge, AD, and landfilling [32]. The FU used in this study is management of one tonne of volatile solids. The environmental impacts associated with biogas production, direct air emissions, energy and material flow, transportation and infrastructure were covered in the system boundary. The ReCipe midpoint method was adopted and impacts in 18 midpoint impact categories were revealed. The LCA results indicated that the most environmentally favorable option was AD of food waste.

Morris et al. investigated the climate change and energy use impacts of four food waste management alternatives, namely AD, aerobic composting, in-sink grinding, and gas-to-energy landfilling [33]. This study revealed 147 relevant LCA studies and selected 28 studies for harmonization using the LCA meta-analysis approach [34–36]. AD was revealed to be the best alternative when considering the average climate impact, while in-sink grinding presented statistically significant lower energy use than the other alternatives. The study also assessed the impacts of soil productivity of the alternatives by ranking four other impact indicators, including soil carbon increase, fertilizer replacement, water conservation, and crop yield increase. The results indicated that aerobic composting would favor soil carbon storage and water conservation, thus benefit soil productivity, the most.

Seven food waste management systems in Australia were assessed using the LCA model CML-IA version 4.2 [37]. The functional unit defined in this study was the management of the annual amount of garbage, food, garden waste and sewage sludge collected from curbside collection services and wastewater treatment plants, respectively, by the Melton city council and Sutherland city council. The evaluated food waste handling scenarios included (1) landfilling, (2) anaerobic codigestion with sewage sludge, (3) in-sink maceration and codigestion, (4) centralized composting, (5) home composting, (6) AD, and (7) mechanical biological treatment (i.e., a combination of mechanical sorting, AD, and aeration). Eight environmental impact categories were assessed. The study neither concluded the overall environmental impacts of the scenarios nor recommended a single scenario to be the most favorable option. Instead, the LCA findings revealed that scenarios with digestion process presented lower GWP impact than landfilling and composting, while mechanical biological treatment, anaerobic codigestion, and home composting showed the most favorable performance in at least two of the impact categories.

Besides solely applying LCA to evaluate food waste management options, some researchers used LCA in combination with other analyzing techniques to obtain more comprehensive results.

The study conducted by Vandermeersch et al. integrated LCA with exergy analysis to investigate the material and energy flows, as well as the environmental impacts of food waste handling scenarios [38]. A three-staged LCA has been conducted to evaluate the environmental performance of two food waste handling options in Belgium [38]. The two scenarios evaluated include (1) AD for all food waste, and (2) animal feed production for the bread fraction of food waste and AD for the nonbread fraction. The assessment approach is consisted of three stages: exergy analysis, exergetic life-cycle assessment (ELCA), and a traditional LCA. In the exergy analysis, material and energy flows were investigated for evaluating the physical and chemical exergy of the systems. Then the exergy analysis was expanded, for revealing the hotspots of the processes, to an ELCA using the cumulative exergy extraction from the natural environment (CEENE) v2.0 method [39]. The third stage was a traditional LCA conducted using the ReCipe Endpoint model, which revealed the environmental impacts in 18 midpoint categories, three endpoint categories and a final aggregated score. The exergy analysis and the ELCA focus more on the resource efficiency, while the traditional LCA focus on the overall environmental performance. The results showed that scenario 2 was 10% more efficient than scenario 1 in the exergy analysis and generated 32% and 26% lower impacts as revealed by the ELCA and LCA, respectively. Therefore, the study concluded that producing animal feeds the bread waste and treating nonbread food waste by AD was a more environmentally favorable option.

Another study evaluated the environmental and economic performance of food waste management scenarios in Hong Kong through the integration of LCA with cost-benefit analysis (CBA) [40]. The study evaluated six food waste management scenarios for the Hong Kong International Airport, including landfilling, dewatering followed by landfilling, on-site/centralized incineration, and on-site/centralized organic waste treatment (AD, dewatering and composting). Using an integrated LCA approach with CBA, the study concluded the on-site incineration to be the most sustainable option. The details of the study are provided in the next section.

LCA has been applied, in combination with data envelopment analysis (DEA), which is a technique in operations research, to evaluate the environmental performance of six food waste management options in the study conducted by Cristóbal et al. [41]. The management options include AD, composting, AD followed by composting, incineration, landfilling with electricity production, and landfilling with gas flaring. The Product

Environmental Footprint (PEF) [42], which is an LCA-based methodology, has been adopted in this study to investigate 12 impact categories of the options. Cristóbal et al. obtained the normalized environmental performance of the scenarios in the 12 impact categories, which were then used as the inputs of the DEA [41]. The final results revealed that four scenarios, including composting, AD followed by composting, incineration and landfilling with gas flaring, were efficient food waste management options.

The information of the above-reviewed LCA studies of food waste management is summarized in Table 18.1.

However, the studies on conventional food waste management strategies as described above showed inconsistence in evaluated technologies, functional units, and LCA models. The two most commonly evaluated food waste treatment options are composting and AD, which make up 71% and 65% of the reviewed studies, respectively. The traditional disposal approaches, namely landfilling and incineration, are also commonly covered in LCAs studies. Due to the inconsistence of the functional units, LCA scopes and LCA models, the results of the studies could hardly be directly comparable. Yet, for most of the LCA studies that included AD of food waste, AD was revealed to be the best option or one of the best options in most of the cases, mainly due to its capability of generating biogas for energy recovery.

18.2.3 LCA on food waste bioconversion and valorization

Food waste contains high organic contents, which could be the raw materials for producing value-added products, such as compost, animal feeds and chemicals. Different from the traditional approach, biological and physiochemical processes for converting food waste into value-added products are emerging in recent years. Numerous published research papers focus on the feasibility, optimized conditions and product yields of the conversions processes, yet investigation on the environmental performance of such processes is limited.

Food waste valorization methods to recycle food waste into useful materials, such as compost and animal feed, through bioconversion process by insects have been studied. LCA has been demonstrated as a useful tool to investigate the environmental performance of such techniques. Salomone et al. has conducted an LCA on food waste bioconversion into compost and aquaculture feed using the insect species *Hermetia illucens* [43]. Three functional units were defined, including one tonne of biodigested food waste, 1.0 kg of proteins and 1.0 kg of lipids for evaluating and comparing the environmental performance of food waste bioconversion against processes with different purposes. For the LCA based on one tonne of biodigested food waste, the environmental impacts and benefits of the inputs and outputs of the bioconversion process were investigated. The bioconversion process produced larvae manure which was used as compost, and dried larvae which was

Table 18.1: Summary of reviewed food waste management LCA studies.

References	Technologies							FU	LCA model	Number of impact categories	Favorable options	Remarks
	AD	Co-digestion	Composting	Incineration	Landfilling	Animal feed	Others					
Bernstad and la Cour Jansen [24]	●		●	●				24.9 kg organic waste/person year	EASEWASTE	5	- AD (biogas for substitution of petrol in light vehicles; digestate for substitution of commercial fertilizers)	
Cristóbal et al. [41]	●		●	●	●			Management of 1 tonne of food waste	PEF	12	- AD - AD + composting - Incineration - Landfilling with gas flaring	LCA + DEA
Edwards et al. [37]	●	●	●		●		●	Collect, manage and treat 1 years' worth of municipal curbside collected garbage, food, garden waste and sewage sludge of the studied jurisdictions	CML-IA	8	- Digestion-based systems (GWP) - Mechanical biological treatment, anaerobic co-digestion, and home composting (lowest impacts for two or more impact categories)	Others: Maceration

Eriksson et al. [31]								1 kg of food waste from the supermarket	specified)		- Donation	Donation only covered carbon footprint
Khoo et al. [21]	●		●	●				570,000 tons of food waste/year	EDIP	5	- Recycling through AD and composting; incineration and aerobic composting for the rest of the food waste	
Kim et al. [27]	●	●		●			●	1 tonne of food waste from households	Total 3.0	6	- Dryer-incineration	Others: Drying
Kim and Kim [23]			●		●	●		1 tonne of food wastes	Total 3.0	1	- Wet/dry feeding	
Lam et al. [40]	●		●	●	●		●	Management of 1 tonne of food waste	ReCipe Endpoint	18	- On-site incineration	LCA + CBA others: Dewatering
Lee et al. [20]			●	●	●	●		Treatment of 1 tonne of food waste	USES-LCA	5	N.A.	The most effective way is to reduce waste generation
Lundie and Peters [18]			●		●		●	Management of food waste produced by a Sydney household in 1 year	(Not specified)	8	- Aerobic home composting	Others: In sinker food waste processor
Morris et al. [33]	●		●		●		●	Treatment of 1 kg of food waste	LCA meta-analysis	6	- AD (climate change) - In-sink grinder (energy use) - Aerobic composting (soil productivity)	Others: In-sink grinder LCA harmonization

Continued

Table 18.1: Summary of reviewed food waste management LCA studies.—cont'd

References	Techno-logies							FU	LCA model	Number of impact categories	Favorable options	Remarks
Nakakubo et al. [26]		●		●				100,000 people receiving food waste and wastewater treatment services	(Not specified)	3	- Co-digestion	Two food waste treatment options combined with six sewage sludge treatment technologies have been evaluated
Ogino et al. [19]				●		●		1 kg dry matter of produced feed with a fixed metabolizable energy content	IPCC	1	- Production of liquid FFR by steriliza-tion with heat	
Saer et al. [28]			●					1 tonne of compost	TRACI 2	9	N.A.	
Vandermeersch et al. [38]	●					●		Valorization of 1000 tonne of food waste with 100 tonne of bread waste	CEENE v2.0 ReCipe Endpoint H/A	18	- Bread frac-tion: production of animal feed; non-bread fraction: AD	
Xu et al. [32]	●	●			●			Management of 1 tonne volatile solids	ReCipe	18	- AD	
Zhao and Deng [30]	●		●		●			Management of 3584 tonne of food waste per day	EASEWASTE	12	- Combined AD and composting	

used as feed for fish. The impacts of transportation and bioconversion, as well as the benefits from the avoided production processes of fertilizers and aquaculture feed, were evaluated in the LCA. As the dried larvae served the same function of protein provision as conventional fish feed, the environmental profiles of fish feed produced from food waste and soybean meal were compared, which was a widely used fish feed, using the FU of 1.0 kg of protein. The environmental profiles of biodiesel production from dried larvae and from rapeseed were also compared using the FU of 1.0 kg of lipids. The study concluded the energy consumption, which was mainly originated from the drying process, to be the most significant contributor to environmental impacts. By avoiding the conventional production process of feed and biodiesel, the food waste bioconversion technique contributed to environmental benefits mainly in the land use aspect.

Numerous studies have shown that food waste could be a reliable feedstock for the production of bio-based chemicals, such as succinic acid (SA) and Hydroxymethylfurfural (HMF). Food waste valorization to valued-added chemicals could be achieved via bioconversion or physiochemical means. The environmental performance of such processes could be assessed by LCA.

Brunklaus et al. conducted an LCA to compare the environmental performance of food waste to biogas, food waste to SA and corn to SA [44]. The FU was defined as one dry tonne of food waste and one tonne of SA crystal for different comparisons. The food waste to biogas was the existing food waste handling option in Sweden, which included the processes of pretreatment, hygenization and AD. Food waste to SA included processes of pretreatment, bacterial (*E. coli*) fermentation and purification, while the conversion of corn to SA involved processes of corn and dextrose production, pretreatment, yeast fermentation and purification. Six environmental impact categories, namely GWP, acidification potential, the eutrophication potential, and human toxicity potential, nonrenewable energy use and renewable energy use, were evaluated. Food waste to biogas was more preferable than conversion to SA from the perspective of food waste treatment, while food waste to SA was more favorable when considering food waste to be an alternative feedstock compared to corn.

The environmental performance of various physiochemical food waste conversion processes to HMF were evaluated using LCA [45]. The food waste-to-HMF process studied was a catalytic conversion approach which used organic solvents, water medium, metal chloride catalysts and microwave heating. The conversion scenario converting bread waste substrate using an acetone-water medium, $AlCl_3$ as the catalyst, reaction temperature of 140°C and reaction time of 30 min was revealed to be the most environmentally friendly. The details of the study are provided in the next section.

The unconventional food waste recycling alternatives, which are mainly biological and physiochemical waste valorization approaches, show notable difference between one another, such as types of food waste inputs, conversion mechanisms and final products. For example, Salomone et al. used mixed food waste as the feedstock for conversion [43]; Brunklaus et al. compared two feedstocks including mixed food waste and corn [44]; and Lam et al. compared the use of bread, rice and kiwi as the food waste substrate [45]. The valorization mechanisms, such as bioconversion using the insect species *Hermetia illucens* [43], bacterial fermentation using *E. coli* [44] and physiochemical conversion using solvents, catalysts and heating, covered in the reviewed studies are also different. Valorization products including compost, animal feeds and value-added chemicals were yielded from food waste conversion. As the LCAs on food waste valorization alternatives are very case-specific, no direct comparison could be made between the findings of such studies. Such studies have successfully demonstrated the suitability of adopting LCA as an appropriate tool for guiding the selection of the most environmentally favorable food waste valorization approach.

18.3 Case studies on LCA application on large-scale conventional food waste management and laboratory-scale food waste valorization scenarios

Two case studies of LCA application for decision-guiding on the selection of food waste recycling strategies are introduced in detail in this section. The first study focused on the selection among large-scale traditional food waste handling options, while the second study evaluated the food waste valorization options to produce value-added chemicals.

18.3.1 Life-cycle cost-benefit analysis on sustainable food waste management in the Hong Kong International Airport

The first detailed case study discussed in this section is a life-cycle cost-benefit analysis (LC-CBA) of the conventional large-scale food waste management alternatives for the Hong Kong International Airport (HKIA). To echo with the principal of reducing solid wastes, including food waste, announced by the Hong Kong government, the HKIA had been developing new strategies to cope with the large amount of food waste collected. Aiming to reduce food waste disposal into landfills, which will be full soon and require extension, other food waste recycling alternatives, including AD, composting, and incineration, were evaluated in the LC-CBA study. The environmental performance of the alternatives was evaluated using the LCA approach to reveal the environmental impacts and benefits, thus facilitate informed selection between the alternatives. In addition to the environmental performance, the economic aspect is also an essential consideration for sustainable waste management. The economic favorability varies with the treatment technology and infrastructure capacity. Therefore, LCA has been integrated with CBA, which is a widely adopted economic evaluation tool, to establish an LC-CBA framework for an all-inclusive evaluation of food waste recycling alternatives.

The LC-CBA framework developed in this study is capable of (1) evaluating the environmental performance (impacts and benefits) of food waste treatment scenarios via LCA, (2) quantifying the environmental performance through economic valuation of emissions via CBA, and (3) providing final indicators in monetary terms for reflecting the sustainability of the food waste management scenarios.

18.3.1.1 Life-cycle cost-benefit analysis methodology

Fig. 18.1 illustrates the integrated LC-CBA framework. The environmental performance of the food waste management scenarios is assessed by the LCA approach. The overall sustainability of the scenarios is evaluated using the CBA approach. The final results would be the net costs of the scenarios given in monetary terms which could be easily understood and effectively guide the decision-making process.

18.3.1.1.1 Step 1: goal and scope definition

The goal of the LCA was to evaluate the environmental performance of different recycling alternatives for handling food waste in the HKIA. The FU was defined as one tonne of food waste. The system boundary of the LCA covered the transportation, treatment, and disposal of the food waste generated from the HKIA. Six food waste management

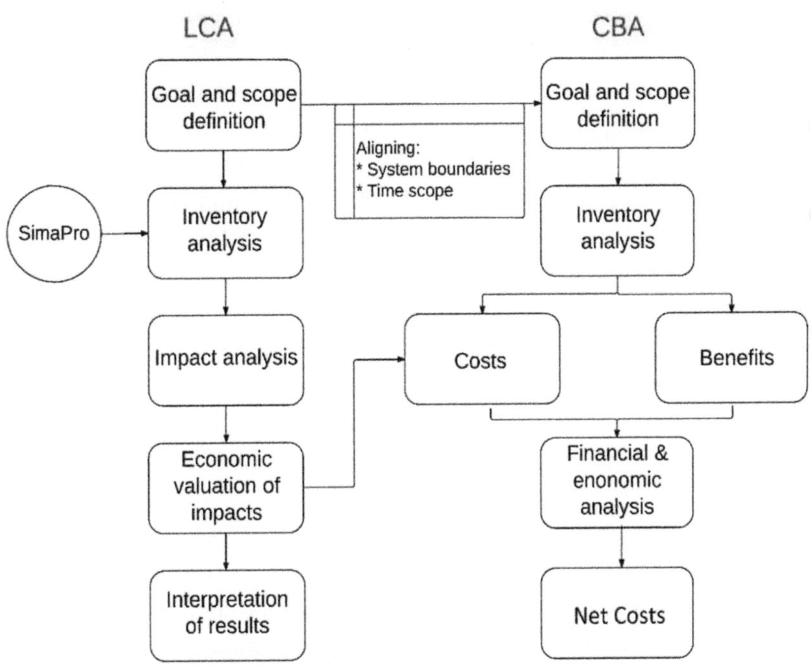

Figure 18.1
Integrated LC-CBA framework.

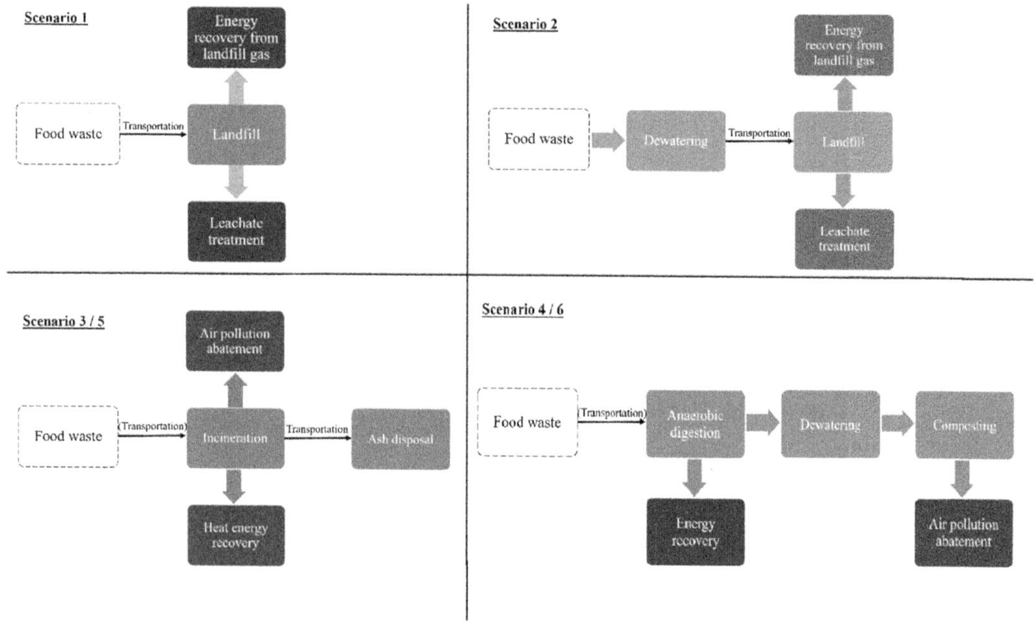

Figure 18.2
Food waste treatment scenarios.

scenarios (Fig. 18.2) were defined based on the existing treatment practice and the proposed treatment methods in Hong Kong. Scenario 1 (S1) is the direct landfill disposal of food waste, while scenario 2 (S2) includes the dewatering of food waste before landfill disposal. Scenario 3 (S3) adopts the centralized incineration of wastes with energy recovery, and the ash is disposed of at the landfill. Scenario 4 (S4) adopts centralized organic waste treatment processes including AD, dewatering and composting. Scenario 5 and 6 (S5 and S6) apply the same treatment processes as in S3 and S4, respectively, yet the food waste is treated by on-site infrastructures.

18.3.1.1.2 Step 2: life-cycle inventory analysis

A process based LCA approach has been adopted for the study in which the LCI was established based on the information of the inputs and outputs involved in the specific processes in each scenario. Table 18.2 lists the items included in the LCI for each scenario.

18.3.1.1.3 Step 3: life-cycle impact assessment

The LCIA was conducted using the ReCipe Endpoint method and the LCA software used was SimaPro 8.3 [46].For the estimation of transportation emissions, the distances of transportation were estimated by the measurement on the map. The EMission FACtors (EMFAC) model version 3.3 [47] is used to estimate the emissions from transportation.

Table 18.2: LCI items for the six scenarios.

Inventory items	S1	S2	S3/S5	S4/S6
Transportation	· From HKIA to landfill	· From HKIA to landfill	· From HKIA to incinerator[a] · From incinerator to landfill	· From HKIA to organic waste treatment facility[a]
Treatment/ disposal process	· Organic waste degradation in landfill	· FW dewatering · Organic waste degradation in landfill	· Incineration	· AD · Dewatering · Composting
Energy recovery	· Energy from landfill gas (LFG)	· Energy from LFG	· Energy from FW incineration	· Energy from biogas in AD
Air pollution control	N.A.	N.A.	· Activated carbon, selective non-catalytic reduction (SNCR) and scrubber	· Odor treatment unit · Air pollution control unit
Destination of byproduct	· Leachate treatment · LFG flaring	· Leachate treatment · LFG flaring	N.A.	· Biogas flaring
Destination of end-product	N.A.	N.A.	· Solidification of fly ash · Ash disposal in landfill	· Compost application on landscaping in facilities

[a]The transportation processes are excluded for on-site scenarios (S5 and S6).

The First Order Decay (FOD) model for solid waste disposal sites (SWDS) developed by the Intergovernmental Panel on Climate Change (IPCC) was used for estimation of landfill gas (LFG) produced by the landfilling of food waste [48]. The parameters used for calculating the CH_4 emissions from landfill disposal of food waste are listed in Table 18.3.

The environmental impacts originated from the FW dewatering process were considered to be the indirect emissions from the electricity consumption of the dewatering machines.

Table 18.3: Inputs for estimation of CH_4 from FW landfilling.

W (Gg FW/year)	1.15
DOC (Gg carbon/Gg FW)[a]	0.15
DOC_f[a]	0.5
MCF[a]	1
DDOCm (Gg carbon/year)	0.08625
k[a]	0.4

[a]Ref. [48].

Table 18.4: Inputs for LCIA on FW dewatering.

Treatment capacity[a] (kg/h)	294.84
Motor power[a] (kW)	2.61
Amount of food waste (kg/day)	3150.68
Operating hour (h/day)	10.69
Electricity consumption (kWh/year)	10,180.08
Electricity consumption (kWh/tonne FW)	8.85

[a]Ref. [48a].

The power and the operating duration of the dewatering machine were used to estimate the electricity consumption (Table 18.4).

Table 18.5 shows the input data for food waste incineration. The stack emissions, fuel consumption and ash production from incineration were based on the data inventory in the LCA study on food waste and sewage sludge treatment in Macau [49]. Information on most of the air emissions, fuel consumption, and ash production was based on the field survey in Macau [50]. The GHG emission data were based on the Intergovernmental Panel on Climate Change (IPCC) model, while emissions including volatile organic compounds (VOCs) and respirable suspended particulate (RSP) were based on the estimation in the previous literature [48,51]. The amount of electricity consumption and material requirements for the ash treatment using cement solidification and the air pollution control technologies, including activated carbon injection, selective noncatalytic reduction (SNCR) and scrubber were also covered in the LCI [50,52].

Table 18.5: Inputs for LCIA of FW incineration (per tonne FW).

SO_2 (kg)[a]	1.00×10^{-2}	Bottom ash (tonne)[a]	1.80×10^{-1}
HCl (kg)[a]	8.90×10^{-3}	Diesel consumption (MJ)[a]	1.57×10^{1}
NO_x (kg)[a]	2.40×10^{-1}	Gasoline consumption (MJ)[a]	1.86×10^{1}
NH_3 (kg)[a]	1.10×10^{-2}	Electricity consumption (kWh)[b]	8.65×10^{1}
CO (kg)[a]	3.00×10^{-2}	Iron (kg)[d]	2.60×10^{-1}
VOCs (kg)[a]	1.32×10^{-1}	Wastewater (m^3)[d]	8.50×10^{-2}
HF (kg)[a]	2.65×10^{-2}	Activated carbon (kg)[d]	2.10×10^{-1}
Dioxin and furans (kg)[a]	6.60×10^{-10}	Aqueous ammonia (kg)[d]	7.30×10^{-1}
PM_{10} (kg)[a]	1.98×10^{-1}	Slaked lime (kg)[d]	7.86
CH_4 (kg)[a]	2.00×10^{-4}	Cement (kg)[d]	1.51
N_2O (kg)[a]	5.00×10^{-2}	Electricity consumption for SNCR (kWh)[c]	1.80×10^{-1}
Fly ash (tonne)[a]	3.60×10^{-2}	Electricity recovered (kWh)[d]	7.07×10^{2}

[a]Ref. [49].
[b]Ref. [50].
[c]Ref. [52].
[d]Calculated based on the heating value of FW (21 MJ/dry kg FW), a moisture content of 79.8% of the FW in Hong Kong and the average combined heat and power (CHP) system efficiency of 60% [49,52a].

The emissions from the AD process, the combined heat and power (CHP) system, flaring system and the odor treatment unit were estimated mainly based on the Environmental Impact Assessment report for the OWTF in Hong Kong [53]. The leakage of CH_4 and CO_2, which are the major constituents of the biogas, from the anaerobic digestor was also considered, and the fraction of leakage was assumed to be 5% [48]. The energy recovery from AD was estimated based on the energy content of CH_4 and the generator efficiency [49]. The environmental impacts originated from dewatering were evaluated using the method introduced above. The GHG emissions from composting, such as CH_4 and N_2O, were included in the LCI. The environmental impacts of the air pollution control technique using biofilter were considered [53,54]. The input data for organic waste treatment processes are summarized in Table 18.6.

18.3.1.1.4 Step 4: life-cycle cost-benefit analysis

The economic costs and benefits, which are already in monetary terms, could be included in the CBA directly, while a valuation process should be conducted to convert the environmental and social performance into external costs and benefits for the inclusion in

Table 18.6: Inputs for LCIA of organic waste treatment processes for FW (per tonne FW).

AD	
SO_2 (kg)[a]	1.85E-02
HCl (kg)[a]	3.71E-03
NO_x (kg)[a]	1.11E-01
CO (kg)[a]	2.41E-01
VOCs (kg)[a]	3.05E-01
HF (kg)[a]	3.71E-04
PM_{10} (kg)[a]	8.85E-02
N_2O (kg)[a]	1.38E-02
CH_4 leakage (kg)	2.45E+00
CO_2 leakage (kg)	4.50E+00
Electricity recovery (kWh)	2.78E+02
Heat recovery (MJ)	1.23E+03
Dewatering	
Electricity consumption (kWh)[b]	4.43
Composting	
CH_4 (kg)[c]	1.83E+00
N_2O (kg)[c]	7.50E-02
NH_3 (kg)[c]	4.06E-01

[a]Refs. [49,53].
[b]Ref. [48a].
[c]Ref. [28].

the CBA. The total cost and the total benefit in the CBA are the summation of economic, environmental and social costs and benefits, respectively.

18.3.1.1.4.1 Economic costs and benefits The economic costs and benefits assessed include capital, operation, and transportation costs. The capital and the operation and maintenance (O&M) costs of the waste treatment and disposal facilities were referred to government documents and previous literature [55–57]. The operation costs of the dewatering facilities were estimated based on the labor and the electricity costs [58,59]. The economic benefit from energy recovery was estimated as the avoided costs for electricity.
The economic value of the compost produced from food waste recycling was included as economic benefit [60]. The transportation costs were estimated based on the traveling distances, diesel consumption and the diesel price [61]. The major inputs for analyzing the economic costs and benefits are a list in Table 18.7.

18.3.1.1.4.2 Environmental costs and benefits The linkage between LCA and CBA is the monetary valuation of the environmental impacts [62]. The information on the included emissions with their economic costs listed in Table 18.8 was extracted from previous local LCA studies on waste treatment strategies [63]. The avoided emissions from energy recovery in forms of electricity and heat were included as the environmental benefits which brought external economic savings to the scenarios.

Table 18.7: Inputs for economic costs and benefits of different processes.

Facilities	Items	Costs (HKD)
WENT landfill extension	Capital cost[a]	10.53 billion
	O&M per tonne of wastes[b]	237.4
	Waste charge[c]	400.0
Dewatering	Labor wage per hour[d]	32.5
	Electricity cost per kWh[e]	0.987
Incineration	Capital cost[f]	12.27 billion
	Annual O&M[f]	434.80 million
Organic waste treatment facilities	Capital cost[g]	1.25 billion
	Annual O&M[g]	78.31 million
	Compost value per tonne[h]	516.94
Transportation	Diesel price per liter[i]	10.96

[a]Ref. [61a].
[b]Ref. [57].
[c]Ref. [61b].
[d]Ref. [59].
[e]Ref. [58].
[f]Ref. [61c].
[g]Ref. [56].
[h]Ref. [60].
[i]Ref. [61].

Table 18.8: External environmental costs of air emissions [63].

Category	Air pollutant compound	HKD/kg emission compound[a]	HKD/kg emission compound (Present value in 2016)[b]
Waste transport	CO_2	0.10	0.11
	NO_x	38.51	41.65
	SO_2	61.95	67.00
	Respirable suspended particulate (RSP)	2040.00	2206.46
Urban pollution	CO_2	0.10	0.11
	NO_x	18.19	19.67
	SO_2	31.105	33.64
	RSP	397.37	429.79
	Total of nine heavy metals[c]	3370.00	3644.99
	Mercury	118,000.00	127,628.80
	Total cadmium and thallium	576.78	623.85
	Dioxins and furans	273,600,000.00	295,925,760,000.00

[a]Adopted from Ref. [63]. Base year for the values is 2014.
[b]Converted to present values in year 2016, using formula $F = P(1 + i)n$, where F denotes future value, P denotes present value, i denotes discount rate (4%), and n denotes number of years.
[c]The heavy metals include As, Co, Cr, Cu, Mn, Ni, Pb, Sb and V.

18.3.1.1.4.3 Social costs and benefits There are two major external social costs generated by the MSW facilities in Hong Kong, namely the opportunity cost of land and the disamenity cost [57]. To account for the significant opportunity cost of land utilization in densely populated cities, such as Hong Kong, the local premium cost of suburban land for recreational purposes was used in the estimation [63,64]. The disamenity cost was represented by the reduction of housing prices in the surrounding areas near the MSW management facilities [63].

18.3.1.2 Life-cycle cost-benefit analysis results

Fig. 18.3 shows the single score LCA results of the six scenarios. The incineration scenarios (S3 and S5) and AD scenarios (S4 and S6) had significant environmental merits due to energy recovery. The incineration scenarios presented the most favorable option, which recovered 82.13 kWh/tonne more energy and presented lower environmental impacts related to air pollution control systems than the AD system. However, the results revealed that the on- and off-site infrastructure had a similar environmental implication (S3 vs. S5 for incineration and S4 vs. S6 for AD) in the case of HKIA due to short transportation distance.

The LC-CBA results are summarized in Table 18.9. The on-site incineration scenario (S5) was the most sustainable food waste management approach in view of the lowest net cost of HKD 462/tonne FW, followed by landfill disposal after dewatering (S2; HKD 711/

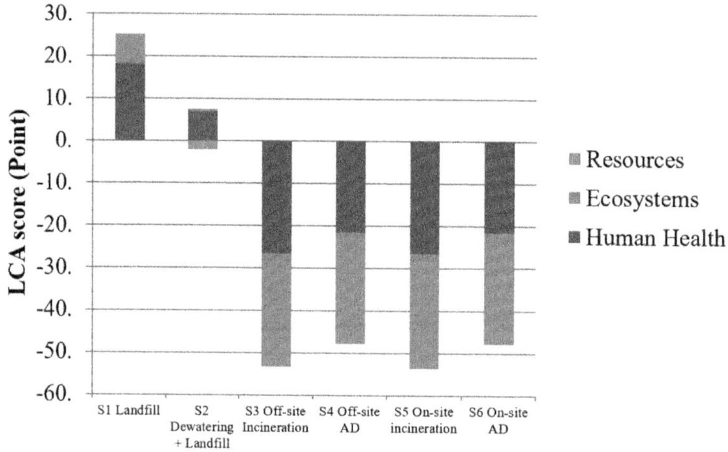

Figure 18.3
Life-cycle assessment results: Environmental impact of different food waste management scenarios (Pt/tonne).

tonne FW), and then off-site incineration (S3). Fig. 18.4 presents the costs and the benefits of the scenarios. The good performance of scenario S5 was attributed to the efficient energy recovery from FW incineration, avoidance of disamenity cost and the relatively low capital and O&M costs. S5 presented a total economic and environmental savings of HKD 1170/tonne from energy recovery. The on-site scenarios (S5 and S6) tended to be more sustainable than the off-site centralized treatments (S3 and S4) in the case of HKIA. Although the off-site incineration (S3) recovered a significant amount of energy, it incurred the high capital cost in building the artificial island in the adjacent

Table 18.9: Life-cycle cost-benefit analysis results.

HKD/tonne FW	S1	S2	S3	S4	S5	S6
Economic						
Capital cost	341.17	154.04	1032.06	1546.22	568.87	1306.13
O&M cost	256.77	180.91	453.00	1124.30	453.00	1124.30
Transportation cost	48.92	20.17	17.60	40.30	0.00	0.00
Waste charge	400.00	85.92	87.11	0	87.11	0
Energy recovery	−195.72	−195.72	−697.81	−611.82	−697.81	−611.82
Compost	0.00	0.00	0.00	−51.69	0.00	−51.69
Environmental						
Env. cost	420.55	219.48	454.14	118.70	451.61	118.70
Env. benefit	−131.81	−131.34	−470.96	−412.92	−470.96	−412.92
Social						
Disamenity cost	254.80	254.80	48.88	566.80	0.00	0.00
Land cost	297.73	122.75	29.77	120.43	69.91	120.43
Net cost	**1692.42**	**711.00**	**953.78**	**2440.32**	**461.73**	**1593.13**

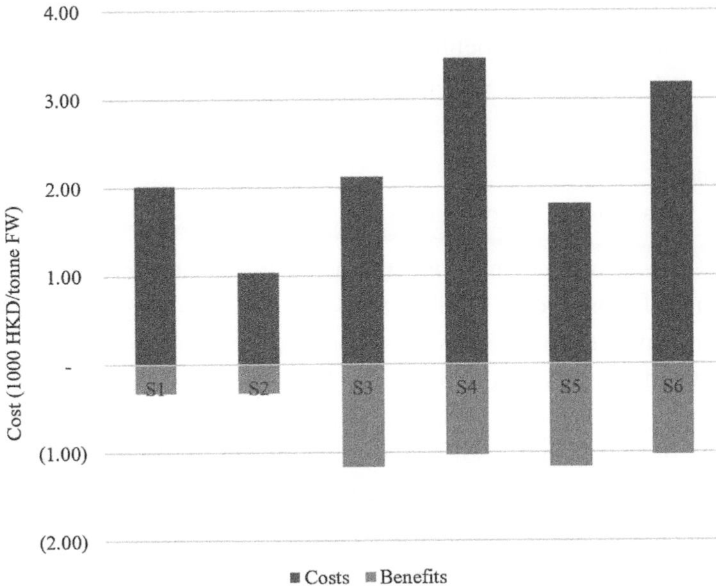

Figure 18.4
Life-cycle costs and benefits of the scenarios (HKD/tonne FW).

area to Shek Kwu Chau in Hong Kong, which could be avoided in the on-site scenario (S5) [56].

In this study, an integrated LC-CBA framework to assist decision-making on sustainable food waste management was developed and demonstrated in a case study as a successful and suitable tool for achieving sustainability. The contributions of this study include the development of the LC-CBA tool, with the following innovative features: (1) the inclusive coverage of the economic, environmental, and social costs and benefits originated from food waste management; (2) the clear and easily understood final indicator in monetary terms with the external environmental and social costs integrated; and (3) the wide applicability of the LC-CBA tool for sustainable decision-making on food waste management worldwide.

18.3.2 Life-cycle assessment on food waste valorization to value-added products

The second detailed case study is on the LCA on food waste valorization options to produce hydroxymethylfurfural (HMF). The valorisation of biomass to HMF has been extensively studied as it is a versatile platform chemical, which has been listed as one of the top 10 bio-based chemicals by the US Department of Energy [65–67]. In this study, food waste is chosen as a representative of waste biomass to produce HMF through catalytic conversion approaches [68,69], yet their environmental performances have not yet been evaluated. Therefore, the primary aim of this study is to develop an LCA framework to assess the environmental significance of various system components (solvents, catalysts, reaction temperature, reaction time, etc.) in the food waste-to-HMF process, by comparing the environmental impacts arising from eight laboratory-scale conversion systems, to inform decision-makers in long-term upscale of the food waste valorization systems.

18.3.2.1 Methodology

18.3.2.1.1 Step 1: goal and scope definition

The goal of this LCA was to assess and compare the environmental performance of eight experimental methods for producing HMF from food waste, which significantly differed in terms of the operating parameters (i.e., process inputs) and product yields (i.e., process outputs). The FU was defined as the conversion of 1.0 g of food waste substrates. The scope of this LCA covered eight scenarios of catalytic conversion of food waste-to-HMF, including processes of the use of solvent and cosolvents, the addition of catalysts, heating, and yielding of HMF. Water was used as the solvent in all scenarios, while various organic solvents were used as the cosolvents. Either tin (IV) chloride ($SnCl_4$) or aluminum chloride ($AlCl_3$) were used as the catalysts. Microwave reactor was used for the heating process. The environmental consequences related to such processes were included in the system boundary, which determines what processes and activities are included in the LCA (Fig. 18.5).

18.3.2.1.2 Life-cycle inventory analysis

The details of the conditions of the laboratory conversion process of food waste-to-HMF were organized and used to build the LCI (Table 18.10). The electricity consumption for heating the reaction mixture was calculated based on the power of the microwave reactor and the duration of heating using equation $E = P \times t$ (E denotes energy, P denotes power and t denotes time). The amounts of solvent, cosolvents, catalysts, and HMF were measured from the experiments.

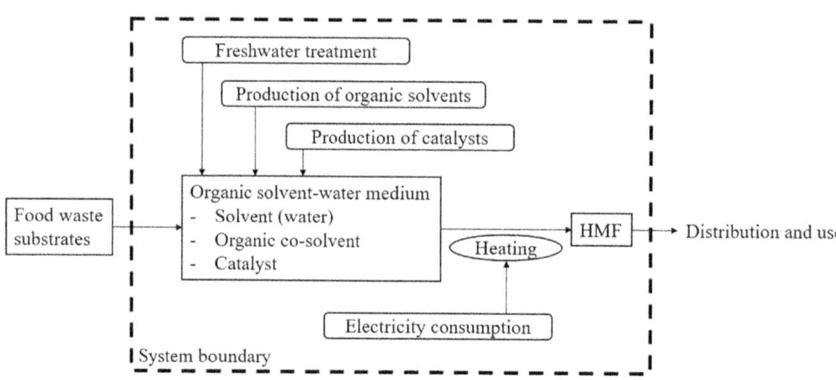

Figure 18.5
System boundary of food waste valorisation LCA.

Table 18.10: LCI of food waste valorization.

	Process input					Process output
	Food waste (1 g)	Solvent (10 mL)	Co-solvent (10 mL)	Catalyst	Electricity (Wh)	HMF yield (g)
S1	Bread waste	Water	DMSO	$SnCl_4$; 0.289 g	100.0	0.214
S2	Bread waste	Water	DMSO	$SnCl_4$; 0.289 g	50.0	0.126
S3	Bread waste	Water	DMSO	$SnCl_4$; 0.289 g	150.0	0.199
S4	Bread waste	Water	THF	$SnCl_4$; 0.289 g	400.0	0.109
S5	Bread waste	Water	Acetone	$SnCl_4$; 0.289 g	33.3	0.191
S6	Bread waste	Water	Acetone	$AlCl_3$; 0.148 g	100.0	0.203
S7	Rice waste	Water	DMSO	$SnCl_4$; 0.289 g	100.0	0.227
S8	Fruit waste	Water	DMSO	$SnCl_4$; 0.289 g	50.0	0.137

18.3.2.1.3 Life-cycle impact assessment

The ReCipe Endpoint method was adopted for conducting the LCIA. Eighteen midpoint indicators and three endpoint indicators are analyzed in the ReCipe Endpoint method. The software SimaPro 8.3.0.0, which is a widely recognized LCA tool, was used in this study. The solvents and electricity were produced through relatively common processes, thus the information on the associated environmental emissions is available in databases. Such information was adopted from the EcoInvent database in this study. The environmental emissions originated from the production process of $SnCl_4$ and $AlCl_3$ have not yet been documented in the databases, thus such emissions were estimated according to the method used by the EcoInvent for building life-cycle inventories of chemicals in order to ensure the consistency [70].

The conventional approach to produce HMF was neither documented in the previous LCA studies nor the databases. In this study, the reaction between sugar syrup (fructose) and sulfuric acid at a temperature of 166°C was assumed [71]. Based on such study, 2.55 g of sugar syrup, 0.07 g of sulfuric acid and 0.21 kWh of electricity are required for producing 1 g of HMF [71]. The environmental impacts of sugar production and sulfuric acid were obtained from the Agri-footprint [72,73] and the U.S. LCI [74] databases, respectively. The Hong Kong fuel mix for electricity generation was considered during the estimation of impacts associated with electricity consumption. The avoidance of starting materials (sugar syrup and sulfuric acid) usage and energy consumption for heating via food waste valorization were considered in the evaluation of environmental performance.

18.3.2.2 Life-cycle assessment results

The single score LCA results of the eight scenarios of food waste valorization are shown in Fig. 18.6. The environmental impacts are presented in milli-points (mPt), which reveal the overall impacts of the scenarios. The results indicate that S6 is the most environmentally friendly option, while S4 is the most polluting scenario.

The environmental impacts were categorized into the human health, Ecosystems and resources aspects. The impacts in resources aspect were the highest, contributing to 74%

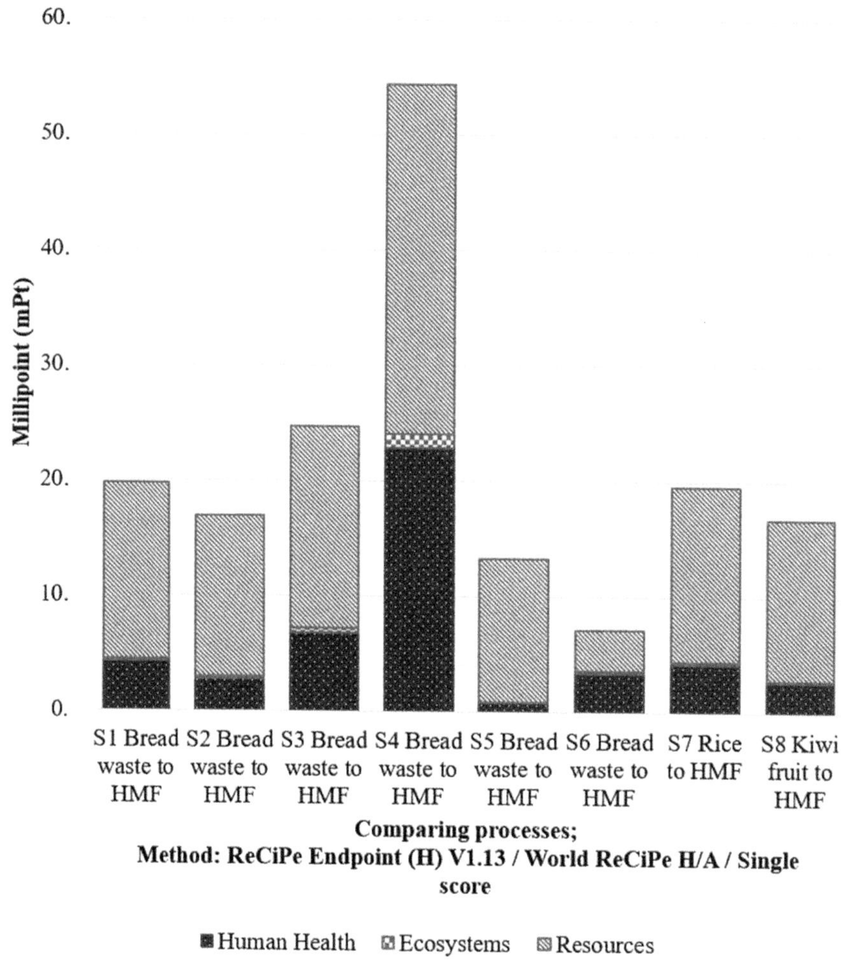

Figure 18.6
Single score LCA results.

to the overall impacts on average (Fig. 18.6). The high environmental stress on resources depletion is mainly attributed to the use of the relatively limited tin resources for producing the metal chloride catalyst. Human health impacts ranked after impacts on resources with an average contribution of 25% to the overall impacts. The production of organic solvents, especially THF, and the metal mining process caused adverse impacts to human health. As mentioned in Section 3.2, exposure to organic solvents through different routes has been revealed to cause human health threats, such as cancer and fatality. The excavation activity of tin mining and the disposal of tailings change the radionuclide compositions in soil, thus increasing the chance of radiological exposure of mine workers and nearby residents [75,76]. During bauxite mining for $AlCl_3$ catalyst production, the excavation activities release air pollutants, such as dust and fine particulate matters, that harm the respiratory and cardiovascular systems after inhalation [77]. Drinking water could also be polluted by the discharge of bauxite washing water. Chronic ingestion of metal-containing water may increase cancer risks. The environmental emissions from energy consumption and the production processes of organic solvents and metal catalysts caused eutrophication, toxicity, and climate change impacts to the ecosystem. The scenarios in this study had relatively low impacts on the Ecosystems aspect (0.27%–2.78% of the overall environmental impacts).

To investigate the significance of different system parameters (i.e., reaction temperature, reaction time, solvents, etc.) in determining the total environmental impacts, the single score LCA results are presented to illustrate the individual process contributions (Fig. 18.7). The processes involved in the conversion of food waste-to-HMF included the

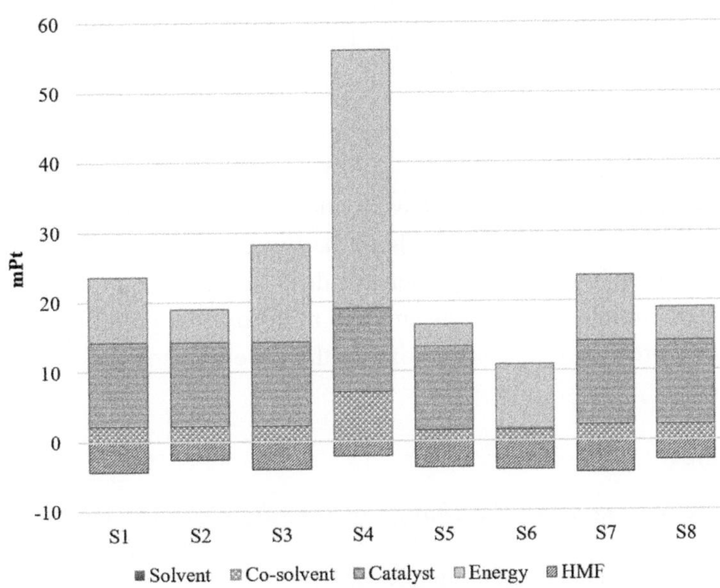

Figure 18.7
Process contributions to LCA results.

utilization of solvent, cosolvents, catalysts, energy, and the yield of HMF. The use of solvents, catalysts, and energy contributed to the adverse environmental impacts, while the production of HMF recovers untapped value from food waste to synthesize high-value products and presents an alternative to a petroleum refinery, thus providing environmental benefits that should be properly recognized and quantified. In all the scenarios, the use of water as a solvent only contributed to trivial environmental impacts (only accounted for 0.003% of the overall impacts on average). The water use was assumed to obtain from conventional potable water treatment methods, which presented significantly lower environmental impacts compared to other processes, such as the production of organic solvents and metal chlorides. It should be noted that in industrial applications, water solvent should come from the indigenous water content of food waste where additional water demand can be avoided or minimized.

To inform the decision-making on the selection of the best food waste valorization option, this study developed an LCA framework for evaluating the environmental performance of the food waste valorization scenarios by including the major processes of the utilization of solvent, cosolvents, catalysts, energy and the recovery of HMF. The LCA conducted in this study assessed eight scenarios of food waste valorization via catalytic conversion and concluded that S4 is the most polluting scenario while S6 is the most environmentally favorable option. The use of a less polluting catalyst ($AlCl_3$) and cosolvent (acetone), as well as the relatively high yield of HMF (27.9 Cmol%), provided S6 the superior environmental performance. Metal depletion impacts, which were attributed mainly to the production of metal chlorides catalyst, were the highest among the categories, followed by the toxicity impacts (marine ecotoxicity, freshwater toxicity, and human toxicity) which were contributed mostly by the production of organic cosolvents. The energy and $SnCl_4$ catalyst consumptions were the most dominant factors of the environmental impacts in most of the scenarios.

To keep the consistence of the framework while the detailed economic information about catalyst recycling was unavailable, only the environmental aspect was considered in this study. However, when the development of such valorization process become more mature, and the information is more readily available in the future, the inclusion of the economic aspect is expected, so that a more comprehensive decision-supporting tool could be developed. The LCA in this study acts as an early milestone for guiding the selection of the best valorization process, thus contributing to the development of the waste valorization systems.

18.4 Challenges

18.4.1 Use of LCA to address the change of paradigm in food waste management

Conventional food waste management approaches viewed food waste as a type of organic waste and mainly focused on proper disposal in landfills, energy recovery through AD or incineration, and material recovery through composting or production of animal feed. In recent years, more research studies have been conducted on food waste conversion to biobased products such as biodiesel and chemicals. Such paradigm shift from treating food waste as unwanted materials to considering it as a valuable feedstock for producing value-added products could be addressed by LCA for identifying key processes to improve environmental performance at the laboratory-testing stage or selecting the most favorable technologies for full-scale application.

18.4.2 Adaptation of LCA framework to emerging technologies

The emerging technologies of food waste valorization to value-added products involved different techniques, such as bacterial fermentation and catalytic conversion. The inputs and outputs related to such techniques should be fully covered for comprehensive LCA evaluations. The inclusion of material and energy flows could be achieved by the conventional procedure of setting up LCI in LCA. However, for more complicated biological or biochemical processes, such as bacterial fermentation and digestion, relevant models could be integrated into the LCA framework for more reliable modeling of the processes.

The outputs of the emerging valorization technologies are products which have added market values. Yet, the valorization processes are often costly. The economic performance is also an essential aspect to be considered for evaluating the overall sustainability of the options. Therefore, economic evaluation tools, such as CBA and life-cycle costing, are recommended for integration into the LCA framework for a more inclusive life-cycle sustainability framework in future study.

18.4.3 Standardization of food waste management LCA framework

Although the methodology and procedures of LCA has been standardized for more than 10 years, the flexibility of the LCA framework requires the practitioners to defines a number of elements, such as FU and scope of study, when conducting an LCA. For example, a diverse selection of functional units and LCIA models has been observed in the reviewed studies. Such differences hinder comparison of food waste management alternative evaluated in different research studies. The benchmarking of food waste handling practices among different service-provider, jurisdictions or countries would also be impeded. Standardization of food waste management LCA framework is recommended to facilitate benchmarking of handling options worldwide and comparing emerging technologies with convention approaches, which in turn favor the selection of sustainable food waste management strategies.

18.5 Conclusions and perspectives

LCA application on food waste management and recycling has been developed rapidly in the past 15 years. Published research studies reviewed in this chapter have evidenced the suitability of using LCA as a decision-supporting tool for guiding food waste management toward sustainability. The flexibility of LCA allows it to be adjusted to cover the environmental impacts and benefits of the changing food waste management approaches, from waste treatment to recycling via conversion and valorization processes. Observing the paradigm shift in food waste management, integration of relevant biological or biochemical models, as well as life-cycle costing models, with LCA is recommended for more comprehensive sustainability evaluation. To enhance comparability, standardization of LCA framework is also required to improve consistence between LCA studies on food waste management.

References

[1] Gustavsson J, Cederberg C, Sonesson U. Global food losses and food waste: extent, causes and prevention; study conducted for the International Congress Save Food! at Interpack 2011, [16–17 May], Düsseldorf, Germany. Rome: Food and Agriculture Organization of the United Nations; 2011.
[2] FAO. Food wastage footprint – impacts on natural resources – summary report. 2013.
[3] Guinée J, editor. Handbook on life cycle assessment: operational guide to the ISO standards. Springer Netherlands; 2002.
[4] Gonzalez-Garcia S, Gullón B, Moreira MT. Environmental assessment of biorefinery processes for the valorization of lignocellulosic wastes into oligosaccharides. Journal of Cleaner Production 2018;172:4066–73.
[5] Pennington DW, et al. Life cycle assessment part 2: current impact assessment practice. Environment International 2004;30(5):721–39.
[6] Hauschild M, Rosenbaum RK, Olsen S, editors. Life cycle assessment: theory and practice. Springer International Publishing; 2018.
[7] Pajula T, Behm K, Vatanen S, Saarivuori E. Managing the life cycle to reduce environmental impacts. In: Grösser SN, Reyes-Lecuona A, Granholm G, editors. Dynamics of long-life assets: from technology adaptation to upgrading the business model. Cham: Springer International Publishing; 2017. p. 93–113.
[8] Yang Y, Campbell JE. Improving attributional life cycle assessment for decision support: the case of local food in sustainable design. Journal of Cleaner Production 2017;145:361–6.
[9] ISO 14040. Environmental management – life cycle assessment – principles and framework: international standard 14040. Geneva: International Standards Organisation; 2006.
[10] ISO 14044. Environmental management – life cycle assessment – requirements and guidelines. Geneva: International Standards Organisation; 2006.
[11] Bruijn H, Duin R, Huijbregts MAJ. Handbook on life cycle assessment. 2002.
[12] Rebitzer G, et al. Life cycle assessment part 1: framework, goal and scope definition, inventory analysis, and applications. Environment International 2004;30(5):701–20.
[13] Morris J. Recycling versus incineration: an energy conservation analysis. Journal of Hazardous Materials 1996;47(1):277–93.
[14] Dalemo M, et al. ORWARE – a simulation model for organic waste handling systems. Part 1: model description. Resources, Conservation & Recycling 1997;21(1):17–37.
[15] Harrison KW, Dumas RD, Barlaz MA, Nishtala SR. A life-cycle inventory model of municipal solid waste combustion. Journal of the Air & Waste Management Association 2000;50(6):993–1003.

[16] Powell J. The potential for using life cycle inventory analysis in local authority waste management decision making. Journal of Environmental Planning and Management 2000;43(3):351–67.

[17] Ohlsson T. Food waste management by life cycle assessment of the food chain. Journal of Food Science 2004;69(3):CRH107–9.

[18] Lundie S, Peters GM. Life cycle assessment of food waste management options. Journal of Cleaner Production 2005;13(3):275–86.

[19] Ogino A, et al. Environmental impact evaluation of feeds prepared from food residues using life cycle assessment. Journal of Environmental Quality; Madison 2007;36(4):1061–8.

[20] Lee S-H, Choi K-I, Osako M, Dong J-I. Evaluation of environmental burdens caused by changes of food waste management systems in Seoul, Korea. The Science of the Total Environment 2007;387(1):42–53.

[21] Khoo HH, Lim TZ, Tan RBH. Food waste conversion options in Singapore: environmental impacts based on an LCA perspective. The Science of the Total Environment 2010;408(6):1367–73.

[22] Hauschild M, Potting J. Spatial differentiation in life cycle impact assessment — the EDIP 2003 methodology. Institute for Product Development, Technical University of Denmark; 2003.

[23] Kim M-H, Kim J-W. Comparison through a LCA evaluation analysis of food waste disposal options from the perspective of global warming and resource recovery. The Science of the Total Environment 2010;408(19):3998–4006.

[24] Bernstad A, la Cour Jansen J. A life cycle approach to the management of household food waste – a Swedish full-scale case study. Waste Management 2011;31(8):1879–96.

[25] DTU. EASEWASTE user manual. Technical University of Denmark; 2012.

[26] Nakakubo T, Tokai A, Ohno K. Comparative assessment of technological systems for recycling sludge and food waste aimed at greenhouse gas emissions reduction and phosphorus recovery. Journal of Cleaner Production 2012;32:157–72.

[27] Kim MH, Song HB, Song Y, Jeong IT, Kim JW. Evaluation of food waste disposal options in terms of global warming and energy recovery: Korea. The International Journal of Energy and Environmental Engineering 2013;4(1):1.

[28] Saer A, Lansing S, Davitt NH, Graves RE. Life cycle assessment of a food waste composting system: environmental impact hotspots. Journal of Cleaner Production 2013;52:234–44.

[29] Bare J. Developing a consistent decision-making framework by using the U.S. EPA's TRACI. US EPA; 2002.

[30] Zhao Y, Deng W. Environmental impacts of different food waste resource technologies and the effects of energy mix. Resources, Conservation and Recycling 2014;92:214–21.

[31] Eriksson M, Strid I, Hansson P-A. Carbon footprint of food waste management options in the waste hierarchy – a Swedish case study. Journal of Cleaner Production 2015;93:115–25.

[32] Xu C, Shi W, Hong J, Zhang F, Chen W. Life cycle assessment of food waste-based biogas generation. Renewable and Sustainable Energy Reviews 2015;49:169–77.

[33] Morris J, Brown S, Cotton M, Matthews HS. Life-cycle assessment harmonization and soil science ranking results on food-waste management methods. Environmental Science & Technology 2017;51(10):5360–7.

[34] Brandao M, Heath G, Cooper J. What can meta-analyses tell us about the reliability of life cycle assessment for decision support? Journal of Industrial Ecology 2012;16(S1):S3–7.

[35] Heath GA, Mann MK. Background and reflections on the life cycle assessment harmonization project. Journal of Industrial Ecology 2012;16(S1):S8–11.

[36] Zumsteg JM, Cooper JS, Noon MS. Systematic review checklist. Journal of Industrial Ecology 2012;16(Suppl. 1):S12–21.

[37] Edwards J, Othman M, Crossin E, Burn S. Life cycle assessment to compare the environmental impact of seven contemporary food waste management systems. Bioresource Technology 2018;248:156–73.

[38] Vandermeersch T, Alvarenga RAF, Ragaert P, Dewulf J. Environmental sustainability assessment of food waste valorization options. Resources, Conservation & Recycling 2014;87:57–64.

[39] Alvarenga RAF, Dewulf J, Van Langenhove H, Huijbregts MAJ. Exergy-based accounting for land as a natural resource in life cycle assessment. International Journal of Life Cycle Assessment 2013;18(5):939–47.

[40] Lam C-M, Yu IKM, Medel F, Tsang DCW, Hsu S-C, Poon CS. Life-cycle cost-benefit analysis on sustainable food waste management: the case of Hong Kong International Airport. Journal of Cleaner Production 2018;187:751–62.

[41] Cristóbal J, Limleamthong P, Manfredi S, Guillén-Gosálbez G. Methodology for combined use of data envelopment analysis and life cycle assessment applied to food waste management. Journal of Cleaner Production 2016;135:158–68.

[42] Manfredi S, et al. Improving sustainability and circularity of European food waste management with a life cycle approach. Publications Office of the European Union; 2015.

[43] Salomone R, Saija G, Mondello G, Giannetto A, Fasulo S, Savastano D. Environmental impact of food waste bioconversion by insects: application of life cycle assessment to process using Hermetia illucens. Journal of Cleaner Production 2017;140:890–905.

[44] Brunklaus B, Rex E, Carlsson E, Berlin J. The future of Swedish food waste: an environmental assessment of existing and prospective valorization techniques. Journal of Cleaner Production 2018;202:1–10.

[45] Lam C-M, Yu IKM, Hsu S-C, Tsang DCW. Life-cycle assessment on food waste valorisation to value-added products. Journal of Cleaner Production 2018;199:840–8.

[46] PRé Sustainability. About SimaPro. 2017 [Online]. Available from: https://simapro.com/about/.

[47] HK EPD. EMFAC-HK user's guide: calculating emission inventories for vehicles in Hong Kong. Hong Kong SAR Government; 2017.

[48] IPCC. 2006 IPCC guidelines for national greenhouse gas inventories (vol. 5) – waste. 2006.

[48a] Vincent Corporation. "Vincent KP Screw Press." Accessed July 22, 2016. http://www.vincentcorp.com/sites/all/files/KP_Screw_Press_1M-09_15.pdf.

[49] Chiu SLH, Lo IMC, Woon KS, Yan DYS. Life cycle assessment of waste treatment strategy for sewage sludge and food waste in Macau: perspectives on environmental and energy production performance. International Journal of Life Cycle Assessment 2016;21(2):176–89.

[50] Song Q, Wang Z, Li J. Environmental performance of municipal solid waste strategies based on LCA method: a case study of Macau. Journal of Cleaner Production 2013;57:92–100.

[51] Woon KS, Lo IMC. Analyzing environmental hotspots of proposed landfill extension and advanced incineration facility in Hong Kong using life cycle assessment. Journal of Cleaner Production 2014;75:64–74.

[52] Møller J, Munk B, Crillesen K, Christensen TH. Life cycle assessment of selective non-catalytic reduction (SNCR) of nitrous oxides in a full-scale municipal solid waste incinerator. Waste Management 2011;31(6):1184–93.

[52a] US EPA, OAR. "Methods for Calculating CHP Efficiency.". Reports and Assessments. US EPA. August 23, 2015. https://www.epa.gov/chp/methods-calculating-chp-efficiency.

[53] HK EPD. AEIAR-180/2013 – development of organic waste treatment facilities, phase 2. 2013. HKSAR.

[54] Bindra N, Dubey B, Dutta A. Technological and life cycle assessment of organics processing odour control technologies. The Science of the Total Environment 2015;527:401–12.

[55] HK LegCo. HEAD 705 – CIVIL ENGINEERING environmental protection – refuse disposal 172DR –organic waste treatment facilities phase 1. April 08, 2014. Public Works Subcommittee of Finance Committee, Legislative Council.

[56] HK LegCo. Head 705 – CIVIL ENGINEERING environmental protection – refuse disposal 177DR – development of integrated waste management facilities phase 1. Public Works Subcommittee of Finance Committee, Legislative Council; April 16, 2014.

[57] Woon KS, Zhou SW. A life cycle eco-efficiency analysis of the proposed landfill extension and advanced incineration facility in Hong Kong. 2015. Hong Kong.

[58] CLP. Electricity tariff. China Light and Power; 2016.

[59] HK LD. "Statutory minimum wage," labour department – employees' rights & benefits. 2017 [Online]. Available from: http://www.labour.gov.hk/eng/news/mwo.htm.

[60] Chen Y-T. A cost analysis of food waste composting in Taiwan. Sustainability 2016;8(11):1210.

[61] Shell. "Price Board," Price Board – Shell Hong Kong and Macau. 2016 [Online]. Available from: http://www.shell.com.hk/en_hk/motorists/shell-fuels/price-board.html#vanity-aHR0cDovL3d3dy5zaGVsbC5jb20uaGsvZW4vcHJvZHVjdHMtc2VydmljZXMvb24td GhlLXJvYWQvZnVlbHMvcHJpY2UtYm9hcmQuaHRtbA.

[61a] HK LegCo. "Environmental Infrastructure Projects." The Legislative Council Panel on Environmental Affairs. May 27, 2013. http://www.legco.gov.hk/yr12-13/english/panels/ea/papers/ea0527cb1-1079-1-e.pdf.

[61b] HK ACE. Municipal Solid Waste Charging: Latest Development. Advisory Council on the Environment; 2015. http://www.epd.gov.hk/epd/sites/default/files/epd/english/boards/advisory_council/files/ACE_Paper_6_2015.pdf.

[61c] HK LegCo. "Head 705 – CIVIL ENGINEERING Environmental Protection – Refuse Disposal 177DR – Development of Integrated Waste Management Facilities Phase 1. Public Works Subcommittee of Finance Committee, Legislative Council. April 16, 2014. http://www.legco.gov.hk/yr13-14/english/fc/pwsc/papers/p14-07e.pdf.

[62] Powell JC, Pearce DW, Brisson I. Valuation for life cycle assessment of waste management options. CSERGE, School of Environmental Sciences, University of East Anglia; 1995.

[63] Woon KS, Lo IMC. An integrated life cycle costing and human health impact analysis of municipal solid waste management options in Hong Kong using modified eco-efficiency indicator. Resources, Conservation and Recycling 2016;107:104–14.

[64] HK LegCo. Background brief on Hong Kong disneyland: LC paper No. CB(1)1063/04-05. 2005. HKSAR.

[65] Rout PK, Nannaware AD, Prakash O, Kalra A, Rajasekharan R. Synthesis of hydroxymethylfurfural from cellulose using green processes: a promising biochemical and biofuel feedstock. Chemical Engineering Science 2016;142:318–46.

[66] Chen SS, Maneerung T, Tsang DCW, Ok YS, Wang C-H. Valorization of biomass to hydroxymethylfurfural, levulinic acid, and fatty acid methyl ester by heterogeneous catalysts. Chemical Engineering Journal 2017;328:246–73.

[67] Yu IKM, Tsang DCW. Conversion of biomass to hydroxymethylfurfural: a review of catalytic systems and underlying mechanisms. Bioresource Technology 2017;238:716–32.

[68] Yu IKM, et al. Polar aprotic solvent-water mixture as the medium for catalytic production of hydroxymethylfurfural (HMF) from bread waste. Bioresource Technology 2017;245:456–62.

[69] Yu IKM, Tsang DCW, Yip ACK, Chen SS, Ok YS, Poon CS. Valorization of food waste into hydroxymethylfurfural: dual role of metal ions in successive conversion steps. Bioresource Technology 2016;219:338–47.

[70] Althaus H-J, et al. Life cycle inventories of chemicals, vol. 8; 2007. Dubendorf.

[71] Fachri BA, Abdilla RM, van de Bovenkamp HH, Rasrendra CB, Heeres HJ. Experimental and kinetic modeling studies on the sulfuric acid catalyzed conversion of D-fructose to 5-hydroxymethylfurfural and levulinic acid in water. ACS Sustainable Chemistry & Engineering 2015;3(12):3024–34.

[72] Blonk Consultants. Agri-footprint 2.0 – part 1: methodology and basic principles. Blonk Agri-footprint BV; 2015.

[73] Blonk Consultants. Agri-footprint 2.0 – part 2: description of data. Blonk Agri-footprint BV; 2015.

[74] Franklin Associates. Cradle-to-gate life cycle inventory of nine plastic resins and four polyurethane precursors. 2011.

[75] Arogunjo AM, et al. Uranium and thorium in soils, mineral sands, water and food samples in a tin mining area in Nigeria with elevated activity. Journal of Environmental Radioactivity 2009;100(3):232–40.

[76] Aliyu AS, Mousseau TA, Ramli AT, Bununu YA. Radioecological impacts of tin mining. Ambio 2015;44(8):778–87.

[77] Abdullah NH, Mohamed N, Sulaiman LH, Zakaria TA, Rahim DA. Potential health impacts of bauxite mining in Kuantan. Malaysian Journal of Medical Sciences 2016;23(3):1–8.

CHAPTER 19

Determining key issues in life-cycle assessment of waste biorefineries

Homa Hosseinzadeh-Bandbafha[1], Meisam Tabatabaei[2,3,4,5], Mortaza Aghbashlo[1], Mohammad Rehan[6], Abdul-Sattar Nizami[6]

[1]Department of Mechanical Engineering of Agricultural Machinery, Faculty of Agricultural Engineering and Technology, College of Agriculture and Natural Resources, University of Tehran, Karaj, Alborz, Iran; [2]Faculty of Plantation and Agrotechnology, Universiti Teknologi MARA (UiTM), Shah Alam, Selangor, Malaysia; [3]Microbial Biotechnology Department, Agricultural Biotechnology Research Institute of Iran (ABRII), Agricultural Research, Education, and Extension Organization (AREEO), Karaj, Alborz, Iran; [4]Biofuel Research Team (BRTeam), Karaj, Alborz, Iran; [5]Faculty of Mechanical Engineering, Ho Chi Minh City University of Transport, Ho Chi Minh City, Vietnam; [6]Center of Excellence in Environmental Studies (CEES), King Abdulaziz University, Jeddah, Makkah Province, Saudi Arabia

19.1 Introduction

The exponential increase in the human population as well as the technological advances observed since the industrial revolution have collectively led to increased energy consumption. This has, in turn, posed tremendous pressure on the environment, including greenhouse gas (GHG) emissions, global warming (GW), and climate change [1,2]. Therefore, a large number of researchers have investigated the connection between urbanization, energy consumption, and GHG emissions from various perspectives striving to offer solutions to address these challenges [1]. These studies have highlighted urbanization as one of the major factors leading to increased energy consumption. Moreover, since 99% of the total energy consumption is currently dependent on fossil-oriented energy carriers, urbanization could be regarded as one of the main sources of GHG emissions [1,3–5]. In light of that, increasing energy efficiency, implementing energy saving projects, sustainable management of urban resources, and outsourcing energy infrastructure have been offered as solutions to address the challenges associated with urbanization and its unfavorable impacts on energy consumption and GHG emissions [5].

In addition to energy-related issues and concerns, the amount of waste produced globally is on the rise endangering both human health (public and occupational health) and

Waste Biorefinery. https://doi.org/10.1016/B978-0-12-818228-4.00019-8

ecosystem quality [6,7]. This necessitates the implementation of efficient, and sustainable waste management strategies to minimize such growing detrimental impacts [8]. It should be noted that appropriate handling of waste is not only critical from a hygienic perspective but also from the economic and social point of view considering the high potential of wastes as a cheap and affordable feedstock for bioenergy and value-added bioproducts. For instance, Table 19.1 presents the energy content and electricity production potential of the different fractions typically found in the municipal solid waste (MSW).

Various techniques have been explored for energy recovery from waste feedstock through modern waste management platforms [10]. It is worthy to mention that extraction of multiple products from waste biomass could increase the economic viability and profit margin of these systems. In this context, concepts such as "waste biorefinery" (analogous to conventional oil refinery; Fig. 19.1) has created stretching aspirations toward integrating various conversion technologies in the field of waste management with an ambition to generate a spectrum of different types of bioproducts like fuels, energy and chemicals [12–14]. The implementation of such a closed-loop bioprocesses cascade could enable an invaluable transformation in the direction of achieving a circular and low-carbon bioeconomy [15]. For instance, Table 19.2 highlights such an economical evaluation of different biorefinery technologies. Overall, waste biorefineries encompassing a number of favorable attributes including high-energy efficiency, zero-discharge nature, as well as low-carbon and water footprints on one hand and higher economic viability on the other could serve as promising platforms for simultaneous waste management and renewable energies/materials production [18–20].

In spite of the advantageous features of waste biorefineries, it is still essential to scrutinize both direct and indirect environmental, economic and social impacts of these systems versus existing waste treatment options, using holistic and systematic methods like life-cycle assessment (LCA) [21]. LCA is an efficient approach to analyze the environmental

Table 19.1: Comparison of different municipal solid waste (MSW) fractions for their energy content and electricity production potential.

Source	Energy content (Btu/lb)	kWh/kg
Food	2,400	1.55
Glass	0.00	0.00
Paper	6,800	4.39
Plastic	14,000	9.05
Textiles	8,100	5.20
Wood	7,300	4.73
Others	5,200	3.36

From Ouda OKM, Cekirge HM, Raza SAR. An assessment of the potential contribution from waste-to-energy facilities to electricity demand in Saudi Arabia. Energy Conversion and Management 2013;75:402–6. With permission from Elsevier. Copyright© 2013.

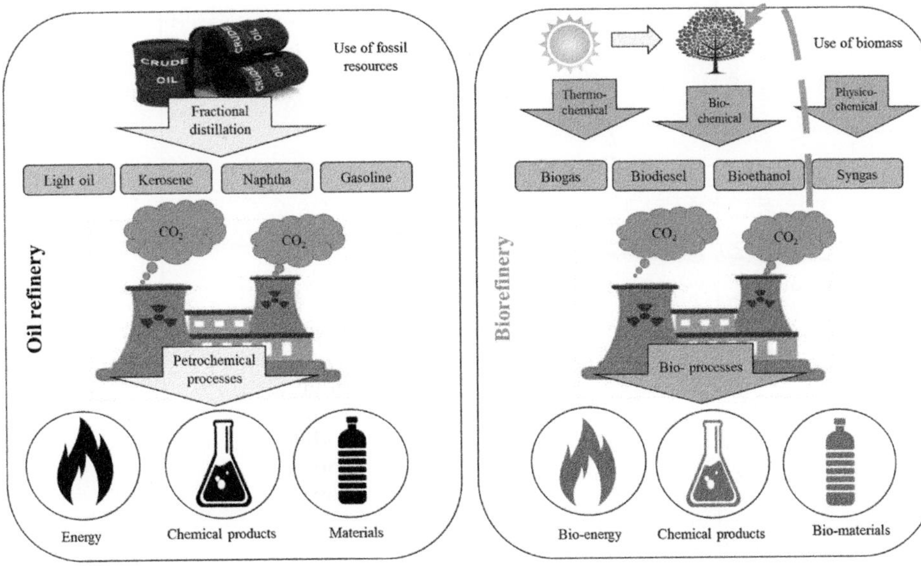

Figure 19.1

Comparison of the biorefinery and the oil refinery frameworks for their production chains. *Adopted from Hülsey MJ. Shell biorefinery: a comprehensive introduction. Green Energy & Environment 2018;3:318–27.*

effects and economic and sustainability aspects of a product, process, and service from the cradle to the grave. Accordingly, these findings can be used practically in decision-making processes through life-cycle thinking (LCT). In fact, an overall, holistic assessment of a system could be defined as LCT. In better words, LCT goes beyond the traditional focus on systems by including the environmental, social, and economic impacts of the (production/service) system under investigation over its entire life-cycle from raw material extraction, processing and production, distribution and transportation, consumption, recycling, to disposal. It should be noted that LCT is a philosophical approach while LCA, life-cycle management (LCM), life-cycle costing (LCC), etc. are in fact scientific approaches enabling such line of thinking.

Considering that, the present chapter is aimed at presenting and discussing the main principles of waste biorefineries and the role of LCA in determining the key issues associated with waste biorefineries. Subsequently, guidelines for efficient use of LCA in investigating waste biorefinery frameworks are also provided.

19.2 Biorefinery: definition and perspectives

According to the international energy agency (IEA), "biorefining is the sustainable synergetic processing of biomass into a spectrum of marketable food and feed ingredients,

Table 19.2: Comparison of biorefinery technologies from an economic viewpoint [16,17].

Biorefinery technologies	Suitable waste type	Annual capital cost per ton waste	Net operational cost per ton waste	Daily power generation (MW per ton waste)	Annual capital cost of daily power generation (per MW)	Net operational cost of daily power generation (per MW)	Total cost of daily power generation (per MW)
Combustion (incineration)	General waste stream	14.5–22 US $	1.5–2.5 US $	0.01–0.02[a]	710–2200 US $	75–250 US $	782–2450 US $
Pyrolysis	Organic and inorganic waste	17–25 US $	2–3 US $	0.01–0.014[a]	1214–2500 US $	142.8–300 US $	1356.8–2800 US $
Gasification	Organic and inorganic waste	19.5–30 US $	2.5–4 US $	0.04–0.045[a]	433.5–666.67 US $	55.55–100 US $	489.05–766.67 US $
Refuse derived fuel	General waste stream	7.5–11.3[b] US $	0.30–0.55[b] US $	0.01–0.014[a]	535.7–1130 US $	21.42–55 US $	557.12–1185 US $
Anaerobic digestion (AD)	Organic waste	0.1–0.14 US $	Minimal	0.015–0.02[c]	5–9.33 US $	Minimal	5–9.33 US $
Natural gas combined cycle							*1608 US $*[d]
Coal convent'l							*2402 US $*[d]

[a]Daily power production.
[b]Refuse derived fuel (RDF) pellet production cost.
[c]Power production spread over the lifespan of the biomethanation plant.
[d]For 24 h daily.

products (chemicals, materials) and energy (fuels, power, heat)" [22]. Therefore, a biorefinery could be translated to a processing unit or a group of processing units integrating various stages (upstream, midstream, and downstream) of biomass valorization, as well as to biomass processing in general. For example, the generation of renewable biological feedstocks/wastes and their subsequent valorization into a wide spectrum of high-value products [23–25].

Biorefineries can be divided into the first- or second-generation biorefineries depending on the type of waste/biomass used as raw materials [26]:

(1) first-generation (1G) biorefineries are based on conventional agricultural commodity crops (food materials) like corn starch and edible vegetable oil.
(2) second-generation (2G) biorefineries are based on nonedible feedstocks that are principally rich in lignocelluloses, like agricultural residues, energy crops, and woody materials.

In addition to these, third-generation (3G) biorefineries have also been introduced which use algal biomass as feedstock [27].

Despite the existing conflicts over the allocation of food crops to the renewable energy generation sector, the most common type of biorefineries currently in use is 1G [28]. This is ascribed to the fact that there are still many technical or economic barriers faced hindering the application of 2G and 3G biorefineries. For example, efforts to initiate commercial production of cellulosic ethanol date back to 2014 [29], whereas ethanol biorefineries are processing US corn and Brazilian sugarcane presently contributes 58% and 25% of global ethanol production, respectively [30]. From a different perspective, crops are not available all year round and hence, access to noncrop feedstock ensuring year-round supplies would be critical to the sustainability of biorefineries [31–33]. In better words, the use of grain resources in biorefineries does not seem logical in the long run. It should also be noted that the feedstock cost contributes to about 40%–60% of the total operating costs of a typical biorefinery [34] and this further highlights the challenging task of choosing the appropriate feedstock for biorefineries. As mentioned earlier, waste production globally is on the rise, and hence, channeling these widely available and economically feasible resources into biorefineries would bring along considerable environmental, economic, and social benefits. Examples of such feedstocks include food waste, green waste biomass, plastics, paper, rubber, metal, wood, etc.

19.2.1 Biorefinery feedstock (residues/wastes)

19.2.1.1 Lignocellulosic materials

Biorefineries for bioethanol production, which are already in use in many countries to produce bioethanol as the potential substitute for gasoline, use three groups of feedstocks including lignocellulosic biomass, starchy crops, sugar crops and byproducts of sugar industries. Lignocellulosic biomass is considered comparatively advantageous as this materials do not endanger food security and could be obtained at stable and low prices, and contain high carbohydrates contents [35]. In addition to these, in comparison with the 1G ethanol production, the produced bioethanol leads to lower net GHG emissions and as a result could reduce the environmental burdens associated with ethanol production [36]. Complete recycling of lignocellulosic wastes could take place in the 2G biorefinery platform, where through integrated and sustainable processes, bioenergy (e.g., bioethanol) and bioproducts (e.g., paper) could be produced [37].

Lignocellulosic materials are composed of carbohydrates such as cellulose (38%–50%), hemicelluloses (23%–32%) and other extraneous components such as lignin (15%–25%) as well as trace amounts of proteins and inorganic substances, which are intensely intermeshed and chemically bound through "covalent" or "noncovalent" forces [38,39]. Cellulose is physically associated with hemicellulose, while both physically and chemically associated with lignin [40].

Lignocellulosic biomass used in bioethanol biorefineries could be classified into three major categories depending on the source of waste [41]:

(1) Woody waste biomass,
(2) Agricultural residues (barley, wheat, and rice straws as well as sugarcane, bagasse, corn stover, etc.),
(3) Various types of cellulosic wastes (lumber mill wastes, MSW, and pulp mill waste).

More details on the composition of these resources are grouped in Table 19.3.

It should be noted that since lignocelluloses are in fact the structural materials in plants, they are inherently recalcitrant against enzymatic attacks. Therefore, to achieve a successful enzymatic saccharification, implementation of a single or a combination of pretreatment methods is required. More specifically, pre-treatments have been proved to improve the bioconversion process of lignocellulose biomass into useful small molecules by disrupting the naturally resistant lignin shield and by reducing the crystallinity index of cellulose [42,43]. Different pretreatment methods used for lignocellulosic biomass are presented in Fig. 19.2.

19.2.1.2 Oils and fats

As mentioned earlier, biorefineries are analogous to current oil refineries, while biodiesel is produced from bio-oils and bio-fats instead of fossil oil [45,46]. Despite the advantageous characteristics of biodiesel over conventional diesel fuel like indigenous availability, higher cetane number, renewability, lower aromatic and sulfur contents, higher efficiency, more favorable safety features, and better emission profile [47,48], the high

Table 19.3: Compositions of the selected lignocellulosic biomass (dry-base percentage).

Type of feedstock	Cellulose	Hemicellulose	Lignin
Bagasse	41	23	18
Bamboo	26–43	15–26	21–31
Banana waste	13	15	14
Barley straw	32	26	23
Coffee pulp	35	46	19
Corn cobs	45	35	15
Corn stalks	43	24	17
Corn stover	40	22	18
Grasses	25–40	25–50	10–30
Hardwood bark	22–40	20–38	30–55
Hardwood stem	40–50	24–40	18–25
Leaves	15–20	80–85	–
Millet husk	33	27	14
Newspaper waste	40–55	25–40	18–30
Nutshells	25–30	25–30	30–40
Pinewood	39	24	20
Poplar wood	35	17	26
Rice husk	31	24	14
Rice straw	37	23	14
Rye straw	33–35	27–30	16–19
Ryegrass (early leaf)	21	16	3
Ryegrass (seed setting)	27	26	7
Softwood stem	45–50	25–35	25–35
Solid cattle manure	1.6–4.7	28	–
Sorted plant refuse	60	20	20
Sweet sorghum bagasse	45	25	18
Swine waste	6	28	–
Switchgrass	45	31	12
Waste papers from chemical pulps	60–70	10–20	5–10
Wheat straw	39	24	16

From Parajuli R, Dalgaard T, Jørgensen U, Adamsen APS, Knudsen MT, Birkved M, Gylling M, Schjørring JK. Biorefining in the prevailing energy and materials crisis: a review of sustainable pathways for biorefinery value chains and sustainability assessment methodologies. Renewable and Sustainable Energy Reviews 2015;43:244–63. With permission from Elsevier. Copyright© 2015.

production cost of biodiesel is one of its drawbacks, hindering its widespread applications. According to Balat [49], the cost of raw materials accounts for 70%–95% of the total cost of biodiesel production and negatively impact the economic competitiveness of biodiesel production from food and crops oils when compared with conventional diesel fuel. Therefore, it is necessary to explore new and economically feasible oil feedstocks for biodiesel production in biorefineries such as inexpensive waste cooking oils (WCOs), waste animal fats (WAFs), nonedible oils, and waste-oily byproducts generated in edible-oil refineries [50,51]. Such a strategy could substantially improve biodiesel production from the sustainability and productivity viewpoints [52].

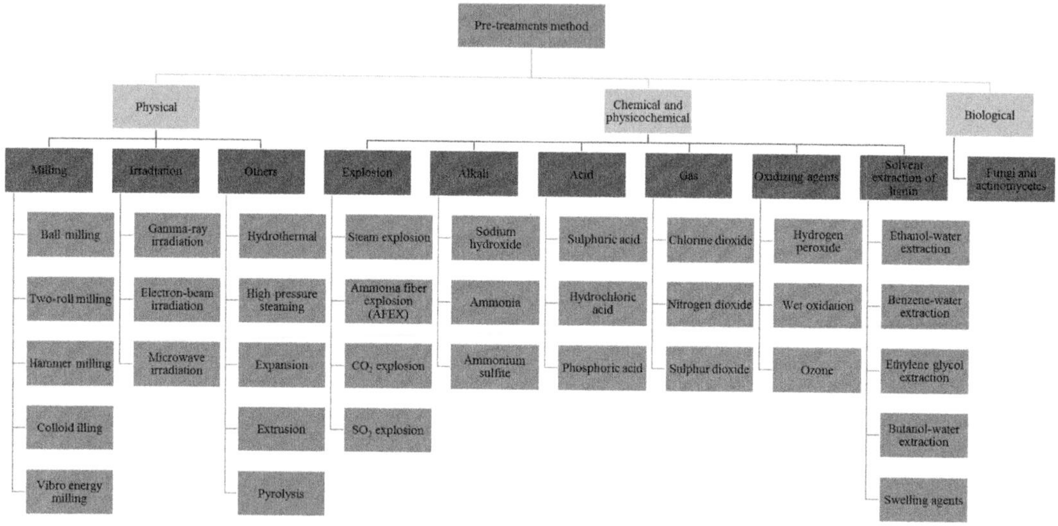

Figure 19.2

Pretreatments methods used for lignocellulosic materials. *Source: Taherzadeh M, Karimi K. Pretreatment of lignocellulosic wastes to improve ethanol and biogas production: a review. International Journal of Molecular Sciences 2008;9:1621–51.*

Waste oils are usually converted into methyl esters (biodiesel) and glycerol through the transesterification reaction between triglycerides and light alcohols such as methanol in the presence of an alkaline catalyst such as sodium hydroxide. Although waste fats and oils could be reacted directly, their generally high free fatty acid (FFA) content might necessitate the implementation of a pretreatment (esterification) step. Upon the completion of the transesterification reaction, the surplus methanol could be recycled and reused in order to improve both the environmental and economic features of the process (Fig. 19.3).

19.2.1.3 Other waste feedstock for the biorefinery

There are various other wastes different from conventional lignocellulosic biomass and oils/fats, which can be used as primary inputs to waste biorefineries for the production of bioenergy and biomaterials such as heat, electricity, chemicals, etc. (Fig. 19.4).

19.2.2 Biorefinery products

19.2.2.1 Energy products

To address the concerns over the sustainability of the energy sector, deployment of renewable energy carriers like renewable heat and electricity in the energy market in various economic regions of the world is increasing [18]. As elaborated earlier, waste-oriented biorefineries could serve as cost-effective solutions to the energy crisis faced as well as to waste disposal problems [54]. The three primary technological pathways

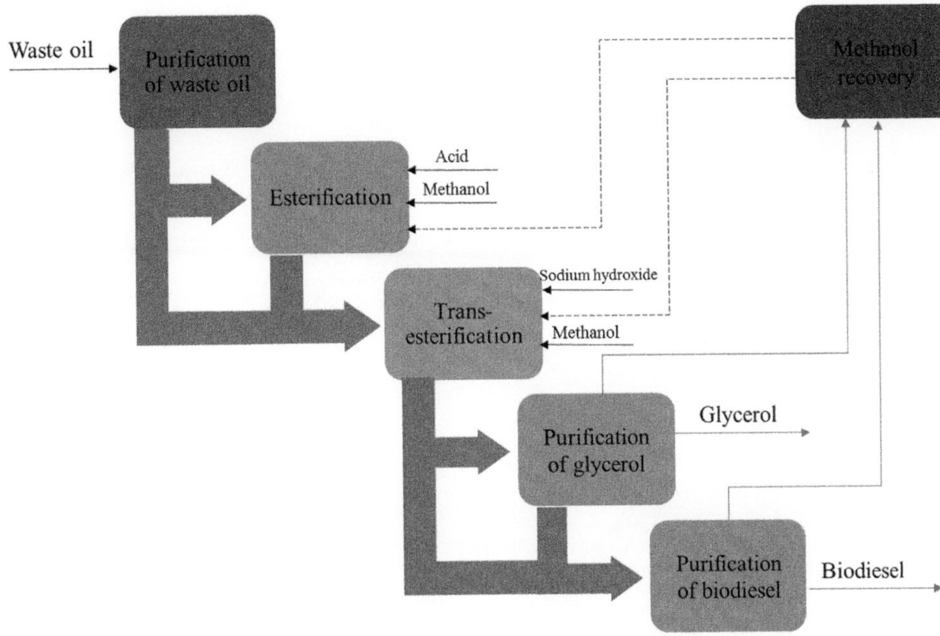

Figure 19.3
Schematic presentation of biodiesel production process from waste oils including methanol recovery.

associated with biorefineries for energy production include thermochemical, biochemical, and physicochemical processes (Fig. 19.5) which are described partially in Section 19.2.3.

- Thermochemical technologies: transform the waste feedstock to energy in the form of fuel, heat, electricity, and value-added products at elevated temperatures through four different routes including incineration, pyrolysis, gasification, and refuse derived fuel (RDF) [55,56].
- Biochemical technologies: transform organic wastes into liquid or gaseous fuels through biomethanation and fermentation using biological agents. Whereas coproducts can be used in agriculture, cosmetics, and cardboards industries [57].
- Physicochemical technologies: transform organic wastes into liquid fuels by chemical agents through transesterification as the most common physicochemical conversion technology [58].

19.2.2.2 Biomaterials

In addition to energy generation, biorefineries could provide an array of biochemicals such as cleaning compounds, adhesives, dielectric fluids, detergents, dyes, inks, hydraulic fluids, packaging materials, lubricants, paper, and boxboard paints and coatings, plastic fillers, polymers, solvents, and sorbents [59]. Such products are estimated to contribute a turnover

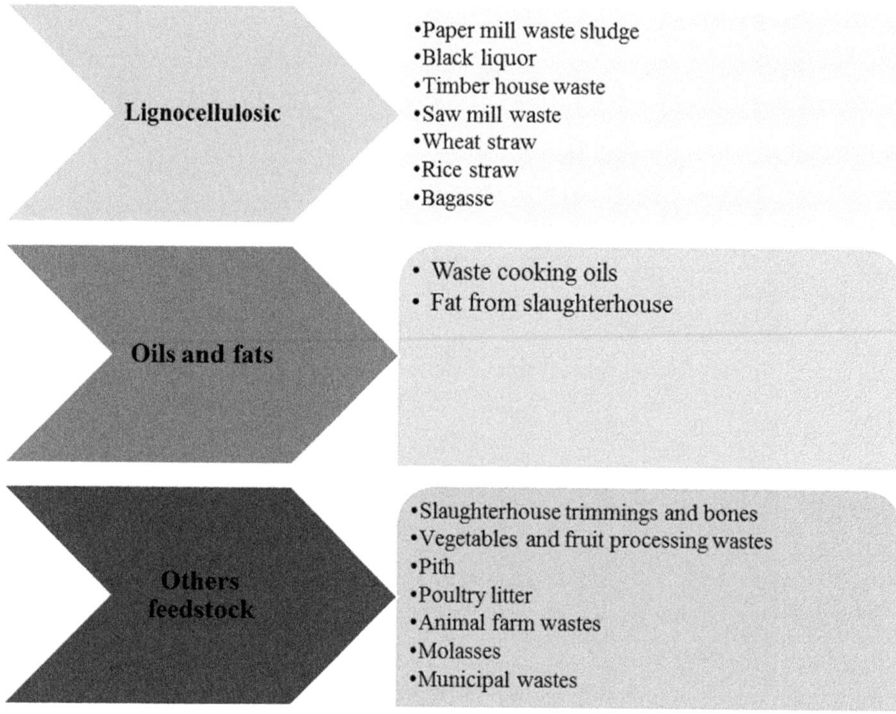

Figure 19.4
Waste feedstocks used in sustainable biorefineries for biomaterials/bioenergy production. *Source: Ghatak HR. Biorefineries from the perspective of sustainability: feedstocks, products, and processes. Renewable and Sustainable Energy Reviews 2011;15:4042–52.*

of €2 trillion to the European Union economy every year [60]. In Fig. 19.6, an illustrative list of biomaterials typically produced in biorefineries is presented. Recovering waste products in biorefineries could also improve supply chain security and lead to cost savings [61]. In spite of that, the majority of research works is focused on biofuels production in biorefineries, and therefore, it seems necessary to further elucidate the improving role of biomaterials from the economic and environmental perspectives.

19.2.3 Energy production pathways in biorefineries

19.2.3.1 Thermochemical

Thermochemical conversion pathways (combustion or incineration) are the most common methods to harness energy from biomass. However, these methods are not the most efficient ones, and more importantly, these methods contradict the multiproduct approach as a basic principle in biorefineries. Hence, other thermochemical conversion processes like pyrolysis, gasification, and RDF should be considered to produce energy and materials in waste biorefineries [56].

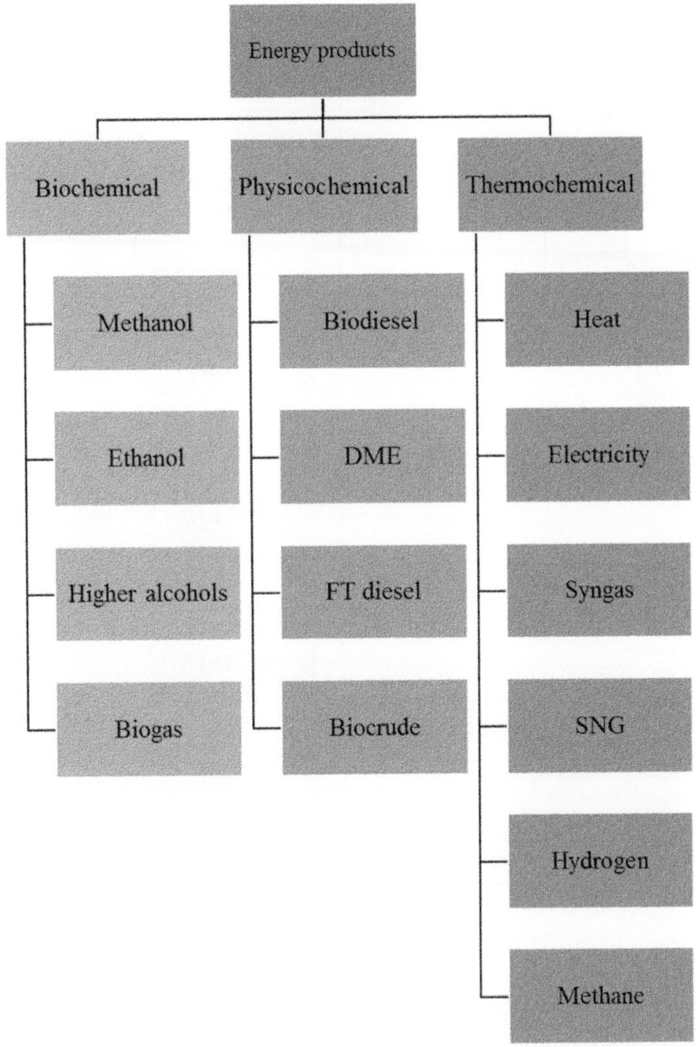

Figure 19.5

Different energy products generated in waste biorefineries. *Source: Ghatak HR. Biorefineries from the perspective of sustainability: feedstocks, products, and processes. Renewable and Sustainable Energy Reviews 2011;15:4042–52.*

Pyrolysis is a thermochemical process that decomposes a feedstock at a high temperature (about 300–1000°C) in an oxygen-depleted environment. Nowadays, this process has been applied to transform waste into useful products like bio-oil, biosyngas, and biochar. There has not been a consensus yet on the main pyrolytic product since various observations reported by researchers are different depending on waste composition and experimental conditions (e.g., vapor residence time, reaction temperature, and heating rates) [62].

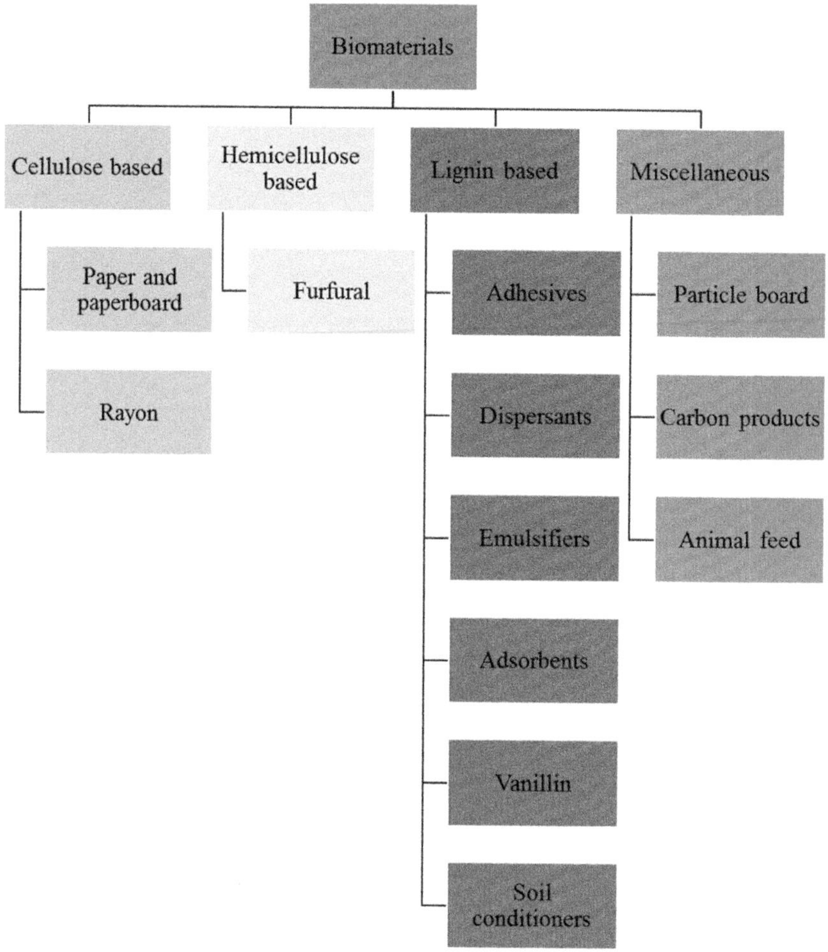

Figure 19.6
List of biomaterials produced in waste biorefineries. *Source: Ghatak HR. Biorefineries from the perspective of sustainability: feedstocks, products, and processes. Renewable and Sustainable Energy Reviews 2011;15:4042–52.*

Generally, pyrolysis can be applied as a feedstock recovery method for hydrocarbon resources, where the wastes are cleaved to produce hydrocarbon oils, gases, and char. Pyrolysis process has been used for various feedstocks. For instance, waste oil treatment using pyrolysis to produce energy-dense products has been reported [63]. Pyrolytic conversion of sewage sludge into tar-free fuel gas and polyaromatic hydrocarbons was also reported by Zhang et al. [64] and Dai et al. [65], respectively. There are other pyrolysis studies focusing on catechol, plastic wastes, tires, ethylene, and acetylene, as well as copyrolysis of the scrap tires with waste lubricating oil [66–69].

Gasification of solid waste is a complex process taking place at temperatures generally higher than 600°C (without combustion). It includes a number of chemical and physical reactions through which solid wastes are converted into a synthetic gas that could subsequently be used to produce electricity and other bioproducts. Syngas content can vary depending on gasification technique, gasification material, reactor type, the residence time of materials in the reactor, reactor temperature, supplied gas type, and supplied gas rate. Oxidation medium is the basis of classification of various types of waste gasification processes. Accordingly, these processes can be categorized into partial oxidation by air, pure oxygen, or oxygen-enriched air by plasma gasification, and by steam gasification [70].

RDF is a solid fuel produced from a mixture of different waste streams such as municipal and industrial wastes, construction and demolition wastes, commercial wastes, and sewage sludge. RDF is produced when the recyclable fraction of the waste streams like glass, metal, and plastics have been removed. The purpose of RDF production is to create a fuel easily burnt in a combustion chamber and to divert materials from landfills. The main steps in producing RDF from waste streams include sorting, shredding, drying, and densification [71].

19.2.3.2 Biochemical

Fermentation is a metabolic process that in the absence of oxygen can convert complex biomass feedstocks such as sugars, into organic acids or alcohol, and gases through the action of microorganisms (yeast and bacteria). Typical reaction involved in ethanolic fermentation is presented by Eq. (19.1) [72].

$$\underset{\text{Glucose}}{C_6H_{12}O_6} \xrightarrow{\text{yeast}} \underset{\text{Ethanol}}{2CH_3CH_2OH} + 2CO_2 \tag{19.1}$$

Ethanolic fermentation of sugars is a commercially viable technology. Using enzymatic hydrolysis, both cellulose and starch can be transformed into fermentable sugars. Enzymatic depolymerization of cellulose into glucose monomers is the first step of bioethanol production from cellulose through biochemical transformation [73] followed by typical fermentation. Given the significant contribution of hemicelluloses to lignocellulosic structures, their fermentation is also considered necessary for the inclusive application of lignocellulosic biomass [74]. Therefore, wastes rich in hemicelluloses such as wheat straw, bagasse, and rice straw could be fermented into bioalcohols as well. However, starch liquefaction is the commercial method used to hydrolyze starch into glucose syrup at a relatively high temperature of about 140–180°C by amylase enzymes [75].

Anaerobic digestion (AD) is one of important components of the waste biorefineries. In this process, organic matters are decomposed under oxygen-free or anaerobic conditions with the aid of a variety of anaerobic microorganisms. The final products of AD mainly include biogas (containing 60%–70% methane (CH_4) and 30%–40% CO_2) and an organic residue rich in nitrogen. This technique has been successfully employed for lowering the biochemical oxygen demand (BOD) and chemical oxygen demand (COD) of different waste streams such as food wastes, wastewater sludge, and agricultural wastes [76]. AD processes are classified based on reactor designs, feedstocks, and operating parameters. For instance, based on operation continuity, into batch versus continuous; based on operating temperature, into thermophilic, temperature, psychrophilic, and mesophilic; based on reactor design, into complete-mix, plug-flow, covered lagoons, etc.; and based on solid content, into dry versus wet [77].

From the mechanism point of view, AD consists of several successive stages of chemical and biochemical reactions including hydrolysis, acidogenesis, acetogenesis, and methanogenesis [78]. In hydrolysis stage, complex compounds such as carbohydrates, proteins, and lipids are converted into their monomeric constituents such as glucose, amino acids, fatty acids, *etc*. In the acidogenesis stage, the acidogenic bacteria convert the resultant monomers into volatile fatty acids (VFAs) (i.e., propionic acid, butyric acid, and acetic acid), alcohols, and CO_2. At the acetogenesis stage, acetogenic bacterial generate acetic acid, CO_2, and hydrogen (H_2) using the products of the preceding stage. Finally, at the methanogenesis stage, methanogens convert the acetic acid, CO_2, and H_2 to CH_4. Typical reactions taking place during AD are presented in Eqs. (19.2)–(19.5) [79].

$$\underset{\text{Organic compounds}}{C_6H_{12}O_6} \rightarrow \underset{\text{Ethanol}}{2C_2H_5OH} + 2CO_2 \tag{19.2}$$

$$\underset{\text{Ethanol}}{C_2H_5OH} + CO_2 \rightarrow CH_4 + \underset{\text{Acetic acid}}{2CH_3COOH} \tag{19.3}$$

$$\underset{\text{Acetic acid}}{CH_3COOH} \rightarrow CH_4 + CO_2 \tag{19.4}$$

$$CO_2 + 4H_2 \rightarrow CH_4 + 2H_2O \tag{19.5}$$

19.2.3.3 Physicochemical

Waste oils are byproducts of restaurants, vegetable oil refineries, animal slaughterhouses, and trapped grease from treatment plants, and are generated at huge amounts. These waste streams such as WCO could be economically used as a biodiesel feedstock. WCO is primarily composed of lipids or triglycerides (i.e., three fatty acid molecules attached to a glycerol backbone) and to a lesser extent, monoglycerides, and diglycerides [80]. Various techniques have been developed to convert these feedstocks into biodiesel. Transesterification has been the most common method used at industrial scale due to its simplicity and widespread application. Transesterification is, in fact, the reversible reaction of long-chain fatty acids contained in oils or fats with light alcohol like methanol or ethanol in the presence of a strong base (NaOH or KOH) [81,82] or acid catalyst [83]. Through this reaction, fatty acids are converted into their corresponding alkyl esters. According to Yaakob et al. [84], "the transesterification process begins with a sequence of three consecutive reversible reactions, wherein triglycerides are converted to diglycerides, diglycerides are converted to monoglycerides, and monoglycerides are converted to glycerol". In each step of the process, an ester is produced, ultimately leading to the production of three ester molecules from one triglyceride molecule [85]. The general transesterification reaction is shown in Eq. (19.6) [86].

$$\begin{array}{lcll} CH_2-OCOR^1 & & CH_2OH & R^1COOCH_3 \\ | & & | & \\ CH-OCOR^2 + 3CH_3OH & \xleftrightarrow{catalyst} & CHOH + & R^2COOCH_3 \\ | & & | & \\ CH_2-OCOR^3 & & CH_2OH & R^3COOCH_3 \\ \text{Triglyceride} \quad \text{Methanol} & & \text{Glycerol} & \text{Methyl ester} \end{array} \tag{19.6}$$

Various waste to energy (WTE) technologies classified based on their conversion processes such as thermochemical, biochemical, and physiochemical are shown in Fig. 19.7.

19.3 Life-cycle approach

According to the ISO 14044, a product's life-cycle initiates from raw material extraction to final disposal, including the production and use phases and waste management [87]. In better words, product's life-cycle starts from the raw material (minerals, water, fossil fuels, *etc.*) extraction and continues with transportation, construction, and consumption of the product, and ends with the waste disposal or waste management [88]. Such a life-cycle approach can also be considered as a "cradle-to-grave" approach [89] where system boundary includes the extraction and processing of raw materials, production, storage, transportation, consumption and final disposal stages of a product. While another approach could also be considered like "cradle-to-cradle" where system boundary, in addition to the phases mentioned above, also covers reuse, recycling and/or recovery to produce parts of a product or the whole product [90]. The latter is the life-cycle approach typically considered in waste biorefineries.

19.3.1 Life-cycle assessment (LCA)

Environmental studies in which "climate change, human health, ecosystem quality, and resource depletion" using the "cradle-to-grave" or "cradle-to-cradle" approach investigated

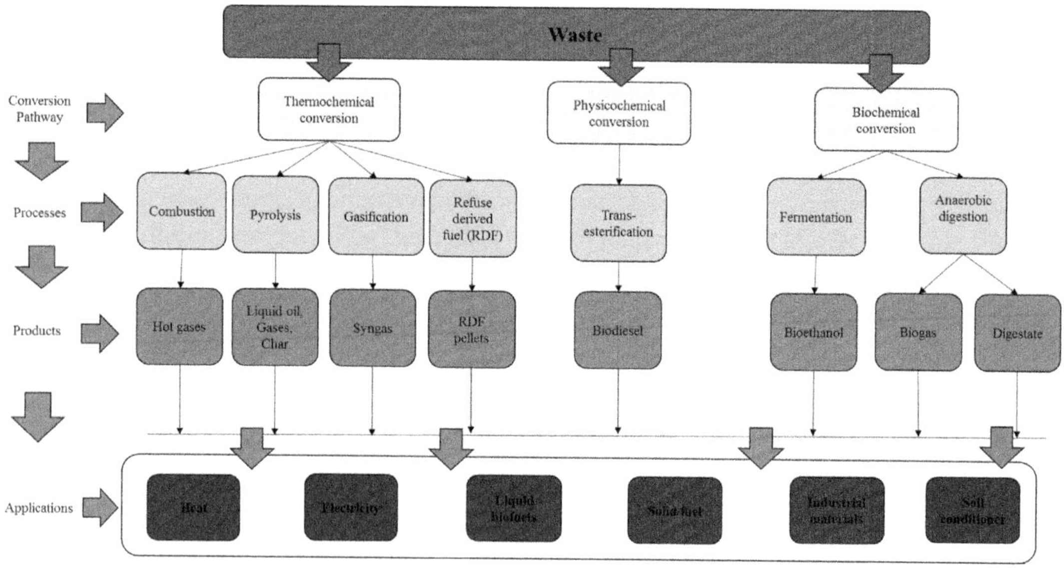

Figure 19.7

Classification of different WTE technologies based on their conversion process. *Adopted from Ouda OKM, Raza SA, Nizami AS, Rehan M, Al-Waked R, Korres NE. Waste to energy potential: a case study of Saudi Arabia. Renewable and Sustainable Energy Reviews 2016;61:328–40. With permission from Elsevier. Copyright© 2016.*

are also known as LCA. LCA assesses all steps of the life-cycle of a product and estimates cumulative environmental and economic burden associated with these steps, and as a result, the most environmentally sound paths or processes could be selected. Accordingly, LCA helps researchers, managers, policymakers, and decision-makers to design and implement the merchandises, technologies, processes, and/or services leading to minimal unfavorable impacts on the surroundings. The diagram conferred in Fig. 19.8 shows the main life-cycle stages taken into consideration during the execution of an LCA study.

As depicted in Fig. 19.8, any products or technologies would need some inputs in the form of energy and raw materials throughout its life-cycle steps, from acquisition to production, application, and eventually disposal stage. All the mentioned life-cycle phases might result in emissions as well as the generation of wastewaters, and/or solid wastes. This is simply because the conversion efficiency of the energy and material is rarely 100%. Moreover, there are sometimes byproducts generated as well that would also end up in waste streams. LCA assists with keeping track of all these favorable and unfavorable outcomes. The scheme delineated in Fig. 19.9 provides guidelines for LCA mapping. A typical LCA project set up includes the following main stages [87]:

- Goal and scope definition: identifies reasons for conducting a given LCA study while also describes system boundaries and a functional unit (FU).

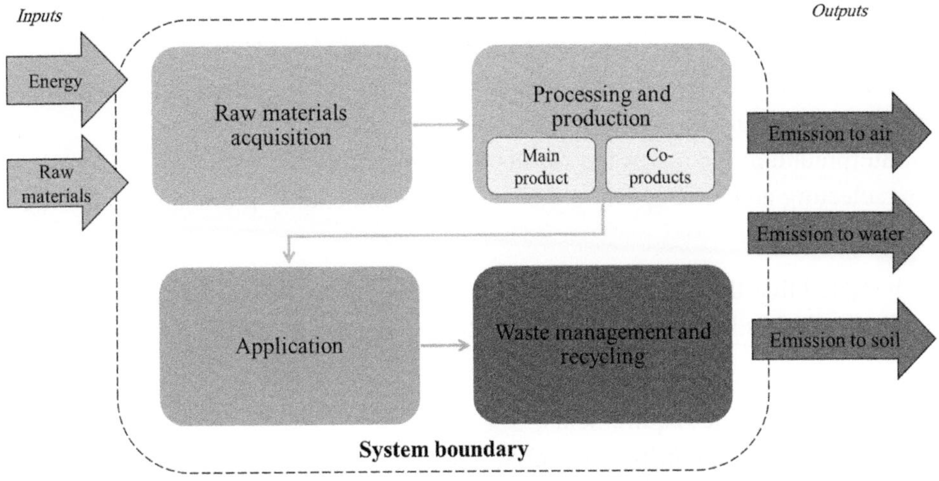

Figure 19.8

The primary inputs and outputs flow in an LCA study. *Adopted from ISO, 14044 International Standard. Environmental management — life cycle assessment — principles and framework. Geneva, Switzerland: International Organisation for Standardization; 2006.*

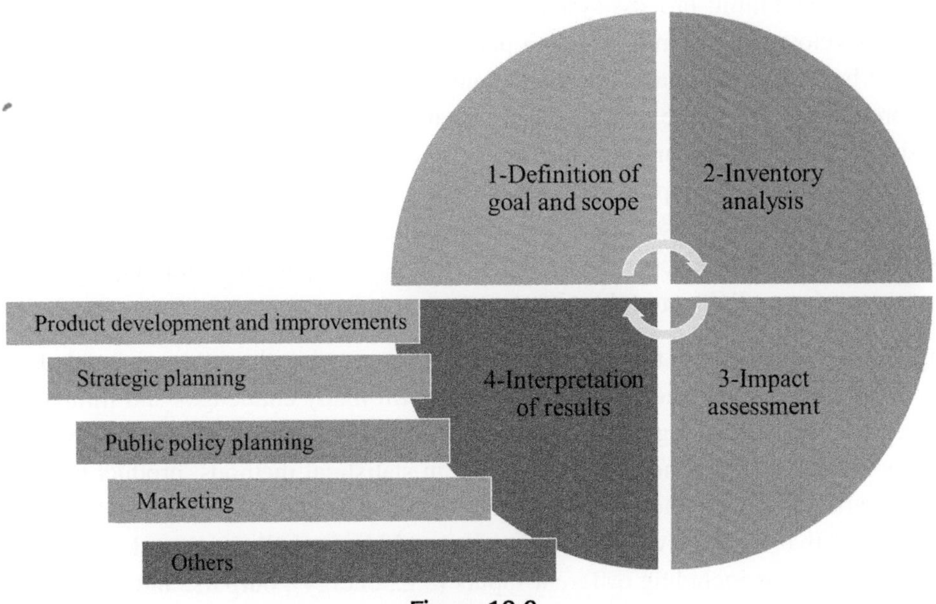

Figure 19.9

LCA methodology phases. *Adopted from ISO, 14044 International Standard. Environmental management — life cycle assessment — principles and framework. Geneva, Switzerland: International Organisation for Standardization; 2006.*

- Inventory analysis: identifies and quantifies materials, water, and energy as inputs and environmental releases as outputs and waste generated per FU.
- Impact assessment: assesses the potential human and ecological effects, quantify metrics and other environmental impacts.
- Data interpretation: compares the results obtained through the two proceeding steps before selecting and/or suggesting the most environmentally promising technology, process, or product.

It should be noted that the standard ISO 14044 is also known as "requirements and guidelines" in LCA studies. The highlights of this standard are as follows:

(1) Provides detailed guidelines for each step of LCA,
(2) Elaborates on specific requirements,
(3) Provides guidelines for critical reviews,
(4) Provides examples of applications.

Fig. 19.10 shows a sample checklist for performing an LCA investigation according to the ISO 14044.

19.3.2 LCA of waste biorefineries

The waste biorefinery principle for single and multiple waste streams/bioresources is based on multiproduct and multi-process systems. To ensure that the very principles of sustainable development are met, and as a part of the detailed design feasibility study, an LCA should be performed to assess the environmental and economic impacts associated with the production and waste management scenarios included in each waste biorefinery [91]. It should be noted that unlike stand-alone processes in which the "cradle-to-grave" LCA concept is generally employed, in waste biorefineries, the "cradle-to-cradle" is used as a powerful technique to assess the environmental and economic effects of the bioprocesses and/or bioproducts. Within this platform, all inputs and outputs (biomaterials and bioenergy) during LCA are taken into account.

Although LCA can be of substantial assistance with improving the environmental and economic features of waste biorefineries, there are still many challenges and limitations associated with its methodologies and implementation, including lack of accurate data availability, rigid system boundaries, differences in statistical methods, product type selectivity, variations in product usability, as well as local conditions and environment [92]. In light of that, LCA of waste biorefineries could be a challenging task requiring various considerations including different inputs (e.g., energy), outputs, and emissions for given biorefinery technologies, waste quantity, and characterization, spectra of biorefinery products considered, local practices and conditions, etc. To address the above-mentioned challenges and to achieve better results, Mohan et al. [15] proposed the integration of

Goal and scope definition
- Specifying of the system boundary
- Definition of the functional unit

Data collection
- Direct collecting
- Using of literature
- Using of databases

Data validation

Relating data to unit process and functional unit

Inventory analysis
- Calculation of emissions to water
- Calculation of emissions to soil
- Calculation of emissions to air

Impact assessment
- Selection of the evaluation method
- Characterization of impacts
- Damage assessment of impacts
- Normalization of impacts
- Weighting of impacts

Interpretation
- Conclusion
- Recommendations
- Identifying of constraints

Reporting

Application

Figure 19.10
A sample checklist for performing an LCA investigation according to the ISO 14044 standard.

socio-economic evaluation, life-cycle sustainability assessment (LCSA), ecological based LCA, economic input-output LCA (EIOLCA), and LCC with the LCA of waste biorefineries.

19.3.2.1 Goal and scope definition in LCA of waste biorefineries

The goal of the analysis is to assess the environmental impacts and/or profits of a waste biorefinery. As mentioned earlier, waste biorefineries present an integrated production of bioenergies and biomaterials and, therefore, LCA of the waste biorefinery is aimed at identifying the critical features influencing resource performance, environmental efficiency, and sustainability of the whole system. Such findings would assist waste biorefinery industries with the right decision-making process.

As mentioned earlier, the definition of goal and scope of the system is an integral part of conducting an LCA, which explains the purpose of the study in its system boundary according to the defined FU. The FU is a quantified performance of goods or services for use as a reference unit for inventory analysis of the inputs (resources) and the outputs (emissions). In the face of the goals mentioned, four FUs could be considered in waste biorefinery:

(1) Mass (kg): this FU is used for comparison of the production of an energy carrier and/or a biomaterial by a biorefinery pathway with its petrochemical counterpart in an oil refinery (mass of output). Moreover, in some studies, the mass of waste used in a biorefinery is selected as FU (mass of input). In these cases, the allocation is required to investigate scenarios of waste biorefineries, due to multi-output processes. The selection of the allocation method in LCAs of waste biorefineries can considerably affect the results and thus, the final decision-making. In the LCA studies of multiproduct systems, allocation of the environmental burdens associated with the products through only energy, economic, or mass value would not suffice the target, and thus multi-allocation methods should be used. For example, in some biorefineries, besides bioenergy, nutrients in the form of biochar and sludge are also generated whose allocation requires mass or economic values while energy value-based allocation should be used for bioenergy. Allocation issues may also arise when it is not possible to separate the multifunctional processes into subprocesses connected to specific products. Under such circumstances, the allocation can be handled by partitioning or by system expansion. According to Heijungs and Guinée [93], the partitioning method should be based on "the artificial splitting up of a multifunctional process into a number of independently operating mono-functional processes," that are "mathematical constructions" and do not exist as real cases. In better words, the environmental impacts of the multi-process systems are distributed among the coproducts by factors such as mass or energy content. In the partitioning method based on mass, the mass ratio of coproducts to the total products are the basis of calculations. However, cases, where the coproducts are based on

energy such as heat, application of mass allocation, is not appropriate. Under such circumstances that are widely common in LCAs of bioenergy systems, the energy content of coproducts is considered as the allocation criterion. Nevertheless, it should be noted that the application of the energy allocation for coproducts, that are not produced for their energy content (e.g., chemicals), would not be leading to accurate results. An exergy content-based allocation to address the shortcoming associated with the partitioning method based on mass and energy content has been proposed [94]. This has been supported by the fact that both material flows and energy could be accounted for by using the exergy-based unit. The partitioning coefficients for allocation are calculated using the following equations (Eqs. 19.7 and 19.8) [95]:

$$u_i y_i = a_i W_{tot} \tag{19.7}$$

$$a_i = \frac{y_i c_i}{\sum_i^n y_i c_i} \tag{19.8}$$

where y_i refers to the flow of the coproduct quantified in energy, mass, or other terms and c_i is its specific factor used for partitioning, which is related to unit of y_i (e.g., MJ/unit or €/unit), a_i is the partitioning coefficient that is defined between 0 and 1, and their sum is equal to 1), W_{tot} is the total environmental impacts in a biorefinery, and u_i is environmental impacts of the products in a biorefinery. A schematic of the partitioning method is shown in Fig. 19.11A.

On the other hand, when system expansion is used, the allocation method is based on expanding the system boundaries as far as alternative production systems of the external functions are included in the system boundaries. In better words, the environmental effects of alternative production systems based on common practices in the area are attributed to the coproducts and are subtracted from the total effects of the current systems. Finally, the effect is charged to the main product. It should be noted that in this allocation method, accurate results of LCA can be achieved only when accurate information is available for exported functions [93]. The system expansion coefficient (allocation coefficient) for the main product could be computed as follows (Eq. 19.9) [95]:

$$a_1 = 1 - \frac{\sum_{i \neq 1}^n u_i^* y_i}{W_{tot}} \tag{19.9}$$

The system expansion coefficient (allocation coefficient) for the coproducts are also calculated using Eq. (19.10) [95]:

$$a_{i \neq 1} = \frac{u_{i \neq 1}^* y_{i \neq 1}}{W_{tot}} \tag{19.10}$$

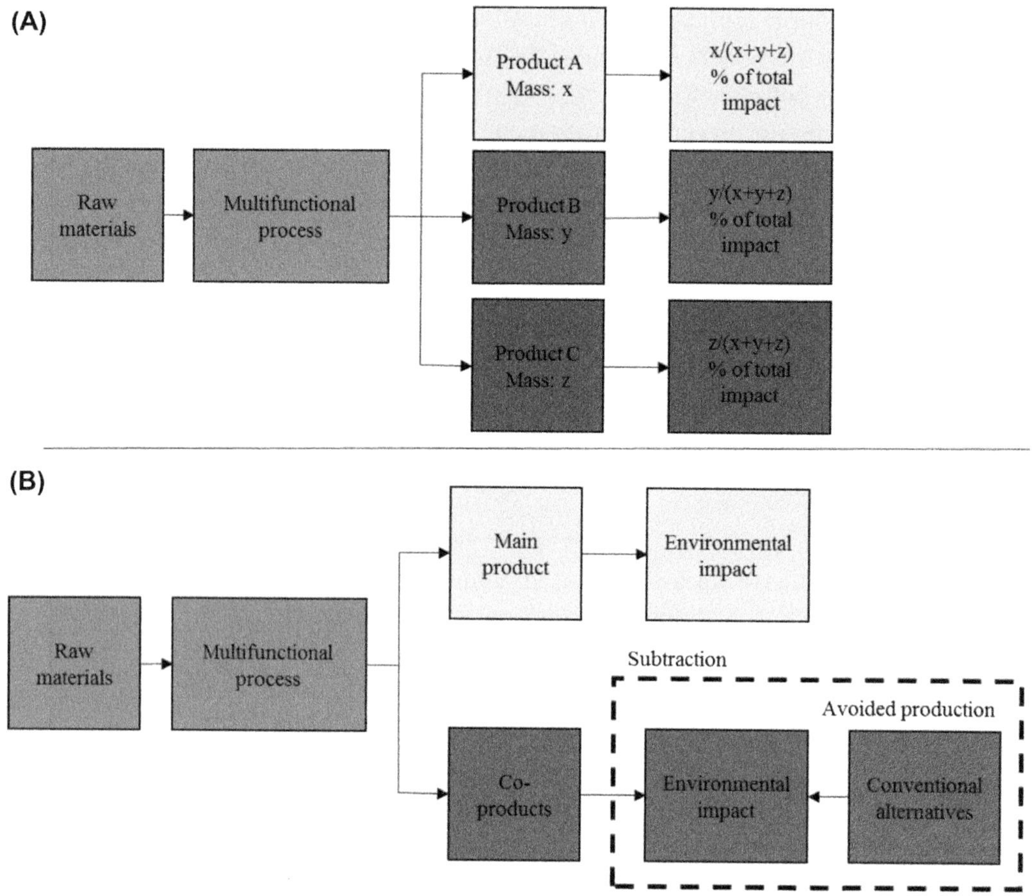

Figure 19.11

A schematic example for partitioning method (A) and system expansion (B). *Adopted from Cherubini F, Strømman AH, Ulgiati S. Influence of allocation methods on the environmental performance of biorefinery products — a case study. Resources, Conservation and Recycling 2011;55:1070–7. With permission from Elsevier. Copyright© 2011.*

where $i = 1$ is the main product, $i \neq 1$ is defined as the exported functions, and u_i^* denotes the environmental impacts of the coproducts for conventional fuels (fossil fuels). In Fig. 19.11B, a schematic example of system expansion is presented. The selection of allocation method can be based on ISO 14044 [87] or the international reference life-cycle data system (ILCD) handbook [96], while it can also be based on the preferences of LCA practitioners or request of the study's commissioner/s for confirming the environmental profits of a certain product [97].

(2) Revenue (US$ earned): this FU allows the comparison of different technological scenarios in waste biorefineries in terms of revenue.

(3) Distance (km run by a vehicle using biofuels): this FU is used with the goal of providing a comparison of environmental effects associated with the utilization of biofuels and fossil fuel. The system boundaries for this evaluation are extended until the use stage of fuels in vehicles.
(4) Energy (MJ): this FU is used for biofuels produced in a waste biorefinery.

As explained earlier, in waste biorefineries various technological routes including thermochemical, biochemical, and physicochemical methods are used to convert waste/biomass into beneficial end products. The generalized system boundary for the bioenergies (or biomaterials) generated in such systems includes three major steps (Fig. 19.12): (1) collection, separation, and transportation, (2) operation in plant site and upgrading of primary products where necessary, and (3) recycling and demolition of the plant.

19.3.2.2 Inventory analysis in LCA of waste biorefineries

In the life-cycle inventory (LCI), an inventory from the emissions and wastes released and the materials and energies used during the operation of the waste biorefinery is collected and calculated. In waste biorefineries, LCI includes the total materials and energies used and the emissions released for biomass collection and transportation. Moreover, it covers total materials and energies used and emissions released for establishing the biorefinery,

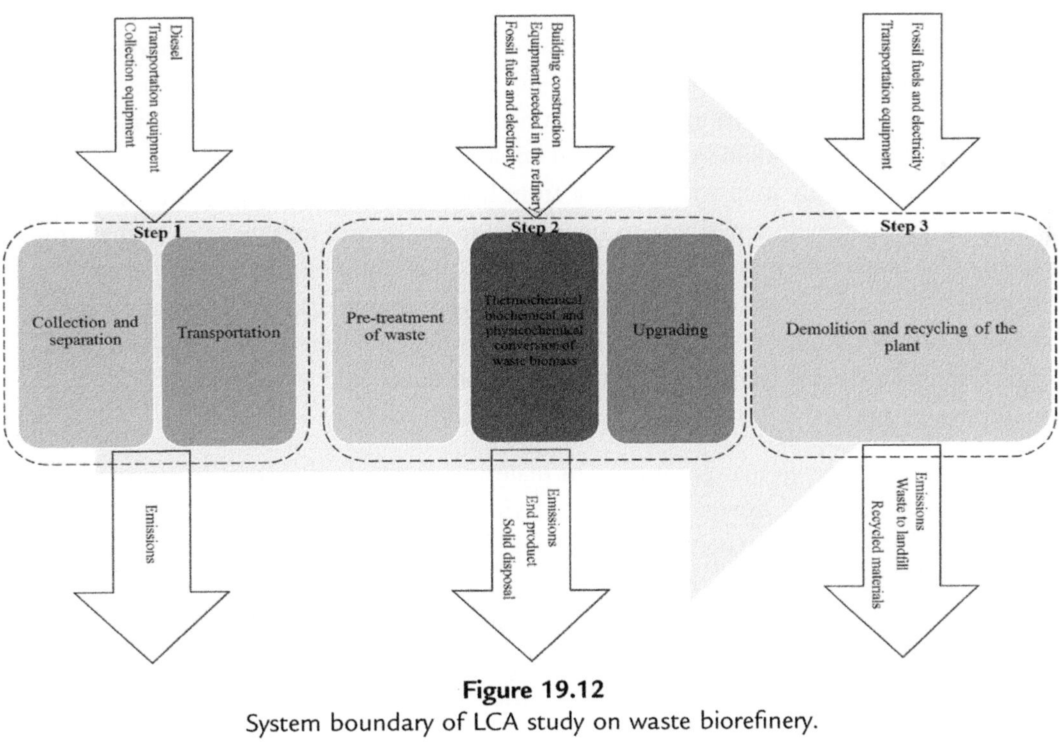

Figure 19.12
System boundary of LCA study on waste biorefinery.

i.e., during the generation and distribution of the natural gas, electricity, steam, heat, *etc.* In addition to these, LCI encompasses the emissions released during the operation of the waste biorefinery in the defined system boundary per the FU as well as the amount of generated bioenergies and biomaterials. If the data is not fully available, LCI can retrieve them from the standard databases like EcoInvent. This stage is the most important phase of the LCA of waste biorefineries, and if at this stage, the accuracy of data could be increased, the results of the LCA could be more accurate as well.

19.3.2.3 Life-cycle impact assessment (LCIA) in LCA of waste biorefineries

Life-cycle impact assessment (LCIA) is the third phase of an LCA as defined by the International Organization for Standardization [87]. This step is implemented after collecting the data on raw material extractions and substance emissions associated with a product's life-cycle. In LCIA, the potential environmental impacts are first identified, and subsequently, their quantity and importance are evaluated through a number of category indicators. These indicators vary for different stages of LCIA including characterization; classification; normalization (optional); grouping (optional); weighting (optional), and data quality analysis (optional) [98]. ISO 14044 provides a full description of the different elements of LCIA [87].

19.3.3 Summary of LCA studies with a focus on waste biorefinery

As revealed by the results of different LCA studies focused on waste biorefineries, developing waste biorefineries through more advanced and sustainable biorefinery technologies as well as switching from the consumption of fossil fuels to renewable and green energy resources would be considered advantageous. More specifically, such an approach could not only help with shifting from linear economies toward circular economies but also could contribute to improving public health and the environment by reducing GHGs emissions and their adverse impacts including climate change [20]. Other environmental advantages associated with waste biorefineries include reduced landfilling and mitigating its detrimental impacts on the environment and public health, production of renewable energies and other green products, and advancement of the agriculture sector.

Many studies have shown that production of materials and energy carriers in waste biorefineries could reduce some or all of the environmental impacts. For example, Cherubini and Ulgiati [59] reported GHGs (CO_2 and CH_4) savings in the range of about 50% using crop residues as raw materials in waste biorefinery systems for production of bioenergy as a replacement for fossil fuels (gasoline and natural gas). Researchers attributed the largest fraction of total GHGs savings to gasoline replacement (81% when using corn stover and 84% when using wheat straw), followed by replacing electricity from natural gas (10% and 3%, respectively), and heat from natural gas (7% and 11%, respectively). Their findings of the role of CO_2 in reducing global warming potential (GWP) in waste biorefineries were consistent with those of Ramachandran et al. [99]. Both research teams stated that the primary reasons for the GHG emissions reduction in biorefineries were (1) CO_2 emissions from biogenic carbon sources has no GWP, and (2) lower fossil fuel consumption in waste biorefineries.

Similar results were also reported by Fan et al. [100] who investigated GHG emissions for electricity generated from waste resources through pyrolysis-based processing. In their study, life-cycle GHG savings of 77%–99% were estimated for electricity generation through the combustion of pyrolysis oil in comparison with fossil fuels combustion, depending on the type of waste and combustion technologies used. Similar results were also obtained in a different study conducted by Iribarren et al. [101]. Their findings showed that GHG emissions savings of biofuels were about 82% as compared to conventional fossil fuels. The employed biofuel production system included a circulating fluidized bed reactor followed by bio-oil upgrading through hydrotreating and hydrocracking. In conclusion, the overall effects of waste biorefineries on GHG emissions were reportedly favorable, the magnitude of such favorable impacts was less considerable in some studies though. For instance, in a study conducted by Sebastião et al. [102] on an ethanol plant converting 5400 tons of dry sludge/year, two biorefinery scenarios (i.e., the reduced HCl scenario and the cofermentation scenario) were considered. Compared with most of the existing literature on waste biorefineries, the results obtained revealed lower reductions in the GWP impact category, i.e., 23% and 15%, respectively. Another example of achieving lower positive impacts on GHG footprint was the study performed by Boldrin et al. [103] on using rapeseed straws-based biorefinery for energy generation to replace fossil rescores. Researchers reported impact reductions ranging from 9% to 29% depending on the adopted conversion process.

These differences could be attributed to a number of reasons such as the type of waste and conversion technology. Nizami et al. [104] studied the impact of different conversion technology for energy generation in waste biorefineries on GWP. Their findings revealed that the highest environmental value (reduction in GWP) caused by reducing fossil fuel consumption for energy generation, was attributed to AD (505,000 Mt CO_2 eq.) in comparison with RDF, pyrolysis, and transesterification (227,100, 199,700, and 75,700 Mt CO_2 eq., respectively). Moreover, the authors reported that AD had the highest CH_4 emission reduction potential (20,200 tons) in comparison with RDF pyrolysis, and transesterification (9100, 7900, and 3000 tons, respectively). Dong et al. [105] reported different reduction rates in GWP impact category depending on the technology used; in descending order, gasification, incineration, and gasification-melting. Accordingly, Dong et al. concluded that further improvements in biorefinery technology should be targeted in order to improve its respective GW impact category further. For example, modern incineration could fulfill the criteria for an environmentally sound technology and therefore, could be a better option as compared to pyrolysis and gasification-melting. Combination of different technologies has also been proposed as a means to achieve further reductions in GHG emissions in a waste biorefinery. For instance, Sebastião et al. [102] claimed that the combination of HCl scenario and cofermentation scenario in a biorefinery-oriented bioethanol plant could more effectively reduce GHG emissions (i.e., 38% vs. <25% for individual biorefinery scenarios).

The type of waste used in biorefineries could play a vital role in the overall performance of the system in terms of GWP. An interesting observation made by Fan et al. [100] who investigated different feedstock types including waste wood, poplar, and willow for pyrolysis oil production. Researchers indicated that the least GHG emissions were associated with the pyrolysis oil produced from waste wood because the feedstock itself did not introduce any GHG emissions. In another study, Chang et al. [106] compared to rice straw (agricultural waste), Napier grass (energy crop), and *Eucalyptus* spp. (short rotation coppice) as bioethanol feedstock for electricity generation in a biorefinery using LCA. Chang et al. argued that in the waste biorefinery based on rice straw as a feedstock, CH_4 emissions from paddy fields led to more climate change impacts in comparison with the biorefineries based on Napier grass and *Eucalyptus* spp.

In addition to the studies focused on GHG emissions and climate change, studies are investigating the effects of waste biorefineries on ecosystem quality. Unlike climate change, ecosystem quality impact may have an incremental trend in waste biorefineries. Cherubini and Ulgiati [59] reported increases in the eutrophication (EU) impact category potentials (effective on ecosystem quality) in waste biorefinery systems aimed at producing heat, electricity, and biomaterials from crop residues versus fossil reference systems (0.39–0.52 kt PO_4-eq./year vs. 0.15–0.17 kt PO_4-eq./year, respectively). Researchers attributed these results to nitrogen fertilization and the consequent leaching of nitrates to groundwater. The unfavorable effects of waste biorefineries on ecosystem quality as compared to fossil systems were also found by Kimming et al. [107] who developed a platform suitable for small-scale combined heat and power (CHP) generation plants in rural areas based on agriculture biomass. The authors highlighted a quite considerable increment in acidifying air pollution (effective on ecosystem quality) in biomass-based scenarios (ley, straw, and willow) versus the fossil fuel–based scenarios; caused by diesel consumption during ley and willow harvest. More specifically, use of diesel in agricultural machinery resulted in more sulfur dioxide (SO_2) and nitrogen oxides (NO_x) emissions to the atmosphere; both are effective on acidification (AC) [108]. Therefore, it should be noted that although waste biorefineries could result in reduced GHG emissions and fossil energy consumption, their unfavorable impacts on the AC and EU impact categories should not be overlooked, even if climate change mitigation is of major concern by policymakers.

Nevertheless, the impact of using waste in biorefineries on ecosystem quality and related impact categories is largely controversial because of the lack of a consensus among the published literature. For example, contrary to the perspective taken up by the authors of the reports mentioned above, utilization of rice straw for bioenergy or biomaterial production in waste biorefineries has been marked as favorable by some other researchers. More specifically, researchers have argued that utilization of this feedstock could avoid the impacts on AC and EU potentials due to avoiding open biomass burning (taking place in some geographical regions) that is also a significant contributor to AC and EU potentials [109]. Shafie et al. [110] also highlighted the favorable attributes of waste biorefineries considering AC and EU impact categories. The authors reported that cofiring of rice straw and coal for power generation could result in 55% less sulfur oxides (SO_x) emissions and consequently reduced acid rain potential as compared with coal firing alone. These findings were in line with those of Sebastião et al. [102] who also associated a waste biorefinery (producing second-generation ethanol from pulp and paper sludge) to positive effects on the AC and EU impact categories. When compared with the base case scenario (fossil fuel), AC and EU impact categories were calculated 12% and 15% loss for the AC impact category and 3% and 16% loss for the EU impact category, respectively.

As for the human health damage category, a considerable deal of investigation carried out so far support the idea that unlike oil refineries, waste are advantageous due to their less toxic gas emissions [59,102,106,110–112]. Moreover, by using WTE technologies, long-term pollution potentials of landfills can be decreased improving human health. For instance, the impacts of waste biorefineries (considering electricity recovery, heat recovery, non-Fe recovery, and Fe recovery) on reducing human toxicity category as compared with landfilling wastes was found very impressive (−6.07E+06, −2.58E+06, −1.01E+08, and −1.30E+06 vs. 1.53E+07 kg 1,4-DB eq., respectively) [112]. Moretti et al. [111] also quantified the environmental effects of AD of organic MSW and pig manure by LCA for two scenarios; biorefinery-based and nonbiorefinery-based. Scenario 1 was a stand-alone treatment of the biomass resources using AD, while scenario 2 was a biorefinery system based on AD. In the latter, all the biogas generated by anaerobic digesters were converted into renewable energy carriers, i.e., electricity and heat, using a CHP. Researchers argued that although, both scenarios, because of energy recovery and the prevention of landfilling were useful for human health, but the biorefinery performed more favorably in terms of human toxicity (−60 vs. −0.72 CTU_h/year, respectively).

Avoiding the depletion of natural resources could probably be considered as the critical factor differentiating waste biorefineries from oil refineries [113]. Although fossil fuels and the electricity generated from fossil resources might be used as a source of power in waste biorefineries, their amounts are considerably lower in waste biorefineries. Sebastião et al. [102] reported a reduction of 15%–20% in the abiotic depletion of fossil fuels in a waste biorefinery (ethanol plant) than the reference fossil system producing gasoline and diesel. This observation could be ascribed to the production of bioethanol from sludge requiring less fossil resources compared to the production of gasoline and diesel. This finding was in line with those of Cherubini and Ulgiati [59], Ripa et al. [112], and Chang et al. [106] who also found waste biorefineries superior in terms of resource damage category than fossil systems. Such favorable effects are intensified when there is a recycling of materials such as ferrous or nonferrous metals [112], and paper [106]. It should be noted that the use of alternative fuels produced in waste biorefineries as a source of power could reduce resource depletion due to reducing fossil fuel consumption [101]. Table 19.4 tabulates LCA of waste biorefineries in which different conversion technologies were used.

Overall, it could be deduced that waste biorefineries are a way forward not only to attain sustainable waste management but also to generate positive environmental effects including reduced GHG emissions in comparison with the current disposal practices and protection of natural resources such as groundwater, soil, land, and energy. Nevertheless, the decision to choose types of waste biorefineries requires detailed technical, social and economic as well as environmental assessment through LCA. Finally, in LCA of waste biorefineries, a number of important limitations and ambiguities should also be considered, particularly when collecting high-credibility data as well as when selecting appropriate allocation methods.

19.4 Conclusions and perspectives

Waste biorefineries could generate significant economic and environmental values by the recovery of energy and valuable products, landfill cost savings, land savings, and creating new business opportunities. Waste biorefineries by saving raw materials and energy could lead to environmental benefits such as reduction of environmental impacts caused by the current disposal practices as well as protection of natural resources such as groundwater, land, and soil. Waste biorefineries could also contribute to the sustainability of industrial and agricultural supply chains, the creation of green jobs and service and help to achieve circular economies. However, the decision to choose among the different types of waste biorefineries requires thorough sustainability assessment of their various features using advanced environmental assessment tools such as LCA.

From an LCA methodological perspective, the critical aspects affecting the coherence and transparency of the results of the environmental impacts of waste biorefineries are the FU definition, the allocation principle, and/or system boundary expansions. It should be noted that for waste-based bioenergy and biomaterial generation in biorefineries, the data used could significantly affect the results obtained. In better words, the incorrect or inappropriate use of data may lead to wrong conclusions. Environmental and economic hotspots of waste biorefinery identified through pursuing different LCIA steps including

Table 19.4: Summary of the LCA of different conversion technologies.

Conversion technologies	Process	Type of waste	End product/s	FU	Environmental impact categories	Comment	References
Biochemical	Fermentation	Pulp and paper sludge	Bioethanol	1 MJ	All impact categories using the CML-IA method.	All the LCA impacts were reduced.	[102]
Biochemical	AD	MSW	Heat, electricity, compost	1 year	Climate change, AC, HT, EU.	The biorefinery was found favorable in terms of climate change and human toxicity impact categories.	[111]
Biochemical	Fermentation	1 Rice straw, Napier grass (energy crop), and *Eucalyptus* spp. 2 Virgin pulp, recycled newspaper, imported coals	Electricity, molded pulp, pellet fuel	1 1 kg bioethanol 2 1 dried kg of waste	All impact and damage categories using the IMPACT 2002+ method	Fermentation-based waste utilization was found suitable for regions requiring high-energy imports.	[106]
Biochemical and thermochemical	Fermentation, AD, pyrolysis	Corn stover, wheat straw	Bioethanol, heat, electricity, phenols	ha	All impact categories using the CML-IA method.	Use of crop residues in a biorefinery led to GHG emissions savings and reduced fossil energy demands but was also accompanied with higher eutrophication potentials versus fossil reference systems.	[59]

Continued

Table 19.4: Summary of the LCA of different conversion technologies.—cont'd

Conversion technologies	Process	Type of waste	End product/s	FU	Environmental impact categories	Comment	References
Biochemical and thermochemical	Mechanical and biological treatment	MSW	Heat, electricity, metal	100,000 tons of MSW	All impact categories using the ReCiPe (H) method.	The recovery of energy and metals resulted in significant environmental benefits.	[112]
Physicochemical and biochemical	Transesterification, fermentation, AD	Rapeseed and its straw	Electricity	1 MJ	GW, AC, terrestrial eutrophication.	Reducing GHGs emission in the range of 9%–29% due to the utilization of rapeseed straw, depending on the considered alternative.	[103]
Physicochemical, biochemical and thermochemical	Transesterification, AD, pyrolysis, RDF, and recycling	MSW	Electricity, recycled materials	1 GWh	GWP	GWP reduction of 1.15 million Mt CO_2 eq. could be achieved.	[104]
Thermochemical	Cofiring	Rice straw	Power	1 MWh	AC, GWP, EU, HT.	Depending on the cofiring ratio, the CO_2 emission was reduced by up to 77% and SO_X by up to 55%, leading to a significant reduction in impact categories.	[110]
Thermochemical	Cofiring	Willow	Electricity	1 MWh	Net energy ratio and net GWP.	Net energy ratio was increased by 9%, and net GWP was decreased by 7%–10% at 10% cofiring. GWP: 910 kg CO_2 eq. per FU.	[114]
Thermochemical	Cofiring	Wood residue	Electricity	1 kWh	GWP	GWP: 894.3 g CO_2 eq. per FU, i.e., an 18.2% redaction for 15% cofiring 1002.9 g CO_2 eq. per FU, i.e., a 5.4% redaction for 5% cofiring	[115]

Thermochemical	Cofiring	Wheat straw	Electricity	1 TJ	GWP	GWP: 298 t CO_2 eq. per FU, 10% Direct cofiring GWP: 300 t CO_2 eq. per FU, 10% indirect cofiring Coal boiler efficiency and biomass treatment were identified as important parameters.	[116]
Thermochemical	Pyrolysis	Corn stover	Biogasoline	1 ha	GWP	GWP: 7.65 t CO_2 eq. per FU. Corn stover removal rate was the sensitive parameter influencing the biochar and bio-oil yields.	[117]
Thermochemical	Pyrolysis	Short rotation poplar wood chips	Gasoline, diesel, and char	1 MJ	CED, GWP, OLD, POFP, LC, AC, EU.	GWP: −50.54 kg CO_2 eq. per FU Biomass pretreatment, pyrolysis, and steam reforming were found as the major contributors to the environmental impact categories.	[101]
Thermochemical	Pyrolysis	Forest residue	Gasoline and diesel	1 km	GWP and NEV.	GWP: 98−117 g CO_2 eq. per FU NEV: 0.92−1.09 MJ per FU GWP and NEV are lower than the conventional gasoline and diesel.	[118]
Thermochemical	Pyrolysis	Corn stover	Gasoline	1 MJ	GWP	Reforming fuel gas/natural gas into H_2 could reduce well-to-wheel GHG emissions by 60% (range of 55%	[119]

Continued

Table 19.4: Summary of the LCA of different conversion technologies.—cont'd

Conversion technologies	Process	Type of waste	End product/s	FU	Environmental impact categories	Comment	References
						−64%) compared to the mean of petroleum fuels.	
Thermochemical	Pyrolysis	Logging residue, hybrid poplar, willow, waste wood	Electricity	1 kWh	GWP	Depending on the type of feedstock (pyrolysis oil to fossil fuel combustion), life-cycle GHG savings of 77% −99% were estimated for power generation.	[100]
Thermochemical	Pyrolysis	Wood chip	Electricity	1 kWh	GWP, ODP, POCP, ACP, EUP.	Production of 1 kWh in waste biorefinery led to significant differences for GWP_{100}: −87.7%, eutrophication potential (EUP: +23.2%), acidification potential (ACP: −44.6%), photochemical ozone creation potential (POCP: −63.6%), and ozone depletion potential (ODP: −92.4%)	[120]
Thermochemical	Pyrolysis	Corn stover, switch grass	Biochar	1 ton of dry biomass	GWP	GWP for corn stover: −864 kg CO_2 eq. per FU GWP for Switchgrass: +36 kg CO_2 eq. per FU	[121]
Thermochemical	Pyrolysis	Wood waste	Biofuel and power	h	GWP, OLD, photochemical smog, AC, EU, HT, toxicity.	Emission from the combustion of bio-oil was effective on GWP, AC, HT, and EU.	[122]

Thermochemical	Gasification	Willow biomass	Heat and power	1 MWh	Fossil energy requirement, primary energy requirement, LU, GWP, AC.	Significant reduction in GHG emissions was observed from willow biomass versus the fossil fuel reference systems, but the biomass-based systems led to higher acidifying emissions.	[107]
Thermochemical	Gasification	Forest residue	Heat and power	1 MJ	GWP, ACP, abiotic depletion potential (ADP), EUP, photochemical oxidation potential (POP), ozone layer depletion potential (OLDP), human toxicity potential (HTP), freshwater aquatic toxicity potential (FWTP), marine aquatic toxicity potential (MTP), and terrestrial toxicity potential (TTP).	GWP: 8.8–10.5 g CO_2 eq. per FU Environmental impacts were found significant for both biomass procurement and plant operation.	[123]
Thermochemical	Gasification	Different agricultural wastes	Hydrogen	1 MJ	GHG, winter smog, summer smog, AC, EU, carcinogenesis, heavy metals.	Biomass-gasification-electricity-electrolysis route was found to possess a more favorable environmental performance.	[124]
Thermochemical	Combustion	Birch wood	Heat	1 kWh	GWP, PCOP, AC, EU.	GWP: 80–110 g CO_2 eq. per FU	[125]
Thermochemical	Combustion	Rice husk	Electricity	1 MWh	GWP, AC, EU, ecotoxicity.	GWP: 217.33 kg CO_2 eq. per FU	[126]

Continued

Table 19.4: Summary of the LCA of different conversion technologies.—cont'd

Conversion technologies	Process	Type of waste	End product/s	FU	Environmental impact categories	Comment	References
Thermochemical	Combustion	Forest residue	Power	1 kWh	GWP	GWP: 11–14 g CO_2 eq. per FU Chipping at landing used less energy and led to less GHG emissions in comparison with chipping at the power plant.	[127]
Thermochemical	Combustion	Wood waste	Electricity	1 MJ	Climate change, respiratory effect, POFP, AC, EU,	Inventory data collection was found as the major factor for LCA analysis.	[128]
Thermochemical	Gasification	Sewage sludge, woody biomass	Biochars	0.2 kg sewage sludge and 0.8 kg of woody biomass, i.e., 1 kg of sewage sludge and woody biomass mixture	Climate change mitigation.	Potentially, a reduction of 137.0–164.1 kt CO_2 eq. per year could be achieved.	[99]
Thermochemical	Pyrolysis, gasification, and incineration	MSW	Electricity	1 ton of MSW	GW, AC, terrestrial eutrophication (TEU), photochemical ozone formation to human health (POFh), human toxicity via air (HT a) and solid (HTs), and ecotoxicity via solid (ETs).	In comparison with direct incineration, pyrolysis and gasification were found more efficient in decreasing the environmental impacts of TEU, POFh, HTa, and ETs, while it incremented the burdens of GW and HTs.	[105]

classification, characterization, normalization grouping, and weighting could provide decision and policymakers with important information toward enhancing sustainable aspects of waste biorefineries in the future. In addition, LCA results of waste biorefineries could supplement economic evaluations leading to a better understanding of the most environmentally sound management policies and decisions.

Acknowledgments

The authors would like to extend their appreciation to Biofuel Research Team (BRTeam), Universiti Teknologi MARA (UiTM), Agricultural Biotechnology Research Institute of Iran (ABRII), University of Tehran, and Iranian Biofuel Society (IBS) for supporting this work.

References

[1] Zhang C, Lin Y. Panel estimation for urbanization, energy consumption and CO_2 emissions: a regional analysis in China. Energy Policy 2012;49:488–98.
[2] Wang Q, Zeng Y, Wu B-w. Exploring the relationship between urbanization, energy consumption, and CO_2 emissions in different provinces of China. Renewable and Sustainable Energy Reviews 2016;54:1563–79.
[3] Poumanyvong P, Kaneko S. Does urbanization lead to less energy use and lower CO_2 emissions? A cross-country analysis. Ecological Economics 2010;70:434–44.
[4] Zhou J, Lin J, Cui S, Qiu Q, Zhao Q. Exploring the relationship between urban transportation energy consumption and transition of settlement morphology: a case study on Xiamen Island, China. Habitat International 2013;37:70–9.
[5] Al-mulali U, Fereidouni HG, Lee JYM, Sab CNBC. Exploring the relationship between urbanization, energy consumption, and CO_2 emission in MENA countries. Renewable and Sustainable Energy Reviews 2013;23:107–12.
[6] Ripa M, Fiorentino G, Vacca V, Ulgiati S. The relevance of site-specific data in Life Cycle Assessment (LCA). The case of the municipal solid waste management in the metropolitan city of Naples (Italy). Journal of Cleaner Production 2017;142:445–60.
[7] Vergara SE, Tchobanoglous G. Municipal solid waste and the environment: a global perspective. Annual Review of Environment and Resources 2012;37:277–309.
[8] Hoornweg D, Bhada-Tata P. What a waste: a global review of solid waste management. Washington, DC: World Bank; 2012. http://hdl.handle.net/10986/17388.
[9] Ouda OKM, Cekirge HM, Raza SAR. An assessment of the potential contribution from waste-to-energy facilities to electricity demand in Saudi Arabia. Energy Conversion and Management 2013;75:402–6.
[10] Astrup TF, Tonini D, Turconi R, Boldrin A. Life cycle assessment of thermal waste-to-energy technologies: review and recommendations. Waste Management 2015;37:104–15.
[11] Hülsey MJ. Shell biorefinery: a comprehensive introduction. Green Energy & Environment 2018;3:318–27.
[12] Mohan SV. Reorienting waste remediation towards harnessing bioenergy: a paradigm shift. In: Industrial wastewater treatment, recycling and reuse. Butterworth-Heinemann Oxford; 2014. p. 235–81.
[13] Aresta M, Dibenedetto A, Dumeignil F. Biorefinery: from biomass to chemicals and fuels. Walter de Gruyter; 2012.
[14] Cherubini F. The biorefinery concept: using biomass instead of oil for producing energy and chemicals. Energy Conversion and Management 2010;51:1412–21.

[15] Mohan SV, Butti SK, Amulya K, Dahiya S, Modestra JA. Waste biorefinery: a new paradigm for a sustainable bioelectro economy. Trends in Biotechnology 2016;34:852–5.
[16] Ouda OKM, Raza SA, Nizami AS, Rehan M, Al-Waked R, Korres NE. Waste to energy potential: a case study of Saudi Arabia. Renewable and Sustainable Energy Reviews 2016;61:328–40.
[17] US EIA. Levelized cost of new generation resources in the annual energy outlook 2013. Washington, DC: US Energy Inf. Adm.; 2013. https://www.eia.gov/outlooks/aeo/electricity_generation.php.
[18] Ouda OKM, Raza SA, Al-Waked R, Al-Asad JF, Nizami AS. Waste-to-energy potential in the Western Province of Saudi Arabia. Journal of King Saud University Science 2017;29:212–20.
[19] Ouda OKM, Cekirge HM. Potential environmental values of waste-to-energy facilities in Saudi Arabia. Arabian Journal for Science and Engineering 2014;39:7525–33.
[20] Mohan SV, Nikhil GN, Chiranjeevi P, Reddy CN, Rohit MV, Kumar AN, Sarkar O. Waste biorefinery models towards sustainable circular bioeconomy: critical review and future perspectives. Bioresource Technology 2016;215:2–12.
[21] Tonini D, Astrup T. Life-cycle assessment of a waste refinery process for enzymatic treatment of municipal solid waste. Waste Management 2012;32:165–76.
[22] IEA Bioenergy. IEA Bioenergy Task 37-country reports summary 2014. France: IEA Bioenergy, International Energy Agency Paris; 2015.
[23] Parada MP, Osseweijer P, Duque JAP. Sustainable biorefineries, an analysis of practices for incorporating sustainability in biorefinery design. Industrial Crops and Products 2017;106:105–23.
[24] Hassan SS, Williams GA, Jaiswal AK. Lignocellulosic biorefineries in Europe: current state and prospects. Trends in Biotechnology 2018;37:231–4.
[25] de Jong E, Jungmeier G. Biorefinery concepts in comparison to petrochemical refineries. In: Industrial biorefineries and white biotechnology. Elsevier; 2015. p. 3–33.
[26] Guo M, Song W, Buhain J. Bioenergy and biofuels: history, status, and perspective. Renewable and Sustainable Energy Reviews 2015;42:712–25.
[27] Parajuli R, Dalgaard T, Jørgensen U, Adamsen APS, Knudsen MT, Birkved M, Gylling M, Schjørring JK. Biorefining in the prevailing energy and materials crisis: a review of sustainable pathways for biorefinery value chains and sustainability assessment methodologies. Renewable and Sustainable Energy Reviews 2015;43:244–63.
[28] Gerssen-Gondelach SJ, Saygin D, Wicke B, Patel MK, Faaij APC. Competing uses of biomass: assessment and comparison of the performance of bio-based heat, power, fuels and materials. Renewable and Sustainable Energy Reviews 2014;40:964–98.
[29] Service RF. Renewable Energy. Cellulosic ethanol at last? Science 2014;345:1111.
[30] Renewable Fuels Association. Going global: 2015 Ethanol industry outlook. Washington, DC. 2015.
[31] Sims REH, Mabee W, Saddler JN, Taylor M. An overview of second generation biofuel technologies. Bioresource Technology 2010;101:1570–80.
[32] Stephen JD, Sokhansanj S, Bi X, Sowlati T, Kloeck T, Townley-Smith L, Stumborg MA. The impact of agricultural residue yield range on the delivered cost to a biorefinery in the Peace River region of Alberta, Canada. Biosystems Engineering 2010;105:298–305.
[33] Hess JR, Wright CT, Kenney KL. Cellulosic biomass feedstocks and logistics for ethanol production. Biofuels, Bioproducts and Biorefining 2007;1:181–90.
[34] Caputo AC, Palumbo M, Pelagagge PM, Scacchia F. Economics of biomass energy utilization in combustion and gasification plants: effects of logistic variables. Biomass and Bioenergy 2005;28:35–51.
[35] Zhao J, Xia L. Ethanol production from corn stover hemicellulosic hydrolysate using immobilized recombinant yeast cells. Biochemical Engineering Journal 2010;49:28–32.
[36] Hahn-Hägerdal B, Galbe M, Gorwa-Grauslund M-F, Lidén G, Zacchi G. Bio-ethanol-the fuel of tomorrow from the residues of today. Trends in Biotechnology 2006;24:549–56.
[37] Liguori R, Amore A, Faraco V. Waste valorization by biotechnological conversion into added value products. Applied Microbiology and Biotechnology 2013;97:6129–47.

[38] McKendry P. Energy production from biomass (part 1): overview of biomass. Bioresource Technology 2002;83:37–46.
[39] Sun S, Sun S, Cao X, Sun R. The role of pretreatment in improving the enzymatic hydrolysis of lignocellulosic materials. Bioresource Technology 2016;199:49–58.
[40] Mussatto SI, Fernandes M, Milagres AMF, Roberto IC. Effect of hemicellulose and lignin on enzymatic hydrolysis of cellulose from brewer's spent grain. Enzyme and Microbial Technology 2008;43:124–9.
[41] Kim JS, Lee YY, Kim TH. A review on alkaline pretreatment technology for bioconversion of lignocellulosic biomass. Bioresource Technology 2016;199:42–8.
[42] Yunus R, Salleh SF, Abdullah N, Biak DRA. Effect of ultrasonic pre-treatment on low temperature acid hydrolysis of oil palm empty fruit bunch. Bioresource Technology 2010;101:9792–6.
[43] Shrotri A, Kobayashi H, Fukuoka A. Catalytic conversion of structural carbohydrates and lignin to chemicals. In: Advances in catalysis. Elsevier; 2017. p. 59–123.
[44] Taherzadeh M, Karimi K. Pretreatment of lignocellulosic wastes to improve ethanol and biogas production: a review. International Journal of Molecular Sciences 2008;9:1621–51.
[45] Aghbashlo M, Tabatabaei M, Rastegari H, Ghaziaskar HS. Exergy-based sustainability analysis of acetins synthesis through continuous esterification of glycerol in acetic acid using Amberlyst® 36 as catalyst. Journal of Cleaner Production 2018;183:1265–75.
[46] Aghbashlo M, Tabatabaei M, Jazini H, Ghaziaskar HS. Exergy-based optimization of a continuous reactor applied to produce value-added chemicals from glycerol through esterification with acetic acid. Energy 2018;150:351–62.
[47] Hosseinzadeh-Bandbafha H, Tabatabaei M, Aghbashlo M, Khanali M, Demirbas A. A comprehensive review on the environmental impacts of diesel/biodiesel additives. Energy Conversion and Management 2018;174:579–614.
[48] Aghbashlo M, Tabatabaei M, Khalife E, Shojaei TR, Dadak A. Exergoeconomic analysis of a DI diesel engine fueled with diesel/biodiesel (B5) emulsions containing aqueous nano cerium oxide. Energy 2018;149:967–78.
[49] Balat M. Potential alternatives to edible oils for biodiesel production – a review of current work. Energy Conversion and Management 2011;52:1479–92.
[50] Aghbashlo M, Hosseinpour S, Tabatabaei M, Dadak A. Fuzzy modeling and optimization of the synthesis of biodiesel from waste cooking oil (WCO) by a low power, high frequency piezo-ultrasonic reactor. Energy 2017;132:65–78.
[51] Aghbashlo M, Tabatabaei M, Hosseinpour S. On the exergoeconomic and exergoenvironmental evaluation and optimization of biodiesel synthesis from waste cooking oil (WCO) using a low power, high frequency ultrasonic reactor. Energy Conversion and Management 2018;164:385–98.
[52] Bankovi-Ili IB, Stojkovi IJ, Stamenkovi OS, Veljkovic VB, Hung Y-T. Waste animal fats as feedstocks for biodiesel production. Renewable and Sustainable Energy Reviews 2014;32:238–54.
[53] Ghatak HR. Biorefineries from the perspective of sustainability: feedstocks, products, and processes. Renewable and Sustainable Energy Reviews 2011;15:4042–52.
[54] Nizami A-S, Rehan M, Ouda OKM, Shahzad K, Sadef Y, Iqbal T, Ismail IMI. An argument for developing waste-to-energy technologies in Saudi Arabia. Chemical Engineering Transactions 2015;45:337–42.
[55] Tozlu A, Özahi E, Abuşoğlu A. Waste to energy technologies for municipal solid waste management in Gaziantep. Renewable and Sustainable Energy Reviews 2016;54:809–15.
[56] Aluri S, Syed A, Flick DW, Muzzy JD, Sievers C, Agrawal PK. Pyrolysis and gasification studies of model refuse derived fuel (RDF) using thermogravimetric analysis. Fuel Processing Technology 2018;179:154–66.
[57] Esen M, Yuksel T. Experimental evaluation of using various renewable energy sources for heating a greenhouse. Energy and Buildings 2013;65:340–51.
[58] Pavlas M, Touš M, Bébar L, Stehlík P. Waste to energy – an evaluation of the environmental impact. Applied Thermal Engineering 2010;30:2326–32.

[59] Cherubini F, Ulgiati S. Crop residues as raw materials for biorefinery systems – a LCA case study. Applied Energy 2010;87:47–57.
[60] Glew D, Stringer LC, McQueen-Mason S. Achieving sustainable biomaterials by maximising waste recovery. Waste Management 2013;33:1499–508.
[61] Sacramento Rivero JC. A methodology for evaluating the sustainability of biorefineries: framework and indicators. Biofuels, Bioproducts and Biorefining 2012;6:32–44.
[62] Bridgwater AV. Review of fast pyrolysis of biomass and product upgrading. Biomass and Bioenergy 2012;38:68–94.
[63] Fernández Y, Arenillas A, Menéndez JÁ. Microwave heating applied to pyrolysis. In: Advances in induction and microwave heating of mineral and organic materials. InTech; 2011.
[64] Zhang L, Xiao B, Hu Z, Liu S, Cheng G, He P, Sun L. Tar-free fuel gas production from high temperature pyrolysis of sewage sludge. Waste Management 2014;34:180–4.
[65] Dai Q, Jiang X, Jiang Y, Jin Y, Wang F, Chi Y, Yan J. Formation of PAHs during the pyrolysis of dry sewage sludge. Fuel 2014;130:92–9.
[66] Sanchez NE, Callejas A, Millera A, Bilbao R, Alzueta MU. Polycyclic aromatic hydrocarbon (PAH) and soot formation in the pyrolysis of acetylene and ethylene: effect of the reaction temperature. Energy & Fuels 2012;26:4823–9.
[67] Scheirs J, Kaminsky W. Feedstock recycling and pyrolysis of waste plastics. Chichester, UK: John Wiley & Sons; 2006.
[68] Thomas S, Wornat MJ. The effects of oxygen on the yields of polycyclic aromatic hydrocarbons formed during the pyrolysis and fuel-rich oxidation of catechol. Fuel 2008;87:768–81.
[69] Uçar S, Karagöz S, Yanik J, Saglam M, Yuksel M. Copyrolysis of scrap tires with waste lubricant oil. Fuel Processing Technology 2005;87:53–8.
[70] Arena U. Process and technological aspects of municipal solid waste gasification. A review. Waste Management 2012;32:625–39.
[71] Robinson T, Bronson B, Gogolek P, Mehrani P. Sample preparation for thermo-gravimetric determination and thermo-gravimetric characterization of refuse derived fuel. Waste Management 2016;48:265–74.
[72] Fan J. The microwave activation of cellulose. 2013.
[73] Wising U, Stuart P. Identifying the Canadian forest biorefinery. Pulp and Paper Canada 2006;107:25–30.
[74] Pan X, Gilkes N, Saddler JN. Effect of acetyl groups on enzymatic hydrolysis of cellulosic substrates. Holzforschung 2006;60:398–401.
[75] Hamelinck CN, Van Hooijdonk G, Faaij APC. Ethanol from lignocellulosic biomass: techno-economic performance in short-, middle-and long-term. Biomass and Bioenergy 2005;28:384–410.
[76] Kwietniewska E, Tys J. Process characteristics, inhibition factors and methane yields of anaerobic digestion process, with particular focus on microalgal biomass fermentation. Renewable and Sustainable Energy Reviews 2014;34:491–500.
[77] Zhang W, Wei Q, Wu S, Qi D, Li W, Zuo Z, Dong R. Batch anaerobic co-digestion of pig manure with dewatered sewage sludge under mesophilic conditions. Applied Energy 2014;128:175–83.
[78] Qian M, Li Y, Zhang Y, Sun Z, Wang Y, Feng J, Yao Z, Zhao L. Efficient acetogenesis of anaerobic co-digestion of food waste and maize straw in a HSAD reactor. Bioresource Technology 2019;283:221–8.
[79] Themelis NJ, Verma S. The better option – anaerobic digestion of organic wastes in MSW. Waste Management World 2004:41–8.
[80] Rabu RA, Janajreh I, Honnery D. Transesterification of waste cooking oil: process optimization and conversion rate evaluation. Energy Conversion and Management 2013;65:764–9.
[81] Aransiola EF, Ojumu TV, Oyekola OO, Madzimbamuto TF, Ikhu-Omoregbe DIO. A review of current technology for biodiesel production: state of the art. Biomass and Bioenergy 2014;61:276–97.
[82] Salam KA, Velasquez-Orta SB, Harvey AP. A sustainable integrated in situ transesterification of microalgae for biodiesel production and associated co-product – a review. Renewable and Sustainable Energy Reviews 2016;65:1179–98.

[83] Gardy J, Hassanpour A, Lai X, Ahmed MH, Rehan M. Biodiesel production from used cooking oil using a novel surface functionalised TiO_2 nano-catalyst. Applied Catalysis B: Environmental 2017;207:297–310.

[84] Yaakob Z, Mohammad M, Alherbawi M, Alam Z, Sopian K. Overview of the production of biodiesel from waste cooking oil. Renewable and Sustainable Energy Reviews 2013;18:184–93.

[85] Sharma YC, Singh B. Development of biodiesel from karanja, a tree found in rural India. Fuel 2008;87:1740–2.

[86] Meher LC, Sagar DV, Naik SN. Technical aspects of biodiesel production by transesterification – a review. Renewable and Sustainable Energy Reviews 2006;10:248–68.

[87] ISO, 14044 International Standard. Environmental management – life cycle assessment – principles and framework. Geneva, Switzerland: International Organisation for Standardization; 2006.

[88] Sanaei S, Janssen M, Stuart PR. LCA-based environmental evaluation of biorefinery projects. In: Integrated biorefineries: design, analysis, and optimization. Boca Raton: CRC; 2013. p. 793–817.

[89] Eskandarpour M, Dejax P, Miemczyk J, Péton O. Sustainable supply chain network design: an optimization-oriented review. Omega 2015;54:11–32.

[90] Silvestre JD, De Brito J, Pinheiro MD. Environmental impacts and benefits of the end-of-life of building materials-calculation rules, results and contribution to a -cradle to cradle- life cycle. Journal of Cleaner Production 2014;66:37–45.

[91] Nizami AS, Rehan M, Waqas M, Naqvi M, Ouda OKM, Shahzad K, Miandad R, Khan MZ, Syamsiro M, Ismail IMI. Waste biorefineries: enabling circular economies in developing countries. Bioresource Technology 2017;241:1101–17.

[92] Shahzad K, Rehan M, Ismail IMI, Sagir M, Tahir MS, Bertok B, Nizami A-S. Comparative life cycle analysis of different lighting devices. Chemical Engineering Transactions 2015;45:631–6.

[93] Heijungs R, Guinée JB. Allocation and 'what-if'scenarios in life cycle assessment of waste management systems. Waste Management 2007;27:997–1005.

[94] Cherubini F, Bargigli S, Ulgiati S. Life cycle assessment of urban waste management: energy performances and environmental impacts. The case of Rome, Italy. Waste Management 2008;28:2552–64.

[95] Cherubini F, Strømman AH, Ulgiati S. Influence of allocation methods on the environmental performance of biorefinery products – a case study. Resources, Conservation and Recycling 2011;55:1070–7.

[96] Wolf M-A, Pant R, Chomkhamsri K, Sala S, Pennington D. The International Reference Life Cycle Data system (ILCD) handbook – JRC reference reports. 2012.

[97] Sandin G, Røyne F, Berlin J, Peters GM, Svanström M. Allocation in LCAs of biorefinery products: implications for results and decision-making. Journal of Cleaner Production 2015;93:213–21.

[98] Mimoso AF, Carvalho A, Mendes AN, Matos HA. Roadmap for environmental impact retrofit in chemical processes through the application of life cycle assessment methods. Journal of Cleaner Production 2015;90:128–41.

[99] Ramachandran S, Yao Z, You S, Massier T, Stimming U, Wang C-H. Life cycle assessment of a sewage sludge and woody biomass co-gasification system. Energy 2017;137:369–76.

[100] Fan J, Kalnes TN, Alward M, Klinger J, Sadehvandi A, Shonnard DR. Life cycle assessment of electricity generation using fast pyrolysis bio-oil. Renewable Energy 2011;36:632–41.

[101] Iribarren D, Peters JF, Dufour J. Life cycle assessment of transportation fuels from biomass pyrolysis. Fuel 2012;97:812–21.

[102] Sebastião D, Gonçalves MS, Marques S, Fonseca C, Gírio F, Oliveira AC, Matos CT. Life cycle assessment of advanced bioethanol production from pulp and paper sludge. Bioresource Technology 2016;208:100–9.

[103] Boldrin A, Balzan A, Astrup T. Energy and environmental analysis of a rapeseed biorefinery conversion process. Biomass Conversion and Biorefinery 2013;3:127–41.

[104] Nizami AS, Shahzad K, Rehan M, Ouda OKM, Khan MZ, Ismail IMI, Almeelbi T, Basahi JM, Demirbas A. Developing waste biorefinery in Makkah: a way forward to convert urban waste into renewable energy. Applied Energy 2017;186:189–96.

[105] Dong J, Tang Y, Nzihou A, Chi Y, Weiss-Hortala E, Ni M. Life cycle assessment of pyrolysis, gasification and incineration waste-to-energy technologies: theoretical analysis and case study of commercial plants. The Science of the Total Environment 2018;626:744–53.

[106] Chang FC, Lin LD, Ko CH, Hsieh HC, Yang BY, Chen WH, Hwang WS. Life cycle assessment of bioethanol production from three feedstocks and two fermentation waste reutilization schemes. Journal of Cleaner Production 2017;143:973–9.

[107] Kimming M, Sundberg C, Nordberg Å, Baky A, Bernesson S, Norén O, Hansson P-A. Biomass from agriculture in small-scale combined heat and power plants – a comparative life cycle assessment. Biomass and Bioenergy 2011;35:1572–81.

[108] González-García S, Moreira MT, Feijoo G. Comparative environmental performance of lignocellulosic ethanol from different feedstocks. Renewable and Sustainable Energy Reviews 2010;14:2077–85.

[109] Silalertruksa T, Gheewala SH. A comparative LCA of rice straw utilization for fuels and fertilizer in Thailand. Bioresource Technology 2013;150:412–9.

[110] Shafie SM, Mahlia TMI, Masjuki HH. Life cycle assessment of rice straw co-firing with coal power generation in Malaysia. Energy 2013;57:284–94.

[111] Moretti M, Van Dael M, Malina R, Van Passel S. Environmental assessment of waste feedstock mono-dimensional and bio-refinery systems: combining manure co-digestion and municipal waste anaerobic digestion. Journal of Cleaner Production 2018;171:954–61.

[112] Ripa M, Fiorentino G, Giani H, Clausen A, Ulgiati S. Refuse recovered biomass fuel from municipal solid waste. A life cycle assessment. Applied Energy 2017;186:211–25.

[113] Cornelissen RL, Hirs GG. The value of the exergetic life cycle assessment besides the LCA. Energy Conversion and Management 2002;43:1417–24.

[114] Heller MC, Keoleian GA, Mann MK, Volk TA. Life cycle energy and environmental benefits of generating electricity from willow biomass. Renewable Energy 2004;29:1023–42.

[115] Mann M, Spath P. A life cycle assessment of biomass cofiring in a coal-fired power plant. Clean Products and Processes 2001;3:81–91.

[116] Sebastián F, Royo J, Gómez M. Cofiring versus biomass-fired power plants: GHG (Greenhouse Gases) emissions savings comparison by means of LCA (Life Cycle Assessment) methodology. Energy 2011;36:2029–37.

[117] Kauffman N, Hayes D, Brown R. A life cycle assessment of advanced biofuel production from a hectare of corn. Fuel 2011;90:3306–14.

[118] Hsu DD. Life cycle assessment of gasoline and diesel produced via fast pyrolysis and hydroprocessing. Biomass and Bioenergy 2012;45:41–7.

[119] Han J, Elgowainy A, Dunn JB, Wang MQ. Life cycle analysis of fuel production from fast pyrolysis of biomass. Bioresource Technology 2013;133:421–8.

[120] Faix A, Schweinle J, Schöll S, Becker G, Meier D. (GTI tcbiomass) life cycle assessment of the BTO® process (biomass to oil) with combined heat and power generation. Environmental Progress & Sustainable Energy 2010;29:193–202.

[121] Roberts KG, Gloy BA, Joseph S, Scott NR, Lehmann J. Life cycle assessment of biochar systems: estimating the energetic, economic, and climate change potential. Environmental Science & Technology 2009;44:827–33.

[122] Zhong ZW, Song B, Zaki MBM. Life-cycle assessment of flash pyrolysis of wood waste. Journal of Cleaner Production 2010;18:1177–83.

[123] Guest G, Bright RM, Cherubini F, Michelsen O, Strømman AH. Life cycle assessment of biomass based combined heat and power plants: centralized versus decentralized deployment strategies. Journal of Industrial Ecology 2011;15:908–21.

[124] Koroneos C, Dompros A, Roumbas G. Hydrogen production via biomass gasification-a life cycle assessment approach. Chemical Engineering and Processing: Process Intensification 2008;47:1261–8.
[125] Solli C, Reenaas M, Strømman AH, Hertwich EG. Life cycle assessment of wood-based heating in Norway. International Journal of Life Cycle Assessment 2009;14:517–28.
[126] Shafie SM, Mahlia TMI, Masjuki HH, Rismanchi B. Life cycle assessment (LCA) of electricity generation from rice husk in Malaysia. Energy Procedia 2012;14:499–504.
[127] Thakur A, Canter CE, Kumar A. Life-cycle energy and emission analysis of power generation from forest biomass. Applied Energy 2014;128:246–53.
[128] Périlhon C, Alkadee D, Descombes G, Lacour S. Life cycle assessment applied to electricity generation from renewable biomass. Energy Procedia 2012;18:165–76.

SECTION G

System dynamics and carbon footprints

CHAPTER 20

System dynamics on wood and yard waste management

Tiffany M.W. Mak[1], Lei Wang[1,2], Daniel C.W. Tsang[1]
[1]Department of Civil and Environmental Engineering, The Hong Kong Polytechnic University, Kowloon, Hong Kong, China; [2]Department of Materials Science and Engineering, The University of Sheffield, Sheffield, United Kingdom

20.1 Introduction

20.1.1 Holistic review on municipal solid waste around the globe

Municipal solid waste (MSW) consists of waste generated from residential, commercial, institutional, and industrial sources, including wood, yard trimmings, durable and nondurable goods, food waste, and inorganic waste. In 2015, over 262 million tonnes of MSW were generated in the United States [1], about 26% of which is recovered for recycling from processes such as composting. Among different types of MSW, there are two types which contain solid wood, namely "wood" and "yard trimmings." Because of the wood component, the type of "wood" can further be categorized into items such as wooden furniture and cabinets, wooden panels, wood formwork, and wood from manufacturing facilities. However, round wood, unprocessed wood, repaired wood, or recycled pallets are not included. Meanwhile, "yard trimmings" refers to leaves and grass clippings, brush, and tree trimmings [1]. In 2015, approximately 16.3 million tonnes of MSW wood waste were produced, with a low recovery rate at 16%. In MSW, the total wood waste is approximately 6.2%. However, the quantity of wood waste varies by countries. At the same time, over 30 million tonnes of yard trimmings were generated, comprising of 55% wood and 45% herbaceous material [2,3]. Yard trimmings waste has a higher recovery rate of 58% compared with that of wood waste [1].

Construction industry influences the socioeconomic development of all regions and the industry has been rapidly growing due to the increase of living standard, demands of infrastructure projects to meet the growing population, and reshaping of consumption traits. Such growth is associated with the waste generation that causes severe social and environmental problems globally [4–6]. Construction and demolition (C&D) wastes which are originated from different types of activities and have various characteristics, and differ

Waste Biorefinery. https://doi.org/10.1016/B978-0-12-818228-4.00020-4

in separation, recovery and recyclability processes. C&D waste is defined as one of the waste streams in MSW. Construction waste is generated from the process of construction, repair and remodeling of residential and nonresidential structures while demolition waste is produced when structures are demolished, which is often contaminated with paints, adhesives, wall covering materials. C&D waste is one of the heaviest and most voluminous waste streams generated worldwide [7], which accounts for over 30% of global waste [8]. C&D waste even accounts for over 70%, 50%, 40%, 35%, and 30% of the total waste in Spain, United Kingdom, Australia, Japan and Italy, respectively [9]. In China, C&D waste generation accounts for 40% of the total waste [10]. Countries such as Japan imposed guidelines in managing C&D waste. For instance, production of construction waste at construction site is avoided, reuse and proper handling of construction waste are encouraged. In Holland, regulations have been set up the to restrict waste disposal and enforce recycling by adopting quality-control systems [9]. In the past decades, environmental pollution, resource overdepletion, and increasing land price have been caused by improper C&D waste treatment and disposal, exerting great threat to the living environment [11]. Previous researchers indicated that waste is generated from planning, design, procurement, and construction stage [12–14], and it affects the economical dynamics of community and poses impacts on the environment [15]. Over 540 million tonnes of C&D waste were generated in the United States in 2015, which contained 7% of wood. In particular, over 90% of wood debris was from demolition activities while the rest was from construction activities. C&D waste generation and recovery rates are affected by various economic drivers such as housing completions and the changing population [2,3]. In the United States, wood waste generation is significantly influenced by the general economic recession, specifically new residential construction. Over 35 million tonnes of C&D wood generated, nearly 48% of waste was recoverable. The percentage of wood varied from around 25%–55%, which the total amount of wood in the C&D waste stream is about 30% [16]. Construction waste can be classified into two groups, namely physical and nonphysical waste. Physical waste contributes significantly to landfill. Previous studies reported that the construction industry generates more than 50% of waste material ending up in landfills [17]. On the other hand, nonphysical construction wastes refer to intangible expenses such as time and expense incurred in completion delay in construction projects [18,19].

20.1.2 Development of system dynamics model

A model is defined as "an external and explicit representation of part of reality as seen by the people who wish to use that model to understand, to change, to manage and to control that part of reality" [20]. The development process of a model emphasizes a part of reality that requires deeper understanding and better management when creating an external and specific representation of this reality [21]. In different scenarios, decision makers encounter a social system that is complex and highly interacting with communities. Development of models offers a platform to decision makers to understand their communities as an interconnected system, which can consequently test the effect of policy interventions. Researcher provides a classification on the types of model that are available to predict future behavior. Firstly, it is a model that provides absolute and precise predictions, for example, when and where the next solar eclipse would be observable. Secondly, it provides conditional and precise predictions, for example, what the maximum pressure exerted on the reactor's containment vessel would be if the cooling system in a nuclear plant failed. Lastly, it provides conditional but imprecise projections on dynamic behavior, for example, what the possible future burden of demand on intensive care facilities would be after 1 month from the spread of an infectious disease [22]. System dynamics (SD) primarily provides modeling for economic and policy simulations [23], and it can be categorized into the third class of model which provides conditional but imprecise projections on dynamic behavior. This may due to the characteristics that social and economic systems can never be predicted in absolute accuracy [22]. Since researcher proposes that most of the models are in fact wrong [24] that they are unable to produce precise point-predictions on future social events, it is challenging to create beneficial models through extensive validation of projected results with data from the real world [25].

The environmental problem is known as a complex and dynamic issue that involves various disciplines to address the issue [26]. SD approach offers a suitable tool to capture and model the key components of environmental systems, which has a wide range of economic and environmental applications. Although SD can be applied to model restoration, it has been limited to either wetland or watershed issue (e.g., Refs. [27–29]). Other SD applications are widespread and increasingly prominent in resolving water, agricultural and other environmental problems, etc. For instance, researchers develop a model to understand the restoration of mountain fynbos ecosystems South Africa [30]. Some researchers focus on modeling wetland management in the Limpopo river basin [31] and estuaries [32]. Other studies include the modeling of the Indonesian agroforestry sector [33] and the aquaculture sector in China [34].

Systems includes four main components: (1) State variables, which act as reservoirs, control variables for referring to ongoing processes within the system to decide reservoir. (2) Converters which are system variables, often influence the rates of operating processes. Balancing or reinforcing relationships represent the connections of all components within the system [35]. In SD model, the four main components to develop various system scenarios include stocks (state variables); flows (processes entering and exiting the state variable), auxiliary variables (converter) and information flows (connectors) [36]. In particular, a stock acts as the foundation of any system [37], which distinguishes the state of the system under investigation and provide the required information to assist in further decisions and actions [38]. Stocks can only be changed by flows, which are the quantities inflowing, or outflowing from stock over time [21]. There are three steps in forming the SD model. Firstly, develop connections of the generic structures of system or consider structural variables that may result in causing conflicts. Secondly, analyze results from a casual loop diagram to envisage the relationship of the problematic system. Lastly, provide answers for future improvement.

Components within the system are envisaged to interact with other recognized variables as feedback processes [36]. The feedback mechanism defines SD [39], and it is indispensable in model development to recognize feedback loops within social systems in high complexity. A feedback loop is a series of circular causal connections, which a stock influences a flow and consequently changes another stock [40]. Two types of feedback loops are identified by researchers: reinforcing feedback and balancing feedback. Reinforcing feedback counteracts the direction of change while balancing feedback drives exponential growth of stocks [21]. The merit of the SD model is to form complex feedback mechanisms effectively by integrating changes from small to large parameters that are incidental or continual. It is easier to analyze and manage complex environmental systems by developing interconnections across various sectors of a process structure, such as the waste-management system. Time-lag refers to the response time of a stimulus to the system. Variation, which occurs when the control variables do not depend on other variables linearly, has arisen in the feedback mechanism due to the presence of nonlinear relationships [40]. SD has the following distinctive methodological principles to facilitate the combination of various kinds of data sources via recognizing numerous interactions and significant time delays [41]. Firstly, SD models represent the system structure. Since the behavior of a system originates from its structure, it also represents the behavior of a system. Secondly, SD models capture disequilibrium. Thirdly, SD emphasizes on a broad boundary, and considers the feedback and the delayed impacts of decisions on its system. Lastly, interactions among system components are consistent with the real world since they are captured when modelers test the models with grounded methods [41].

This chapter aims to provide a holistic literature review on the management of wood and yard waste from the MSW and C&D waste streams. This chapter gives a brief overview of global solid waste composition, then introduces the SD approach and its development, applications of SD on various environmental issues, comprehensive literature review on wood and yard waste management in the stream of MSW and C&D, and the current implementation of SD on MSW and C&D waste management.

20.2 Literature review on the application of SD model

SD can be applied to numerous areas. For instance, SD has been applied in social studies, such as business administration, physical and social sciences, mathematics, law, medicine, and education [42]. Business administration is often the starting point and ending up in resolving some environmental problems to predict individual's behavior or search for solutions in more complex issues. Regarding business administration, previous researches made use of SD to investigate production cycles of hogs, chickens and cattle and the generation of business cycle for demonstrating production scheduling and workforce management policies. Studies find that business cycles are caused by "capital investment policies that fail to account for delays in acquiring long-lead time plant and equipment" [43]. Other applications of SD such as management and the use of marketing models on corporate planning, policy design, strategy support models and organizational learning are emphasized. For instance, SD was adopted in taxation policy that integrated the approach of public health in the tobacco industry in New Zealand. Study analyzed the relationships of customs service outputs to desired government outcomes, in relation to the collection of tobacco excise duties and cigarette smoking. Policy experiments with the model investigated the impacts of changes in excise duties and the impacts of a price change on tobacco related behaviors. Such results offered valuable insights into the customs and health-related activities associated with the supply and consumption of tobacco products in New Zealand [44]. The use of SD also extends from business administration to dealing with environmental problems. For instance, studies on climate change and solid waste management [45], and mining problem [46] are prominent.

Improving organizational performance is a general approach of the SD application. SD, can be applied in multidiscipline, aims to simulate the reality in order to achieve improvements and. In terms of Industrial Dynamics, which is one of the most common application areas of SD. In the context of enterprise design approach (EDA), the problem managed is first recognized that the modeler is responsible for creating the structure system of the problem definition. Results from EDA helps develop SD model that only contains the factors affecting the problem. This state-of-the-art approach is that the interactions of variables in the structure are also under consideration when SD facilitates the development of the organization. Model development provides basic knowledge on the system structure and testing of improvements on the model provides more knowledge that lead to findings of possible improvements. The process may be perhaps simple, but it is time-consuming and requires effort in understanding and hypothesizing relationships of variables within one system. The complexity of developing a valid SD model and the characteristics of problem depends on the types of organizational application that modeler chooses. Theoretically, there is no limitation on the area of the organization that SD can be applied. SD simulation is a possible organizational application to understand the characteristic behavior of the actual system on any dilemma that can be quantified and validated.

SD has been considered as a quantitative method, which all relationships among variables are assigned with values. Previous literature discovers that the outcome of qualitative analysis is not measurable and cannot be validated. SD founder, Forrester, believed that simulation was the only solution to reveal systems' behavior since they are too complicated. SD is a method with great capacity and the system behavior can be further understood when making a model for perception of the system [47]. At the same time, systems thinking comes from the application of SD approach but not from the simulation. In particular, SD applications on some environmental problems are further investigated, such as water management, energy policy and solid waste management.

20.2.1 Literature review on SD application in water management

An SD model is developed for determining the most suitable utilization scenario for water resources system in China, which initiates a dynamic hypothesis by separating the regional water resources system into subsystems and then into different departments. In order to illustrate the overall feedback mechanism of the water resources system in China, researchers define the causal feedback structure of water supply and demand [48]. Previous literature emphasizes on the demand-sided approach by developing a set of mathematical equations to project water demand, water supply and the effect of water price on demand in Yulin city in China. It is more effective to resolve the issue of water scarcity, when compared to supply-sided approach [49]. Moreover, the supply and demand balances trend of water resources in Shandong Province in China is predicted by establishing an SD model to analyze the regression equation of the growing population [50]. In the global context, a participatory SD model is initiated to establish a model for sustainable water resource management for the Palouse basin in the United States. Researchers perform an iterative process that includes workshops and model building between workshops to propose a dynamic hypothesis [51]. Workshops including surveys indicated that most participants strongly agreed that the process of building a participatory model was a good method for providing social input into water management decisions and an effective way to communicate personal values. There was also concern that such a process would end up political and have the potential to be detrimental to relationships that were already fragile. The above study demonstrated a positive attitude toward how participatory SD modeling may be used to assist communities with sustainable water resource management. In order to improve the effectiveness of irrigation water systems in Australia, an SD model is developed to assist irrigators, water policy decision makers and water supply authorities [52]. While in Florida, the domestic water demand for Manatee County is estimated by an SD model consisting of three submodels, namely socioeconomic, population, and water demand. Researchers have successfully quantified the interrelationships among socioeconomic variables of these submodels by collecting population data for population dynamics submodel from the US Census Bureau and the historical mean annual wage in Florida from the US Department of Labor and other relevant literature to formulate regression and empirical equations [53].

SD is often applied in policy analysis, which aims to investigate how specific change in a parameter in the model impacts its response. It allows modelers to recognize the policy levers that would have the desired effects on the proposed model [54]. Ten different policies and scenarios are proposed in previous research on the exploration of water supply and demand scenarios to examine the sustainability of the water resources in China. The targeted outcome of the above policy analysis aims to provide decision makers with viable options of higher contributions to sustain water supply in the Yellow River Basin. Each scenario consists of various interconnected factors and inputs from different socioeconomic sectors [55]. Other researchers compare the effect of different policy options such as low flow appliances, xeriscaping and pricing with the status quo scenario over a time horizon of 25 years to reduce municipal water demands. Results reveal that a maximum possible reduction in the municipal water demands can be achieved by integrating the use of different policies [56]. In Canada, researchers conduct policy analysis to uncover the significant relationships between the future development and a sustainable and acceptable quality of water resources and restraining demand. An integrated water resource management SD model is developed by classifying 12 scenarios into four groups that are population, water, economy and energy. Each scenario contains multiple policy variables from different sectors [57]. Previous literature creates five alternatives from a different combination of four policy options on the aquifer management plan in Idaho to evaluate system reliability. The four policy options include (1) a conservation plan for groundwater and surface water (i.e., CON); (2) managed aquifer recharge to maximize outcomes for fish and wildlife, surface and groundwater quality, hydropower, and recreation (i.e., MAR); (3) demand reduction via irrigation efficiency and groundwater curtailment during drought (i.e., GWC); and (4) weather modification program to increase winter snowpack and augment surface-water flow (i.e., WMO). Five alternatives created from the above four policy options can be found in Table 20.1. The system reliability which is at 97% is highly satisfactory in simulating the

Table 20.1: Five alternatives from four policy options.

Alternatives	Policy options
1	CON
	CON + MAR
2	CON
	CON + MAR
	CON + MAR + WMO
3	GWC
4	GWC + WMO
5	CON
	CON + MAR
	CON + MAR + WMO
	CON + MAR + WMO + GWC

outcomes of the designed policy options stipulated for Snake River basin [58]. In order to determine the maximum possible growth in variables such as crop yield, net-farm income, population, agricultural, domestic and industrial, water demand by 2050, three different policy scenarios, including the development of water infrastructure, cropland expansion and dry conditions, are designed for the sustainable water resources management in the Volta River basin, Ghana [59].

20.2.2 Literature review on SD application in energy policy formulation

Besides water management, energy policy analysis is another popular application of SD model. Three categories can be classified in the context of energy policy formulation, which are strategic, tactical and operational problems. To better understand the use of SD model on energy system policy, problems are categorized into energy–economy–environment (3Es) including energy demand–supply management problem, new product innovation problem, capacity management problem, and energy pricing problem [60,61].

The 3Es concept is desired to be incorporated in energy policies by decision makers [61,62], that models developed emphasize on fulfilling both economic profit focused objectives and the sustainability of energy resources, and environmental health [63]. For instance, polluter pays principle is promoted by the implementation of the emissions tax policy [64]. Firstly, energy demand and energy supply related factors are investigated by researchers and energy policy makers by establishing energy systems models [60,61]. Important dynamic factors, such as time delays and nonlinear relationships among variables within system, are considered [64,65]. Previous literature recognized two limitations to the growth of demand–supply related factors, which are energy resources and supply capacity. Demand–supply related factors in this study refer to variables such as economic growth, status of energy resources, industrial activity, propensity for investment, societal development, customer pressure, and technological supply capacity [60,63]. Triple-bottom-line concept was adopted in previous literature to investigate the above-mentioned environmental, economic and social factors. Data were collected from over 670 manufacturing plants over 19 countries. Findings suggested that internal environmental program has a positive impact, whereas internal social initiatives have a positive impact on only social and environmental performance [63].

Secondly, new strategies for innovative energy technologies are crucial to meet the growing global energy demand in an economically and environmentally competitive manner [60,61,63]. It is vital to develop new and clean energy sources, which can be commercialized and adopted by users to benefit our communities. Nevertheless, it is complex to establish innovative energy systems, which are time-consuming and with high failure rate to successfully implement all energy innovation projects. Extensive support from a combination of different operators, commercial disciplines, public and private entities that are in partnerships and investors are necessary to commercialize new energy technology [61]. Thirdly, capacity adjustment decisions can resolve capacity management problem in light of energy demand-supply factors [60] such as availability of energy resources. Policy makers would like to forecast the possible impacts of their decisions during the implementation of energy policies, such problem would be affected by investment opportunities from the commercial companies [66–68]. Fourthly, energy pricing problem can be resolved with the consideration of the dynamic relationships between energy demand, supply and pricing, which are nonlinear [69]. Energy systems constantly move toward the point of equilibrium when the market prices are adjusted by dynamic changes of energy supply. SD model is a suitable approach to evaluate nonlinearities interactions among dynamic variables by formulating multiple feedback loops and time delays that lead to unexpected system behavior [61].

20.2.3 Literature review of on wood and yard waste management

USEPA defines MSW wood waste which include two types—wood and yard trimmings. "Wood" includes wood generated from construction and demolition processes such as scrap lumber, cabinets, and wooden furniture, and others such as wood containers and pallets. On the other hand, yard trimmings refer to brush, leaves, grass clippings, tree trimmings, and removals [70,71]. However, it appears that the major regulatory agency dealing with waste has a different definition for urban wood waste. They define urban wood waste as landfill disposal of yard trimmings, wood waste generated from C&D projects, site removals, pallets, furniture, packaging, and other commercial or household wood waste [72]. Also, some regions put more emphasis on residential and community yard waste than C&D waste. For instance, urban wood waste is described as "yard waste" in the regulations of North Carolina, which yard waste is categorized into two classes: (1) wood waste generated from land-clearing debris and (2) trash that contains yard. During the construction of households, infrastructure, and other commercial buildings, land-clearing debris including trees and other vegetations are generated. On the other hand, plantations and shrub branches, logs, wood from landscaping, and debris from natural disasters are referred to as yard trash [73]. Urban tree removals are identified as a crucial part of the urban wood waste stream in the wood recycling campaign in the San Francisco Bay area [74]. The above examples illustrate that the emphasis is put on trees and yard waste as urban wood waste, while wood from MSW stream and C&D debris are excluded.

Previous literature intend to increase the market value of some low-value products, such as mulch or firewood by conducting urban wood utilization studies and examining the utilization of urban trees and woody yard residues from municipalities [75–77]. Some of these studies discover that plantation accounts for the major constitution of the urban wood waste stream [75,77]. Previous studies emphasize the utilization of plantation from urban wood waste and neglect the composition of urban wood waste. International organizations, such as The Urban Forest Products Alliance and The Tree Care Industry Association, define urban wood as wood from felled urban trees and trimmings [78] and describe a wood product produced from an urban or community tree harvested from residential or public lands as "urban forest product" [79], respectively. The Urban Forest Products Alliance, which consists of representatives from industry and public authorities, aims at promoting sustainable recovery and uses of products from urban forests. The Tree Care Industry Association is a group of leading corporates that work in trade association for commercial tree care. Similarly, corporates consider urban trees and woody debris as urban wood waste [80]. The above section provides discussion on the various definitions of wood and yard waste globally. In short, wood and yard waste can be considered as C&D waste within the MSW stream.

As a tool to manage waste and reduce pollution, industrial symbiosis has been adopted in European countries [81]. Industrial symbiosis refers to a waste that can be considered as a resource for another industry [82]. It is possible to exchange waste within the vicinity of industrial estate. Industries can therefore be benefited financially by exchanging such type of materials [83]. For instance, there are some EU support networks for industrial symbiosis and European Innovation Partnerships such as National Programmes (e.g., NISP [UK], Finnish Industrial SymbioSis System [FISS]), regional initiatives (e.g., Sotenäs municipality [Sweden]) [84]. In this way industrial symbiosis plays a vital role in collaboration among industries for pollution abetment [85]. On the other hand, developing countries such as South Africa, implement a variety of waste regulations, policies and legislation, such as the National Environmental Management: Waste Act 59, the Municipal Systems Act, the Mineral and Petroleum Resources Development Act, and the Air Quality Act [86]. Also, China has passed the National Sword Policy and tightened its restrictions on recyclable waste that only 1% of impurities of waste would be accepted for import in 2018 [87].

In order to conduct a comprehensive literature review on both wood and yard waste in general, studies of the implementation of the SD model on MSW and C&D waste are investigated.

20.2.3.1 Implementation of SD on MSW waste management

The causal interactions of different variables are studied by conducting simulation on the performance of a closed-loop chain [88]. The complexity of the waste generation and management process are incorporated by combining some simpler subprocesses to develop the dynamic system. Parameters that influence waste generation are identified and input into the model for further interpolation with the consideration of a specific prosperity level, depending on the income level of individuals. The dynamic characteristics of the changing environment of a city are captured. Over the past decades, a wide range of disciplines have been studied with the applications of SD, such as the modeling of MSW management systems in the Netherlands that draw attention on the limitations of adopting SD such as the optimum amount of soft variables included in the system of concern and their quantification [89]; environmental sustainability [90]; strategic management [91]; systems to assist decision-making processes [92], and environmental impact assessment [93]. At different prosperity levels, the generation of total MSW is estimated by formulating the suitable regression equations derived through time-series and cross-sectional data analysis for the assessment year [36]. In order to further process the consequences and deepen the understanding of authorities in the complex interactions, it is indispensable to predict and consider recyclables, organic waste, and other discards including mixed and un-separated residual waste. Estimations should also consider the relationships between socioeconomic and demographic conditions and the waste generation rate. Significant parameters in affecting MSW generation are discovered in previous studies, including gross domestic product per capita, infant mortality rates, population distribution, size of household, life expectancy and labor force in agriculture sector [94]. The above six indicators are recognized as strong economic outputs and related process to influence the waste generation. According to established principles and relationships, the basic model structure for city-scale waste generation is developed in Newark city, which is one of the larger urban centers in the New Jersey [94]. A 10-year assessment horizon, the year 2003–13, is selected for modeling. The projections of total waste generation and the collected waste fraction are performed independently. Paper and cardboard, metals, plastic, glass, hazardous waste and organic waste are the modeled waste fraction, which data is collected from the municipal recycling tonnage records at the Essex County Utilities Authority [95]. The state variables include total annual MSW generated per person as stock, the total MSW generation in the assessment period and the available space of landfill with and without respective waste prevention measures. Other studies also investigate factors of waste generation [96].

Economic factors such as gross domestic products are often emphasized in previous waste generation studies. Waste-to-energy technology such as incineration has been implemented in many countries to manage MSW. For instance, previous research in Sweden identified economic benefits was always an issue while designing waste treatment technologies. Landfill tax and waste management treatment cost are also as key economic drivers for Sweden. However, earlier literature proposed other additional factors such as purchasing power [97]. On the other hand, Social indicators identified as potential drivers for technological development of the waste sector in Sweden, are population, the volume of waste generation, people behavior, local waste management practices and the process of urbanization. Population and the volume of waste generation are vital for designing waste management systems. Socio-political drivers such as local and international rules and regulations are also important in the development of waste treatment technology. Regulations have been acting as a supporting tool for promoting, developing or restricting a system [98]. Furthermore, interactions between these additional factors and MSW generation are conceptualized due to an increasing economic growth on regions. This leads to higher spending per capita and results in more MSW generation [99]. Qualitative aspects of waste generation and the separation of recyclable waste at source are analyzed by SD model with their respective impacts of transition from landfill to other waste disposal options [100]. Such model is utilized as a basis to project MSW generation and formulate sustainable waste policies in developing countries [101]. At the same time, an integrated MSW waste management system is established with the emphasis on the collection, transport means and its associated economic and environmental impacts [102]. Poor MSW management also create social impacts, including the unpleasant odor when garbage is left uncollected and the unpleasant odor due to landfill site, the dirty surroundings, breeding of mosquito, worms, insects and flies due to the landfill site and the uncollected garbage and the release of smoke and poisonous gases that cause safety problems. These impacts are also referred to as disamenities due to MSW [103]. Five different SD models are simulated to investigate the associated impacts on the future generated quantities of MSW in terms of site selection, cost assessment, and capacity planning of MSW [104]. Other potential applications of SD model include the assessment of MSW treatment facilities to achieve a desired improvement on the environmental quality by developing MSW generation models [105].

20.2.3.2 Implementation of SD on C&D waste management

General studies on C&D waste management are further explored to provide a broader review to address the problems caused by C&D waste. Over the past 2 decades, C&D waste management have been emphasized and people aim to reduce its amount and impacts on the environment [106,107]. The generation of waste influences the economic, environmental and social aspects in undertaking construction projects either positively or negatively. Therefore, the attempts on its management are of utmost importance to improve C&D waste practices [107]. To investigate the economic benefits of suitable C&D recycling techniques, many studies are conducted on the economic suitability of CDW recycling plants. A comparison is conducted in the implementation of a fixed recycling C&D plant between mobile recycling stations in China and the recycling centers with mobile stations in the Netherlands. It aims to investigate the success factors influencing the viability of a recycling plant. Results demonstrate that both fixed and mobile recycling centers with used equipment have higher commercial viability than centers equipped with new apparatus. This further explains that a higher profit margin is resulted for recycling centers with used equipment and location advantage while recycling costs are reduced with the scale of economy in fixed centers. Furthermore, investment risks can be reduced by introducing economic and political instruments [108]. The importance of creating an economic system to allow reusable and recyclable material flow is emphasized, that is in line with a critical principle of industrial ecology [109]. In the global context, studies on evaluating the economic suitability of a large-scale recycling plant in a densely populated urban region in Portugal are conducted. The recycling plant can achieve a high-profit potential in about 2 years with the return of invested capital despite the absence of regulatory government policy and high initial investment [110]. The authors further conduct a life cycle assessment on the above large-scale recycling plant in Portugal about the primary energy consumption and carbon dioxide equivalent emissions. The results show that the plant has a capacity of 350 tonnes per hour and 60-year operating lifespan. This means the performance in using recycled materials in the plant was good in saving energy and carbon dioxide equivalent emissions are lower during the operating lifespan [111]. The above studies affirm the commercial viability of operating a recycling center of C&D waste when different conditions are considered.

Another research endeavors to minimize C&D waste by the implementation of various management measures. A wide range of aspects such as building design, on-site management, handling and storage of raw materials [112], and the transportation, recycling and disposal of C&D waste are examined [113]. For instance, survey is conducted among large Singapore contractors to gather information on project design-related waste sources and to assess building design [114]. At the same time, in China, major determinants influencing the implementation of on-site management are explored for improving the efficiency [115]. The generation of construction waste in developing countries is grievous due to the following reasons. On one hand, a vast amount of construction waste is resulted from increasing large-scale construction activities in these countries due to urbanization and infrastructure development [115]. On the other hand, decision makers neglect the impacts of increasing construction activities on the environment and focus on conventional project objectives such as cost, duration, quality, and safety [116]. Therefore, the regulatory environment is immature and the application of waste management practices is insufficient to achieve a high level of C&D waste management [117]. Social impacts of poor management on C&D waste are seldomly investigated. Previous literature suggests that the collective development of economic, environmental, and social aspects can achieve sustainable construction in the long run [118].

The unique characteristics of SD approach are proved to be an effective tool for simulating the effects on policy implementation [119]. SD improves the soundness and effectiveness of the decision-making process, which makes it a prevailing technique for modeling construction project and waste management [120]. While each construction activity involves various stakeholders [107], SD approach provides decision makers with analytical hierarchy processes to model sustainable waste management measures [121] and predicts the materials flow of concrete waste by simulation to reduce C&D waste [122]. SD also interrelates subsystems within a complex system and provide knowledge on the dynamic interactions and interdependencies of the key areas in C&D management [102]. To help decision makers and various stakeholders better grasp and understand the architecture involved in C&D waste management, SD approach for strategic planning of construction waste management is applied to explain the complex information from different perspectives. Quantitative studies are conducted to assess the social performance of C&D waste management in China. Results indicate that "physical working environment," "safety of operatives" and "practitioners long-term health" significantly contribute to poor social performance. Also, scenarios integrating various management measures can maximize the effect on enhancing the social performance of C&D waste management [123]. Besides quantifying social impacts on different C&D management strategies, studies which analyze the cost-effectiveness of implementing C&D waste management define the construction waste chain in Shenzhen, China. Conducting C&D waste management is beneficial, while a higher landfill charge, which gives higher net benefit, an earlier realization. Moreover, a higher environmental cost is resulted from illegal dumping when the general public is required to pay a higher landfill charge. Results also reveal some key characteristics of the dynamic system, including various elements involved, such as waste generation, reduction, reuse, recycling, and disposal. The dynamic system is different from the conventional researches that solely focus on the system from a static point of view [124]. Lastly, SD models compare and evaluate alternatives for a better operation of C&D recycling centers with the consideration of different policies and economic environments. Three major determinants, including profit, unit of recycling costs and extra revenue from location advantage contribute significantly to the economic feasibility of C&D recycling and the ratio of savings to costs. And the optimum ratio of savings to costs, the design of recycling centers and selection of governmental instruments can be achieved at a low level of the above three determinants upon the comparison between the ratios of public and private sectors [98].

20.3 Conclusions and perspectives

In summary, this chapter provides a holistic review of SD approach on wood and yard waste, including a brief review of global solid waste composition, an introduction to SD approach and its development, applications of SD on various environmental issues, comprehensive literature review on wood and yard waste management in the stream of MSW and C&D, and the current implementation of SD on MSW and C&D waste management. Since the above-mentioned studies mainly focus on regional context, future researches can be conducted by broadening the scope to cross-national study. Countries with similar wood and yard waste-management systems can be compared and analyzed by formulating SD models. Besides, the use of SD approach draws concern on its application. It is widely accepted that the model does not perfectly mimic the real world as the structure is proposed and governed by modelers' understanding toward the real situation. Therefore, objective justification on the proposed system structure should be verified by third parties, such as related experts and academics to affirm the system structure. Lastly, the application of top-down SD approach can be supplemented by the use of other modeling approaches, such as Agent-based modeling, which is a bottom-up method taking individual's behavior into account. Such model goes beyond the limitation of SD model that more dynamics can be captured even without the knowledge on global interdependencies. The hybrid model can resolve more complex issues by allowing a deeper understanding on both the global interdependencies of variables within a system structure and individuals' behavior.

Acknowledgments

The authors appreciate the financial support of the Hong Kong Research Grants Council (E-PolyU503/17 and PolyU 15223517) for this study.

References

[1] USEPA. Advancing sustainable materials management: 2015 fact sheet. Washington: U.S. Environmental Protection Agency; 2015. https://www.epa.gov/sites/production/files/2018-07/documents/2015_smm_msw_factsheet_07242018_fnl_508_002.pdf.

[2] Falk RH, McKeever DB. Recovering wood for reuse and recycling: a United States perspective. Greece: University Studio Press; 2004.

[3] McKeever DB. Inventories of woody residues and solid wood waste in the United States, 2002. 2004. Canada.

[4] Katz A, Baum H. A novel methodology to estimate the evolution of construction waste in construction sites. Waste Management 2011;31:353–8.

[5] Begum RA, Satari SK, Pereira JJ. Waste generation and recycling: comparison of conventional and industrialized building systems. American Journal of Environmental Sciences 2010;6:383–8.

[6] Nagapan S, Ismail AR, Ade A. A review of construction waste cause factors,. In: Proceedings of the Asian Conference of Real Estate: sustainable growth managing challenges (ACRE), Johor, Malaysia; 2011.

[7] Ng S, Engelsen CJ. Construction and demolition wastes. Amsterdam, The Netherlands: Elsevier; 2018.

[8] Llatas C. A model for quantifying construction waste in projects according to the European waste list. Waste Management 2011;31:1261–76.

[9] Poon CS, Yu ATW, Wong A, Yip R. Quantifying the impact of construction waste charging scheme on construction waste management in Hong Kong. Journal of Construction Engineering and Management 2013;139:466–79.

[10] Wang R, Li S. Talking about the production and disposing of construction waste from the view of sustainable development. In: Proceeding power and energy engineering conference. IEEE; 2011. p. 1–4.

[11] Chen J, Shen L, Song X, Shi Q, Li S. An empirical study on the CO_2 emissions in the Chinese construction industry. Journal of Cleaner Production 2017;168:645–54.

[12] Yunpeng H. Minimization management of construction waste. IEEE International Symposium of Water Resource and Environmental Protection 2011:2769–72.

[13] Bossink AG, Brouwers HJH. Construction waste: quantification and source evaluation. Journal of Construction Engineering and Management 1996;122:55–60.

[14] Ekanayake LL, Ofori G. Construction material waste source evaluation. In: Proceedings: strategies for a sustainable built environment; 2000.
[15] Kralj D. Innovative systemic approach for promoting sustainable innovation for zero construction waste. Kybernetes 2011;40:275–89.
[16] Staley BF, Barlaz MA. Composition of municipal solid waste in the United States and implications for carbon sequestration and methane yield. Journal of Environmental Engineering 2009;135:901–9.
[17] Hwang BG, Yeo ZB. Perception on benefits of construction waste management in the Singapore construction industry. Journal of Engineering, Construction and Architectural Management 2011;18:394–406.
[18] Alwi S, Hampson K, Mohamed S. Waste in the Indonesian construction projects. In: Proceedings 1st international conference of CIB W107 – creating a sustainable construction industry in developing countries, South Africa; 2002.
[19] Formoso CT, Isatto EL, Hirota EH. Method for waste control in the building industry. In: Proceedings of the seventh annual conference of the international group for lean construction, Berkeley, USA; 1999.
[20] Pidd M. Tools for thinking; modeling in management sciences. Journal of the Operational Research Society 1997;48:1150.
[21] Duggan J. An introduction to system dynamics. United States: Springer International Publishing; 2016.
[22] Meadows DL, Behrens WW, Meadows DH, Naill RF, Randers J, Zahn E. Dynamics of growth in a finite world. Cambridge: Wright-Allen Press; 1974.
[23] Homer J. Models that matter: selected writings on system dynamics, 1985–2010. Grapeseed Press; 2012.
[24] Box GE. Science and statistics. Journal of the American Statistical Association 1976;71:791–9.
[25] Sterman JD. All models are wrong: reflections on becoming a systems scientist. System Dynamics Review 2002;18:501–31.
[26] Aronson J, Milton SJ, Blignaut JN. Restoring natural capital: business, science, and practice. Washington, DC: Island Press; 2007. p. 3–8.
[27] Arquitt S, Johnstone R. Use of system dynamics modeling in design of an environmental restoration banking institution. Ecological Economics 2008;65:63–75.
[28] Liu Y, Guo H, Yu Y, Dai Y, Zhou F. Ecological-economic modeling as a tool for watershed management: a case study of Lake Qionghai watershed, China. Limnologica 2008;38:89–104.
[29] Bendor T. A dynamic analysis of the wetland mitigation process and its effects on no net loss policy. Landscape and Urban Planning 2009;89:17–27.
[30] Higgins SI, Turpie JK, Costanza R, Cowling RM, Le Maitre DC, Marais C, Midgley GF. An ecological economic simulation model of mountain fynbos ecosystems: dynamics, valuation and management. Ecological Economics 1997;22:155–69.
[31] Jogo W, Hassan R. Balancing the use of wetlands for economic well-being and ecological security: the case of the Limpopo wetland in Southern Africa. Ecological Economics 2010;69:1569–79.
[32] Turpie JK, Clark BM, Cowley P, Bornman T, Terörde A. Integrated ecological-economic modeling as an estuarine management tool: a case study of the east Kleinemonde estuary. In: Model construction, evaluation and user manual; 2008. Pretoria.
[33] Wise RM, Cacho OJ. Tree-crop interactions and their environmental and economic implications in the presence of carbon-sequestration payments. Environmental Modelling & Software 2005;20:1139–48.
[34] Nobre AM, Musango JK, De Wit MP, Ferreira JG. A dynamic ecological economic modeling approach for aquaculture management. Ecological Economics 2009;68:3007–17.
[35] Deaton ML, Winebrake JJ. Dynamic modeling of environmental systems. New York: Springer Science-Business Media; 2000.
[36] Kollikkathara N, Feng H, Yu DL. A system dynamic modeling approach for evaluating municipal solid waste generation, landfill capacity and related cost management issues. Waste Management 2010;30:2194–203.
[37] Meadows DH. Thinking in systems: a primer. Vermont: Chelsea Green Publishing; 2008.

[38] Sterman JD. Business dynamics: systems thinking and modeling for a complex world. Boston: McGraw-Hill; 2000.

[39] Lane DC. IFORS' operational research hall of fame Jay Wright Forrester. International Transactions in Operational Research 2006;13:483–92.

[40] Hannon B, Ruth M. Dynamic modeling. 2nd ed. New York, USA: Springer Science-Business Media; 2001.

[41] Sterman JD, Oliva R, Linderman K, Bendely E. System dynamics perspectives and modeling opportunities for research in operations management. Journal of Operations Management 2015;39–40:1–5.

[42] Alessi S. The application of system dynamics modeling in elementary and secondary school curricula. University of Iowa; 2000.

[43] Sastry A, Sterman J. Desert island dynamics: an annotated survey of the essential system dynamics literature. Cambridge, USA. 1992.

[44] Robert YC, Leslie VC. Demonstrating the utility of system dynamics for public policy analysis in New Zealand: the case of excise tax policy on tobacco. System Dynamics Review 2006;22:321–48.

[45] Aboyade A. The potential for climate change mitigation in the nigerian solid waste disposal sector: a case study from Lagos. Lund: Centre for Sustainability Studies, Lund University; 2004. p. 47.

[46] Coyle G. The practice of system dynamics: milestones, lessons and ideas from 30 years' experience. System Dynamics Review 1998;14:343–65.

[47] Forrester JW. Industrial dynamics. Cambridge: Wright-Allen Press; 1961.

[48] Gao Y, Liu C. Research on simulation of optimal decision making for a regional water resources system. International Journal of Water Resources Development 1997;13:123–34.

[49] Wang XJ, Zhang JY, Liu JF, Wang G, He R, Elmahdi A, Elsawah S. Water resources planning and management based on system dynamics – a case study of Yulin city. Environment, Development and Sustainability 2011;13:331–51.

[50] Huang L, Yin L. Supply and demand analysis of water resources based on system dynamics model. Journal of Engineering and Technological Sciences 2017;49:705–20.

[51] Beall A, Fiedler F, Boll J, Cosens B. Sustainable water resource management and participatory system dynamics. Case study: developing the palouse basin participatory model. Sustainability 2011;3:720–42.

[52] Elmahdi A, Malano H, Etchells T. Using system dynamics to model water-reallocation. The Environmentalist 2007;27:3–12.

[53] Qi C, Chang NB. System dynamics modeling for municipal water demand estimation in an urban region under uncertain economic impacts. Journal of Environmental Management 2011;92:1628–41.

[54] Saleh M, Oliva R, Kampmann CE, Davidsen PI. A comprehensive analytical approach for policy analysis of system dynamics models. European Journal of Operational Research 2010;203:673–83.

[55] Xu ZX, Takeuchi K, Ishidaira H, Zhang XW. Sustainability analysis for Yellow River water resources using the system dynamics approach. Water Resources Management 2002;16:239–61.

[56] Ahmad S, Prashar D. Evaluating municipal water conservation policies using a dynamic simulation model. Water Resources Management 2010;24:3371–95.

[57] Simonvic SP, Rajasekaram V. Integrated analyses of Canada's Water Resources: a system dynamics approach. Canadian Water Resources Journal 2004;29:223–50.

[58] Ryu JH, Contor B, Johnson G, Allen R, Tracy J. System dynamics to sustainable water resources management in the Eastern Snake Plain Aquifer under water supply uncertainty. Journal of the American Water Resources Association 2012;48:1204–20.

[59] Kotir JH, Smith K, Brown G, Marshall N, Johnstone R. A system dynamics simulation model for sustainable water resources management and agricultural development in Volta River Basin, Ghana. The Science of the Total Environment 2016;573:444–57.

[60] Jebaraj S, Iniyan AS. A review of energy models. Renewable and Sustainable Energy Reviews 2006;10:281–311.

[61] Mutingi M, Mbohwa C. Fuzzy system dynamics: a framework for modelling renewable energy policies. In: Quadrat Ullah H, editor. Energy policy modelling in the 21st century. London: Springer; 2013. p. 31–47.
[62] Qudrat-Ullah H. Understanding the dynamics of electricity generation capacity in Canada: a system dynamics approach. Energy 2013;59:285–94.
[63] Gimenez C, Sierra V, Rodon J. Sustainable operations: their impact on the triple bottom line. International Journal of Production Economics 2012;140:149–59.
[64] Shafiei E, Davidsdottir B, Leaver J, Stefansson H, Asgeirsson EI. Simulation of alternative fuel markets using integrated system dynamics model of energy system. Procedia Computer Science 2015;51:513–21.
[65] Sisodia GS, Sahay M, Singh P. System dynamics methodology for the energy demand fulfillment in India: a preliminary study. Energy Procedia 2016;95:429–34.
[66] Braun W. The system archetypes. System **27**. 2002.
[67] Wolstenholme EF. Towards the definition and use of a core set of archetypal structures in system dynamics. System Dynamics Review 2003;19:7–26.
[68] Wolstenholme EF. Using generic system archetypes to support thinking and modelling. System Dynamics Review 2004;20:341–56.
[69] Liuguo S, Shijing Z, Jianbai H. Pricing simulation platform based on system dynamics. Systems Engineering Procedia 2012;5:445–53.
[70] Falk B, McKeever D. Generation and recovery of solid wood wastes in the US. Biocycle 2012;53:30–2.
[71] US Environmental Protection Agency. Yard and wood waste. Washington, DC. 2013.
[72] US Environmental Protection Agency. Biomass combined heat and power catalog of technologies. Washington, DC. 2007.
[73] Heinen K, Lawler M, McHale M, Peterson MN. Urban wood waste: a guide to managing your community's resource. North Carolina Cooperative Extension Service; 2012.
[74] Fairchild JK. Utilizing urban forests. Finding highest uses for urban wood waste. Biocycle 2003;44:59–60.
[75] Bratkovich SM. Utilizing municipal trees: ideas from across the county. St. Paul, Minnesota: USDA Forest Service, Northeastern Area State and Private Forestry; 2001.
[76] Cesa ET, Lempicki EA, Knotts JH. Recycling municipal trees: a guide for marketing sawlogs from street tree removals in municipalities. Morgantown, West Virginia: USDA Forest Service, Northeastern Area State & Private Forestry; 2003.
[77] McKeever DB, Skog KE. Urban tree and woody yard residue another wood resource. Madison, Wisconsin: USDA Forest Service, Forest Products Laboratory; 2003.
[78] Urban Forest Products Alliance. FSC and urban wood. 2013.
[79] Tree Care Industry Association, Inc. Harvesting a. evaluation, removal, recovery – working document. Londonderry, New Hampshire: Tree Care Industry Association, Inc.; 2012.
[80] Little D, Makra E, Greenwood R, Beaulieux M. Feasibility study for a community wood reclamation system. 2011 [Online], http://www.illinoisurbanwood.org/documents/RobbinswoodFinal_12_21_2011.pdf.
[81] Baina A, Shenoy M, Ashton W, Chertow M. Industrial symbiosis and waste recovery in an Indian industrial area. Resources, Conservation and Recycling 2010;54:1278–87.
[82] Puente RMC, Arozamena ER, Evans S. Industrial symbiosis opportunities for small and medium sized enterprises: preliminary study in the Besaya region (Cantabria, Northern Spain). Journal of Cleaner Production 2015;87:357–74.
[83] Zhang Y, Zhenga H, Fath BD. Ecological network analysis of an industrial symbiosis system: a case study of the Shandong Lubei Eco-Industrial Park. Ecological Modelling 2014:1–11.
[84] Bonnet F, Courtois M, Koulouri A, Munoz de la Torre DH, Yilmaz O, Salihcavusoglu K, Yontem E, Ozdogru ZG. Identification of best practices and lessons learnt in industrial symbiosis. 2016 [Online], http://fissacproject.eu/wp-content/uploads/2018/06/FISSAC-D1.2-Best-practices-and-lessons-learnt-in-IS-Summary.pdf.

[85] Leigh M, Li X. Industrial ecology, industrial symbiosis and supply chain environmental sustainability: a case study of a large UK distributor. Journal of Cleaner Production 2008:1–12.
[86] Kubanza NS, Simatele MD. Sustainable solid waste management in developing countries: a study of institutional strengthening for solid waste management in Johannesburg, South Africa. Journal of Environmental Planning and Management 2019. https://doi.org/10.1080/09640568.2019.1576510.
[87] Johnson J. China's National Sword to cut post-industrial plastic scrap. 2018 [Online], https://global-factiva-com.ezproxy.lb.polyu.edu.hk/ga/default.aspx.
[88] Ilgin MA, Gupta SM. Environmentally conscious manufacturing and product recovery: a review of the state of the art. Journal of Environmental Management 2010;91:563–91.
[89] Yücel G, Miluska C, Meza C. Studying transition dynamics via focusing on underlying feedback interactions modelling the Dutch waste management transition. Computational & Mathematical Organization Theory 2008;14:320–49.
[90] Kunsch P, Springael J. Simulation with system dynamics and fuzzy reasoning of a tax policy to reduce CO_2 emissions in the residential sector. European Journal of Operational Research 2008;185:1285–99.
[91] Warren K. Improving strategic management with the fundamental principles of system dynamics. System Dynamics Review 2005;21:329–50.
[92] Nail RF, Gelanger S, Klinger A, Peterson E. An analysis of cost effectiveness of US energy policies to mitigate global warming. System Dynamics Review 1992;8:111–8.
[93] Vizayakumar K, Mohapatra PKJ. Modeling and simulation of environmental impacts of a coalfield: system dynamic approach. Journal of Environmental Management 1993;42:59–73.
[94] Beigl P, Wassermann G, Schneider F, Salhofer S, Mackow I, Mrowinski P, Sebastian M. The use of life cycle assessment tool for the development of waste management strategies for cities and regions with rapid growing economies. Vienna, Austria. 2003.
[95] ECUA. Municipal recycling tonnage and rates reports. Fairfield, New Jersey, USA: ECUA; 2003.
[96] Liu Z, Li H. Extreme learning machine based spatiotemporal modelling of lithium-ion battery thermal dynamics. Journal of Power Sources 2015;277:228–38.
[97] Lam W, Chan K. How authoritarianism intensifies punctuated equilibrium: the dynamics of policy attention in HK. Governance 2014;28:212–23.
[98] Zaman AU. Identification of waste management development drivers and potential emerging waste treatment technologies. International Journal of Environmental Science and Technology 2013;10:455–64.
[99] Hong J, Li X, Zhaojie C. Life cycle assessment of four municipal solid waste management scenarios in China. Waste Management 2010;30:2362–9.
[100] Mashayekhi AN. Transition in the New York State solid waste system: a dynamic analysis. System Dynamics Review 1993;9:23–47.
[101] Hao JL, Hills MJ, Huang T. A simulation model using system dynamic method for construction and demolition waste management in Hong Kong. Construction Innovation 2007;7:7–21.
[102] Wang FS. Deterministic and stochastic simulations for solid waste collection systems – a SWIM approach. Environmental Modeling & Assessment 2001;6:249–60.
[103] Blore I, Nuan F. Living with waste: public valuation of solid waste impacts in Bangkok, papers in the administration of development, vol. 57. Development Administration Group, University of Birmingham; 1996.
[104] Dyson B, Chang NB. Forecasting municipal solid waste generation in a fast growing urban region with system dynamics modeling. Waste Management 2005;25:669–79.
[105] Sufian MA, Bala BK. Modeling of urban solid waste management system: the case of Dhaka city. Waste Management 2007;27:858–68.
[106] Faniran OO, Caban G. Minimizing waste on construction project sites. Engineering, Construction and Architectural Management 1998;5:182–8.
[107] Yuan HP, Shen LY. Trend of the research on construction and demolition waste management. Waste Management 2011;31:670–9.

[108] Zhao W, Ren H, Rotter VS. A system dynamics model for evaluating the alternative of type in construction and demolition waste recycling center—the case of Chongqing, China. Resources, Conservation and Recycling 2011;55:933–44.
[109] Allenby BR. Industrial ecology: policy framework and implementation. Upper Saddle River, New Jersey: Prentice Hall; 1999.
[110] Coelho A, de Brito J. Economic viability analysis of a construction and demolition waste recycling plant in Portugal – part I: location, materials, technology and economic analysis. Journal of Cleaner Production 2013;39:338–52.
[111] Coelho A, de Brito J. Environmental analysis of a construction and demolition waste recycling plant in Portugal – part II: environmental sensitivity analysis. Waste Management 2013;33:147–61.
[112] Chen Z, Li H, Wong CTC. An application of bar-code system for reducing construction wastes. Automation in Construction 2002;11:521–33.
[113] Hu MM, Voet E, Huppes G. Dynamic material flow analysis for strategic construction and demolition waste management in Beijing. Journal of Industrial Ecology 2010;14:440–56.
[114] Ekanayake LL, Ofori G. Building waste assessment score: deign-based tool. Building and Environment 2004;39:851–61.
[115] Wang JY, Yuan HP, Kang XP, Lu WS. Critical success factors for onsite sorting of construction waste: a China study. Resources, Conservation and Recycling 2010;54:931–6.
[116] Shen LY, Yao H, Alan G. Improving environmental performance by means of empowerment of contractors. Management of Environmental Quality: An International Journal 2006;17:242–57.
[117] Lu WS, Yuan HP. Exploring critical success factors for waste management in construction projects of China. Resources, Conservation and Recycling 2010;55:201–8.
[118] Yao H. A dynamic approach for evaluating the sustainability performance of infrastructure projects. Hong Kong: The Hong Kong Polytechnic University; 2009.
[119] Tam WY, Li J, Cai H. System dynamic modeling on construction waste management in Shenzhen, China. Waste Management & Research 2014;32:441–53.
[120] Hao JL, Hills MJ, Huang T. A simulation model using system dynamic method for construction and demolition waste management in Hong Kong. Construction Innovation: Information, Process, Management 2007;7:7–21.
[121] Rong L. Using system dynamics in decision support for sustainable waste management. Singapore: National University of Singapore; 2004.
[122] Hsiao TY, Huang YT, Yu YH, Wernick IK. Modeling materials flow of waste concrete from construction and demolition wastes in Taiwan. Resources Policy 2002;28:39–47.
[123] Yuan H. A model for evaluating the social performance of construction waste management. Waste Management 2012;32:1218–28.
[124] Yuan HP, Shen LY, Hao JL, Lu WS. A model for cost-benefit analysis of construction and demolition waste management throughout the waste chain. Resources, Conservation and Recycling 2011;55:604–12.

CHAPTER 21

Waste-to-biofuel and carbon footprints

Yize Li, Asam Ahmed, Ian Watson, Siming You
Division of Systems, Power & Energy, James Watt School of Engineering, University of Glasgow, Glasgow, United Kingdom

21.1 Introduction

A large amount of 2.12 billion tons of waste is generated each year due to rapid global urbanization and economic development, which imposes a huge burden on achieving the United Nations' Sustainable Development Goals [1]. The waste collection rate in some low- and middle-income countries is still low (<70%) with about half of the waste being directly discharged into streets and drains, which causes various environmental and health problems in addition to the wastage of a potential energy resource. It is expected that the urban population explosion and development will double the urban solid waste generation rate, incurring an annual waste management costs of $375 billion globally [2]. Hence, waste management is not only an environmental issue but also a national economic issue that is critical to the sustainable development of a country and the world.

Waste-to-bioenergy development is closely related to the sustainable development goals (SDGs) proposed by the United Nations (UN) in 2015, i.e., the goals of ensuring access to affordable, reliable, sustainable and modern energy for all and make cities and human settlements inclusive, safe, resilient, and sustainable. Waste-to-bioenergy can contribute to the fulfillment of the specific targets corresponding to the goals, e.g., ensuring universal access to affordable, reliable and modern energy services and reducing the adverse environmental impact of cities by focusing on the problems of air quality and municipal and other waste management [3]. Bioenergy recovery from waste serves as an attractive solution for the SDGs, as it serves to tackle the challenges of climate change, fossil fuel depletion, and sustainable waste management simultaneously. The UN plans to achieve each goal and specific targets by 2030 which, however, calls for significant technology advancement and implementation.

Conventional waste management relies heavily on landfill and incineration both of which suffer from a variety of issues. For example, noxious gases (e.g., nonmethane organic compounds, mercury, tritium) and chemicals (e.g., benzene, toluene, vinyl chloride, carbon

Waste Biorefinery. https://doi.org/10.1016/B978-0-12-818228-4.00021-6

tetrachloride, dioxins, and furans) can be formed during the degradation process of landfilled waste, which pose a threat to the surrounding environmental and ecological systems (e.g., soil and water contamination). The primary gas product, methane, is a greenhouse gas (GHG), with a global warming potential of around 28 times higher than carbon dioxide on a 100-year time scale [4]. Incineration of waste suffers from problems such as emissions of NOx and dioxin, high off-gas flow rates, residual ash disposal obligation, etc. [5]. Environmentally friendly waste management practices are highly demanded.

Recent studies have shown that waste can be converted into various value-added biofuels (e.g., biohydrogen, biomethane, biodiesel, and bioethanol) with potentially high economic values [6–8]. Converting waste into the biofuels can potentially increase the economic viability of relevant waste management processes and relevant knowledge serves as the basis for developing accurate waste utilization schemes that make the best use of the different compositions in waste.

Waste-to-biofuel conversion technologies and systems vary in the carbon footprints which affect their environmental sustainability and practical implementation. Low carbon footprints are always expected and desirable for waste-to-biofuel systems, which do not necessarily hold when relevant upstream (e.g., waste generation, transport, and pretreatment) and downstream processes (e.g., product upgrading and transport) are taken into consideration, requiring a lifecycle assessment-based carbon evaluation. It is important to compare the footprints of different technologies and systems and identify relevant footprint "hotspots" to shape the priorities of future research and development.

21.2 Biofuel classification

First-generation biofuels (e.g., bioethanol, biogas, and biodiesel) refer to the ones directly produced from crops. Some of the typical feedstocks include wheat and sugar for bioethanol production and oilseed rape for biodiesel production [9]. However, first-generation biofuels suffer from two major problems. First, the carbon abatement potential of first-generation biofuels may be debilitated due to the fact that relevant production processes incur considerable net energy input and carbon emissions offsetting the low-carbon benefits from biomass use. For example, 17–420 times more carbon dioxide might be released by biofuel production based on rainforests, peatlands, savannas in Brazil, and Southeast Asia than the corresponding GHG reductions resulted by fossil fuel displacement, leading to a net biofuel carbon debt [10]. Second, the increasing production of first-generation biofuels implies a risk chance of diverting farmland or crops away from food production, resulting in food versus fuel dilemma and an overall rise in food prices [11].

Second-generation biofuels (e.g., biomethane, biohydrogen, and bio-DMF) are derived from nonfood biomass such as agricultural waste, organic waste and food crop waste. The nonagricultural source dependence of second-generation biofuels suggests a reduced impact on food security, and thus they can potentially be more environmentally friendly than first-generation biofuels [12]. Second-generation biofuels could be as price competitive as fossil fuels. However, to be a sustainable source, the energy input for biomass production still needs to be less than the energy output achieved from the biomass [12]. Lifecycle assessment (LCA) showed that the production of second-generation biofuels had a net energy gain and a desirable carbon abatement potential [13]. The second generation of biofuels from waste will be the main focus of this chapter.

Third-Generation biofuels refer to the ones produced from particularly engineered energy crops like algae. Algae serve as an energy-intensive, low cost, and renewable feedstock for energy production. They have the potential to create more energy per acre compared to traditional crops [14]. 2000–5000 gallons of fuels per acre could be produced by algae. it can grow on noncrop or marginal and land. It can be growing algae using water and land which are not for food production; therefore, they lessen the straining on already exhausted land and water sources. Algae have less environmental impacts compared with other sources of biomass used for biofuels [10].

Photobiological solar fuels and electrofuels are referred to as fourth-generation biofuels that are based on designer photosynthetic microorganisms and a combination of photovoltaics and microbial fuel production, respectively [15]. Synthetic biology technologies play a central role in the development of fourth-generation biofuel processes that depend on relevant biological systems to utilize solar energy and economic and abundantly available feedstock for the production of high-value chemicals and biofuels. The production and commercial uses of fourth-generation biofuels are still at an early stage.

21.3 Waste-to-biofuel

A waste-to-biofuel system normally consists of a waste pretreatment unit, a waste-to-biofuel reactor, a cleaning/purification unit, and an upgrading unit. However, actual system configurations are highly contingent upon the types of waste feedstocks, conversion technologies (i.e., chemical, biological, and thermochemical), and targeted biofuels. This section will differentiate the waste-to-biofuel generation based on four major types of biofuels, i.e., bioethanol, biohydrogen, biomethane, and biodiesel.

21.3.1 Waste-to-bioethanol

First-generation bioethanol can be produced from cellulose and starch-rich crops such as potato, rice, and sugar cane. As mentioned earlier, the first-generation bioethanol production competes with food to satisfy the increasing fuel demand and have a negative impact on biodiversity [16]. Lignocellulosic wastes (e.g., agricultural and forestry residues like corn stover, straw, cotton stem and sugarcane bagasse, and municipal solid wastes like waste papers, waste wood and spent grains), as an abundant and cheap feedstock, have been exploited as alternative substrates for bioethanol production (second-generation bioethanol). The lignocellulosic feedstock constitutes the richest biosphere renewable organic ingredients [17].

Lignocellulose is a readily available substrate with a productivity of 2×10^{11} tons per year [18]. The application of lignocellulosic substances for ethanol production increases the complexity of the process and energy-intensive pretreatment steps are required to open the stubborn structure of lignocellulose to obtain its polymer [19]. There are physical, physicochemical, solvent fractionation and biological pretreatment methods [17]. The bioethanol yield of cellulose hydrolysis can exceed 90% after pretreatment compared with 20% for the case without pretreatment. Enzymatic pretreatment can be carried out simultaneously with the fermentation that requires microorganisms to degrade sugars into alcohols.

Conventional lignocellulose-to-bioethanol processes generally include several steps like pretreatment, enzymatic hydrolysis, fermentation and product recovery (distillation) [20]. In these processes, significant costs are incurred by the purification of enzymes that need to be produced exogenously. Consolidated bioprocessing methods that combine enzyme production, saccharification, and multisugar fermentation have been proposed to achieve cost-effective and efficient bioethanol production, and have such benefits as pretreatment avoidance, few reactor requirements, and own enzyme generation by processing organisms [20]. It is still unclear, however, a single organism or community of organisms is a better choice for achieving a higher efficiency for the process [21].

Saccharomyces cerevisiae has been commonly used to produce a high yield of ethanol from sugar (90% of theoretical yield) [22]. Separate hydrolysis and fermentation (SHF) is a traditional method based on *S. cerevisiae* for bioethanol production. However, *S. cerevisiae* ferments hexose sugar only in this process. Simultaneous saccharification and fermentation (SSF) with a combined hydrolysis and fermentation bioprocess in a single reactor has the benefits of reducing costs and avoiding a large number of inhibitory compounds that are derived from the side reactions of pretreatment and inhibit downstream biochemical processes.

The process of bioconversion of lignocellulose to produce ethanol, which mainly involves three main steps: (1) pretreatment, to break down the recalcitrant structures of lignocellulose; (2) enzymatic hydrolysis, to hydrolyze polysaccharides (e.g., cellulose, hemicellulose) into fermentable sugars; (3) fermentation, to convert sugars into ethanol [23].

The organic-rich nature of food waste makes it a good resource for the production of bioethanol [24]. Food waste mainly consists of carbohydrates (starch, cellulose, and hemicellulose), proteins, and lipids. The cleavage of glycosidic bonds in food waste releases the polysaccharide as an oligosaccharide and a monosaccharide by hydrolysis of carbohydrates, which is easier to ferment. The total sugar and protein contents in food waste are in the range of 35.5%–69% and 3.9%–21.9%, respectively [14].

21.3.2 Waste-to-biohydrogen

Organic waste can be converted into hydrogen through biological (e.g., fermentation) and thermochemical (e.g., gasification and pyrolysis) processes [25]. Fermentation-based hydrogen production methods can be classified into photofermentation, dark fermentation, biophotolysis, and microbial electrolysis cells [26]. Thermochemical methods can be divided into supercritical water gasification, steam gasification, and combined pyrolysis and gasification. Additionally, hybrid processes such as combined slow pyrolysis and steam gasification have been developed to improve the yield and quality of hydrogen production [27].

Food waste serves as a good source for biological process-based hydrogen production. During fermentation, hydrogen is produced when acetate and butyrate are major by-products [28]. Dark fermentation-based biohydrogen production has two main advantages: mild operating condition requirements and residual biomass reutilization [29]. During dark fermentation, fermenting bacteria convert organic substrates into hydrogen, volatile fatty acids (VFA) such as butyric and acetic acids, and carbon dioxide in the absence of light. Carbohydrates are a preferred carbon source for this process [30]. Typically, Eqs. (21.1) and (21.2) are the main reactions for the bioconversion of organic waste-to-biohydrogen [26]. Batch, semicontinuous, continuous, single or multi-stage fermentation systems have been developed to convert food waste into hydrogen-enriched product gas [31].

$$C_6H_{12}O_6 + 2H_2O \rightarrow 4H_2 + 2CH_3COOH + 2CO_2 \tag{21.1}$$

$$C_6H_{12}O_6 + 2H_2O \rightarrow 2H_2 + CH_3CH_2CH_2COOH + 2CO_2 \tag{21.2}$$

The dark fermentation process working condition operating under mesophilic (25–40°C), thermophilic (40–65°C), or hyperthermophilic (>80°C) temperature. Under the higher temperature, thermophilic or hyperthermophilic could get higher biohydrogen production rates and yields, due to the expected thermal activation of chemical reactions [32].

Supercritical water gasification has been used to produce hydrogen-rich product gases from carbonaceous waste. Water can be present in three material states, namely solid, liquid, and steam. The gas phase and liquid phase of water will coexist with oxygen as an oxidizing agent under the supercritical conditions of 22.1 MPa and 374°C. When a carbonaceous material reacts with supercritical water, the carbon particles are oxidized to release carbon monoxide. Part of the formed carbon monoxide is further oxidized to carbon dioxide, while hydrogen element present in the biomass and water are released as hydrogen. This method is very suitable for raw materials with high water content [27].

Steam gasification is a special type of gasification process where steam is used as the gasifying agent. Since pyrolysis is faster than gasification, a pyrolysis stage occurs prior to gasification during a full gasification process. The volatiles are removed by pyrolysis and the formed solid residue, biochar, will react with steam to produce hydrogen, carbon monoxide and carbon dioxide in steam gasification. High yields (80%–99%) of high-purity hydrogen (>99.9 vol%) are produced by catalytic steam gasification of chestnut wood chips in a combined down-flow fluidized bed and fixed bed reactor [33]. The steam facilitates the water-gas shift reaction ($CO + H_2O \rightarrow CO_2 + H_2$), and only a smaller amount of coke and tar is produced during the steam gasification process. The hydrogen production cost of supercritical water gasification is normally higher than steam gasification due to the requirement of the high-pressure condition [34].

The biochar produced by pyrolysis serves as a good feedstock for gasification to produce hydrogen. Combining slow pyrolysis with steam gasification can improve syngas (a mixture of hydrogen, carbon monoxide, and methane) quality, hydrogen yield, and conversion efficiency, reduce tar yield, and achieve better control of product composition, however, the combined process has the disadvantage of high operating energy and long residence time requirements [27]. Optimum temperature and residence time for maxim hydrogen yields were reported 800°C and 30 min for combined slow pyrolysis and steam gasification of a variety of waste biomass such as sugarcane bagasse, coir pith, rice husk, groundnut shell, sawdust, and casuarina leaves, and the resultant hydrogen generation ranged from 36.87% for sawdust to 57.87% for coir pith [35]. It is worth noting that studies have shown that biochar can be used as a cost-effective catalyst to promote the hydrogen production of gasification [36].

21.3.3 Waste-to-biomethane

Biomethane production through anaerobic digestion is a widely accepted organic waste management solution because of its low cost and the production of valuable digestate which can be used for soil conditioning. Depending on the use of substrate, anaerobic digestion can be classified as monodigestion with a single substrate and/or codigestion with two or more substrates, or dry digestion with total solids $\geq$15% and wet digestion with total solids $\leq$15% [37,38]. Depending on the digesting temperature, it can be classified as psychrophilic digestion (<25°C), mesophilic digestion (25–40°C), and thermophilic digestion (>45°C), with the first one being less efficient for biomethane production than the two others [39].

Single-stage and two-stage methods have been developed for anaerobic digestion-based biomethane production. All reactions (hydrolysis, acid production, acetogenic, and methanogenic) occur simultaneously in a single-stage reactor and the corresponding systems generally encounter lower frequency of technical failures leading to lower investment costs [40]. Two-stage anaerobic digestion is typically used to produce hydrogen and methane in two separate reactors compared to single-stage anaerobic digestion. Rapidly growing acid and hydrogen-producing microorganisms are enriched in the first stage for the production of hydrogen and VFA. In the second phase, slow-growing acetogens and methanogens are established, and VFA is converted to methane and carbon dioxide.

The hydrolysis or liquefaction reactions mainly include four stages: (1) lipids to fatty acids; (2) polysaccharides to monosaccharides; (3) protein to amino acids; (4) nucleic acids to purines and pyrimidines. The principal acids produced include acetic acid (CH_3COOH), propionic acid (CH_3CH_2COOH), butyric acid ($CH_3CH_2CH_2COOH$), and ethanol (C_2H_5OH). At the third stage, bacteria called methane formers (methanogens) produce methane either by cleaving acetic acid molecules to generate carbon dioxide and methane, or by reduction of carbon dioxide with hydrogen [30]. The reaction equations are:

$$C_6H_{12}O_6 \rightarrow 2C_2H_5OH + 2CO_2 \tag{21.3}$$

$$CH_3COOH \rightarrow CH_4 + CO_2 \tag{21.4}$$

$$2C_2H_5OH + CO_2 \rightarrow CH_4 + 2CH_3COOH \tag{21.5}$$

$$CO_2 + 4H_2 \rightarrow CH_4 + 2H_2O \tag{21.6}$$

Ref. [41] compared the hydrogen yields and overall energy outputs of two-stage and single-stage processing of food waste and wheat feed. They showed that higher hydrogen yields but lower overall energy output were resultant during food waste treatment by two-stage batch processing than single-stage processing. However, for wheat feed, lower hydrogen yields but higher overall energy output were achieved by two-stage processing. Ref. [42] compared the methane production by the anaerobic digestion of potato waste in two different two-stage systems. One system connected the solid bed reactor to an up-flow anaerobic sludge blanket methanogenesis reactor, while for the other system, a methanogenic reactor equipped with a straw biofilm carrier was connected to a solid bed reactor. Similar methane yields (0.39 $m^3/kg\ VS_{added}$) and cumulative methane production are achieved from both systems, however, the latter achieved a higher waste degradation speed.

Packed bed reactors or fixed bed systems have been developed to achieve high loads, the immobilization of microbial consortia and stabilization of methanogenesis [43]. A fixed bed reactor using hydrogenophilic methanogens has the potential to efficiently convert carbon dioxide to methane with a conversion of 100% and a retention time of 3.8 h [44].

Biohythane (hydrogen + methane) production has attracted increasing attention because it serves as a high-value solution for waste reutilization and biohythane can be potentially used as a transport fuel in place of fossil-based hythane [14]. Main VFA components produced during the hydrogen fermentation stage include acetic and butyric acids with concentrations ranging from 10 to 25 mmol/L [45]. During the second stages, a small amount of VFA was accumulated in the methane fermentation. The total energy recovery of the second-stage process was 18% higher than that of the first stage. An integrated system consisting of continuous stirred tank reactor (CSTR) and anaerobic fixed bed reactor (AFBR) has been proposed to continuously produce a mixture of hydrogen and methane from food waste [46]. AFBR exhibited stable operation and excellent performance, and the AFBR effluent was recycled to the CSTR to effectively provide alkalinity, maintaining the pH in an optimal range (5.0–5.3) for hydrogen-producing bacteria. Two types of two-stage systems have been developed to convert food waste and sewage sludge into hydrogen and methane [47]. The first type consists of the first stage of dark fermentative hydrogen production and the second stage of an anaerobic sequencing batch reactor, and the second type consists of the same first stage but the different second stage of an up-flow anaerobic sludge blanket reactor. The first type system led to a higher biogas conversion (78.6%), while the second type achieved a higher biogas production rate (2.03 L H_2/L_{system}/d, 1.96 L CH_4/L_{system}/d).

21.3.4 Waste-to-biodiesel

Biodiesel is generally more expensive than fossil fuels due to its higher raw material and production costs. The use of low-cost raw materials like waste could make biodiesel costs close to conventional diesel [48]. Biodiesel can be produced from a variety of waste (e.g., food waste, waste cooking oil, sewage sludge, grease trap waste, animal fat, etc.) by esterification/transesterification processes which can be homogeneous, heterogeneous, enzymatic or noncatalytic [49]. For example, biodiesel can be produced from waste vegetable oil and animal fat via sulfuric acid-catalyzed esterification and sodium hydroxide-catalyzed transesterification, and via sewage via sulfuric acid-catalyzed in situ transesterification [50].

Lipid-containing grease trap waste has been a great environmental burden that is costly and difficult to dispose of. The use of grease trap waste as a feedstock for biodiesel production has the potential to facilitate waste management while reducing the cost of biodiesel production. Ref. [51] studied the environmental impacts of a grease trap waste-based biodiesel system that consists of components of lipid extraction, lipid conversion, crude biodiesel washing and vacuum distillation-based purification. The core lipid conversion technology was based on the sulfuric acid (0.5% w/w) catalyzed esterification under atmospheric pressure and 120°C with methanol being recycled. The grease trap waste-to-biodiesel system considered by Ref. [52] includes a stage of fats, oils, and greases (FOG) separation (based on moderate heating and gravity settling), followed by acid esterification-based pretreatment and alkaline transesterification-based biodiesel production. An additional anaerobic digestion unit was considered to generate biogas for electricity and heat generation by utilizing the solid residue from the pretreatment stage.

The vegetable oils or fats with methanol or ethanol in the presence of a suitable catalyst, which can through transesterification to produce fatty acid alkyl esters. The stoichiometry of the transesterification reaction requires 3 mol of methanol (or ethanol) and 1 mol of triglyceride to give 3 mol of fatty acid methyl (or ethyl) ester and 1 mol of glycerol [53].

Biodiesel can be generated from waste cooking oil through a hydrogenation method that consists of a series of steps: waste cooking oil is firstly degraded at 400–500°C to form organic acids that are subsequently converted into hydrocarbons; crude biodiesel corresponding to the oils with an intermediate-boiling point in the hydrocarbons is separated; after acid removal, the biodiesel is enhanced regarding stability using hydrogenation at 150–250°C in hydrogen [54]. Sludge, concentrates, and scum in sewage treatment plants have been converted into biodiesel via a multistage process that includes pretreatment, glycerol hydrolysis, base catalyzed transesterification and fractionation [55]. Sludge can be converted into biodiesel through methods of hydrothermal dehydration, drying, and pyrolysis [56]. Dewatered sludge produces an internal waste stream, which is called concentrate. The conversion process from concentrate to biodiesel consists of five steps: algae cultivation, harvesting, dehydration, drying, and pyrolysis. By far the most effective method is pyrolysis, especially microwave-assisted pyrolysis, where process conditions were shown to be economically advantageous [57].

21.4 Carbon footprints

21.4.1 Lifecycle assessment method

Lifecycle assessment (LCA) is an integrated approach that is used to quantify the environmental impacts and resource use of processes and systems from a whole lifecycle perspective (i.e., from the raw material extraction to production, use, management of environmental impact during production, disposal and recycling) for reliable decision-making [58]. It can also help to identify the key "hotspots" for improving the environmental impacts of a product, process or system [59]. Since the late 1990s, ISO has been working to coordinate the LCA process, and developed ISO 14040 series standards. LCA consists of four phases: goal and scope definition, lifecycle inventory (LCI), lifecycle impact assessment (LCIA), and lifecycle interpretation [60].

The phase of goal and scope definition defines the functional unit of a study and describes the benefits of relevant products or systems. All products or technologies being compared and their effects, as well as different scenarios of the same product system, need to be extended to the same functional unit to ensure comparability [61]. This phase also needs to specify the system boundary, the input and output assignment procedures, impact categories considered, data quality requirements including collection time and geographical area, as well as the source of the data collected [62]. The establishment of the system boundaries needs to ensure that the same product and energy services are delivered not only by the bioenergy study but also by the fossil energy reference systems. The LCI records all substances and energy that flow into or out of the system being evaluated. While some of the most important and focused steps of the LCI data lifecycle are directly measured and collected, the large amount of data that is added or subtracted during the lifecycle comes from the generic LCI database. In the LCIA phase, LCI records are assessed based on their contribution to certain categories of environmental impacts, such as global warming potential, eutrophication, or resource depletion. An impact assessment method is used to evaluate the environmental impacts of the process or product. Two commonly used impact assessment methods are mid-point end-point approaches. For the former, all materials from the LCI are combined into impact categories based on the common characteristics of causality. A variety of impact categories such as global warming potential ($CO_{2\text{-eq}}$ emission), acidification potential ($SO_{2\text{-eq}}$ emission), eutrophication potential ($PO_{4\text{-eq}}$ emission), etc, are available, while this chapter focuses on global warming potential. For the latter, the environmental impacts at the end of the cause-effect chain are addressed and the severity of environmental damage is characterized [63]. The interpretation phase evaluates the results and reveals the uncertainty and deficiencies of the previous steps by combining the information in the three previous phases [64].

There are various commercial LCA software such as Ecoinvent, Gabi, SimaPro, Umberto, etc. They normally differ in terms of information credibility, datasets understandability, the ease of looking for data orthe accessible breadth of processes. There is a big difference in the extent of systems and processes accessible within the software. In many cases, data availability for a particular region or multiple regions and countries of interest is an important criterion for selecting LCA software [62].

21.4.2 LCA carbon footprints

The GHG emission of waste-to-biofuel can be represented in a unit of grams carbon dioxide -equivalents per megajoule of energy (g $CO_{2\text{-eq}}$/MJ), or kilograms carbon dioxide -equivalents per kilogram or gallon of biofuel (kg $CO_{2\text{-eq}}$/kg or kg $CO_{2\text{-eq}}$/gal). The use of

MJ rather than volumetric units facilitates the comparison of ethanol with the existing gasoline system.

21.4.2.1 Waste-to-bioethanol

Ref. [65] carried out LCA to compare the environmental impacts of production and use of bioethanol blends, i.e., E10 (10 vol.% bioethanol with 90 vol.% gasoline) and E85 (85 vol.% bioethanol with 15 vol.% gasoline) with that of conventional gasoline. Bioethanol was produced from wheat straw and corn stover via three steps: pretreatment and conditioning where hemicellulose sugars were decomposed by treatment with sulfuric acid and steam, saccharification and fermentation, and distillation and dehydration to achieve 99.5% bioethanol. It was shown that E10 and E85 reduced the GHG emissions by 4.3% and 47%, respectively, on a 1 km driving distance. Ref. [66] analyzed that the lifecycle GHG emissions associated with the processes of food waste conversion into bioethanol with coproduction of compost and animal feed. This study was based on an SSF process at room temperature with a grinding pretreatment. The lifetime GHG emission associated with the ethanol production process is 1458 g $CO_{2\text{-eq}}$/L bioethanol.

Ref. [67] summarized that the GHG emissions of bioethanol production are highly contingent upon the types of feedstock and the relative proportion of bioethanol in the gasoline–bioethanol blend. For example, for an E100 blend, bioethanol production from agricultural residues (corn stover and wheat straw), switchgrass, and wood led to the GHG emission reductions of 82%–91%, 53% - 93% and 50%–62%, respectively. as compared with conventional gasoline per unit driving distance. For an E10 blend, the GHG emission was lower than 10% as compared to more than 40% for E85 and upper blends. When switching from conventional gasoline to E10, GHG savings were found to range from 4% to 15%, E85 from 12% to 96%, and E100 from 46% to 90% [68].

Ref. [69] analyzed the lifecycle environmental impacts of bioethanol production from cattle manure via an SHF process. The GHG emission was found to be 1.7 g $CO_{2\text{-eq}}$/MJ with major contributions from energy use, drying milling, acid pretreatment, buffer use during hydrolysis, and sodium phosphate use during fermentation. Ref. [70] analyzed the GHG emissions of bioethanol production from cassavia via a five-step process, i.e., milling, mixing and liquefaction, saccharification and fermentation, distillation (stillage was digested to produce biogas and subsequently steam used by the process), and dehydration, and from molasses via a three-step process, i.e., alcohol generation by fermentation, distillation (stillage digested to produce biogas for energy generation), and dehydration. The former had an average GHG emission of 37.3 g $CO_{2\text{-eq}}$/MJ with major contributions from coal combustion for steam production followed by fertilizer application during cassava cultivation, while the latter had a GHG emission of 25.7–39.0 g $CO_{2\text{-eq}}$/MJ with major contributions from fertilizer application during sugarcane cultivation and sugarcane burning before harvesting. High environmental performance can be achieved in bioethanol production from sludge from the bleached pulp process, with the lowest amount of carbonate added [71]. The bleaching process allows for maximum sugar content in the sludge, which can then be added using the proposed bioethanol production system.

21.4.2.2 Waste-to-biomethane

Ref. [72] compared GHG emissions associated with the production of ethanol, biomethane, limonene and digestion products from citrus waste, a byproduct of the citrus processing industry. For large biorefineries, bioethanol used as E85 in light vehicles resulted in a 134% reduction in GHG emissions compared to gasoline-fueled vehicles when applying systems expansion methods. For small biorefineries, Biomethane replaces natural gas to generate electricity, limonene replaces acetone in a solvent, and digestion in anaerobic digestion replaces synthetic fertilizer. when biomethane is used instead of natural gas, GHG emissions were reduced by 77%.

A comprehensive LCA was carried out in the EU-funded demonstration project "Industrial-scale biofuel production sustainable algae cultivation demonstration". The results of LCA showed that the algae biorefineries offered significant benefits in protecting the climate, fossil resources and ozone compared to conventional wastewater treatment and the use of biomethane instead of compressed natural gas as a carrier fuel [73]. Ref. [74] used LCA to evaluate the environmental footprints of three biogas upgrading technologies, which are high-pressure water washing (HPWS) and alkaline regeneration (AwR) and bottom ash upgrading (BABIU). It was determined that the AwR process had an 84% higher impact on all LCA categories, primarily due to energy-intensive production of alkaline reactants. Even with the other five carbon dioxide capture technologies on the market, the BABIU process has the least impact on most categories. As AwR, it was determined that the use of NaOH instead of KOH improved its environmental performance by 34%. For the BABIU process, the use of renewable energy improved its impact as it accounted for 55% of the impact.

Ref. [75] evaluated the LCA of biomethane produced by lignocellulosic biomass as a biofuel. This paper describes a case study of grass biomethane produced by anaerobic digestion of grass silage and used as a transportation fuel. Biomethane production as a transport fuel through grass silage can achieve a GHG emission savings of 89%. The study showed that cumulative GHG emission savings under the various sensitivity analysis scenarios were up to 89.4%. It also showed that processing emissions accounted for the largest GHG emission (38.13 g $CO_{2\text{-eq}}$/MJ) followed by emissions associated with plant construction (12.64 g $CO_{2\text{-eq}}$/MJ), product upgrading (12.64 g $CO_{2\text{-eq}}$/MJ) and biogas leakage (10.82 g $CO_{2\text{-eq}}$/MJ).

21.4.2.3 Waste-to-biohydrogen

Ref. [76] evaluated the lifecycle environmental impacts (GHG emissions, acidification potential, and fossil energy demand) of biomass hydrogen production for transport use. The raw feedstock materials included woody biomass from forestry or short rotation coppice, energy crops, straw, bio-waste, and organic by-products. The technologies included gasification of biomass, steam reforming of biomethane from fermentation, pyroreforming of glycerol from biodiesel production, and alkaline water electrolysis supported by biomass cogeneration. The results showed that the GHG emissions of hydrogen production from the considered routes ranged between 124.1 and 28.3 g $CO_{2\text{-eq}}$/MJ, with the lowest emission (28.3 g $CO_{2\text{-eq}}$/MJ) corresponding to the gasification of forest residues followed by the gasification of short rotation coppice wood (29.7 g $CO_{2\text{-eq}}$/MJ). The study also showed that the GHG emissions of gasification-based routes were mainly determined by the electricity demand that was considered to be 0.159 kWh/MJ.

Biohydrogen production from autothermal reforming of waste poultry fat was studied by [77]. It was shown that the GHG emission corresponding to 1 kg of hydrogen production was 9.57 kg $CO_{2\text{-eq}}$, with thermal energy requirement being the dominant GHG contributor (8.15 kg $CO_{2\text{-eq}}$) followed by transport and electricity requirement. [80] explored biohydrogen production from the gasification of forestry wood chips and fermentation of maize silage, manure, and organic waste. They showed that the gasification-based method had a GHG emission of around 4.08 kg $CO_{2\text{-eq}}$/kg H_2, while the fermentation-based method had a GHG emission of around 5.28 kg $CO_{2\text{-eq}}$/kg H_2, with the distribution of biohydrogen contributing more emissions than waste biomass supply and biohydrogen conversion combined.

21.4.2.4 Waste-to-biodiesel

Numerous LCA studies have been conducted to calculate the GHG emissions of waste-to-biodiesel generation. Ref. [54] compared the environmental impacts of biodiesel derived from waste cooking oil using a hydrogenation method with that of the one using a catalysis method. The route of the catalysis-based biodiesel had a GHG emission of −150 tons $CO_{2\text{-eq}}$/year as compared to 547 tons $CO_{2\text{-eq}}$/year for the route of the hydrogenated biodiesel.

Ref. [51] compared the GHG emissions of grease trap waste-based biodiesel production with that of existing grease trap waste disposal, soybean-based biodiesel production and low-sulfur diesel. Depending on the relevant waste management consideration, the average GHG emissions of grease trap waste-based biodiesel production ranged from 22 to 37 g $CO_{2\text{-eq}}$/MJ, which was better than the existing waste disposal practice. If the produced biodiesel was used to displace low-sulfur diesel, this led to a reduction of GHG emissions by 20%–75%. The environmental impacts of grease trap waste-based biodiesel production were comparable to that of soybean-based biodiesel production when lipid concentrations in waste were greater than 10%.

Ref. [50] found that the GHG emissions of four biodiesel production routes, i.e., esterification–transesterification of waste vegetable oil, beef tallow, and poultry fat, and

Table 21.1: The carbon footprints of the different waste-to-biofuel generation.

Production	Technology involved	Feedstock	GHG emissions	References
Bioethanol	Saccharification and fermentation	Cane molasses	25 g $CO_{2\text{-eq}}$/MJ	[78]
	Saccharification and fermentation	Cattle manure	1.7 g $CO_{2\text{-eq}}$/MJ	[69]
	Liquefaction and fermentation	Cassava	65.5–73.8 g $CO_{2\text{-eq}}$/MJ	[14]
	Saccharification and fermentation	Cassavia	37.3 g $CO_{2\text{-eq}}$/MJ	[70]
	Fermentation and dehydration	Malasse	25.7–39.0 g $CO_{2\text{-eq}}$/MJ	[70]
Biomethane	Anaerobic digestion	Grass silage	69.74 g $CO_{2\text{-eq}}$/MJ	[75]
	Anaerobic digestion	Straw	20–50 g CO_2-eq/MJ	[79]
	Anaerobic digestion	Manures	−104–44 g $CO_{2\text{-eq}}$/MJ	[79]
Biohydrogen	Gasification	Forest residues	28.3–124.1 g $CO_{2\text{-eq}}$/MJ	[76]
	Gasification	Wood chips	29.7 g $CO_{2\text{-eq}}$/MJ	[76]
	Fermentation	Maize silage, manure, and organic waste	5.28 kg $CO_{2\text{-eq}}$/kg H_2 (44 g $CO_{2\text{-eq}}$/MJ)	[80]
	Gasification	Forestry wood chips	4.08 kg CO_2-eq/kg H_2 (34 g $CO_{2\text{-eq}}$/MJ)	[80]
	Autothermal reforming	Waste poultry fat	9.57 kg CO_2-eq/kg H_2 (80 g $CO_{2\text{-eq}}$/MJ)	[77]
Biodiesel	Acid-catalyzed esterification + alkali-catalyzed transesterification	Waste vegetable oil	16.97 g $CO_{2\text{-eq}}$/MJ	[50]
	Acid-catalyzed esterification + alkali-catalyzed transesterification	Beef tallow	23.32 g $CO_{2\text{-eq}}$/MJ	[50]
	Acid-catalyzed esterification + alkali-catalyzed transesterification	Poultry fat	23.55 g $CO_{2\text{-eq}}$/MJ	[50]
	Acid-catalyzed in situ transesterification	Sewage sludge	20.84 g $CO_{2\text{-eq}}$/MJ	[50]
	Transesterification	Grease trap waste	−55.49 g $CO_{2\text{-eq}}$/gal	[52]
	Catalytic hydrotreatment	Waste cooking oil	12.15 g $CO_{2\text{-eq}}$/MJ	[81]

transesterification of sewage sludge, were 16.97, 23.32, 23.55, and 20.84 g $CO_{2\text{-eq}}$/MJ, respectively. They found that electric and thermal energy use accounted for the major factor for GHG emissions. For example, over 30% and 20% of GHG emissions were contributed by thermal energy requirements in biodiesel production from beef tallow and poultry, and waste vegetable oil, respectively. However, for biodiesel production from sewage sludge, over 50% of GHG emissions were accounted for by the significant use of methanol to promote the transesterification of sewage sludge.

Ref. [52] applied LCA to assess energy consumption and GHG emissions of biodiesel production from grease trap waste. In their system, an anaerobic digestion unit was used to utilize the solid residue after the FOG pretreatment for energy generation, that significantly reduced the GHG emission to −55.49 g $CO_{2\text{-eq}}$/gal. When the solid was considered as waste for disposal instead, the GHG emission was 5735.22 g $CO_{2\text{-eq}}$/gal. Esterification and transportation accounted for GHG emissions of 2179.03 and 2410.11 g $CO_{2\text{-eq}}$/gal, respectively and were two major GHG emission contributors followed by FOG separation (873.35 g $CO_{2\text{-eq}}$/gal) and transesterification (234.06 g $CO_{2\text{-eq}}$/gal). The carbon footprints of the different waste-to-biofuel generation are summarized in Table 21.1.

21.5 Conclusions and perspectives

Second-generation biofuel production based on waste serves as a good solution for fulfilling the demands of low carbon fuel and sustainable waste management. Different technology routes are available, however, a systematic database about the optimal configuration and design need to be developed to guide practical implementation. Lifecycle assessment serves an important tool for identifying the configurations and designs with better environmental sustainability. Depending on the system boundary, technology, design, feedstock, the carbon footprints of waste-to-biofuel generation vary significantly with an average value of tens of g $CO_{2\text{-eq}}$/MJ.

References

[1] Karmee SK. Liquid biofuels from food waste: current trends, prospect and limitation. Renewable & Sustainable Energy Reviews 2016;53:945–53.

[2] ISWA. Globalization and waste management. International Solid Waste Association; 2012.

[3] Nations U. Sustainable development goals. Sustainable Development 2015;11(1).

[4] Cubasch U, et al. Climate Change 2013 the Physical science Basis Working Group I contribution to the fifth assessment report of the Intergovernmental Panel on climate change introduction. Climate Change 2013: The Physical Science Basis. 2014. p. 119–58.

[5] Safa S, Soucy G. Liquid and solution treatment by thermal plasma: a review. International Journal of Environmental Science and Technology 2014;11(4):1165–88.

[6] Li S, Yang X. Biofuel production from food wastes. In: Handbook of biofuels production: processes and technologies. 2nd ed., vol. 98; 2016. p. 617–53.

[7] de Wild PJ, Huijgen WJJ, Heeres HJ. Pyrolysis of wheat straw-derived organosolv lignin. Journal of Analytical and Applied Pyrolysis 2012;93:95–103.
[8] Myglovets M, et al. Preparation of carbon adsorbents from lignosulfonate by phosphoric acid activation for the adsorption of metal ions. Carbon 2014;80:771–83.
[9] Hammond GP, Li B. Environmental and resource burdens associated with world biofuel production out to 2050: footprint components from carbon emissions and land use to waste arisings and water consumption. Global Change Biology Bioenergy 2016;8(5):894–908.
[10] Gomiero T. Are biofuels an effective and viable energy strategy for industrialized societies? A reasoned overview of potentials and limits. Sustainability 2015;7(7):8491–521.
[11] Taghizadeh-Alisaraei A, et al. Biofuel production from citrus wastes: a feasibility study in Iran. Renewable & Sustainable Energy Reviews 2017;69:1100–12.
[12] Mohammadshirazi A, et al. Energy and cost analyses of biodiesel production from waste cooking oil. Renewable & Sustainable Energy Reviews 2014;33:44–9.
[13] Zhou ZY, et al. Lignocellulosic biomass to biofuels and biochemicals: a comprehensive review with a focus on ethanol organosolv pretreatment technology. Biotechnology and Bioengineering 2018;115(11):2683–702.
[14] Kiran EU, et al. Bioconversion of food waste to energy: a review. Fuel 2014;134:389–99.
[15] Aro E-M. From first generation biofuels to advanced solar biofuels. Ambio 2016;45(1):24–31.
[16] Bertrand E, et al. First generation bioethanol. In: Green fuels technology: biofuels; 2016. p. 175–212.
[17] Saini JK, Saini R, Tewari L. Lignocellulosic agriculture wastes as biomass feedstocks for second-generation bioethanol production: concepts and recent developments. 3 Biotech 2015;5(4):337–53.
[18] Chen H. Lignocellulose biorefinery engineering: principles and applications. Woodhead Publishing; 2015.
[19] Bhatia L, Johri S, Ahmad R. An economic and ecological perspective of ethanol production from renewable agro waste: a review. AMB Express 2012;2.
[20] Schuster BG, Chinn MS. Consolidated bioprocessing of lignocellulosic feedstocks for ethanol fuel production. Bioenergy Research 2013;6(2):416–35.
[21] Zuroff TR, Curtis WR. Developing symbiotic consortia for lignocellulosic biofuel production. Applied Microbiology and Biotechnology 2012;93(4):1423–35.
[22] Spyridon A, Euverink W, Jan G. Consolidated briefing of biochemical ethanol production from lignocellulosic biomass. Electronic Journal of Biotechnology 2016;19(5):44–53.
[23] Huang RL, et al. Bioconversion of lignocellulose into bioethanol: process intensification and mechanism research. Bioenergy Research 2011;4(4):225–45.
[24] Guo Y, et al. Hydrogen-rich gas production by steam gasification of hydrochar derived from sewage sludge. International Journal of Hydrogen Energy 2014;41(5):3363–72.
[25] You S, et al. Towards practical application of gasification: a critical review from syngas and biochar perspectives. Critical Reviews in Environmental Science and Technology 2018;48(22–24):1165–213.
[26] Hay JXW, et al. Biohydrogen production through photo fermentation or dark fermentation using waste as a substrate: overview, economics, and future prospects of hydrogen usage. Biofuels Bioproducts & Biorefining 2013;7(3):334–52.
[27] Parthasarathy P, Sheeba KN. Combined slow pyrolysis and steam gasification of biomass for hydrogen generation – a review. International Journal of Energy Research 2015;39(2):147–64.
[28] Dhar BR, et al. Hydrogen production from sugar beet juice using an integrated biohydrogen process of dark fermentation and microbial electrolysis cell. Bioresource Technology 2015;198:223–30.
[29] Gomez X, et al. Hydrogen production: two stage processes for waste degradation. Bioresource Technology 2011;102(18):8621–7.
[30] Molino A, et al. Biomethane production by anaerobic digestion of organic waste. Fuel 2013;103:1003–9.
[31] Gai C, et al. Hydrogen-rich gas production by steam gasification of hydrochar derived from sewage sludge. International Journal of Hydrogen Energy 2016;41(5):3363–72.
[32] Lee DJ, Show KY, Su A. Dark fermentation on biohydrogen production: pure culture. Bioresource Technology 2011;102(18):8393–402.

[33] Fermoso J, Rubiera F, Chen D. Sorption enhanced catalytic steam gasification process: a direct route from lignocellulosic biomass to high purity hydrogen. Energy & Environmental Science 2012;5(4):6358–67.
[34] Vamvuka D. Bio-oil, solid and gaseous biofuels from biomass pyrolysis processes – an overview. International Journal of Energy Research 2011;35(10):835–62.
[35] Parthasarathy P, et al. Optimization of parameters for the generation of hydrogen in combined slow pyrolysis and steam gasification of biomass. Energy & Fuels 2017;31(12):13692–704.
[36] You S, et al. A critical review on sustainable biochar system through gasification: energy and environmental applications. Bioresource Technology 2017;246:242–53.
[37] Zhang Y, et al. Computational fluid dynamics study on mixing mode and power consumption in anaerobic mono-and co-digestion. Bioresource Technology 2016;203:166–72.
[38] Jiang Y, et al. Inhibition of volatile fatty acids on methane production kinetics during dry co-digestion of food waste and pig manure. Waste Management 2018;79:302–11.
[39] Li H, et al. Performance, granule conductivity and microbial community analysis of upflow anaerobic sludge blanket (UASB) reactors from mesophilic to thermophilic operation. Biochemical Engineering Journal 2018;133:59–65.
[40] Nasir IM, Ghazi TIM, Omar R. Production of biogas from solid organic wastes through anaerobic digestion: a review. Applied Microbiology and Biotechnology 2012;95(2):321–9.
[41] Patterson T, et al. Life cycle assessment of biohydrogen and biomethane production and utilisation as a vehicle fuel. Bioresource Technology 2013;131:235–45.
[42] Parawira W, et al. Profile of hydrolases and biogas production during two-stage mesophilic anaerobic digestion of solid potato waste. Process Biochemistry 2005;40(9):2945–52.
[43] Kastner V, Somitsch W, Schnitzhofer W. The anaerobic fermentation of food waste: a comparison of two bioreactor systems. Journal of Cleaner Production 2012;34:82–90.
[44] Lee JC, et al. Biological conversion of CO_2 to CH_4 using hydrogenotrophic methanogen in a fixed bed reactor. Journal of Chemical Technology and Biotechnology 2012;87(6):844–7.
[45] Nathao C, Sirisukpoka U, Pisutpaisal N. Production of hydrogen and methane by one and two stage fermentation of food waste. International Journal of Hydrogen Energy 2013;38(35):15764–9.
[46] Yeshanew MM, et al. Production of biohythane from food waste via an integrated system of continuously stirred tank and anaerobic fixed bed reactors. Bioresource Technology 2016;220:312–22.
[47] Jung KW, et al. Conversion of organic solid waste to hydrogen and methane by two-stage fermentation system with reuse of methane fermenter effluent as diluting water in hydrogen fermentation. Bioresource Technology 2013;139:120–7.
[48] Bankovic-Ilie IB, et al. Waste animal fats as feedstocks for biodiesel production. Renewable & Sustainable Energy Reviews 2014;32:238–54.
[49] Yaakob Z, et al. Overview of the production of biodiesel from waste cooking oil. Renewable & Sustainable Energy Reviews 2013;18:184–93.
[50] Dufour J, Iribarren D. Life cycle assessment of biodiesel production from free fatty acid-rich wastes. Renewable Energy 2012;38(1):155–62.
[51] Hums ME, Cairncross RA, Spatari S. Life-cycle assessment of biodiesel produced from grease trap waste. Environmental Science & Technology 2016;50(5):2718–26.
[52] Tu QS, McDonnell BE. Monte Carlo analysis of life cycle energy consumption and greenhouse gas (GHG) emission for biodiesel production from trap grease. Journal of Cleaner Production 2016;112:2674–83.
[53] Georgogianni KG, et al. Biodiesel production: reaction and process parameters of alkali-catalyzed transesterification of waste frying oils. Energy & Fuels 2007;21(5):3023–7.
[54] Yano J, et al. Life cycle assessment of hydrogenated biodiesel production from waste cooking oil using the catalytic cracking and hydrogenation method. Waste Management 2015;38:409–23.
[55] Mu DY, et al. A life cycle assessment and economic analysis of the Scum-to-Biodiesel technology in wastewater treatment plants. Bioresource Technology 2016;204:89–97.

[56] Xin CH, et al. Waste-to-biofuel integrated system and its comprehensive techno-economic assessment in wastewater treatment plants. Bioresource Technology 2018;250:523–31.
[57] Capodaglio AG, Callegari A. Feedstock and process influence on biodiesel produced from waste sewage sludge. Journal of Environmental Management 2018;216:176–82.
[58] Hauschild MZ, Huijbregts MAJ. Introducing life cycle impact assessment. In: Life cycle impact assessment; 2015. p. 1–16.
[59] Pandey D, Agrawal M, Pandey JS. Carbon footprint: current methods of estimation. Environmental Monitoring and Assessment 2011;178(1–4):135–60.
[60] Chang J, et al. Biohydrogen production: current perspectives and the way forward. International Journal of Hydrogen Energy 2014;37(20):15616–31.
[61] Van den Heede P, De Belie N. Environmental impact and life cycle assessment (LCA) of traditional and 'green' concretes: literature review and theoretical calculations. Cement and Concrete Composites 2012;34(4):431–42.
[62] Vieira DR, Calmon JL, Coelho FZ. Life cycle assessment (LCA) applied to the manufacturing of common and ecological concrete: a review. Construction and Building Materials 2016;124:656–66.
[63] Cavalett O, et al. Comparative LCA of ethanol versus gasoline in Brazil using different LCIA methods. International Journal of Life Cycle Assessment 2013;18(3):647–58.
[64] Schebek L, et al. Land-use change and CO_2 emissions associated with oil palm expansion in Indonesia by 2020. In: From science to society: new trends in environmental informatics; 2018. p. 49–59.
[65] Daylan B, Ciliz N. Life cycle assessment and environmental life cycle costing analysis of lignocellulosic bioethanol as an alternative transportation fuel. Renewable Energy 2016;89:578–87.
[66] Ebner J, et al. Life cycle greenhouse gas (GHG) impacts of a novel process for converting food waste to ethanol and co-products. Applied Energy 2014;130:86–93.
[67] Morales M, et al. Life cycle assessment of lignocellulosic bioethanol: environmental impacts and energy balance. Renewable & Sustainable Energy Reviews 2015;42:1349–61.
[68] Borrion AL, McManus MC, Hammond GP. Environmental life cycle assessment of lignocellulosic conversion to ethanol: a review. Renewable & Sustainable Energy Reviews 2012;16(7):4638–50.
[69] de Azevedo A, et al. Life cycle assessment of bioethanol production from cattle manure. Journal of Cleaner Production 2017;162:1021–30.
[70] Papong S, et al. Environmental life cycle assessment and social impacts of bioethanol production in Thailand. Journal of Cleaner Production 2017;157:254–66.
[71] Sebastiao D, et al. Life cycle assessment of advanced bioethanol production from pulp and paper sludge. Bioresource Technology 2016;208:100–9.
[72] Pourbafrani M, et al. Life cycle greenhouse gas impacts of ethanol, biomethane and limonene production from citrus waste. Environmental Research Letters 2013;8(1).
[73] Maga D. Life cycle assessment of biomethane produced from microalgae grown in municipal waste water. Biomass Conversion and Biorefinery 2017;7(1):1–10.
[74] Starr K, et al. Life cycle assessment of biogas upgrading technologies. Waste Management 2012;32(5):991–9.
[75] Nizami AS, Ismail IM. Life-cycle assessment of biomethane from lignocellulosic biomass. In: Life cycle assessment of renewable energy sources; 2013. p. 79–94.
[76] Wulf C, Kaltschmitt M. Life cycle assessment of biohydrogen production as a transportation fuel in Germany. Bioresource Technology 2013;150:466–75.
[77] Hajjaji N, Houas A, Pons MN. Thermodynamic feasibility and life cycle assessment of hydrogen production via reforming of poultry fat. Journal of Cleaner Production 2016;134:600–12.
[78] Khatiwada D, et al. Energy and GHG balances of ethanol production from cane molasses in Indonesia. Applied Energy 2016;164:756–68.
[79] Tonini D, et al. GHG emission factors for bioelectricity, biomethane, and bioethanol quantified for 24 biomass substrates with consequential life-cycle assessment. Bioresource Technology 2016;208:123–33.

[80] Zech K, et al. Technical, economic and environmental assessment of technologies for the production of biohydrogen and its distribution results of the Hy-NOW study. International Journal of Hydrogen Energy 2015;40(15):5487–95.
[81] Bezergianni S, Chrysikou LP. Sustainability assessment of fuels production via hydrotreating waste lipids and co-processing waste lipids with petroleum fractions. In: Energy, transportation and global warming; 2016. p. 387–400.

SECTION H

Country specific case studies

CHAPTER 22

Biorefineries in Germany

Maria Alexandri[1], Francesca Demichelis[2], Silvia Fiore[2], Mette Lübeck[3], Daniel Pleissner[4]

[1]Leibniz Institute for Agricultural Engineering and Bioeconomy Potsdam, Potsdam, Germany; [2]DIATI, Politecnico di Torino, Torino, Italy; [3]Department of Chemistry and Bioscience - Section for Sustainable Biotechnology, Denmark; [4]Sustainable Chemistry (Resource Efficiency), Institute of Sustainable and Environmental Chemistry, Leuphana University of Lüneburg, Lüneburg, Germany

22.1 Introduction

Integrated biorefineries are a combination of material and energy production in accordance with the principle of cascade use. An integrated biorefinery can be defined as "an integrated production plant using biomass or biomass-derived feedstocks to produce a range of value-added products and energy" [1]. Using physical and mechanical, thermochemical, chemical and biotechnological processes, different platform compounds obtained from various biogenic feedstocks are converted in products belonging to the classes: Food, feed, materials, chemicals and energy. The process steps can be divided into pretreatment and separation of biomass components (primary refining) as well as conversion of biomass components into products (secondary refining) [2].

During feedstock processing in biorefineries, residual materials and waste materials are created. The integration of different processes resulted from the necessity to exploit the potential of all material streams for product formation (Table 22.1, Fig. 22.1).

Integrated biorefineries are not only more efficient in terms of feedstock utilization, but also in terms of costs. For instance, the production of lactic acid as single product from food waste might be cost-inefficient. The economy, however, can significantly be improved when lactic acid production is combined with anaerobic digestion of the residues for biogas and consequently energy generation [3]. Generated energy can successively be used to cover the energy demand of biorefineries and consequently reduce operation costs.

The integration of different processes for feedstock utilization was not properly considered in the past. Energy-rich compounds like bioethanol and biodiesel, for instance, were the major products to be obtained from biorefineries. The focus on biofuels was a consequence of the limitation in fossil-fuel and its climate and environmental effects. The

Waste Biorefinery. https://doi.org/10.1016/B978-0-12-818228-4.00022-8

Table 22.1: Elements of a biorefinery.

Feedstock	Process	Platform compounds	Products
Agricultural biomass -Oil crops -Starch crops -Sugar crops -Grasses -Wood and woody biomass Aquatic biomass -Algae Biogenic residual- & waste materials -Agricultural and forestry residues -Biogenic residual materials (e.g., pulp, stillage, spent grains) -Biogenic waste materials (e.g., yellow grease, waste wood)	-Physical and mechanical -Thermochemical -Chemical -Biotechnological	-Low molecular weight carbohydrates -Polymeric carbohydrates -Lignocellulose -Proteins -Fibers -Vegetable oils and lipids -Press juice	Food and feed -Proteins -Pigments -Vegetable oils and lipids Materials -Chemicals -Materials Bioenergy -Solid, liquid and gaseous fuels -Electricity -Heat

Modified from BMELV, BMBF, BMU, BMWI. Biorefineries roadmap; 2012.

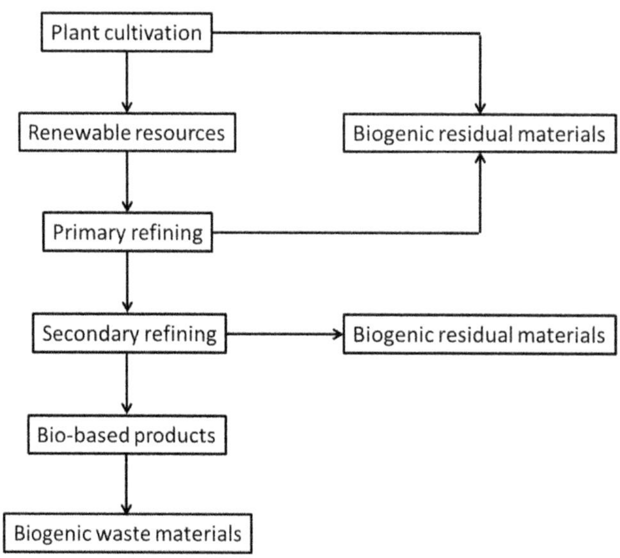

Figure 22.1
Process scheme of an integrated biorefinery. *Modified from BMELV, BMBF, BMU, BMWI. Biorefineries roadmap; 2012.*

development of biorefineries is a continuous process oriented on the demand for products, and thus more and more "plug-in" solutions are under investigation to increase to product portfolio (Table 22.1).

Biorefineries are dependent on the availability of substrates and social acceptance. While the focus was on the production of energy-rich compounds using first generation biorefinery concepts in the past, the focus of research activities is on the use of residual biomass streams in second generation biorefinery nowadays. The switch from first generation to second generation was caused by the controversial discussion whether food and feed should serve as sources of energy.

In 2017, nearly 2.65 million ha of arable land were associated with the production of renewable resources in Germany [4]. Most of the renewable resources (2.35 million ha) were occupied for energy plants, such as plants for bioethanol, biodiesel, biogas, and solid fuels. The production of biomass and the associated assimilation of carbon dioxide as well as fixation of carbon in long-life products positively affect the carbon balance. It seems that the positive effect on carbon balance makes the production of biomass sustainable. However, biomass production requires water, arable land, fertilizers and energy. With an increasing contribution of the bioeconomy to the total economy, the amount of resources needed to cover the demand increases accordingly and this may negatively affect the overall sustainability. Therefore, integrated biorefineries should not only utilize the whole potential of biomass, but also focus on recovery and recirculation of fertilizers, such as phosphate, to arable land to keep biomass production continuing.

A successful biorefinery constitutes first of all a sustainable business plan, giving a clear value proportion on site, meaning that all units including raw material, intermediate products and end-products are somehow connected. A good example of a successful biorefinery is the one based on sugarcane utilization. Sugarcane biorefinery is composed of a sugar-production unit, an ethanol-production unit and a second generation (2G) ethanol unit. The 2G ethanol unit operates using the sugarcane bagasse, exploiting at the same time the surplus steam and water. Thus, a biorefinery producing 2G ethanol or other bio-based products can take advantage of the sugar and 1G ethanol-production units, leading to an economically competitive integrated biorefinery [5]. A biorefinery should be energy efficient and exploit its "wastes" either as coproducts or utilize them in the bioconversion processes [6]. A successful biorefinery is integrated in a traditional refining process, like the example of sugar and ethanol production from sugarcane. A very important aspect of a biorefinery is found in its flexibility. Flexibility gives the opportunity to choose among different processes, to select optimal operational conditions and adapt them according to the specific features of the available feedstock to be processed. In this way, the production of a broad spectrum of marketable products is feasible [6].

This chapter gives an overview to bioeconomy and biorefineries based on various substrates, such as biowaste, oil/fat and sugar/starch as well as green biomass, currently operated in Germany. The use of each substrate will be investigated for availability, process and scale as well as products. Furthermore, future challenges of currently operated processes are discussed.

22.2 Bioeconomy and biorefineries in Germany

In 2010, five million people were employed in the bioeconomy sector with a gross value added of 140 billion Euros [5]. The German government understands bioeconomy as a "... knowledge-based production und utilization of renewable resources, in order to provide products, processes and services in all economic sectors within the context of a future-capable economic system." Many research funding programs have been initiated by the Federal Ministry of Education and Research as well as the Federal Ministry of Food and Agriculture, which provided 2.4 billion Euros by 2016 [6]. Since then, the bioeconomy sector in Germany has been continuously growing and consequently also the number of biorefineries that started to operate. In 2017, from the 224 biorefineries operating in Europe, 58 were located in Germany dealing with the processing of various substrates. Fig. 22.2 illustrates the different biorefineries working with different substrates and operating in Europe and particularly Germany. Biorefineries are split over the whole country with more than 1.1 million employees, but there is a trend that the operation is based on locally available substrates. A predominant number of biorefineries is oil/fat-based for biodiesel and oleochemicals production. There are further a couple of biorefineries which are biowaste-, sugar/starch-, or green biomass-based (Table 22.1, Fig. 22.2, [1]). Those processes left the research and development stage behind and are operated by enterprises for value creation. It is expected that more biorefineries will be established to utilize locally available biomass streams, such as lignocellulosic materials, in the future.

22.2.1 Biowaste-based biorefinery

Biowaste in urban areas is essentially made of two streams: A solid stream known as organic fraction of municipal solid waste (OFMSW), including organic waste from households, restaurants, markets, bakeries and tertiary service organic waste, and a liquid stream known as civil wastewater and sludge deriving from wastewater treatment. Currently, in Germany urban biowaste management is carried out by 48% recycling, 35% incineration with energy recovery and 17% anaerobic digestion and composting [7,8], while in EU-28 urban biowaste management is carried out by 44% recycling, 29% incineration and 25% composting [8]. To achieve social, economic and environmental sustainability, Germany is promoting the decoupling of resource consumption and

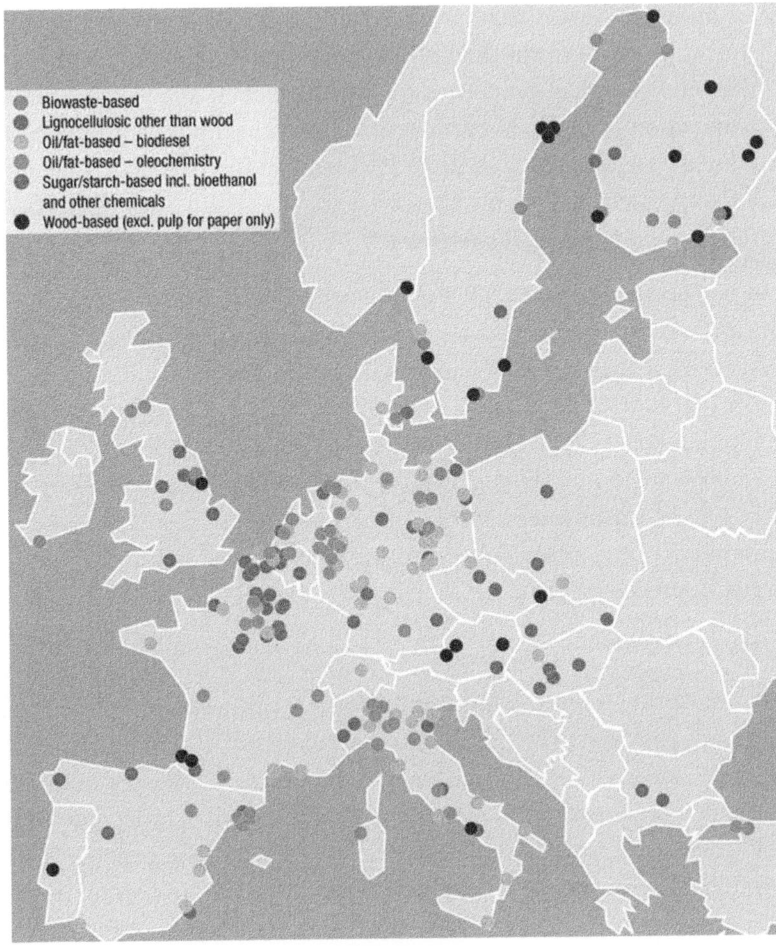

Figure 22.2
Operating biorefineries in Europe in 2017. *Modified from Bio-based Industries Consortium, NovaInstitute. Biorefineries in Europe 2017; 2018.*

economic development with national programs, such as ProgRess II [9], adopted on March 2, 2016, and the German Closed Cycle Management Act [10] to boost the valorization of biowaste as secondary raw material for the production of platform chemicals and energy, and to contribute to sustainable production by means of recycling and recovery actions, to reduce and preserve raw materials and primary energy consumption [2].

The use of biowaste as feedstock in biorefinery processes has two main benefits: (1) it is an alternative to petroleum-based refinery; and (2) it is a nonfood competitive biomass [11]. Among waste biomasses and organic solid wastes, biowaste is the most abundant carbon source for the production of platform chemicals and bioenergy [12].

The application of biowaste as feedstock in biorefineries is based on the knowledge of its physical and chemical properties. Physical properties include pH, total solids (TS) and volatile solids (VS), and chemical properties include elemental analysis (C, H, N, S, and O) and macro-composition in terms of carbohydrate, protein, lipid, and lignocellulose. OFMSW and civil wastewater have high organic contents (65%–70% VS, *w/w*), however, OFMSW and civil wastewater have with 1%–2% (*w/w*) [13] and 18%–24% (*w/w*) [14], respectively, different TS contents.

From a chemical perspective, biowaste is a complex and heterogeneous feedstock, mainly depending on local food-diet trends and wastewater management infrastructures [15]. The elemental analysis proved that: OFMSW and civil wastewater have carbon contents over 50% TS (*w/w*) [13,14]. In detail, OFMSW has around 74% nutrients directly soluble in water without pretreatments [12]. OFMSW is made of (*w/w*) 55.25% $\pm$ 13.08 carbohydrates, 13.00% $\pm$ 3.54 proteins, 3.60% $\pm$ 0.85 oil/fat and 8.0% $\pm$ 1.08 lignocellulose [14]. Civil wastewater consists of (*w/w*) 43.1% $\pm$ 3.77 carbohydrates and 13.7% $\pm$ 8.49 proteins [13]. Carbohydrates and proteins are major carbon and nitrogen sources in biorefinery processes [16].

22.2.1.1 Substrate availability

Based on Eurostat database, the availability of biowaste was assessed both as annual total production (Mt/y) and per capita (kg/pc) within a period from 2012 to 2016 (see Table 22.2). OFMSW and civil wastewater were classified as: W100-03: household and similar waste (bakery, restaurant, etc.), mixed and indifferent materials, and W033: liquid waste from wastewater treatment, respectively, which considers the generation of not hazardous waste. In Germany, the total and per capita biowaste production in the considered time period were 13.02 $\pm$ 0.24 Mt and 160 $\pm$ 1.51 kg/pc of OFMSW, and 0.09 $\pm$ 0.02 Mt and 1.33 $\pm$ 0.58 kg/pc of civil wastewater, respectively, [17]. The availability of biowaste is not only season dependent but varies with urbanization level and population growth, and this is the key strength of a biowaste-based biorefinery system and process designed for scale-up and full-scale applications.

22.2.1.2 Processes and scale

Due to chemical composition, biowaste is a versatile and suitable feedstock for thermochemical and biological biorefinery processes. Among thermochemical processes the most studied and implemented are: Gasification, thermo-valorization and pyrolysis. The three main goals of thermochemical treatments of biowaste are: Energy generation, biowaste sanitation and volume reduction. Gasification converts carbon-rich feedstocks into H_2, CO, CO_2, and CH_4 by means of gasification agents, such as steam, oxygen and air, and catalytic materials. Currently, the bottleneck of biowaste gasification is the low and nonconstant syngas quality due to high amount of char and tar. To solve this problem,

Table 22.2: Biowaste availability [8].

	Wastewater			OFMSW		
	EU 28 (Mt)	Germany (Mt)	German contribution in EU 28 (%)	EU 28 (Mt)	Germany (Mt/y)	German contribution in EU 28 (%)
2012	8.34	0.09	1.02	81.56	12.75	15.64
2014	10.17	0.07	0.73	84.69	13.06	15.42
2016	11.10	0.12	1.08	88.36	13.24	14.98
Average of five years	9.87	0.09	0.94	84.87	13.02	15.34
Standard deviation	1.40	0.02	0.19	3.41	0.24	0.33
	EU 28 (kg/pc)	**Germany (kg/pc)**	**German contribution in EU 28 (%)**	**EU 28 (kg/pc)**	**Germany (kg/pc)**	**German contribution in EU 28 (%)**
2012	17.00	1.00	5.88	161.70	158.70	98.14
2014	20.00	1.00	5.00	166.80	161.10	96.58
2016	27.00	2.00	7.41	165.60	160.80	97.10
Avarage of five years	21.33	1.33	6.10	164.70	160.20	97.28
Standard deviation	5.13	0.58	1.22	2.67	1.31	0.80

steam is employed as gasification agent. Gasification, performed between 750 and 900°C with steam, provides syngas consisting of 28$\%_{v/v}$ H_2, 21$\%_{v/v}$ CH_4, 16.5 $\%_{v/v}$ CO and 17.5$\%_{v/v}$ CO_2, with a low heating value of 15.0 MJ/Nm^3, 7.9$\%_{v/v}$ char yield and 0.2$\%_{v/v}$ tar yield [18]. The steam agents benefits, compared with oxygen and air agents, are: A H_2 content of 82$\%_{v/v}$ and a reduction in tar as well as char yields by 40$\%_{v/v}$–50$\%_{v/v}$ [18]. The valorization of biowaste is economically and environmentally sustainable after drying due to high water contents (OFMSW 80%–82% (*w/w*) and civil wastewater 96%–98% (*w/w*)). Thermo-valorization process can generate both electric energy and combined heat and electric power (CHP). Optimized thermo-valorization processes (temperature between 200 and 400°C for 3 h) reached high heating values (HHV) equal to MJ/kg and a reduction of the amount of biowaste by 86% (*w/w*) [19].

Among thermochemical processes, pyrolysis is able to treat biowaste without predrying. In detail, pyrolysis of biowaste with a moisture content of around 80%–99% results in the production of 36$\%_{v/v}$ H_2 and a reduction in the yield of solid residues due to the steam gasification and steam reforming reactions [20]. The fast evaporation of water creates a steam-rich atmosphere and the local pressure breaks the biowaste from the inside to outside [21]. Comparing the three above-mentioned thermochemical processes: Gasification provides gas with a high energy content, requires external heat for steam generation and has a low energy efficiency for the endothermic nature of the process [22], thermo-valorization produces variable electric energy amounts depending on biowaste moisture and composition [23], and pyrolysis exhibits high quality gas without predrying and requires high capital cost as well as low operational costs [24]. In Germany, thermochemical processes of biowaste are implemented mostly at pilot-scale and seldom at full-scale [9].

Biological biowaste valorization includes: Fermentation for platform chemicals formation and anaerobic digestion (AD) for biogas production. In a biowaste-based biorefinery, the biological path consists of three steps: Upstream, conversion and downstream. Each of these steps depends on: (1) composition of the biowaste, (2) biowaste availability according to current industrial infrastructure as well as (3) quality and type of the required product. The potential to produce the appropriate platform chemical depends on the chemical structure of the biowaste (i.e., simple or double carbon bonds, amino groups, hydroxyl groups, carboxyl groups, etc.) and on the performances of upstream, fermentation and downstream, which can affect each other. Usually, hydrolysis of biowaste is carried out as upstreaming in order to make carbon and nitrogen compounds available for growth and metabolism of microorganism. At pilot- and full-scales, the most adopted upstream configurations are acid-, alkali-, thermal-, and enzymatic hydrolyses [25].

Depending on the composition of biowaste applied, fermentative platform chemical production may require addition of external nutrients, such as yeast extract, agar and minerals, and an inoculum made up of selective and/or engineered microorganisms [26]. For instance, microorganisms used for ethanol, lactic acid, propionic acid and succinic acid productions are *Saccharomyces cerevisiae*, *Streptococcus* sp. and *Escherichia coli* [27]. Temperature and pH conditions are set according to the employed microorganism. Batch or semicontinuous feeding mode are defined in accordance with the available industrial infrastructure [15]. Different types of bioreactors are utilized according to the desired product. For examples, high hydrogen yields have been reported with continuously stirred tank reactors (CSTR) whereas, high methane yields have been achieved in anaerobic fixed bed reactors [15]. The quality of upstream steps influences the conversion yield and by-products formation in fermentative process. Biological conversion processes are followed by downstream, to reach the purity grade required by market. Downstreaming is expensive both by economic and environmental perspectives [3]. Downstream efficacy highly depends on the efficient optimization of fermentation processes (e.g., low byproduct formation, high concentration of desired product and low concentration of remaining nutrients). The most adopted downstream scheme includes centrifugation of fermentation broth, filtration, ion-exchange and condensation [28].

AD is a mature technology implemented at industrial scale, which produces biogas as energy resource and digestate as nutrient-rich substrate used as soil amendment in addition to mineral fertilizers. AD has two main advantages: Stabilization of putrescible matter with energy production and valorization of organic matter and COD of 60%–65% [29]. Currently at industrial scale, AD of biowaste is carried out both in mesophilic and thermophilic conditions with a substrate-to-inoculum-ratio ranging from 0.5:1 to 1:1, with a carbon-nitrogen-phosphorous ratio equal to 100–150:5:1 and in semicontinuous and continuous feed mode [15,29]. In Germany, biogas production from biowaste is implemented at full-scale with Vesta Biofuel and 3B Biofuels.

22.2.1.3 Integration in other processes

In a biowaste-based biorefinery, integration can be achieved by: (1) integration of feedstock and (2) integration of processes. The aims of integration are to maximize the biowaste conversion into product with the advantage of enhancing the revenue and minimizing the amount of waste which is in agreement with the circular economy and bioeconomy principles.

Currently, the integrated biorefinery configuration is biowaste-based biorefinery for platform chemical production and sequential bioenergy generation. This configuration is demonstrated in Refs. [3,14] for sequential lactic acid and biogas productions by means of biological paths. According to Refs. [29,30], biogas plays a key-role, both in single and integrated biorefineries, which makes the processes more versatile and resilient, and results in two products with market values and demands. On the other side, sequential production of platform chemicals or bioenergies is not economic profitable [31]. However, such an approach to utilize OFMSW materially and energetically is currently neither running at pilot- nor at full-scale.

22.2.1.4 Products

Biorefinery converts biowaste into added-value products according to the following hierarchy: First platform chemicals and then biofuels and bioenergy. The products deriving from biorefinery routes depend on the composition of biowaste and on process configuration. According to Ref. [32] the classification of added-value products generated by biowaste-based biorefineries is: (1) Platform chemicals, which are of high-value and low-volume products, due to limited quantities for high-technology applications, as: Bioethanol, lactic acid, propionic acid, succinic acid, butyric acid, bioplastics, etc.; (2) biofuels, classified as medium-value, and volume products; and (3) compost and animal feed, considered low-value and high-volume products. In Germany, the platform chemicals produced at pilot and full-scales by means of biological processes are: Bioethanol 0.44–0.46 g/g biowaste [27], lactic acid 0.29–0.33 g/g biowaste, propionic acid 0.45 g/g biowaste [33], succinic acid 0.57–1.13 g/g biowaste [27,34] and butyric acid 0.37–0.45 g/g biowaste [27]. Among biofuels and energies, biogas yield ranges between 0.5 and 0.9 Nm^3/kg_{vs} [35,36], bio-oil yield is between 13.0 and 38.5 g/g biowaste with an HHV equal to 27.6−35.8 MJ/kg biowaste. For the last group of products, digestate valorization for soil amendment and fertilizer production is 30%−40% (*w/w*) of the fed biowaste [37].

22.2.2 Oil/fat-based

Oil/fat-based biorefineries are predominantly known for the formation of biodiesel. Saturated and mono-unsaturated long-chain fatty acids, such as palmitic acid (C16), stearic acid (C18), or oleic acid (C18:1) present in form of triglycerides, are transesterified with methanol (Fig. 22.3) and the resulting fatty acid methyl ester applied as biodiesel.

The application of glycerol, a side-product from biodiesel formation, as carbon source in fermentation processes thereby allows the establishment of integrated processes (Fig. 22.4).

Even though biodiesel is produced in many biorefineries, the product portfolio obtainable from oil/fat is large. To mention are products for surfactants, cosmetics, lubricants, dyes, and plasticizers, but also the direct use of vegetable oils as solvents is possible. With those more specialized products a higher value gain can be achieved compared with biodiesel. Despite a strong research regarding the use of vegetable oils, globally acting companies, well-defined primary refining technologies and the experiences and know-how on chemical and biotechnological conversions of vegetable oils in Germany, there is a weak secondary refining by integration of other processes [2].

22.2.2.1 Substrate availability

Major feedstocks used in oil/fat-based biorefineries in Germany are sunflower, linseed and rapeseed oils. Furthermore, wasted vegetable oils from food sector and algal lipids can find application. From the 2.65 million ha of arable land associated to the production of renewable resources in Germany, around 0.143 million ha were used to grow vegetable oil-rich plants (0.131, 0.008 and 0.004 million ha for rapeseed, sunflower and linseed, respectively) for material use and 0.713 million ha for rapeseed cultivation for energetic use in 2017 [4]. Alone in 2015, German companies processed 13.1 million t of oilseed,

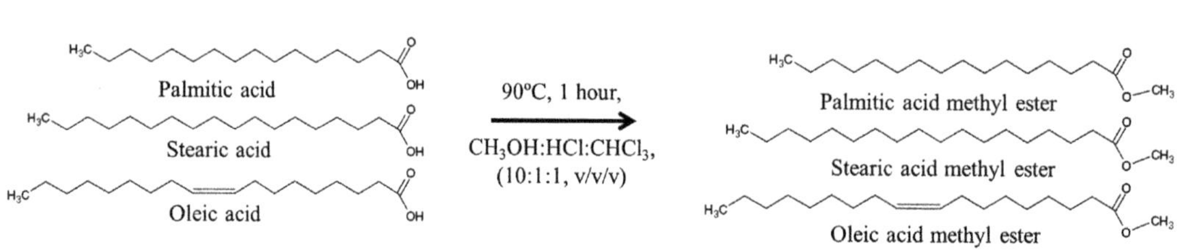

Figure 22.3
Acid catalyzed reaction of fatty acids in presence of methanol for the formation of fatty acid methyl ester (biodiesel).

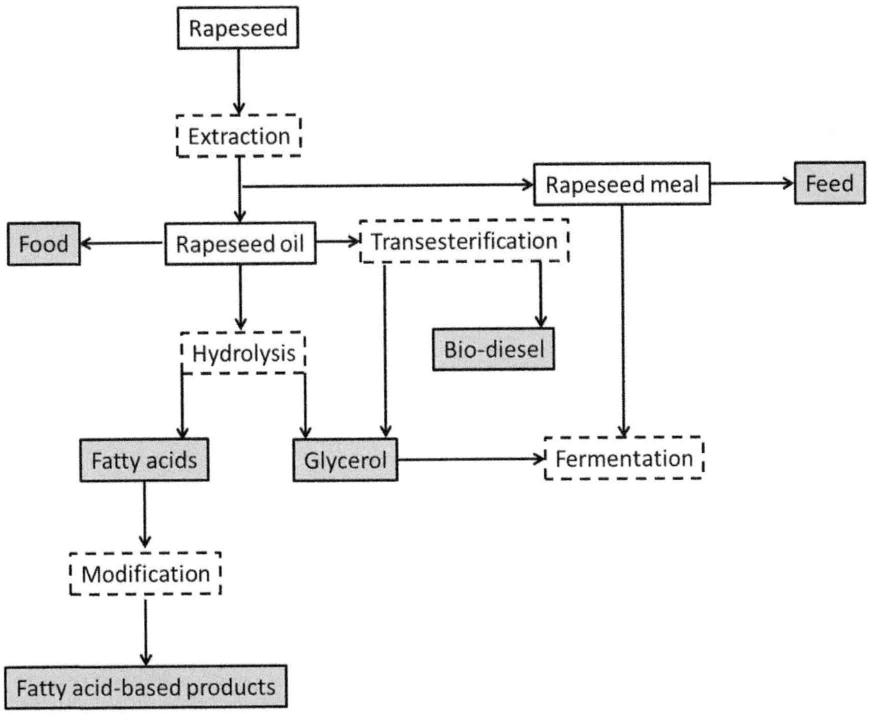

Figure 22.4
Integrated biorefinery for rapeseed processing. Processes are illustrated by a *dashed line* and products are shown in gray.

consisting 75% of rapeseed [38]. Most of the fatty acids present in rapeseed and sunflower are unsaturated or monosaturated (Table 22.3), and thus a production of biodiesel seems favorable due to better properties of fatty acid methyl esters in terms of density, viscosity, cetane number, HHV, iodine, and saponification values and cold filter plugging point [39].

The use of wasted vegetable oil for the generation of biodiesel exceeded the use of rapeseed oil for the first time in 2017. 0.87 million t of biodiesel were waste-based and 0.86 million t were rapeseed oil-based [40]. In addition to the "home grown" plants, 1.168 million t palm oil were imported to Germany in 2015. Almost 43% of the imported palm oil was used for energy generation [41].

Algal biomass has been discussed since decades as new feedstock for biorefineries and in particular as source of lipids. Despite intensive research activities in the past, there is no significant contribution of algal biomass to the bioeconomy. Most operated processes are still at research and development level. An advantage of algal biomass to rapeseed and sunflower is the presence of omega-3 fatty acids, such as docosahexaenoic acid (C22:6) and eicosapentenoic acid (C20:5) [42]. Only linseed (*Linum usitatissimum*) contains with

Table 22.3: Fatty acid weight percentage in rapeseed, sunflower and linseed [43].

Fatty acid (%, *w/w*)	Rapeseed (*Arachis hypogaea*)	Sunflower (*Helianthus annus*)	Linseed (*Linum usitatissimum*)
C12:0	-	-	0.03
C14:0	0.04	0.04	0.04
C16:0	4.06	6.35	5.18
C16:1	0.23	0.07	0.10
C18	1.54	3.92	3.26
C18:1	62.29	20.91	19.04
C18:2	20.65	67.58	16.12
C18:3	8.71	0.17	54.59
C20:0	0.87	0.22	0.09
C20:1	1.09	0.11	0.07
C22:0	0.27	0.66	0.10
C22:1	0.77	-	0.20
C24:0	0.04	0.26	0.03

54.59% (*w/w*) of all fatty acids a significant amount of the omega-3 fatty acid alpha-linolenic acid [43]. A high degree of polyunsaturation of fatty acids allows the formation of special products, such as biobased plasticizer [44].

22.2.2.2 Process and products

At least 35 biorefineries are oil/fat-based and use predominantly rapeseed as feedstock in Germany. An example of a rapeseed-based biorefinery is shown in Fig. 22.4.

The first step is the extraction of oil from seeds (primary refining) using pressing or solvent extraction. The remaining press-cake is rich in proteins and can be fed to animals. When press-cake is used as feed [45] or food [46], a solvent needs to be chosen which can be completely removed and is not harmful to environment and organisms. For instance, it has been suggested to substitute hexane by a harmless solvent [47,48]. A promising substitute is isopropanol, which in combination with ultrasound treatment reaches a similar extraction efficiency (79%) as hexane [49]. In recently developed material utilization processes, press-cake has been hydrolyzed and the nitrogen-rich hydrolysate applied as nutrient source in fermentation. Products formed from rapeseed meal are: poly(3-hydroxybutyrate) [50], phenolic compounds [51], succinic acid [52] or free amino acids and sugars [53]. Rapeseed oil can further undergo secondary refining or used as vegetable oil for food purpose.

The secondary refining of extracted oil depends on the products to be produced. When the focus is on biodiesel production then a transesterification (Fig. 22.3) is carried out and formed fatty acids methyl esters are separated from glycerol. Glycerol finds then application as carbon source for various microorganisms in fermentations [54,55].

More advanced is the material use of fatty acids for the production of surfactants, cosmetics, lubricants, dyes and plasticizers. Hydrolysis ensures the separation of glycerol and fatty acids. Fatty acids are successively chemically and/or biochemically converted into products of interest. The illustrated process from rapeseed to products is also applicable to other oil-rich feedstocks (Fig. 22.4).

Fig. 22.4 illustrates the possibility of integrating various utilization processes in order to exploit as most of the potential of residual and waste materials. Such holistic approach is currently not operating by one single enterprise in Germany. Excluding those plants focusing predominantly on biodiesel production, most operating biorefineries provide oils and glycerol for secondary refining to partners that are specialized in the chemical or biochemical modification. ADM Hamburg AG provides oils from primary refining for food and feed purpose, for technical as well as energetic use. The production of oleochemicals requires knowledge and expertise. An example for a specialist in secondary refining is the Baerlocher GmbH who produces tailor-made metal soaps from oil, a know-how which can hardly be provided by biorefinery operators.

Oleochemicals are literally chemicals derived from oil. Surfactants are used in detergents (soaps) and consist of hydrophobic and hydrophilic parts are well-known examples of oleochemicals [44]. The hydrophobic part is formed by a long-chain fatty acid. Other oleochemicals are fatty amines used as flotation agents, anticaking agents, corrosion inhibitors, dispersants, emulsifiers and additives as well chemical intermediates. The German company Ecogreen Oleochemicals GmbH produces fatty amines from a couple of feedstocks, such as cocos and tallow. The challenge associated with the formation of oleochemicals is the supply of reactants. Irrespective whether vegetable oils were obtained from rapeseed, sunflower, linseed or algal biomass, from palm oil or wasted oils the diversity of fatty acids requires a separation. Separation of fatty acids is possible using chromatographic methods and the principle of fatty acid quantification. At industrial scale when complex feedstocks are applied large-scale chromatographic units can be used. For instance the separation of docosahexaenoic acid and docosapentanoic acid has been carried out after esterification with ethanol using two octadecylsilica-packed columns with an inner diameter of 400 mm and a length of 1000 mm [56]. Another possibility to separate und purify fatty acids is distillation [57]. Appling high vacuum, effective heating, and short contact times can thereby minimize the modification of unsaturated fatty acids by high temperatures or side-reaction.

22.2.3 Sugar/starch-based biorefineries

Sugar and starch biorefineries are based on the conversion of sugar or starch-rich crops into biotechnological products, with bioethanol being the predominant one. This type of biorefinery is also called "first generation" and even though, there is a lot of discussion due to competition with food and feed, up to date sugar and starch crops are important raw materials for the production of not only biobased fuels but also chemicals [58].

A sugar-based biorefinery utilizes sugar beet (*Beta vulgaris*) or sugar cane (*Saccharum officinarum*) for the production of organic acids, vitamins, biofuels and other fermentation products. Sugar production in Europe is dependent mainly on sugar beets, but also EU is one of the main importers of sugar from sugar cane. The estimated sugar production (sugar beet based) in 2016 was 110,119,913 t, from which 12,683,383.81 t were used for biofuels [59].

Cereal grains and tubers are the raw materials of a starch-based biorefinery. Considering the availability and global production yield, the most important crops are wheat (*Triticum* spp.) and corn (*Zea mays*), followed by rice (*Oryza sativa*), potato (*Solanum tuberosum*), and cassava (*Manihot esculenta*). Wheat is the predominant crop worldwide, having a starch content of approximately 58%–70% of the total dry weight [58]. Starch content (*w/w*) represents almost the 72% of the kernel weight in corn, while there are some varieties containing even 80% starch [58]. Potatoes are another significant starch source, as starch could constitute to 65%–80% of the total dry weight [58]. Cereal production in EU for the year 2016 was more than 303 million t, placing EU as one of the largest producers worldwide [59]. From the total production, 68,165,214.2 t were used for food, whereas 14,334,294.93 t were employed for biofuels [59].

A successful biorefinery should stand on feedstock and product flexibility, in order to reduce the risks related to raw material availability and the final product's market demand [58]. There are very good examples of sugar and starch biorefineries that are processing different feedstocks for the production of multiple products. The most characteristic one is the complex "Le Sohettes" in France, which consists of a sugar and wheat refinery plant combined with a straw-based paper unit, producing bioethanol, succinic acid and paper. Even though, there is not yet such an example in Germany, many companies are trying to exploit the available resources to the fullest, either by locating the biorefinery at the same site as the raw material production, or by fully utilizing the crop for more than one end-product.

22.2.3.1 Substrate availability

From the aforementioned crops, sugar beets, wheat and corn are the main raw materials utilized in Germany for in sugar/starch-based biorefineries. Bioethanol for fuel use is the main product from this type of biorefineries, whereas other biobased products are still in R&D, pilot or demo-stage phase [58].

In Germany, for the year 2016, the estimated area harvested for cereals was 6,316,000 ha, and the total production reached 45,364,400 t. The main grains used for biorefineries were rye, wheat and triticale. In 2016, the harvested area for these crops was 570,900 ha for rye, 396,100 ha for triticale and 3,201,700 ha for wheat with a production of 3,173,800 t,

2,397,300 t and 2.4×10^7 t, respectively, [60]. Potatoes are another very important crop for Germany. In 2016, 1.1×10^7 t was produced on 242,500 ha [60].

Sugar beet ranks number seven commodity worldwide, with an annual production of about 270 million t. Germany is one of the leading countries on sugar beet production, with a total harvested area in 2016 of 334,500 ha, and a production yield reaching 25,497,200 t (almost 10% of the global production) [61]. In addition to "domestic" crops, Germany also imports sugar cane, presenting a market share on global imports of 11.7% [62].

Besides edible crops, algae have also been proposed as alternative feedstocks for bioethanol production (third-generation biofuels), as well as sago palm (*Metroxylon sagu*). However, up to date, there use is limited at lab-scale [58].

22.2.3.2 Processes during primary and secondary refining

22.2.3.2.1 Sugar biorefinery

In Europe, France and Germany are the leading countries in sugar beet cultivation. Sugar beet is a perennial plant, growing from spring to September and harvested from September to November. Sucrose is accumulated in the roots of sugar beets during winter, and depending on the cultivation conditions, sugar content could reach 12%–21% (*w/w*). Sugar extraction (primary refining) is carried out following a number of processing steps including diffusion, juice purification, thickening and finally crystallization. More specifically, initially, the beets are washed and sliced. Sugar is then extracted through the process of reverse osmosis, by applying hot water (up to 70°C) in diffusion towers. The remaining solids are the so-called sugar beet pulp, and they are dried, pelletized and sold as animal feed. The raw juice is subjected to further purification with lime, in order to remove nonsugar components. These components are precipitated together with the excess of calcium carbonate and sold as fertilizer. With this process, the thin juice (~60% sugar content) is produced, with a clear, pale yellow color, which is subsequently introduced to a multi-stage evaporation system until a dry matter content of 70%–75% (syrup formation). Crystallization occurs in steam-heated evaporation crystallizers, and the sugar crystals are separated from the syrup via centrifugation. White sugar is afterward dried, fined, and coarsed. The residual syrup is crystallized two more times to give molasses as byproduct. All the intermediate stages (raw juice, thin juice, syrup) as well as the refined sugar, could be utilized for biobased products or bioenergy [2].

22.2.3.2.2 Starch biorefinery

The primary refining in a starch biorefinery differs depending on the initial raw material. If cereal grains are used as starch source, they have to be soaked and expanded. Then, the germ must be separated, grinded and sieved. When potatoes are employed, they have to be initially cleaned and mashed. The next steps are similar regardless of the initial raw material used and involve the separation of fibers and proteins and starch dissolution. After cleaning and drying of the final slurry, pure starch is produced.

Products of the secondary refining of native starch are used in food and chemical industries, either as starch modifications, or for the manufacturing of paper, adhesives or tires. Starch modifications are used in the food industry as thickening agents, or as additives in paper and cosmetic industries. Maltodextrins, glucose and dextrose syrups are products of different hydrolysis degree of starch [63], and could be utilized for different applications either directly or after processing. For example, many fermentation products like lactic acid, gluconic acid, citric acid and amino acids are derived from the glucose syrup [2].

These two types of biorefineries could be combined in one facility for constant bioethanol productions (or other biobased chemicals) regardless of substrate availability. Fig. 22.5 illustrates all the products and processes of a starch and sugar biorefinery.

22.2.3.3 Products

The main product of all the sugar/starch-based biorefineries in Germany is bioethanol, directed mainly for fuel use. FNR (Fachagentur Nachwachsende Rohstoffe e.V) with the support of the Federal Ministry of Agriculture published in 2017 a report entitled "Bioenergy in Germany: Facts and Figures." Table 22.4 presents the results regarding the main raw materials used in Germany for bioethanol production as well as the efficiency of each crop. Starch-based feedstocks yield in higher bioethanol production per ton of

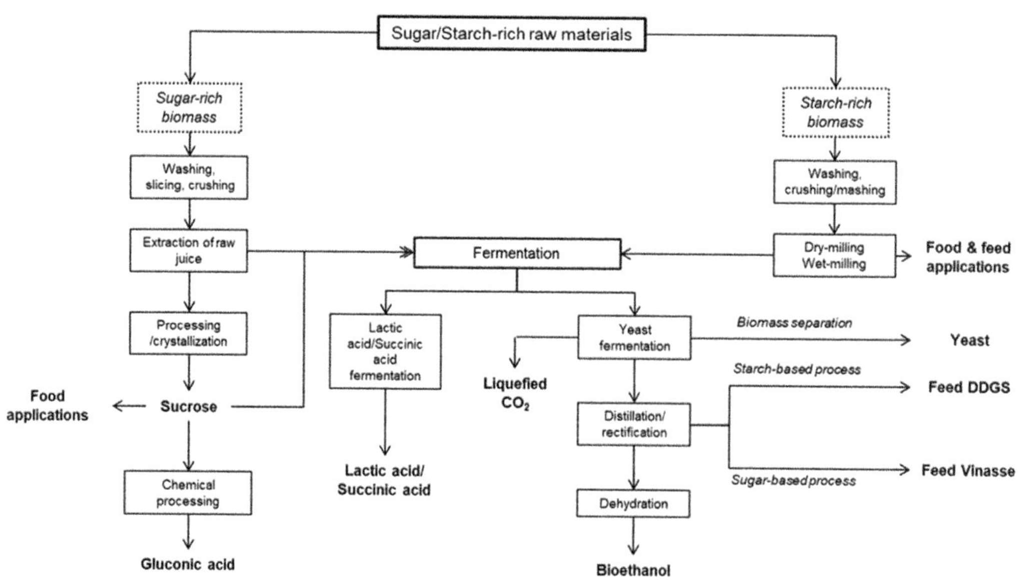

Figure 22.5
Sugar and starch-based biorefineries. *Adapted from BMELV, BMBF, BMU, BMWI. Biorefineries roadmap; 2012.*

Table 22.4: Raw materials for bioethanol production, the biomass (BM) yield, bioethanol yield as well as the required biomass for the production of 1 L of fuel.

Raw materials	Biomass yield (in wet basis) (t/ha)	Bioethanol yield (L/t BM)	Required BM per liter of fuel (kg/L)
Grain maize	9.9	400	2.5
Wheat	7.7	380	2.6
Rye	5.4	420	2.4
Sugar beets	70.0	110	9.1
Sugar cane	73.0	88	11.4

Taken by Bioenergy in Germany, Facts and Figures 2017. http://www.fnr.de/fileadmin/allgemein/pdf/broschueren/broschuere_basisdaten_bioenergie_2017_engl_web.pdf.

biomass, in comparison to sugar beets and sugar cane. Besides its use as biofuel, ethanol is also a precursor for the production of olefin and ethylene via dehydration [58].

Besides bioethanol, other biobased products from sugar/starch biorefineries include succinic and lactic acids. Succinic acid is an important platform chemical presenting the fastest growing market, since it can be used in food industry, cosmetics, pharmaceuticals and for polymer production like polybutylene succinate (PBS) and polyester polyols among others [64,65]. Lactic acid is another platform chemical with a growing market, mainly induced by its established use in food industry, chemicals and pharmaceuticals and also for its utilization in polylactic acid (PLA) production. PLA is a biodegradable polymer used in packaging, textiles, in automobile industry and due to its biocompatibility also in biomedical applications [65]. A list of the operating biorefineries using sugar and starch crops is presented in Table 22.5. A multi-purpose fermentation plant with an annual capacity of 1 kt, operating by the US company Myriant together with ThyssenKrupp Industrial Solutions AG is located in Leuna, and converts glucose and/or sucrose (using sugar cane, sugar beet, corn, cassava, cellulosic materials) to succinic or lactic acid [58]. Depending on the feedstock, pretreatment might be required prior to fermentation. For example, when sorghum grains are employed, hydrolysis is carried out by mixing the milled grains with hot water and sulfuric acid. Then enzymes are added in order to produce a syrup rich in sugars. An industrial-scale pilot-plant producing PLA was launched in 2011 by the company Uhde Inventa-Fischer (part of ThyssenKrupp) in Guben. The company has also developed a cost-efficient downstream process in order to obtain high purity lactic acid. The process produces ammonium sulfate, which is used as fertilizer [66].

Lactic acid can also be applied as the chemical industry as a raw material for the production of lactate ester, propylene glycol, 2,3-pentanedione, propanoic acid, acrylic acid, acetaldehyde, and di-lactide [67]. Biobased lactic acid is therefore an important bulk chemical and has a growing market not least due to the increased awareness of sustainable production.

Table 22.5: Industrial facilities based on sugar/starch biorefinery located in Germany.

Company	Capacity	Raw material	Product	References
Myriant, ThyssenKrupp	1000 t/y	Sugar cane, sugar beet, corn, cassava, cellulosic materials (sucrose/glucose)	Lactic acid, succinic acid, ammonium sulfate	[65,66]
Uhde Inventa-Fischer (ThyssenKrupp)	100,000 t/y	Sugar cane, sugar beet, corn, cassava, cellulosic materials (sucrose/glucose)	Lactic acid, PLA, ammonium sulfate	[66]
ADM Hamburg AG	-	Corn	Bioethanol, animal feed	[68]
CropEnergies (Südzucker)	400,000 m^3/y	Sugar syrups (sugar beet), grains	Bioethanol, animal feed (DDGS), liquefied CO_2	[63]
Barby (Cargill)	50,000 m^3/y	Wheat	Bioethanol	[69]
Nordzucker	130,000 m^3/y	Sugar beets (raw juice, thick juice, molasses)	Bioethanol, animal feed vinasse	[70]
KWST	80,000 m^3/y	Grain, sugar beets, sugar cane	Bioethanol	[71]
VERBIO ethanol Zörbig	60,000 t/y 240 GWatt-h/y	Rye, triticale, wheat	Bioethanol, biomethane, organic fertilizers	[72]
VERBIO ethanol Schwedt	170,000 t/y 360 GWatt-h/y	Rye	Bioethanol, biomethane, organic fertilizers	[73]

ADM Hamburg AG produces ethanol from corn, and animal feed as a secondary product of the fermentation process [68]. Cargill's Barby starch factory in Saxony-Anhalt operates since 2016 and produces high-grade alcohol via fermentation from wheat, which comes mainly from the region of Magdeburg [69]. One of the largest bioethanol plants in Europe is operating in Zeitz since 2005. The company CropEnergies Bioethanol GmbH (parent organization is Südzucker) produces approximately 400,000 m^3 bioethanol annually by processing up to 750,000 t of sugar syrups and grains from more than 1,000,000 t of sugar beets. The plant also produces animal feed from DDGS (Distillers' Dried Grains with Solubles) under the brand name ProtiGrain and liquefied dioxide, which is captured and purified from the fermentation process. Since 2010, the plant has been certified as fully sustainable [63]. More specifically, the main sources of starch utilized by the company are potatoes, wheat and maize and besides bioethanol, other biobased products are also developed directed to pharmaceuticals, cosmetics or construction chemicals industries among others.

The bioethanol plant in Klein Wanzleben/Sachsen-Anhalt of the company Nordzucker, started in 2007, and its annual capacity is 130,000 m^3. Bioethanol is produced via fermentation of the raw juice, the thick juice and the molasses derived from sugar beets. The nonsugar components like pectins and cellulose are processed for the production of protein-rich animal feed vinasse [70]. The company KWST produces highly purified neutral ethanol from grain, sugar beets and sugar cane, with a capacity of 80,000 m^3. The company disposes five continuously operating distillation plants and it is located in Hannover [71].

The plant VERBIO Ethanol Zörbig GmbH & Co. KG was the first bioethanol plant in Germany for fuel use. Operating since 2004 in Zörbig (Saxony-Anhalt) and fermenting annually more than 270,000 t of rye, triticale and wheat to bioethanol. In 2010, the company developed a process enabling the exploitation of the entire crop in a closed loop, combining the production of bioethanol with biomethane. The by-products of the process are used as organic fertilizers. In total, every year the plant has a capacity of 60,000 t bioethanol and 240 GW-hours biomethane [72]. The company launched another bioethanol plant in 2005 in Schwedt (Brandenburg) which processes mainly rye. As for the facilities in Zörbig, the company developed a biorefinery in 2010, by adding biomethane production to the process. The annual capacity of the plant involves 170,000 t of bioethanol, 360 GW-hours of bioethane and the by-products are sold as organic fertilizers [73]. Germany's market of biomethane is shown in Fig. 22.6.

22.2.4 Green biomass-based

Green biorefineries are based on the utilization of green crops i.e., grasses, legumes (e.g., alfalfa and clover) and catch crops or the green part of crops (e.g., beet and carrot leaves),

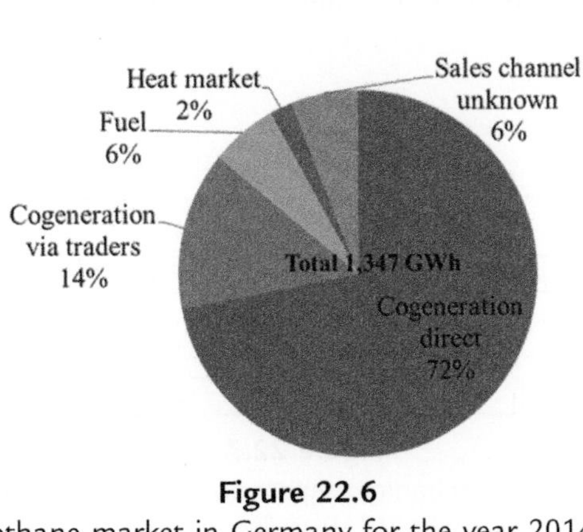

Figure 22.6
Biomethane market in Germany for the year 2014 [74].

which can be fresh or ensiled [75–77]. As all plants are green, the difference from other feedstock biomasses in the green biomass-based biorefineries is defined as the use of the aerial parts of plants that usually are in the growth phase performing photosynthesis or are ensilaged and the biomass is not dried prior to use. Grasses, clover and alfalfa are perennials and can be harvested several times during the growth season. Mechanical fractionation of the wet fresh or ensiled green biomass into a liquid (green juice) and a solid fraction (press-cake) is essential in the green biorefinery (Fig. 22.7). These two fractions can be used for a broad range of products including feed, food, chemicals, materials and biofuels (Figs. 22.8 and 22.10). There is an increased interest in green biorefineries due to the potential for protein recovery (reviewed in Ref. [78]), which fits well with the current focus in EU on local protein production.

Green biorefineries have been studied for numerous years but despite the experiences and know-how on many of the various chemical and biotechnological conversion routes, there are only very few commercial production units in Europe. In Germany, there are two identified companies which have a very different focus as Biowert produces insulation material, fiber-plastic granulates, biogas and fertilizer while Biofabrik mainly produces amino acids [79]. These production units produce only a small portion on what could be applied in a full green biorefinery, and in principle are complementary, as Biowert primarily utilize the press-cake and Biofabrik the juice fraction. Integration of several processes e.g., focusing on producing high-value compounds such as pharmaceuticals, pigments, proteins and amino acids for human consumption may enable profitable production [76]. The vision for utilizing the green biomasses for production of food grade protein and amino acid for human consumption is appealing but needs to consider food safety aspects in the production and the EU-regulations on novel food.

22.2.4.1 Substrate availability

There is a range of different crops that can be utilized as feedstocks, including forage grasses and perennial legumes (e.g., clover and alfalfa), including grasses on marginal

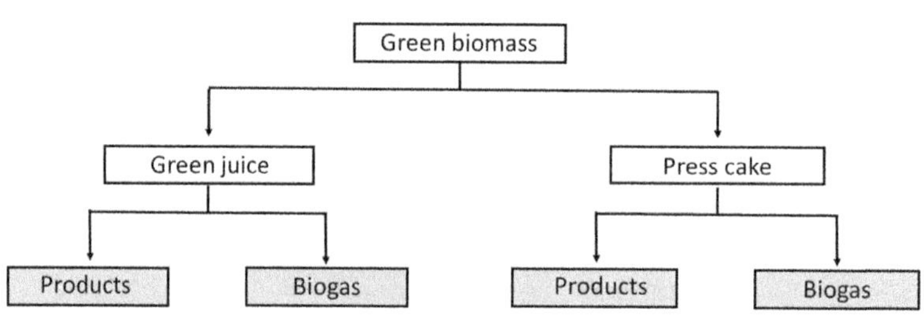

Figure 22.7
Simplified schematic overview of mechanical separation of green biomass *Modified from BMELV, BMBF, BMU, BMWI. Biorefineries roadmap; 2012.*

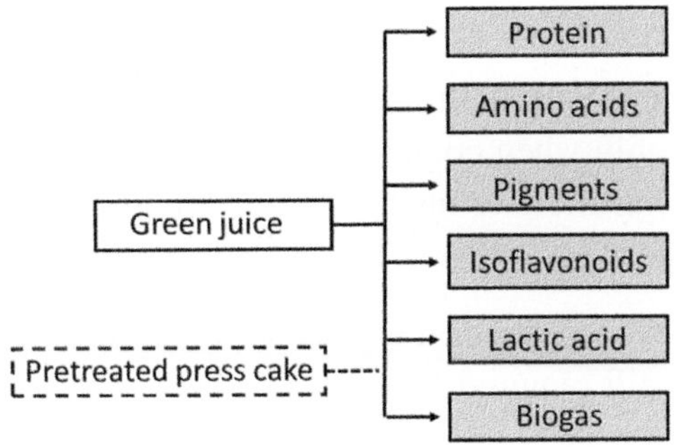

Figure 22.8
Simplified schematic overview of potential products from green juice. Silage juice can mainly generate amino acids, lactic acid and biogas. Sugars from pretreated and hydrolyzed press-cake can be added to juice for continuous fermentation purposes.

land. The green crops especially perennials can be regarded as more sustainable compared with many other crops due to less demand for pesticides, and to their ability to reduce nitrate leaching. In a study comparing different agricultural cropping systems with perennial grasses, they outperformed the other systems by doubling biomass N and reducing nitrate leaching by 70%–80% [80]. Due to their nitrogen fixing abilities, legumes such as clover and alfalfa have the advantage of reducing the need for nitrogen fertilizer. The crops are valuable in crop rotations as they can deliver nitrogen to the proceeding crop which especially in organic farming is important. Traditionally, the use of the crops is limited to cattle feed eventually in the form of silage or feed pellets produced in drying plants. In some of the European countries such as Ireland, Belgium and the Netherlands, most of the pasture area is used for animal production, whereas in other countries including Germany only approx. half of the area is used for grazing [81]. As also catch crops, beet and carrot leaves can be utilized, the substrate availability is already high and could be extended. In future, if the processes for protein recovery combined with integrated biorefining processes become economic feasible, other feed crops (e.g., cereals) can be substituted with grasses, including clover and alfalfa.

One of the main challenges in green biorefineries is the utilization of freshly harvested crops that are processed quickly after harvest, unless they are ensiled prior to use. The silage process is a well-known preservation method but the ensiled biomass has different applications than the fractions from freshly harvested biomass and is not usable for some of the applications [81]. The freshly harvested biomass needs urgent treatment in order to avoid contamination with unwanted microbes that may result in uncontrolled fermentation processes and low juice quality [82]. As fresh feedstock is seasonal and only available during a part of the year, a way for storage the whole plant or the fractionated press-cake and juice for later use is by using lactic acid bacteria for preservation [83,84].

Briefly, the main drawbacks of green biorefineries are the following [77]:

- Applicable only in regions with high grassland
- Economically feasible only when combined with a biogas plant
- The obtained products have low or insufficient quality to meet high-value applications
- Seasonality of the biomass

More studies on the technical aspects of the green biorefineries, in respect to both cost and environmental impacts are still required [77].

A biorefinery with a system capacity of 18.000 t dry matter per year of green biomass and/or silage would need approx. 2300 ha of grassland for raw material supply [2].

22.2.4.2 Processes and products

The biorefining of green crops starts with a mechanical separation into green juice and press-cake. The fractions have several different possible applications as outlined in Figs. 22.8 and 22.9.

The green juice contains proteins, free amino acids, fibers, sugars, vitamins and minerals and can be used for production of lactic acid by microbial fermentation [83,85]. Efficient lactic acid production requires a fermentation media containing nitrogen in the form of amino acids or peptides, simple sugars and vitamins for growth due to the nutrient requirements of lactic acid bacteria. For developing cost-effective large-scale production, the green juice is a very valuable low-cost nutrient medium compared with commercially available expensive protein extracts such as yeast extract and peptone [82,86]. Due to the high nutrient content, the green juice has also proven to be applicable for microbial fermentation to produce lysine as a feed additive [82].

A significant portion of the proteins from leaves of freshly harvested green crops is soluble and will be present in the green juice. The protein content varies among different plant species, and e.g., alfalfa and clover can have more than 20% protein content in their dry matter. Proteins from green juice can be recovered by heat-coagulation, alkali- or acid-precipitation followed by harvesting of the precipitated proteins by sedimentation, filtration or centrifugation [78]. Protein extraction can also be performed using lactic acid fermentation of the green juice resulting in lowering the pH due to the conversion of carbohydrates into lactic acid [87,88]. The use of bacterial fermentation is an interesting and more sustainable alternative to heat-treatment or the addition of chemicals for protein recovery. Proteins derived from green biomass represent an attractive solution to the increasing demand for protein-rich animal feed while decreasing the dependency on soybean imports from China or South America. Many plant resources cultivated in Europe including legumes such as fava beans and peas do not have a balanced amino acid profile compared to soybeans. In organic farming, it is not allowed to add synthetic amino acids and it was found that the protein products recovered from different green crops presented balanced amino acid composition compared to soybeans [88]. Especially there is a high content of methionine, which is the limiting amino acid in poultry production and therefore the protein product is very promising as a feed ingredient for poultry.

Biofabrik Green Biorefinery was founded in 2012 and focuses on creating a business based on the Austrian experiences from Green Biorefinery Upper Austria with extraction of amino acids from the juice of grass silage [89]. Biofabrik's concept consists of a decentralized plant (primary process unit), which produces concentrates and then a central plant that refines the product to amino acids and a fertilizer product. Biofabrik currently has two production facilities: a decentralized plant producing the intermediate product located in the Czech Republic and the central plant that refines the intermediate into amino acids, located in Germany (https://biofabrik.com; [79]). However, the company awaits approval for the use of the product as substance in dietary supplements for human consumption. The production capacity of the decentralized plant is 3000 t dry matter per year.

The press-cake will have an increased dry matter from approx. 15%–20% in the fresh crop to 30%–40%, which is the normal dry matter for silage processes. It consists mainly of plant cell wall material with fibers as the main components which consist of cellulose, hemicellulose, pectin and lignin. In addition, the press-cake contain fiber-bound proteins and may still contain up to 70% of the total protein content of the plants depending on the efficiency of the mechanically separation [88]. Due to the protein and carbohydrate content, the press-cake can be used as feed for ruminants, either ensiled or as feed pellets. Ensiled press-cake or the production of feed pellets from the press-cake have good nutritional value for ruminants. It was recently found that a higher milk production was obtained in the cows fed with ensiled press-cake than control cows fed with traditional silage from the same field [90]. This result was surprising as proteins were removed from the screw-pressed material.

The press-cake can also serve as a lignocellulosic feedstock for production of chemicals, fibrous composite materials (e.g., for insulation and building materials) or for biogas production [75,76]. Compared with many other lignocellulosic plant materials such as wheat straw, the press-cake is less recalcitrant due to the mechanical fractionation. Silage of the press-cake will also be advantageous prior to pretreatment for generation of a sugar platform for further conversion into biofuels or biochemicals. Extraction and enzymatic hydrolysis are commonly employed for the saccharification of the press-cake or low-moisture anhydrous ammonia pretreatment [77].

Biowert's current production focus is primarily fiber-plastic granules for the plastic industry, especially in injection molding. Fiber granules are developed in close cooperation with the buyers, which are typically plastic manufacturer's injection molding technology.

The fiber-plastic granules contain between 25% and 75% grass fibers as well as they also contain polypropylene, polyethylene and PLA, typically in the form of residues from plastic manufacturers. Biowert also produces insulation material and as their production is coupled with biogas, the digestates are sold as a fertilizer (https://biowert.com/company; [79]). The raw material is ensiled grass, and the capacity of Biowert is 7000 t dry matter per year.

As described, there are several fundamentally different green biorefinery concepts, as also evident from the two German commercial biorefining units. In general, green biorefinery concepts are typically coupled with biogas production as also seen in Fig. 22.9. The remaining material after separation of the various different products can be utilized as substrates for AD. Santamaría-Fernandéz et al. analyzed the different residue fractions; press-cake and residue juice after protein recovery from the green juice [91]. During the AD process, organic matter is converted into energy-rich biogas and plant nutrient-rich digestate [92]. The generated biogas can add value as heat generated during the biogas process can be used in the biorefining processes. Furthermore, the plant nutrient-rich digestate contains mineralized organic matter (NH_4^+-N, P, and K) and can serve as a fertilizer with high value for the green crops.

Green biorefineries has a wide range of possible products with high-priced sales opportunities. Providing the quality of protein products from green crops are comparable with soybean for animal feed in terms of digestibility and no major issues with antinutritional factors, there are huge market opportunities especially within the organic sector. There are still major technical and economic uncertainties in establishing these biorefineries in terms of attainable revenues. Techno-economic analysis showed that maximizing product yield of protein and cascade utilization of the different platform

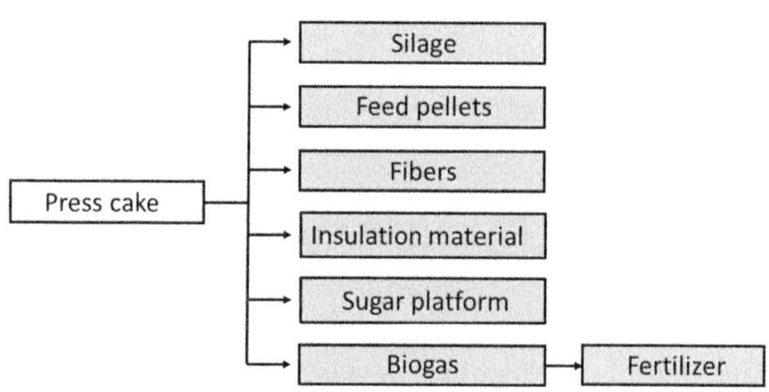

Figure 22.9
Simplified schematic overview of potential products from press-cake. The sugar platform can generated by pretreatment and enzyme hydrolysis, and can be utilized in fermentation for production of various products such as organic acids (biochemical platform molecules), ethanol or other biofuels.

products are the most important environmental optimization parameters for the green biorefinery, even more important than reducing the energy consumption of the biorefinery [93].

22.3 Conclusions and future perspectives

Germany has an established bioeconomy, but biorefinery processes are currently limited to only one or two products: Material and/or energy. It is assumed that future biorefineries in Germany will focus on the simultaneous production of food and feed, materials and energy in accordance to a cascade use of biogenic feedstocks as recommended by the German Bioeconomy Society. This is expected to improve the economic feasibility as different value-added products can be formed. The focus on the production of more than one or two products is also expected to increase flexibility to changing market prices. The production of one or more platform chemicals is beneficial since both the upstream and downstream processes can be similar, providing the desired flexibility to the biorefinery. Succinic acid and lactic acid are two good examples, already produced in Germany. The market price of biobased succinic acid is approximately 2.5 $/kg, while for lactic acid ranges between 1.30 and 2.30 $/kg, depending on the application. In addition to the production of various value added compounds the recovery of particularly fertilizers from waste streams should be envisaged in order to recycle at least parts of the initial applied resources to biomass production. Furthermore, it may release pressure from the competition of the three sectors: Food and feed, materials and energy, for feedstocks in biorefineries. The transition to multi-products biorefineries from second generation biomass still requires extensive research. Its heterogeneity and recalcitrant structure make an adaption of process steps necessary.

References

[1] Bio-based Industries Consortium, NovaInstitute. Biorefineries in Europe 2017. 2018.
[2] BMELV, BMBF, BMU, BMWI. Biorefineries roadmap. 2012.
[3] Demichelis F, Fiore S, Pleissner D, Venus J. Technical and economic assessment of food waste valorization through a biorefinery chain. Renewable and Sustainable Energy Reviews 2018;94:38–48.
[4] FNR, BMELV. Anbau nachwachsender Rohstoffe in Deutschland 2015–17. 2018.
[5] Efken J, Dirksmeyer W, Kreins P, Knecht M. Measuring the importance of the bioeconomy in Germany: concept and illustration. NJAS – Wageningen Journal of Life Sciences 2016;77:9–17.
[6] BMBF, BMEL. Bioeconomy in Germany. 2015.
[7] Nelles M, Grünes J, Morscheck G. Waste management in Germany – development to a sustainable circular economy? Procedia Environmental Sciences 2016;35:6–14.
[8] http://ec.europa.eu/eurostat/web/products-datasets/-/cei_pc031.
[9] https://www.bmu.de/fileadmin/Daten_BMU/Pools/Broschueren/abfallwirtschaft_2018_en_bf.pdf.
[10] Schüch A, Morscheck G, Lemke A, Nelles M. Bio-waste recycling in Germany – further challenges. Procedia Environmental Sciences 2016;35:308–18.

[11] Cristóbal J, Caldeira C, Corrado S, Sala S. Techno-economic and profitability analysis of food waste biorefineries at European level. Bioresource Technology 2018;259:244–52.
[12] Carmona-Cabello M, Garcia IL, Leiva-Candia D, Dorado MP. Valorization of food waste based on its composition through the concept of biorefinery. Current Opinion in Green and Sustainable Chemistry 2018;14:67–79.
[13] Appels L, Baeyens J, Degrève J, Dewil R. Principles and potential of the anaerobic digestion of waste-activated sludge. Progress in Energy and Combustion Science 2008;34(6):755–81.
[14] Pleissner D, Demichelis F, Mariano S, Fiore S, Navarro Gutiérrez IM, Schneider R, et al. Direct production of lactic acid based on simultaneous saccharification and fermentation of mixed restaurant food waste. Journal of Cleaner Production 2017;143:615–23.
[15] Dahiya S, Kumar AN, Shanthi Sravan J, Chatterjee S, Sarkar O, Mohan SV. Food waste biorefinery: sustainable strategy for circular bioeconomy. Bioresource Technology 2018;248:2–12.
[16] Barik S, Paul KK. Potential reuse of kitchen food waste. Journal of Environmental Chemical Engineering 2017;5(1):196–204.
[17] http://appsso.eurostat.ec.europa.eu/nui/show.do?dataset=env_ww_spd&lang=en.
[18] Watson J, Zhang Y, Si B, Chen W-T, de Souza R. Gasification of biowaste: a critical review and outlooks. Renewable and Sustainable Energy Reviews 2018;83:1–17.
[19] Vakalis S, Sotiropoulos A, Moustakas K, Malamis D, Vekkos K, Baratieri M. Thermochemical valorization and characterization of household biowaste. Journal of Environmental Management 2017;203:648–54.
[20] Xiong S, Zhuo J, Zhang B, Yao Q. Effect of moisture content on the characterization of products from the pyrolysis of sewage sludge. Journal of Analytical and Applied Pyrolysis 2013;104:632–9.
[21] Zhu J, Wan L, Chen H, Zhu Y, Yang L. Pilot test of co-pyrolysis characteristics of wet sewage sludge and sawdust in an external heating moving bed. Energy Procedia 2017;105:570–5.
[22] Hu M, Gao L, Chen Z, Ma C, Zhou Y, Chen J, et al. Syngas production by catalytic in-situ steam co-gasification of wet sewage sludge and pine sawdust. Energy Conversion and Management 2016;111:409–16.
[23] Singh R, Krishna BB, Mishra G, Kumar J, Bhaskar T. Strategies for selection of thermo-chemical processes for the valorisation of biomass. Renewable Energy 2016;98:226–37.
[24] Farooq MZ, Zeeshan M, Iqbal S, Ahmed N, Shah SAY. Influence of waste tire addition on wheat straw pyrolysis yield and oil quality. Energy 2018;144:200–6.
[25] Maity SK. Opportunities, recent trends and challenges of integrated biorefinery: part II. Renewable and Sustainable Energy Reviews 2015;43:1446–66.
[26] Nguyen QA, Cho E, Trinh LTP, Jeong J-S, Bae H-J. Development of an integrated process to produce d-mannose and bioethanol from coffee residue waste. Bioresource Technology 2017;244:1039–48.
[27] Koutinas AA, Vlysidis A, Pleissner D, Kopsahelis N, Lopez Garcia I, Kookos IK, et al. Valorization of industrial waste and by-product streams via fermentation for the production of chemicals and biopolymers. Chemical Society Reviews 2014;43(8):2587–627.
[28] Alves de Oliveira R, Komesu A, Vaz Rossell CE, Maciel Filho R. Challenges and opportunities in lactic acid bioprocess design—from economic to production aspects. Biochemical Engineering Journal 2018;133:219–39.
[29] Hagman L, Blumenthal A, Eklund M, Svensson N. The role of biogas solutions in sustainable biorefineries. Journal of Cleaner Production 2018;172:3982–9.
[30] Kim M-S, Na J-G, Lee M-K, Ryu H, Chang Y-K, Triolo JM, et al. More value from food waste: lactic acid and biogas recovery. Water Research 2016;96:208–16.
[31] Narra M, Macwan K, Vyas B, Harijan MR, Shah D, Balasubramanian V, et al. A bio-refinery concept for production of bio-methane and bio-ethanol from nitric acid pre-treated corncob and recovery of a high value fuel from a waste stream. Renewable Energy 2018;127:1–10.
[32] Jong E, Higson A, Walsh P, Wellisch M. Biobased chemicals - value added products from biorefineries. IEA Bioenergy-Task 2011;36.

[33] Wang Z, Lin M, Wang L, Ammar EM, Yang S-T. Metabolic engineering of *Propionibacterium freudenreichii* subsp. shermanii for enhanced propionic acid fermentation: effects of overexpressing three biotin-dependent carboxylases. Process Biochemistry 2015;50(2):194–204.

[34] Nghiem N, Kleff S, Schwegmann S. Succinic acid: technology development and commercialization. Fermentation 2017;3(2):26.

[35] Arnò P, Fiore S, Verda V. Assessment of anaerobic co-digestion in areas with heterogeneous waste production densities. Energy 2017;122:221–36.

[36] Ardolino F, Parrillo F, Arena U. Biowaste-to-biomethane or biowaste-to-energy? An LCA study on anaerobic digestion of organic waste. Journal of Cleaner Production 2018;174:462–76.

[37] Cerda A, Mejias L, Rodríguez P, Rodríguez A, Artola A, Font X, et al. Valorisation of digestate from biowaste through solid-state fermentation to obtain value added bioproducts: a first approach. Bioresource Technology 2019;271:409–16.

[38] UFOP. Supply report 2016/2017. 2017.

[39] Martínez G, Sánchez N, Encinar JM, González JF. Fuel properties of biodiesel from vegetable oils and oil mixtures. Influence of methyl esters distribution. Biomass and Bioenergy 2014;63:22–32.

[40] UFOP. Absatzentwicklung biodiesel in Deutschland; 2018. Available from: https://www.ufop.de/presse/aktuelle-pressemitteilungen/erstmals-mehr-abfalloel-als-rapsoel-zur-herstellung-von-biodiesel-verwendet/.

[41] UFOP. Biodiesel 2016/2017; 2017.

[42] Breuer G, Evers WAC, de Vree JH, Kleinegris DMM, Martens DE, Wijffels RH, et al. Analysis of fatty acid content and composition in microalgae. Journal of Visualized Experiments: JoVE 2013;(80):50628.

[43] Giakoumis EG. Analysis of 22 vegetable oils' physico-chemical properties and fatty acid composition on a statistical basis, and correlation with the degree of unsaturation. Renewable Energy 2018;126:403–19.

[44] Pleissner D, Lau KY, Zhang C, Lin CSK. Plasticizer and surfactant formation from food-waste- and algal biomass-derived lipids. ChemSusChem 2015;8(10):1686–91.

[45] Zhu LP, Wang JP, Ding XM, Bai SP, Zeng QF, Su ZW, et al. Effects of dietary rapeseed meal on laying performance, egg quality, apparent metabolic energy, and nutrient digestibility in laying hens. Livestock Science 2018;214:265–71.

[46] Ivanova P, Chalova V, Uzunova G, Koleva L, Manolov I. Biochemical characterization of industrially produced rapeseed meal as a protein source in food industry. Agriculture and Agricultural Science Procedia 2016;10:55–62.

[47] Byrne FP, Jin S, Paggiola G, Petchey THM, Clark JH, Farmer TJ, et al. Tools and techniques for solvent selection: green solvent selection guides. Sustainable Chemical Processes 2016;4(1):7.

[48] Li Y, Fine F, Fabiano-Tixier A-S, Abert-Vian M, Carre P, Pages X, et al. Evaluation of alternative solvents for improvement of oil extraction from rapeseeds. Comptes Rendus Chimie 2014;17(3):242–51.

[49] Perrier A, Delsart C, Boussetta N, Grimi N, Citeau M, Vorobiev E. Effect of ultrasound and green solvents addition on the oil extraction efficiency from rapeseed flakes. Ultrasonics Sonochemistry 2017;39:58–65.

[50] Salakkam A, Webb C. Production of poly(3-hydroxybutyrate) from a complete feedstock derived from biodiesel by-products (crude glycerol and rapeseed meal). Biochemical Engineering Journal 2018;137:358–64.

[51] Laguna O, Barakat A, Alhamada H, Durand E, Baréa B, Fine F, et al. Production of proteins and phenolic compounds enriched fractions from rapeseed and sunflower meals by dry fractionation processes. Industrial Crops and Products 2018;118:160–72.

[52] Chen K, Zhang H, Miao Y, Wei P, Chen J. Simultaneous saccharification and fermentation of acid-pretreated rapeseed meal for succinic acid production using *Actinobacillus succinogenes*. Enzyme and Microbial Technology 2011;48(4):339–44.

[53] Wang R, Shaarani SM, Godoy LC, Melikoglu M, Vergara CS, Koutinas A, et al. Bioconversion of rapeseed meal for the production of a generic microbial feedstock. Enzyme and Microbial Technology 2010;47(3):77–83.

[54] Sanguanchaipaiwong V, Leksawasdi N. Using glycerol as a sole carbon source for *Clostridium beijerinckii* fermentation. Energy Procedia 2017;138:1105–9.
[55] Mohandas SP, Balan L, Jayanath G, Anoop BS, Philip R, Cubelio SS, et al. Biosynthesis and characterization of polyhydroxyalkanoate from marine *Bacillus cereus* MCCB 281 utilizing glycerol as carbon source. International Journal of Biological Macromolecules 2018;119:380–92.
[56] Yamamura R, Shimomura Y. Industrial high-performance liquid chromatography purification of docosahexaenoic acid ethyl ester and docosapentaenoic acid ethyl ester from single-cell oil. Journal of the American Oil Chemists' Society 1997;74(11):1435–40.
[57] Cermak S, Evangelista R, Kenar J. Distillation of natural fatty acids and their chemical derivatives. 2012.
[58] Höfer R. In: Pandey A, Höfer R, Taherzadeh M, Madhavan Nampoothiri K, Larroche C, editors. Sugar- and starch-based biorefineries. Amsterdam, Oxford, Waltham: Elsevier; 2015. p. 157–235.
[59] OECD-FAO. Agricultural outlook 2018–27.
[60] http://www.fao.org/faostat/en/#data/QC.
[61] http://faostat.fao.org/site/339/default.aspx.
[62] https://www.tridge.com/intelligences/sugarcane/DE/import.
[63] http://www.cropenergies.com/en/Unternehmen/Standorte/Zeitz/.
[64] Pateraki C, Patsalou M, Vlysidis A, Kopsahelis N, Webb C, Koutinas AA, et al. *Actinobacillus succinogenes*: advances on succinic acid production and prospects for development of integrated biorefineries. Biochemical Engineering Journal 2016;112:285–303.
[65] From the sugar platform to biofuels and biochemicals. 2015.. http://www.e4tech.com/project/from-the-sugar-platform-to-biofuels-and-biochemical/.
[66] https://www.thyssenkrupp-industrial-solutions.com/media/products_services/chemical_plants_processes/polymer_plants/brochure_pla.pdf.
[67] Wee Y-J, Kim J-N, Ryu H-W. Biotechnological production of lactic acid. Food Technology and Biotechnology 2006;44:163–72.
[68] https://www.adm.com/adm-worldwide/europe/germany.
[69] https://www.cargill.de/en/barby-location.
[70] https://www.nordzucker.com/en/company/about-nordzucker/locations/klein-wanzleben-bioethanol.html.
[71] http://www.kwst.com/en/ethanol.html.
[72] https://www.verbio.de/en/group/corporate-structure/verbio-ethanol-zoerbig/.
[73] https://www.verbio.de/en/group/corporate-structure/verbio-ethanol-schwedt/.
[74] http://www.fnr.de/fileadmin/allgemein/pdf/broschueren/broschuere_basisdaten_bioenergie_2017_engl_web.pdf.
[75] Kamm B, Kamm M. Principles of biorefineries. Applied Microbiology and Biotechnology 2004;64(2):137–45.
[76] Kamm B, Schönicke P, Hille C. Green biorefinery – industrial implementation. Food Chemistry 2016;197:1341–5.
[77] Mandl MG. Status of green biorefining in Europe. Biofuels, Bioproducts and Biorefining 2010;4(3):268–74.
[78] Santamaría-Fernández M. A novel green biorefinery concept: protein refining by lactic acid fermentation and biogas production from green biomass. Aalborg University; 2018.
[79] DEA. Report on the business potential for biorefining based on organic farming. Danish Environmental Agency; 2018.
[80] Manevski K, Lærke P, Olesen J, Jørgensen U. Nitrogen balances of innovative cropping systems for feedstock production to future biorefineries. The Science of the Total Environment 2018;633:372–90.
[81] Sharma S, Mandle M. Green biorefinery. CRC Press, Taylor-Francis; 2014. p. 535–64.
[82] Thomsen MH, Alassali A, Cybulska I, Yousef AF, Brown JJ, Andersen M, et al. Microorganisms for biorefining of green biomass. In: Kamm B, editor. Microorganisms in biorefineries. Berlin, Heidelberg: Springer Berlin Heidelberg; 2015. p. 157–81.

[83] Papendiek F, Venus J. Cultivation and fractionation of leguminose biomass for lactic acid production. Chemical and Biochemical Engineering Quarterly 2014;28:375–82.
[84] Weinberg ZG, Szakacs G. Extraction of bioactive plant materials by simultaneous enzyme treatment and lactic acid fermentation – a mini-review. Israel Journal of Plant Sciences 2018;65(01–02):31–7.
[85] Dietz D, Schneider R, Papendiek F, Venus J. Leguminose green juice as an efficient nutrient for l(+)-lactic acid production. Journal of Biotechnology 2016;236:26–34.
[86] Pleissner D, Venus J. Utilization of protein-rich residues in biotechnological processes. Applied Microbiology and Biotechnology 2016;100(5):2133–40.
[87] Ajibola OO. Physical, chemical and microbiological changes during green plant juice fermentation. Food Chemistry 1984;13(3):181–92.
[88] Santamaría-Fernández M, Molinuevo-Salces B, Kiel P, Steenfeldt S, Uellendahl H, Lübeck M. Lactic acid fermentation for refining proteins from green crops and obtaining a high quality feed product for monogastric animals. Journal of Cleaner Production 2017;162:875–81.
[89] Schaffenberger M, Ecker J, Koschuh W, Essl R, Mandl M, Boechzelt H, et al. Green biorefinery – production of amino acids from grass silage juice using an ion exchanger device at pilot scale. Chemical Engineering Transactions 2012;29:505–10.
[90] Damborg V, Stødkilde L, Jensen S, Weisbjerg M. Protein value and degradation characteristics of pulp fibre fractions from screw pressed grass, clover, and Lucerne. 2018.
[91] Santamaría-Fernández M, Molinuevo-Salces B, Lübeck M, Uellendahl H. Biogas potential of green biomass after protein extraction in an organic biorefinery concept for feed, fuel and fertilizer production. Renewable Energy 2018;129:769–75.
[92] Risberg K, Cederlund H, Pell M, Arthurson V, Schnürer A. Comparative characterization of digestate versus pig slurry and cow manure – chemical composition and effects on soil microbial activity. Waste Management 2017;61:529–38.
[93] Corona A, Ambye-Jensen M, Vega GC, Hauschild MZ, Birkved M. Techno-environmental assessment of the green biorefinery concept: combining process simulation and life cycle assessment at an early design stage. The Science of the Total Environment 2018;635:100–11.

CHAPTER 23

Integrated biorefinery concept for Indian paper and pulp industry

Megha Sailwal[1,2], Ayan Banerjee[1,2], Thallada Bhaskar[1,2], Debashish Ghosh[1,2]
[1]Material Resource Efficiency Division, CSIR-Indian Institute of Petroleum, Dehradun, Uttarakhand, India; [2]Academy of Scientific and Innovative Research (AcSIR), New Delhi, India

23.1 Introduction

Sustainability is not a tailor-made term for the entire world; it's a global need for the betterment of the future generations. The industrial development and population explosion has put a high demand on the natural resources even exceeding the limit beyond their rate of replenishment. The development is dimensionally technological, and industries are the basis for this advancement. The paper and pulp industry is a significant consumer of forest resource [1]. Increasing demand for paper has led to clearing of hectares of green coverage reducing the existing natural carbon sinks [2]. This consumption of resources not only lead to resource depletion but it generates a large proportion of waste, which are entitled to be carried to the existing natural systems already functioning at limited capacity [3].

Industrial waste exists as one of the fatiguing tasks to handle. The industrial waste is consistently designated as toxic and hazardous waste which makes its treatment the necessity before disposal. The presence of metal chippings (arsenic, lead, cadmium, mercury, and nickel), wastewater, refinery sludge, fly ash, solvents in the industrial waste leads to toxication of the environment where it is released [4]. The untreated discharge from industries not only contributes to pollution but has an acute negative impact on the whole environmental health. Industrial waste such as oil industry waste, pulp and paper industry waste, mining industry waste, textile industry waste, and municipal solid waste are considered as some of the primary collaborators of the environmental pollution load [5]. Mining industry wastes includes acid generating tailings from processing of sulfide ore, dangerous substances from physical and chemical processing of metalliferous and nonmetalliferous minerals, as well as drilling muds and other drilling wastes containing oil [6]. Textile industry waste comprises metals, acids, alkalis, hydrogen peroxide, starch, surfactants dispersing agents or chlorides, dyes, and soaps of metals [7].

Waste Biorefinery. https://doi.org/10.1016/B978-0-12-818228-4.00023-X

The wastes from the paper and pulp industry is a growing concern as being present amid the natural environment it directly releases the waste to self-sustaining natural ecosystems. This industry is also having some of the high energy demanding processes which brings in the concern for energy crisis [8]. The waste generated in the paper and pulp industries when treated for material and energy recovery will aid in achieving waste minimization, resource efficiency and improved environmental health.

23.1.1 Wastes from the paper and pulp industry: current status

The paper and pulp industry affects the environment in a two-fold way. The paper and pulp processing waste is in massive quantity and contains toxic components such as adsorbable organically bound halides (AOX), mercury, sulfides, and chlorides. Secondly, the processes of this industry depend on wood as a primary source of raw material. This dependence creates vulnerability to the existing forest reserves. The global production of paper is expected to reach 490 million tons by 2020 [9].

The paper and pulp sector is one of the major industries of the world. The United States, China, and Japan share the largest portion of the total paper produced worldwide. They jointly cover over nearly half of the global production of paper [10,11]. The output of the pulp and paper industry is rapidly growing; at the same time, the amount of wastes produced is also becoming a difficult task to handle. This requires a new approach to balance and manage both of the major issue with this industry.

23.1.2 Biorefinery: an approach toward circular economy

Biorefineries are the linking hub for the sustainable production of renewable fuel, energy and value-added chemicals. Biorefineries primarily are of various types that include agricultural biorefinery, cereal biorefinery, oilseed biorefinery, green biorefinery, lingo-cellulosic (biomass) biorefinery, forestry biorefinery, and industrial waste biorefinery [12,13]. The industrial waste biorefinery is based on the conversion processes to produce particular fuels and chemicals from a particular waste stream [14]. Considering that all the waste has a specific property of having an integral net positive energy, this energy can be regained and reused by the biochemical and thermochemical reactions in a closed loop. It permits the industry to shift from a high carbon utilizing economy toward a circular and low-carbon bioeconomy [15]. Integrated waste biorefinery is the modified and efficient form of biorefineries. It utilizes waste from multiple sources as raw material for the production different fuels, chemical commodities and energy [16]. This opens an option of reducing the cost for industries in waste management while producing added products in the existing industrial premises.

23.1.3 The necessity of paper and pulp waste biorefinery

The waste generated in the paper and pulp industry undergoes treatment before discharge. The waste treatment procedures reduce the pollution load of the industrial effluents while increasing the capital cost of the whole industry without providing any other useful material. Because the pulp and paper industry is a capital-intensive industry, material recovery from the waste stream is a viable solution to this problem [17]. The paper and pulp industry waste is the preeminent source for various chemical and biological commodities. Considering the presence of high lignin and hemicellulose content in the black liquor generated from paper mills, integration of biorefinery for utilization of this lignocellulosic waste is a promising concept [18,19]. Designing the integrated biorefinery depends on the characteristics and the amount of waste generated by the paper and pulp industry that vary between geographical regions. An interdisciplinary knowledge of the pulp and paper industry processes and the characteristics of its waste streams are essential before designing an integrated biorefinery. Therefore, this chapter first provides an outlook of the functioning of the pulp and paper industry of two contrasting economies, i.e., the Indian and the European economy before going to the concept of integrated paper and pulp waste biorefinery. The understanding of the dynamics and resourcefulness of the pulp and paper industry makes the modeling of biorefinery easier.

23.2 Indian paper and pulp industry

The pulp and paper industry in India is considered one of the earliest systemized industry compared to other industries in the country [20]. This industry grew from 17 mills in 1951 to 850 mills in 2019 despite having a low R&D input in this industry [21,22]. The Indian paper industry rapidly grew after 1914 when the use of bamboo for making pulp emerged as an option against hardwood. After 1970, Eucalyptus wood and agricultural residues originate as another major option for raw material in the Indian paper and pulp production [23].

In 2018, India reported exiguous portion around 4% out of the global paper production [20]. According to Indian Paper Manufacture Association (IPMA) in 2018, the Indian paper and pulp industry contributed to human resource development by employing 5 lakh people and derivatively benefiting over 15 lakh peoples. In 2017, annual turnover of this industry was about INR 50,000 crores [24]. The Indian paper and pulp industry arose as an industry with the world's highest growth rate. As proclaimed, in 2018, the Indian paper industry is presently growing at an annual rate of 6%–7% (CAGR) [25]. According to Associated Chambers Of Commerce And Industry (ASSOCHAM), the Indian paper and pulp industry is prepared to expand and ready to hit the target of 25 million tons (2019–20) from 20.3 million tons (2017–18) at a rate of 10% per annum soon [26].

23.2.1 Structure of the Indian paper industry

The Indian paper industries are structurally distributed as depicted in Fig. 23.1. The paper industry is systematically distributed based on its production capacity as large, medium and small scale [27]. Large scale industry (also known as a large integrated mill) uses bamboo, hardwood, and recycled paper. The production capacity of large mills is more than 33,000 tons per annum. Medium-sized industries rely on the agricultural wastes and the recovered paper while the small-scale industries depend on recycled fibers only as the raw material [28]. Medium-sized mills have production capacity in between 10,000–33,000 tons per annum. Small-sized mills production capacity is less than 10,000 TPA [29].

As India is an agrobased country, a vast portion of the paper industries in India use agrowaste as the raw material. Other than agrowaste, the paper and pulp industry also relies on wastepaper and wood biomass as the raw material for pulp formation. Concerning the deforestation issues, the Indian government has strict legislation about the use of forest land, due to which the industry confronted a scarcity of wood as raw material in India for the paper production. This restriction makes the industry to rely majorly on the agricultural and waste residues. That's why only 30%–35% of pulp production is based on wood in India. Waste paper accounts for 40%–45%, and agrowaste (contains bassage, wheat straw, rice husk, rice straw, etc.) shares 20%–22% in pulp production [30].

Based on the finished product, this industry is again divided into four sectors. Industrial paper (packaging and paper board) sector cover nearly 47% of the domestic industry. The industrial paper sector is the highest growing sector as well as the largest sector of the paper and pulp industry. This sector is further divided into two parts: tertiary packaging

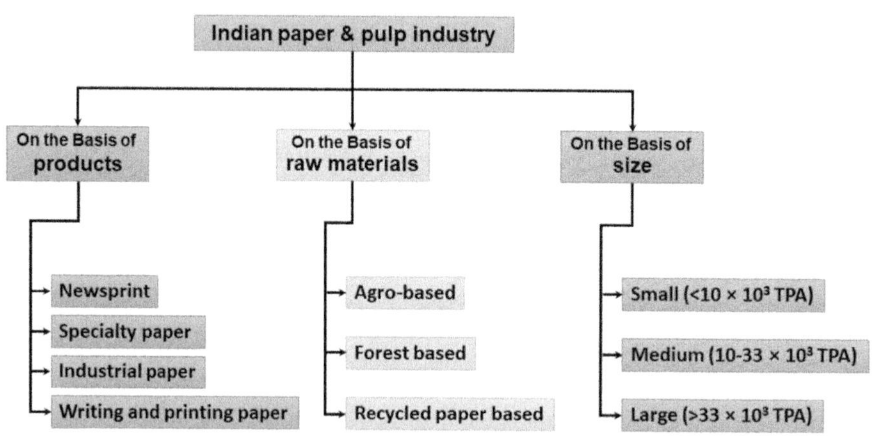

Figure 23.1
Indian pulp and paper industry: classification.

(kraft) and consumer packaging (greyback/whiteback and folding/solid boxboard/others). This sector produces packaging material for poster, pharmaceuticals, kraft paper, food and beverage, fast-moving consumer goods (FMCG), etc. [31,32].

Printing and writing sector is the second largest sector. It comprises nearly 31% of the domestic paper industry. This sector is further divided into four sections viz. Coated Wood-Free (CWF), Uncoated Wood-Free (UWF), Coated Mechanical (CM) and Uncoated Mechanical (UM). Newsprint segment nearly comprises 15% of the total industry in India. Specialty papers shares 4% of the sector and majorly deals with the gift wrapping and tissue papers [33].

23.2.2 Processes in Indian paper industry

The size and raw material used govern the process and the functioning of the Indian pulp and paper industry. However, the basic pathway followed for the production of paper in the Indian pulp and paper industry consists of five subprocesses, these includes preparation of raw material, pulping, washing and bleaching, recovery of chemicals, and papermaking [34]. These processes are explained in this chapter to provide insight into the paper and pulp industry. From the preparation of raw material to papermaking, the process of the paper industry is vital to understand the functioning of the industry.

23.2.2.1 Preparation of raw material

This is the first step toward paper production. Slightly different methods are implemented for raw material preparation based on the disparate raw material used in the pulping industry. In the case of the wood-based industry, procurement of wood plays an important role. After procuring, wood is harvested usually during the winter season and transported to the industries [35]. All of the Indian paper and pulp industry are situated in the vicinity of the source of water as well as wood (wood-based industries). Therefore, transportation seems to be less energy intensive. In some cases, transportation is long distances by road and railway.

Reception of wood at the target site follows quality assurance, documentation, and scaling of wood. Woodyards are used to store the wood to protect them from the climatic effects. Next step is processing of wood that includes debarking (manually or by machine) and chipping [36].

In the case of agrowaste, different kinds of procedures can be used depending on the type of waste such as wheat straw, rice straw, and bassage. Straws are opened by carrying off the wrapping wires and then the straws are cut down. Dust and loose fines are removed in the cyclone separator to enhance the quality of pulp. Leaves are removed because they have a high content of silica that is undesirable in the pulp. Silica makes the chemical recovery process difficult. Then the straws are charged in the hydra pulpers where the process of pulping begins [37]. Also recovered paper is treated before pulping. This consists of cutting down of the paper, its dust and ink removal processes. After the framing up, the raw material is on the brink of pulping.

23.2.2.2 Pulping

Pulp production is the process of separating the fibers from the protective lignin layer. This process dissociates the three main components, i.e., cellulose, lignin, and hemicelluloses, of wood and agricultural residues. Approximately, 26% of the total energy is used in pulping [38].

Pulping can be done in various ways (Fig. 23.2) depending on the pulp production process, raw material availability, and the type digester used (batch and continuous) [39].

The main pulping process used worldwide is the chemical pulping process. The chemical pulping process is carried out at a high temperature and pressure [40]. The chemical method can be used for both the wood and agrobased industries. The chemicals are used to dissolve the lignin content and free the cellulose and hemicelluloses part. Based on the chemicals used, the chemical pulping process is further split into three branches. The first is the soda pulping in the chemical pulping processes. In this, the cooking of wood chips in caustic soda solution is done.

Next in this order is the kraft pulping (sulfate pulping) [41]. The chemicals used in this process are sodium hydroxide and sodium sulfide (white liquor) [42]. This alkaline liquid causes the lignin to disintegrate. Cellulose becomes free from the lignin. As the lignin components are acidic, the pH of the solution decreases. This is the most common process

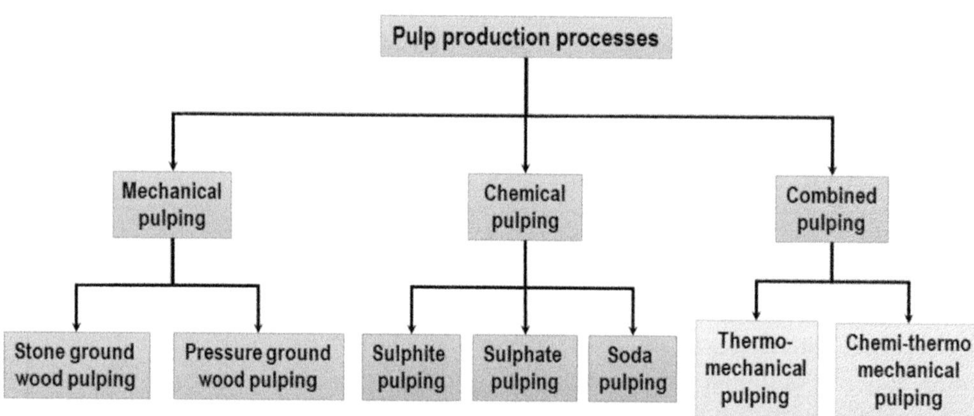

Figure 23.2
Different types of pulping processes.

used in Indian pulp industry to produce brown pulp. During this, lignin is degraded by α-aryl ether and β-aryl ether bonds cleavage of phenolic group of lignin [43].

$$\text{Lignin} + \text{Sodium hydroxide} \rightarrow \text{Sodium salt of lignin molecule} + \text{Alcohol}$$
$$\text{Lignin} + \text{Sodium sulfide} \rightarrow \text{Mercaptans}$$

One more method is also practiced, the sulfite pulping. In this, sulfuric acid is used to cook the wood chips. This process is very less used in Indian pulp and paper industries.

Despite the presence of chemical pulping, the three other processes, i.e., the mechanical pulping, the combined-pulping, and the hydropulping processes are furthermore used to a certain extent [44]. During the mechanical pulping, physical forces are adopted to break the wood into separate fibers by breaking the bond between them. Wood is pressed against the wet rotating grindstones (stone groundwood pulping and pressure groundwood pulping). This technique provides lower quality fibers due to the damage of fiber caused by mechanical grinding. Only 45% of the raw material is converted into the pulp by this process [45]. Even after this, the mechanical pulping process has few advantages over the other pulping procedures. This process gives good optical property and paper-surface properties to the pulp. Wood-based industry usually relies on this process for pulping [46].

In the combined-pulping process, the wood chips passes through the counterrevolving grooved metal disks or refiners with the increasing temperature (thermomechanical pulping or refiner mechanical pulping). The quality of the produced pulp depends on the temperature of the process. This treatment provides entire long fibers giving strength to the product made from this pulp. In chemo-thermo-mechanical pulping, lignin is tenderized by chemical action (sulfur-based chemicals, i.e., sodium sulfite) and further work is performed by mechanical action.

In the case of wastepaper, hydro-pulping is the only way for pulp production. In the dust-free recovered papers, water is added to make the slurry. This slurry is defibrinated that represents the hydro-pulp. This pulp can be deinked or can be left nondeinked based on the paper mill demand. Deinking of this slurry is done by froth flotation and wash deinking methods. Froth flotation process is more commonly used. In this process, at the bottom of the tank, air bubbles are formed which move to the top by carrying the ink with them to form the froth. This froth is then removed carefully and disposed of. The pulp produced at the end of the process is purified. Generated waste in recovered paper pulping technique is more compared to other raw materials (wood and agroresidues) [47,48].

A substantial amount of heat is produced during all the pulping processes. This heat is restored from the pulp slurry and used to preheat the wood chips before pulping. After the pulp production, the next footstep is the washing and bleaching. The pulp is washed due to the presence of unwanted residues in it.

23.2.2.3 Washing and bleaching

After chemical pulping, the pulp is washed three to four times to remove the chemicals and undigested blocks of wood chips (black liquor). Then the pulp is screened to remove the residual undigested particles. Consequent of the pulping processes, bleaching of the immature or brown pulp is done to produce the white pulp. The energy used in bleaching is nearly 7% of the total energy used in paper production.

Alkali solutions and oxidizing chemicals are used to bleach the chemical pulp [49]. Chlorine dioxide, chlorine, sodium hypochlorite, oxygen, ozone, and hydrogen peroxide are used for bleaching accordingly to the pulp-making process, cost and availability of chemicals and environmental protection laws. Chlorine produces dioxins after bleaching that is quite undesirable and dangerous for the environment.

Bleaching process releases toxic chemicals such AOX in the wastewater. These chemicals make their way into wastewater (contains high COD value). This wastewater is treated by various methods such as physicochemicals (electrocoagulation, flocculation, ozonation) and biological anaerobic and aerobic processes as well as combination of these processes [50]. Physicals methods can only separate the different harmful components while biological method has the potential to degrade these pollutants.

Caustic soda is also used during bleaching of pulp, is a chloralkali process product. This process leads to the release of mercury which is also a toxic element [51]. These show that bleaching and recovery of chemicals are the steps in paper production that affect the environment maximally.

23.2.2.4 Recovery of chemicals

Recovering chemicals maintain the process economy as well as protect the environment by not dumping the chemicals directly into nature. The chemical pulping process produces effluent characterized by dark brown color effluent known as black liquor [52]. The black liquor consists most of the chemicals used during the pulp production. The weak black liquor passes into the evaporators to produce the steam and heavy black liquor. This concentrated heavy black liquor is passed into the recovery boilers where inorganic solids are heated to get the energy.

The remaining chemicals (mainly sulfides, sulfates, sodium chloride, and potassium chloride) known as smelt in the recovery boilers are then mixed with water that forms the green liquor (Na_2CO_3). This green liquor is placed in causticizers where lime is added to form the white liquor containing NaOH [50].

The chemical reactions involved during smelting and causticizing are as follows.

$$2NaR\ (lignin) + O_2\ (air) \rightarrow Na_2CO_3 + CO_2\ (lignin)$$

$$\text{Sodium sulfate} + 2\text{C (Lignin)} \rightarrow \text{Sodium sulfide} + \text{Carbon dioxide}$$

$$\text{Green liquor (aq)} + \text{Calcium hydroxide (s)} \rightarrow \text{White liquor (aq)} + \text{Calcium carbonate (s)}$$

$$CaCO_3 \rightarrow CaO + CO_2$$

$$CaO + H_2O \rightarrow Ca(OH)_2$$

The waste lime sludge is burned in a lime kiln to get lime again. After all of the above steps, the mature pulp is produced that can be sold as the market pulp or can be utilized in the same industry to make the paper [53].

23.2.2.5 Papermaking

Papermaking consists of the preparation of pulp, pressing and drying. In the pulp preparation process, water and additives are added into the beater containing pulp. Beater separates the fibers to get uniformity in the pulp. Water is removed from the pulp and pulp is spread onto the screen to form the web of the fibers. The rollers on the mat (made of nylon and polyester) are used to compress the paper to remove the water content to about 50%. In the last step, fibers are dried thermally to have paper having 2%–6% water. Depending on the product formed, paper undergoes sizing and calendering procedures. Lastly, the paper is woven into rolls and ready for transport. During the whole process of paper production, a lot of waste is generated. This waste is treated to reduce its quantity and to remove the harmful contaminants from it (Fig. 23.3) [54].

23.2.3 Introduction of treatment processes

Several treatment methods have been used to reduce the solid and liquid fraction of the waste produced by the pulp and paper industry. Primary methods (physicochemical treatment) used are screening, settling/clarification, and flotation. The primary treatment removes a major portion of suspended solids from the wastewater [55]. Secondary treatment procedures (biological treatment procedures) include aerobic and anaerobic digestion. Aerobic and anaerobic methods lead to degraded pollutants, present in the sludge produced after primary treatment, by the action of microorganisms. Activated sludge produced after the biological treatment is then secondarily clarified [56].

The tertiary treatment procedure is used to remove nitrogen, phosphorus, additional suspended solids, refractory organics, or dissolved solids. The tertiary method includes filtration assisted crystallization technology, multifosoftening technology, electrodialysis technique among the others [57].

23.3 Paper industries of the west

The Confederation of European Paper Industries (CEPI) is already working for a low-carbon economy. In this reference, CEPI Forest Fiber Industry 2050 Roadmap

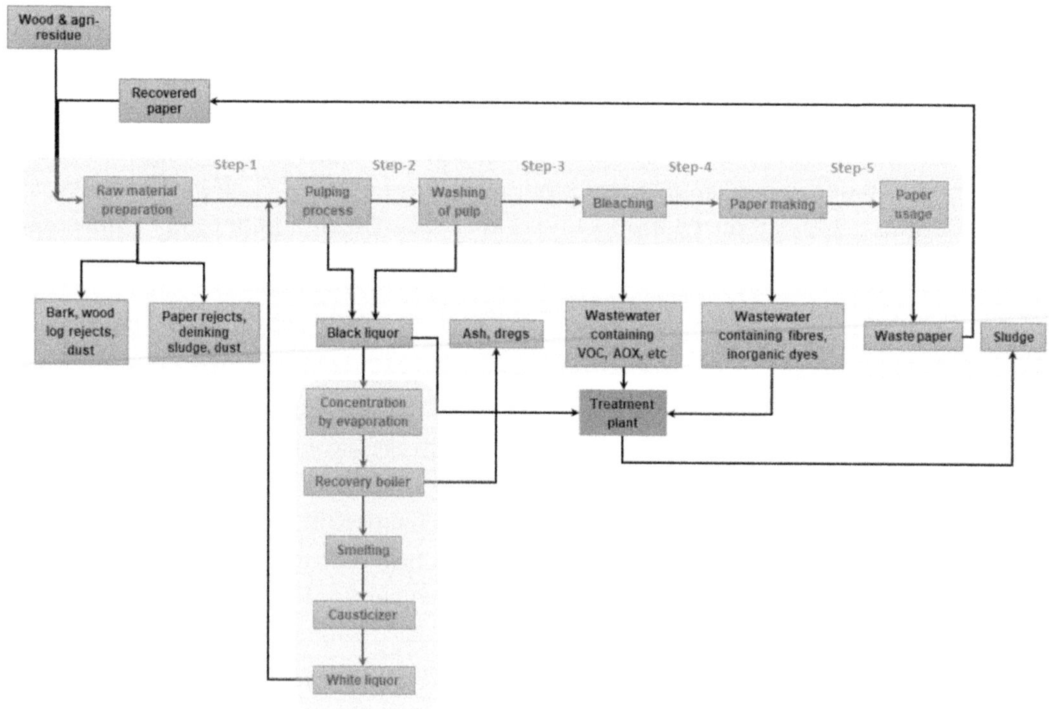

Figure 23.3
Paper-making processes and generated wastes and waste streams.

program was launched in 2011. Pulping techniques using Deep Eutectic Solvent (DES) are developed to produce pulp at ambient temperature and pressure. DES is the organics produced by the plants during its metabolism. DES could be a great option to recover cellulose from the waste [58]. Pulping through the use of DES produces pure form lignin. Similarly, best available techniques (BAT) are also enabled. These programs explore the opportunities to reduce 80% carbon dioxide emission and to produce 50% value-added products simultaneously. BAT also works to have new and more efficient machinery to reduce the pollution load. About 89% of the total production capacity inside the CEPI countries is certified or registered under the Environment Management Standards ISO 14001 and Eco-Management and Audit Scheme (EMAS) [59].

23.3.1 Structure of the Western paper industry

CEPI is the group of 18 nations of the western hemisphere that represents more than 900 paper and pulp mills. CEPI member countries in 2018 include Belgium, Finland, Austria, France, Norway, Czech Republic, France, Germany, Hungary, Italy, Norway, United

Kingdom, Poland, Spain, Portugal, Slovakia, Netherlands, and Slovenia. In 2005, the total production of paper in Europe was around 99.3 million tons, and in 2018, CEPI's total pulp and paper production decreased to 92.2 million tons. The production of paper is nearly constant in the CEPI countries since a few years because they are already established industries whereas pulp and paper industries of the developing nations shows faster growth rate [60].

23.3.2 Operation of the Western paper industry

The main operations are quite the similar as the Indian pulp and paper industry. CEPI claims to recycle 93% of the water in acceptable quality. Meanwhile, European mills are also working to revamp the paper and pulp industry waste to produce fuels and various chemicals used in the pharmaceuticals industry, in food products, and cosmetics [61]. Production of chemicals and fuels in a high quantity requires familiarity with the waste streams in the paper industry.

23.4 Wastes generated in paper and pulp industry

The paper and pulp industry is sixth globally in terms of polluting the environment [62]. This industry utilizes a very significant amount of water and chemicals during the production process that leads to generating a lot of wastewater and sludge. The waste produced by this industry is huge in amount and hazardous to nature. Because of this, paper and pulp industry confront various environment-related issues.

Central Pollution Control Board (CPCB) in India puts the paper and pulp industry in the group of 17 most polluting industries [63]. CBCB have notified the standard limits of waste discharge from the pulp and paper industry (Table 23.1). On adjacently, the European pulp and paper industry every year generates 11 million tons of wastes, reported by CEPI in 2005 [64]. The industry generates wastes in three forms: solid, liquid, and gas. Solid and liquid wastes are the predominant waste and are present in the slurry form that is quite difficult to manage. Solid waste is separated from the slurry by primary and secondary treatment methods [65].

23.4.1 Liquid waste

Liquid waste of the pulp and paper industry is reported as one of the biggest environment enemy [66]. Only one ton of paper produced can generate 60 m^3 of wastewater globally [67]. In India, the paper and pulp industry still generates 162–380 m^3 of wastewater per ton paper produced [68,69]. This is in huge amount as compared to the European paper industry. Currently, nearly 14 million tons of paper is produced in India that generates massive quantity of wastewater. In India, the pulp and paper industry wastewater has high

Table 23.1: Standard discharge limits of effluent of the paper and pulp industry.

Discharge limits	BIS standards	CPCB standards	
Parameter	**I.S. 3307**	**Small pulp and paper industry**	**Large pulp and paper industry**
Volume of wastewater (m^3 per ton)	Not specified	Agrobased: 200 Waste paper-based: 75	150–200
pH	5.5–9.0	5.5–9.0	7.0–8.5
COD (mg/L)	250	Not specified	250
Suspended solids (mg/L)	Not specified	100	50
BOD	500	30 (discharge into inland surface water) 100 (disposal on land)	30
Total dissolved solids (mg/L)	2100	Not specified	Not specified
Sodium adsorption ratio	Not specified	26	Not specified
Absorbable Organic Halogens (AOX) in effluent discharge (kg/ton of paper produced)	Not specified	2.00	1.00
Total residual chlorine (mg/L)	1.00	Not specified	Not specified
Emissions of particulate matter (mg/m^3)	Not specified	Not specified	250
Emissions of H_2S (mg/m^3)	Not specified	Not specified	10

COD (5.7 kg/m^3) and BOD value [70]. Treatment of the wastewater is essential if this high amount of COD and BOD reduction is planned before discharge. As India is already dealing with the water scarcity, wastewater treatment is a necessity for the paper as well as other industries.

Depending on the pulping process, bleaching process, the raw material used, reuse of water and chemicals recovered (from chemically recovery process), different amount and composition of effluent are produced in this industry. As the fresh water consumption is based on the raw material used, agrobased industry, wood-based and recycled paper-based industry consumes nearly 125–200, 125–225, and 75–100 m^3 water respectively per ton of the paper produced [71]. The wastewater contains both organic and inorganic substances. The composition of organic matter in the wastewater depends on the raw material used for the pulping process. Inorganic material comprises inorganic components of black liquor (such as salts of sodium, sulfates, and calcium), washing and bleaching wastes such as chlorinated compounds [72].

23.4.1.1 Organic fraction

Nearly 40% of the organics produced in the chemical pulping process are of low biodegradation capability [73]. Black liquor consists of a significant portion of the organic fraction. Some compounds such as lignin present in the black liquor if left untreated persists in the environment for several years due to its low biodegradability. Lignin and its derivatives cause the color of effluent dark brownish. Approximately 100 kg of colored compounds waste is generated per ton paper produced [74]. Degradation of lignin, during the paper production process, leads to the production of high, medium and low weight lignin compounds. These compounds can be chlorinated or nonchlorinated. High molecular weight lignin derivatives are recalcitrant to degradation and if they are directly released into the water bodies present nearby, they can severely pollute the environment [75].

Other than lignin, the effluent of the pulp industry consists of hazardous compounds such as AOX. AOX present in pulp and paper industry waste contain trichlorophenol, trichloroguicol, tetrachloroguicol, dichloro-phenol, dichoroguicol, and pentachlorophenol. These are produced by the reaction of lignin and its derivatives with the chlorinated compounds (added for bleaching) [76]. Bleaching process effluent consists of high BOD and COD (1000–7000 mg/L) values. Biodegradability ratio value is also very low for bleaching effluent (0.02–0.07) [77]. The chlorinated organic compounds released during bleaching are also recalcitrant and toxic to nature. Approximately 2–4 kg of chlorinated organic compounds is generated per ton of the paper [78]. Bleaching wastewater is treated by various methods such as electrochemical, electrochemical advanced oxidation, biological fungal treatment, chemical precipitation, ultrasonication, and electrooxidation treatment to remove toxic AOX and COD of it [79].

Nonchlorinated compounds of the organic fraction consist of resin acids, fatty acids, sterols, diterpene alcohols, and tannins. These compounds are also hard to biodegrade and persist in the environment for a long period.

23.4.1.2 Inorganic fraction

Inorganic fraction of the pulp and paper industry effluent contains salt cake, sodium hydroxide, sodium sulfides, bisulfites, sodium carbonate, calcium carbonates, sulfates, HCl, calcium oxide, chlorinated inorganic compounds. Chlorine forms high molecular weight and low molecular weight chlorinated compounds on reacting with the other constituents of the effluent. These include chlorinated cymene, chlorinated phenols, and many others. During the pulping and bleaching processes, polychlorinated dibenzo-p-dioxins and polychlorinated dibenzofurans are also formed which are recognized as the highly toxic chemical compounds [80]. Chelating agents such as ethylene-diamine-tetraacetic acid (EDTA) and diethylene-triamine-pentaacetic acid (DTPA) are used during

Table 23.2: Characteristics of wastewater generated during various stages of paper and pulp mill.

	Parameters					
	pH	TS	SS	BOD	COD	Color
Process	–	ppm	ppm	ppm	ppm	
Raw material processing	7	1,160	600	250	1275	Dark brown
Pulping	10	1,309	256	360	–	Dark brown
Bleaching	2.5	2,285	216	352	–	–
Paper-making	6.5	645	760	641	1116	Black
Large mill (India)	11	5,250	1233	983	2530	Black
Small mill (India)	12.3	15,120	4890	2628	6145	Dark brown

BOD, biochemical oxygen demand; *COD*, chemical oxygen demand; *SS*, suspended solid; *TS*, total solid.

the ozone and peroxide bleaching process are nonbiodegradable to a larger extent and are present as an inorganic fraction in the paper and pulp wastewater (Table 23.2) [81,82].

23.4.2 Solid waste

Solid waste generated by the paper and pulp industry is composed of the sludge removed by the primary and secondary treatment of the wastewater, the causticizing plant wastes (lime slaker grits, lime mud, etc.) and the boiler and furnace ash. Nearly 40–50 kg of dried sludge is produced per ton of paper production [83]. Solid waste is present in two forms namely suspended solids and dissolved solids.

23.4.2.1 Suspended solids

Suspended solids are defined as the solids that do not pass through a 0.45-micron filter [84]. Suspended solids of the paper and pulp industry are removed from the liquid waste by the physicochemical treatment methods.

Paper and pulp industry wastewater contains solid particles such as fine bark particles, pith (from bagasse pulping) and silt. These particles are burned in bark boiler for energy recovery in the form of steam. In the recovery boiler, ash is generated due to the burning of organic fraction of the wastewater. Suspended solid wastes are easily removed from the wastewater in this settling and filtration methods.

23.4.2.2 Dissolved solids

Dissolved solids are extracted from the sludge generated by the secondary and tertiary treatment plants. Scrubber sludge is also a part of the dissolved solids. Dissolved solids fraction consists of the remaining organic (wood fibers, scrubber sludge) and inorganic compounds (such as calcium carbonate) [85].

23.4.3 Gaseous waste

This waste stream is commonly neglected in the pulp and paper industry. It is the primary source of air pollution from the paper and pulp industries. Gas wastes exuded by the paper and pulp industry consist of reduced sulfur compounds (hydrogen sulfide, methyl mercaptan, dimethyl sulfide, particulate matter, SO_2, and NO_x) are either primary pollutants or are highly toxic as methyl mercaptan [86]. The chemicals such as H_2S, CH_3SH, $(CH_3)_2S$, and $(CH_3)_2S_2$ are released from the digestion, washing and evaporation plants in the sulfate mills while sulfur dioxide is released from the sulfite mills digestion and evaporation plants. Chlorine dioxide gas is released during the bleaching of pulp [87]. Occupational safety and health administration (OSHA) time-weighted average limit (usually based on 8 h workday) of dimethyl sulfide is 10 ppm, dimethyl disulfide is 0.5 ppm, hydrogen sulfide is 10 ppm and methyl mercaptan is 0.5 ppm. Above this, these gases can cause serious damage. Ceiling limit for hydrogen sulfide and methyl mercaptan are 20 and 10 ppm respectively, increasing this limit during any part of the work experience can lead to serious health damage [88]. Gaseous waste is not usually considered for recycling approach as the technology for trapping and chemical processing of the gases is costly and seldom installed in paper and pulp processing (Table 23.3) [89,90].

23.5 Integrated biorefinery concept

Wastes generated by the pulp and paper industry gives a direct negative impact on the environment with losing resources present in the waste. Current research is focusing on the integrated biorefinery approach to tackle these issues. So, an overall process can be developed that can dig out cost benefits from this industrial waste with simultaneously protecting the environment. The pulp and paper industry waste can be treated through the zero-waste biorefinery route to utilize every component of the waste.

However, implementing biorefineries can be done by two routes in any industry. The first route is the presence of integrated biorefinery at the same premises in the industry. While the second model of integrated biorefinery in the field of pulp and paper industry is its presence somewhere near the industry, through this, the waste of more than one pulp and paper industry can be together collected and used. This model will create a new industrial sector for the pulp and paper industry waste usage. Apart from this, it will also give a pathway for increasing the employment ratio in developing countries like India. The disadvantage in this route is only the transportation cost that will increase the capital expenditure. The integrated biorefinery approach should maintain the economic viability of the system and provide profit to the industry. Therefore, the optimization of the structure, area, raw material use, and products needed in the biorefinery would be an interesting

Table 23.3: Characteristic compounds of solid, liquid and gaseous effluent of the paper and pulp industry.

Types of wastes		
Waste phase	**Fraction/solids**	**Effluent**
Liquid	Inorganic	• Salt cake, sodium hydroxide, sodium sulfides, bisulfites • Sodium carbonate, calcium carbonates and sulfates • HCl • Calcium oxide • Inorganic chlorine compounds (TOXIC) such as chlorate, elemental chlorine, chlorine dioxide, phosphate, nitrate, silicates, calcium carbonate, kaolin clay, inorganic dyes
	Organic	• Black liquor consisting lignin, hemicellulose, cellulose, extractives, starch, cellulose fibers (suspended solids), resins, chlorinated resin acids, fatty acids, high BOD and COD, tannins and sulfur compounds • Chlorinated organic compounds or AOX: furans, dioxins, chlorophenols, guaiacols, catechols, veratroles, aromatic chloroethers, cymenenes, chlorinated hydrocarbons, polychlorodibenzofurans, alkyl polychlorobiphenyls, alkyl polychlorophenanthrenes, polychlorinated dibenzothiophenes • VOC (terpenes, diterpene alcohols, alcohols, phenol, methanol, acetone, chloroform, methyl chloride, carbon disulfide, chloromethane, trichloroethane)
Solid	Dissolved	• Sludge from the treatment plant, scrubber sludge, etc.
	Suspended	• Remaining bark particles, soil, dirt, fibers silt, clay, chalk, lime mud, lime slaker grits, green liquor dregs, boiler and furnace ash, scrubber sludges

Table 23.3: Characteristic compounds of solid, liquid and gaseous effluent of the paper and pulp industry.—cont'd

Types of wastes		
Waste phase	Fraction/solids	Effluent
Gaseous	–	• Poisonous gases such as total reduced sulfur gases (TRS) such as hydrogen sulfide, methyl mercaptan, dimethyl sulfide, sulfur oxides, steam, NO_x, particulates from recovery boiler, CO

topic that needs to be addressed. The conversion of kraft mill biorefineries into integrated biorefineries has been reported for the production of ethanol [91]. Other than ethanol, many valuable commodities can be obtained from the waste of pulp and paper industry. For this, the different processes are described below that can be used to maximize the use of the resources available in the wastes (Fig. 23.4).

For integrated biorefinery in the pulp and paper industry, two main processes are needed to be in operation. These are the biochemical and thermochemical processes [92]. As a high amount of organics are present in the pulp and paper industry waste, biochemical pathways give the first vision to achieve some beneficial product following this pathway. In biochemical route, biofuels and fine chemicals are produced by the processes such as fermentation and hydrolysis. Hemicelluloses in the organic fraction can be extracted by the process known as value before the pulping process [93]. Processes such as dilute acid pretreatment, alkaline treatment, ammonia fiber/freeze explosion, dilute acid steam explosion are some of the methods that can be used to extract hemicelluloses from the lignocellulosic biomass. The hemicelluloses then can be used for fermentation, hydrolysis, and saccharification to produce fine chemicals like ethanol, succinic acid, lactic acid, furfural, acetic acid, among the others [94].

The biochemical methods include three processes namely separate hydrolysis and fermentation, simultaneous saccharification and fermentation, and simultaneous saccharification and co-fermentation [95]. Hydrolysis and saccharification are carried out to free the C5 and C6 sugars from the lignocellulosic biomass [96]. Fermentation of these free sugars can be carried out by bacterial and yeast strains such as *Saccharomyces cerevisiae, Kluyveromyces lactis, Zymomonas mobilis, Candida guilliermondii, Actinobacillus succinogenes,* and *Clostridium beijerinckii* depending on the product

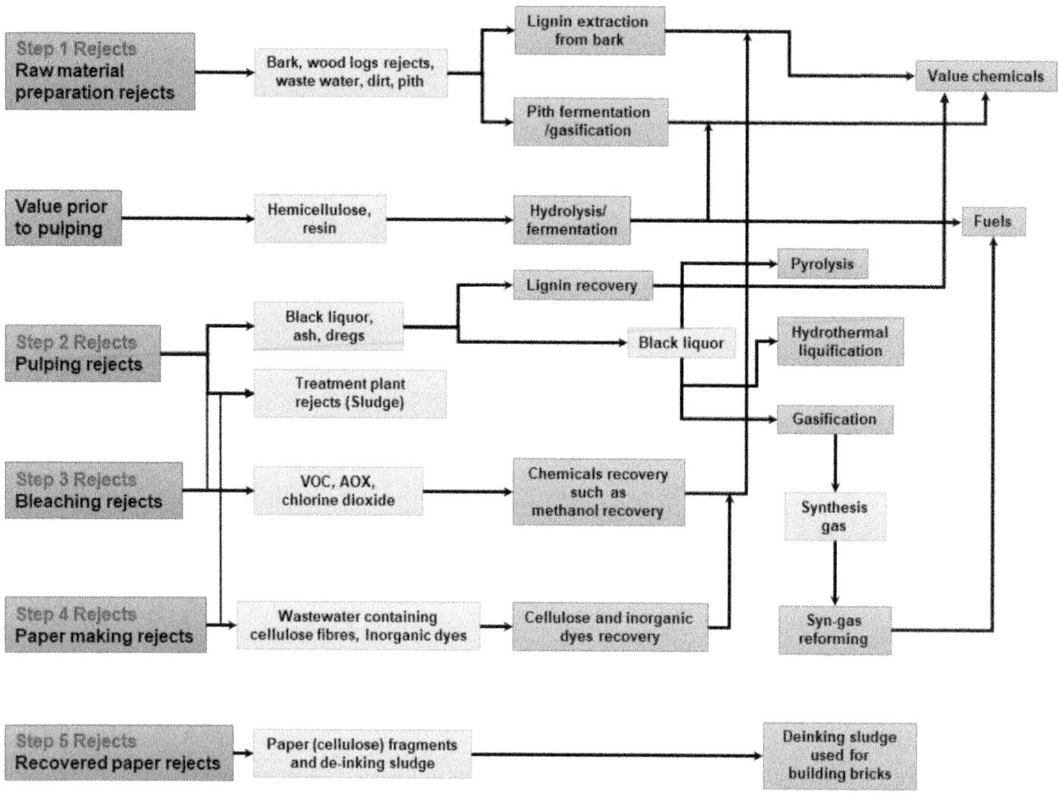

Figure 23.4
Biorefinery concept for pulp and paper industry.

require [97]. In simultaneous saccharification and fermentation, hydrolysis step is carried out first, then saccharification and fermentation are done together of the C5 sugars to generate biofuels [98]. Simultaneous saccharification and cofermentation process is similar to simultaneous saccharification and fermentation except for C5 and C6 sugars are fermented together in this process. This will also generate supplementary earnings in the industry. Hemicellulose extraction is commercially being utilized during dissolving pulp production [99]. The use of grits (unreacted slaker $CaCO_3$ and CaO) and dregs (smelt) that are present in the organic fraction is reported in cement clinker production [100]. The use of C6 and C5 sugar present in the prehydrolyzate has been reported for production of lactic acid by a chemical process known as Plaxica's Versala, the technique which utilizes prior to pulp process for the production of lactic acid along with the dissolving pulp production. Before pulping, the biomass is prehydrolyzed to extract the C5/C6 sugars. This C5/C6 stream contains the hemicelluloses that are further converted into lactic acid, an industrially important chemical. Lignin and acetic acid are isolated from this stream before lactic acid production [101].

The thermochemical processing other than biochemical processing, remains the preferred route for the production of the desired chemical, with the application of high temperature and pressure. Gasification, pyrolysis, and hydrothermal liquefaction are the major valorization methods used for conversion of black liquor into the valuable compounds [102].

The high amount of lignin fraction, a natural resource present in the black liquor can be derived to produce energy and several chemical commodities. However, the extraction of lignin from black liquor is a tiresome job. The structure of lignin is complicated; it is a cross-linked three-dimensional amorphous phenylpropanoid polymer [103]. The sulfur present in the black liquor reacts with the lignin and forms recalcitrant compounds known as lignosulfonates, and makes it further difficult to reach the lignin for further chemical reactions. To tackle this, many techniques are available to extract lignin in the purified form before feeding black liquor to the recovery boiler. Few of these techniques are the acid precipitation, ultrafiltration, and ion-exchange [104]. Acid precipitation includes the protonation of the hydrophilic ionized phenolic group of lignin. This leads to the reduction of electrostatic repulsive forces present between the lignin molecules. This makes the lignin molecules less hydrophilic leading to the precipitation of lignin [105]. This method is commonly used because of its simplicity as only a strong acid is needed to separate lignin from black liquor. But has a disadvantage of colloids formation that makes separation difficult. Second process, ultrafiltration of lignin, is the membrane-based (usually ceramic membrane of 5, 10, and 15 kDa) technique which requires the filtration of black liquor solution to extract lignin from it. This method is also a simple process that leads to obtain lignin fractions of defined molecular weight. The cons associated with this system are in-service life and fouling and cleaning cycle of the membrane [106]. Ion-exchange resins are used to remove dissolved organic matter from the kraft mill effluent. Cation-exchange membranes are used to precipitate the lignin by modulating the pH by electrochemical reactions. As solubility of lignin in black liquor is pH dependent, changing it can significantly lead to lignin extraction [107]. The only con attached with ion-exchange is the high cost of the system. LignoBoost, LignoForce System, and Sequential Liquid-Lignin Recovery and Purification (SLRP) named processes are specifically used to extract pure lignin from the black liquor are also available. In the LignoBoost process, lignin is precipitated at temperature 55–70°C by acidifying the black liquor using CO_2; the precipitate is dewatered and acidified by adding H_2SO_4 [108]. The extracted lignin is washed at the end to get the purified product. Domtar's Plymouth mill in North Carolina and Stora Enso's Sunila mill in Finland are the commercial plants are that already operating on LignoBoost technique for extraction of lignin fraction [109]. LignoForce System is also being operated commercially at West Fraser pulp mill in Hinton [110].

The purified lignin readily can be transformed into many derivates such as phenol, biooil, activated carbon, carbon fibers, BTX, and other aromatic hydrocarbons, vanillin, carbon fiber, and other lignin-based compounds [111]. About 36%–42% of theoretical yield BTX was reported to be produced from lignin [112]. Specific use of lignin is reported in the manufacturing of a novel N-doped fused carbon fibrous mat that is made of 9:1 blend of lignin-polyethylene oxide. It is used as high-performance anode material in lithium-ion batteries [113]. Lignosulfonates can also be utilized as concrete plasticizers [114].

The black liquor can be made to undergo gasification to produce energy and synthesis gas. Gasification of black liquor can be done at a lower temperature (<700°C) or a higher temperature (>700°C) under the reducing conditions [115]. This process produces synthesis gas (H_2 and CO). The synthesis gas formed can further be reacted to produce many other chemicals such as Fischer-Tropsch liquid transportation fuel, alcohol, dimethyl ether and can be used to produce heat by combustion [116]. Syn-gas functions also as an energy source for gas turbine and a steam turbine to produce electricity and steam. These fuels and chemicals can be the alternatives to the depleting nonrenewable fuels. The production of these fuels can provide significant revenues to the pulp and paper industry. Methanol production by the black liquor gasification has been reported by two processes, oxygen-blown pressurized thermal BLG and dry BLG with direct causticization [117].

In pyrolysis, the black liquor is heated in the absence of oxygen to produce biochar, volatile organics, and gaseous compounds at a lower temperature (400–650°C) than gasification [118]. In the pulp and paper industry, sawdust or wood residues are the major raw material for pyrolysis. The retention time for the reaction can be more or less that makes the pyrolysis process further divided into slow and fast pyrolysis. The substances present in the gaseous phase generated by the reaction produces pyrolysis oil (biooil). The pyrolysis oil can be used in place of lime kiln fuel partly [119]. It also has the potential to be used as road transportation fuels [120]. Other than gasification and pyrolysis, liquid phase thermal treatment (a method of liquification) is the process for obtaining high viscosity liquids (that are not soluble in water) mainly from lignin by reacting it under high temperature (300–350°C), high pressure (20 Mpa) and reducing environment (CO or H_2). Apart from lignin, extractives such as turpentine and tall oil are also present in the black liquor can be extracted to provide a way for revenue generation. Due to the density difference, these extractives are easily removed from the concentrated black liquor. Also, the deinking sludge can be separated. It is reported to be used in building bricks. Almost all of the waste produced in the paper and pulp industry is reported to produce energy and valuable chemicals [121].

23.6 Research needs and directions

The paper and pulp industry is an established industry. The records and future development in this industry is clear to the global community. The new element that must be included in the future development must consider the resource inventory and recovery option. The necessity of material flow analysis in the processing of pulp and paper is clear from the discussion that the majority of the organic parts of feedstock goes to the waste stream. Without the understanding of the existing material, environment impact and energy loss and its recovery after installing the biorefinery, an economic and viable solution for optimum recovery of material and production is difficult. Therefore, environmental impact assessment and life cycle assessment of pulp and paper industry are being analyzed to know the impact of this industry gives onto the environment throughout the life cycle of paper from production to reuse or disposal. These assessments can also help in providing new pathway to develop better waste utilization processes of this industry along with the minimal amount of load on the environment. Environmental impact assessment of soda-anthraquinone pulping process has been reported. This report is based on the nonwood fibers production i.e., pulp production from hemp and flax biomass. It shows that the environmental impact is majorly caused by the production of chemicals used in the paper mills, electricity production if purchased from the grit, production of fibers (cultivation of agricultural biomass) [122]. Another report shows the life cycle assessment of the integrated biorefinery producing dissolving pulp, ethanol and lignosulfonates. This report shows that the production of chemicals, wastewater treatment plant (sludge disposal) and cogeneration units (recovery boiler unit) have more negative environment effect than the cooking and bleaching (close-loop total chlorine free bleach) steps [123]. Both of these reports are from cradle-to-gate perspective. More information on cradle-to-grave or cradle-to-cradle perspective will help to find newer processes in integrated pulp and paper biorefinery that will not or less harm to the environment. Considering the previous reports, the design of paper and pulp biorefinery must be produced by an interdisciplinary approach with the consideration of principles of circular economy, the green chemistry and industrial ecology (Fig. 23.5).

23.7 Conclusions and perspectives

Paper is a necessary commodity that cannot be limited in supply when its demand grows consistently. The impact that this industry has put on the environment, though not severe but due to the excess quantity of waste, is alarming. The paper and pulp industry generating black liquor and discharge from the bleaching unit puts the local environment

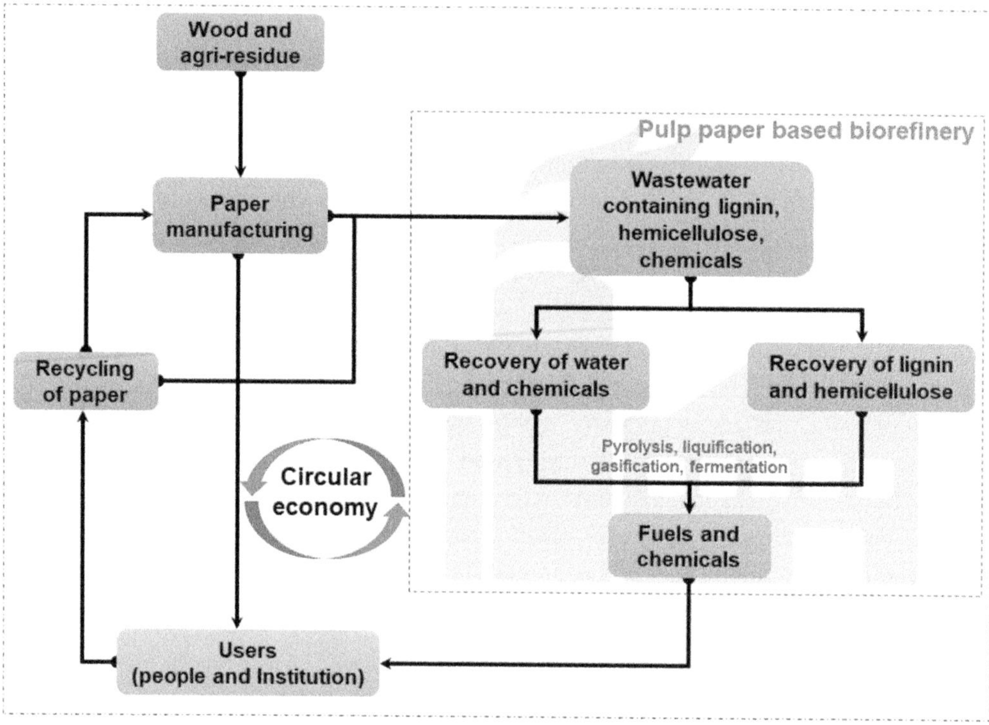

Figure 23.5
Circular economy concept for pulp and paper industry.

in a vulnerable position for contamination. The suspended organic load in the waste stream and AOX, if not removed may directly hinder the aquatic ecosystem at the area of waste discharge. In the preview of environmental protection, the regulation for discharge is documented to be limited but malfunctioning of a single pollution abatement unit leads to failure in pollution control as a whole. The biorefinery is preferable because the removal of the contaminants from the waste stream is not enough, their fate after treatment must be environmentally sound. In a biorefinery especially with the target of achieving a circular economy, the materials and energy recovery is optimized to form a sustainable system. The cost involved in the treatment and disposal of paper and pulp processing waste is a major concern for the industry as it incurs high expenditure. Processing the waste biochemically or thermochemically leads to recovery of material, production of valuable chemicals and fuel. This can compensate for the expenditures of the industry and reduce the pollution load over the local environment. The holistic approach to obtain the best from the waste needs proper identification of the problems existing within the industry. The possible options to select process in the biorefinery to obtain the products with a high market value are important target is generating economic returns.

References

[1] Hetemäki L, Hurmekoski E. Forest products markets under change: review and research implications. Current Forestry Reports 2016;2(3):177–88.

[2] Report on "key issues in the forest sector today: part II". http://www.fao.org/3/y0900e/y0900e05.html.

[3] Tietenberg T, Lewis L. Environmental & natural resource economics. 10th ed. New York, USA: Routledge; Taylor and Francis; 2016.

[4] Artiola JF. Chapter 21: industrial waste and municipal solid waste treatment and disposal. Remediation, restoration, treatment, and reuse. In: Environmental and pollution science. Academic Press; 2019. p. 377–91.

[5] Abdel-Shafy HI, Mansour MSM. Solid waste issue: sources, composition, disposal, recycling, and valorization. Egyptian Journal of Petroleum 2018;27(4):1275–90.

[6] Szczepanska J, Twardowska I. Mining waste (III.6). In: Solid waste: assessment, monitoring and remediation. Elsevier B.V.; 2004.

[7] Holkar CR, Jadhav AJ, Pinjari DV, Mahamuni NM, Pandit AB. A critical review on textile wastewater treatments: possible approaches. Journal of Environmental Management 2016;182:351–66.

[8] Shabbir I, Shabbir I, Mirzaeian M, Mooney J, Anvari F. Energy efficiency improvement potentials of pulp and paper sector through energy benchmarking and cogeneration. In: 29th international conference on efficiency, cost, optimization, simulation and environmental impact of energy systems, Portorož, Slovenia; 2016.

[9] Bajpai P. Introduction; chapter 1. In: Recycling and deinking of recovered paper. Elsevier Book Series; 2013. p. 1–19.

[10] Report on "the global paper market – current review". 2018. PG Paper, https://www.pgpaper.com/wp-content/uploads/2018/07/Final-The-Global-Paper-Industry-Today-2018.pdf.

[11] Mixer Direct Report Blog on "the top 10 paper producing countries in the world". 2017.

[12] Mongkhonsiri G, Gani R, Malakul P, Assabumrungrat S. Integration of the biorefinery concept for the development of sustainable processes for pulp and paper industry. Computers & Chemical Engineering 2018;119:70–84.

[13] Monte MC, Fuente E, Blanco A, Negro C. Waste management from pulp and paper production in the European Union. Waste Management 2009;29:293–308.

[14] Robus CLL, Gottumukkala LD, van Rensburg E, Gorgens JF. Feasible process development and techno-economic evaluation of paper sludge to bioethanol conversion: South African paper mills scenario. Renewable Energy 2016;92:333–45.

[15] Mohan SV, Butti SK, Amulya K, Dahiya S, Modestra JA. Waste biorefinery: a new paradigm for a sustainable bioelectro economy. Trends in Biotechnology 2016;34(11):1–3.

[16] Arevalo-Gallegos A, Ahmad Z, Asgher M, Parra-Saldivar R, Iqbal HM. Lignocellulose: a sustainable material to produce value-added products with a zero-waste approach – a review. International Journal of Biological Macromolecules 2017;99:308–18.

[17] Posch A, Brudermann T, Braschel N, Gabriel M. Strategic energy management in energy-intensive enterprises: a quantitative analysis of relevant factors in the Austrian paper and pulp industry. Journal of Cleaner Production 2015;90:291–9.

[18] Thakur VK, Thakur MK, Raghavan P, Kessler MR. Progress in green polymer composites from lignin for multifunctional applications: a review. ACS Sustainable Chemistry & Engineering 2014;2(5):1072–92.

[19] Holladay JE, White JF, Bozell JJ, Johnson D. Top value-added chemicals from biomass-volume II – results of screening for potential candidates from biorefinery lignin (No. PNNL-16983). Richland, WA, USA: Pacific Northwest National Lab (PNNL); 2007.

[20] Madan SM, Bhalerao S, Mishra M. Indian paper industry – moving out of the woods? Industry research, care ratings professional risk opinion. 2018.

[21] Thapliyal BP. Wastewater treatment technologies and practices in Indian pulp & paper industries. Asia Wastewater Management Seminar. Tokyo: United Nations Industrial Development Organization. Investment and Technology Promotion Office; 2017.
[22] Report on "national & international best practices manual pulp & paper industry vol. II". India: IPMA; Confederation of Indian Industry, Sohrabji Godrej Green Business Centre Hyderabad; 2009.
[23] Report on "ecological & environmental factors in paper industry vol. II". Development Council for Pulp Paper & Allied Industries. CESS. Ministry of Commerce and Industry, Govt. of India. http://www.dcpulppaper.org.
[24] Report on "Indian paper industry pre-budget memorandum 2017–18". New Delhi, India: IPMA.
[25] Tripathi JG. Study of India's paper industry – potential and growth in 21st century. Indian Journal of Applied Research 2014;5(4):112–5.
[26] Report on "Indian paper industry poised touch 25 million tonnes in 2019–20". http://timesofindia.indiatimes.com/articleshow/63096462.cms?utm_source=contentofinterest&utm_medium=text&utm_campaign=cppst.
[27] Report on "technology compendium on energy saving opportunities – pulp & paper". Hyderabad, India: Shakti Sustainable Energy Foundation (SSEF) & Confederation of Indian Industry (CII); 2013.
[28] Venkatesh R. Origin and growth of paper industries in Indian context; chapter II. In: A study of employee's perception on the human resource management practices in select paper industries in Tamil Nadu; 2012. http://shodhganga.inflibnet.ac.in/bitstream/10603/20387/5/08_chapter%202.pdf.
[29] Report on "technical EIA guidance manual for pulp and paper industries". IL&FS Ecosmart Limited, Hyderabad, Ministry of Environment & Forest, Govt. of India; 2010.
[30] Report on "global competitiveness of the Indian paper industry". Saharanpur, India: Jaakko Poyry Consulting. Final report draft for CPPRI; 2002. https://dipp.gov.in/sites/default/files/Final_9092002.pdf.
[31] Kathuria V. Draft 1: comments welcome industrial policy, structural change and technology gap: a study of Indian pulp and paper industry. 2014. p. 1–32.
[32] Kirwan MJ. Paper and paperboard packaging technology. Blackwell Publishing Ltd; 2011.
[33] Saxena N. Newsprint demand in India expected to reach 2.8-mn tonnes in FY20. 2018. https://www.exchange4media.com/media-print-news.
[34] Schumacher K. India's pulp and paper industry: productivity and energy efficiency. Berkeley, CA 94720, USA: Energy Analysis Program, Environmental Energy Technologies Division, Lawrence Berkeley National Laboratory; 1999.
[35] Onali H. The wood procurement in Indian pulp and paper industry and its challenges [Master's thesis], Forest Technology and Logistics. Department of Forest Sciences, Faculty of Agriculture and Forestry, University of Helsinki; 2017.
[36] Avşar E, Demirer GN. Cleaner production opportunity assessment study in SEKA Balikesir pulp and paper mill. Journal of Cleaner Production 2008;16(4):422–31.
[37] Fahmy Y, Fahmy TYA, Mobarak F, El-Sakhawy M, Fadl MH. Agricultural residues (wastes) for manufacture of paper, board, and miscellaneous products: background overview and future prospects. International Journal of ChemTech Research 2017;10:424–48.
[38] Ray BK, Reddy BS. Understanding industrial energy use: physical energy intensity changes in Indian manufacturing sector. Energy Policy 2011;39(11):7234–43.
[39] Dahlquist E. Process simulation for pulp and paper industries: current practice and future trend. Chemical Product and Process Modeling 2008;3(1):e1–18. Article 18.
[40] Wanrosli WD, Zainuddin Z, Law KN, Asro R. Pulp from oil palm fronds by chemical processes. Industrial Crops and Products 2007;25(1):89–94.
[41] Hu J, Zhang Q, Lee D-J. Kraft lignin biorefinery: a perspective. Bioresource Technology 2018;247:1181–3.
[42] Namazi AB, Allen DG, Jia CQ. Effect of chemical components in white liquor on carbon activation. Canadian Journal of Chemical Engineering 2015;93(10):1705–12.

[43] Report on "lecture 26: preparation of wood pulp by sulfate (kraft) process". NPTEL – Chemical – Chemical Technology II. Joint initiative of IITs and IISc – funded by MHRD.
[44] Stephenson RJ, Duff SJ. Coagulation and precipitation of a mechanical pulping effluent – I. Removal of carbon, colour and turbidity. Water Research 1996;30(4):781–92.
[45] Roth S, Zetterberg L, Acworth W, Kangas H, Neuhoff K, Zipperer V. The pulp and paper overview paper: sector analysis for the climate strategies project on inclusion of consumption in carbon pricing. Climate Strategies. 2016.
[46] Li B, Li H, Zha Q, Bandekar R, Alsaggaf A, Ni Y. Effects of wood quality and refining process on TMP pulp and paper quality. BioResources 2011;6(3):3569–84.
[47] Wahaab RA, Moawad AK, Taleb EA. Wastewater management and resource recovery in hard paper industry: a case study. World Applied Sciences Journal 2010;8(4):470–6.
[48] Tiku DK, Kumar A, Sawhney S, Singh VP, Kumar R. Effectiveness of treatment technologies for wastewater pollution generated by Indian pulp mills. Environmental Monitoring and Assessment 2007;132(1–3):453–66.
[49] Beyer M, Koch H, Fischer K. Role of hemicelluloses in the formation of chromophores during heat treatment of bleached chemical pulps. Macromolecular Symposia 2005;232(1):98–106. Weinheim: WILEY-VCH Verlag.
[50] Deshmukh NS, Lapsiya KL, Savant DV, Chiplonkar SA, Yeole TY, Dhakephalkar PK, Ranade DR. Upflow anaerobic filter for the degradation of adsorbable organic halides (AOX) from bleach composite wastewater of pulp and paper industry. Chemosphere 2009;75(9):1179–85.
[51] Jaishankar M, Tseten T, Anbalagan N, Mathew BB, Beeregowda KN. Toxicity, mechanism and health effects of some heavy metals. Interdisciplinary Toxicology 2014;7(2):60–72.
[52] Irfan M, Butt T, Imtiaz N, Abbas N, Khan RA, Shafique A. The removal of COD, TSS and colour of black liquor by coagulation–flocculation process at optimized pH, settling and dosing rate. Arabian Journal of Chemistry 2017;10:S2307–18.
[53] Armstrong AD, Bentley KM, Galeano SF, Olszewski RJ, Smith GA, Smith Jr JR. The pulp and paper industry. In: The ecology of industry. Washington, DC, USA: National Academy Press; 1998. p. 101–41.
[54] Pokhrel D, Viraraghavan T. Treatment of pulp and paper mill wastewater—a review. The Science of the Total Environment 2004;333(1–3):37–58.
[55] Dilek FB, Gökçay CF. Treatment of effluents from hemp-based pulp and paper industry. Waste characterization and physico-chemical treatability. Water Science and Technology 1994;29(9):161–3.
[56] Pere J, Alen R, Viikari L, Eriksson L. Characterization and dewatering of activated sludge from the pulp and paper industry. Water Science and Technology 1993;28(1):193–201.
[57] Tsai JH, Macedonio F, Drioli E, Giorno L, Chou CY, Hu FC, Li CL, Chuang CJ, Tung KL. Membrane-based zero liquid discharge: myth or reality? Journal of the Taiwan Institute of Chemical Engineers 2017;80:192–202.
[58] Report on "unfold the future – a two team project". Belgium: Confederation of European Paper Industries; 2013.
[59] Report on "CEPI sustainability report". Belgium: Confederation of European Paper Industries; 2017.
[60] Report on "preliminary statistics 2018". Belgium: Confederation of European Paper Industries; 2018.
[61] European paper industry – advancing the bioeconomy Report on "CEPI sustainability report". Belgium: Confederation of European Paper Industries; 2013.
[62] Tewari PK, Batra VS, Balakrishnan M. Efficient water use in industries: cases from the Indian agro-based pulp and paper mills. Journal of Environmental Management 2009;90(1):265–73.
[63] Gupta S, Goldar B. Do stock markets penalize environment-unfriendly behaviour? Evidence from India. Ecological Economics 2005;52(1):81–95.
[64] Bajpai P. Chapter II: generation of waste in pulp and paper mills. In: Management of pulp and paper mill waste. Switzerland: Springer International Publishing; 2015. p. 9–17.
[65] Dey A, Gupta BS. Pollution abatement in the Indian pulp and paper industry. Environmentalist 1992;12(2):123–9.

[66] Latha A, Arivukarasi M, Keerthana CM, Subashri R, Priya V. Paper and pulp industry manufacturing and treatment processes – a review. International Journal of Engineering Research and Technology 2018;6(2). PECTEAM 2K18 Conference Proceedings.
[67] Manuja S, Kumar A, Pandey S. Greenhouse gas emissions and reduction stratagems from waste sector in India. 2018. p. 17–26.
[68] Madan S, Sachan P, Singh U. A review on bioremediation of pulp and paper mill effluent – an alternative to conventional remedial technologies. Journal of Applied and Natural Science 2018;10(1):367–74.
[69] Pradesh A, Reddy MA. Assessment of waste generation and industry – a case study from Rajahmundry. Andhra Pradesh 2016;6(8):628–30.
[70] Kamali M, Alavi-Borazjani SA, Khodaparast Z, Khalaj M, Jahanshahi A, Costa E, Capela I. Additive and additive-free treatment technologies for pulp and paper mill effluents: advances, challenges and opportunities. Water Resources and Industry 2019;21:100–9.
[71] Saadia A, Ashfaq A. Environmental management in pulp and paper industry. Indian Journal of Industrial Pollution Control 2010;26(1):71–7.
[72] Gopal PM, Sivaram NM, Barik D. Paper industry wastes and energy generation from wastes. In: Energy from toxic organic waste for heat and power generation; 2019. p. 86–7.
[73] Hubbe MA, Metts JR, Hermosilla D, Blanco M, Yerushalmi L, Haghighat F, Lindholm-Lehto P, Khodaparast Z, Kamali M, Elliott A. Wastewater treatment and reclamation: a review of pulp and paper industry practices and opportunities. BioResources 2016;11(3):7953–8091.
[74] Subashini LM. Review on biological treatment processes of pulp and paper industry waste water. International Journal of Innovative Research in Science, Engineering and Technology 2015;4(5). 2319-8753.
[75] Ko JJ, Shimizu Y, Ikeda K, Kim SK, Park CH, Matsui S. Biodegradation of high molecular weight lignin under sulfate reducing conditions: lignin degradability and degradation by-products. Bioresource Technology 2009;100(4):1622–7.
[76] Ali M, Sreekrishnan TR. Aquatic toxicity from pulp and paper mill effluents: a review. Advances in Environmental Research 2001;5(2):175–96.
[77] Yeber MC, Rodriguez J, Baeza J, Freer J, Zaror C, Duran N, Mansilla HD. Toxicity abatement and biodegradability enhancement of pulp mill bleaching effluent by advanced chemical oxidation. Water Science and Technology 1999;40(11–12):337–42.
[78] Gavrilescu D, Puitel1 AC, Dutuc G, Craciun G. Environmental impact of pulp and paper mills. Environmental Engineering and Management Journal 2012;11(1):81–5.
[79] Sridhar R, Sivakumar V, Immanuel VP, Maran JP. Treatment of pulp and paper industry bleaching effluent by electrocoagulant process. Journal of Hazardous Materials 2011;186(2–3):1495–502.
[80] Yamashita N, Kannan K, Imagawa T, Villeneuve DL, Hashimoto S, Miyazaki A, Giesy JP. Vertical profile of polychlorinated dibenzo-p-dioxins, dibenzofurans, naphthalenes, biphenyls, polycyclic aromatic hydrocarbons, and alkylphenols in a sediment core from Tokyo Bay, Japan. Environmental Science & Technology 2000;34(17):3560–7.
[81] Gupta A, Gupta R. Treatment and recycling of wastewater from pulp and paper mill. In: Advances in biological treatment of industrial waste water and their recycling for a sustainable future. Singapore: Springer; 2019. p. 13–49.
[82] Hinck ML, Ferguson J, Puhaakka J. Resistance of EDTA and DTPA to aerobic biodegradation. Water Science and Technology 1997;35(2–3):25–31.
[83] Krishna KV, Sarkar O, Venkata Mohan S. Bioelectrochemical treatment of paper and pulp wastewater in comparison with anaerobic process: integrating chemical coagulation with simultaneous power production. Bioresource Technology 2014;174:142–51.
[84] Woodard F. Chapter 5: waste characterization. In: Industrial waste treatment handbook. 2nd ed. Elsevier; 2006. p. 107.

[85] Badar S, Farooqi IH. Pulp and paper industry – manufacturing process, wastewater generation and treatment. In: Environmental protection strategies for sustainable development. Dordrecht, the Netherlands: Springer; 2012. p. 397–436.

[86] Das TK, Houtman C. Evaluating chemical-, mechanical-, and bio-pulping processes and their sustainability characterization using life-cycle assessment. Environmental Progress 2004;23(4):347–57.

[87] Kangas J, Jäppinen P, Savolainen H. Exposure to hydrogen sulfide, mercaptans and sulfur dioxide in pulp industry. American Industrial Hygiene Association Journal 1984;45(12):787–90.

[88] Deshmukh S, Jana A, Bhattacharyya N, Bandyopadhyay R, Pandey RA. Quantitative determination of pulp and paper industry emissions and associated odor intensity in methyl mercaptan equivalent using electronic nose. Atmospheric Environment 2014;82:401–9.

[89] Dunn RF, El-Halwagi MM. Optimal recycle/reuse policies for minimizing the wastes of pulp and paper plants. Journal of Environmental Science & Health Part A 1993;28(1):217–34.

[90] Chakraborty D, Dahiya S, Amulya K, Srivastav V, Mohan SV. Valorization of paper and pulp waste: opportunities and prospects of biorefinery. In: Industrial and municipal sludge. Butterworth-Heinemann; 2019. p. 623–56.

[91] Sarks CJ, Johnson AC, Slupska MM, Karl ZJ, Eichmann MA. Methods and systems for propagation of a microorganism using a pulp mill and/or a paper mill waste by-product, and related methods and systems. 2019. U.S. Patent No. US20190071696A1.

[92] García-Velásquez CA, Cardona CA. Comparison of the biochemical and thermochemical routes for bioenergy production: a techno-economic (TEA), energetic and environmental assessment. Energy 2019;172:232–42.

[93] Penín L, Peleteiro S, Santos V, Alonso JL, Parajó JC. Selective fractionation and enzymatic hydrolysis of *Eucalyptus nitens* wood. Cellulose 2019;26(2):1125–39.

[94] Lee H. Development of lactic and succinic acid biorefinery configurations for integration into a thermomechanical pulp mill [Doctoral dissertation]. École Polytechnique de Montréal; 2015. Available from: https://publications.polymtl.ca/1750/.

[95] Guan W, Shi S, Tu M, Lee YY. Acetone-butanol-ethanol production from Kraft paper mill sludge by simultaneous saccharification and fermentation. Bioresource Technology 2016;200:713–21.

[96] Branco R, Serafim L, Xavier A. Second generation bioethanol production: on the use of pulp and paper industry wastes as feedstock. Fermentation 2019;5(1):4–34.

[97] Hahn-Hägerdal B, Karhumaa K, Fonseca C, Spencer-Martins I, Gorwa-Grauslund MF. Towards industrial pentose-fermenting yeast strains. Applied Microbiology and Biotechnology 2007;74(5):937–53.

[98] Zhao T, Yasuda K, Tashiro Y, Darmayanti RF, Sakai K, Sonomoto K. Semi-hydrolysate of paper pulp without pretreatment enables a consolidated fermentation system with in situ product recovery for the production of butanol. Bioresource Technology 2019;278:57–65.

[99] Kim CH, Lee J, Treasure T, Skotty J, Floyd T, Kelley SS, Park S. Alkaline extraction and characterization of residual hemicellulose in dissolving pulp. Cellulose 2019;26(2):1323–33.

[100] Frías M, Rodríguez O, Sánchez de Rojas MI. Paper sludge, an environmentally sound alternative source of MK-based cementitious materials. A review. Construction and Building Materials 2015;74:37–48.

[101] Report on "NWBC". In: The 6th Nordic wood biorefinery conference, Helsinki, Finland, 20–22 October, 2015; 2015.

[102] Bustos G, Torre N, Moldes AB, Cruz JM, Domínguez JM. Revalorization of hemicellulosic trimming vine shoots hydrolyzates through continuous production of lactic acid and biosurfactants by L. pentosus. Journal of Food Engineering 2007;78(2):405–12.

[103] Kohli K, Prajapati R, Sharma BK. Bio-based chemicals from renewable biomass for integrated biorefineries. Energies 2019;12(2):233.

[104] Kasurinen JA. Developing a tool for techno-economic analysis of pulp mill integrated biorefineries [MS thesis]. Finland: Programme in Environmental and Energy Engineering; Tampere University of Technology; 2016.

[105] Zhu W, Theliander H. Precipitation of lignin from softwood black liquor: an investigation of the equilibrium and molecular properties of lignin. BioResources 2015;10(1):1696–714.
[106] Toledano A, García A, Mondragon I, Labidi J. Lignin separation and fractionation by ultrafiltration. Separation and Purification Technology 2010;71(1):38–43.
[107] Jin W, Tolba R, Wen J, Li K, Chen A. Efficient extraction of lignin from black liquor via a novel membrane-assisted electrochemical approach. Electrochimica Acta 2013;107:611–8.
[108] Patankar SC, Liu LY, Ji L, Ayakar S, Yadav V, Renneckar S. Isolation of phenolic monomers from kraft lignin using a magnetically recyclable TEMPO nanocatalyst. Green Chemistry 2019;21(4):785–91.
[109] Belkheiri T, Andersson S, Mattsson C, Olausson L, Theliander H, Vamling L. Hydrothermal liquefaction of kraft lignin in sub-critical water: the influence of the sodium and potassium fraction. Biomass Conversion & Biorefinery 2018;8:585–95.
[110] Johnson MA, Hart PW. Integrating a biorefinery into an operating kraft mill. BioResources 2016;11(4):10677–710.
[111] Cesari L, Canabady-Rochelle L, Mutelet F. Separation of phenols from lignin pyrolysis oil using ionic liquid. Separation and Purification Technology 2019;209:528–34.
[112] Tribot A, Amer G, Alio MA, de Baynast H, Delattre C, Pons A, Mathias JD, Callois JM, Vial C, Michaud P, Dussap CG. Wood-lignin: supply, extraction processes and use as bio-based material. European Polymer Journal 2019;112:228–40.
[113] Wang SX, Yang L, Stubbs LP, Li X, He C. Lignin-derived fused electrospun carbon fibrous mats as high-performance anode materials for lithium ion batteries. ACS Applied Materials & Interfaces 2013;5(23):12275–82.
[114] Koutinas AA, Vlysidis A, Pleissner D, Kopsahelis N, Garcia IL, Kookos IK, Papanikolaou S, Kwan TH, Lin CS. Valorization of industrial waste and by-product streams via fermentation for the production of chemicals and biopolymers. Chemical Society Reviews 2014;43(8):2587–627.
[115] Farmer M, Sinquefield S. Final report. An external benefits study of black liquor gasification. Georgia Institute of Technology; 2003. 15.
[116] Clark J, Li Q, Lilik G, Reed N, Wang C. Novel design of an integrated pulp mill biorefinery for the production of biofuels for transportation. EGEE 2007;580:105.
[117] Naqvi M, Yan J, Dahlquist E. Bio-refinery system in a pulp mill for methanol production with comparison of pressurized black liquor gasification and dry gasification using direct causticization. Applied Energy 2012;90(1):24–31.
[118] Attia M, Farag S, Habibzadeh S, Hamzehlouia S, Chaouki J. Fast pyrolysis of lignocellulosic biomass for the production of energy and chemicals: a critical review. Current Organic Chemistry 2016;20(23):2458–79.
[119] Spath PL, Dayton DC. Preliminary screening – technical and economic assessment of synthesis gas to fuels and chemicals with emphasis on the potential for biomass-derived syngas (No. NREL/TP-510-34929). Golden, CO (US): National Renewable Energy Lab; 2003.
[120] Zhang Z, Matharu AS. Thermochemical valorisation of paper deinking residue through microwave-assisted pyrolysis. Waste Biorefinery 2018:671–92.
[121] Alen R. Pulp mills and wood-based biorefineries. In: Industrial biorefineries & white biotechnology. Elsevier; 2015. p. 91–126.
[122] Gonzalez-Garcia S, Moreira MT, Artal G, Maldonado L, Feijoo G. Environmental impact assessment of non-wood based pulp production by soda-anthraquinone pulping process. Journal of Cleaner Production 2010;18(2):137–45.
[123] Gonzalez-Garcia S, Hospido A, Agnemo R, Svensson P, Selling E, Moreira MT, Feijoo G. Environmental life cycle assessment of a Swedish dissolving pulp mill integrated biorefinery. Journal of Industrial Ecology 2011;15(4):568–83.

CHAPTER 24

Integration of biorefineries for waste valorization in Ulsan Eco-Industrial Park, Korea

Izhar Hussain Shah[1,2], Shishir Kumar Behera[3], Eldon R. Rene[4], Hung-Suck Park[1]

[1]Department of Civil and Environmental Engineering, University of Ulsan, Ulsan, Republic of Korea; [2]Institute of Environmental Sciences and Engineering, School of Civil and Environmental Engineering, National University of Sciences and Technology, Islamabad, Pakistan; [3]Industrial Ecology Research Group, School of Chemical Engineering, Vellore Institute of Technology, Vellore, Tamil Nadu, India; [4]Department of Environmental Engineering and Water Technology, IHE Delft, Institute for Water Education, Delft, The Netherlands

24.1 Introduction

Contrary to the traditional perception of waste as an economic and environmental burden, waste valorization is being promoted as an alternative for enhanced resource recovery. Waste valorization usually involves the recovery of valuable resources and bioproducts that can be used as a feedstock in energy generation and manufacturing and process industries [1,2]. Industry-scale extraction of energy and biobased products (chemicals, biofuels, biocommodities etc.) from waste is a promising approach for waste valorization [3]. Waste valorization, considered second to waste reduction and recycling, involves the valorization of waste residues through coproduction of materials and energy. Waste valorization is mostly applied to combustible organic waste which possesses: limited recycling opportunity, lower value of recovered materials, potential for waste contamination, or is preferred when avoidance of land disposal of waste materials is required [4].

A waste biorefinery can be termed as a bioprocess used to extract biobased materials and energy from renewable waste resources through sustainable biotechnology and can be seen as an integration of remediation and material recovery [3,5]. According to the International Energy Agency (IEA), biorefining is the sustainable processing of biomass feedstocks into a range of biobased products and bioenergy including food, feed,

Waste Biorefinery. https://doi.org/10.1016/B978-0-12-818228-4.00024-1

chemicals, materials, biofuels, power and heat [6]. Among several available technologies used in biorefineries such as anaerobic digestion, fermentation, incineration, etc., the use of any technology or a combination of technologies depends on the type of feedstock, its availability, characteristics and market demand [4,7]. Moreover, the ability of biorefineries to process diverse organic feedstocks from agriculture, forests, municipalities, and industries [8,9] makes it a cost-effective method for waste management and energy recovery especially in the form of biodiesel, ethanol, methane, hydrogen, steam and heat etc. [7]. Among other technologies, the transition from petroleum refinery to waste biorefinery [9,10] seems to be among the best available options to achieve resource sustainability and climate neutrality—both at the same time [11].

The most widely cited case with established resource sustainability pathways is Kalundborg in Denmark [12]. In Kalundborg, a variety of resource exchange networks have existed since 1970s and have evolved into a complex and adaptive industrial ecosystem. Among the various symbiotic exchange networks, Inbicon Biomass Refinery (or Inbicon biorefinery) was inaugurated in 2009 as a demonstration biomass refinery in Kalundborg. Inbicon biorefinery has a capacity to produce 5.4 million liters of bioethanol annually from 30 kilotons of wheat straw along with the cogeneration of lignin pellets (13 kilotons) and C5-molasses (11 kilotons), with biorefinery outputs subject to increase in the scale-up phases [13]. The bioproducts from this biorefinery are used in a variety of ways including ethanol used in transport (fuel replacement), lignin used in power and heat generation (coal replacement), and molasses used in food production and chemical production (chemical replacement). Thus, various bioproducts were able to substitute virgin resources following the concept of integrated biorefineries. In fact, biorefineries can very well meet all of their internal energy demands and export rest of the bioproducts for commercial purposes—steps leading to the substitution of fossil fuels and mitigation of environmental impacts [14]. For developing countries, based on the above described biorefinery application, a transition from fossil-based energy consumption to biobased energy use can partly contribute to climate change mitigation and resource efficiency improvements. This becomes more relevant when high shares of biodegradable waste remain untapped or untreated in developing countries [15,16]. Nevertheless, in spite of low-income levels in developing economies, progress toward biorefinery implementation may come sooner given their rising climate vulnerabilities [17,18]. Thus, sustained policies and actions are necessary for scaling-up the application of integrated biorefineries in the long run.

24.1.1 Waste valorization: Korean context

Biorefineries, mainly classified depending on the type of feedstock sources, can be used to process both industrial and municipal wastes [10]. Both of these wastes are highly relevant to waste management practices in Korea. With rapid urban and industrial development, increasing waste generation coupled with limited land space has put Korea in pursuit of alternative methods of sustainable waste management policies and valorization technologies. Under the national government, the Ministry of Environment (MoE) and Ministry of Trade, Industry and Energy (MOTIE) have been continuously looking for ways to valorize waste generated in the country. This includes implementation of multiple policies and laws such as "waste deposit-refund system in 1991," "Act on promotion of saving and recycling of resources in 1992," "volume based waste fee system in 1995," "extended producer responsibility in 2003," "food waste separation in 2005," "Act on the promotion of the conversion into environment-friendly industrial structure in 2006," and "low carbon green growth vision in 2008." Since 2010, a transition toward high-value, material-frugal, and technology-based manufacturing has risen significantly and the share of service sectors is also on the rise [19]. Changes in industrial structure have been promoted through the governmental support under "low carbon green growth strategy (2009)" and the five-year plan for green growth (until 2013)" [20,21]. More recently, alternate energy policies are being actively pursued since 2017 in an effort to reduce fossil fuel consumption and switch to cleaner energy resources including biobased energy [22]. Overall, efforts at the national level have resulted in reduced municipal waste generation (0.40 tons/capita/year in 1995 compared to 0.36 tons/capita/year in 2009 [23]) in spite of population and economic growth, and has promoted waste valorization, material reuse, material conservation, and energy efficiency (recovery) in all economic sectors including industries.

Industrial sector in Korea in energy-intensive and is responsible for consuming most of the energy resources [22]. As of 2017, the final energy consumed by the industrial sector amounted to about 144 million tons of oil equivalent (toe) which represents a share of about 62% in total energy consumption in the country [24]. Similarly, as of 2016, greenhouse gas (GHG) emissions from industrial sector (including processing and manufacturing but excluding energy production industry) amounted to about 236 million tons CO_2-equivalent, representing a share of about 34% in national GHG emissions [25]. This is an understated figure as most of the processed energy ends up being consumed by the industrial sector in terms of electricity, steam etc. Nevertheless, this situation has pushed successive governments to act upon rising climate consequences from large-scale energy consumption by domestic industries. Efforts have been, therefore, extensively carried out to mitigate fossil fuel consumption and reduce consequent GHG emissions [21]. From this perspective, energy efficiency and GHG mitigation have been specially promoted at both the national and sectoral levels through green growth initiatives and ecoindustrial development strategies [19]. At the sectoral level, industries have been encouraged to employ smart grid storage systems [26], develop offshore wind farms [27], use low carbon power and renewables [19], and invest in energy research and development projects [28] to improve energy efficiency at an individual firm or an industrial park level. Similarly, national level strategies are also complementing regional plans in light of ecoindustrial development, as discussed in the following paragraphs.

Following the Rio Earth Summit (1992), Korean government was harnessing ideas to restructure local industries and businesses with cleaner production practices and industrial ecology tools in order to improve their economic, environmental and social performance [29]. Active governmental involvement toward resource efficiency enhancement paved the way for the 15-year, 3-stage national EIP program that was initiated in 2005. The Korean EIP program was initiated by the Korea National Cleaner Production Center (KNCPC) in collaboration with the MOTIE under the title "Eco-industrial Park: Construction for establishing infrastructure of cleaner production in Korea" [29]. An EIP is a community of firms and businesses pursuing enhanced environmental, economic, and social performance by mutual collaboration in order to conserve natural resources and energy; increase productivity, improve industrial efficiency, promote worker health and public image, and provide economic benefits from the use and sale of waste materials and/or by-products [30,31]. An EIP may consist of a group of firms and companies that seek higher economic benefits and improved environmental performance by mutual collaboration and resource connectivity, thus, making collective benefits larger than the sum of individual benefits each firm would accrue if they are improving individually [32]. Industries and governments usually employ industrial ecology tools, including industrial symbiosis, for reduced overall carbon footprint of industrial ecosystems. The carbon footprint of an EIP can be a measure of the total carbon dioxide [33] or sum of all GHG [34] emissions directly associated with the functioning of an EIP, though incorporating indirect emissions are subject to interpretation. With this definition in mind, EIP development—whether planned or unplanned—creates innovative pathways for higher resource efficiency at the intrafirm, interfirm, and regional levels. With higher resource efficiency, use of virgin raw materials is reduced, and waste resources are optimized within the industrial ecosystem, thus mimicking the principles of a natural ecosystem. Fig. 24.1 presents the basic concept of EIPs in the context of resource efficiency and waste (by-product) valorization.

According to this concept, only the resources flowing inside (from outside) and wastes flowing outside (from inside) are considered when environmental impacts of EIPs are to be quantified (13). The external resources (R_E) are received by firms inside an EIP along with internal resources (R_i) in the form of exchanged wastes and/or by-products. As shown in Fig. 24.1, total resource consumption is equal to the sum of R_E^1, R_E^2, and R_E^3 whereas total waste discharges from the EIP are equal to the sum of W_E^1, W_E^2, and W_E^3. Under this system, the total waste from the EIP is reduced by a quantity equal to the quantity of materials exchanged inside. Therefore, both the consumption of external resources and generation of external wastes is reduced proportionally to the level of internal resources exchanged between the firms.

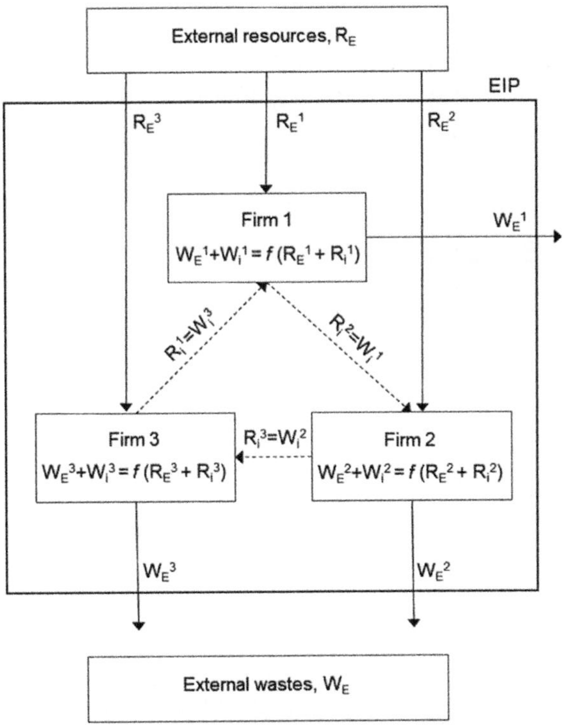

Figure 24.1

Concept of internal versus external resource consumption and waste circulation in the context of EIPs. RE = external resources; WE = external wastes; Wi and Ri = internal wastes and resources.

As EIPs integrate both environmental and economic benefits, all participating firms take part in resource sharing networks to increase their market competitiveness and public image through locally developed business models. Following this approach in the Korean EIP program, several waste valorization projects through the regional EIPs were successfully materialized producing significant benefits [35,36]. Table 24.1 presents the summary of economic and environmental benefits of the Korean EIP program.

24.1.2 Waste valorization under Ulsan EIP

In Ulsan, the transition of industrial complexes into EIPs was an evolving phenomenon that has been systematically accelerated by national level policies on ecoindustrial development. The city of Ulsan, well known as the industrial capital of Korea, is one of the nation's seven metropolitan cities and has been Korea's most significant heavy industrial region for more than 40 years. It has a total area of 1061 km^2 and population of 1.19 million [37]. For the built-up of urban development, 11% of the land area is allocated while the remaining 89% is allocated as forests and agricultural land. The industrial base

Table 24.1: Economic and environmental benefits from Korea EIP program, as of 2016.

Regional center	Projects	Economic benefits, billion KRW			Environmental benefits[a]			
		Investment	Cost reduction	Revenue	Waste, ton	Water, m^3	Energy, toe	CO_2, ton
Kyeonggi	22	125	14.2	17.4	86,216	887,085	11,192	56,724
Ulsan	36	212	80.0	63.0	40,042	79,388	279,761	665,712
Busan	17	11.3	7.2	10.5	22,961	2890	22,946	95,669
Chonbuk	25	36.1	6.0	32.6	150,236	—	—	204,139
Chonnam	42	93.9	30.6	87.8	791,784	10,980	—	530,173
Kyeongbuk	40	210.4	47.0	85.8	436,016	36,571,514	55,053	351,834
Daegu	24	10.1	7.2	40.2	68,220	600	50	35,176
Choongbuk	19	44.9	5.1	25.6	40,007	36,860	20,521	73,462
Choongnam	6	15	0.1	11.1	636	—	—	42,337
DaeJeon	1	0.6	0.2	1.5	322	—	—	—
Incheon	3	1.5	—	1.3	10	—	1200	2869
Cumulative	235	760.8	197.6	376.8	1,636,450	37,589,317	390,723	2,058,095

[a]Environmental benefits are reported as (1) waste and byproducts reduction/exchange, (2) water saved, (3) energy savings, and (4) CO_2 reduction.

includes two majors EIPs comprising petrochemical, chemical, nonferrous, automobile, and shipbuilding industries with multiple material and energy symbiosis networks [35,38,39]. Under the national EIP program, an EIP center in Ulsan was institutionalized to systemically engineer and cultivate symbiotic exchanges of waste materials and process by-products in order to achieve resource optimization, energy recovery and waste valorization.

In 2016 alone, total energy consumption in Ulsan was equal to 27.1 million toe from which 89% (24.2 million toe) was attributed to the industrial use. The overwhelming demand for energy resources had put immense pressure on city's government to promote resource and energy recovery programs. The transition from conventional landfilling of waste during 1990s to energy/resource recovery during 2000s was, therefore, a major achievement that is partly attributable to spontaneous energy efficiency activities. However, the waste valorization approach, under Ulsan EIP program, has systematically helped in meeting the rising energy demand from the industrial sector while reducing GHG emissions at the same time.

From a transitional perspective, during the 1990s, the organic waste generated within the city was traditionally sent to sanitary landfills with no value creation or energy recovery, although, consciousness about waste reduction was emerging. This approach transformed, during early 2000s, toward resource recovery mostly in terms of landfill gas extraction and use. This practice further evolved when multiple technologies including recycling, landfill gas collection, anaerobic digestion, and incineration etc. were combined and successful waste valorization began. Under this waste valorization approach, biorefineries became integrated with EIPs and energy sharing networks were developed.

Since Ulsan has an energy-intensive industrial base, most of the focus has been on providing energy from clean and alternative resources, of which, bioenergy is considered to be carbon-neutral, if not carbon-negative [40]. Therefore, most of the biorefineries in Ulsan focus on providing cleaner bioenergy to EIPs in terms of biogas and steam (generated from biogas) that mitigates the overall carbon footprint of EIPs. The EIPs themselves showcase the successful sharing of waste resources, energy products, infrastructure and communication networks to the mutual benefit of participating companies. This symbiotic approach adopted under the Ulsan EIP program has greatly reduced city's heavy dependence on fossil fuels, improved its sustainability status, enhanced industrial market competitiveness and provided triple bottom line benefits.

This chapter sheds light on the evolution of biorefinery processes in Ulsan through the integration of biorefineries waste valorization. Strategies from cost reduction to industrial competitiveness and sustainable industrial development are contextualized under the ecoindustrial park/development concept. First, successful cases of waste valorization using landfill gas reclamation and steam production from municipal waste have been introduced.

This will be followed by biogas production from food waste and municipal wastewater treatment sludge and its utilization by a chemical processing company through industrial symbiosis. Lastly, strengthening biorefinery of a paper mill business through steam and CO_2 networking between a zinc smelter and a bioenergy center under industrial symbiosis approach has been discussed. These case studies will provide insights on how biorefinery concept can be adapted into the real field in the context of EIPs. The case studies are followed by summary of triple bottom line benefits from Ulsan EIP along with a discussion on progress made by developing Asian countries in this area.

24.2 Integration of biorefineries in Ulsan EIP

The waste recovery and resource sharing projects successfully executed in Ulsan provide critical understanding of the success factors and strategies involved in upscaling the waste to energy infrastructures and biorefineries.

24.2.1 Landfill gas reclamation and industrial symbiosis

Ulsan metropolitan city implemented a landfill gas (LFG) reclamation project with a cost of 55 billion Korean Won (KRW) at Seongam landfill. The LFG reclamation project was initiated due to several reasons such as large generation rates of organic waste, scarcity of landfill sites, need for landfill stabilization, increasing energy costs and demand from local industries, and most importantly the motivation for resource recovery. The LFG supply began in 2002 at an average rate of 100–230 m^3/ton of landfilled waste with a reported calorific value of 4707 kcal/m^3, with preference given to industrial units in order to supplement their high energy demand. Before the waste valorization project began, extraction of LFG was smooth due to landfill maturity and effective collection pipe network consisting of 49 extraction wells. The gas collection network at the landfill site, having a capacity of 4.45 km^2, was able to collect and then discharge through the gas flare system without any heat/energy recovery. Some of the municipal waste of Ulsan city was also sent to a nearby waste incinerator that was used to produce steam for energy production. Fig. 24.2 shows the schematic diagram of landfill gas reclamation project before and after industrial symbiosis at Ulsan Seongam landfill.

As these facilities were located within the industrial park area, their proximity to industrial units was turned into a business idea through the industrial symbiosis approach. Through the spontaneous symbiosis project, LFG was supplied to (1) Kumho Petrochemicals Co. Ltd. (until 2012) and (2) city waste incinerators to heat selective catalytic reduction (SCR) for NO_x and dioxin control (until 2018). This symbiotic exchange project created six additional jobs to support the operation and maintenance of the energy network. With the help of a gas purification facility, LFG with a methane concentration of 54.4% was

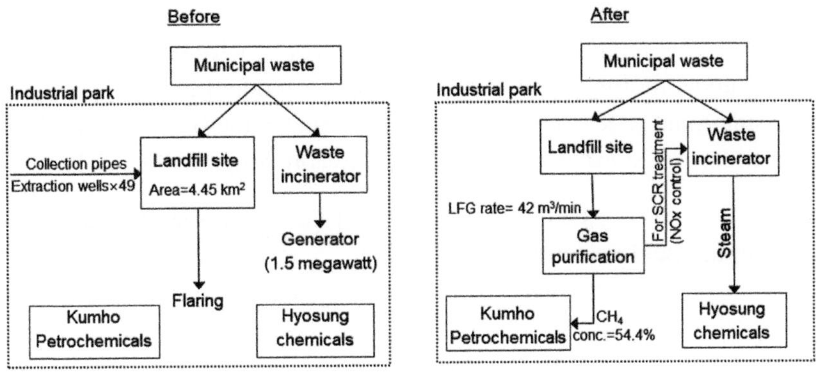

Figure 24.2
Schematic diagram of landfill gas reclamation project at Ulsan Seongam landfill.

supplied to Kumho Petrochemicals Co. Ltd. The methane concentration in LFG usually varies between 55% and 58% throughout the year depending on several parameters such as composition of feedstock and other environmental conditions.

Under this project, the designed LFG supply of 42 m^3/min (average 30–33 m^3/min) was able to generate a total revenue of 13.3 billion KRW during 2003–2016 with an average annual income of 1.77 billion KRW. The LFG supply to the petrochemical company helped them to reduce fossil fuel consumption and their corresponding GHG emissions. This project also supported steam networking between Ulsan municipal solid waste incinerators and nearby Hyosung chemicals. With the help of Ulsan EIP program, these exchange networks helped industries to reduce their boiler operating costs, odor generation, and greenhouse gas emissions, all paving way for an enhanced ecofriendly status of Ulsan EIP.

24.2.2 Biogas sharing network with a chemical plant

A biogas sharing network was developed between a municipal wastewater treatment plant (MWTP) and a chemical company. The integrated MWTP, located at Yongyeon, was established in 2010 with an investment of 197 billion KRW comprising primary, secondary and advanced treatment facilities (treatment capacity = 250,000 m^3/day). Although the plant was initially designed for conventional primary and secondary treatment, however, it was later upgraded with advanced bioreactor processes (tertiary treatment). The primary function of this facility was wastewater treatment and biogas disposal. The MWTP was located within the industrial park and received sewage wastewater from both municipalities and industries. At the treatment plant, the bioreactor comprised of anaerobic processes which provided an opportunity to produce biogas through digestion. Prior to any resource networking, the biogas produced at the digester was sent to the gas storage tank where it was temporarily stored before being sent to the gas flare system (open combustion) with no heat and/or energy recovery. However, with the development of bioenergy sharing network, biogas was collected, pretreated and then sold to a nearby chemical processing plant (SK Chemicals). Fig. 24.3 shows the bioenergy utilization before and after industrial symbiosis project at Yongyeon integrated MWTP.

Owing to the higher bioenergy demand from receiver company, the integrated MWTP was expanded by doubling the digester capacity (7000 m^3 × 2) through adopting ultrasonic technology from "Scandinavian Biogas." With doubled digester capacity, an additional 180 tons per day of food wastes were also added to the processing facility after a redesign that costed 21 billion KRW. Food waste was now added, to the existing processing of 600 tons of sludge, to increase biogas production and utilize the food waste for resource recovery. This enhanced waste treatment capacity along with higher biogas production also provided an annual revenue of 3 billion KRW. The symbiotic biogas sharing project also resulted in the creation of 10 new jobs. The biogas sharing network helped the biogas to be used as a bioenergy resource for SK Chemicals while wastage of biogas through flaring was also avoided. Thus, the substitution of fossil fuels by biomethane at SK Chemicals became a driver of GHG mitigation [41]. With the help of this symbiotic project, successful waste valorization was achieved in which organic wastes were processed to produce biomethane for use in industrial sector—an excellent example of integrating biorefineries with EIPs. Consequently, the whole situation, from biogas flaring to its use as a resource, mutually benefited both partners.

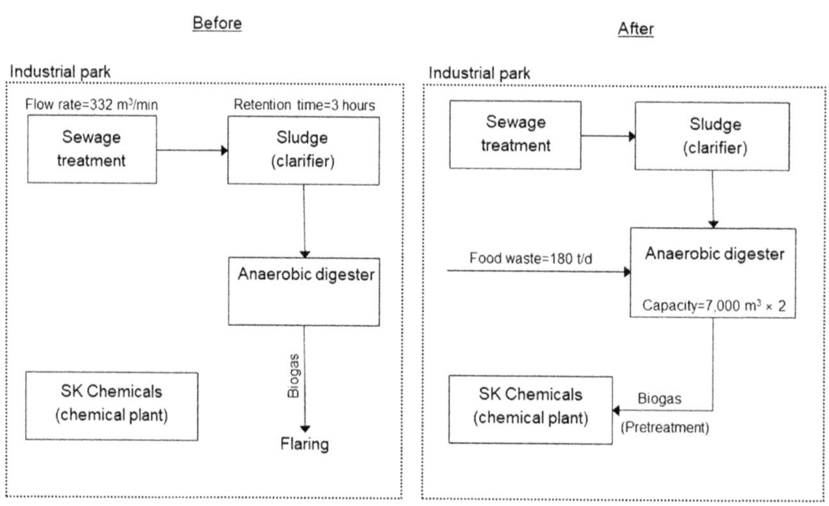

Figure 24.3
Biogas utilization project at Yongyeon integrated wastewater treatment plant.

Among future plans, the MWTP utilization (treatment capacity = 250,000 m^3/day, throughput in 2017 = 217,100 m^3/day) can be increased to treat an additional 32,900 m^3/day of wastewater and integrate more industries through this waste valorization approach.

24.2.3 Biorefinery strengthening and bioenergy networking

This case pertains to the integration of a bioenergy facility with a paper mill and a zinc smelter—all colocated within the EIP. The evolution of such a biorefinery networking provides valuable insights for readers and will be explained in two parts. The first part explains the actual paper mill competitiveness strengthening through steam and CO_2 networking, while the second part describes the development of a bioenergy center that focused on bioenergy production from organic wastes.

24.2.4 Paper mill strengthening through steam and CO_2 networking

Steam and CO_2 sharing network, among two different entities, is also an interesting case of waste valorization in which successful integration of biorefineries and EIPs took place [42]. The project involved a receiver company i.e., Hankook Paper, classified as a stakeholder of the biorefinery business, and a sender company i.e., Korea Zinc, classified as a large zinc smelter—both located 3.8 km apart. Through the resource sharing network, Hankook Paper received steam and CO_2 from Korea Zinc. The investments by Korea Zinc and Hankook Paper were 16.87 billion KRW and 4.16 billion KRW, respectively, mainly for infrastructure and pipeline development. The entire project development and planning was supervised by Ulsan EIP center. Fig. 24.4 shows the schematic diagram of the steam

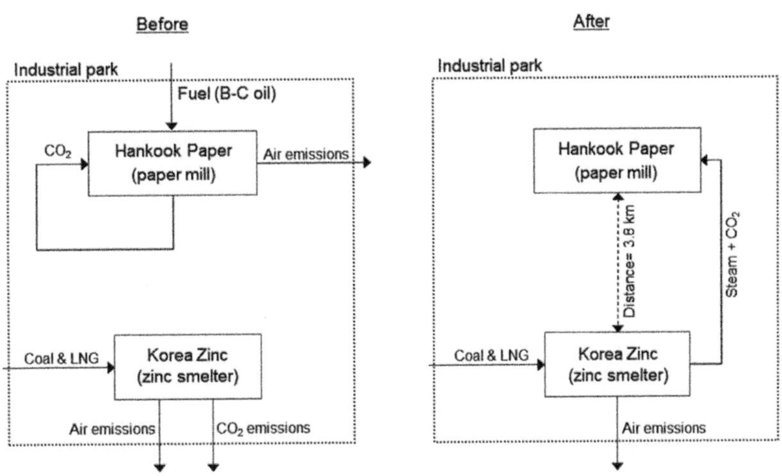

Figure 24.4
Industrial symbiosis between Hankook Paper and Korea Zinc.

and CO_2 networking before and after industrial symbiosis project between Hankook Paper and Korea Zinc.

Before the symbiosis between the two companies took place, Korea Zinc and Hankook Paper did not explore any resource sharing opportunities as both were working independently at the industrial park level. That approach transformed when potential resource networking was identified under the leadership of Ulsan EIP center. Hankook Paper previously operated internal boilers to produce both steam (about 50 tons/hour) and CO_2 (for conversion into calcium carbonate to be used as a filler material). Korea Zinc used to emit large quantities of CO_2 as waste flue gas before the implementation of symbiosis project. However, after the successful implementation of symbiosis project between the two companies in 2011, steam was supplied from the Korea Zinc company amounting to 690,638 tons, while 77.72 million Nm^3 of CO_2 flue gases were also received by Hankook Paper. This multi-faceted symbiosis helped Hankook Paper to shut down its existing boilers that were being run on B–C oil. This helped in fuel cost reductions for Hankook Paper company along with reduced GHG emissions. Through this exchange networking, the annual profits of Korea Zinc and Hankook Paper were 4.19 billion KRW and 2.42 billion KRW, respectively. The project also helped in a net reduction of GHG emissions by 60,522 tons of CO_2 equivalent.

24.2.5 Ulsan Bio Energy Center

Under the policy to promote renewable bioenergy production from organic wastes, energy recovery and to prepare concrete business plans for the environmental-friendly treatment of food waste and livestock manure, "Ulsan Bio Energy Center" in Onsan was established with a cost of 23 billion KRW in 2014. The MoE provided 70% funding while the rest was managed by Ulsan city government. The production of primary products included biogas at a rate of 9000 Nm^3/day, out of which, 5850 Nm^3/day of biomethane was produced after the refining process. The bioenergy center generated 16.5 tons/day of impurities and 8.8 tons/day of digestion sludge. Fig. 24.5 shows the schematic diagram of the Ulsan bioenergy center.

The designed treatment capacity of the bioenergy center was 150 tons/day of waste comprising of both food waste (100 tons) and livestock manure (50 tons). The treatment process was based on the anaerobic digestion method in which the generated biogas was initially used for steam production but was later shared through industrial symbiosis. The bioenergy center had facilities for anaerobic digestion, biogas production, sludge treatment, odor prevention, and wastewater treatment. Anaerobic digestion process comprised of feed-in and pretreatment, and acid and methane fermentation. The biogas production at the bioenergy center was used to produce steam at site which was then supplied to Hankook Paper Co. Ltd. and resulted in significant economic and

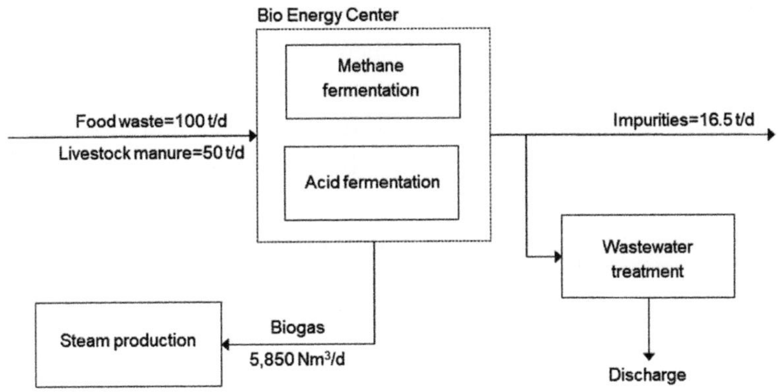

Figure 24.5
Schematic diagram of Ulsan Bio Energy Center.

environmental benefits. Steam generated at this facility created a profit of 700 million KRW per year. For the year 2016, the bioenergy facility revenues through steam supply and waste disposal fee amounted to 2.49 billion KRW along with the creation of 10 new jobs. Since improvements in design have been taken up recently, revenue generation is expected to increase in coming years.

The integration of biorefinery concept in Ulsan EIP is further elaborated in Fig. 24.6 which presents the steam and CO_2 networking before and after industrial symbiosis project involving all three participants i.e., Ulsan bioenergy center, Hankook Paper and Korea Zinc. Prior to the symbiosis project, food and animal waste in Ulsan city was disposed using conventional methods mainly landfilling and ocean dumping. Such a practice required higher resources for waste disposal with no energy recovery or material recycling. Further, Ulsan bioenergy center became operational during 2014 which provided an

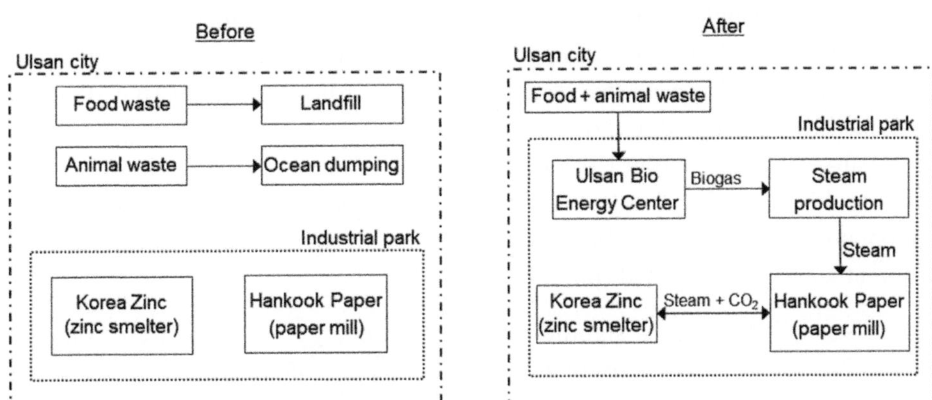

Figure 24.6
Steam and CO_2 networking project in Ulsan.

opportunity to link its steam production with nearby industries. Therefore, this idea further evolved when the bioenergy center was established within the industrial park, thus, increasing its proximity with two participating firms. After the successful integration for waste valorization, the bioenergy center received organic waste and produced steam from biogas and sold it to Hankook Paper which was already receiving steam and CO_2 from Korea Zinc. This case highlights the importance of waste networking which can bring forward significant waste valorization opportunities even when industries are located at a considerable distance from each other. This case also highlights how the integration of biorefineries for waste valorization can help reduce GHG emissions especially at a time when global consensus over climate change mitigation is very strong.

24.3 Ulsan EIP program and waste valorization

Economic, environmental, and social benefits—the basis of sustainable development—were also significant for Ulsan EIP program in the context of waste valorization. Ulsan EIP was also able to greatly contribute to the national economy by improving energy and resource intensity at one of Korea's important industrial cities. The successful commercialization of multiple projects also provided motivation for new industries to implement industrial symbiosis for waste valorization. In total, Ulsan EIP initiative resulted in 77 project proposals from 294 firms (for which feasibility was conducted) that materialized in the implementation of 34 symbiotic projects among 123 firms. The economic benefits calculated as the sum of cost savings (reduced resource procurement, operations, waste management, replaced virgin material with by-products) and revenues (selling of by-products) were highly significant during the first 10 years of the program i.e., from 2005 till 2016. By the end of 2016, government investments totaled to 16.64 billion KRW for project research and development, including EIP center operations. From this government research fund, new income of 73.11 billion KRW per year was generated through selling of by-products and recovery of materials/energy. An additional income of 87.84 billion KRW per year was generated from energy and material savings. Moreover, a total private investment for the construction of industrial symbiosis networking facilities amounted to 276.46 billion KRW and created 195 new jobs, thus, adding a social value to the Ulsan EIP program. The environmental benefits were evaluated in terms of the direct reductions in waste or byproducts, wastewater, energy consumption, and CO_2 emissions. Table 24.2 presents the environmental benefits achieved under different stages of Ulsan EIP program.

During the Ulsan EIP program, a significant amount of energy resources were conserved amounting to 279,761 toe, which resulted in a reduction of 665,712 tons of CO_2 emissions and 4052 tons of air pollutants such as SOx and NOx. In addition, a total of 79,388 m^3 of wastewater generation was avoided, and 40,044 tons of by-products and wastes were either

Table 24.2: Environmental benefits achieved under different stages of Ulsan EIP program, as of 2016.

	Implementation	Waste reduction, ton	Water savings, m^3	Energy savings, toe	CO_2 reduction, ton
Outcomes	Stage 1 (2005–09)	31,719	37,048	51,767	118,377
	Stage 2 (2010–14)	6826	41,975	144,335	369,249
	Stage 3 (2015–19)[a]	1497	365	83,659	178,086
	Cumulative	40,042	79,388	279,761	665,712

[a]The Korean EIP program was ended in 2016 instead of the planned year, i.e., 2019.

reduced or reused. A large share of environmental benefits was attributed to energy symbiosis networks under the Ulsan EIP program [35]. Multiple energy sharing networks were developed among several firms where high-grade heat and waste steam were shared between different symbiotic partners. The symbiotic networks for energy exchange directly reduced combustion of fossil fuels including major fuels, such as B–C oil and other petroleum products, and indirectly caused significant GHG emission reductions. For further details on several symbiotic projects under Ulsan EIP program, we refer to published literature [35,43]. The program also resulted in the creation of 195 jobs during the construction, operation and maintenance of resource sharing networks. Consequently, these outcomes positively helped industrial complexes to improve their public image and enhance their relations with the neighboring local communities. Industries along with local companies and regional government also disseminated outcomes of Ulsan EIP program through televised reports and social media campaigns under corporate social responsibility (CSR) initiatives for positive public relations. Environmental benefits were portrayed as ethical responsibility of industries, thus negating any environmental concerns local communities may have.

Although the national EIP program was initially planned to finish by 2019, however, it was stopped in the year 2016 by the government of Korea mainly due to governmental policy shift. The Korean government considered that the national EIP program had sufficiently achieved its original objectives and the development of industrial symbiosis should be voluntarily continued in regional EIPs.

24.4 Progress on biorefineries: Asian context

The development of biorefineries, especially in the developing world, can be the first step toward upscaling biomass and organic waste valorization—especially at a time when agriculture, municipalities and industries lack adequate waste management technologies. Moreover, waste management and co-generation of bioenergy through biorefinery development has become increasingly pertinent to developing countries, as most of them are facing severe environmental challenges [7]. This argument is strengthened when energy demand is growing nearly three times faster in developing countries than industrialized countries [44]. Therefore, biorefinery development can provide opportunities for both waste valorization and carbon-neutral energy production through circular resource loops in developing countries.

From an ecosystem's perspective, Asian countries also carry huge potential to garner benefits from environmental-friendly waste management, bioenergy production, GHG mitigation via fossil fuel substitution and cogeneration of biobased chemicals for industrial and agricultural purposes—all through the biorefinery development approach. However, progress on biorefinery development in most of the Asian countries is rather modest. According to a report by World Economic Forum (WEF), the policies on biorefineries and biofuels are inconsistently implemented in most Asian countries except China which has invested largely in biomass-derived energy [45]. For instance, in China, starch crops have been used to produce bioethanol at five ethanol based biorefineries with an estimated production of 1.7 million tons in 2009, however, food security concerns have forced these biorefineries to substitute grain-based biomass feedstock with municipal and agroindustrial wastes. India, second largest country by population in the world, is also promoting policies on bioproducts and biofuels since last two decades [46]. Recent efforts have focused on large-scale bioenergy production from both energy crops and waste resources [47] with increasing efforts on valorization of agricultural wastes such as those disclosed in Indian patent application 443/DEL/2003 (recovery of bioactive compounds from mango peels) [48,49].

Similarly, Pakistan, an agricultural dominant economy, has a huge potential for bioenergy production from agricultural biowastes, and governmental support has been increasing [50]. Although waste valorization through biorefineries is yet to be seen at large scales in Pakistan, biogas production from organic waste at community levels, and bioethanol production from a few sugar mills, is taking place [51]. This is in line with the wider integration of sugar mills with biorefineries where cogeneration of ethanol, organic fertilizers, and bioenergy is gaining more interest [11]. In Bangladesh, a densely populated Asian country, energy recovery from organic waste is gaining more attention including biogas cogeneration and LFG reclamation [52]. Huge potential for biorefinery development is available particularly for agricultural wastes [53] which also provide economic integration of farming communities through waste valorization [54].

In rest of the Asian countries, biorefinery approach is also emerging as a sustainable strategy for waste valorization. Examples include biorefinery in Thailand using molasses for bioenergy and biofertilizer with integrated sugar mills [55,56] along with regional-scale implementation of biorefinery approach using organic wastes in Nepal, Vietnam, Cambodia, Laos, etc., where biogas is extracted as an energy resource and bioslurry is coproduced as an organic fertilizer and fish feed [57]. Therefore, a transition away from first generation biorefineries (which mainly use raw biomass and energy crops) is taking place with rising implementation of integrated biorefineries (which mainly use organic by-products and wastes) for waste valorization especially in the developing Asian countries. Nevertheless, with the absence of large-scale commercial biorefineries [47], systematic integration of biorefineries with municipalities and industries is still in the early research and development phase and is likely to mature in the coming decades provided that sustained policies are in place.

24.5 Conclusions and perspectives

Due to the advent of symbiotic networks and successful commercialization of several projects and their inherent benefits obtained through the Ulsan EIP program, waste valorization has become an important tool toward sustainability and urban ecoliving. Moreover, integration of biorefineries with existing industrial symbiosis networks provides an exciting opportunity to tackle organic waste disposal issues and recover biobased products, including bioenergy. A biorefinery, similar to a petroleum refinery, maximizes revenue through the coproduction of value-added bioproducts, including chemicals and energy. Yet, issues such as inconsistent supply of feedstocks, land use competition with food crops, lack of governmental support, and technological limitations may hinder the expansion of biorefineries and need to be addressed appropriately.

In Korea, Ulsan's economy depends heavily on its manufacturing and processing industries and demand for clean energy supply is becoming more important especially in the context of global movement against greenhouse gas emissions. This chapter discussed the emergence of Ulsan EIP program along with integration of biorefineries for waste valorization. Some of the successful biorefinery projects, undertaken in light of Ulsan EIP program, were also discussed. This concept has made considerable contribution to reduce city's carbon footprint and enhance its ecofriendly performance. The Ulsan EIP initiative was based on the industrial symbiosis (IS) research and development into business model. As such, a pilot post-EIP strategy to replicate and mainstream waste valorization and biorefinery concepts through ecoindustrial development and promote a biobased economy for energy security is highly recommended for future applications. This is expected to help establish the basis for self-reliance in EIP development and increase business awareness and motivation for further opportunities.

Acknowledgments

The authors acknowledge the support from Brain Korea 21plus (BK-21 plus) program from the Ministry of Education, Science, and Technology through the Environmental Engineering Program and the Basic Science Research Program, National Research Foundation (NRF), Ministry of Science and ICT, Korea with grant number (NRF-2017R1A2B4011978) at the University of Ulsan.

References

[1] Poinern GEJ, Fawcett D. Food waste valorization: new manufacturing processes for long-term sustainability. Encyclopedia of Food Security and Sustainability 2019;1:429–33.
[2] Ferronato N, Rada EC, Portillo MA, Cioca LI, Ragazzi M, Torretta V. Introduction of the circular economy within developing regions: a comparative analysis of advantages and opportunities for waste valorization. Journal of Environmental Management 2019;230:366–78.
[3] Mohan VS, Butti SK, Amulya K, Dahiya S, Modestra JA. Waste biorefinery: a new paradigm for a sustainable bioelectro economy. Trends in Biotechnology 2016;34:852–5.
[4] Mata TM, Martins AA, Caetano NS. Bio-refinery approach for spent coffee grounds valorization. Bioresource Technology 2018;247:1077–84.
[5] Dahiya S, Kumar AN, Sravan J, Chatterjee S, Sarkar O, Mohan SV. Food waste biorefinery: sustainable strategy for circular bioeconomy. Bioresource Technology 2018;215:2–12.
[6] International Energy Agency (IEA). Biorefining in a future BioEconomy. 2018. Available from: www.ieabioenergy.com/task/biorefining-sustainable-processing-of-biomass-into -a-spectrum-of-marketable-biobased-products-and-bioenergy/.
[7] Nizami AS, Rehan M, Waqas M, et al. Waste biorefineries: enabling circular economies in developing countries. Bioresource Technology 2017;241:1101–17.
[8] Naik SN, Goud VV, Rout PK, Dalai AK. Production of first and second generation biofuels: a comprehensive review. Renewable and Sustainable Energy Reviews 2010;14:578–97.
[9] Mohan VS, Nikhil GN, Chiranjeevi P, et al. Waste biorefinery models towards sustainable circular bioeconomy: critical review and future perspectives. Bioresource Technology 2016;215:2–12.
[10] Kamm B, Kamm M. Principles of biorefineries. Applied Microbiology and Biotechnology 2004;64:137–45.
[11] Nguyen TLT, Gheewala SH, Sagisaka M. Greenhouse gas savings potential of sugar cane bio-energy systems. Journal of Cleaner Production 2010;18:412–8.
[12] Chertow M, Park J. Scholarship and practice in industrial symbiosis: 1989–2014. In: Clift R, Druckman A, editors. Taking stock of industrial ecology. AG Switzerland: Springer International Publishing; 2015. p. 87–116.
[13] European Biofuels Technology Platform (ETIP). Inbicon demonstration plant. 2019. Available from: http://www.etipbioenergy.eu/.
[14] Martin M, Eklund M. Improving the environmental performance of biofuels with industrial symbiosis. Biomass and Bioenergy 2011;35:1747–55.
[15] World Bank. Solid waste management. 2019. Available from: http://www.worldbank.org/en/topic/urbandevelopment/brief/solid-waste-management.
[16] Kawai K, Tasaki T. Revisiting estimates of municipal solid waste generation per capita and their reliability. Journal of Material Cycles and Waste Management 2016;18:1–13.
[17] Sarkodie SA, Strezov V. Economic, social and governance adaptation readiness for mitigation of climate change vulnerability: evidence from 192 countries. The Science of the Total Environment 2019;656:150–64.
[18] Choudhary BK, Tripathi AK, Rai J. Can 'poor' cities breathe: responses to climate change in low-income countries. Urban Climate 2019;27:403–11.

[19] Sonnenschein J, Mundaca L. Decarbonization under green growth strategies? The case of South Korea. Journal of Cleaner Production 2016;123:180–93.

[20] Jung S, An KJ, Dodbiba G, Fujita T. Regional energy-related carbon emission characteristics and potential mitigation in eco-industrial parks in South Korea: logarithmic mean divisia index analysis based on the Kaya identity. Energy 2012;46:231–41.

[21] Sasikala SP, Jeong GH, Yun T, Kim SO. A perspective on R&D status of energy storage systems in South Korea. Energy Storage Materials 2019;23:154–8.

[22] Park J, Kim B. An analysis of South Korea's energy transition policy with regards to offshore wind power development. Renewable and Sustainable Energy Reviews 2019;109:71–84.

[23] Seo Y. Current MSW management and waste-to-energy status in the Republic of Korea [Masters thesis]. Department of Earth and Environmental Engineering, Columbia University; 2014.

[24] Korea Energy Economics Institute (KEEI). Yearbook of regional energy statistics of 2017. Seoul, Republic of Korea: Ministry of Trade, Industry, and Energy, Government of Korea; 2017.

[25] Greenhouse Gas Inventory and Research Center of Korea (GGIRC). National energy statistics comprehensive information system: emissions from industrial processes. 2016 [Seoul, Republic of Korea].

[26] Lee SI. Policy issue paper by Korea Energy Economics Institute. 2015 [Seoul, Republic of Korea].

[27] Ministry of Trade, Industry and Energy (MOTIE). Offshore wind development task force. 2010. Available from: http://motie.go.kr/.

[28] Lee SK, Mogi G. Relative efficiency of energy technologies in the Korean mid-term strategic energy technology development plan. Renewable and Sustainable Energy Reviews 2018;91:472–82.

[29] Park HS, Rene ER, Choi SM, Chiu ASF. Strategies for sustainable development of industrial park in Ulsan, South Korea – from spontaneous evolution to systematic expansion of industrial symbiosis. Journal of Environmental Management 2008;87(1):1–13.

[30] Lowe EA. Creating by-product resource exchanges: strategies for eco-industrial parks. Journal of Cleaner Production 1997;5:57–65.

[31] Côté R, Hall J. Industrial parks as ecosystems. Journal of Cleaner Production 1995;3:41–6.

[32] Doyle B, Martin SA, Weitz KA, et al. Eco-industrial parks: a case study and analysis of economic, environmental, technical, and regulatory issues. Office of Policy, Planning and Evaluation. Washington, DC: USEPA; 1996.

[33] Wiedmann T, Minx J. A definition of 'carbon footprint'. In: Pertsova CC, editor. Ecological economics research trends. New York: Nova Science Publishers; 2007. p. 1–11.

[34] Johnson E. Disagreement over carbon footprints: a comparison of electric and LPG forklifts. Energy Policy 2008;36:1569–73.

[35] Kim HW, Dong L, Choi AES, Fujii M, Fujita T, Park HS. Co-benefit potential of industrial and urban symbiosis using waste heat from industrial park in Ulsan, Korea. Resources, Conservation and Recycling 2018;135:225–34.

[36] Yedla S, Park HS. Eco-industrial networking for sustainable development: review of issues and development strategies. Clean Technologies and Environmental Policy 2017;19(2):391–402.

[37] Ulsan Metropolitan City. Ulsan Metropolitan City statistical database. 2018. Available from: http://www.ulsan.go.kr/rep/index.

[38] Park JY, Park JM, Park HS. Scaling-up of industrial symbiosis in the Korean national eco-industrial park program: examining its evolution over the 10 years between 2005-2014. Journal of Industrial Ecology 2018;23:197–207.

[39] Nguyen TNT, Jung KS, Son JM, Kwon HO, Choi SD. Seasonal variation, phase distribution, and source identification of atmospheric polycyclic aromatic hydrocarbons at a semi-rural site in Ulsan, South Korea. Environmental Pollution 2018;236:529–39.

[40] Bui M, Fajardy M, Mac-Dowell N. Bio-energy with carbon capture and storage (BECCS): opportunities for performance improvement. Fuel 2018;213:164–75.

[41] Singlitico A, Goggins J, Monaghan RFD. The role of life cycle assessment in the sustainable transition to a decarbonised gas network through green gas production. Renewable and Sustainable Energy Reviews 2019;99:16–28.
[42] Kim HW, Ohnishi S, Fujii M, Fujita T, Park HS. Evaluation and allocation of greenhouse gas reductions in industrial symbiosis. Journal of Industrial Ecology 2017;22:275–87.
[43] Behera SK, Kim J, Lee S, Suh S, Park HS. Evolution of 'designed' industrial symbiosis networks in the Ulsan Eco-industrial Park: 'research and development into business' as the enabling framework. Journal of Cleaner Production 2012;29–30:103–12.
[44] Keho Y. What drives energy consumption in developing countries? The experience of selected African countries. Energy Policy 2016;91:233–46.
[45] World Economic Forum (WEF). The future of industrial biorefineries: a white paper. 2010. Available from: https://www.iwbio.de/fileadmin/Publikationen/IWBio-Publikationen/WEF_ Biorefineries_Report_ 2010.pdf.
[46] Saravanan AP, Mathimani T, Deviram G, Rajendran K, Pugazhendhi A. Biofuel policy in India: a review of policy barriers in sustainable marketing of biofuel. Journal of Cleaner Production 2018;193:734–47.
[47] Naqvi M, Yan J. Biorefinery: production of biofuel, heat, and power utilizing biomass. In: Yan J, editor. Handbook of clean energy systems. Chichester: John Wiley and Sons; 2015. p. 169–88.
[48] Rao P, Gururaja S, Matheyam AC, Mahadevamma. Indian Patent No. 443/DEL/2003. 2008. Retrieved from: http://www.allindianpatents.com/patents/225497-a-process-for-extraction-of-bioactive-compounds-polyphenols-polyphenol-oxidase-peroxidase-dietary-fibre-from-mango-peel.
[49] Banerjee J, Singh R, Vijayaraghavan R, MacFarlane D, Patti AF, Arora A. A hydrocolloid based biorefinery approach to the valorisation of mango peel waste. Food Hydrocolloids 2018;77:142–51.
[50] Khan NA, el Dessouky H. Prospect of biodiesel in Pakistan. Renewable and Sustainable Energy Reviews 2009;13:1576–83.
[51] Shah SH, Raja IA, Rizwan M, et al. Potential of microalgal biodiesel production and its sustainability perspectives in Pakistan. Renewable and Sustainable Energy Reviews 2018;81:76–92.
[52] Hossain ZHM, Hossain QH, Monir MM, Ahmed MT. Municipal solid waste (MSW) as a source of renewable energy in Bangladesh: Revisited. Renewable and Sustainable Energy Reviews 2014;39:35–41.
[53] Jahan MS, Sultana N, Rahman M, Quaiyyum A. An integrated biorefinery initiative in producing dissolving pulp from agricultural wastes. Biomass Conversion and Biorefinery 2013;3:179–85.
[54] Jahan MS, Uddin MN, Akhtaruzzaman AFM. An approach for the use of agricultural by-products through a biorefinery in Bangladesh. The Forestry Chronicle 2016;92(4):447–52.
[55] Gheewala SH, Bonnet S, Prueksakorn K, et al. Sustainability assessment of a biorefinery complex in Thailand. Sustainability 2011;3:518–30.
[56] Silalertruksa T, Gheewala SH, Pongpat P. Sustainability assessment of sugarcane biorefinery and molasses ethanol production in Thailand using eco-efficiency indicator. Applied Energy 2015;160:603–9.
[57] Sawatdeenarunat C, Nguyen D, Surendra KC, et al. Anaerobic biorefinery: current status, challenges, and opportunities. Bioresource Technology 2016;215:304–13.

CHAPTER 25

Tannery wastewater treatment and resource recovery options

Hassan Sawalha[1], Maher Al-Jabari[1], Amer Elhamouz[2], Abdelrahim Abusafa[2], Eldon R. Rene[3]

[1]Renewable Energy and Environment Research Unit, Mechanical Engineering Department, Palestine Polytechnic University, Hebron, Palestine; [2]Chemical Engineering Department, An-Najah National University, Nablus, Palestine; [3]Department of Environmental Engineering and Water Technology, IHE Delft Institute for Water Education, Delft, The Netherlands

25.1 Introduction

The production of leather from rawhides, which is called tanning, has been considered as one of the most important industrial processes since ancient times. For centuries, leather was one of the few available materials for the production of high durability garments and footwear. Nowadays, leather is still one of the leading materials for clothing and footwear production due to its unique properties [1].

In tanneries, the wastewater steam is often characterized by high concentrations of pollutants with low biodegradability and it is a major challenge both technologically and environmentally [2–8]. Tanning industry wastes cause deleterious effects on the water, terrestrial, and atmospheric systems due to its high oxygen demand, discoloration, and toxic chemical constituents in liquid, solid and gaseous phases. Many toxic chemicals such as chromium, titanium, caustic soda, ammonium sulfate, sodium sulfide, lime, formic acid, sulfuric acid, enzymes, dyes, sodium formate, sodium bicarbonate, soap, and detergents are used in the tanneries. Researchers have proposed new technologies for the treatment of waste streams and recover valuable resources from these wastes [9–12]. Researchers have also proposed strategies for substituting the more toxic and valuable materials used in the tanning process with the help of environmental friendly materials [13].

Wastewater from the tanning industry contains chromium, a toxic heavy metal, high chemical oxygen demand (COD), chlorides, and sulfides [8]. From a legislative view point, Hassen and Woldeamanuale [7] reported that the wastewater quality from various tanning processes does not comply with the discharge limits of water that can be discharged to the sewer networks. Reusing the treated wastewater in the tanning processes

Waste Biorefinery. https://doi.org/10.1016/B978-0-12-818228-4.00025-3

has also been suggested an option as long as the water does not cause damage to the quality of leather [5]. Electrocoagulation with aluminum electrodes was also recommended for the removal of chromium and COD from the wastewater [6]. In that study, duralumin aluminum alloy was found to be more efficient for COD and chromium removal than pure aluminum electrodes. In another study, Abdulla et al. [4] showed that ~98% of the chromium could be recovered from the waste water through the process of chemical precipitation with lime.

From a life-cycle assessment (LCA) perspective, a recent study has shown that the use of a large amount of chemicals as well as the production and transportation of rawhides for the leather tanning process was shown as the main contributor to the environmental impact in the leather industry [14]. In other studies, environmental impact of the leather tanning industry was shown to increase with the use of end-of-pipe treatment technologies because large amount of chemicals and energy is being continuously used for waste treatment. Besides, the feasibility of chromium recovery was also shown to increase when the concentration of the metals were high in the tannery sludge and when resource recovery based options are applied at the industrial scale [15,16]. In this chapter, some of the most recent techniques, tools and technologies used to remove, recycle or replace tannery waste chemicals, and in particular, chromium and sodium sulfide, are discussed. Moreover, options for composting the wastes and recovery of the enzymes and other value-added products such as protein and fats have been discussed from an application viewpoint.

25.2 Tannery waste characterization

The modern leather industry is based on hides which are a byproduct of the meat industry. In this aspect, tanneries reuse waste materials from other process industries. On the other hand, the tanning processes generate even greater quantities of byproducts and wastes than those of the finished leather. Tannery waste typically contains a complex mixture of both organic and inorganic chemicals. The major public concern over tanneries has traditionally been about the generation of odors and water pollution from untreated discharges. The most important pollutants associated with the tanning industry include chlorides, tannins, chromium, sulfate, and sulfides and other trace organic chemicals [2,17–19]. In terms of quantity, an average of 30–35 m^3 of wastewater is produced per ton of rawhide processed. However, wastewater production varies in a rather wide range (10–100 m^3 per ton hide) depending on the raw material, the finishing product, and the production processes [2].

Tables 25.1 and 25.2 show the amount of released waste and the main wastewater characteristics from various leather making processes for two tanneries located in the cities of Hebron and Nablus in Palestine. The main pollution characteristics of wastewater released from the two local tanneries including chemical oxygen demand (COD), total solids (TS), pH, concentrations of chloride, and total chromium were determined by

Table 25.1: The leather manufacturing chemical processes showing the operational time, wastes generated and main wastewater characteristics for a tanner in Hebron, Palestine [20].[a]

Processing time	Wastes generated	Wastewater characteristics	
Soaking	≈1.5 m^3 WW/ton	COD (×10^3)	29
(24–48 h)	(Salt, dirt, fats, soap)	TS (×10^3)	125
		pH	6.33
		Chloride (×10^3)	200
		Total chromium	0
Hair removal and liming	≈1.2 m^3 WW/ton	COD (×10^3)	167
(48 h for goat skin and 18 h	(Sulfide, very toxic [7])	TS (×10^3)	140
for cow hides)	Lime and hair	pH	12.41
		Chloride (×10^3)	42.5
		Total chromium	0
Deliming	1 m^3 WW/ton from each stage	COD (×10^3)	10.4
(3.17 h)		TS (×10^3)	37.4
Stage wise:		pH	9.8
40, 90, and then 60 min		Chloride (×10^3)	10.5
		Total chromium	0
Pickling	Zero waste	COD (×10^3)	8.98
(2.7 h mixing)		TS(×10^3)	105
Then		pH	4.65
Hides are left in the drum		Chloride (×10^3)	35
overnight: pH 2.5–2.8		Total chromium	0
Tanning	0.8 m^3 WW/ton	COD (×10^3)	7.39
(8 h mixing)	(Chromium)	TS (×10^3)	77.6
Then		pH	3.65
Hides are left in the drum for		Chloride (×10^3)	27.5
24 h		Total chromium	3506

[a]All values, except for pH, are expressed in mg/L.

collecting representative wastewater samples from the tanneries. Characterization of such processes effluents assists in identifying the waste generation rates and discharges and for suggesting cleaner production options. Interestingly, the amount of wastewater produced in

Table 25.2: Wastewater characteristics for a tannery in Hebron, Palestine.

Process/parameter	Soaking	Liming	Deliming	Pickling	Tanning	Combined in the pool
pH	6.73	12.37	10.89	<2.00	4.6	12
COD (mg/L)	10,870	32,425	3,800	3,130		12,600
BOD_5 (mg/L)	3,560	1,510	750	480		4,050
Cl^- (mg/L)	17,750	13,500	1,500	0		13,250
SO_4^{2-} (mg/L)	545	3,100	1,240	1,396		436
SS (mg/L)	2,885	5,093	572	522		4,955
Ammonia			60	80		
Total chromium (mg/L)					3,600	280

the local Palestinian tanneries is much lower than those produced worldwide, however, it is more concentrated with a wide variety of toxic pollutants (Table 25.3). The liming process has the highest COD and the highest pH value, while the tanning process releases wastewater highly concentrated with chromium.

According to the FAO (Food and Agriculture Organization) statistics, the worldwide annual production of bovine hides and skins amounts to ~6 million ton (wet salted weight). Sheepskins, lambskins, goatskins and kidskins accounts for ~600,000 ton (dry weight). The world leather production generates about 3.0–3.5 million tons of solid wastes. One ton of wet salted hides yields only ~200 kg of leather. The rest, ~800 kg becomes a waste, including tanned solid waste (~250 kg), nontanned waste (~350 kg) and waste lost together with wastewater (~200 kg) [22]. The amount of water required for the processing of 1000 kg of hides amounts to ~45–50 m^3. Typically, for processing ~1000 kg of hides, ~400 kg of chemicals is required, including sodium chloride, lime, sodium sulfide, sulfuric acid, basic chromium sulfate, and others [23].

These complex characteristics of wastewater requires appropriate treatment in order to meet the legislative environmental discharge standards. The chemical composition of untreated hide or skin waste (fleshing, trimming, splits) depends mainly on the type and quality of the raw material used, the processing steps and the process operating conditions. The main components are proteins and fat contributing to ~10.5% (*w/w*) for both groups, while the water content is ~60%. These wastes also contain small amounts of mineral substances, 2%–6% (*w/w*) [1]. The tanned leather wastes are mainly useless splits, shavings and trimmings. These waste groups differ mostly in size and shape, while the chemical composition is usually comparable. They contain 3%–6% (*w/w*) of fat and ~15% (*w/w*) of mineral components, including 3.5%–4.5% (*w/w*) of chromium in the form of Cr_2O_3 [1].

Moreover, tanneries emit odors and other volatile organic compounds during the tanning processes [24], as well as from the biological (natural) decomposition processes that occurs during the storage of rawhides, wastewater, among others. Pollutants as ammonia, hydrogen sulfide, volatile hydrocarbons, amines, and aldehydes are emitted from tanneries. Landfilling of tannery wastes have also shown to pose serious threats to the environment [25]. The nontanned waste undergoes biological degradation, which may be the main source of pathogenic bacteria and volatile organic compounds emission [1]. For example, the release of chlorinated phenols and chromium were found to be closely associated with tannery wastes.

Chromium, as an inorganic pollutant, is a transition metal and it exists in several oxidation states. The trivalent (Cr^{3+}) and hexavalent (Cr^{6+}) species are the most common forms. Other pollutants of concern within the tanning industry include azo dyes, cadmium

Table 25.3: Average composition of tannery effluent bath.

Parameters	Soaking		Unhairing liming		Bating deliming		Pickling	Chrome tanning	Re-tanning	
	Min	Max	Min	Max	Min	Max	Min	Max	Min	Max
pH	6	10	12.5	13	6	11	4	3.2	4	10
	7.7		11.9		8.6		3.6		5	
T [°C]	10	30	10	25	20	35	—	—	20	60
BOD_5 [mg/L]	2,000	5,000	5,000	20,000	1,000	4,000	100	250	6,000	15,000
COD [mg/L]	3,000	6,000	—	—	—	—	1,000	300	—	—
	5,000	11,800	20,000	40,000	2,500	7,000	800	400	15,000	75,000
	31,000		58,000		5,325		2,900		4,365	
TSS [mg/L]	25,000	40,000	—	—	—	—	30,000	70,000	—	—
	2,300	6,700	6,700	25,000	2,500	10,000	—	—	—	—
TDS [mg/L]	22,000	33,000	—	—	—	—	29,000	67,000	—	—
Cl^- [mg/L]	15,000	30,000	—	—	—	—	20,000	30,000	—	—
	17,000	50,000	3,300	25,000	2,500	15,000	8,950	2,000	5,000	10,000
Sulfides [mg/L]	0	700	2,000	3,300	25	250	—	—	—	—
	—		2,650		134		—		—	
Cr(III) [mg/L]	—	—	—	—	—	—	—	4,100	0	3,000
NH_3—N	850		380		3,800		670			530

Adopted from Lofrano G, Meric S, Belgiorno V. Tannery wastewater treatment by advanced oxidation processes. In: Water, wastewater and soil treatment by advanced oxidation processes (AOPs); 2011. p. 197 [chapter 13].

compounds, cobalt, copper, antimony, barium, lead, selenium, mercury, zinc, arsenic, polychlorinated biphenyls (PCB), nickel, formaldehyde-based resins, and several pesticide residues [26].

25.3 Tanning process

The tanning process consists of a series of several mechanical and chemical steps/stages as shown in Fig. 25.1. They can be further divided into three fundamental subprocesses, namely the preparatory stage, tanning and crusting. The first subprocess includes preservation, soaking, liming, unhairing, fleshing, splitting, bating, degreasing, and pickling. After that, the pretreated raw material is tanned [27]. During the tanning process, the protein of the rawhide or skin is converted into a stable nonputrefactive material. Chromium (III) compounds are commonly used as the tanning agent (in ~90% of the tanneries worldwide). Crusting consist of stages such as thinning, retanning, lubrication, and coloring steps. Thereafter, the tanned leather is subjected to a finishing step, which gives it the required pattern, gloss or waterproof qualities. The finishing operations includes oiling, brushing, impregnation, polishing, embossing, glazing, and tumbling [1].

25.4 Tannery waste treatment options

The treatment of tannery waste is rather difficult compared to other industrial waste and there is no standalone method for its comprehensive utilization. Usually, the nontanned waste is used as a raw material for the production of glue, gelatin, protein sheaths, and fertilizers. This waste is also suitable as a substrate for biogas production in anaerobic digesters. Recently, numerous treatment technologies have been proposed by researchers to treat the toxic effluents generated from tanneries. For example, raw vegetable tannery wastewater was treated following the coagulation–flocculation method with poly aluminum chloride (PACL) as a standalone process and in combination with a coagulant [28]. In that report, the removal efficiency of COD, TSS, and color was low when PACL was used as a standalone strategy, whereas the COD, TSS, and color removal significantly improved when PACL was used in combination with a flocculent [28]. In another study, bittern was used as coagulant, ozone for oxidation and activated carbon for adsorption. The combined process resulted in the reduction of suspended solids (SS), color, turbidity, chromium, and nitrogen from wastewater. Apart from these methods, the conventional activated sludge process (ASP), both as a standalone treatment step or in combination with several physicochemical technologies, has also been tested for tannery wastewater treatment and very high removals of BOD and COD has been reported [29].

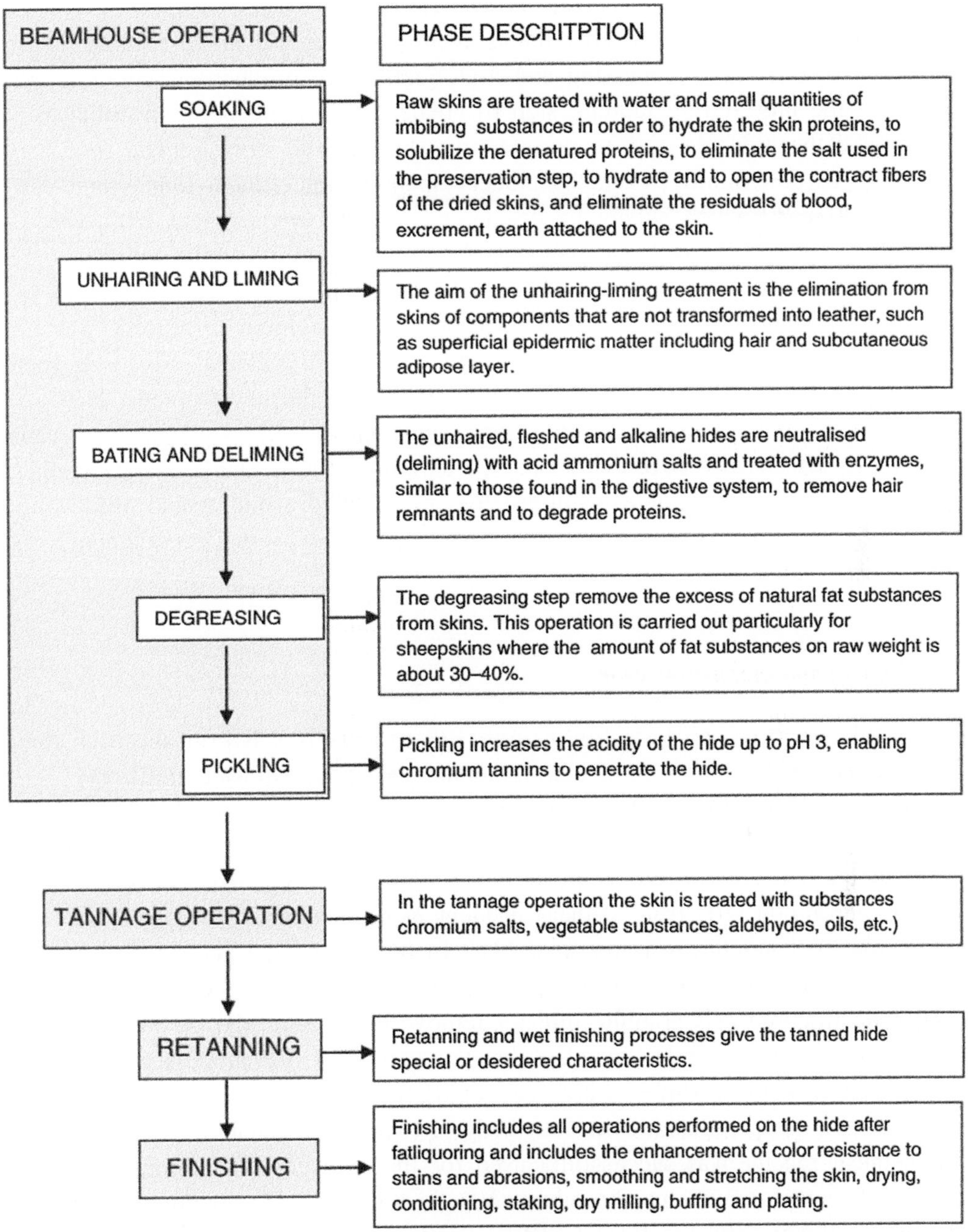

Figure 25.1

Steps involved in the leather tanning process. *Adopted from Lofrano G, et al. Chemical and biological treatment technologies for leather tannery chemicals and wastewaters: a review. The Science of the Total Environment 2013;461:265–81.*

25.5 Chromium removal and recovery

Chromium (III) is the most utilized tanning agent and it is crucial to tanning industries because of easy processing and high-quality end products. Chromium salts are added to the hide during tanning step to create crosslinking between collagen fibers to make the hide resistant to putrefaction, durable and flexible. The concentration of chromium (III) in the tannery spent liquor effluent ranges between 2000 and 5000 mg/L which cannot be discharged into the environment [30]. The tanned waste contains up to 4.5% (*w/w*) of chromium, mostly as relatively nontoxic to living organisms. It can, however, undergo oxidation to chromium (VI) which is known for its carcinogenic properties. The disposal of chromium in tannery waste is significant, about 30,000 ton per year, worldwide. The environmental effects of such proceedings go far beyond the potential contamination of the environment. The chronic exposure to chromium via inhalation, ingestion, or dermal contact may cause adverse health effects. Moreover, the reuse of chromium from tannery waste could help save energy and financial expenditure for chromium ore output and processing. Therefore, the residual chromium (III) should be recovered by the environment-friendly and sustainable methods [30].

25.5.1 Membrane electroflotation

Electrocoagulation (EC) with electroflotation (EF) process is a hybrid approach that has been tested for the removal of heavy metals from polluted water [30]. EC process coagulates the dissolved heavy metals using soluble iron or aluminum as an anode and the coagulated heavy metals are floated to the surface by the hydrogen/oxygen bubbles that are evolved from the electrodes during the EF process. The removal of chromium (III) from tannery effluent using different anodes such as titanium, iron, and graphite was reported [31]. In that study, the authors have reported that soluble iron anode resulted in better efficiency (95% removal) by EC, and insoluble titanium and graphite did not remove chromium (III) by the standalone EF process. It is noteworthy to mention that the hybrid EC-EF process generates iron/aluminum containing sludge which cannot be reused, thereby requiring further treatment of the sludge. Besides, in that study, a two-compartment membrane electrochemical reactor was successfully used for the recovery of chromium (III), without oxidizing chromium to chromium (VI), in high chloride containing tannery spent liquor effluent collected from CSIR-CLRI (Central Leather Research Institute). Fig. 25.2 shows the removal rate of chromium (III) and the changes in the pH of the spent liquor effluent during the EF process in a two-compartment electrochemical membrane cell.

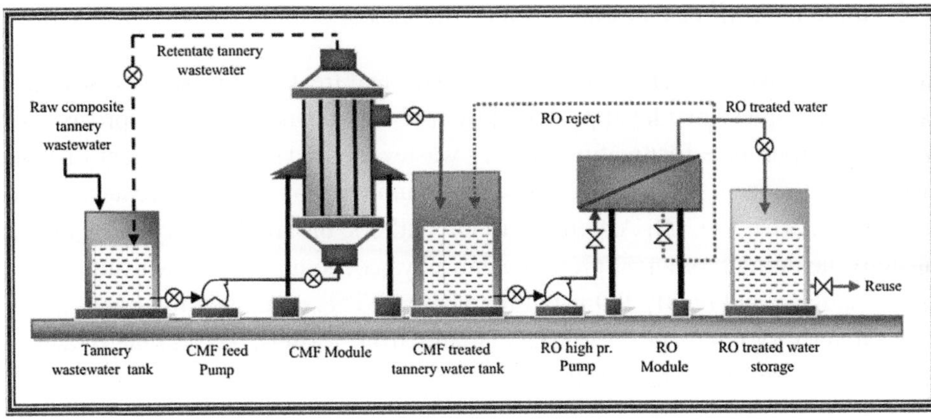

Figure 25.2

Schematic of the combined membrane filtration (CMF) and reverse osmosis (RO) process. *Adopted from Bhattacharya P, Ghosh S, Mukhopadhyay A. Combination technology of ceramic microfiltration and biosorbent for treatment and reuse of tannery effluent from different streams: response of defence system in Euphorbia sp. International Journal of Recycling of Organic Waste in Agriculture 2013;2(1):19.*

In the single compartment, oxidation of chromium (III) to chromium (VI) was observed [31]. The anodic evolution of chlorine in the anode leads to the accumulation of active chlorine species (HOCl and OCl-) in the effluent. These active chlorine species leads to the oxidation of chromium (III) to chromium (VI), which is highly toxic. Furthermore, the authors also prepared chrome tanning agent from the recovered chromium hydroxide in order to compare it with the commercially available basic chrome sulfate (BCS). The chromium content and shrinkage temperature of the wet blue leathers were ascertained in order to test the tanning efficiency. The chrome tanned leathers were converted into crust leathers. Table 25.4 shows the measured organoleptic characteristics of crust leathers. In summary, $Cr(OH)_3$ recovered from the spent liquor effluent by an EF process can be directly used without any pretreatment.

Table 25.4: Organoleptic parameters of crust leathers made from the recovered chromium (scale 0–10).

	Control	Experiment
Smoothness	9	9
Tightness	8	8.5
Fullness	8.5	8
Overall	9	9

Adopted from Selvaraj R, et al. A membrane electroflotation process for recovery of recyclable chromium (III) from tannery spent liquor effluent. Journal of Hazardous Materials 2018;346:133–9.

25.5.2 Ceramic microfiltration and reverse osmosis

To overcome the drawbacks associated with conventional treatment processes, several new and cleaner technologies have been proposed by researchers and membrane processes have recently gained importance for treating tannery wastewater [31]. The treatment of tannery wastewater collected from beam house units with nanofiltration (NF) and reverse osmosis (RO) has been reported [29,32]. According to the authors, the treatment consists of different steps including biological pretreatment, physicochemical process using a polymeric coagulant, and finally RO using membranes. This combined treatment method resulted in satisfactory removal of the contaminants from wastewater, and the treated water was found to be suitable for reuse.

Nevertheless, the commonly used polymeric membranes have drawbacks such as lower thermal and acid-alkali resistance. Ceramic membrane can operate at a wide range of temperature and are stable even under harsh operating environments of acid and alkali. They have narrow and well-defined pore size distribution and excellent chemical resistance against strong cleaning agents which ensures the stability of membrane performance for a longer period of time. Ceramic membranes have been effectively used for the treatment of industrial and domestic wastewater. Fig. 25.2 illustrates the schematic of a combined MF and RO process. The sludge produced from the MF is subjected to RO as the second step of treatment. Pretreatment is essential for RO process since fouling occurs in the RO membrane due to hard scales and soft amorphous complexes present in wastewater [33].

According to Bhattacharya et al. [29], ~91% reduction in COD and BOD, 62% reduction in TOC, and complete removal of sulfide can be obtained in the direct microfiltration process, while the turbidity can be reduced to below 1 NTU (Nephelometric Turbidity Units). The RO process results in ~99% reduction of TOC and 82% reduction of sodium, while the turbidity value was ~0.025 NTU. Compared to the several stages in the conventional effluent treatment process, the proposed two-stage process is able to successfully reduce the COD, BOD, and heavy metal content in wastewater. The authors also compared the physical properties between leathers produced by fresh water and the treated water [30]. Table 25.5 shows the characteristics of the leather produced with fresh and treated water. Interestingly, the leather produced using treated water was of better quality compared to that of the control (i.e., fresh water) in terms of the physical properties. Dye uptake by leather was more in case of leather samples prepared with the treated effluent. The tensile strength of this leather was ~19% more than that of leather tanned using fresh water. The elongation was ~38% compared to the control (31.8%). Full crack in conventional leather occurred at 845 kg/cm (fresh water), while the crack occurred at 918 kg/cm in the case of the leather tanned using treated water.

Table 25.5: Physical properties of the tanned leather using fresh and treated water.

Leather	Tensile strength (kg/cm^2)	Stitch tear strength (kg/cm)	Grain crackness (kg/cm)		Percent elongation (%)	Extension at grain crack (%)	Dye uptake (λ_{360})
			Just crack	Full crack			
Conventional method	558	135	279	845	31.8	12.6	867
RO permeate	691	217	384	918	38	13.9	2.839

Adopted from Bhattacharya P, Ghosh S, Mukhopadhyay A. Combination technology of ceramic microfiltration and biosorbent for treatment and reuse of tannery effluent from different streams: response of defence system in Euphorbia sp. International Journal of Recycling of Organic Waste in Agriculture 2013;2(1):19.

Accordingly, the proposed process reduces the cost of treatment compared to that of the conventional treatment. The process may be up-scaled further for proper implementation in industries. This technology would be of great help to reduce the freshwater consumption that is used in large quantities in tanneries, i.e., about 100,000 ton, and also wasted in large quantities (~80,000 tons).

25.5.3 Biological treatment

The main advantages of biological treatment methods are: (1) low capital and operating costs compared to conventional chemical oxidation processes; (2) complete mineralization of organics, versus mere phase separation, such as with air stripping or activated carbon based adsorption; (3) oxidation of a wide variety of organic compounds; (4) removal of reduced inorganic compounds such as sulfides and ammonia; (5) total nitrogen removal through denitrification; (6) operational flexibility to handle a wide range of flow rates and wastewater characteristics; and (7) reduction of aquatic toxicity [17].

25.5.3.1 Coagulation and flocculation

Coagulation–flocculation (CF) of tannery wastewater has been reported in the literature using inorganic coagulants such as aluminum sulfate, ferric chloride, and ferrous sulfate to reduce the organic load (COD) and suspended solids (SS) as well as to remove the toxic heavy metals, e.g., chromium, before biological treatment [2]. According to the authors, total chromium was also effectively removed by alum, while it was completely removed using ferric chloride. Moreover, 40%–70% removal of COD and >99% of total chromium from leather tanning wastewater using $FeSO_4$, $FeCl_3$, and alum was achieved [2]. In summary, 30%–37% of total COD removal, 74%–99% of chromium removal and

38%–46% of suspended solids removal was achieved using 800 mg/L of alum, at pH 7.5 for presettled tannery wastewater containing 260 mg/L of SS, 16.8 mg/L of chromium, and 3300 mg/L of COD, at pH 9.2.

25.5.3.2 Bioleaching

Bioleaching has been developed as a successful and cost-effective way to remove Cr (III) from tannery sludge [34,35]. According to Ma et al. [36], Cr (III) can be solubilized by tannery sludge acidification through both direct and indirect mechanisms driven by *Acidithiobacillus* species, mainly *Acidithiobacillus ferrooxidans* and *Acidithiobacillus thiooxidans*. The authors observed that bioleaching of Cr (III) from tannery sludge using the mixture of indigenous iron- and sulfur-oxidizing bacteria could result in 100% oxidation of Fe^{2+} to Fe^{3+} in the bioleachate, while the dissolved chromium concentration reached its maximum removal of 95.6%. Tannery sludge usually contains high concentrations of chromium, iron, some suspended solids, and soluble organic matter. The chromium recovered from the bioleaching process could be reused in the tanning process, but the bioleachate containing large amount of iron would affect chromium absorption by wet blue to a certain extent. At present, in the tanneries, obtaining chromium from the bioleachate through alkali precipitation makes the chromium mud unrecyclable because the addition of alkali may precipitate Fe (III) and Cr (III) at the same time. Therefore, it is worthwhile to reuse Fe (III) and Cr (III) in the tanning process rather than to separate them from the bioleachate.

Ma et al. [36] suggested that Cr (III) and Fe (III) can be separated by adjusting the pH of the bioleachate as the Cr (III) precipitates at a pH of 4.60, and complete precipitation can be achieved at a pH of 5.6. On the other hand, Fe (III) begins to precipitate at a pH of 1.81 and it can be completely precipitated at a pH of 2.81 (Fig. 25.3). Therefore, Cr (III) and Fe (III), theoretically, can be separated by adding alkaline to the solution. However, Cr (III) and Fe (III) cannot be effectively separated by directly adding hydroxide to regulate the pH.

Ma et al. [36] compared chromium tanned leather with chromium-iron tanned leather in terms of its physical, mechanical and sanitation properties. According to the information shown in Table 25.6, the chromium-iron tanned leather offered better properties, except for the slightly lower water vapor permeability. The better performance in tensile strength, tear strength and break load revealed that the presence of iron content in the prepared chromium-iron tanning agent could enhance toughening of leather collagen.

25.6 Sodium sulfide recovery and removal

The most versatile and commonly used depilation agent in the leather industry is sodium sulfide. During the depilation process, sodium sulfide can either be used alone or in

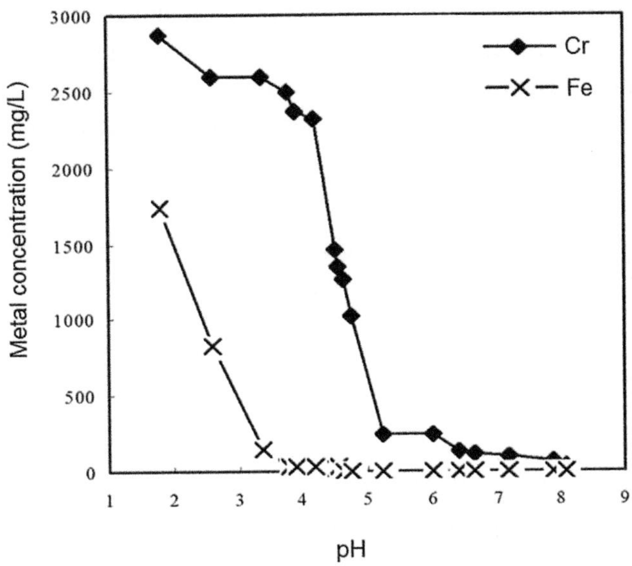

Figure 25.3

Variation of the concentration of Cr (III) and Fe (III) at different pH values. *Adopted from Ma H, et al. Chromium recovery from tannery sludge by bioleaching and its reuse in tanning process. Journal of Cleaner Production 2017;142:2752–60.*

combination with $Ca(OH)_2$. This process using sodium sulfide and $Ca(OH)_2$ is typically responsible for 84% of the biochemical oxygen demand (BOD), 75% of the chemical oxygen demand (COD) and 92% of the suspended solids of pretanning effluent. Moreover,

Table 25.6: Physical and mechanical properties of crust leathers.

Parameters		Chromium	Chromium-iron
Tensile strength	N/mm^2	15.4	21.4
Elongation under specific load	%	17	14
Elongation at break	%	40	41
Tear strength	N/mm	67.9	84.6
Break load	kg	10	21
Break height	mm	9.8	12
Water vapor permeability	mL/ (10 cm^2 24 h)	271.2	230.5
Air permeability	mL/ (10 cm^2 24 h)	7.26	50.57
Substance increase	%	25.4	42.3

Adopted from Ma H, et al. Chromium recovery from tannery sludge by bioleaching and its reuse in tanning process. Journal of Cleaner Production 2017;142:2752–60.

the use of lime during the unhairing process requires its removal, and this is usually done by the addition of ammonium salts. These salts contribute to high amounts of nitrogen in the wastewater. From a cleaner production view point, a replacement of this process using a less polluting process is required in order to sustainably operate the leather manufacturing process [37].

25.6.1 Enzymatic unhairing

Among the different new technologies, the use of proteolytic enzymes has been experimentally tested for the unhairing process [38]. Despite being consolidated for other industrial applications, the use of enzymes in the leather industry is not usually common. In the tanneries operating in south Brazil, only 22% use a safer unhairing process, substituting sulfides for other chemicals such as amines, while 78% use sulfides and none of the companies use enzymes [39]. The enzymatic unhairing is a very special case of the application of enzymes in the beam house, wherein proteolytic enzymes attack the hair roots and the epidermis. In this process, the amount of chemicals added to significantly less compared to the conventional unhairing process [40].

Dettmer et al. [39] obtained the enzymatic preparations from cultures of *Bacillus subtilis*, a bacterium isolated from local tannery sludge. The authors performed kinetic studies to determine the time required for enzymatic unhairing. The results showed that, after 6 h it is possible to obtain hides successfully unhaired, without causing damages on the grain and with satisfactory removal of interfibrillary proteins. The processing time for the entire unhairing procedure using enzymes varies from 24 h for cattle hides, 18 h for goat skins, and up to 12 h for pig skins. In this enzymatic unhairing procedure, there is no requirement of mechanical force, or the assistance of knifes or blades. Fig. 25.4 shows the photographs of the skin after enzymatic unhairing [39]. The authors also measured the chromium content in the leather, along with the tensile strength and elongation, and the tear strength of all the leather samples. The results of their measurements are shown in Table 25.7.

25.6.2 Aqueous ionic liquid solution

Several alternative reagents have been studied as a means to carry out the reductive cleavage of sulfide linkages that is required. Enzymes are also sensitive to storage conditions and the temperature at which they are employed and such variations can yield inconsistent results. These limitations have prompted research into oxidative processes using hydrogen peroxide. Even though the depilation obtained by this method has been satisfactory, the damages caused to the finished product is unacceptable. There have also been reports on using alkaline calcium peroxide (at a pH of 13.5 and temperature of 45°C) to loosen the hair (Table 25.8) [42]. However, the process suffers from the generation of

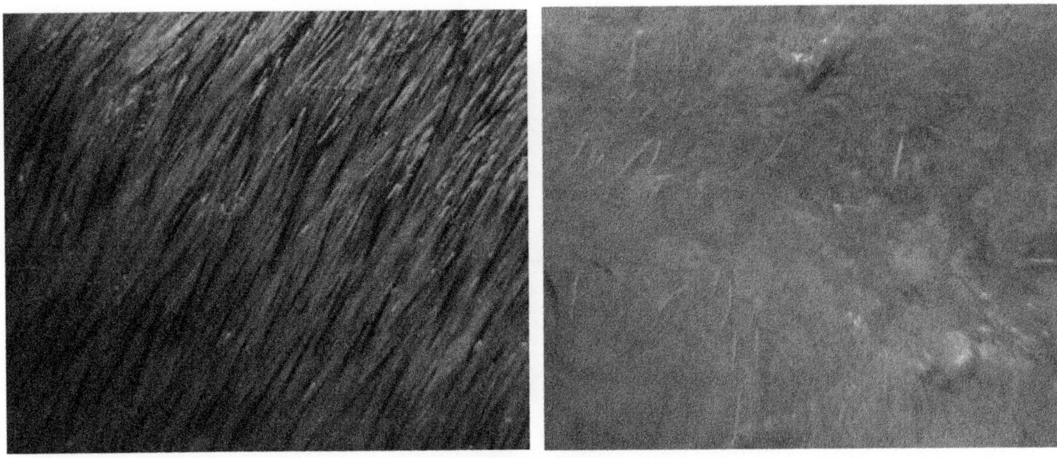

Figure 25.4
Optical microscopy images of enzymatic unhairing: left—control and right—after 6 h of enzymatic treatment. *Adopted from Dettmer A, et al. Environmentally friendly hide unhairing: enzymatic hide processing for the replacement of sodium sulfide and delimig. Journal of Cleaner Production 2013;47:11–8.*

Table 25.7: Comparison of physical and mechanical proprieties, chromium content in leather, and shrinkage temperature of chromium tanned leather.

	Direction	Conventional process	Enzymatic process
Tensile strength (MPa)	Along	78.53 ± 9.95	70.10 ± 1.02
	Across	75.25 ± 4.49	67.52 ± 8.94
Elongation at break (%)	Along	42.11 ± 22.32	73.68 ± 29.77
	Across	39.47 ± 20.09	76.32 ± 26.05
Tear strength (N/mm)	Along	35.83 ± 2.76	39.72 ± 1.96
	Across	35.25 ± 4.03	33.75 ± 2.53
% Chrome		3.03 ± 0.01	3.02 ± 0.04
Shrinkage temperature (°C)		96 ± 1.41	96 ± 0.71

Adopted from Dettmer A, et al. Environmentally friendly hide unhairing: enzymatic hide processing for the replacement of sodium sulfide and delimig. Journal of Cleaner Production 2013;47:11–8.

toxic products. Recently, the use of ozone as an oxidizing agent has been studied. There is, however, the need for specialized equipment's for the production of ozone which adds to the investment and operational cost of the process.

In another work [41], a wider range of "reducing" ionic liquids of aprotic cations, including salts of the thioglycolate anion and the dihydrogen phosphite anion was tested. The aim of that work was to reduce the S–S linkages in the depilation process and thereby use them as the active agent in aqueous solution for the removal of keratinous

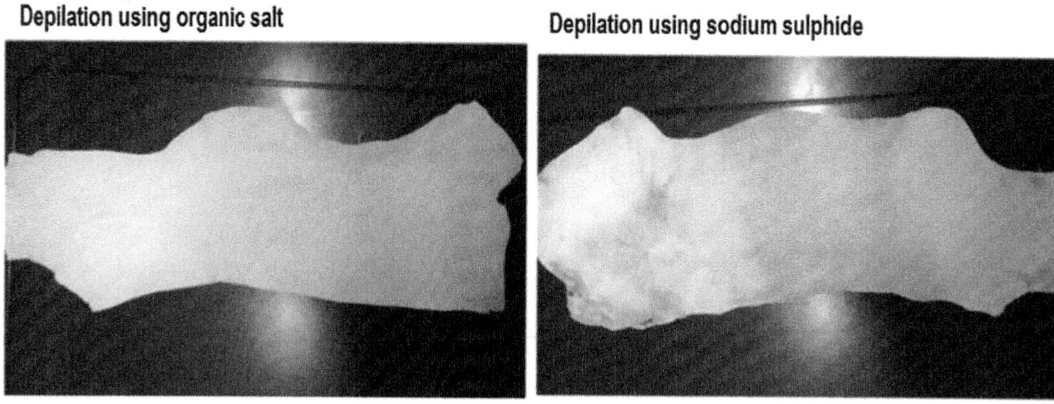

Figure 25.5
Comparison of depilation processes using choline thioglycolate and sodium sulfide. *Adopted from Vijayaraghavan R, et al. Aqueous ionic liquid solutions as alternatives for sulphide-free leather processing. Green Chemistry 2015;17(2):1001–7.*

materials. The depilation process, as shown in Fig. 25.5, is the same as the conventional process, except that the active agent is an ionic liquid solution instead of sodium sulfide. It can be seen from Table 25.8 that the solutions based on the thioglycolate anion outperforms the dihydrogen phosphite, as well as the sodium sulfide and sodium thioglycolate, in term of its depilation ability.

Table 25.8: Depilation using different ionic liquid solutions and $Ca(OH)_2$.

Sample	Depilation ability (assessed by leather experts)
$Ca(OH)_2$ + [TBA][Phos]	1
$Ca(OH)_2$ + [Cho][TG]	5
$Ca(OH)_2$ + [TBA][TG]	5
$Ca(OH)_2$ + [Chol][Phos]	2
$Ca(OH)_2$ + [TIBA][Phos]	1
$Ca(OH)_2$ + sulfide	4
$Ca(OH)_2$ + sodium thioglycolate	3
$Ca(OH)_2$ (without sulfide)	1

Note: [a] Depilation ability: 1 = very poor; 2 = poor; 3 = fair; 4 = good; 5 = excellent. [b] Concentration of ionic liquids used in the experiments: [TBA][TG], [TBA][Phos], [TIBA][Phos] and [TIBA][TG] = 0.1 mol/kg (hide).
Adopted from Vijayaraghavan R, et al. Aqueous ionic liquid solutions as alternatives for sulphide-free leather processing. Green Chemistry 2015;17(2):1001–7.

25.7 Composting of wastes

Hair waste is one of the most important solid waste generated from tanneries. Keratin is a fibrous protein structural protein which offers high resistance to degradation. Generally, unhairing has been done by dissolving the hair waste with chemicals, i.e., Na_2S, which becomes a wastewater pollutant and thus it requires further treatment [42]. Dumping or landfilling have been conventionally used for the disposal of tannery solid waste; however, it causes severe environmental problems, i.e., due to the slow degradation process, and it requires a large area to prevent the leaching of toxic compounds [43]. In addition to the solid hair waste, tannery sludge contains high chromium and other chemicals, e.g., salts and carbonate, which can cause pollution problems when the sludge is applied to the soil. Tannery sludge is also problematic if the disposal does not have a good management practice, particularly leachate and toxic gases from landfills. The applications of composting of tannery sludge shown to improve the soil quality [44]. Composting is one of the alternative treatment options due to its inexpensive operating cost and production of valuable organic fertilizers. Besides, the composting of hair contains high organic matter and nitrogen content which is preferable to the soil [45]. Composting provides the required conditions for tannery solid waste degradation including thermophilic temperature (40–50°C), proper moisture content (55%–65%) and adequate carbon to nitrogen ration (C/N = 20:1–35:1) [46].

In order to have a good quality compost for soil application, the tannery waste is usually cocomposted with other organic wastes, i.e., food waste [47], raw sludge from municipal waste and deink sludge [45]. As the solid waste from tannery contains high polymeric material such as protein, lipid, and carbohydrate, the hydrolysis process of tannery solid waste is occasionally provided before the composting process. During hydrolysis, some mixed bacterial culture is added to accelerate the reaction, leading to an effective composting process [48].

Composting is a sustainable and effective technology for solid residues. Composting tannery sludge for soil application offers the following advantages [49]:

- To improve the soil moisture content
- To increase the soil resistance to pets and diseases
- To prevent soil erosion

25.7.1 Case studies

25.7.1.1 Aerated composting in MAHK & Sons, Ranipet, India

This tannery operates using wet salting for buffalo hides and chrome tanning to finish the leather product [50]. Composting is being carried under a shelter (Fig. 25.6). The area for

Figure 25.6
Composting under the shelter in MAKH & Sons, Ranipet, India. *Adopted from Sampathkumar S, Buljan J. Composting of tannery sludge. United Nations Industrial Development Organization; 2001.*

composting is 1200 m^2 under the shelter, therefore sieving the harvested compost and storage can be done in the same area. The composting composition consists of fleshing waste, paddy straw, green biomass, and cow dung.

25.7.1.2 Aerated composting in Shafeeq Shameel & Co. (SSC), Ambur, India

This tannery processes raw skins and chrome tanning of cow and goat skins [50]. The factory also has a chrome liquor recovery unit. The composting is carried in the open area. The composting piles are set up on the unused evaporation pans and it is made of granite and cudappah stones which maintains a proper temperature and moisture content to the composting piles. The composting composition consisted of fleshing waste, coir pith, paddy straw, green biomass and cow dung.

25.7.1.3 Pilot scale composter

This case study uses the cocomposting of hair waste generated during the unhairing process with wastewater treatment sludge [45]. The composting is carried out in a composter (Fig. 25.7). The attractive point of this composting system is the use of static respiration indices (SRI) to determine the stability of the compost. Besides, temperature probes are also fitted inside the composter to monitor the temperature fluctuations during the day and night (Fig. 25.7).

25.7.1.4 Recovery of enzymes and other value-added products

Tannery produces large amount of fleshing waste containing mainly lipids and proteins which has potential for various industrial applications [51]. Table 25.9 shows the value-added products obtained from tannery wastes that can be used in other industries.

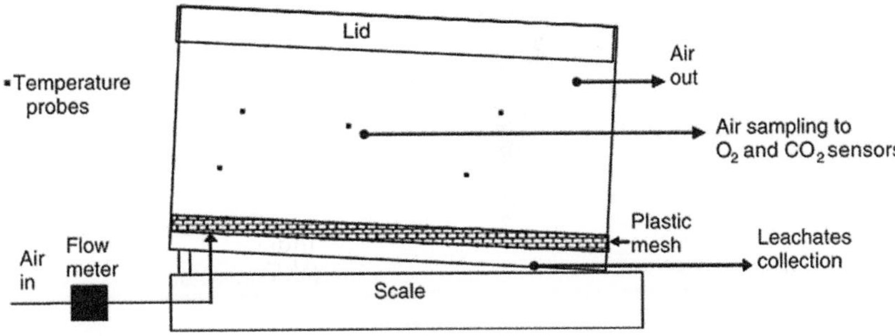

Figure 25.7

Schematic of a pilot scale composter for cocomposting of tannery solid waste and municipal sludge. *Adopted from Barrena R, et al. Co-composting of hair waste from the tanning industry with de-inking and municipal wastewater sludges. Biodegradation 2007;18(3):257–68.*

Table 25.9: The applications of value-added products obtained from tannery wastes.

Products	Applications
Collagen	Drug carrier
Collagen gels	Clinical application
Collagen fibers	Paper making
Fat	Animal feed, fertilizer
Keratin, collagen hydrolysates	Glue
Polypeptides	Cosmetic and chemical industry
Protein	Additive for concrete and ceramic
Shredded trimmings	Biodiesel, biomethane
Keratin hydrolysate	Cosmetics, animal feed, fertilizers

Adopted from Sundar VJ, et al. Recovery and utilization of proteinous wastes of leather making: a review. Reviews in Environmental Science and Bio/Technology 2011;10(2):151–63.

25.7.2 Recovery of fat

Fats/lipids can be recovered during after lime fleshing. Recovery of fat has been achieved using enzymatic process which could yield ~92% of soluble fat [53]. The recovered fat can be used for firing the boiler used for generating the steam [54,55]. Fat recovery in tannery has many advantages, i.e., it decreases the disposal costs and is attractive for the fuel markets.

Recently, several techniques are used in the industries for fat recovery. A company in Germany, Flottweg, has introduced the technology of centrifuge for limed fleshing coupled to a fat recovery unit (Fig. 25.8). Flottweg carried out animal fat recovery using a Tricanter processing limed fleshing. Subsequently, the recovered fat was used as a fuel for

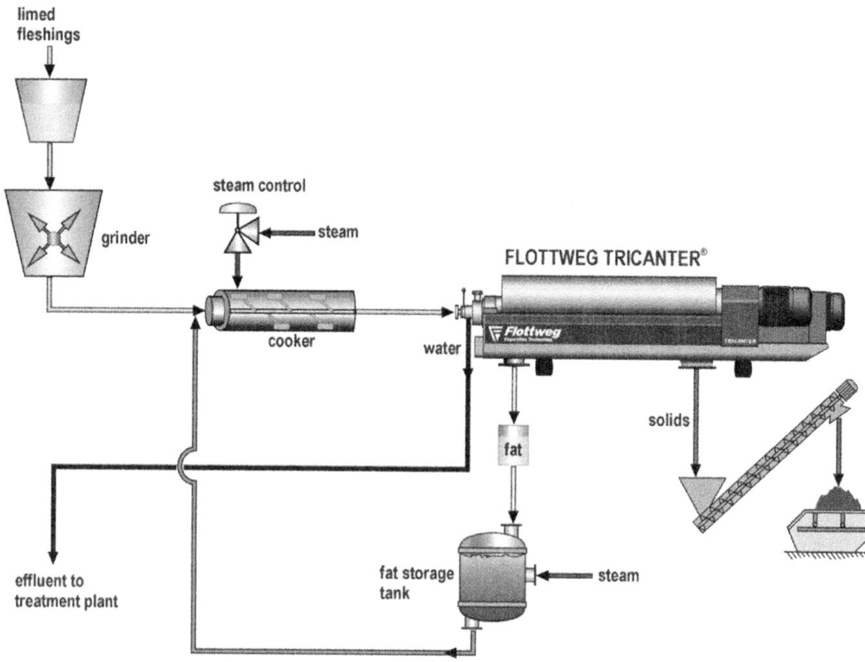

Figure 25.8
The centrifuge used for processing limed fleshing coupled to the recovery of fats, located at Flottweg, Germany.

a cooker located in the industry. The recovery process has proven to be reliable, efficient and successful for reducing the disposal costs and earning a profit from the fat recovery system.

25.7.3 Protein

Proteins during tannery processing can be recovered from wastewater. This liquid waste usually contains degraded products of proteoglycans and fibrous proteins, i.e., collagenous and keratin. However, the presence of these proteins significantly increases the biochemical oxygen demand and COD concentrations in wastewater. Therefore, the recovery of protein can be an effective option in order to reduce the contaminants in wastewater and obtain profit from the recovered proteins which has potential applications in the food and biopharmaceutical industries. Protein extraction techniques have been previously done using filtration, precipitation, centrifugation and electrophoresis [52].

25.7.3.1 Precipitation

Magnesium ammonium phosphate (MAP) precipitation has been successfully tested for protein recovery from tannery effluent [56]. It is achieved by the addition of metal ions

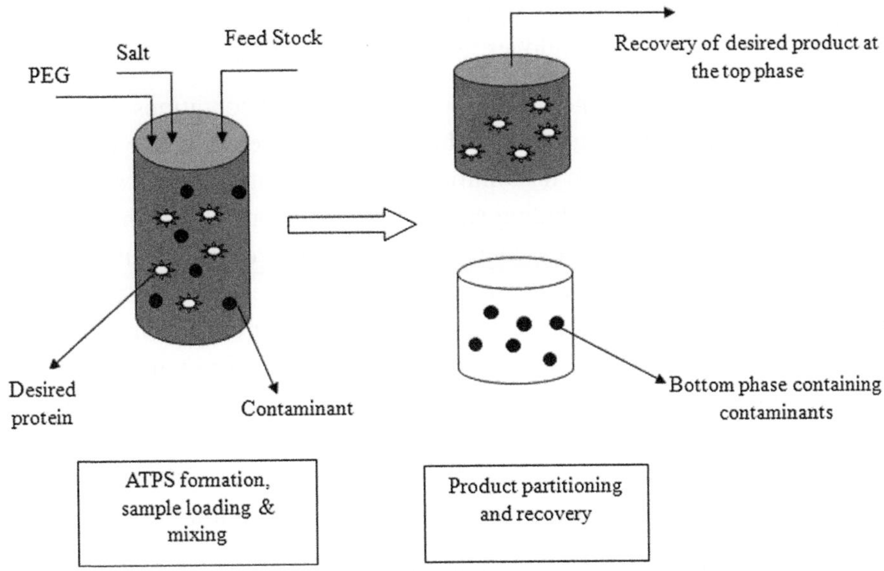

Figure 25.9
Aqueous two-phase systems using PEG-salt. *Adopted from Raja S, Murty VR. Development and evaluation of environmentally benign aqueous two phase systems for the recovery of proteins from tannery waste water. ISRN Chemical Engineering 2012;2012.*

(i.e., Fe^{3+}, Zn^{2+}, Cd^{2+}, or Cu^{2+}) and acidification methods which could recover ~40–90% of the protein. Additionally, using this technology for tannery wastewater could also remove nitrogen containing pollutants such as ammonia by 85% [56].

25.7.3.2 Aqueous two-phase system

Aqueous two-phase system (ATPS) is a liquid–liquid extraction method [57] and it is considered as an attractive alternative technique compared to the conventional techniques (e.g., precipitation and filtration). This technique is cost-effective, it requires short processing time and environmentally friendly [58]. Fig. 25.9 shows the schematic of the mechanism involved in compound separation. The extraction is easily completed using a low-speed centrifugation unit. Polyethylene glycol (PEG) and salts are used as the two-phase liquid. Eventually, the ATPS separates the soluble protein from the contaminants.

25.8 Health and safety aspects

On the basis of human exposure to toxic wastes, tanning industry are considered as the world dirtiest manufacturing sites. Even under the best practicing circumstances, it can be a dangerous working place. The factory setting is almost vomit-inducing, a combination of garbage, rotting animal hides and toxic chemicals, and it lacks an effective management system. The requirements for making a good quality leather is not a clean and white-collar job. Workers must be provided with gloves, goggles, respirator masks, and boots.

Chemical exposure leads to both short- and long-term medical conditions, lung disease, mainly asthma, bronchitis, lung cancer, urine bladder cancer, reproductive tract infection, and also other diseases like stomach discomfort or gastroenteritis [59]. In addition to the workers, residents living nearby, including small children encounter the chemicals released from tannery wastewater [60]. With the increasing complexity of industrial processes, the knowledge of hazards has also increased among the workers and the industrial managers. The combination of an unorganized industrial and labor structure, subhuman conditions at the workplace such as unguarded machines, improper handling of raw materials, chemical leather dust, wet floors, heavy noise, among others, cause different types of health hazards. According to a recent report, the number of accidents and illness rate is five times higher in tanneries than other industries [61]. The different types of hazards that tannery workers are exposed to and the possible reasons can be summarized as follows:

(A) Accident hazard: (1) Slips, trips and falls on the level, especially on wet, slippery or cluttered floors, while moving heavy loads such as containers of chemicals, bundles of hides, skin, leather etc.; (2) falls into unguarded tanning vats and pits; (3) electronic shocks caused by contract with defective and inadequate electric installations; (4) blows and crushing injuries caused by unguarded rotating or moving parts of machinery; (5) burns caused by contact with hot surfaces or splashes of hot solutions; (6) cuts and stabs caused by flying particles from rotary buffing machines; and (7) poisoning of confined spaces, in particular during the cleaning of vats or tanning baths or removal of clogging in draining pipes.

(B) Physical hazard: (1) Exposure to high noise levels from mechanical equipment (particularly drums, reverse settling machines, through-feed staking machines); (2) callosities on hands caused by continuous strenuous work with hands tools; and (3) eye strain due to poor illumination levels in the tannery.

(C) Chemical hazard: (1) Skin rashes and dermatitis as a result of exposure to cleaners, solvents, disinfectants, pesticides, leather-processing chemicals, etc.; and (2) allergies-contact and systematic-caused by many of the chemicals used in tanneries.

(D) Biological hazard: (1) Rawhides and skins may be contaminated with a variety of bacteria, molds, yeasts, etc., and various diseases may be transmitted to the tanners; and (2) large quantities of dust produced during buffering operations would normally be contaminated with disease-bearing microorganism and putrefaction products.

(E) Ergonomic, psychosocial and organizational factors: (1) Acute musculoskeletal injuries caused by physical overexertion and awkward posture while moving heavy or bulky loads, in particular bundles of hides, skins, and leather; (2) lower back pain due to prolonged working in a standing or semibending posture; and (3) heating stress, when working on warm days in premises lacking good ventilation or air conditioning.

Besides, the workers in the tanneries are also exposed to different unsafe conditions. Some of the commonly reported situations can be summarized as follows:

(1) Machinery and installation is unprotected with moving machine parts.
(2) Rolling, sliding, thrown parts, hot and cold surfaces, liquids, steam, electrical current, etc.
(3) Workplace slippery floors, unprotected floor, opening or elevated locations.
(4) Chemical and biological agents, hazardous chemicals, expired chemicals, nonlabeled bottles and containers.
(5) Contamination due to bacteria, fungus, parasites, virus, mold and yeast.
(6) Working environment, temperature and humidity and illumination level.

25.9 Standards and regulation related to the leather tanning industry

Several international and national agencies have formulated standards and regulations for the leather industry in terms of the product quality, mechanical, chemical, and physical characterization tests, as well as environmental and occupational health and safety regulations. For instance, the International Organization for Standardization (ISO) has established two standards, one that focuses on rawhides and skins, including pickled pelts, while the other focuses on tanned leather. Table 25.10 provide some examples of these agencies and the relevant standards that is related to the tannery industry.

25.10 Conclusions and perspectives

Due to the use of undeveloped and conventional methods, the tannery industry releases large quantities of toxic chemicals to the environment. Different options have been recommended by researchers for recovering and recycling the valuable resources from liquid waste streams, especially, chromium and sodium sulfide, which are used commonly in this industry. As most of the tanneries in the world are located in the developing countries, implementing new methods and technologies for cleaner production needs a lot of efforts and investment, as most of the methods reported in the literature are still at the lab scale and therefore, they require more research and attention in order to scale up to the pilot and semiindustrial scale.

From a future perspective, more research and developmental initiatives should be directed toward the recovery of chemicals and byproducts from the tannery's effluent. Materials recovery projects from the tanning effluent should be implemented at the pilot and industrial levels. Utilization of leather tanning byproducts in other industries should be facilitated through industrial symbioses and ecoindustrial clusters or parks. Improving the environmental and economic performance of the leather tanning sector requires integrated

Table 25.10: International agencies and organizations that have formulated regulations for the leather industry.

Agency/organization	Related standards/regulation
ISO	ISO/TC 120/SC 1: Rawhides and skins, including pickled pelts ISO/TC 120/SC 2: Tanned leather
American Society for Testing and Materials (ASTM)	Leather appeal standard tests Leather chemical analysis standard tests Leather physical properties standard tests
Occupational Safety and Health Administration (OSHA)	Occupational safety and health standards and regulations at tanneries
UNIDO	Environmental regulation
EEA	Environmental regulation
National standard institutes i.e., Bureau of Indian Standards (BIS)	Quality standards and environmental regulations

cleaner production measures, of which resource recovery is a vital option for achieving zero-waste discharge from this industry.

Acknowledgments

The authors would like to thank the Palestinian-Dutch Academic Cooperation Program (PADUCO) for funding this research project: "Managing heavy metals contaminated industrial WW from inorganic chemical industries in the West Bank: Implementing cleaner production for sustainability". The authors also thank the project partners: Palestinian Environment Quality Authority, the Leather and Shoes Association in Palestine and Al-Waleed Leather Company, for their cooperation. The authors also thank the MSc Students from IHE Delft for their support during the literature review stage.

References

[1] Famielec S, Wieczorek-Ciurowa K. Waste from leather industry. Threats to the environment. Czasopismo Techniczne – Chemia 2011;108(1-Ch):43–8.
[2] Lofrano G, et al. Chemical and biological treatment technologies for leather tannery chemicals and wastewaters: a review. The Science of the Total Environment 2013;461:265–81.
[3] Mwinyihija M. Ecotoxicological diagnosis in the tanning industry. Springer Science & Business Media; 2010.
[4] Abdulla HM, et al. Chromium removal from tannery wastewater using chemical and biological techniques aiming zero discharge of pollution. In: Proceeding of fifth scientific environmental conference; 2010.
[5] de Aquim PM, Hansen É, Gutterres M. Water reuse: an alternative to minimize the environmental impact on the leather industry. Journal of Environmental Management 2019;230:456–63.

[6] Elabbas S, et al. Treatment of highly concentrated tannery wastewater using electrocoagulation: influence of the quality of aluminium used for the electrode. Journal of Hazardous Materials 2016;319:69–77.
[7] Hassen AS, Woldeamanuale TB. Evaluation and characterization of tannery wastewater in each process at Batu and Modjo tannery, Ethiopia. International Journal of Rural Development, Environment and Health Research 2017;1(3):1–10.
[8] Islam B, et al. Evaluation and characterization of tannery wastewater. Journal of Forest Products and Industries 2014;3(3):141–50.
[9] Ortolano L, et al. Cleaner production in Pakistan's leather and textile sectors. Journal of Cleaner Production 2014;68:121–9.
[10] Ayalew A. Cleaner production options for solid waste management in the leather industry. Addis Ababa University; 2005.
[11] Alexander K, Donohue V. Cleaner technologies in the tanning industry. Northampton, England: British Leather Confederation; 1990.
[12] Chen W, et al. Resourceful utilization of tannery wastewater. Chinese Journal of Population Resources and Environment 2012;10(1):116–21.
[13] Wu J, et al. Recent progress in cleaner preservation of hides and skins. Journal of Cleaner Production 2017;148:158–73.
[14] Daddi T, et al. Enhancing the adoption of life cycle assessment by small and medium enterprises grouped in an industrial cluster: a case study of the tanning cluster in Tuscany (Italy). Journal of Industrial Ecology 2016;20(5):1199–211.
[15] Kiliç E, et al. Environmental optimization of chromium recovery from tannery sludge using a life cycle assessment approach. Journal of Hazardous Materials 2011;192(1):393–401.
[16] Tasca AL, Puccini M. Leather tanning: life cycle assessment of retanning, fatliquoring and dyeing. Journal of Cleaner Production 2019;226:720–9.
[17] Jahan M, et al. Characterization of tannery wastewater and its treatment by aquatic macrophytes and algae. Bangladesh Journal of Scientific & Industrial Research 2014;49(4):233–42.
[18] Sabur M, Rahman M, Safiullah S. Treatment of tannery effluent by locally available commercial grade lime. Journal of Scientific Research 2013;5(1).
[19] Pantazopoulou E, Zouboulis A. Chemical toxicity and ecotoxicity evaluation of tannery sludge stabilized with ladle furnace slag. Journal of Environmental Management 2018;216:257–62.
[20] Sawalha H, et al. Wastewater from leather tanning and processing in Palestine: Characterization and management aspects. Journal of Environmental Management 2019;251:109596.
[21] Lofrano G, Meric S, Belgiorno V. Tannery wastewater treatment by advanced oxidation processes. In: Water, wastewater and soil treatment by advanced oxidation processes (AOPs); 2011. p. 197 [chapter 13].
[22] Kanagaraj J, et al. Solid wastes generation in the leather industry and its utilization for cleaner environment – a review. Journal of Scientific and Industrial Research 2006;65:541–8.
[23] Joseph K, Nithya N. Material flows in the life cycle of leather. Journal of Cleaner Production 2009;17(7):676–82.
[24] Fela K, et al. Present and prospective leather industry waste disposal. Polish Journal of Chemical Technology 2011;13(3):53–5.
[25] Celary P, Sobik-Szołtysek J. Vitrification as an alternative to landfilling of tannery sewage sludge. Waste Management 2014;34(12):2520–7.
[26] Mwinyihija M, et al. Ecological risk assessment of the Kenyan tanning industry. Journal of the American Leather Chemists Association 2005;100:466.
[27] Covington AD. Tanning chemistry: the science of leather. Royal Society of Chemistry; 2009.
[28] Borchate SS, Kulkarni GSS, Kore VV, Kore S. Application of coagulation–flocculation for vegetable tannery wastewater. International Journal of Engineering Science and Technology 2012;4(5):1944–8.
[29] Bhattacharya P, Ghosh S, Mukhopadhyay A. Combination technology of ceramic microfiltration and biosorbent for treatment and reuse of tannery effluent from different streams: response of defence system in Euphorbia sp. International Journal of Recycling of Organic Waste in Agriculture 2013;2(1):19.

[30] Selvaraj R, et al. A membrane electroflotation process for recovery of recyclable chromium (III) from tannery spent liquor effluent. Journal of Hazardous Materials 2018;346:133–9.
[31] Suthanthararajan R, et al. Membrane application for recovery and reuse of water from treated tannery wastewater. Desalination 2004;164(2):151–6.
[32] De Gisi S, Galasso M, De Feo G. Treatment of tannery wastewater through the combination of a conventional activated sludge process and reverse osmosis with a plane membrane. Desalination 2009;249(1):337–42.
[33] Das C, et al. Treatment of tanning effluent using nanofiltration followed by reverse osmosis. Separation and Purification Technology 2006;50(3):291–9.
[34] Fang D, Zhou L-X. Enhanced Cr bioleaching efficiency from tannery sludge with coinoculation of *Acidithiobacillus thiooxidans* TS6 and Brettanomyces B65 in an air-lift reactor. Chemosphere 2007;69(2):303–10.
[35] Zheng G, Zhou L, Wang S. An acid-tolerant heterotrophic microorganism role in improving tannery sludge bioleaching conducted in successive multibatch reaction systems. Environmental Science & Technology 2009;43(11):4151–6.
[36] Ma H, et al. Chromium recovery from tannery sludge by bioleaching and its reuse in tanning process. Journal of Cleaner Production 2017;142:2752–60.
[37] Andrioli E, Petry L, Gutterres M. Environmentally friendly hide unhairing: enzymatic-oxidative unhairing as an alternative to use of lime and sodium sulfide. Process Safety and Environmental Protection 2015;93:9–17.
[38] Anzani C, et al. Towards environmentally friendly skin unhairing process: a comparison between enzymatic and oxidative methods and analysis of the protein fraction of the related wastewaters. Journal of Cleaner Production 2017;164:1446–54.
[39] Dettmer A, et al. Environmentally friendly hide unhairing: enzymatic hide processing for the replacement of sodium sulfide and delimig. Journal of Cleaner Production 2013;47:11–8.
[40] Murugappan G, et al. A novel approach to enzymatic unhairing and fiber opening of skin using enzymes immobilized on magnetite nanoparticles. ACS Sustainable Chemistry & Engineering 2016;4(3):828–34.
[41] Vijayaraghavan R, et al. Aqueous ionic liquid solutions as alternatives for sulphide-free leather processing. Green Chemistry 2015;17(2):1001–7.
[42] Vidal G, et al. Unhairing effluents treated by an activated sludge system. Journal of Hazardous Materials 2004;112(1–2):143–9.
[43] Santos C, Rodriguez E. Review on some emerging endpoints of chromium (VI) and lead phytotoxicity. In: Botany. IntechOpen; 2012.
[44] Patel A, Patra D. Influence of heavy metal rich tannery sludge on soil enzymes vis-à-vis growth of *Tagetes minuta*, an essential oil bearing crop. Chemosphere 2014;112:323–32.
[45] Barrena R, et al. Co-composting of hair waste from the tanning industry with de-inking and municipal wastewater sludges. Biodegradation 2007;18(3):257–68.
[46] Onyuka A, et al. Parameters for composting tannery hair waste. Journal of the American Leather Chemists Association 2012;107(5):159–66.
[47] Zuriaga-Agustí E, et al. Pollution reduction in an eco-friendly chrome-free tanning and evaluation of the biodegradation by composting of the tanned leather wastes. Journal of Cleaner Production 2015;87:874–81.
[48] Ravindran B, Sekaran G. Bacterial composting of animal fleshing generated from tannery industries. Waste Management 2010;30(12):2622–30.
[49] Araujo ASF, et al. Soil microbial properties after 5 years of consecutive amendment with composted tannery sludge. Environmental Monitoring and Assessment 2015;187(1):4153.
[50] Sampathkumar S, Buljan J. Composting of tannery sludge. United Nations Industrial Development Organization; 2001.
[51] Puhazhselvan P, et al. Enzyme based cleaner process for enhanced recovery of lipids from tannery fleshing waste. Journal of Cleaner Production 2017;144:187–91.

[52] Sundar VJ, et al. Recovery and utilization of proteinous wastes of leather making: a review. Reviews in Environmental Science and Bio/Technology 2011;10(2):151–63.
[53] Dayanandan A, et al. Enhanced production of Aspergillus tamarii lipase for recovery of fat from tannery fleshings. Brazilian Journal of Microbiology 2013;44(4):1089–95.
[54] Pop E, et al. Energetic characteristics of animal fats waste from tannery for energy production. In: 4th international conference on thermal equipment, renewable energy and rural development, TE-RE-RD; 2015.
[55] Bondrea DA, et al. Energy recovering from tanneries by biodiesel production. In: 4th international conference on thermal equipment, renewable energy and rural development, TE-RE-RD 2015, proceeding; 2015.
[56] Kabdaşli I, Ölmez T, Tünay O. Nitrogen removal from tannery wastewater by protein recovery. Water Science and Technology 2003;48(1):215–23.
[57] Raja S, Murty VR. Development and evaluation of environmentally benign aqueous two phase systems for the recovery of proteins from tannery waste water. ISRN Chemical Engineering 2012;2012.
[58] Saravanan S, et al. Partition of tannery wastewater proteins in aqueous two-phase poly (ethylene glycol)-magnesium sulfate systems: effects of molecular weights and pH. Chemical Engineering Science 2007;62(4):969–78.
[59] Rastogi SK, Pandey A, Tripathi S. Occupational health risks among the workers employed in leather tanneries at Kanpur. Indian Journal of Occupational and Environmental Medicine 2008;12(3):132.
[60] Tinni SH, et al. Impact of tanneries waste disposal on environment in some selected areas of Dhaka City Corporation. Journal of Environmental Science and Natural Resources 2014;7(1):149–56.
[61] Islam MS, et al. Heavy metals in the industrial sludge and their ecological risk: a case study for a developing country. Journal of Geochemical Exploration 2017;172:41–9.

Index

'*Note*: Page numbers followed by "f" indicate figures and "t" indicate tables.'

B

C

D

E

F

H

I

M

N

O

Q

R

X

Y

Z

CPI Antony Rowe
Eastbourne, UK
November 24, 2021